W0036784

Birkhäuser Advanced Texts Basler Lehrbücher

Series Editors

Steven Krantz, Washington University, St. Louis, USA

Shrawan Kumar, University of North Carolina at Chapel Hill, Chapel Hill, USA

A series of Advanced Textbooks in Mathematics

This series presents, at an advanced level, introductions to some of the fields of current interest in mathematics. Starting with basic concepts, fundamental results and techniques are covered, and important applications and new developments discussed. The textbooks are suitable as an introduction for students and non-specialists, and they can also be used as background material for advanced courses and seminars.

Heath Emerson

An Introduction to C*-Algebras and Noncommutative Geometry

 Birkhäuser

Heath Emerson
Department of Mathematics and Statistics
University of Victoria
Victoria, BC, Canada

ISSN 1019-6242   ISSN 2296-4894   (electronic)
Birkhäuser Advanced Texts Basler Lehrbücher
ISBN 978-3-031-59849-4   ISBN 978-3-031-59850-0   (eBook)
https://doi.org/10.1007/978-3-031-59850-0

© Springer Nature Switzerland AG 2024
This work is subject to copyright. All rights are solely and exclusively licensed by the Publisher, whether
the whole or part of the material is concerned, specifically the rights of translation, reprinting, reuse
of illustrations, recitation, broadcasting, reproduction on microfilms or in any other physical way, and
transmission or information storage and retrieval, electronic adaptation, computer software, or by similar
or dissimilar methodology now known or hereafter developed.
The use of general descriptive names, registered names, trademarks, service marks, etc. in this publication
does not imply, even in the absence of a specific statement, that such names are exempt from the relevant
protective laws and regulations and therefore free for general use.
The publisher, the authors and the editors are safe to assume that the advice and information in this book
are believed to be true and accurate at the date of publication. Neither the publisher nor the authors or
the editors give a warranty, expressed or implied, with respect to the material contained herein or for any
errors or omissions that may have been made. The publisher remains neutral with regard to jurisdictional
claims in published maps and institutional affiliations.

This book is published under the imprint Birkhäuser, www.birkhauser-science.com by the registered
company Springer Nature Switzerland AG
The registered company address is: Gewerbestrasse 11, 6330 Cham, Switzerland

If disposing of this product, please recycle the paper.

Dedicated to Juliette.

Preface

The study of C*-algebras was initiated by physicists working in quantum mechanics like Heisenberg, but was continued by the mathematician Gelfand, especially in connection with representation theory of locally compact groups. If one has two compactly supported, continuous functions on a locally compact group, then their convolution product is the function on the group given by

$$(1) \qquad (f_1 * f_2)(g) = \int_G f_1(h) f_2(h^{-1}g) dh,$$

where dh denotes Haar measure on the group. Using this multiplication, which is something like matrix multiplication, one can construct a C*-algebra $C^*(G)$ out of any locally compact group, and the interest in the construction was that the representation theories of the group and of the C*-algebra are in natural one-to-one correspondence with each other.

A C*-algebra can be defined to be any subalgebra A of the algebra of bounded operators $\mathbb{B}(H)$ on a Hilbert space, which is closed under operator adjoint and topologically closed in the operator norm topology. If X is any compact Hausdorff space, then the collection $C(X)$ of complex-valued continuous functions $f : X \to \mathbb{C}$ vanishing at infinity acts by multiplication operators on $L^2(X, \mu)$ with respect to any Borel measure and forms a C*-algebra where the algebra multiplication operation is pointwise multiplication

$$(f_1 \cdot f_2)(x) = f_1(x) f_2(x),$$

which is a *commutative* operation: $f_1 f_2 = f_2 f_1$ for any f_1, f_2. Gelfand formulated an abstract definition of C*-algebra, and proved that *any* abstract unital commutative C*-algebra is canonically isomorphic to $C(X)$ for X the compact space of characters of the C*-algebra. Gelfand's theorem generalizes the Fourier isomorphism and shows that the category of *commutative* (unital) C*-algebras is contravariantly equivalent to the category of compact Hausdorff spaces, and suggests the idea that a general, possibly noncommutative C*-algebra may be thought of as a kind of "noncommutative space."

The need for such a concept had already been felt in subatomic particle physics, as Heisenberg had observed, because pairs of observables, like position and momentum, in subatomic systems had a strange "noncommutative behavior" that made it seem better to model their behavior not as usual in classical physics, by functions on appropriate topological spaces, but by self-adjoint operators on a Hilbert space, which do not, in general, commute with each other, unlike functions, when one multiplies them pointwise.

In the 1950s and 1960s, seminal results in topology and analysis—the Bott Periodicity and Atiyah-Singer Index theorems—revitalized interest in C*-algebra theory from a more geometric point of view. The content of the Index Theorem is that the Fredholm index of an elliptic operator on a compact manifold may be described purely topologically, in terms of a K-theory invariant of the symbol of the operator. Topological K-theory was developed by Atiyah, Hirzebruch, Segal, and others, to a large extent in connection with various statements and proofs of the Index Theorem. But K-theory actually makes perfect sense for even noncommutative C*-algebras, and determines a canonical homology theory for them.

In the 1980s, the Fields' Medallist Alain Connes initiated a number of innovative reformulations of constructions from Riemannian geometry, topology, and geometric analysis, with the aim of applying them to potentially noncommutative C*-algebras. Among Connes' discoveries were cyclic cohomology, the concept of a spectral triple, and a formula for a noncommutative Chern character linking the two. Connes' work extends the setup of de Rham cohomology and Chern-Weil theory for manifolds X, to the noncommutative situation. In his fascinating book [47] Connes suggested the possibility of a new field of "Noncommutative Geometry."

Concurrently, the Russian mathematician Genadi Kasparov was developing KK-theory, a bivariant generalization of C*-algebra K-theory, and using it to prove cases of the Novikov Conjecture [111]. New and striking connections between the Novikov conjecture, group C*-algebras, and the large-scale geometry of groups developed by Gromov and others emerged in the 1990s and stimulated cross-pollination between C*-algebra K-theory and topology. A number of mathematicians began exploring various classes of C*-algebras attached to various types of dynamical systems, by the "groupoid" construction. By associating a C*-algebra to a groupoid, one can then study the groupoid C*-algebraically, compute its K-theory invariants, representation-theoretic dual, primitive ideal space, tracial state space, etc. The use of this strategy has produced interesting results in connection with representation theory of locally compact groups, hyperbolic dynamical systems, Gromov hyperbolic groups, Penrose tilings, "coarse," or "large-scale" geometry of metric spaces, foliations, index theory on contact manifolds, quantum groups, manifolds with singularities, homeomorphisms of Cantor sets, systems and C*-algebras associated to number fields, to name a few examples from recent years.

The field of noncommutative geometry, at present, offers what seems to be a very enticing program. It is the author's goal to bring the relative novice up to speed on the basic ideas of C*-algebra theory, K-theory, K-homology, Index Theory, and

Connes' Noncommutative Riemannian geometry, and to offer a glimpse at some of the more advanced topics in this fascinating subject.

This book was written with the explicit purpose of giving my students something to read to learn the parts of the subject I have worked on or am working on, and at least the first chapter should be readily comprehensible to a strong student at the third or fourth year undergraduate level in the Canadian system, equipped with basic knowledge of algebra, analysis, and a bit of functional analysis, especially basic Hilbert space theory. As the book progresses, slightly stronger demands are made on the student, as we deal with smooth manifolds, some differential geometry and elliptic operator theory, and K-theory, and the reader may need to occasionally consult other more specialized texts for the occasional detail, or to deepen their understanding, although I have tried to keep the book as self-contained as possible.

All Hilbert spaces and C*-algebras in this book are assumed separable, with a very small number of exceptions, e.g., the C*-algebra $\mathbb{B}(H)$, and all locally compact spaces are tacitly assumed second countable.

The symbol μ in connection with a locally compact group G denotes Haar measure, normalized if G is compact.

Victoria, Canada   Heath Emerson
April, 2022

Contents

Chapter 1
An Introduction to C*-Algebras

C*-algebras were first considered in connection with quantum mechanics to model algebras of physical observables by Werner Heisenberg. Heisenberg called his new mathematics "matrix mechanics," but the existence of the noncommutative algebra of matrices had already been noted by Jordan. Von Neumann made numerous striking discoveries later about various classes of algebras of operators on a Hilbert space. The abstract characterization of C*-algebra (Definition 1.1.7) is due to Gelfand in around 1943. The term "C*-algebra" is due to I.E. Segal in 1947, who used it to describe norm-closed *-subalgebras of bounded operators on a Hilbert space; the "C" stands for "closed."

The introduction to Connes' book [47] contains an excellent explanation of why the experimental results of spectroscopy require noncommuting observables.

This chapter discusses some of the most important examples of C*-algebras: matrix algebras, the algebra of bounded operators on a Hilbert space, the Toeplitz algebra, group C*-algebras, and crossed products, Fredholm operators, and the Fredholm index. We state and prove the Toeplitz Index Theorem—the simplest case of the Atiyah–Singer Index Theorem.

The reader wishing more sources on basic C*-algebra theory is invited to consult the excellent books [126] and [7].

1.1 The Definition of C*-Algebra

An (associative) *algebra* over the complex numbers is a complex vector space A equipped with an associative, bilinear (linear in each variable separately) multiplication operation $A \times A \to A$, $(a, b) \mapsto ab$. An algebra is unital if it contains an element $1 \in A$ such that $1a = a1 = a$ for all $a \in A$.

© Springer Nature Switzerland AG 2024
H. Emerson, *An Introduction to C*-Algebras and Noncommutative Geometry*,
Birkhäuser Advanced Texts Basler Lehrbücher,
https://doi.org/10.1007/978-3-031-59850-0_1

The zero algebra {0} is an algebra, albeit an extremely uninteresting one, as is the complex numbers itself $\mathbb{C}$. Both are unital. There are of course many, many other examples: polynomial algebras $\mathbb{C}[x_1, \ldots, x_n]$ and their quotients, the quaternion algebra, and matrix algebras $M_n(\mathbb{C})$.

For our purposes, it is extremely interesting and important that every compact Hausdorff space X has an (commutative) algebra canonically associated to it: namely the algebra $C(X)$ of continuous complex-valued functions on X. To get the algebra structure we apply the algebra structure of $\mathbb{C}$ pointwise: Thus

$$(\lambda f)(x) := \lambda f(x), \ (f + g)(x) := f(x) + g(x), \ (fg)(x) := f(x)g(x).$$

The constant function 1 is the unit. The Hausdorff condition is to ensure an adequate supply of continuous functions to produce something meaningful from the construction.

The algebra $C(X)$ has further structure, which turns out to be important. Firstly, complex conjugation on the complex numbers $\mathbb{C}$ gives rise to an *involution* on functions $f \in C(X)$ by setting $f^*(x) := \overline{f(x)}$. Secondly, any continuous function on a compact space is bounded. We set

(1.1) $$\|f\| := \sup_{x \in X} |f(x)|, \text{ for } f \in C(X),$$

and call it the norm of f; it satisfies the standard set of conditions for a norm on a linear space:

$$\|\lambda f\| = |\lambda| \|f\|, \quad \|f + g\| \le \|f\| + \|g\|, \quad \|f\| = 0 \iff f = 0.$$

So $C(X)$ gains a topology from the metric $d(f, g) = \|f - g\|$. It is a quite easy enough exercise (see below) to prove that $C(X)$ is also complete with respect to this metric (thus is a Banach space). This follows from completeness of the complex numbers $\mathbb{C}$. Furthermore, the conditions

$$\|fg\| \le \|f\| \|g\|, \quad \|f^*\| = \|f\|,$$

hold, for all f, g, which makes all the algebra operations continuous.

Exercise 1.1.1 Prove that if X is compact Hausdorff, (f_n) is a sequence of continuous functions on X which is Cauchy with respect to the norm (1.1), then (f_n) converges in the same norm, to a continuous function f. That is, show that $C(X)$ is complete with respect to the norm (1.1).

If A is a unital algebra, an element $a \in A$ is *invertible* if there exists $b \in A$ such that $ab = ba = 1$.

Definition 1.1.2 If A is any unital algebra, and $a \in A$, a complex number λ is in the *spectrum* of a if $\lambda 1 - a$ is *not* invertible in A.

As a matter of notation we normally write just $\lambda - a$ instead of $\lambda\, 1 - a$.

The spectrum of an element, therefore, is a purely algebraic invariant of a (it does not make any reference to norms, nor, in fact, to adjoints).

We let $\mathrm{Spec}(a) \subset \mathbb{C}$ be the spectrum of a.

Exercise 1.1.3 If A is a unital algebra, $u \in A$ is an invertible element, and $a \in A$, then the spectrum of uau^{-1} is the same as the spectrum of a.

Exercise 1.1.4 Verify the formula $(\lambda - ba)^{-1} = \lambda^{-1} + \lambda^{-1}b(\lambda - ab)^{-1}a$ for any elements a, b of a unital algebra A and any $\lambda \neq 0$. Deduce that $\mathrm{Spec}(ab) - \{0\} = \mathrm{Spec}(ba) - \{0\}$ for any $a, b \in A$, where a is a unital algebra.

Exercise 1.1.5 If $A \in M_n(\mathbb{C})$, then $\mathrm{Spec}(A)$ is the set of eigenvalues of A.

If $A = C(X)$, for X compact Hausdorff as above, then a function $f \in C(X)$ is invertible if and only if it does not vanish on X. We deduce that the spectrum of such f is precisely the range of f. But the norm $\|f\|$ of f is precisely the modulus of the largest complex number in the range of f. Hence we get

$$(1.2) \qquad\qquad \|f\| = \sup_{\lambda \in \mathrm{Spec}(f)} |\lambda|.$$

This equation relates the *topology* and the *algebra* in a very tight way.

If A and B are two algebras, an *algebra homomorphism* $\alpha \colon A \to B$ is a linear map such that $\alpha(ab) = \alpha(a)\alpha(b)$. We say α is *unital* if $\alpha(1) = 1$.

Proposition 1.1.6 *Let X and Y be compact Hausdorff, $A = C(X)$, and $B = C(Y)$. Then if $\alpha \colon A \to B$ is a unital algebra homomorphism, then α is automatically continuous with respect to the topologies on A, B determined by their norms as in* (1.1).

Proof Firstly, α is a unital algebra homomorphism. Hence it maps invertibles in A to invertibles in B. It follows that $\mathrm{Spec}(\alpha(f)) \subset \mathrm{Spec}(f)$ for any $f \in A$. Since the norm of f is the radius of the range of f, that is, of the spectrum of f, we deduce that $\|\alpha(f)\| \leq \|f\|$ as claimed.   □

The algebra of n-by-n matrices has a similar property to the one appearing in (1.2). Firstly, the correct involution, or "adjoint" to use on this algebra, is the conjugate transpose A^* of a matrix: Thus $A^*_{ij} := \overline{A_{ji}}$. For the norm, we use the *operator norm*, defined for $A \in M_n(\mathbb{C})$ by

$$(1.3) \qquad\qquad \|A\| := \sup_{\xi \in \mathbb{C}^n, \|\xi\|_{\mathbb{C}^n} = 1} \|A\xi\|_{\mathbb{C}^n},$$

with $\|\cdot\|_{\mathbb{C}^n}$ the Hilbert space norm on $\mathbb{C}^n$. It is an easy exercise to check that this defines a norm on $M_n(\mathbb{C})$: that is: $\|\lambda A\| = |\lambda|\|A\|$, $\|A + B\| \leq \|A\| + \|B\|$. Furthermore, $\|AB\| \leq \|A\|\|B\|$, so that the algebra multiplication is continuous in this norm.

The operator norm is easily checked to be invariant under unitary conjugation $\|UAU^*\| = \|A\|$ for all $A \in M_n(\mathbb{C})$. Secondly, if A is any matrix, $\|A\|^2 = \|A^*A\|$ (this is more challenging to prove and is done in the next section), which reduces the problem of finding $\|A\|$ to finding the norm of A^*A. Now A^*A is unitarily

diagonalizable. So it is unitarily conjugate to $\begin{bmatrix} \lambda_1 & 0 & \cdots & 0 \\ 0 & \lambda_2 & & \vdots \\ \vdots & & \ddots & \vdots \\ 0 & \cdots & \cdots & \lambda_n \end{bmatrix}$ where $\lambda_1, \ldots, \lambda_n$

are the eigenvalues of A^*A (repeated according to multiplicity). Hence $\|A^*A\|$ equals the norm of the diagonal matrix. By another routine exercise, the operator norm of a diagonal matrix with entries $\lambda_1, \ldots, \lambda_n$ is equal to the supremum $\max\{|\lambda_1|, \ldots, |\lambda_n|\}$ of the entries. We thus have that $\|A\|$ (for any A) can be characterized as

$$\|A\| = \sup_{\lambda \in \mathrm{Spec}(A^*A)} \sqrt{|\lambda|},$$

where $\mathrm{Spec}(A^*A)$ is the set of eigenvalues of A^*A. And this is consistent with the case of $C(X)$ with a small adjustment, since we can write (1.2) in the fancier but equivalent way

$$\|f\| = \sup_{\lambda \in \mathrm{Spec}(f^*f)} \sqrt{|\lambda|}.$$

Definition 1.1.7 A *C*-algebra* A is an (associative) algebra over the complex numbers equipped with a map $*: A \to A$ (usually called the *adjoint*) and a norm $\|\cdot\|: A \to [0, \infty)$ satisfying:

a) The map $*$ is a conjugate linear, involutative anti-homomorphism, i.e., satisfies:
   - $(\lambda a + b)^* = \bar{\lambda}a^* + b^*$ for all $\lambda \in \mathbb{C}$, $a, b \in A$.
   - $(ab)^* = b^*a^*$ for all $a, b \in A$.
   - $(a^*)^* = a$ for all $a \in A$.

b) With the metric assigning distance $\|a - b\|$ from a to b, A is complete, i.e., $(A, \|\cdot\|)$ is a Banach space.
c) $\|ab\| \leq \|a\|\|b\|$ for all $a, b \in A$.
d) $\|a^*a\| = \|a\|^2$ for all $a \in A$.

A is *unital* if there exists an element $1 \in A$ acting as the identity under multiplication.

The condition d) in Definition 1.1.7 is often called the C*-condition.

As an easy exercise in the definitions, note that $\|a^*\| = \|a\|$ for all $a \in A$, for using c) above, we have that $\|a\|^2 = \|a^*a\| \leq \|a^*\|\|a\|$. The claim follows by switching the roles of a and a^*.

The following exercise is also routine, and it follows from the uniqueness of the unit.

Exercise 1.1.8   The unit 1 in a unital C*-algebra satisfies $1^* = 1$.

As a matter of terminology, an *-*algebra* is an associative complex algebra with an involution satisfying the condition a) above. A *Banach algebra* is an algebra (without necessarily an involution operation) together with a norm with respect to which it is complete, which also satisfies $\|ab\| \leq \|a\| \|b\|$ for all a, b. A *Banach* *-*algebra* is a Banach algebra with an involution $*: A \rightarrow A$ making it also a *-algebra, and as well, the requirement that $\|a^*\| = \|a\|$—a weakening of the C*-condition, as explained above. This weakening has a substantial effect on the resulting different theories.

Exercise 1.1.9   Let A be a C*-algebra (not necessarily unital). Prove that if $a \in A$ and $a^*a = 0$, then $a = 0$. Deduce that if $xa = 0$ for all $x \in A$, then $a = 0$.

A *-*homomorphism* $\varphi: A \rightarrow B$ between two C*-algebras is an algebra homomorphism such that $\varphi(a^*) = \varphi(a)^*$ for all $a \in A$. The notion of isomorphism of C*-algebras is then the obvious one: There must be two *-homomorphisms which compose to the identity.

Exercise 1.1.10   If $\alpha: A \rightarrow B$ is an isomorphism of unital C*-algebras, then α is automatically unital.

A C*-algebra is *commutative* if, of course, its multiplication is commutative, $ab = ba$ for all $a, b \in A$. It is clear from the preceding discussion that $C(X)$ is a commutative C*-algebra for any compact Hausdorff space X.

We will finish in the next section the proof that the *-algebra of n-by-n matrices $M_n(\mathbb{C})$, with the operator norm, is also a C*-algebra, but it is evidently not commutative.

The mathematical principal of quantum mechanics is roughly as follows. In classical (Newtonian) physics, one deals with points in an appropriate space. If a particle moves through space time, its positions with regard to a fixed set of—say, three—coordinate axes as time changes, all of these constitute the points of a 4-dimensional space X. In the language of physics, continuous functions on X are "observables." For example, if one has a system $(x_1, \ldots, x_n)$ of coordinates on X (valued in $\mathbb{R}^n$), then each $x_i: X \rightarrow \mathbb{R}$ is an observable: At a given point in space, and at a given time, one observes the x_ith coordinate of the particle.

At any rate, *observables* stripped down to their mathematical essentials are continuous functions.

However, it was shown experimentally that when one attempts to study electrons within an atom, certain different measurements, namely position and momentum, interfere with each other in such a way as to make the *simultaneous* measurement of them impossible. Heisenberg postulated that the mathematics describing quantum physics should be the mathematics, not of functions on a space, but of linear operators on a Hilbert space, which, taken as an algebra, behaves, algebraically, much like the algebra of continuous functions on a space, but is not commutative.

And the experimental fact noted with position and momentum of electrons would correspond to the failure of two specific operators to commute with each other. The operators are each self-adjoint and diagonalizable, and thus each, in its own right, is, up to unitary equivalence, just a (real-valued) function on a space (the space being the spectrum of the operator, the function being the inclusion of the spectrum in $\mathbb{R}$), but one cannot consider them both *simultaneously* as functions, because they cannot be simultaneously diagonalized, because they do not commute.

Exercise 1.1.11 Let A be a C*-algebra and X be a compact Hausdorff space. Consider $C(X, A)$, the collection of continuous functions $f: X \to A$. Endow $C(X, A)$ with the algebra operations

$$(f_1 + \lambda f_2)(x) := f_1(x) + \lambda f_2(x), \quad (f_1 f_2)(x) := f_1(x) f_2(x),$$

adjoint $(f^*)(x) = f(x)^*$ and norm $\|f\| := \sup_{x \in X} \|f(x)\|$.
    Prove that $C(X, A)$ is a C*-algebra.

Example 1.1.12 An important example of a Banach algebra which is not a C*-algebra is the *disk* algebra $\mathcal{A}(\mathbb{D})$, consisting of all continuous functions $f \in C(\overline{\mathbb{D}})$ on the closed disk, which are analytic in the open disk $\mathbb{D}$.
    The norm on $\mathcal{A}(\mathbb{D})$ may be taken to be the supremum norm on the closed disk, or, equivalently,

$$\|f\|_{\mathcal{A}(\mathbb{D})} = \sup_{z \in \partial \overline{\mathbb{D}} = \mathbb{T}} |f(z)| = \|f|_{\mathbb{T}}\|_{C(\mathbb{T})}.$$

Hence, $\mathcal{A}(\mathbb{D})$ can be regarded as a closed Banach subalgebra of the C*-algebra $C(\mathbb{T})$ (that is, the norm on $\mathcal{A}(\mathbb{D})$ is the restriction of the norm on $C(\mathbb{T})$). In particular it is a Banach algebra. It is not a C*-subalgebra of $C(\mathbb{T})$, however, because the $f^*(z) := \overline{f(z)}$ is not analytic even if f is analytic.

Exercise 1.1.13 Let A and B be two C*-algebras. Their *direct sum* $A \oplus B$ is defined to be the direct sum of A and B as vector spaces, with the algebra structure $(a, b) \cdot (c, d) := (ac, bd)$, adjoint $(a, b)^* := (a^*, b^*)$, and norm $\|(a, b)\| := \max\{\|a\|, \|b\|\}$.
    Prove that $A \oplus B$ is a C*-algebra and that the two projection maps $\pi_1: A \oplus B \to A$ and $\pi_2: A \oplus B \to B$ are *-homomorphisms.
    Similarly one defines the direct sum $A_1 \oplus \cdots \oplus A_n$ of finitely many C*-algebras, or even infinitely many. If I is an index set, and $\{A_i\}_{i \in I}$ is a family of C*-algebras, we let $\oplus_{i \in I} A_i$ be the C*-algebra completion of the collection of I-tuples $(a_i)_{i \in I}$ with $a_i \neq 0$ for at most finitely many $i \in I$, with respect to the norm

$$\|(a_i)_{i \in I}\| := \sup_{i \in I} \|a_i\|.$$

This is a C*-algebra, and the coordinate projections

$$\pi_j \colon \oplus_{i \in I} A_i \to A_j$$

are surjective *-homomorphisms. Of course if I is infinite, the direct sum is non-unital, even if all the A_i are unital.

We close with a basic but important C*-algebraic construction.

Definition 1.1.14 Let A be any C*-algebra, possibly non-unital. Let $A^+ = A \oplus \mathbb{C}$ as a vector space. Equip this vector space with the multiplication and adjoint

$$(a, \lambda)(b, \mu) := (ab + \lambda b + \mu a, \lambda \mu), \quad (a, \lambda)^* := (a^*, \bar{\lambda}).$$

If we let $1 = (0, 1) \in A^+$, we can write elements of A^+ in the form $a + \lambda 1$, or just $a + \lambda$. Then the multiplication becomes the "obvious" one on such symbols

$$(a + \lambda)(b + \mu) = ab + \lambda b + \mu a + \lambda \mu.$$

We can identify A with a *-subalgebra of A^+ by the map $a \mapsto (a, 0)$. Moreover, with this identification, if $a + \lambda \in A^+$ and $b \in A$, then $(a + \lambda)b \in A$. Hence A is an *ideal* in A^+, which is also clearly closed under adjoint.

For the norm, we set

$$\|a + \lambda\| := \max\{ \sup_{\|b\| \leq 1} \|(a + \lambda)b\|, \; |\lambda| \}.$$

Exercise 1.1.15 A^+ with the given norm is a unital C*-algebra. Moreover, if A is already unital, then $A^+ \cong A \oplus \mathbb{C}$, where the direct sum is defined in Exercise 1.1.13.

The C*-algebra A^+ so defined is called the *unitization* of A.

Exercise 1.1.16 Let X be a locally compact Hausdorff space. A continuous function on X *vanishes at infinity* if for all $\varepsilon > 0$ there exists a compact subset $K \subset X$ such that $|f(x)| < \varepsilon$ for all $x \in X \setminus K$. Prove that the collection of continuous functions f on X which vanish at infinity, with the supremum norm, is a C*-algebra. It is denoted $C_0(X)$. (It is non-unital.)

More generally, if A is any C*-algebra, $C_0(X, A)$ denotes continuous functions on X valued in A, which "vanish at infinity" in the above sense. Prove that $C_0(X, A)$ is a C*-algebra with norm $\|f\| := \sup_{x \in X} \|f(x)\|$.

Exercise 1.1.17 If X is a locally compact Hausdorff space, then the 1-point compactification X^+ of X is the disjoint union (we write $X \sqcup \{\infty\}$) of X and an additional point, labelled ∞, topologized with open sets the open subsets of X, together with the sets $U_K \cup \{\infty\}$, where $K \subset X$ is a compact subset and $U_K = X \setminus K$.

Thus, X^+ contains X as an open subset:

a) Prove that X^+ is compact Hausdorff. Notice how the *locally compact* assumption on X is used here.

b) Prove that if X is already compact, then $\{\infty\}$ is an isolated point in X^+.

c) Prove that a continuous map $f : X \to Y$ between locally compact Hausdorff
   spaces extends continuously to a map $X^+ \to Y^+$ mapping the points of infinity
   to each other, if and only if the map $f : X \to Y$ is *proper*, i.e., if and only if
   $f^{-1}(K)$ is compact in X for every compact subset $K \subset Y$.
d) Prove that the C*-algebras $C(X^+)$ and $C_0(X)^+$ are canonically isomorphic.

Exercise 1.1.18 A *projection* $p \in A$ in a C*-algebra is an element such that $p^2 = p$
and $p = p^*$. Thus, it is an idempotent, $p^2 = p$, but is also required to be self-
adjoint. Projections will play an important role in C*-algebra theory, particularly in
K-theory.

Let A be the C*-algebra $C(X)$, where X is compact Hausdorff. Show that
projections in $C(X)$ correspond to connected components of X. (Thus, a Cantor
set has many projections.)

Exercise 1.1.19 Show that an element p of a C*-algebra is a projection if and only
if $p^*p = p$.

1.2   The C*-Algebra of Bounded Operators on a Hilbert Space

A *pre-Hilbert* space is a complex vector space H equipped with a map, called an
inner product

$$\langle \cdot, \cdot \rangle : H \times H \to \mathbb{C},$$

which is linear in the second variable, and the conditions

$$\langle \xi, \xi \rangle \geq 0,$$
$$\langle \xi, \eta \rangle = \overline{\langle \eta, \xi \rangle},$$
$$\langle \xi, \xi \rangle = 0 \Rightarrow \xi = 0,$$

for any $\xi \in H$.

It follows that $\langle \cdot, \cdot \rangle$ is *conjugate linear* in the first variable.

Remark 1.2.1 In Hilbert space theory in mathematics, inner products are conventionally linear in the first variable, conjugate linear in the *second* variable, while
in physics they are conventionally conjugate linear in the first variable, linear in
the second. We follow the physicists' convention in this book because it is also the
convention used in Hilbert module theory.

We set

$$\|\xi\| := \sqrt{\langle \xi, \xi \rangle}.$$

The Cauchy–Schwarz inequality asserts that

(1.4) $$|\langle \xi, \eta \rangle| \leq \|\xi\| \|\eta\|.$$

Exercise 1.2.2 Use the Cauchy–Schwarz inequality (1.4) to prove that

$$\|\xi + \eta\| \leq \|\xi\| + \|\eta\|.$$

By the exercise, a pre-Hilbert space is a special case of a normed linear space, and the formula

$$d(\xi, \eta) := \|\xi - \eta\|.$$

defines a *metric* on H.

A *Hilbert space* is a pre-Hilbert space which is also *complete* with respect to the metric d. A Hilbert space is *separable* if it contains a countable dense set.

In this book, we will work exclusively with separable Hilbert spaces, and accordingly, we will place a blanket assumption throughout that all our Hilbert spaces are separable.

We emphasize that the concept of *orthogonality* is essential in Hilbert space theory. Two vectors ξ, η are orthogonal if $\langle \xi, \eta \rangle = 0$. If $W \subset H$ is a subset of a Hilbert space, then the *orthogonal complement of* W is the closed subspace $W^{\perp} := \{\xi \in H \mid \langle \xi, \eta \rangle = 0 \ \forall \eta \in W\}$. The *Pythagorean Law* asserts that $\|\xi + \eta\|^2 = \|\xi\|^2 + \|\eta\|^2$ for any pair of orthogonal vectors ξ, η.

Exercise 1.2.3 Show that if $W \subset H$ is any subset, then $W^{\perp}$ is a closed subspace of H.

Example 1.2.4 Let $H = \mathbb{C}^n$ with inner product

$$\langle (z_1, \ldots, z_n), (w_1, \ldots, w_n) \rangle := \sum_{i=1}^{n} \overline{z_i} w_i.$$

Then H is a finite-dimensional Hilbert space.

For an infinite-dimensional example, take

$$H := l^2(\mathbb{N}) := \{(a_n)_{n=0}^{\infty} \mid \sum_{n=0}^{\infty} |a_n|^2 < \infty\}$$

with inner product

$$\langle (z_n), (w_n) \rangle := \sum_{n=0}^{\infty} \overline{z_n} w_n.$$

We leave it to the reader to check that $l^2(\mathbb{N})$ is complete and separable.

Exercise 1.2.5 Let H be a Hilbert space. Prove the *Parallelogram Equality*:

$$\|\xi + \eta\|^2 + \|\xi - \eta\|^2 = 2(\|\xi\|^2 + \|\eta\|^2).$$

Let H, K be a pair of Hilbert spaces. A linear operator $T: H \to K$ is *bounded* if

$$(1.5) \qquad\qquad\qquad \sup_{\xi \in H, \|\xi\|_H = 1} \|T\xi\|_K$$

is finite, where we have (exceptionally) subscripted the norms with the Hilbert space to which they are attached. If (1.5) is finite, then we write $\|T\|$ for the supremum. It is called the *operator norm* of T. We let $\mathbb{B}(H, K)$ denote all bounded linear operators $H \to K$.

If $H = K$, we just write $\mathbb{B}(H)$.

Exercise 1.2.6 It is a consequence of Zorn's Lemma that every Hilbert space (indeed, every vector space) has a Hamel basis: a basis in the purely algebraic sense of linear algebra. (Such a basis will be uncountable unless the space is finite-dimensional.)

Use this fact to show there exists a linear operator $T: l^2(\mathbb{N}) \to l^2(\mathbb{N})$ which is not bounded.

An extremely important class of bounded operators on a Hilbert space are *orthogonal projections*. We refer to [54] for the proof of the following, or the reader may attempt the proof themselves using the Parallelogram Equality (Exercise 1.2.5).

Proposition 1.2.7 *If H is a Hilbert space and $W \subset H$ is a closed subspace, then for each $\xi \in H$ there exists a unique vector $P\xi \in W$ such that $\|\xi - P\xi\| = \text{dist}(\xi, W)$.*

Moreover, the map P is linear and bounded, $\|P\| \leq 1$, $P^2 = P$, and $\ker(P) = W^\perp$ and $\text{ran}(P) = W$.

The map P is called the *orthogonal projection to W*.

Exercise 1.2.8 Suppose that $P \in \mathbb{B}(H)$, $P^2 = P$ and $P = P^*$. Let $W = \text{ran}(P)$. Prove that W is closed and P is the orthogonal projection to W and that $1 - P$ is orthogonal projection to $W^\perp$.

Exercise 1.2.9 Prove using projections that if W is a closed subspace of H, then $(W^\perp)^\perp = W$.

An *orthonormal set* $\{e_i\}_{i \in I}$ *of vectors* in a Hilbert space is a set of vectors such that $\langle e_i, e_j \rangle = \delta_{ij}$ for all $i, j \in I$. An *orthonormal basis* for H is a maximal orthonormal set of vectors in H. Any orthonormal set $\{e_i\}_{i \in I}$ in H is a subset of an orthonormal basis, by an application of Zorn's Lemma. Hence every Hilbert space has an orthonormal basis.

Exercise 1.2.10 Prove Bessel's inequality: If $\{e_i\}_{i=1}^{\infty}$ is an orthonormal set of vectors in a Hilbert space H, then $\sum_i |\langle e_i, \xi \rangle|^2 \leq \|\xi\|^2$ for all $\xi \in H$. (*Hint.* Let $\xi_n = \xi - \sum_{i=1}^{n} \langle e_i, \xi \rangle \xi$. Note that ξ_n is orthogonal to e_i for $i \leq n$. Apply the Pythagorean theorem to $\|\xi\|^2 = \|\xi_n + \sum_{i=1}^{n} \langle e_i, \xi \rangle e_i\|^2$.)

We leave the proof, a consequence of Bessel's inequality, to the reader.

Proposition 1.2.11 *If $W \subset H$ is a closed subspace, P the orthogonal projection to W, and $(e_i)_{i \in I}$ is an orthonormal basis for W, then the series $\sum_{i \in I} \langle e_i, \xi \rangle e_i$ converges in H to $P\xi$.*

In particular, if $(e_i)_{i \in I}$ is an orthonormal basis for H, then $\xi = \sum_{\in I} \langle e_i, \xi \rangle e_i$.

An *unitary isomorphism* $U \colon H \to K$ between two Hilbert space is a linear bijection such that $\langle U\xi, U\eta \rangle = \langle \xi, \eta \rangle$ for all $\xi, \eta \in H$.

Corollary 1.2.12 *Let H, K be Hilbert spaces with orthonormal bases $\{e_i\}_{i \in I}$ and $\{e'_j\}_{j \in J}$; then any set bijection $\phi \colon I \to J$ induces a unitary isomorphism $H \to K$ mapping e_i to $e'_{\phi(i)}$.*

Therefore, up to unitary isomorphism, a Hilbert space is completely determined by the cardinality of an orthonormal basis.

Exercise 1.2.13 Suppose that H and K are Hilbert spaces and that $H' \subset H$ is a dense linear subspace, and $K' \subset K$ another dense linear subspace:

a) Suppose that $c \geq 0$ a constant such that $\|T\xi\| \leq c\|\xi\|$ for all $\xi \in H'$. Prove that T extends to a bounded operator $\overline{T} \colon H \to K$, with $\|\overline{T}\| \leq c$.

b) Suppose that $c \geq 0$ and $T \colon H' \to K$ is a linear map, such that

$$\text{(1.6)} \qquad\qquad |\langle T\xi, \eta \rangle| \leq c \|\xi\| \cdot \|\eta\|$$

holds for all $\xi \in H', \eta \in K'$. Prove that T extends to a bounded operator $\overline{T} \colon H \to K$, with $\|\overline{T}\| \leq c$.

Exercise 1.2.14 Prove the following about the operator norm:

a) Prove that the operator norm on $\mathbb{B}(H, K)$ is a norm, i.e., that

$$\|\lambda T\| = |\lambda| \cdot \|T\|, \quad \|S + T\| \leq \|S\| + \|T\|, \quad \forall \lambda \in \mathbb{C}, S, T \in \mathbb{B}(H, K)$$

and $\|T\| = 0$ if and only if $T = 0$.

b) Prove that $\|TS\| \leq \|T\| \cdot \|S\|$ for any $S \in \mathbb{B}(H, K)$ and $T \in \mathbb{B}(K, L)$.

c) Prove that $\mathbb{B}(H, K)$ is complete in the operator norm.

d) Prove that if H and K are *finite-dimensional* Hilbert spaces, then any linear operator $T \colon H \to K$ is automatically bounded.

Exercise 1.2.15 Show that if $H = K = \mathbb{C}^n$ and T is a diagonal operator

$$T = \begin{bmatrix} \lambda_1 & & & \\ & \lambda_2 & & \\ & & \ddots & \\ & & & \lambda_n \end{bmatrix}$$

with respect to the standard orthonormal basis, then $\|T\| = \sup_n |\lambda_n|$.

Deduce from the Spectral Theorem for self-adjoint matrices, that if T is self-adjoint, then $\|T\| = \sup\{|\lambda| \mid \lambda$ is an eigenvalue of $T\}$.

Exercise 1.2.16 If T is the operator on $\mathbb{C}^2$ with matrix $\begin{bmatrix} 1 & 1 \\ 1 & 0 \end{bmatrix}$, then at what point ξ of the unit sphere in $\mathbb{C}^2$ is $\|T\xi\|$ maximized?

A *bounded linear functional* on a Hilbert space H is (by definition) a bounded linear operator $L\colon H \to \mathbb{C}$. For an example of such a functional, let $\xi \in H$ be a vector. Let $L_\xi\colon H \to \mathbb{C}$ be defined $L_\xi(\eta) = \langle \xi, \eta \rangle$. Then L_ξ is linear. By the Cauchy–Schwarz inequality,

$$|L_\xi(\eta)| = |\langle \eta, \xi \rangle| \le \|\eta\|\|\xi\|,$$

whence L_ξ is a bounded operator. Moreover, $\|L_\xi\| = \|\xi\|$, as is easily checked.

The *Riesz representation theorem* asserts that if $L\colon H \to \mathbb{C}$ is a bounded linear functional on a Hilbert space, then there is a unique vector $\xi \in H$ such that $L = L_\xi$ (as it is part of standard Hilbert space material we will not prove it—see [54] for a proof).

If $T\colon H \to K$ is bounded linear, where $H = K$, we usually just speak of a "bounded linear operator on H." The bounded operator on H forms an algebra under composition of operators and is denoted $\mathbb{B}(H)$. The following shows that $\mathbb{B}(H)$ has an important adjoint operation.

Lemma 1.2.17 *For any bounded operator $T\colon H \to K$ between two Hilbert spaces H, K, there is a unique bounded operator $T^*\colon K \to H$ such that*

$$\langle \eta, T\xi \rangle = \langle T^*\eta, \xi \rangle$$

holds for all $\xi \in H, \eta \in K$.

The operator T^* is called the *adjoint* of T.

Exercise 1.2.18 Lemma 1.2.17 is proved (below) using the Riesz Representation Theorem. Show that, conversely, the Lemma immediately implies the Riesz Representation Theorem.

Proof Let T be as in the statement. Let $\eta \in H$ and $T_\eta\colon H \to \mathbb{C}$ be defined $T_\eta(\xi) = \langle \eta, T\xi \rangle$. Clearly T_η is linear. By the Cauchy–Schwarz inequality T_η is bounded.

So by the Riesz Representation Theorem there is a unique vector $T^*(\eta)$ such that $T_\eta(\xi) = \langle T^*(\eta), \xi \rangle$. Thus

$$\langle \eta, T\xi \rangle = \langle T^*(\eta), \xi \rangle.$$

That T^* is linear and bounded is left as an exercise. $\square$

Exercise 1.2.19 Show the following properties of the operator adjoint:

- $*$ is conjugate linear.
- $(TS)^* = S^*T^*$.
- $(T^*)^* = T$.
- If $H = \mathbb{C}^n$, $\mathbb{B}(H) \cong M_n(\mathbb{C})$; under this identification, the adjoint T^* of an operator defined above corresponds to the conjugate transpose of a matrix: $(a^*)_{ij} := \overline{a}_{ji}$.

Hence $\mathbb{B}(H)$ is a *-algebra. We next show that when equipped with the operator norm, it is a C*-algebra.

Theorem 1.2.20 $\mathbb{B}(H)$ *is complete in the operator norm, and* $\|T^*T\| = \|T\|^2$ *for any bounded operator T. Hence $\mathbb{B}(H)$ is a C*-algebra.*

Proof We just verify the C*-identity; the other requirements to be a C*-algebra are checked in Exercise 1.2.14. If $T \in \mathbb{B}(H)$, then
(1.7)
$$\|T\xi\|^2 = \langle T\xi, T\xi \rangle = \langle T^*T\xi, \xi \rangle \leq \|T^*T\xi\| \|\xi\| \leq \|T^*T\| \|\xi\|^2 \leq \|T^*\| \|T\| \|\xi\|^2.$$

So

(1.8) $$\|T\|^2 \leq \|T^*T\| \leq \|T^*\| \|T\|.$$

In particular, $\|T\| \leq \|T^*\|$, and by interchanging the roles of T and T^*, we get that $\|T\| = \|T^*\|$.

Now, returning to (1.8), since $\|T\| = \|T^*\|$, we now have $\|T\|^2 \leq \|T^*T\| \leq \|T\|^2$ so equality holds as required. $\square$

Definition 1.2.21 A bounded operator $T \in \mathbb{B}(H)$ is *invertible* if there exists a bounded operator $S \in \mathbb{B}(H)$ such that $TS = 1_H = ST$.

The *spectrum* of $T \in \mathbb{B}(H)$ is the subset of the complex plane of all λ such that $\lambda - T$ is not invertible.

This definition accords with Definition 1.1.2, where T is understood as an element of the algebra $\mathbb{B}(H)$.

Exercise 1.2.22 Prove that $T \in \mathbb{B}(H)$ is invertible if and only if T is surjective and there exists a constant $C > 0$ such that

$$\|Tv\| \geq C\|v\|, \quad \forall v \in H.$$

Exercise 1.2.23 A bounded operator $T \in \mathbb{B}(H)$ is *normal* if $T^*T = TT^*$. Prove that if T is normal, then $\|T^*v\| = \|Tv\|$ for any $v \in H$.

Exercise 1.2.24 Let $(P_n)_{n=1}^{\infty}$ be a sequence of projections in $\mathbb{B}(H)$ with the property that $\lim_{n\to\infty} P_n\xi = \xi$ for all $\xi \in H$. Prove that

$$\|T\| = \sup_n \|P_n T P_n\|$$

for all $T \in \mathbb{B}(H)$.

It is obvious that a closed C*-subalgebra of $\mathbb{B}(H)$ (or more generally, of any C*-algebra) is in its own right a C*-algebra. Thus, we can find a lot of other examples of C*-algebras based on the fact that $\mathbb{B}(H)$ is one, since, for example, whenever $T \in \mathbb{B}(H)$ is a single bounded linear operator, it generates a C*-algebra.

More generally:

Definition 1.2.25 The C*-algebra *generated* by a family $\{T_\lambda\}$ of operators on a Hilbert space H is the smallest C*-algebra containing all the T_λ's.

Since $\mathbb{B}(H)$ is such a C*-algebra, there is at least one, and since the intersection of C*-subalgebras of $\mathbb{B}(H)$ is also a C*-subalgebra, the C*-algebra generated by a family of operators is the intersection of all C*-subalgebras of $\mathbb{B}(H)$ containing them.

Remark 1.2.26 It may or may not be the case that the C*-algebra generated by a family of operators, or even a single operator T, contains the unit $1 \in \mathbb{B}(H)$. Nor are such C*-algebras necessarily unital in an intrinsic sense. The C*-algebra generated by a projection $p \in \mathbb{B}(H)$ is all scalar multiples of p. This C*-algebra has a unit, namely p, but the C*-algebra generated by p does not contain the unit of $\mathbb{B}(H)$. On the other hand, if T is the operator M_f on $L^2(\mathbb{R})$ of multiplication by $f(x) = \frac{1}{x+i}$, then the C*-algebra generated by M_f is isomorphic to $C_0(\mathbb{R})$ (by the Stone–Weierstrass Theorem) which is not unital.

We will generally be specific about it if we want a C*-algebra defined by generators to be unital or not.

Example 1.2.27 Let $T \in \mathbb{B}(H)$ be a *self-adjoint operator*, $T^* = T$. We denote by $C^*(T)$ the C*-algebra generated by $\{T\}$ and denote by $C^*(1, T)$ the (unital) C*-algebra generated by T and the unit $1 \in \mathbb{B}(H)$. It is clear that $C^*(1, T)$ is the closure in $\mathbb{B}(H)$ of the *-algebra of polynomials

$$\sum_{k=0}^{n} \lambda_k T^k$$

in T, with complex coefficients, where by T^0 we understand the unit $1 \in \mathbb{B}(H)$. It is commutative and unital. And $C^*(T)$ is the closure of the polynomials $\sum_{k=0}^{n} \lambda_k T^k$, where $\lambda_0 = 0$.

The *Spectral Theorem* gives a complete analysis of $C^*(T)$ and $C^*(1, T)$, purely in terms of the spectrum of T. In finite dimensions, this boils down to the standard results on diagonalization.

Since T is self-adjoint, all the eigenvalues of T are real, and there is an orthonormal basis of H consisting of eigenvectors of T. If the eigenvalues of T are $\lambda_1, \ldots, \lambda_n$, we may write T as a diagonal matrix organized into (diagonal) blocks

$$
T = \begin{bmatrix} \lambda_1 & & & \\ & \lambda_2 & & \\ & & \ddots & \\ & & & \lambda_n \end{bmatrix},
$$

where each λ_i is understood to mean the corresponding multiple of the identity operator on the λ_i-eigenspace $\ker(\lambda_i - T)$. The spectral information is more than enough to determine completely the isomorphism class of $C^*(1, T)$. To see this, observe that $\prod_{j \neq i}(T - \lambda_j)$ is in $C^*(1, T)$ (remember that we have by definition included the unit $1 \in \mathbb{B}(H)$ in $C^*(1, T)$, so $\lambda_j := \lambda_j \cdot 1$ is in $C^*(T)$) and is the diagonal matrix with the nonzero scalar $\mu_i := \prod_{j \neq i}(\lambda_i - \lambda_j)$ in the ith block. This shows us that $P_i := \frac{1}{\mu_i} \prod_{j \neq i}(T - \lambda_j)$, the orthogonal projection onto the ith eigenspace, is in $C^*(T)$. Each P_i generates a one-dimensional *-subalgebra of $C^*(1, T)$ (namely, consisting of all scalar multiples of P_i) and $C^*(1, T)$ is the direct sum of these one-dimensional *-subalgebras, each of which, of course, is isomorphic to $\mathbb{C}$.

Hence $C^*(1, T) \cong \mathbb{C}^n$ where n is the number of distinct eigenvalues of T.

Exercise 1.2.28 Two operators $T_1 \in \mathbb{B}(H_1)$ and $T_2 \in \mathbb{B}(H_2)$ on Hilbert spaces H_1 and H_2 are *unitarily conjugate* if there is a unitary operator $u : H_1 \to H_2$ such that $\mathrm{Ad}_u(T) := uT_1u^* = T_2$.

Show that if T_1 and T_2 are unitarily conjugate, then $C^*(1, T_1) \cong C^*(1, T_2)$ by an isomorphism Ad_u taking T_1 to T_2.

Exercise 1.2.29 Let T_1 and T_2 be self-adjoint operators on a pair of finite-dimensional Hilbert spaces, with respective eigenvalue sets $\mathrm{Spec}(T_1)$ and $\mathrm{Spec}(T_2)$:

a) $C^*(T_1) \cong C^*(T_2)$ if and only if T_1 and T_2 have the same *number* of distinct eigenvalues.
b) $C^*(1, T_1) \cong C^*(1, T_2)$ by an isomorphism sending T_1 to T_2 if and only if T_1 and T_2 have the *same* (sets of) eigenvalues.
c) $C^*(1, T_1)$ and $C^*(1, T_2)$ are isomorphic by a unitary conjugacy with $uT_1u^* = T_2$ if and only if $\mathrm{Spec}(T_1) = \mathrm{Spec}(T_2)$ and if $\ker(\lambda - T_1)$ and $\ker(\lambda - T_2)$ have the same dimension for all $\lambda \in \mathrm{Spec}(T_1) = \mathrm{Spec}(T_2)$.

We close this chapter with some remarks on bounded self-adjoint operators. As mentioned above, a bounded operator T is *self-adjoint* if $T = T^*$.

Exercise 1.2.30 Show that if T is a bounded self-adjoint operator, then $\langle T\xi, \xi \rangle \in \mathbb{R}$ for all vectors ξ.

Lemma 1.2.31 *If T is a bounded, self-adjoint operator on a Hilbert space, then*

$$\|T\| = \sup_{\xi \in H, \|\xi\|=1} |\langle T\xi, \xi \rangle|.$$

Proof By the Cauchy–Schwarz inequality,

$$(1.9) \qquad\qquad |\langle T\xi, \xi \rangle| \le \|T\| \|\xi\|^2$$

for any vector $\xi \in H$, so if we let

$$M := \sup_{\xi \in H, \|\xi\|=1} |\langle T\xi, \xi \rangle|,$$

then $M \le \|T\|$, and we need to show that equality holds.

For T self-adjoint, $\langle T\xi, \xi \rangle$ is real, since $\overline{\langle T\xi, \xi \rangle} = \langle \xi, T\xi \rangle = \langle T^*\xi, \xi \rangle = \langle T\xi, \xi \rangle$.

If now ξ and η are two unit vectors in H, then it follows from a little algebra that

$$\langle T(\xi \pm \eta), \xi \pm \eta \rangle = \langle T\xi, \xi \rangle \pm 2\langle T\xi, \eta \rangle + \langle T\eta, \eta \rangle.$$

Subtracting one of these equations from the other and using the fact that the equalities only involve real numbers (by Exercise 1.2.30) give

$$\langle T(\xi + \eta), \xi + \eta \rangle - \langle T(\xi - \eta), \xi - \eta \rangle = 4\mathrm{Re}\langle T\xi, \eta \rangle.$$

By the triangle inequality and the Cauchy–Schwarz inequality, we get

$$|\langle T(\xi+\eta), \xi+\eta \rangle - \langle T(\xi-\eta), \xi-\eta \rangle| \le |\langle T(\xi+\eta), \xi+\eta \rangle| + |\langle T(\xi-\eta), \xi-\eta \rangle|$$
$$\le M\|\xi + \eta\|^2 + M\|\xi - \eta\|^2.$$

The last inequality uses (1.9).

By the Parallelogram Equality (Exercise 1.2.5), $\|\xi + \eta\|^2 + \|\xi - \eta\|^2 = 2\|\xi\|^2 + 2\|\eta\|^2$. Since ξ and η were assumed unit vectors, this gives, putting everything together, that $4|\mathrm{Re}\,\langle T\xi, \eta \rangle| \le 4M$, so

$$(1.10) \qquad\qquad |\mathrm{Re}\,\langle T\xi, \eta \rangle| \le M$$

for any pair of unit vectors $\xi, \eta \in H$.

Now for a suitable complex number, $e^{i\theta}$, we have $e^{i\theta}\langle T\xi, \eta \rangle$ is real and positive. This equals $\langle T\xi, e^{i\theta}\eta \rangle$. Since (1.10) holds for all unit vectors η and in particular for $e^{i\theta}\eta$ for any η, $|\langle T\xi, \eta \rangle| \le M$ holds for any unit vectors ξ, η. If in this expression we put $\eta = \frac{T\xi}{\|T\xi\|}$ for an arbitrary unit vector ξ, we get that $\|T\xi\| \leq M$, whence, taking sup over ξ, we get $\|T\| \leq M$ as required.   □

Exercise 1.2.32 Prove that if $T \in \mathbb{B}(H)$ is a bounded linear operator, then $\ker(T^*)^{\perp} = \overline{\mathrm{ran}}(T)$.

Exercise 1.2.33 This exercise addresses *polar decompositions* of operators $T \in \mathbb{B}(H)$ on a Hilbert space:

a) A *partial isometry* $u : H \to H$ is a bounded operator which is an isometry from $\ker(u)^{\perp}$ to its range. Prove that u is a partial isometry if and only if $p := uu^*$ and $q := u^*u$ are projections (that is, $p^2 = p$ and $p = p^*$, and similarly for q).
b) Prove that if u is a partial isometry, then $\mathrm{ran}(u) = \mathrm{ran}(uu^*)$ and $\ker(u) = \ker(u^*u)$.
c) If T is a bounded operator on a Hilbert space, let $|T| := (T^*T)^{\frac{1}{2}}$. The existence of such "operator square roots" requires some spectral theory, but for purposes of the exercise, the reader may assume for the moment merely that $|T|$ is an operator such that $|T|^2 = T^*T$. Prove that $\|T\xi\| = \||T|\xi\|$ for any $\xi \in H$. (In particular, $\ker(|T|) = \ker(T)$ follows.)
d) Show that the restriction of $|T|$ to $\ker(T)^{\perp}$ maps into $\ker(T)^{\perp}$ and that the range of this restricted operator is dense in $\ker(T)^{\perp}$. (*Hint.* Note first that $\mathrm{ran}(T^*T) \subset \mathrm{ran}((|T|) = \ker(|T|)^{\perp}$, since $|T|^2 = T^*T$. On the other hand $\mathrm{ran}(|T|) = \ker(|T|)^{\perp} = \ker(T)^{\perp}$.)
e) Define $u: \ker(T)^{\perp} \to \mathrm{ran}(T)$ by defining it on the dense subspace $\mathrm{ran}(|T|) \subset \ker(T)^{\perp}$ in the way it has to be defined, i.e., so that $T\xi = U|T|\xi$. Show that this densely defined operator is *isometric*.
f) Complete the proof of the Polar Decomposition theorem: that if $T \in \mathbb{B}(H)$ is any bounded operator on H, then there exists a partial isometry u with initial space $\ker(T)^{\perp}$ and final space $\overline{\mathrm{ran}}(T)$ such that $T = U|T|$.

Exercise 1.2.34 Let $T: H \to H$ be a self-adjoint operator on a finite-dimensional Hilbert space. Let $S \subset H$ be the unit sphere in H. By maximizing the functional

$$f: S \to \mathbb{C}, \quad f(v) = \langle Tv, v \rangle,$$

and Lemma 1.2.31, show that T has at least one eigenvalue.

Many bounded operators (e.g., convolution operators on groups) which arise from geometry are *integral* operators in the sense of the following Exercise.

Exercise 1.2.35 Let $1 \leq p \leq \infty$. Assume that (X, μ) is a σ-finite measure space and let k be a measurable function on $X \times X$. Suppose there exists $C \geq 0$ such that $\int_X |k(x, y)| d\mu(x) \leq C$ for a.e. $y \in X$, and that $\int_X |k(x, y)| d\mu(y) \leq C$ for a.e. $x \in X$.

Then if $f \in L^p(X)$, then the integral $(I_k f)(x) := \int_X k(x, y) f(y) d\mu(y)$ exists a.e. $x \in X$, $I_k f$ lies in $L^p(X)$, and $\|I_k f\|_p \leq C\|f\|_p$.

In particular, I_k defines a bounded operator on the Hilbert space $L^2(X)$.

(*Hint.* If $1 < p < \infty$, write

$$|k(x, y)f(y)| = |k(x, y)|^{\frac{1}{q}} \cdot \left(|k(x, y)|^{\frac{1}{p}}|f(y)|\right)$$

and apply the Hölder inequality.)

1.3   Group C*-Algebras

In this section, we explain an important construction that associates a C*-algebra to a (locally compact) *group*. The C*-algebra's structure reflects the representation theory of the group. We will give the general definition, but for more detailed theorems and proofs we will specialize to particular classes of groups.

In the second part of this section, we discuss the examples where the group is the integers and the circle, respectively, and in Section 1.5 we analyze the case of finite groups. We will defer a discussion of the group $\mathbb{R}$ of real numbers, which involves a lot of other considerations, to Section 1.9.

Any locally compact group has a Borel measure μ which is left translation-invariant, in the sense that the left translation maps $G \to G$ elements of G, are measure-preserving. Such a measure is called Haar measure. It is unique up to a positive scalar multiple. It follows from left translation invariance that $\int_G f(gh)d\mu(h) = \int_G f(h)d\mu(h)$.

If G is compact, we will always assume that μ is normalized so that $\mu(G) = 1$. We frequently use the abbreviated notation $\int_G f(g)dg$ for $\int_G f(g)d\mu(g)$.

Let f be a continuous function on G of compact support on G. Then f determines a *convolution operator* $\lambda(f) \colon C_c(G) \to C_c(G)$,

$$(1.11) \qquad \left(\lambda(f)\xi\right)(g) := \int_G f(h)\xi(h^{-1}g)dh.$$

The integrand is continuous and compactly supported so converges absolutely.

Exercise 1.3.1 Verify that if $f, \xi \in C_c(G)$, then $\lambda(f)\xi$ is continuous, and if supp$(f) = K \subset G$ and supp$(\xi) = L \subset G$, K, L compact, then supp$(\lambda(f)\xi) \subset K \cdot L$. In particular, $\lambda(f)\xi \in C_c(G)$.

Proposition 1.3.2 *If $f \in C_c(G)$, then $\lambda(f)$ extends continuously to a bounded operator $L^2(G) \to L^2(G)$, and $\|\lambda(f)\| \leq \|f\|_{L^1(G)}$.*

Proof Let ξ, η be a pair of compactly supported continuous functions on G. Then by the definitions

$$(1.12) \qquad \langle \lambda(f)\xi, \eta \rangle = \int_G \left(\int_G \overline{f(h)\xi(h^{-1}g)} \, dh\right) \eta(g) \, dg$$

and we can switch the order of integration by Fubini's Theorem, giving

$$(1.13) \qquad = \int_G \overline{f(h)} \left(\int_G \overline{\xi(h^{-1}g)} \eta(g) dg \right) dh.$$

Set $U_h \xi(g) := \xi(h^{-1}g)$. Then $U_h \xi \in L^2(G)$ and $\|U_h \xi\| = \|\xi\|$ (U_h is a *unitary*.) This is because of the left translation invariance of Haar measure. For each $h \in G$, the Cauchy–Schwarz inequality gives

$$\left| \int_G \overline{\xi(h^{-1}g)} \eta(g) dg \right| = |\langle U_h \xi, \eta \rangle| \leq \|U_h \xi\| \cdot \|\eta\| = \|\xi\| \cdot \|\eta\|.$$

Putting everything together, we get

$$(1.14) \quad |\langle \lambda(f)\xi, \eta \rangle| \leq \int_G |f(h)| \cdot |\int_G \overline{\xi(h^{-1}g)} \eta(g) dg| dh$$

$$\leq \|\xi\| \cdot \|\eta\| \cdot \int_G |f(h)| dh = \|f\|_{L^1(G)} \cdot \|\xi\| \cdot \|\eta\|.$$

Now set

$$\eta := \lambda(f)\xi,$$

η is then compactly supported and continuous and so by above

$$(1.15) \qquad \|\lambda(f)\xi\|^2 = \langle \lambda(f)\xi, \lambda(f)\xi \rangle \leq \|f\|_{L^1(G)} \cdot \|\xi\| \cdot \|\lambda(f)\xi\|,$$

divide both sides by $\|\lambda(f)\xi\|$ to get

$$(1.16) \qquad \|\lambda(f)\xi\| \leq \|f\|_{L^1(G)} \cdot \|\xi\|.$$

Since compactly supported ξ are dense in $L^2(G)$, (1.16) implies that $\lambda(f)$ extends continuously a linear map $L^2(G) \to L^2(G)$, with operator norm $\leq \|f\|_{L^1(G)}$, as claimed. □

Definition 1.3.3 The C*-algebra generated by the convolution operators $\lambda(f)$, as f ranges over $C_c(G)$, is the *C*-algebra of G* and is written $C^*(G)$.

The C*-algebra of a group G is a completion of an intrinsically defined *-algebra, namely of $C_c(G)$ under *convolution* of functions (not pointwise multiplication). Convolution is defined for $f_1, f_2 \in C_c(G)$, by

$$(1.17) \qquad (f_1 * f_2)(g) := \int_G f_1(h) f_2(h^{-1}g) \, d\mu(h).$$

It is clear that $f_1 * f_2 \in C_c(G)$ if $f_1, f_2 \in C_c(G)$. It is easy to check that convolution is an associative bilinear operation.

The *modular function* on a locally compact group G is the function $\delta \colon G \to \mathbb{R}_+^*$ defined by the equation

$$\mu(Ag) = \delta(g^{-1}) \cdot \mu(A)$$

for any $A \subset G$ measureable, with μ Haar measure on G. If $f \colon G \to \mathbb{C}$ is an integrable function, then

$$\delta(h) \cdot \int_G f(gh) \, d\mu(g) = \int_G f(g) \, d\mu(g).$$

The map δ is a group homomorphism. A group is *unimodular* if $\delta = 1$. Almost all the groups we will consider in this book are unimodular: For example, discrete groups, compact groups, and abelian groups are all unimodular.

The reason we bring up the modular function is that the operator adjoint on $f \in C_c(G) \subset \mathbb{B}(L^2(G))$ corresponds to the following adjoint, intrinsically defined on $C_c(G)$:

(1.18) $f^*(g) = \delta(g^{-1}) \cdot \overline{f(g^{-1})}.$

Exercise 1.3.4 Check that the convolution formula (1.17) matches composition of operators on $L^2(G)$, i.e., that

$$\lambda(f_1 * f_2) = \lambda(f_1)\lambda(f_2) \in \mathbb{B}(L^2(G)).$$

Check as well that the adjoint (1.18) agrees with the operator adjoint in $\mathbb{B}(L^2(G))$.

The following exercise gives an example of a non-unimodular group.

Exercise 1.3.5 Let G be the upper triangular group

$$G := \{g = \begin{bmatrix} a & b \\ 0 & a^{-1} \end{bmatrix} \mid a, b \in \mathbb{R}, \ a > 0\} \subset \mathbf{SL}_2(\mathbb{R})$$

of matrices in $\mathbf{SL}_2(\mathbb{R})$. It has evident coordinates a, b putting it into bijective correspondence with $\mathbb{R}_+^* \times \mathbb{R}$. Compute the Haar measure on G in these coordinates and compute the modular function

$$\delta \colon G \to \mathbb{R}_+^*.$$

Proceeding with the general theory, the regular representation λ can be viewed as a *-algebra map $C_c(G) \to \mathbb{B}(L^2(G))$. Hence $C^*(G)$ is the completion of the *-algebra $C_c(G)$ (that is, with its intrinsically defined convolution multiplication, and adjoint), with the operator norm via the representation λ.

In the important case where G is a *discrete* group, we can represent an element of $C_c(G)$, often alternatively denoted $\mathbb{C}[G]$, and called the *complex group algebra* of G, in the form

$$f = \sum_{g \in G} a_g[g],$$

where the sum is finite, each $a_g \in \mathbb{C}$, and where $[g]$ means point mass at g: the function, often written δ_g, which is 1 at g and 0 otherwise.

Thus, f is a function on G whose value at g is a_g.

If we are considering f as an operator on $l^2(G)$, by the regular representation λ, then $\lambda([g])$ (or $\lambda(g)$) means the unitary operator of left translation by g, $\lambda([g])\xi(h) = \xi(g^{-1}h)$.

It is easy to check that convolution multiplication in this notation becomes the "obvious" multiplication of such expressions, using the rules that

$$[g] * [h] = [gh] \quad [g]^* = [g^{-1}].$$

Thus, for example

$$\left(\sum_{g \in G} a_g[g]\right) * \left(\sum_g b_g[g]\right) = \sum_{g,h} a_g b_h[gh],$$

and rearranging the sum gives $= \sum_{g \in G} \left(\sum_{h \in G} a_h b_{h^{-1}g}\right)[g]$, which reproduces the convolution formula (1.17).

Similarly, the adjoint is given by $\left(\sum_{g \in G} a_g[g]\right)^* = \sum_{g \in G} \overline{a_g}[g^{-1}]$.

Example 1.3.6 Let G be the finite cyclic group $\mathbb{Z}/n\mathbb{Z}$.

Then the reduced C*-algebra $C^*(G) = \mathbb{C}[G]$ is finite-dimensional, since $\mathbb{C}[G]$ is, and consists of all combinations $\sum_{k=0}^{n-1} \lambda_k[k + n\mathbb{Z}]$.

The regular representation $\lambda \colon C^*(G) \to \mathbb{B}(l^2 G) \cong \mathbb{B}(\mathbb{C}^n) \cong M_n(\mathbb{C})$ maps the generator $1 + n\mathbb{Z}$ to the shift matrix

$$U := \begin{bmatrix} 0 & \cdots & \cdots & \cdots & 1 \\ 1 & 0 & \cdots & & 0 \\ 0 & 1 & 0 & & \vdots \\ \vdots & & \ddots & & \vdots \\ 0 & \cdots & & 1 & 0 \end{bmatrix}.$$

Thus, as a *-subalgebra of $M_n(\mathbb{C})$, $C^*(G)$ is generated by the shift and consists of all operators of the form $\sum_{k=0}^{n-1} \lambda_k U^k$.

To understand the structure of this C*-algebra, note that the shift U acting on $\mathbb{C}^n$ can be diagonalized. Its eigenvalues are precisely the nth roots of unity,

$1, \omega, \omega^2, \ldots \omega^{n-1}$, and the eigenvector v_k corresponding to the eigenvalue ω^k is given by

$$v_k = (\omega^{-k}, \omega^{-2k}, \omega^{-3k}, \ldots, \omega^{-nk}) \in \mathbb{C}^n.$$

Let χ_k be the *character*, that is, group homomorphism to the circle group $\mathbb{T}$, given by

$$\chi_k : \mathbb{Z}/n\mathbb{Z} \to \mathbb{T}, \quad \chi_k(m + n\mathbb{Z}) := \omega^{km}.$$

Since χ_k is a function on the group, it defines an element of $\mathbb{C}[\mathbb{Z}/n\mathbb{Z}] = C^*(\mathbb{Z}/n\mathbb{Z})$. In group algebra notation

$$\chi_k = \sum_m \omega^{km} \cdot [m] \in \mathbb{C}[\mathbb{Z}/n\mathbb{Z}].$$

Exercise 1.3.7 Show that $\lambda(\chi_k)$ acts on $l^2(\mathbb{Z}/n\mathbb{Z})$ by the matrix

$$p_k = \begin{bmatrix} 1 & \omega^{-k} & & & & \ddots \\ \omega^k & 1 & \omega^{-k} & & & \\ \omega^{2k} & \omega^k & 1 & \omega^{-k} & & \\ & \omega^{2k} & \ddots & & & \\ & & \ddots & & & \omega^{-k} \\ & & & \omega^{2k} & \omega^k & 1 \end{bmatrix},$$

(where the ellipses indicate a similar pattern; the matrix has "constant diagonals") and check that the matrix

$$p_k := \frac{1}{n} \cdot \lambda(\chi_k)$$

is orthogonal projection to the eigenspace spanned by v_k.

Thus, the group C*-algebra $C^*(\mathbb{Z}/n)$ contains n orthogonal projections $p_0, p_1, \ldots, p_{n-1}$ which sum to the identity operator. And in this notation $U = \sum_{k=0}^{n-1} \omega^k p_k$ expresses U as a diagonal operator with respect to the basis $v_0, \ldots, v_{n-1}$: With respect to this basis, U is the diagonal matrix

(1.19)
$$U = \begin{bmatrix} 1 & & & \\ & \omega & & \\ & & \omega^2 & \\ & & & \omega^{n-1} \end{bmatrix}.$$

Hence the C*-algebra $C^*(U) = C^*(\mathbb{Z}/n\mathbb{Z})$ generated by U consists of diagonal matrices in this basis. We have shown that this C*-algebra contains, as well, the projections to the elements of this basis, and hence $C^*(\mathbb{Z}/n)$ is isomorphic to the C*-algebra of n-by-n diagonal matrices—i.e., is isomorphic to $\mathbb{C}^n$ as C*-algebras.

The representation theory of the group $\mathbb{Z}/n\mathbb{Z}$ is thus reflected by the structure of the group C*-algebra $C^*(\mathbb{Z}/n\mathbb{Z})$.

Exercise 1.3.8 Let $G = \mathbb{T}$ be the circle:

a) Show that the characters $\chi_n(z) := z^n$, viewed as elements of $C^*(\mathbb{T})$, are projections: $\chi_n = \chi_n^*$ and $\chi_n * \chi_n = \chi_n$, and that $\chi_n * \chi_m = 0$ unless $n = m$.
b) Prove that the infinite sum $\sum_{n \in \mathbb{Z}} \chi_n$ does *not* converge in the norm in $C^*(\mathbb{T})$, but that $\sum_{n \in \mathbb{Z}} f * \chi_n$ *does* converge (in norm) in $C^*(\mathbb{T})$, for any $f \in C^*(\mathbb{T})$, with $f * \chi_n$ convolution multiplication (the multiplication in $C^*(\mathbb{T})$).

Exercise 1.3.9 Let G be a finite group and $f \in \mathbb{C}[G] = C^*(G)$ a function which is constant on conjugacy classes in G. (Such functions are called *class functions*.) Prove that f is in the center of $C^*(G)$, that is, f commutes with all other elements of $C^*(G)$.

Exercise 1.3.10 Let G be any discrete group:

a) In $l^2(G)$, let $\tau : \mathbb{C}[G] \to \mathbb{C}$ be the map $\tau(\sum \lambda_{[g]}) := \lambda_e$. Prove that τ extends to a continuous linear functional $C^*(G) \to \mathbb{C}$, and that τ is a *trace*: $\tau(ab) = \tau(ba)$ for all $a \in C^*(G)$. (*Hint.* Let $e_0 \in l^2(G)$ be the point mass at the identity of the group, and check that $\tau(T) = \langle \lambda(T)e_0, e_0 \rangle$ for all $T \in \mathbb{C}[G]$.)
b) Prove that if $T \neq 0$ is an element of $C^*(G)$ of the form $T = S^*S$, for some $S \in C^*G$, then $\tau(T) > 0$.
c) Prove that the map $\lambda : \mathbb{C}[G] \to C^*(G)$ is an injective map of *-algebras.
d) Suppose that $H \subset G$ is a finite subgroup. Let

$$p := \frac{1}{|H|} \sum_{h \in H} [h].$$

Show that p is a projection in C^*G), and that $\tau(p) = \frac{1}{|H|}$.
e) Show that p_H is projection onto the subspace of $l^2(G)$ consisting of functions $\xi \in l^2(G)$ which are constant on each right H-coset in G (so for example if G is finite and $H = G$, then p_G is projection to the constant functions in $l^2(G)$).

It is a conjecture of Kadison and Kaplansky that if G has no torsion, then τ only takes integral values on projections. The problem is as yet unsolved, with some of the strongest results derived from positive cases of the Baum–Connes conjecture (discussed in the last section of the book).

Exercise 1.3.11 Let G be a discrete group:

a) Show that if $f = \sum f(g)[g] \in \mathbb{C}[G]$, then

$$\langle e_h, \lambda(f)e_k \rangle = f(hk^{-1}).$$

b) If $T \in C^*(G)$, set $\xi_T := T(e_o) \in l^2(G)$, where $e_o \in l^2(G)$ is point mass at the
identity of the group. Show that $\xi_{\lambda(f)} = f$ regarded as a function in $l^2(G)$, for
$f \in C_c(G) = \mathbb{C}[G] \subset C^*(G)$.

c) Show that the map $T \mapsto \xi_T$ is a norm contractive map $C^*(G) \to l^2(G)$, which
recovers f from $\lambda(f)$, for f a finitely supported function on G.

d) Show that if $\xi_T = 0$, then $T = 0$. (*Hint.* The statement is obvious if $T = \lambda(f)$
for $f \in \mathbb{C}[G]$. To prove the stronger statement using limits, supposed $T_n \to T$
in $C^*(G)$, where $T_n = \lambda(f_n)$, $f_n \in \mathbb{C}[G]$. Then $\xi_{T_n} \to \xi_T$ (part c)), so if $\xi_T = 0$,
then $\xi_{T_n} = f_n \to 0$ in $l^2(G)$ and in particular $f_n \to 0$ pointwise on G. By part
a) $\langle e_h, \lambda(f_n))e_k \rangle = f_n(hk^{-1}) \to 0$ for h, k fixed and $n \to \infty$. From Cauchy–
Schwarz $\langle e_h, \lambda(f_n))e_k \rangle \to \langle e_h, Te_k \rangle$, whence $\langle e_h, Te_k \rangle = 0$ for all h, k.)

e) If $T \in C^*(G)$, let $f = \xi_T$. Observe that the convolution $f * \xi$ is well defined for
$\xi \in C_c(G)$ and f *any* function (confirm), producing a function on G. With this
convention, verify that

$$T\eta = f * \eta$$

for all $\eta \in C_c(G)$.

Thus one can describe $C^*(G)$ fairly concretely as consisting of a priori densely
defined convolution operators $\lambda(f)\xi := f * \xi$, $*f \in l^2(G)$, which are defined on
$C_c(G)$, for which the norm estimate

$$\|f * \xi\|_{L^2(G)} \le C\|\xi\|_{L^2(G)}$$

holds for all $\xi \in C_c(G)$. The estimate then implies that $\lambda(f)$ extends to a bounded
operator on $l^2(G)$, even though $f * \xi$ may not actually be defined as a convergent
integral if $\xi \in l^2(G)$ is more general than a finitely supported function.

Exercise 1.3.12 If H is a Hilbert space and $A \subset \mathbb{B}(H)$ is any self-adjoint set of
bounded operators on H, then the *commutant* A' of A is the collection of bounded
operators T on H which commute with all elements of A:

a) Prove that A' is a norm-closed *-subalgebra of $\mathbb{B}(H)$ and hence is a C*-algebra.
b) Check that if G is finite and

$$\rho(g): l^2(G) \to l^2(G), \quad \rho(g)(e_h) := e_{hg^{-1}},$$

is the right regular representation, then $C^*(G) = \{\rho(g) \mid g \in G\}'$ and that
$C^*(G)$ is unitarily conjugate to $C^*(G)'$.

Remark 1.3.13 If G is infinite, the commutant $W^*(G) = \{\rho(g) \mid g \in G\}'$ is called
the *group von Neumann algebra of G*. It contains $C^*(G)$ as a dense subalgebra but
is closed in a much finer topology. The relation between the C*-algebra and the von

Neumann algebra is analogous to the relation between $C_0(X)$ and $L^\infty(X, \mu)$ for a locally compact space with a Borel measure μ.

Exercise 1.3.14 Prove that a bounded operator $T \in \mathbb{B}(l^2 G)$ is in the group von Neumann algebra $L(G)$ if and only if its matrix representation in the canonical basis $\{e_g \mid g \in G\}$ has constant "diagonals" (the "diagonals" are the vectors $(T_{k,gk})_{k \in G}$, one for each g, where $T_{k,h} := \langle T(e_h), e_k \rangle$ as usual).

Exercise 1.3.15 If G is a discrete group, then verify that $C^*(G) \subset \rho(G)'$, where $\rho \colon G \to \mathbf{U}(l^2)$ is the right regular representation.

Exercise 1.3.16 Let G be any locally compact group. An alternative to forming the C*-algebra of G is to form the *Banach algebra* $L^1(G)$.

Use the Fubini–Tonelli Theorem to prove that if f_1, $f_2 \in C_c(G)$, then $f_1 * f_2 \in L^1(G)$ and

$$\|f_1 * f_2\|_{L^1(G)} \leq \|f_1\|_{L^1(G)}\|f_2\|_{L^1(G)}.$$

Deduce that convolution extends continuously to a multiplication on $L^1(G)$ and that with this multiplication and the L^1-norm, $L^1(G)$ is a Banach algebra.

Actually, $L^1(G)$ has a natural adjoint as well, and it is easy to check that $\|f^*\|_{L^1(G)} = \|f\|_{L^1(G)}$. Hence $L^1(G)$ has the structure of a Banach *-algebra. The regular representation λ extends continuously to a contractive map $L^1(G) \to C^*(G)$ of Banach *-algebras.

Exercise 1.3.17 Suppose that $\pi \colon G \to \mathbf{U}(H)$ is a unitary representation of a discrete group G. It extends linearly to a *-algebra map (for which we use the same letter)

(1.20) $$\pi \colon \mathbb{C}[G] \to \mathbb{B}(H),$$

and the question addressed here is whether (1.20) extends to a C*-algebra homomorphism $C^*(G) \to \mathbb{B}(H)$. If it does, we say that π is *weakly contained in the regular representation of G*:

a) Show that the diagonal action of G on the direct sum $\oplus_{i \in I} l^2(G)$ of an arbitrary number of copies of the left regular representation is weakly contained in the regular representation.

b) Let X be a discrete G-space on which G acts freely. Let $\pi \colon G \to \mathbf{U}(l^2(X))$ be the resulting representation

$$(\pi(g)\xi)(x) = \xi(g^{-1}x).$$

Prove that π is weakly contained in the regular representation.

c) Prove that the trivial representation $\varepsilon \colon \mathbb{Z} \to \{1\} \subset \mathbb{T}$ of the integers is weakly contained in the regular representation.

The condition that the trivial representation of a discrete group G is weakly contained in the regular representation is sufficiently important we record it in a definition.

Definition 1.3.18 If G is a countable group, then G is *amenable* if the trivial representation extends continuously to $C^*(G)$, equivalently, if

$$|\varepsilon(f)| \leq \|\lambda(f)\|$$

for all $f \in \mathbb{C}[G]$, where ε is the trivial representation.

The group $\mathbb{Z}$ is amenable, but the free group $\mathbb{F}_2$ on two generators is not, and so the trivial representation of $\mathbb{F}_2$ does not extend to $C^*(\mathbb{F}_2)$.

Exercise 1.3.19 Let G be a discrete, amenable group. Prove that if $S \subset G$, then

$$\left\| \sum_{s \in S} \lambda(s) \right\| = |S|,$$

where λ is the regular representation.

We close this section with the following remark.

What we are calling the C*-algebra of G in this book is usually called the *reduced C*-algebra* of G in the literature. There are other ways of completing $L^1(G)$ to a C*-algebra. Such completions come from injective unitary representations of the group, and these are not all equivalent to each other, that is, they can produce different norms, and when one completes, one can obtain different C*-algebras.

But the "reduced" C*-algebra has a certain concreteness about it and is important for other reasons as well.

1.4   C*-Algebras of the Integers and the Circle

The structure of the C*-algebras of the groups of the integers and the group consisting of the points of the unit circle $\mathbb{T}$ of the complex plane is completely elucidated by the classical *Fourier transform*, which we describe in slightly more general terms as follows.

If G is a locally compact, second countable, abelian group, its *Pontryagin dual* is the group $\widehat{G}$ of continuous group homomorphisms $\chi : G \to \mathbb{T}$ (they are called *characters*). The group structure on $\widehat{G}$ is by pointwise multiplication of characters $(\chi_1 \cdot \chi_2)(g) = \chi_1(g)\chi_2(g) \in \mathbb{T}$. The identity element is the trivial character, which we will usually denote $\varepsilon : G \to \mathbb{T}$, $\varepsilon(g) = 1 \in \mathbb{T}$ for all g. If χ is a character, then $\chi^*(g) := \overline{\chi(g)}$ is also a character, since complex conjugation on the circle is a (continuous) group homomorphism $\mathbb{T} \to \mathbb{T}$. And $\overline{\chi} \cdot \chi = \varepsilon$. Hence characters from a group. It is clearly abelian. We topologize $\widehat{G}$ with the *compact open* topology, with basis $U(K, \varepsilon, \chi_0) := \{\chi \in \widehat{G} \mid |\chi(g) - \chi_0(g)| < \varepsilon \; \forall g \in K\}$, as K range over compact subsets of G, $\chi_0 \in \widehat{G}$, and $\varepsilon > 0$.

Lemma 1.4.1 *If G is compact, then $\widehat{G}$ is discrete. If G is discrete, then $\widehat{G}$ is compact.*

Proof If G is compact, then $\{\chi \in \widehat{G} \mid |\chi(g) - 1| < \frac{1}{2}, \; \forall g \in G\}$ is a neighborhood of the trivial character $\varepsilon \in \widehat{G}$ containing only one point, namely ε itself, since the image of any character is a subgroup of the circle, and there are no subgroups of $\mathbb{T}$ which lie entirely within $\frac{1}{2}$ of 1. This shows that $\widehat{G}$ is discrete if G is compact.

On the other hand, if G is discrete, then the continuity requirement on a character becomes trivial, and it is then easy to check that $\widehat{G}$ embeds continuously as a closed subset of $\prod_G \mathbb{T}$, which is compact by Tychonoff's Theorem. Hence $\widehat{G}$ is compact if G is discrete. □

Exercise 1.4.2 Prove that if G is discrete, then $\widehat{G}$ can be identified with a closed subset of $\prod_G \mathbb{T}$ with the product topology.

Example 1.4.3 Let $G = \mathbb{T}$, the circle. To each integer $n \in \mathbb{Z}$, we associate the character $\chi_n(z) := z^n$. This gives an isomorphism $\widehat{G} \cong \mathbb{Z}$.

If G is locally compact abelian, the *Fourier transform* for G involves the following construction, which can be applied to various classes of functions, with various results. Suppose that $f \in C_c(G)$. We let

$$(1.21) \qquad \hat{f}(\chi) := \int_G f(g)\overline{\chi(g)}d\mu(g),$$

where μ is Haar measure on G. Then it is immediate that

$$|\hat{f}(\chi)| \leq \|f\|_{L^1(G)}.$$

And if $|\chi(g) - \chi_0(g)| < \varepsilon$ on $\mathrm{supp}(f)$, then

$$|\hat{f}(\chi) - \hat{f}(\chi_0)| < \|f\|_{L^1(G)}\varepsilon.$$

Hence $\hat{f}$ is continuous on $\widehat{G}$ if $f \in C_c(G)$.

Now if G is compact, any two distinct characters χ_1, χ_2 of G, viewed as vectors in $L^2(G)$, are *orthogonal*. Indeed, if $h \in G$, then by invariance of Haar measure

$$\langle \chi_1, \chi_2 \rangle = \int_G \chi_1(g)\overline{\chi_2(g)}d\mu(g) = \int_G \chi_1(hg)\overline{\chi_2(hg)}d\mu(g)$$

$$= \chi_1(h)\chi_2(h)^{-1}\int_G \chi_1(g)\overline{\chi_2(g)}d\mu(g) = \chi_1(h)\chi_2(h)^{-1}\langle\chi_1, \chi_2\rangle,$$

which implies that either $\chi_1 = \chi_2$ or the product is zero.

If $f \in L^2(G)$, then by the definition (1.21), we have $\hat{f}(\chi) = \langle f, \chi \rangle$. Hence

$$(1.22) \qquad \sum_{\chi \in \widehat{G}} |\hat{f}(\chi)|^2 \leq \|f\|_{L^2(G)},$$

from Bessel's inequality (1.2.10). In particular, if $\widehat{G}$ is infinite, then $\hat{f}(\chi) \to 0$ as $\chi \to \infty$. This shows that if $f \in C_c(G)$, then $\hat{f} \in C_0(\widehat{G})$, when G is compact. Furthermore, as $\|\hat{f}\|_{C_0(\widehat{G})} \leq \|f\|_{L^1(G)}$, the map $f \mapsto \hat{f}$ extends to a contractive map $L^1(G) \to C_0(\widehat{G})$. The following is an easy exercise left to the reader.

Exercise 1.4.4 If G is any locally compact abelian group and $f_1, f_2 \in L^1(G)$, then $\widehat{f_1 * f_2} = \hat{f}_1 \hat{f}_2$.

Hence $f \mapsto \hat{f}$ is a contractive homomorphism of Banach algebras $L^1(G) \to C_0(\widehat{G})$, when G is compact.

If G is not compact, but is discrete, then $\widehat{G}$ is compact, and the above arguments show that if $f \in L^1(G)$, then $\hat{f} \in C(\widehat{G})$, and that $\|\hat{f}\|_{C(\widehat{G})} \leq \|f\|_{L^1(G)}$. Therefore, putting things together:

Proposition 1.4.5 *If G is compact or discrete, the Fourier transform defines a contractive homomorphism of Banach *-algebras $L^1(G) \to C_0(\widehat{G})$.*

We will see shortly that the Fourier transform extends continuously to an C^*-algebra isomorphism $C^*(G) \to C_0(\widehat{G})$.

Example 1.4.6 If $G = \mathbb{T}$, with Haar measure normalized Lebesgue measure μ, then $\widehat{\mathbb{T}} \cong \mathbb{Z}$ with the integer n corresponding to the character $\chi_n(z) := z^n$ of $\mathbb{T}$. The Fourier transform in this notation is

$$\hat{f}(n) = \int_{\mathbb{T}} f(z) z^{-n} d\mu(z), \quad f \in L^1(\mathbb{T}) \subset C_r^*(\mathbb{T}).$$

If $G = \mathbb{Z}$, then $\widehat{\mathbb{Z}} \cong \mathbb{T}$ with $z \in \mathbb{T}$ corresponding to the character $\chi_z(n) := z^n$. The Fourier transform for the integers is given by

$$\hat{f}(z) = \sum_{n \in \mathbb{Z}} f(n) z^{-n}, \quad f \in l^1(\mathbb{Z}) \subset C_r^*(\mathbb{Z}).$$

If G is a *compact* (abelian) group, then one can show that the characters $\{\chi\}_{\chi \in \widehat{G}}$ form an orthonormal *basis* for $L^2(G)$. Hence the inequality in (1.22) is actually an equality. So when G is compact, the restriction of the Fourier transform to $L^2(G) \subset L^1(G)$ determines an *isometry* between Hilbert spaces which we denote by

$$F_G \colon L^2(G) \to l^2(\widehat{G}).$$

For example,

$$F_{\mathbb{T}} \colon L^2(\mathbb{T}) \to L^2(\widehat{\mathbb{T}}) \cong l^2(\mathbb{Z})$$

maps a function in $L^2(\mathbb{T}) \subset L^1(\mathbb{T})$ to the bi-infinite sequence $(\hat{f}(n))_{n \in \mathbb{Z}}$ of its Fourier coefficients with respect to the standard orthonormal basis $\{z^n\}_{n \in \mathbb{Z}}$ for $L^2(\mathbb{T})$.

When G is discrete (abelian), $\hat{f}(\chi) = \sum_{g \in G} f(g)\chi(g)$ for $f \in C_c(G) = \mathbb{C}[G]$, $\chi \in \widehat{G}$. Thus $\hat{f} = \sum_{g \in G} f(g)\check{g}$ where $\check{g} \colon \widehat{G} \to \mathbb{T}$ is the function $\check{g}(\chi) := \chi(g)$. Since $\check{g}$ so defined is a character of the compact dual group $\widehat{G}$, and all characters appear this way and make an orthonormal basis of $L^2(\widehat{G})$, we again obtain that

$$\|\hat{f}\|_{L^2(\widehat{G})} = \sum_{g \in G} |f(g)|^2 = \|f\|_{l^2(G)}.$$

Thus, again, we see that F_G induces, now by extension by continuity from $l^1(\mathbb{Z}) \subset l^2(\mathbb{Z})$, an isometry $F_G \colon l^2(G) \to L^2(\widehat{G})$, for any *discrete* G. So in either case, Fourier transform induces a unitary isomorphism.

For example, if $G = \mathbb{Z}$, then $F_G \colon l^2(\mathbb{Z}) \to L^2(\mathbb{T})$ maps a sequence $(a_n)_{n \in \mathbb{Z}}$ to the L^2-function $\sum_{n \in \mathbb{Z}} a_n z^n$.

Although one might expect that $F_{\widehat{G}}$ inverts F_G, this *almost* happens, but not quite. The statement is that

$$F_{\widehat{G}} \circ F_G = S_G,$$

where $S_G \colon L^2(G) \to L^2(G)$ is the self-adjoint unitary $(S_G f)(g) = f(g^{-1})$.

Similarly, still assuming G is compact, $F_G \circ F_{\widehat{G}} = S_{\widehat{G}}$. We conclude that if G is either compact or discrete, then since S_G is a self-adjoint unitary isomorphism, $F_G \colon L^2(G) \to L^2(\widehat{G})$ is a unitary isomorphism as well, and $F_G^* = F_G^{-1} = F_{\widehat{G}} \circ S_{\widehat{G}}$.

Lemma 1.4.7 *If $\lambda(f) \in C^*(G) \subset \mathbb{B}(L^2(G))$, then $F_G \lambda(f) F_G^* = M_{\hat{f}} \colon L^2(\widehat{G}) \to L^2(\widehat{G})$, with $M_{\hat{f}}$ the multiplication operator by $\hat{f} \in C_0(\widehat{G})$.*

We leave the proof as an exercise.

Theorem 1.4.8 *If G is any compact or discrete abelian group, then the Fourier transform $L^1(G) \to C_0(\widehat{G})$ of Proposition 1.4.5, extends continuously to a C*-algebra isomorphism $C_r^*(G) \to C_0(\widehat{G})$; moreover, this C*-algebra isomorphism is implemented by unitary conjugation with the Fourier transform, as a unitary $L^2(G) \to L^2(\widehat{G})$.*

The statement contains the important fact that

$$\|\lambda(f)\| = \sup_{\chi \in \widehat{G}} |\hat{f}(\chi)|,$$

for all $\lambda(f) \in C^*(G)$ (in particular for $f \in C_c(G)$, for example).

Remark 1.4.9 Although the space $\widehat{G}$ of characters of G, group homomorphisms $G \to \mathbb{T}$, may be rather trivial, and hence not a useful thing to consider, when G is not abelian, so that $C_0(\widehat{G})$ no longer is a very useful thing to think about, $C^*(G)$ always is defined, and, of course, agrees (by above) with $C_0(\widehat{G})$ in the commutative case. So this is an example where (noncommutative) C*-algebras may be of use—to study representation theory of nonabelian groups.

1.5   C*-Algebras of Finite Groups

Let G be a locally compact group. A *unitary representation* of G is a group homomorphism $\pi: G \to \mathbf{U}(H)$, where $\mathbf{U}(H)$ is the unitary group of a Hilbert space H, such that for every $\xi \in H$, the map $g \mapsto \pi(g)\xi$ is continuous. Sometime we drop the adjective "unitary," since we only consider unitary representations in this book.

A (unitary) representation is *irreducible* if it has no closed, G-invariant subspace. Two representations ρ_1, ρ_2 of G on H_1, H_2 are *equivalent* (or unitarily equivalent) if there is a unitary $U: H_1 \to H_2$ intertwining them: i.e.,

$$U\rho_1(g)U^* = \rho_2(g), \quad \forall g \in G.$$

In this section we are going to briefly describe some general features of the representation theory of finite groups and show how it ties in with the structure of their group C*-algebras.

Exercise 1.5.1 Let $\pi: G \to \mathbf{U}(H)$ be a unitary representation of a finite group G on a Hilbert space H:

a) Prove that if π is irreducible, then H is finite-dimensional. (*Hint.* Any orbit of G acting on H is finite and so spans a finite-dimensional, G-invariant subspace.)
b) Prove that if π is any finite-dimensional unitary representation of G, then π is completely reducible: There exists a decomposition $H = H_1 \oplus \cdots H_n$ and irreducible representations $\pi_i: G \to \mathbf{U}(H_i)$, such that π is unitarily equivalent to the direct sum representation $\pi_1 \oplus \cdots \oplus \pi_n$.

Lemma 1.5.2 (Schur's Lemma) *If $\pi: G \to \mathbf{U}(H)$ is an irreducible representation of G finite, then the only operators T on H commuting with $\pi(G)$ are scalar multiples of the identity operator 1_H.*

Proof Suppose that T commutes with G, that is, $\pi(g)T = T\pi(g)$ for all $g \in G$. Since H is finite-dimensional, T has at least one eigenvalue. The corresponding eigenspace $H' \subset H$ is G-invariant since $\pi(G)$ commutes with T. Since $H' \neq \{0\}$, $H' = H$, and we are done. ☐

Lemma 1.5.3 *Let $\pi: G \to \mathbf{U}(H))$ be a finite-dimensional irreducible unitary representation of the finite group G. Then*

(1.23)
$$\frac{1}{|G|} \sum_{g \in G} \pi(g) u \pi(g)^* = \frac{\text{Trace}(u)}{\dim H} \cdot 1,$$

for any $u \in \mathbb{B}(H)$, where $1 \in \mathbb{B}(H)$ is the identity operator.

Proof The group G acts by unitary conjugation on the C*-algebra $\mathbb{B}(H)$, and the content of Schur's Lemma is that the fixed-point *-subalgebra of this action is linearly spanned by the identity operator. Clearly the operator on the left hand side of (1.23) is G-fixed in this sense. Hence it is a scalar multiple $c(u)$ of 1_H.

Taking the trace of each side of the equation

(1.24)
$$\frac{1}{|G|} \sum_{g \in G} \pi(g) u \pi(g)^* = c(u) \cdot 1_H$$

then gives

$$\text{Trace}(u) = c(u) \cdot \dim H$$

so $c(u) = \frac{\text{Trace}(u)}{\dim H}$, as required.   □

The statement can be generalized to deal with *pairs* of representations.

Suppose that if $\pi_i \colon G \to U(H_i)$ are irreducible representations. And let $u \colon H_1 \to H_2$ be any linear map. Define a linear map $T_u \colon H_1 \to H_2$ by

(1.25)
$$T_u := \frac{1}{|G|} \sum_{g \in G} \pi_2(g) u \pi_1(g)^*.$$

Lemma 1.5.4 *The operator T_u of (1.25) is zero if π_1 and π_2 are inequivalent representations, for any $u \colon H_1 \to H_2$.*

Proof By a quick computation, $\pi_2(g) T_u = T_u \pi_1(g)$. If T_u were invertible, it would give rise to a unitary conjugacy between the representations. Hence T_u must be non-invertible for any u. On the other hand, $\ker(T_u)$ is a $\pi_1(G)$-invariant subspace of H_1, whence is either all of H_1 (making $T_u = 0$ as claimed) or, as we may assume, is the zero subspace. This makes T_u injective. Next, since $\pi_i(g)$ are unitaries, $i = 1, 2$ and all $g \in G$, it follows that

$$\pi_1(g) T_u^* = T_u^* \pi_2(g)$$

for all $g \in G$, from which it follows that $\ker(T_u^*)$ is $\pi_2(G)$-invariant. Reasoning as above, we conclude that T_u^* is either the zero operator, making T_u also zero, as claimed, or that T_u^* is injective. This would make T_u surjective, and hence invertible, a contradiction.   □

In the notation of the Lemma, let $\xi \in H_2, \eta \in H_1$. Define the rank-one operator

(1.26) $\theta_{\xi,\eta} \colon H_1 \to H_2, \quad \theta_{\xi,\eta}(\zeta) := \langle \eta, \zeta \rangle \, \xi.$

Note that

(1.27) $\langle \theta_{\xi,\eta}(\eta'), \xi' \rangle = \overline{\langle \eta, \eta' \rangle} \cdot \langle \xi, \xi' \rangle.$

Furthermore,

$$\pi_2(g)\theta_{\xi,\eta}\pi_1(g)^*(\zeta) = \langle \eta, \pi_1(g)^*\zeta \rangle \cdot \pi_2(g)\xi = \langle \pi_1(g)\eta, \zeta \rangle \cdot \pi_2(g)\xi,$$

and hence

$$\pi_2(g)\theta_{\xi,\eta}\pi_1(g)^* = \theta_{\pi_2(g)\xi,\pi_1(g)\eta}.$$

Hence, applying Lemma 1.5.4 to $u := T_{\xi,\eta}$ gives the operator equation

(1.28) $\dfrac{1}{|G|} \sum_{g \in G} \theta_{\pi_2(g)\xi,\pi_1(g)\eta} = 0.$

Now evaluate the operator on the left hand side at a vector $\eta' \in H_1$, and take the inner product of the result with ξ'. With the above remarks (and recall that our inner products are conjugate linear in the first variable), we obtain

(1.29) $\dfrac{1}{|G|} \sum_{g \in G} \overline{\langle \pi_2(g)\xi, \xi' \rangle} \cdot \langle \pi_1(g)\eta, \eta' \rangle = 0,$

which is an orthogonality statement for two functions on G, thought of as vectors in $l^2(G)$. Such functions, of the form

$$f^{\pi}_{\xi,\xi'} \colon G \to \mathbb{C}, \quad f^{\pi}_{\xi,\xi'}(g) = \langle \pi(g)\xi, \xi' \rangle,$$

are called *matrix coefficients* (of the representation). Our computations have shown that

(1.30) $\langle f^{\pi_2}_{\xi,\xi'}, f^{\pi_1}_{\eta,\eta'} \rangle = 0$

for any pair of inequivalent, irreducible representations π_1, π_2, and any vectors ξ, ξ', η, η', where the inner product is in the Hilbert space $l^2(G)$.

Finally, we return to the case $H_1 = H_2 = H$ and $\pi_1 = \pi_2 = \pi$. Set $u = \theta_{\xi,\eta}$ in Lemma 1.5.3. It is an easy exercise to check that $\mathrm{Trace}(\theta_{\xi,\eta}) = \langle \eta, \xi \rangle$. Hence

$$(1.31) \qquad \frac{1}{|G|} \sum_{g \in G} \theta_{\pi(g)\xi, \pi(g)\eta} = \frac{\langle \eta, \xi \rangle}{\dim H} \cdot 1_H$$

with 1_H the identity operator on H. Applying this operator equation to a vector η', taking product with a vector ξ', and proceeding as above in the case of two representations, we get the identity

$$(1.32) \qquad \frac{1}{|G|} \sum_{g \in G} \overline{f^\pi_{\xi, \xi'}(g)} \cdot f^\pi_{\eta, \eta'}(g) = \frac{\langle \eta, \xi \rangle \langle \xi', \eta' \rangle}{\dim H}.$$

We summarize:

Proposition 1.5.5 *Let G be a finite group:*

a) *Matrix coefficients of any two inequivalent irreducible representations of G are orthogonal to each other as vectors in $l^2(G)$.*

b) *If $\pi : G \to \mathbb{B}(H)$ is an irreducible representation and ξ, ξ', η, η' are vectors in H, then*

$$(1.33) \qquad \langle f^\pi_{\xi, \xi'}, f^\pi_{\eta, \eta'} \rangle = \frac{|G|}{\dim H} \cdot \langle \eta, \xi \rangle \cdot \langle \xi', \eta' \rangle,$$

where the $f^\pi_{\xi, \xi'}$, etc., are the corresponding matrix coefficients, regarded as elements of $l^2(G)$.

The *character* of a finite-dimensional representation (π, H) is the conjugation-invariant function

$$\chi : G \to \mathbb{C}, \qquad \chi_\pi(g) := \mathrm{Trace}\big(\pi(g)\big).$$

If $\xi_1, \ldots, \xi_n$ is an orthonormal basis for H, then

$$\chi_\pi(g) = \sum_i f^\pi_{\xi_i, \xi_i}$$

from the definitions.

If $\pi_i : G \to \mathbf{U}(H_i)$, $i = 1, 2$, are two inequivalent representations, it follows from (1.30) that

$$\langle \chi_{\pi_1}, \chi_{\pi_2} \rangle = 0$$

by setting $\xi = \xi_i = \xi'_i$ and $\eta = \eta_j = \eta'$, for orthonormal bases $\xi_1, \ldots, \xi_n$ of H_2 and $\eta_1, \ldots \eta_m$ of H_1, and summing over i, j.

Similarly, for a single representation apply (1.32) to $\xi = \xi_i = \xi'_i$, and $\eta = \eta_j = \eta'_j$, for two indices i, j, and then sum over i, j to get:

Proposition 1.5.6 *If* $\pi_i : G \rightarrow \mathbf{U}(H_i)$, $i = 1, 2$ *are inequivalent irreducible representations with characters* χ_{π_1} *and* χ_{π_2}, *viewed as vectors in* $l^2(G)$, *then*

$$\langle \chi_{\pi_1}, \chi_{\pi_2} \rangle = 0.$$

If $\pi : G \rightarrow \mathbf{U}(H)$ *is an irreducible representation, then*

$$\|\chi_\pi\|_{l^2(G)}^2 = |G|.$$

Every equivalence class of irreducible representation has a uniquely defined character, since trace is invariant under conjugation. The results above show that the set $\{\chi_\pi \mid [\pi] \in \widehat{G}\}$ forms an orthonormal set of vectors in $l^2(G)$, and hence there are only finitely many of them. In particular, we conclude that a finite group has only finitely many equivalence classes of irreducible representation, *i.e.*, $\widehat{G}$ has only finitely many points.

Suppose they are $[\pi_1], \ldots, [\pi_n]$. Let (H, π) be any finite-dimensional representation. Since it is completely reducible, it can be written as a direct sum

$$\pi \cong \oplus_i n_i \, \pi_i.$$

of the irreducibles. The integers n_i determine π up to isomorphism. Now $\chi_\pi = \sum_i n_i \, \chi_{\pi_i}$. Since $\langle \chi_{\pi_i}, \chi_{\pi_j} \rangle = |G| \cdot \delta_{ij}$,

$$\langle \chi_\pi, \chi_{\pi_i} \rangle = n_i \cdot |G|$$

so that the character χ_π determines the multiplicities n_i and hence the representation, up to isomorphism.

The basic example is the regular representation $(\lambda, l^2(G))$. Note that its character χ_λ is supported at the identity of the group and has the value $|G|$ there.

Proposition 1.5.7 *A finite group has only finitely many unitary equivalence classes of irreducible unitary representations.*

If the irreducible representations of G *are* $[\pi_1], [\pi_2] \ldots$, *then*

$$\lambda \cong \oplus_i \dim(H_{\pi_i}) \cdot \pi_i,$$

where λ *is the regular representation.*

That is, the multiplicity of π_i *in* λ *is* $\dim(H_{\pi_i})$.

Finally,

$$C^*(G) \cong \oplus_i \mathbb{B}(H_{\pi_i}),$$

as C*-algebras, by taking the direct sum of the maps $\pi_i : \mathbb{C}[G] \rightarrow \mathbb{B}(H_{\pi_i})$.

Proof By the discussion above, the multiplicities appearing in the decomposition $\lambda \cong \oplus_i n_i \pi_i$ of the regular representation into irreducibles are given by

$$n_i = \frac{1}{|G|} \cdot \langle \chi_\lambda, \chi_{\pi_i} \rangle = \frac{1}{|G|} \cdot \sum_{g \in G} \chi_\lambda(g) \chi_{\pi_i}(g).$$

Since χ_λ is supported at the identity $e \in G$ and $\chi_\lambda(e) = |G|$, $\chi_{\pi_i}(e) = \dim(\pi_i)$, the first statement follows.

For the second statement, we have shown above that there is a unitary isomorphism of Hilbert spaces

(1.34) $$l^2(G) \cong \oplus_i n_i H_{\pi_i},$$

where $n_i := \dim(H_{\pi_i})$. This unitary conjugates the regular representation λ to the direct sum representation

$$g \mapsto \oplus_i (\pi_i(g) \oplus \cdots \oplus \pi_i(g)),$$

with the term in brackets the direct sum of the unitary $\pi_i(g)$ with itself m_i times, acting on $H_{\pi_i} \oplus \cdots \oplus H_{\pi_i}$ (m_i times). Thus, an element $T \in C^*(G)$, $T = \sum_{g \in G} a_g[g]$, acts on the direct sum $l^2(G) \cong \oplus_i n_i H_{\pi_i}$ by the direct sum operator $\oplus_i \pi_i(T) \oplus \cdots \oplus \pi_i(T)$.

For each $i \in I$, follow this C*-algebra homomorphism by the projection to one of the summands in the ith factor, another C*-algebra homomorphism. Then the composition

(1.35) $$C^*(G) \to \oplus_i \dim(H_{\pi_i}) \cdot \mathbb{B}(H_{\pi_i}) \to \oplus_i \mathbb{B}(H_{\pi_i})$$

is an injective *-homomorphism, because the first map is injective, by construction, and the second is injective because a direct sum operator of the kind $\pi_i(T) \oplus \cdots \oplus \pi_i(T)$ is obviously determined by any of its summands.

The dimension (as a complex vector space) of the C*-algebra $\oplus_{i \in I} \mathbb{B}(H_{\pi_i})$ is

$$\sum_{i \in I} \dim(H_{\pi_i})^2$$

and by the first statement of the Proposition, this equals $\dim l^2(G)$. This equals $|G|$ and equals $\dim C^*(G)$. Therefore the map (1.35) is an isomorphism for dimension reasons.

We have thus argued that

$$C^*(G) \cong \oplus_i \mathbb{B}(H_{\pi_i}),$$

as C*-algebras, by taking the direct sum of the maps $\pi_i : \mathbb{C}[G] \to \mathbb{B}(H_{\pi_i})$. $\qquad\square$

Notice the following interesting Corollary:

Corollary 1.5.8 *If G is a finite group, $\widehat{G}$ the set of equivalence classes of irreducible representations of G, then*

$$|G| = \sum_{[\pi] \in \widehat{G}} \dim(H_\pi)^2.$$

These simple observations lead to the following fact. Since we have shown $C^*(G) \cong \oplus_i \dim(H_{\pi_i}) \cdot \mathbb{B}(H_{\pi_i})$, there must be, for each irreducible representation $\pi \in \widehat{G}$, a projection $e_\pi \in C^*(G)$, which is mapped by this isomorphism to the projection to the ith factor (the i-tuple with zeros everywhere except in the ith spot, and the identity operator on H_{π_i} there). Note that e_π commutes with all elements of $C^*(G)$, i.e., it is in the center of $C^*(G)$. The condition of projecting to the ith factor means that $\rho(e_\pi) = 0$ if ρ is not equivalent to π, and that $\pi(e_\pi) = 1$.

If χ is a function on G, let $\chi^*(g) := \overline{\chi(g)}$. If χ is the character of a unitary representation, then $\chi^*(g) = \chi(g^{-1})$ because of properties of the operator trace (exercise).

Proposition 1.5.9 *If π is an irreducible representation of G, then the induced C^*-algebra representation $C^*(G) \to \mathbb{B}(H_\pi)$ maps the function $\frac{\dim(H_\pi)}{|G|} \cdot \chi_\pi^*$ on G to the projection e_π.*

In particular, $e_\pi = \frac{\dim(H_\pi)}{|G|} \cdot \chi_\pi^* \in \mathbb{C}[G]$ is a projection in the center of $C^*(G)$, and

$$e_\pi \cdot C^*(G) \cong \mathbb{B}(H_\pi).$$

Proof Let π be one of the irreducible representations $\pi_1, \pi_2, \ldots$ of G. Let $e_\pi = \sum_{g \in G} a_g[g] \in \mathbb{C}G$; we solve for a_g, by the following arguments. Firstly, observe that $\text{Trace}(\lambda(e_\pi)) = a_e|G|$. And similarly $\text{Trace}(\lambda([g^{-1}]e_\pi)) = a_g|G|$ for any $g \in G$. Thus, using our direct sum decomposition of $C^*(G)$, based on a spatial decomposition of the Hilbert space $l^2(G)$, we deduce

$$a_g = \frac{1}{|G|}\text{Trace}\left(\lambda([g^{-1}]e_\pi)\right) = \frac{1}{|G|}\sum_i \dim(H_{\pi_i})\,\text{Trace}\left(\pi_i([g^{-1}]e_\pi)\right)$$

$$= \frac{\dim H_\pi}{|G|}\,\text{Trace}\left(\pi_i(g^{-1})\right).$$

Therefore

$$\chi_\pi^*(g) = \frac{|G|}{\dim H_\pi}a_g$$

for all $g \in G$.

We conclude that

$$e_\pi = \frac{\dim(H_\pi)}{|G|} \cdot \chi_\pi^* \in \mathbb{C}[G],$$

i.e., as functions on G.   □

Exercise 1.5.10 Suppose G is a finite *abelian* group. Let $\chi: G \to \mathbb{T}$ be a group character. Prove by direct computation that $\frac{1}{|G|}\chi^*$ (that is, that $\frac{1}{|G|}\sum_{g \in G} \chi(g^{-1})[g])$ is a projection in $C^*(G)$.

For a more challenging exercise, give a more direct proof that the one we have given that the elements

$$e_\pi = \frac{\dim(H_\pi)}{|G|} \cdot \chi_\pi^* \in \mathbb{C}[G]$$

are projections in the center of $C^*(G)$, if π is any irreducible representation.

We summarize what has been proved about C*-algebras of finite groups.

Theorem 1.5.11 *Let G be a finite group, $\widehat{G}$ its set of equivalence classes of irreducible representations. Let $[\pi] \in \widehat{G}$, and χ_π its character. Let $e_\pi = \frac{\dim H_\pi}{|G|}\sum_{g \in G} \overline{\chi_\pi(g)}[g] \in C^*(G)$. Then:*

a) *e_π is a central projection in $C^*(G)$.*
b) *$\pi(e_\pi) = 1$, the identity in $\mathbb{B}(H_\pi)$, and if $[\rho] \in \widehat{G}$ and $[\rho] \neq [\pi]$, then $\rho(e_\pi) = 0$.*
c) *Summing the representations π for $[\pi] \in \widehat{G}$ gives an isomorphism*

$$C^*(G) \cong \oplus_{[\pi] \in \widehat{G}} \, e_\pi C^*(G), \quad \text{and } e_\pi C^*(G) \cong \mathbb{B}(H_\pi).$$

d) *If $\tau: C^*(G) \to \mathbb{C}$ is the trace $\tau(\sum_g a_g[g]) := a_e$ of Exercise 1.3.10 a), then*

$$\tau(e_\pi) = \frac{\dim H_\pi^2}{|G|}$$

for all $[\pi] \in \widehat{G}$.

Exercise 1.5.12 The finite group S_3 has 3 irreducible representations: the trivial representation ε, the sign representation $\sigma(g) = \pm 1$ according as g is an even or odd permutation, and a 2-dimensional representation defined as follows. Let S_3 act on $\mathbb{C}^3$ by permutation matrices. In $\mathbb{C}^3$ let V be the subspace $(1, 1, 1)^\perp = \{(x, y, z) \in \mathbb{C}^3 \mid x + y + w = 0\}$. Then V is S_3-invariant, and the restriction of the S_3 action on $\mathbb{C}^3$ to V is irreducible.

Deduce that $C^*(S_3) \cong \mathbb{C} \oplus \mathbb{C} \oplus M_2(\mathbb{C})$. Exhibit an explicit isomorphism.

1.6 The Compact Operators

In noncommutative geometry, "compact" (operator) tends to suggest "small" in some sense, as in a "small perturbation." Sometimes they are argued to be the quantum physical (or noncommutative) analogue of the "infinitesimals" one meets in calculus, or differential geometry.

Definition 1.6.1 A bounded linear operator $T \in \mathbb{B}(H, K)$ between Hilbert spaces H, K is a *compact* operator if the image $T(B_H)$ of the unit ball $B_H := \{\xi \in H \mid \|\xi\| < 1\}$ in H is pre-compact, that is, if its closure is compact.

Remark 1.6.2 A standard result from point-set topology is that a subspace $A \subset X$ of a complete metric space is pre-compact if and only if it is totally bounded. Hence, a bounded operator $T : H \to K$ is compact if and only if for all $\varepsilon > 0$ there exist vectors $\xi_1, \ldots, \xi_n \in B_H$ such that $T(B_H) \subset \bigcup_i B_\varepsilon(T\xi_i)$.

If H is an infinite-dimensional Hilbert space, then the closed unit ball B_H is noncompact since any infinite orthonormal set in B_H provides a net with no convergent subnet. It follows that the identity operator on H is not compact, whence certainly not all bounded operators are compact.

More generally:

Exercise 1.6.3 Let $T : H \to K$ be a bounded operator for which there exists $c > 0$ such that $\|T\xi\| \geq c\|\xi\|$ for all $\xi \in H$. Prove that if T is compact, then H is finite-dimensional.

Example 1.6.4 Let T be a bounded operator on a Hilbert space whose matrix with respect to an orthonormal basis $\{e_1, e_2, \ldots\}$ is

$$
T = \begin{bmatrix} \lambda_1 & & & \\ & \lambda_2 & & \\ & & \lambda_3 & \\ & & & \ddots \end{bmatrix}
$$

Then T is compact if and only if $\lim_{n\to\infty}|\lambda_n| = 0$. Indeed, if T is compact, then Exercise 1.6.3 implies that for all $\varepsilon > 0$ the set $\{n \mid |\lambda_n| \geq \varepsilon\}$ is finite. Hence $\lambda_n \to 0$ as claimed.

Conversely, suppose that $\lim_{n\to\infty}|\lambda_n| = 0$. Choose $\varepsilon > 0$. Let N such that $|\lambda_n| < \frac{\varepsilon}{2}$ if $n \geq N$. Any vector in $T(B_H)$ can be written in the form $T\xi + T\eta$ where ξ is a linear combination of $e_1, \ldots, e_N$, η is in the closed span of $e_{N+1}, e_{N+2}, \ldots$, with $\|\xi\| \leq 1$ and $\|\eta\| \leq 1$.

By choice of N and Exercise 1.2.15, $\|T\eta\| < \frac{\varepsilon}{2}$, because the restriction of T to the closed span of $e_{N+1}, e_{N+2}, \ldots$ is diagonal, with entries bounded by $\frac{\varepsilon}{2}$.

Hence every vector in $T(B_H)$ is at distance at most ε to a vector in $T(B_{H'})$ where H' is the subspace $\text{span}(e_1, \ldots, e_N)$.

Now since H' is finite-dimensional, $B_{H'}$ is pre-compact, so $T(B_{H'})$ is also pre-compact, whence it is totally bounded, so there are finitely many vectors $\xi_1, \ldots, \xi_k$ in $B_{H'}$ such that $T(B_{H'}) \subset \bigcup_{i=1}^{k} B_{\frac{\varepsilon}{2}}(T(\xi_i))$.

Putting these together gives that $T(B_H) \subset \bigcup_{i=1}^{k} B_{\varepsilon}(T(\xi_i))$, as required.

As observed in the previous proof, an operator $T \in \mathbb{B}(H, K)$ with finite-dimensional range maps B_H into a bounded subset of a finite-dimensional subspace of K. Such operators are said to have *finite rank*. Since a bounded and closed subset of a finite-dimensional Hilbert space is compact, we get the following basic result.

Proposition 1.6.5 *Any bounded finite-rank operator is compact.*

Exercise 1.6.6 Let H be a separable Hilbert space. Fix an orthonormal basis $\{e_1, e_2, \ldots\}$ for H and represent operators by their matrices in the usual way with $T = (T_{ij})$, $T_{ij} = \langle e_i, T(e_j) \rangle$. An operator has a *finitely supported matrix* (with respect to the given orthonormal basis) if it has only finitely many nonzero entries. Prove that:

a) If T is bounded and has finite rank, there is a unitary u and a finitely supported operator S such that $uTu^* = S$.
b) If T is a bounded finite-rank operator and $\varepsilon > 0$, then there exists a finitely supported operator S such that $\|S - T\| < \varepsilon$.
c) If $\xi, \eta \in H$, let

$$T_{\xi, \eta}(v) := \langle \eta, v \rangle \xi.$$

Prove that T is a rank-one linear operator.

Prove that if T is a bounded finite-rank operator on H, then there exist vectors $\xi_1, \ldots, \xi_n, \eta_1, \ldots \eta_n$ in H and $a_1, \ldots, a_n \in \mathbb{C}$ such that $T = \sum_i a_i T_{\xi_i, \eta_i}$.

The following theorem gives the basic tool in showing that operators are compact.

Theorem 1.6.7 *A bounded operator $T: H \to K$ between Hilbert spaces is compact if and only if it is a norm limit of finite-rank operators.*

The proof of Theorem 1.6.7 amounts to the following two Lemmas.

Lemma 1.6.8 *The set of compact operators $\mathcal{K}(H, K)$ from H to K is a closed subset of $\mathbb{B}(H, K)$ in the operator norm topology. That is, any operator norm limit of compact operators is a compact operator.*

Proof See Remark 1.6.2. Let T be a limit point of the set of compact operators from H to K. Choose $\varepsilon > 0$. There exists a compact operator S such that $\|S - T\| < \frac{\varepsilon}{3}$. Since S is compact, there exist finitely many vectors $\xi_1, \ldots, \xi_n \in B_H$ such that $S(B_H) \subset \bigcup_{i=1}^{n} B_{\frac{\varepsilon}{3}}(S\xi_i)$. Now we claim that $T(B_H) \subset \bigcup_{i=1}^{n} B_{\varepsilon}(T\xi_i)$. For if $\xi \in B_H$, choose ξ_i so that $S\xi \in B_{\frac{\varepsilon}{3}}(\xi_i)$, and then by the triangle inequality

$$\|T\xi - T\xi_i\| \le \|T\xi - S\xi\| + \|S\xi - S\xi_i\| + \|S\xi_i - T\xi_i\| < \frac{\varepsilon}{3} + \frac{\varepsilon}{3} + \frac{\varepsilon}{3} = \varepsilon,$$

proving the claim.   □

Lemma 1.6.9 *The finite-rank operators from H to K are dense in* $\mathcal{K}(H, K)$.

Proof We show that if $T : H \to K$ is a compact operator, then there exists a sequence (T_n) of finite-rank operators from H to K such that $T_n \to T$ in operator norm.

Let $e_1, e_2, \ldots$ be an orthonormal basis and let P_n be the orthogonal projection to the span of $e_1, \ldots, e_n$:

$$P_n\xi = \sum_{i=1}^{n} \langle e_n, \xi \rangle e_n.$$

By standard Hilbert space theory $P_n\eta \to \eta$ for all $\eta \in L$. In particular, $P_nT\xi \to T\xi$ for all $\xi \in H$.

We claim that this convergence is actually uniform over B_H, i.e., that $P_nT \to T$ in operator norm, and thus provides the required approximation of T by finite-rank operators.

Choose $\varepsilon > 0$. Choose a finite set of vectors $\xi_1, \ldots, \xi_n$ in B_H such that $T(B_H) \subset \bigcup_i B_\varepsilon(T\xi_i)$. Choose any $\xi \in B_H$. Then $\|T\xi - T\xi_i\| < \varepsilon$ for some i. Since $P_nT \to T$ pointwise as observed above, there exists N such that if $n \ge N$ then $\|P_nT\xi_i - T\xi_i\| < \varepsilon$ for $i = 1, \ldots, n$. So we argue

$$\|P_nT\xi - T\xi\| \le \|P_nT\xi - P_nT\xi_i\| + \|P_nT\xi_i - T\xi_i\| + \|T\xi_i - T\xi\|.$$

The first term is bounded by $\|P_n\| \cdot \|T\xi - T\xi_i\| = \|T\xi - T\xi_i\| < \varepsilon$, and this is why the third term is also $< \varepsilon$. The second term is also $< \varepsilon$ if $n \ge N$.   □

Corollary 1.6.10 *If* $T \in \mathbb{B}(H, K)$ *is a compact operator,* $S \in \mathbb{B}(M, H)$ *and* $R \in \mathbb{B}(K, L)$, *then* TS *and* RT *are compact operators* $M \to K$ *and* $H \to L$, *respectively. The adjoint* $T^* : K \to H$ *of a compact operator* $H \to K$ *is compact.*

Proof All of these statements are clear for finite-rank operators, and they follow by taking norm limits for compact operators, by the density result Lemma 1.6.9.   □

Corollary 1.6.11 *The collection* $\mathcal{K}(H)$ *of compact operators* $H \to H$ *is a C*-subalgebra of* $\mathbb{B}(H)$ *for any Hilbert space* H *and, in particular, is a C*-algebra, for any Hilbert space* H.

Moreover, $\mathcal{K}(H)$ *is an ideal of* $\mathbb{B}(H)$: *If* $T \in \mathbb{B}(H)$ *and* $S \in \mathcal{K}(H)$, *then* $ST, TS \in \mathcal{K}(H)$.

Exercise 1.6.12 Let H be a separable Hilbert space and $T \in \mathcal{K}(H)$ be any compact operator on H. Show that $\|Te_n\| \to 0$ as $n \to \infty$ for any infinite orthonormal set $e_0, e_1, \ldots$ of vectors in H. Give an example of a bounded operator T and an orthonormal basis $e_0, e_1, \ldots$ such that $\|Te_n\| \to 0$ as $n \to \infty$, but T is not compact.

(*Hint.* For the second question, let P be a countable infinite direct sum of the n-by-n

(projection) matrices $\begin{bmatrix} \frac{1}{n} & \cdots & \frac{1}{n} \\ \vdots & & \vdots \\ \frac{1}{n} & \cdots & \frac{1}{n} \end{bmatrix}$.)

Exercise 1.6.13 Let $W, F \subset H$ be two linear subspaces of a Hilbert space H with $\dim F < \infty$. Prove that W is closed if and only if $W + F$ is closed.

Exercise 1.6.14 Let T be a compact operator on a Hilbert space H and $\lambda \neq 0$ a nonzero complex number. Prove:

a) The subspace $H_\lambda := \{\xi \in H \mid T\xi = \lambda\xi\} = \ker(\lambda - T)$ of T is finite-dimensional.

b) If

$$\inf\{\|(\lambda - T)v\| \mid \|v\| = 1\} = 0,$$

then λ is an eigenvalue for T. (*Hint.* Otherwise there exists a sequence (ξ_n) of unit vectors such that $(\lambda - T)\xi_n \to 0$. Use compactness of T to deduce an appropriate convergent subsequence.)

c) If λ is not an eigenvalue of T, then $\lambda - T$ has closed range.

d) As an element of the unital C*-algebra $\mathbb{B}(H)$, the compact operator T has a spectrum: the set of $\lambda \in \mathbb{C}$ such that $\lambda - T$ is invertible as a bounded operator. Let $\lambda \in \mathrm{Spec}(T)$ be nonzero. Show that either λ is an eigenvalue of T or $\bar{\lambda}$ is an eigenvalue of T^*. (*Hint.* Use Exercise 1.2.32.)

Exercise 1.6.15 The following exercise shows that the set of nonzero eigenvalues of a compact operator is always discrete in $\mathbb{C} - \{0\}$. Suppose that T is compact and that $Tv_n = \lambda_n v_n$ for nonzero, distinct elements $\lambda_n \in \mathbb{C}$ and unit vectors v_n:

a) Prove that $\{v_1, \ldots, v_n\}$ is linearly independent.

b) Deduce there exist unit vectors $q_n \in \mathrm{span}\{v_1, \ldots, v_n\}$ such that $q_n \perp v_1, \ldots, v_{n-1}$ for all n.

c) Check that

$$Tq_n = \lambda_n q_n + w_n,$$

where $w_n \in \mathrm{span}\{v_1, \ldots, v_{n-1}\}$, and deduce that for any $n > m$

$$\|Tq_n - Tq_m\| \geq |\lambda_n|.$$

d) Deduce that the sequence (Tq_n) has no convergent subsequence unless $\lambda_n \to 0$.

Exercise 1.6.16 If $T : H \to K$ is a compact operator, then there exists a vector $\xi \in H$ such that $\|\xi\| \leq 1$ and $\|T\xi\| = \|T\|$.

(*Hint*. By compactness of T and the definition of norm, there is a sequence $\xi_1, \xi_2, \ldots$ of unit vectors in H such that $T\xi_n \to \eta$, where $\|\eta\| = \|T\|$. Apply the the parallelogram law to get the estimate

$$\|\xi_n - \xi_m\|^2 = 4 - \|\xi_n + \xi_m\|^2 \le 4 - \frac{\|T\xi_n + T\xi_m\|^2}{\|T\|^2} \to 0$$

as $n, m \to \infty$, and deduce the result.)

Exercise 1.6.17 If $T \in \mathcal{K}(H)$ is self-adjoint, then T has an eigenvalue λ such that $|\lambda| = \|T\|$. (*Hint*. Assume $\|T\| = 1$ without loss of generality. Let (v_n) a sequence of unit vectors with $\lim_{n\to\infty} |\langle Tv_n, v_n \rangle| = 1$ (by Lemma 1.2.31) such a sequence exists. By compactness we may assume $Tv_n \to w$ for some w. Argue $\|w\| = 1$. As in Exercise 1.6.16 argue now that (v_n) converges, say to v. Prove that $|\langle Tv, v \rangle| = 1$, and then deduce that $Tv = \pm v$.)

We now discuss compact operators arising geometrically. Let X be a locally compact Hausdorff space and μ a Borel measure on X. For example, $X = \mathbb{R}$, μ Lebesgue measure, or $X = \mathbb{T}$ with Lebesgue measure. Let $k \in L^2(X \times X, \mu \times \mu)$. If $\xi \in L^2(X, \mu)$, set

$$(1.36) \qquad I_k\xi(x) := \int_X k(x, y)\xi(y)d\mu(y).$$

Since k is in $L^2(X \times X)$, it follows from the Fubini Theorem that for a.e. $x \in X$,

$$\int_X |k(x, y)|^2 d\mu(y) < \infty$$

so that $k(x, \cdot) \in L^2(X)$ for a.e. $x \in X$. By the Cauchy–Schwarz inequality, the integral (1.36) converges absolutely. By the Cauchy–Schwarz inequality $|I_k\xi(x)|^2 \le \|k(x, \cdot)\|^2_{L^2(X)}\|\xi\|^2_{L^2(X)}$. Integrating this over X gives that I_k is a bounded operator and $\|I_k\| \le \|k\|_{L^2(X \times X)}$.

Proposition 1.6.18 I_k *is a compact operator for all* $k \in L^2(X \times X)$ *and* $\|I_k\| \le \|k\|_{L^2(X \times X)}$.

Proof We proved the first assertion above. For the second, measurable functions of the form $r(x, y) = \sum_i f_i(x)g_i(y)$, where the sum is finite, and the f_i's and g_i's are in $L^2(X)$, are dense in $L^2(X \times X, \mu \times \mu)$. For such a function r, by the definitions,

$$I_r\xi = \sum_i \left[\int_X g_j(y)\xi(y)d\mu(y) \right] f_i.$$

In particular $\text{ran}(I_k) \subset \text{span}\{f_i\}$, which is finite-dimensional. Hence I_r has finite rank.

Now if $k \in L^2(X \times X)$, let $k_n \to k$ with k_n of the form of r above. Then $\|I_{k_n} - I_k\| \leq \|k_n - k\|_{L^2(X \times X)} \to 0$ so I_k is a compact operator. $\qquad\qquad\square$

Example 1.6.19 If G is a compact group and $f \in C(G)$, $\lambda(f)$ the corresponding convolution operator on $L^2(G)$, then $\lambda(f)$ is a special case of a compact integral operator, for we can write

$$\lambda(f)\xi \ (g) = \int_G f(h)\xi(h^{-1}g)d\mu(h) = \int_G f(gh^{-1})\xi(h)d\mu(h) = I_k\xi \ (g),$$

where I_k is the integral operator with kernel $k(g, h) = f(gh^{-1})$ (μ normalized Haar measure over G as usual).

Compactness of G is needed here to ensure that $k \in L^2(G \times G)$.

In particular, convolution operators $\lambda(f)$ with $f \in C(G)$ are compact operators. Since it is an immediate consequence of Theorem 1.6.7 that operator norm limits of compact operators are also compact, it follows that the C*-algebra $C^*(G)$ of G consists entirely of compact operators on $L^2(G)$, when G is a compact group.

Exercise 1.6.20 If $G = \mathbb{T}$ and $f(z) = \sum_{k=-n}^{m} a_k z^k$ is a trigonometric polynomial in $C(\mathbb{T}) \subset C^*(\mathbb{T})$, then the convolution operator $\lambda(f): L^2(\mathbb{T}) \to L^2(\mathbb{T})$ has rank at most $n + m$.

Exercise 1.6.21 If G is a compact group and $\lambda(f) \in C^*(G)$ is convolution by a continuous function f on G, then f then $\lambda(f)$ is a compact operator on $L^2(G)$. By Theorem 1.6.23 (and Exercise 1.6.15) $\lambda(f)$ has a countable collection of nonzero eigenvalues. What are they if G is abelian? What about G finite?

Exercise 1.6.22 Let G be a locally compact group, $f \in C_c(G)$ and $h \in C_c(G)$, and let $\lambda(f) \in \mathbb{B}(L^2(G))$ be convolution with f and M_h be multiplication by h. We have already noted that $\lambda(f)$ is compact if G is compact; it is clear that M_h is compact if G is discrete.

Prove that $\lambda(f)M_h$ is a compact operator *for any locally compact group* G. (*Hint.* Show that $\lambda(f)M_h = I_k$, an integral operator, with appropriate compactly supported kernel.)

We conclude the general discussion with a Spectral Theorem for compact operators.

Theorem 1.6.23 *Let T be any self-adjoint compact operator on a separable Hilbert space H and $\Lambda \subset \mathbb{R}$ denote the set of eigenvalues of T. If $\lambda \in \Lambda$, let $H_\lambda = \ker(\lambda - T)$ be the corresponding eigenspace.*

Then Λ is at most a countable set, H_λ is finite-dimensional for all nonzero λ, H_λ is orthogonal to $H_{\lambda'}$ if $\lambda \neq \lambda'$, and $H = \bigoplus_{\lambda \in \Lambda} H_\lambda$.

Proof The proofs that T self-adjoint implies that all eigenvalues are real and that eigenspaces for distinct eigenvalues are orthogonal, are left to the reader—they are identical to the arguments used to prove these facts in finite-dimensional linear algebra. The existence of an eigenvalue λ such that $\|T\| = |\lambda|$ is shown in Exercise 1.6.17.

Let $H' = \left(\oplus_{\lambda \in \Lambda} H_\lambda \right)^\perp$ and T' be the restriction of T to T'; then since the restriction of a compact operator is compact, T' is both compact and injective, so H' is finite-dimensional. Furthermore, T' clearly has no eigenvalues, as they would also be eigenvalues for T, whence $H' = 0$.   □

Exercise 1.6.24 A compact operator T is *positive* if it is self-adjoint and all its eigenvalues are positive. Show that if T is a positive compact operator, then there is a unique positive compact operator $\sqrt{T}$ such that $\sqrt{T}^2 = T$. Deduce that

$$\langle T\xi, \xi \rangle \geq 0$$

for all $\xi \in H$, and every positive compact operator.

Exercise 1.6.25 Let T be a compact operator. Prove that T^*T is positive.

1.7   Inductive Limits of C*-Algebras

We start by discussing the category-theoretic idea of a direct limit. Let C be a category and I a directed set: a set I equipped with a reflexive and transitive relation such that any two elements of I have an upper bound in I.

A *directed system* of objects of C is a family $\{A_i \mid i \in I\}$ of objects of the category and a family of morphisms $\varphi_{ji} \colon A_i \to A_j$ for all $i \leq j$ such that $\varphi_{ii} = \mathrm{id}_{A_i}$, the identity morphism $A_i \to A_i$, for all i, and such that

$$\varphi_{kj} \circ \varphi_{ji} = \varphi_{ki}, \quad \forall i \leq j \leq k.$$

Remark 1.7.1 If the directed set is just the natural numbers $1, 2, \ldots$ with its usual ordering, then the directed system is often written simply in the form

$$A_1 \xrightarrow{\varphi_1} A_2 \xrightarrow{\varphi_2} \to \cdots$$

with $\varphi_n \colon A_n \to A_{n+1}$ the maps between the adjacent objects of the system.

Definition 1.7.2 A *direct limit* of a directed system $\{A_i \mid i \in I\}, \{\varphi_{ji} \colon A_i \to A_j \mid i \leq j\}$ in a category C is an object A of C together with a family $\varphi_i \colon A_i \to A$ of morphisms that satisfies the following universal property.

If B is any object of C and $\{\psi_i \colon A_i \to B \mid i \in I\}$ is a family of morphisms which is coherent in the sense that $\psi_j \circ \varphi_{ji} = \psi_i$ for all $i \leq j$, then there is a unique morphism $\psi \colon A \to B$ such that $\psi \circ \varphi_i = \psi_i$ for all $i \in I$.

In general, direct limits may or may not exist in a category. In many familiar categories, however, like the categories of groups, rings, modules over a ring, and so on, they exist and are defined by a construction similar to the case of groups, which we explain first.

Let $\{G_i \mid i \in I\}$, $\{\varphi_{ij} \mid i \leq j\}$ be a directed family of groups. On the disjoint union $\sqcup G_i$ let $\sim$ be the equivalence relation generated by $g_j \sim \varphi_{ji}(g_i)$ if $i \leq j$. Thus, if $g \in G_i$, we identify g with any of its images $\varphi_{ji}(g) \in G_j$.

We endow $\sqcup G_i / \sim$ with the following group operation: If $g \in G_i$ and $h \in G_j$, then choose any $k \geq i, j$, push g and h into the same G_k using the structure maps φ_{ki} and φ_{kj}, respectively, and multiply them in G_k. Thus, if $[g]$ denotes the equivalence class of $g \in G_i$ in $\sqcup G_i / \sim$, and similarly for h, then $[g] \cdot [h] := [\varphi_{ki}(g)\varphi_{kj}(h)]$. This is easily seen to be a group operation.

The morphisms $\varphi_i \colon G_i \to \varinjlim G_i$ are the evident maps; G_i embeds firstly into the disjoint union, as a set, and then by the quotient map into the direct limit, and this is clearly a group homomorphism. We leave it to the reader to check the universal property.

The following exercise gives practise in dealing with inductive limits (of abelian groups). These examples appear later as K_0-groups of certain inductive limit C*-algebras.

Example 1.7.3 With the usual ordering on the natural numbers, making it a directed set, for $n \leq m$, let $\varphi_{m,n} \colon \mathbb{Z} \to \mathbb{Z}$ be the group homomorphism of multiplication by 2^{m-n}.

The corresponding direct limit G of groups is $\mathbb{Z}[\frac{1}{2}]$, the subgroup of $\mathbb{Q}$ generated by $\mathbb{Z}$ and the numbers $\frac{1}{2^n} \in \mathbb{Q}$, $n = 1, 2, \ldots$.

To see this, note that a typical element of the inductive limit is the equivalence class $[(n, m)]$ of a pair (n, m). The equivalence relation is that $(n, m) \sim (n+k, 2^k m)$ for $k = 1, 2, \ldots$. The group operation at the level of pairs is

$$[(n, m)] + [(r, s)] := [(n + r, 2^r m + 2^n s)].$$

The map $\phi \colon \varinjlim \mathbb{Z} \to \mathbb{Z}[\frac{1}{2}]$ by

$$\phi([(n, m)]) := \frac{m}{2^n}$$

is a well-defined group isomorphism.

The same works for any positive integer d, not just $d = 2$.

Exercise 1.7.4 Let $\mathbb{N}$ be made into a directed set by letting $n \leq m$ if $n \mid m$.

For each $n \mid m$ let $\varphi_{mn} \colon \mathbb{Z} \to \mathbb{Z}$ be the group homomorphism of multiplication by $\frac{m}{n}$ (an integer).

Show that $\varinjlim \mathbb{Z} \cong \mathbb{Q}$. *Hint.* We denote elements of the direct limit as classes $[(n, m)]$ of pairs $(n, m) \in \mathbb{N} \times \mathbb{Z}$ of integers, where the equivalence relation is that $(n, m) \sim (nk, mk)$. Identify $[(n, m)]$ with the fraction $\frac{n}{m}$.

Proposition 1.7.5 *Direct limits exist in the category of C*-algebras and C*-algebra homomorphisms.*

For the proof, which we will give only in the situation where the structure maps φ_{ij} of the system are injective, we will need the following Lemma; the (easy) result is the content of Corollary 3.1.18 of Chapter 3, and we refer the interested reader to the proof presented there.

Lemma 1.7.6 *If A and B are C*-algebras and $\varphi: A \to B$ is a *-homomorphism, then φ is norm contractive:*

$$\|\varphi(a)\| \leq \|a\|, \quad \forall a \in A.$$

If φ is injective, then it is isometric.

$$\|\varphi(a)\| = \|a\|, \quad \forall a \in A.$$

Proof of Proposition 1.7.5 We will assume for simplicity that the structure maps $\varphi_{ij}: A_j \to A_i$ are all injective. See Example 4.3.13 for the general case.

As in the example of a direct limit of groups, we start by defining the algebra direct limit as $\mathcal{A} := \sqcup A_i / \sim$ where $\sim$ is the equivalence relation generated by identifying $a \in A_i$ with $\varphi_{ji}(a_i) \in A_j$ for any $i \leq j$. Note that this results in the zero elements $0 \in A_i$ all being identified (similarly the identity elements of the groups G_i are all identified in the construction of the direct limit of groups). Let $\varphi_i: A_i \to \mathcal{A}$ be the evident maps of A_i into $\mathcal{A}$.

As in the case of groups, we endow $\mathcal{A}$ with the structure of a *-algebra. If we wish to multiply $a \in A_i$ and $b \in A_j$ (or more precisely, if we want to multiply $\varphi_i(a)$ and $\varphi_j(b)$ in $\mathcal{A}$, we instead choose $k \geq i, j$ and define the product to be the class in $\mathcal{A}$ of

$$\varphi_{ki}\big(\varphi_i(a)\big) \cdot \varphi_{kj}\big(\varphi_j(b)\big).$$

Similarly, we define the sum of two elements. The adjoint may be defined in the obvious way. We obtain a *-algebra $\mathcal{A}$. If $a \in A_i$, we set $\|\varphi_i(a)\| := \lim_{j \to \infty} \|\varphi_{ji}(a)\|$. The limit exists because it is a decreasing net of positive real numbers, because *-homomorphisms are automatically contractive, by Lemma 1.7.6. Now, if the φ_{ij} are all injective, then by the same Lemma, they are isometric, and the norm defined above on $\mathcal{A}$ is actually a norm. In this case we can $\varinjlim A_i$ to be the completion of $\mathcal{A}$ with respect to this norm. It is easy to see that this results in a C*-algebra.

In the general case (if the φ_{ij} are not all injective), we obtain a pre-C*-algebra $(\mathcal{A}, \|\cdot\|)$ in the sense of Definition 4.3.1 of Chapter 3, and we define $\varinjlim A_i$ to be its completion in the sense of completions of pre-C*-algebras as discussed in Chapter 3. We will be largely focusing on examples where the structure maps are all injective.

We leave it as an exercise to check the universal property.   □

Exercise 1.7.7 Show that if the structure maps $\varphi_{ij} : A_j \to A_i$, $i \geq j$ of a directed system are all injective, then the induced inclusions $\varphi_j : A_i \to \varinjlim A_i$ into the direct limit are also injective, so that case $\varinjlim A_i$ is effectively the closure of the *union* of the A_i's in this case.

Example 1.7.8 Infinite direct sums $\oplus_{i \in I} A_i$ of a family $\{A_i\}_{i \in I}$ of C*-algebras are inductive limits of finite direct sums.

Indeed, let $\mathcal{F}$ be the directed set of finite subsets of I under inclusion. If $F \in \mathcal{F}$, let $A_F := \oplus_{i \in F} A_i$. If $F_1 \leq F_2$, there is an associated injective C*-algebra homomorphism $\varphi_{F_2, F_1} : A_{F_1} \to A_{F_2}$ by adding zeros to an F_1-tuple until one gets an F_2-tuple. We obtain a directed system of C*-algebras.

For each finite subset F, let $\psi_F : A_F \to B := \oplus_{i \in I} A_i$ be the map which adds zeros to an F-tuple to get an I-tuple. Evidently if $F_1 \leq F_2$, then $\psi_{F_2} \circ \varphi_{F_2, F_1} = \psi_{F_1}$, so we obtain a *-homomorphism

$$\varinjlim A_F \to \oplus_{i \in I} A_i.$$

Thus, we recover infinite direct sums by combining inductive limits and finite sums.

Example 1.7.9 There is a natural C*-algebraic analogue of the inductive system of Example 1.7.3. The directed system is $\mathbb{N}$ with its usual ordering.

For $n = 1, 2, \ldots$ we set $A_n := M_{2^n}(\mathbb{C})$. If $m \geq n$, the map $\varphi_{m,n} : M_{2^n}(\mathbb{C}) \to M_{2^m}(\mathbb{C})$ places 2^{m-n} copies of a matrix $A \in M_{2^n}(\mathbb{C})$ along the diagonal, to make a matrix in $M_{2^m}(\mathbb{C})$.

For example,

$$\varphi_{1,2}\left(\begin{bmatrix} 1 & 2 \\ 3 & 4 \end{bmatrix}\right) = \begin{bmatrix} 1 & 2 & 0 & 0 \\ 3 & 4 & 0 & 0 \\ 0 & 0 & 1 & 2 \\ 0 & 0 & 3 & 4 \end{bmatrix}.$$

The inductive limit is a special kind, called a UHF algebra, and this particular one is usually denoted $U(2^\infty)$ in the literature.

Example 1.7.10 The inductive system of groups of 1.7.4 also has a kind of C*-algebraic analogue. Here the directed system is the natural numbers with the relation $n \leq m$ if and only if $n | m$.

If A is a k-by-k matrix, then we can place l copies of A along the diagonal of a kl-by-kl matrix

$$\varphi_{kl,k}(A) = \begin{bmatrix} A & & & \\ & A & & \\ & & \ddots & \\ & & & A \end{bmatrix}.$$

This procedure defines a *-homomorphism $\varphi_{lk,l}: M_l(\mathbb{C}) \rightarrow M_{kl}(\mathbb{C})$; put otherwise, if n and m are positive integers and $n|m$, then by the procedure just explained we obtain a canonical unital *-homomorphism $M_n(\mathbb{C}) \rightarrow M_m(\mathbb{C})$.

We set

$$\mathcal{N} = \varinjlim_n M_n(\mathbb{C})$$

to be the corresponding direct limit.

Example 1.7.11 The *Bunce–Deddens* algebra is defined by the following inductive system. For each natural number n, let $\phi_{n+1,n}: C(S^1, M_{2^n}(\mathbb{C})) \rightarrow C(S^1, M_{2^{n+1}}(\mathbb{C}))$ be the *-homomorphism $\phi_{n+1,n}(f)(z) := \begin{bmatrix} f(z^2) & 0 \\ 0 & f(z^2) \end{bmatrix}$. The Bunce–Deddens algebra B_{2^∞} is then defined by

$$B_{2^\infty} := \varinjlim C(S^1, M_{2^n}(\mathbb{C})).$$

One similarly can define B_{d^∞} for any natural number d.

Exercise 1.7.12 Let X be a locally compact Hausdorff topological space. Let $\mathcal{U}$ be the directed set of all pre-compact open subsets of X, under the inclusion relation.

If $U \subset V$ and $f \in C_0(U)$, then by extending f to zero on $V \setminus U$, we obtain a continuous function $\varphi_{V,U}(f) \in C_0(V)$. Check that this describes an inductive system $\{\varphi_{V,U}: C_0(U) \rightarrow C_0(V) \mid U \subset V\}$, and prove that the associated direct limit satisfies $C_0(X) \cong \varinjlim C_0(U)$.

Exercise 1.7.13 An important example of a C*-algebra connected to topological dynamics (of topological Markov chains) is the *Cuntz–Krieger* algebra associated to an n-by-n matrix A of 0's and 1's. Such a matrix can be interpreted as the adjacency matrix of a directed graph on n vertices, with an edge from vertex i to vertex j iff $A_{ij} = 1$.

The Cuntz–Krieger algebra O_A is the C*-algebra generated by n partial isometries $s_1, \ldots, s_n$ and satisfying

$$s_i^* s_i = \sum_{j=1}^n A_{ij} s_j s_j^*.$$

If A is a primitive matrix, then O_A is uniquely defined in this way and is simple. Cuntz–Krieger algebras were invented by J. Cuntz and W. Krieger [58]. If $\mu = (i_1, \ldots, i_k)$ is a word in $\{1, \ldots, n\}$, let $s_\mu := s_{i_1} \cdots s_{i_k}$. Prove that the subalgebra of O_A generated by the elements $s_\mu s_\nu^*$ with μ and ν words of equal length is an inductive limit of finite-dimensional C*-algebras.

1.8 Crossed Product C*-Algebras

In dynamical systems, one is generally interested in the long-term or asymptotic properties of a group action, such as, for instance, the action of the group of integers induced by a single homeomorphism $\varphi\colon X \to X$ of (usually) a compact space.

We might be interested, for example, in the case of such an integer action, in the orbit of a single point: how the orbit $\{\varphi^n(x)\}_{n\in\mathbb{Z}}$ wanders around the space.

One might also be interested in parameterizing the *set* of all the orbits. In fact, this set, the quotient of X by the equivalence relation $x \sim y$ if $y = \varphi^n(x)$ for some $n \in \mathbb{Z}$, has a natural topology: the quotient topology. One might therefore hope that one could study the *space* of orbits and acquire information about the dynamics or geometry by computing the standard invariants of algebraic topology of this space.

However, as the following example shows, in examples which are interesting from a dynamical system's point of view, the quotient space of a space by an interesting action is rarely any good as a topological space.

Example 1.8.1 Let $\omega = e^{2\pi i\theta} \in \mathbb{T}$ with θ irrational and let $R_\theta\colon \mathbb{T} \to \mathbb{T}$ be group multiplication by ω (in other words, rotation by the angle θ). Then every orbit of R_θ is dense in $\mathbb{T}$, and the quotient topology on the quotient space $\mathbb{Z}\backslash\mathbb{T}$ is trivial. This is because there are no nonempty proper open subsets of $\mathbb{T}$ invariant under R_θ.

The crossed product construction makes a C*-algebra which is a substitute in a certain sense for the (C*-algebra of continuous functions on the) space of orbits of the action, but which is *noncommutative*. This crossed product C*-algebra carries a great amount of interesting information about the action—unlike, in general, the quotient space. The C*-algebra $C(\mathbb{Z}\backslash\mathbb{T})$ of continuous functions on the quotient space of Example 1.8.1, with the quotient topology, is simply isomorphic to $\mathbb{C}$, the constant functions on $\mathbb{Z}\backslash\mathbb{T}$. But the C*-algebraic crossed product $C(\mathbb{T}) \rtimes_\theta \mathbb{Z}$ associated to the action of the integers generated by group translation by θ, called the *irrational rotation algebra*, is an extremely interesting, noncommutative C*-algebra, containing a great deal of fine geometric information about arithmetic properties of $\theta \in \mathbb{R}/\mathbb{Z}$, and the dynamics of the action.

Let G be a (countable) discrete group.

An *action* of G on a locally compact Hausdorff space X is a group homomorphism $G \to \mathrm{Homeo}(X)$ of G into the group of homeomorphisms of X. We say that X is a *G-space*.

If G acts on X, then G acts by C*-algebra automorphisms of the C*-algebra $C_0(X)$ by $f \mapsto f \circ g^{-1}$. More generally, a *group action on a C*-algebra* A is a group homomorphism $G \to \mathrm{Aut}(A)$, where $\mathrm{Aut}(A)$ is the group of *-automorphisms of A.

We call A a *G-C*-algebra*.

Definition 1.8.2 Let G be a discrete group and A be a G-C*-algebra. The *twisted group algebra* $A[G]$ is the vector space $C_c(G, A)$ of finitely supported functions from G to A, with multiplication and involution defined as follows.

We write an element of $C_c(G, A)$ in the form $\sum_{g \in G} a_g[g]$, with $a_g \in A$ being the value of the function at g, and where it is to be understood that $a_g = 0$ for all but finitely many $g \in G$. With this convenient notation, we equip $C_c(G, A)$ with an algebra multiplication and an involution:

(1.37)
$$\left(\sum_{g \in G} a_g[g]\right) * \left(\sum_{g \in G} b_g[g]\right) = \sum_{g,h} a_g g(b_h)[gh], \quad \left(\sum_{g \in G} a_g[g]\right)^* := \sum_{g \in G} g^{-1}(a_g^*)[g^{-1}].$$

Exercise 1.8.3 If $A = \mathbb{C}$, then $A[G]$ with twisted convolution is the same as the group algebra $\mathbb{C}[G]$ with convolution, as in (1.17). In particular, $\mathbb{C}[G]$ is a *-subalgebra of $A[G]$ if A is *unital*.

The algebra multiplication (1.37) is a kind of twisted version of the convolution operation (1.17) on scalar-valued functions on groups. It can be rewritten

(1.38) $(f_1 \star f_2)(g) = \sum_{h \in G} f_1(h)h\left[f_2(h^{-1}g)\right], \quad f_1, f_2 \in C_c(G, A).$

We generally prefer the group algebra notation, as it involves fewer brackets and seems more transparent.

Exercise 1.8.4 Answer the following questions about the twisted group algebra construction:

a) Prove that the map $A \to A[G]$, $a \mapsto a[e]$, with $e \in G$ the identity, is a *-homomorphism of *-algebras. Thus, A can be viewed as a *-subalgebra of $A[G]$, consisting of functions supported at the identity of the group.
b) Prove that if A is unital, then the elements $[g] \in A[G]$ are unitaries: $[g]^* = [g^{-1}] = [g]^{-1}$, that the resulting copy of the group G as unitaries in $A[G]$ satisfies $[g]a[g]^* = g(a)$ for all $g \in G$ and $a \in A$.
   Thus, $A[G]$ is a larger algebra than A, in which the original action of G on A becomes one by *inner automorphisms*.
c) If G acts on C*-algebras A and B, and if there is a G-equivariant *-isomorphism $\alpha\colon A \to B$, then α induces a canonical *-homomorphism of *-algebras $A[G] \to B[G]$.
d) More generally, let A and B be C*-algebras, G and G' be two discrete groups, with G acting on A and G' acting on A'. Suppose that $\alpha\colon A \to A'$ is a *-homomorphism, $\varphi\colon G \to G'$ is a group homomorphism, and that

(1.39) $\varphi(g)\big(\alpha(a)\big) = \alpha\big(g(a)\big).$

   Check that the map $A[G] \to A'[G']$ mapping $\sum_{g \in G} a_g[g]$ to $\sum_{g \in G} \alpha(a_g)[\varphi(g)]$ is a *-algebra homomorphism.
e) Suppose that B is a unital C*-algebra, $\alpha\colon A \to B$ is a *-homomorphism, and $\varphi\colon G \to \mathbf{U}(B)$ is a group homomorphism from G into the group of unitaries in B, such that the *covariance condition*

(1.40) $$\varphi(g)\alpha(a)\varphi(g)^* = \alpha\big(g(a)\big)$$

holds for all $a \in A$, $g \in G$. Then α and φ combine to make a *-homomorphism $A[G] \to B$, mapping $\sum_{g \in G} a_g[g]$ to $\sum_{g \in G} \alpha(a_g)\varphi(g)$.

The pair (α, φ) is called a *covariant pair*. Prove that any unital *-homomorphism $A[G] \to B$ to a unital C*-algebra arises from a covariant pair.

Example 1.8.5 Let $X = \{1, 2\}$ be the 2-point space, $A = C(X) \cong \mathbb{C} \oplus \mathbb{C}$, and $G = \mathbb{Z}/2$ act with the generator u flipping the points. Since the group has two elements, every element of $C(X)[G]$ can be written $f + g[u]$. Associate to $f + g[u]$ the matrix $\begin{bmatrix} f(1) & g(1) \\ g(2) & f(2) \end{bmatrix}$.

Under twisted convolution we have

$$(f + g[u])(f' + g'[u]) = ff' + gu(g') + \big(gu(f') + fg'\big)[u],$$

which is easily checked to correspond to the product of the two matrices.

The adjoint: $(f + g[u])^* = f^* + u(g^*)[u]$ (since $u = u^{-1}$) corresponds to the adjoint on 2-by-2 matrices.

Hence $C(\{1, 2\})[\mathbb{Z}/2] \cong M_2(\mathbb{C})$ as *-algebras.

Exercise 1.8.6 Generalize the above and prove that $C(X)[\mathbb{Z}/n] \cong M_n(\mathbb{C})$ if X is the n-point space $\{1, 2, \ldots, n\}$ and $\mathbb{Z}/n$ acts on X by shifting.

An important concept in noncommutative geometry is that of *Morita equivalence*. Matrix algebras $M_n(\mathbb{C})$ are all Morita equivalent to each other, and thus to $\mathbb{C}$. The above calculations show (with a bit more work) that if G is a finite group acting freely and transitively on a finite set, then the twisted group algebra $C(X)[G]$ is a matrix algebra $M_n(\mathbb{C})$ and hence is Morita equivalent to $\mathbb{C}$, or, in other words, to the algebra of continuous functions on a point. This is because the quotient $G \backslash X$ by the action is a single point in this case. As we will see, if one allows a variety of orbits, with stabilizers, then $C(X)[G]$ decomposes as a direct sum over the set of orbits, and each term in the sum is Morita equivalent to the group C*-algebra of the isotropy group of the orbit. These matters will be pursued later.

We proceed to define a C*-algebra $A \rtimes G$, by suitably completing $A[G]$.

Definition 1.8.7 Let the discrete group G act on the C*-algebra A by automorphisms, and suppose that $\pi : A \to \mathbb{B}(H)$ is an injective and unital *-homomorphism representing A on a Hilbert space H. Let $l^2(G, H)$ be the Hilbert space of L^2-functions on G valued in H, that is, the completion of $C_c(G, H)$ under the inner product $\langle f_1, f_2 \rangle := \sum_{g \in G} \langle f_1(g), f_2(g) \rangle$.

Define a *covariant pair*, in the sense of Exercise 1.8.4, e), and induced *-homomorphism which we denote by

(1.41) $$\mathrm{Ind}(\pi): A[G] \to \mathbb{B}\big(l^2(G, H)\big)$$

by

$$(g\xi)(h) = \xi(g^{-1}h), \quad (a \cdot \xi)(h) = \pi\big(h^{-1}(a)\xi(h)\big).$$

The *crossed product* $A \rtimes G$ is the completion of $A[G]$ in the norm

(1.42) $$\left\|\sum a_g[g]\right\| := \left\|\mathrm{Ind}(\pi)(\sum a_g[g])\right\|,$$

the norm on the right hand side being the operator norm on $l^2(G, H)$.

By definition $A \rtimes G$ comes equipped with a natural, injective representation, which we continue to denote by $\mathrm{Ind}(\pi)\colon A \rtimes G \to \mathbb{B}\big(l^2(G, H)\big)$, by extending the homomorphism (1.41).

Exercise 1.8.8 If G is a finite group, acting on X, then $C(X)[G]$ is already complete with respect to the norm defined in Definition 1.8.7, for any choice of π, and hence $A \rtimes G = A[G] = C(X)[G]$.

Note that if G acts trivially on the C*-algebra $\mathbb{C}$, then $\mathbb{C} \rtimes G$ is exactly the same thing as the reduced C*-algebra $C^*(G)$ of Definition 1.3.3, where, of course, for an injective representation of $\mathbb{C}$ we use the identity map.

Example 1.8.9 Let G act on X locally compact Hausdorff by homeomorphisms, with induced action by C*-algebra automorphisms on $C_0(X)$ by $g(f) := f \circ g^{-1}$. Let μ be a Borel measure on X of full support, so that $f \in C_c(X)$ $f \neq 0$ and $f \geq 0$ imply $\int f d\mu > 0$. Let $\pi_\mu \colon C_0(X) \to \mathbb{B}\big(L^2(X, \mu)\big)$ be the representation by multiplication operators.

Then $\mathrm{Ind}(\pi_\mu) =: \lambda_\mu$ is an injective representation of $C_0(X) \rtimes G$ on $l^2\big(G, L^2(X, \mu)\big)$. It is injective because π_μ is, because μ has full support.

The formulas for the covariant pair are given by

$$(g\xi)(h) = \xi(g^{-1}h), \quad (f \cdot \xi)(h) = (f \circ h) \cdot \xi(h).$$

Example 1.8.10 Let $G = \mathbb{Z}/2$ acting on $X := [-1, 1]$ by letting the generator u of G act by $u(x) = -x$. Then

$$C(X) \rtimes G \cong \{f\colon [0, 1] \to M_2(\mathbb{C}) \mid f \text{ is continuous, and } f(0) \text{ is a diagonal matrix}\}.$$

As G is finite, $C(X) \rtimes G = C(X)[G]$, i.e., no completion is involved in forming the crossed product. So there is no need to locate an injective representation of $C(X)$. Instead, to understand the C*-algebra better, define a covariant pair and induced *-homomorphism $C(X)[G] \to M_2(\mathbb{C})$ by letting the group generator u map to the constant (unitary) matrix-valued function $\begin{bmatrix} 0 & 1 \\ 1 & 0 \end{bmatrix}$ on $[0, 1]$. If $f \in C([-1, 1])$, we map f to the matrix-valued function on $[0, 1]$ (note the change of domain) given by

$$\tilde{f}(x) = \begin{bmatrix} f(x) & 0 \\ 0 & f(-x) \end{bmatrix}.$$

This is clearly a covariant pair (exercise). It is injective on functions because if both $f(x)$ and $f(-x)$ vanish on $[0, 1]$, then f vanishes on $[-1, 1]$. An element $f + g[u] \in C(X)[G]$ is mapped to the function T with

$$T(x) = \begin{bmatrix} f(x) & g(x) \\ g(-x) & f(-x) \end{bmatrix}$$

at $x \in [0, 1]$. Fix any $x \neq 0$. The collection of matrices $T(x)$ obtained from some f, g is then $M_2(\mathbb{C})$. To see this, note that the range is automatically a *-subalgebra of $M_2(\mathbb{C})$. Since we can find a continuous function f with value 1 at x and value 0 at $-x$, and setting $g = 0$, we obtain the matrix

$$T = \begin{bmatrix} f(x) & 0 \\ 0 & f(-x) \end{bmatrix} = \begin{bmatrix} 1 & 0 \\ 0 & 0 \end{bmatrix}.$$

Since the image under our *-homomorphism is $\begin{bmatrix} 0 & 1 \\ 1 & 0 \end{bmatrix}$, and these two matrices generate $M_2(\mathbb{C})$ as an algebra, we obtain that the possible set of values of $T(x)$ is all of $M_2(\mathbb{C})$, if $x \neq 0$.

If $x = 0$, consider the collection of matrices of the form

$$\begin{bmatrix} f(0) & g(0) \\ g(0) & f(0) \end{bmatrix},$$

for some continuous f, g on $[-1, 1]$. This is just the *-algebra of matrices

$$\begin{bmatrix} a & b \\ b & a \end{bmatrix},$$

and this collection is simultaneously diagonalizable using the eigenvectors $\frac{1}{\sqrt{2}}(1, 1)$ and $\frac{1}{\sqrt{2}}(1, -1)$. From this we conclude, as is easily checked by the definitions, that the algebra obtained at $x = 0$ is isomorphic to $\mathbb{C} \oplus \mathbb{C}$ by the map

$$\begin{bmatrix} a & b \\ b & a \end{bmatrix} \mapsto (a + b, a - b).$$

We obtain therefore a sort of "picture" of this C*-algebra, as the space of sections of a continuous field of C*-algebras over $[0, 1]$ equal to $M_2(\mathbb{C})$ for $x \neq 0$, and equal to $\mathbb{C} \oplus \mathbb{C}$ at $= 0$.

Actually, the contribution at $x = 0$ is the representation ring Rep($\mathbb{Z}/2$) of the isotropy group at that point. The relation of $[0, 1]$ to the original space $[-1, 1]$ on which the group acts is that $[0, 1]$ is naturally homeomorphic to the quotient space of $[-1, 1]$ by the $\mathbb{Z}/2$-action.

Before discussing more examples of crossed products, we discuss an important construction. We start with the following exercise. It deals with whether or not it makes sense to speak of infinite "expansions" $a = \sum_h a_h[h]$ in $A \rtimes G$, as one does for the twisted group algebra $A[G]$.

Exercise 1.8.11 Let $a = \sum a_h[h] \in A[G]$. Show that the operator $\mathrm{Ind}(\pi)([g^{-1}]a])$ on $l^2(G, H) = \oplus_{h \in G} H$ leaves the summand of the direct sum over $g = e$, a copy of H, invariant, and acts by the operator $\pi(a_g)$. Thus, the coefficients a_g of an element of the twisted group algebra, or at least their images $\pi(g)$ under the injective representation π, have an interpretation purely in terms of the operator $\mathrm{Ind}(\pi)(a)$.

Suppose that we declare for $a \in A \rtimes G$ to have the "expansion" $a = \sum_h a_h[h]$, if $a_g = \mathrm{Ind}(\pi)([g^{-1}a])$ for all $g \in G$. Check that (from the previous paragraph) this agrees with the usual expansions of elements of $A[G]$, and if $a_n \in A[G], a \in A \rtimes G$, and $a_n \to a$ in the crossed product, then the coefficients $(a_n)_g$ of a_n converge in A to the coefficients a_g of a.

Exercise 1.8.12 Let $a_e \in A$ be the coefficient of e in an expansion $a = \sum_{h \in G} a_h[h] \in A[G]$ of an element of the twisted group algebra:

a) Show that $A[G] \to A, a \mapsto a_e$ extends continuously to a norm contractive linear map $E \colon A \rtimes G \to A$.
b) Prove that if $a \in A \rtimes G$ has expansion $a = \sum_h a_h[h]$ in the sense of Exercise 1.8.11, then $E(a) = a_e$.
c) Prove that $E(a^*a) \geq 0$, so E is positive.
d) Prove that if $\mu \colon A \to \mathbb{C}$ is a state, then $\tau_\mu := \mu \circ E$ is a state on $A \rtimes G$, and that if μ is G-invariant in the sense that $\mu(g(a)) = \mu(a)$ for all $g \in G, a \in A$, then τ_μ is a *trace*: $\tau_\mu(ab) = \tau_\mu(ba)$.
e) Show that if G acts by homeomorphisms on X compact preserving a probability measure μ, then $\tau_\mu(\sum_h f_h[h]) := \int_X f_e d\mu$ defines a tracial state on $C(X) \rtimes G$.
f) Prove that if $\tau \colon C(X) \rtimes G \to \mathbb{C}$ is a tracial state, where X is compact, then $\tau = \tau_\mu$ for some G-invariant probability measure μ.

There is of course an enormous variety of (discrete) group actions. It is useful sometimes to conceptually classify them into three types. These correspond approximately to the types, in the theory of von Neumann algebras, of the von Neumann closures of their C*-algebras.

Type I examples are proper actions, whose C*-algebras are discussed in the following chapters. Example 1.8.10 is of this kind. Up to Morita equivalence, these examples are nearly commutative, although not quite. They are important in topology.

Type II examples involve actions $G \times X \rightarrow X$ of compact spaces where the action preserves a probability measure. See Exercise exercise:dsoifjsdjflskdf e). The invariant measure determines trace $\tau_\mu \colon C(X) \rtimes G \rightarrow \mathbb{C}$. The irrational rotation algebra discussed below involves an interesting dynamics of the circle by the integers, leaving Lebesgue measure invariant, and is an excellent example of a Type II action.

Example 1.8.13 (Rotation by an Irrational Angle) Let $\omega = e^{2\pi i\theta} \in \mathbb{T}$ with θ irrational. The corresponding homeomorphism $R_\theta \colon \mathbb{T} \rightarrow \mathbb{T}$ of group multiplication by ω, or rotation by θ, has infinite order, since ω has infinite order (an easy exercise). We obtain an action of the integers $\mathbb{Z}$ on $\mathbb{T}$ and on $C(\mathbb{T})$ with the integer n acting by $f \mapsto f \circ R_\theta^{-n}$.

The corresponding C*-algebra crossed product, which we frequently denote in the form $C(\mathbb{T}) \rtimes_\theta \mathbb{Z}$, is called the *irrational rotation algebra* and is denoted A_θ.

It has some remarkable properties, among which is its *simplicity*: It has no proper, nonzero closed ideals.

As an injective representation of $C(\mathbb{T})$ to build the crossed product, we can take $H := L^2(\mathbb{T})$ with $C(\mathbb{T})$ acting by multiplication operators. We then complete the twisted group algebra $C(\mathbb{T})[\mathbb{Z}]$ by letting it act as operators on $l^2(\mathbb{Z}, L^2(\mathbb{T}))$ as described above. The group algebra $C(\mathbb{T})[\mathbb{Z}]$ consists of all finite sums $\sum_{n=-N}^{N} f_n[n]$ with $f_n \in C(\mathbb{T})$. If we require f to be a trigonometric polynomial, we obtain a smaller *-subalgebra consisting of all finite double sums $\sum a_{n,m} z^n [m]$, and these act on the Hilbert space $l^2(\mathbb{Z}, L^2(\mathbb{T}))$ by operators of the form

$$(1.43) \qquad \sum a_{n,m} u^n v^m,$$

where u is the operator on $l^2(\mathbb{Z}, L^2(\mathbb{T}))$ corresponding to the function $f(z) = z \in C(\mathbb{T}) \subset C(\mathbb{T})[\mathbb{Z}]$ and v the operator corresponding to the unitary group generator $[1] \in \mathbb{C}[\mathbb{Z}] \subset C(\mathbb{T})[\mathbb{Z}]$ of the integers $\mathbb{Z}$.

If we use the standard identification of $L^2(\mathbb{T})$ as $l^2(\mathbb{Z})$, we can identify $l^2(\mathbb{Z}, L^2(\mathbb{T}))$ with $l^2(\mathbb{Z} \oplus \mathbb{Z})$ with orthonormal basis $e_{n,m}$, with $e_{n,m}$ corresponding to $z^n[m] \in l^2(\mathbb{Z}, L^2(\mathbb{T}))$.

Exercise 1.8.14 In this notation,

$$u(e_{n,m}) = e_{n,m+1}, \quad v(e_{n,m}) = \omega^n \cdot e_{n,m+1},$$

so that u by a vertical shift and v by a horizontal weighted shift.

Verify the relation

$$(1.44) \qquad\qquad uv = \omega \cdot vu$$

for this pair of unitary operators. Of course this follows from a similar relation in the group algebra $C(\mathbb{T})[\mathbb{Z}]$. If $f(z) = z$, then $(f \circ R_\theta^{-1})(z) = \bar{\omega} z$. Hence in group algebra notation,

$$[1]z[1]^* = \bar{\omega} \cdot z.$$

Remark 1.8.15 It turns out that A_θ is the unique C*-algebra, up to canonical isomorphism, generated by a pair u, v of unitaries satisfying $uv = \omega \cdot vu$, i.e., satisfying (1.44).

Roughly, Type III examples of actions are those which leave invariant *no* probability measure. In this case, the C*-algebras $C(X) \rtimes G$ therefore have no traces, by Exercise 1.8.12 f). Some interesting examples of Type III examples involving group actions on boundaries are discussed at several points in this book. The following is a simple instance of them.

Example 1.8.16 Let $G = \mathbb{F}_2$ be the free group on 2-generators a, b. Elements of $\mathbb{F}_2$ may be written uniquely as reduced words $s_1 \cdots s_n$ with s_i in the generating set $S := \{a, a^{-1}, b, b^{-1}\}$, where a word is *reduced* if it contains no occurrence of an $s_i s_i^{-1}$.

The collection of *infinite* reduced words $s_1 s_2 \cdots$ is a subspace of the product space $\prod_{n=1}^\infty S$. It consists of those sequences $s_1, s_2, \cdots$ for which no term s_i is followed by s_i^{-1}. It is an easy exercise to prove that this is a closed subspace of the product space. We let $\partial \mathbb{F}_2$ be the collection of infinite such words, with the subspace topology. It is a Cantor set. If w is a reduced word in the generators, we let $U_w \subset \partial \mathbb{F}_2$ be the set of infinite words $s_1 s_2 \cdots$ which begin with w. Then the U_w's form a basis for the topology.

The left translation action of $\mathbb{F}_2$ on itself extends to an action of $\mathbb{F}_2$ on $\partial \mathbb{F}_2$ by left group multiplication on infinite reduced words. For example the group element $g = ab^{-2} \in \mathbb{F}_2$ maps the boundary point $\xi = baba^{-1}bbb \cdots \in \partial \mathbb{F}_2$ to $g(\xi) = ab^{-1}aba^{-1}bbb \cdots$.

The dynamics of this group action is very interesting, and the associated crossed product C*-algebra has some special properties not possessed by the irrational rotation algebra A_θ.

Exercise 1.8.17 Let $i \in S$ be a generator, and $U_i \subset \partial \mathbb{F}_2$ the clopen subset of all infinite reduced words beginning in i. Let $\chi_i := \chi_{U_i} \in C(\partial \mathbb{F}_2)$ and

$$s_i := \chi_s[i] \in C(\partial \mathbb{F}_2)[\mathbb{F}_2] \subset C(\partial \mathbb{F}_2) \rtimes \mathbb{F}_2.$$

a) Prove that the s_i's, for $i \in S$ are partial isometries and that

(1.45) $$\sum_{i \in S} s_i s_i^* = 1, \quad \text{and} \quad s_j^* s_j = \sum_{i \neq j^{-1}} s_i s_i^*.$$

b) If $g = i_1 \cdots i_k$ is a reduced word in $\mathbb{F}_2$, let $s_g := s_{i_1} \cdots s_{i_k}$. Prove that

$$s_g s_g^* = \chi_{U_g},$$

where U_g is all infinite reduced words in $\partial \mathbb{F}_2$ which begin with g.

c) Prove that $\{s_i \mid i \in S\}$ generates $C(\partial \mathbb{F}_2) \rtimes \mathbb{F}_2$. This is an example of a *Cuntz–Krieger algebra*. Such algebras occur in connection with topological Markov chains and are among the most important basic examples of C*-algebras. See [58].

The following two exercises give two proofs that $C(\partial \mathbb{F}_2) \rtimes \mathbb{F}_2$ has no traces—is "Type III."

Exercise 1.8.18 Deduce from (1.45) that there is no nonzero trace $\tau \colon C(\partial \mathbb{F}_2) \rtimes \mathbb{F}_2 \to \mathbb{C}$. (*Hint.* In the notation of the displayed equation, τ was a trace and then $\tau(s_i s_i^*) = \tau(s_i^* s_i)$ for all i.)

Exercise 1.8.19 Let $s \in \mathbb{F}_2$ be one of the generators.

Prove that if μ were any (s)-invariant probability measure on $\partial \mathbb{F}_2$, where (s) is the subgroup generated by s, then μ is supported at the two fixed points of s.

Deduce that $\partial \mathbb{F}_2$ has no probability measure invariant under $\mathbb{F}_2$.

The action of $\mathbb{F}_2$ on $\partial \mathbb{F}_2$ is a special case of a more general construction which applies to the class of *Gromov hyperbolic groups* G, for which Gromov has defined a natural geometric boundary ∂G, on which G acts by homeomorphisms. See Section 10.4. The boundary determines a compactification $\overline{G} = G \cup \partial G$, all invariant under the group action. The dynamics of G on its boundary has many interesting and important features (see [91]). Example 1.8.20 is of this kind, involving the (elementary) hyperbolic group $\mathbb{Z}$. The absence of traces and the important *purely infinite* property are verified for these examples in [150] and [4].

Example 1.8.20 Let $\overline{\mathbb{Z}} := \mathbb{Z} \cup \{\pm\infty\}$ the usual 2-point compactification of the integers. Let the group of integers act on this set by translation, fixing the points at $\pm\infty$. This induces an action of $\mathbb{Z}$ on the C*-algebra $C(\overline{\mathbb{Z}})$.

The C*-algebra $C(\overline{\mathbb{Z}})$ is represented on $l^2(\mathbb{Z})$ by multiplication operators. Together with the left regular representation of $\mathbb{Z}$ on the same Hilbert space, we obtain a covariant pair and *-homomorphism

$$C(\overline{\mathbb{Z}})[\mathbb{Z}] \to \mathbb{B}\big(l^2(\mathbb{Z})\big).$$

For the following exercise, the reader may assume the fact (which is not obvious) that this representation extends continuously to an injective representation of $C(\overline{\mathbb{Z}}) \rtimes \mathbb{Z}$ on $l^2(\mathbb{Z})$.

Exercise 1.8.21 A *corner* of a C*-algebra A is a sub-C*-algebra of the form pAp, where $p \in A$ is a projection. A corner is *full* if ApA (the ideal generated by p) is dense in A. Prove that the Toeplitz algebra is a corner of the crossed product C*-algebra $C(\overline{\mathbb{Z}}) \rtimes \mathbb{Z}$, but that it is not full. What is the ideal generated by p?

Example 1.8.22 (Rational Rotations of the Circle) Let $\mathbb{Z}/n$, the cyclic group of order n, be realized as the corresponding subgroup of roots of unity in the circle $\mathbb{T}$. This subgroup then acts by group multiplication on $\mathbb{T}$. Taking a primitive nth root of unity $\omega = \exp(\frac{2\pi i}{n}) \in \mathbb{T}$ representing the generator of $\mathbb{Z}/n$, its action on $\mathbb{T}$ is rotation by $\frac{2\pi}{n}$ radians.

For a faithful representation of $A = C(\mathbb{T})$, we use the representation by multiplication operators on $L^2(\mathbb{T})$. The Hilbert space $l^2(\mathbb{Z}/n, L^2(S^1))$ may be identified with $L^2(\mathbb{T}, \mathbb{C}^n)$. Let $f \in C(\mathbb{T})$, then by the definitions:

- f acts by the $M_n(\mathbb{C})$-valued function

$$\tilde{f}(z) = \begin{bmatrix} f(z) & & & \\ & f(\omega z) & & \\ & & \cdots & \\ & & & f(\omega^{n-1} z) \end{bmatrix},$$

  where we let such matrix-valued functions act on $L^2(\mathbb{T}, \mathbb{C}^n)$ in the obvious way.
- The group $\mathbb{Z}/n$ acts on $L^2(\mathbb{T}, \mathbb{C}^n)$ by the unitary representation implemented by sending the generator ω to the shift

$$U := \begin{bmatrix} 0 & 0 & \cdots & 1 \\ 1 & 0 & \cdots & 0 \\ 0 & 1 & \cdots & \cdots \\ \cdots & \cdots & \cdots & \cdots \\ 0 & \cdots & 1 & 0 \end{bmatrix}.$$

A change of coordinates makes things more clear. The Hilbert space $\mathbb{C}^n$ with $\mathbb{Z}/n$ acting by the shift is the regular representation of $\mathbb{Z}/n$, and it decomposes into character spaces $V_k = \{v \in \mathbb{C}^n \mid Uv = \omega^k v\}$ for the set of characters of $\mathbb{Z}/n$, which may be identified with the set of nth roots of unity, i.e., the points $1, \omega, \omega^2, \ldots, \omega^{n-1}$ (each power ω^k determines a character of $\mathbb{Z}/n$ by sending the generator ω to ω^k). So we have another orthonormal basis $v_0 \ldots, v_{n-1}$ for $\mathbb{C}^n$ with, explicitly, v_k the vector $\frac{1}{\sqrt{n}}(\omega^{-k}, \omega^{-2k}, \omega^{-3k}, \ldots, \omega^{-nk})$. With respect to this orthonormal basis, U acts by the (constant) diagonal matrix-valued function

$$U(z) = \begin{bmatrix} 1 & & & \\ & \omega & & \\ & & \cdots & \\ & & & \omega^{n-1} \end{bmatrix},$$

while the function $f(z) = z$ on the circle acts by the weighted shift

$$V(z) = z \begin{bmatrix} 0 & 0 & \cdots & \omega^{n-1} \\ \omega & 0 & \cdots & 0 \\ 0 & \omega^2 & \cdots & \cdots \\ \cdots & \cdots & \cdots & \cdots \\ 0 & \cdots & 1 & 0 \end{bmatrix}.$$

Let P_0 be the diagonal matrix $\begin{bmatrix} 1 & & & \\ & 0 & & \\ & & \cdots & \\ & & & 0 \end{bmatrix}$. It is projection onto the span of

the trivial character 1 (it is the matrix representation in this basis of the element $\frac{1}{n} \sum_{k=0}^{n-1} \omega^k \in C_r^*(\mathbb{Z}/n) \subset C(\mathbb{T}) \rtimes \mathbb{Z}/n$). An easy computation shows that $V(z)^i P_0 V(z)^{-j} = z^{i-j} \omega^{i-j} E_{ij}$ where E_{ij} is the matrix with a 1 in the (i, j)th entry and has zeros in all other entries.

As the crossed product also contains a copy of $C(\mathbb{T})$, identified with continuous functions valued in scalar multiples of the identity operator on $\mathbb{C}^n$, it follows that $C(\mathbb{T}) \rtimes \mathbb{Z}/n$ contains all matrix-valued functions of the form $f(z) E_{ij}$, for any $f \in C(\mathbb{T})$. Therefore it contains every element of $C(\mathbb{T}, M_n(\mathbb{C}))$.

This argument shows therefore that

Proposition 1.8.23 $C(\mathbb{T}) \rtimes \mathbb{Z}/n \cong C(\mathbb{T}, M_n(\mathbb{C}))$.

Exercise 1.8.24 The Bunce–Deddens algebra B_{2^∞} appears in Example 1.7.11. Let G be the group of diadic rationals in the circle $G = \cup_{n=0}^\infty \{\omega \in \mathbb{T} \mid \omega^{2^n} = 1\}$. Prove that $C(\mathbb{T}) \rtimes G \cong B_{2^\infty}$ by constructing an appropriate inductive system.

Exercise 1.8.25 Prove that the *-algebras $A_\theta := C(\mathbb{T}) \rtimes_\theta \mathbb{Z}$ and $A_{-\theta} := C(\mathbb{T}) \rtimes_{-\theta} \mathbb{Z}$ are isomorphic, for any $\theta \in \mathbb{R}$ irrational, by constructing an appropriate covariant pair (see Exercise 1.8.4).

We close this section with the definition of crossed products by general (not necessarily discrete) locally compact groups. We will not work much with this level of generality in this book, so we will be brief. C*-algebras of locally compact groups were discussed to some extent in Section 1.3.

If G is a locally compact group, a *G-C*-algebra* is a C*-algebra A equipped with a group homomorphism $G \to \mathrm{Aut}(A)$ satisfying $g \mapsto g(a)$ is continuous from G to A, for every $a \in A$.

In this situation, the formula (1.38) is generalized by defining the twisted convolution of $f_1, f_2 \in C_c(G, A)$ by

$$(f_1 * f_2)(g) := \int_G f_1(h) h \left(f_2(h^{-1} g) \right) d\mu(h).$$

By a routine exercise, $f_1 * f_2 \in C_c(G, A)$. We can make $C_c(G, A)$ into a *-algebra by defining

$$f^*(g) := g\left(f(g^{-1})^*\right) \cdot \delta(g^{-1}),$$

where δ is the modular function of G.

Now fix an injective representation $\pi : A \to \mathbb{B}(H)$, and as in the discrete case we represent $C_c(G, A)$ on $L^2(G, H)$ by the covariant pair

$$(\rho(a)\xi)(g) := \pi\left(g^{-1}(a)\right)\xi(g), \qquad (\rho(g)\xi)(h) := \xi(g^{-1}h).$$

This induces a representation $\rho = \mathrm{Ind}(\pi)$ of the *-algebra $C_c(G, A)$ on the Hilbert space
$L^2(G, H)$ by "integrating" this covariant pair:

$$(\rho(f)\xi)(g) := \int_G \pi\left(f(h)\right)\xi(h^{-1}g)d\mu(h), \quad f \in C_c(G, A).$$

Then the crossed product $A \rtimes G$ is by definition the completion of $C_c(G, A)$ with respect to the norm $\|f\| := \|\pi(f)\|$.

An important family of examples are sometimes called *time evolutions*, which refer to actions of $\mathbb{R}$ on C*-algebras. The C*-algebra $C^*(\mathbb{R})$ of the real line is discussed in the next section. This is the crossed product $\mathbb{C} \times \mathbb{R}$ of $\mathbb{R}$ acting on $\mathbb{C}$ (trivially, of course).

The following two exercises give other interesting examples of $\mathbb{R}$-C*-algebras.

Example 1.8.26 Let $\mathbb{T}^2 = \mathbb{R}^2/\mathbb{Z}^2$ be the 2-torus, and let $v \in \mathbb{R}^2$ be a vector. For $f \in C(\mathbb{T}^2)$, define $\alpha_t(f)(w) := f(w + tv)$, for $w \in \mathbb{T}^2$. This is a time evolution on $C(\mathbb{T}^2)$. It underlies the flow $w \mapsto w + tv$ on $\mathbb{T}^2$, which is especially interesting when v has irrational slope, and is called *Kronecker flow*.

As we show later, the C*-algebra $C(\mathbb{T}^2) \rtimes \mathbb{R}$ is Morita equivalent to the irrational rotation algebra $A_\hbar$, where $\hbar$ is the slope of v.

Example 1.8.27 This example continues Example 1.8.16. On the boundary $\partial\mathbb{F}_2$ of the free group, define a probability measure as follows. If w is a reduced word, let U_w be the basic clopen set of all infinite reduced words which begin in w. There are 4 generators a, a^{-1}, b, b^{-1}, and for each such generator s we define $\mu(U_s) := \frac{1}{4}$. For $w = s_1 \cdots s_k$ a reduced word of length > 1, we let $\mu(U_w) := \frac{3}{4}3^{-k}$, so that the sum $\sum_{|w|=k} \mu(U_w) = 1$.

This defines a probability measure on $\partial\mathbb{F}_2$, as one may verify.

The measure μ is not $\mathbb{F}_2$-invariant, but its measure class is left invariant, and if $g \in \mathbb{F}_2$, then

$$\int_{\partial\mathbb{F}_2} f(g\xi)d\mu(\xi) = \int_{\partial\mathbb{F}_2} f(\xi)\frac{d(g_*\mu)}{d\mu}d\mu.$$

The functions

$$\sigma_g := \frac{d(g_*\mu)}{d\mu} d\mu$$

are discrete Gaussian probability distributions on the boundary and are in particular continuous. See Lemma 10.4.14.

We now define a time evolution on the crossed product $C(\partial\mathbb{F}_2) \rtimes \mathbb{F}_2$ by setting

$$\sigma_t\left(\sum f_g[g]\right) := \sum \sigma_g^{it} f_g[g].$$

Exercise 1.8.28 Verify that σ satisfies the "Chain Rule" $\sigma_{gh} = (\sigma_h \circ g^{-1}) \cdot \sigma_g$. Deduce that $\sigma_t : C(\partial\mathbb{F}_2)[\mathbb{F}_2] \to C(\partial\mathbb{F}_2)[\mathbb{F}_2]$ is a *-automorphism, for all $t \in \mathbb{R}$. Show that σ_t extends continuously to an automorphism of $C(\partial\mathbb{F}_2) \rtimes \mathbb{F}_2$ and that $C(\partial\mathbb{F}_2) \rtimes \mathbb{F}_2$ is an $\mathbb{R}$-C*-algebra with this action.

The crossed product $C(\partial\mathbb{F}_2) \rtimes \mathbb{F}_2 \rtimes_\sigma \mathbb{R}$ of the "Type III" example $C(\partial\mathbb{F}_2) \rtimes \mathbb{F}_2$ is of a "Type II" character: It has a (densely defined) trace.

Exercise 1.8.29 Let $\tau\left(\sum f_g[g]\right) := \int_{\partial\mathbb{F}_2} f d\mu$:

a) Prove that τ extends continuously to a bounded linear functional $\tau : C(\partial\mathbb{F}_2) \rtimes \mathbb{F}_2 \to \mathbb{C}$.

b) Let $(\sigma_t)_{t\in\mathbb{R}}$ be the time evolution of Exercise 1.8.28. Show that for elements $a \in C(\partial\mathbb{F}_2)[\mathbb{F}_2]$ in the group algebra, the function $t \mapsto \sigma_t(a)$ analytically extends to $\mathbb{C}$. Let $\sigma_z(a)$ denote the value of this extension to $z \in \mathbb{C}$. Show that for each $z \in \mathbb{C}$, $\sigma_z : C(\partial G)[G] \to C(\partial G)[G]$ is an algebra homomorphism, but that it does not extend to a C*-algebra automorphism unless $z \in \mathbb{R}[i]$.

c) Show that

$$\tau(\sigma_{-i}(a)b) = \tau(ba).$$

This is called the KMS$_\beta$ condition.

d) Define, for $f \in C_c(\mathbb{R}, C(\partial\mathbb{F}_2) \rtimes \mathbb{F}_2)$,

$$\tilde{\tau}(f) = \tau\left(f(0)\right).$$

Show that $\tilde{\tau} : C_c(\mathbb{R}, C(\partial\mathbb{F}_2) \rtimes \mathbb{F}_2) \to \mathbb{C}$ has the tracial property: $\tilde{\tau}(ab) = \tilde{\tau}(ba)$. Thus, the crossed product $C(\partial\mathbb{F}_2) \rtimes \mathbb{F}_2 \rtimes_\alpha \mathbb{R}$ has a densely defined trace.

We have for the most part avoided dealing with the more general class of locally compact group crossed products in this book—with a few exceptions.

In fact, many of the interesting actions of locally compact groups appearing in dynamics, especially foliation theory, are equivalent in a certain sense (Morita equivalent) to discrete group actions. Therefore, from a certain point of view, crossed products by *discrete* groups are really the essential examples.

The C*-algebra of the real line $\mathbb{R}$ is important for other reasons and is discussed in more detail in the next section.

1.9   The C*-Algebra and Fourier Transform for the Group of Real Numbers

The structure of the group C*-algebra $C^*(\mathbb{R})$ of the topological group of real numbers is, as with compact or discrete abelian groups, determined by the Fourier transform for $\mathbb{R}$, see [83] for a good exposition.

Let $f \in C_c(\mathbb{R})$, $\lambda(f)\colon L^2(\mathbb{R}) \to L^2(\mathbb{R})$ be the operator of convolution by f:

$$(1.46) \qquad (\lambda(f)u)(x) := (f * u)(x) := \int_{\mathbb{R}} f(y)u(x - y)dy.$$

The integral converges absolutely, by the Cauchy–Schwarz inequality, and using translation invariance of Lebesgue measure $|(f * g)(x)|m \leq \|f\|_2 \cdot \|u\|_2$ where $\|\cdot\|_2$ denotes the L^2-norm. More generally, we have:

Exercise 1.9.1 Let $1 \leq p \leq \infty$ and $f \in L^1(\mathbb{R})$ and $g \in L^p(\mathbb{R})$. Then the integral (1.46) defining $f * g$ exists for a.e. $x \in \mathbb{R}$, and $\|f * g\|_p \leq \|f\|_1 \cdot \|g\|_p$. (*Hint.* Follows from Exercise 1.2.35 with $k(x, y) = f(y - x)$.)

In particular, $f * u \in L^2(\mathbb{R})$ if $f \in L^1(\mathbb{R})$ and $u \in L^2(\mathbb{R})$ and $\|f * u\|_2 \leq \|f\|_1 \cdot \|u\|_2$ so we get a representation $\lambda\colon L^1(\mathbb{R}) \to \mathbb{B}(L^2(\mathbb{R}))$ of the Banach *-algebra $L^1(\mathbb{R})$ (see Exercise 1.3.16).

The C*-algebra of $\mathbb{R}$ is the completion of this Banach *-algebra to a C*-algebra, using the operator norm on $L^2(\mathbb{R})$.

The following heuristic is sometimes helpful, if one wants to remember how convolution works.

Let $U_t\colon L^2(\mathbb{R}) \to L^2(\mathbb{R})$ by the unitary translation operator by t, $U_t v(x) = v(x - t)$. Then, if $f \in L^1(\mathbb{R})$, then

$$(1.47) \qquad \lambda(f) = \int f(t)U_t dt \in C^*(\mathbb{R}),$$

which is an integral version of the group algebra notation $\sum a_g[g]$ we were using before, for discrete group C*-algebras. The integral converges absolutely since $f \in L^1(\mathbb{R})$.

If one applies the operator-valued integral (1.47), somewhat formally, to an L^2-function v, and evaluate at $x \in \mathbb{R}$, one gets the formula (1.46). Thus:

$$\left(\int f(t)U_t v \, dt \right)(x) = \int f(t)(U_t v)(x) \, dt = \int f(t)v(x - t)dt = (f * v)(x).$$

Remark 1.9.2 Note that the unitaries $U_t \in \mathbb{B}(L^2\mathbb{R})$ are not in $C^*(\mathbb{R})$, but they are *multipliers* of $C^*(\mathbb{R})$, see Section 5.3 for this concept. That is, left multiplication by the unitary operator U_t maps $C^*(\mathbb{R}) \subset \mathbb{B}(L^2\mathbb{R})$ to itself. Indeed, if $f \in C_c(\mathbb{R})$, then $U_t \cdot \lambda(f) = \lambda(f_t)$ by an easy computation, where $f_t(x) = f(x - t)$.

Exercise 1.9.3 Prove that:

a) The map $\mathbb{R} \to U(L^2(\mathbb{R}))$, $t \mapsto U_t$, is continuous as a map from $\mathbb{R}$ with its standard topology, and $\mathbb{B}(L^2\mathbb{R})$ with the strong operator topology.
b) The integral (1.47) converges in the norm topology of $C^*(\mathbb{R})$ if $f \in L^1(\mathbb{R}) \subset C^*(\mathbb{R})$.

Exercise 1.9.4 Suppose that $(k_t))_{t>0}$ is a family of continuous functions on $\mathbb{R}$ such that for some $C \geq 0$, $\int |k_t(x)| dx \leq C$ for all $t > 0$, $\int k_t(x) dx = 1$, and for all $\delta > 0$, $\lim_{t \to 0} \int_{|x| \geq \delta} |k_t(x)| dx = 0$.

Then if $f \in C_b(\mathbb{R})$, then the convolution integral $(f * k_t)(x) := \int f(y) k_t(x - y) dy$ is absolutely convergent and $f * k_t \to f$ uniformly on compact subsets of $\mathbb{R}$.

If $\xi \in \mathbb{R}$ is a real number, then $\chi_\xi(t) := e^{2\pi i t \xi}$ is a character $\chi_\xi : \mathbb{R} \to \mathbb{T}$ of the group of real numbers, and, conversely, all characters arise in this way.

Exercise 1.9.5 Let $\chi : \mathbb{R} \to \mathbb{T}$ be a continuous group homomorphism. Let $a > 0$ be suitable. Use the identity

$$\chi(s) \cdot \int_0^a \chi(t) dt = \int_s^{s+a} \chi(t) dt$$

to deduce that χ satisfies a differential equation $\chi'(s) = C\chi(s)$ and deduce that $\chi(s) = e^{2\pi i s \xi}$ for some $\xi \in \mathbb{R}$.

Thus, $\mathbb{R}$ may be canonically identified with its own Pontryagin dual $\widehat{\mathbb{R}}$. With this identification, the Fourier transform as defined in (1.21) has the form

$$(1.48) \qquad \hat{u}(\xi) := (F_{\mathbb{R}} u)(\xi) := \int u(x) e^{-2\pi i x \xi} dx,$$

where for initial purposes, we can take $u \in C_c^\infty(\mathbb{R})$, but the integral converges absolutely if merely $u \in L^1(\mathbb{R})$.

The following two features of the Fourier transform are key: Let u be any measurable function on $\mathbb{R}$:

a) If u has *rapid decay*, i.e., if $p(x)u(x)$ is bounded for every polynomial $p(x)$ (so that in particular $u \in L^1(\mathbb{R})$), then the integral defining $\hat{u}$ converges absolutely at each point and $\hat{u}$ is an infinitely differentiable function.
b) If u is infinitely differentiable with all derivatives in $L^1(\mathbb{R})$, then $\hat{u}$ has rapid decay, i.e., $\hat{u}(\xi) p(\xi)$ is bounded for every polynomial $p(\xi)$.

To prove a), use the Dominated Convergence theorem to prove that $\hat{u}$ is differentiable everywhere with

$$\hat{u}'(\xi) = -2\pi i \int u(x) x e^{-2\pi i x} dx,$$

i.e., differentiate under the integral sign. Thus,

(1.49) $$\hat{u}'(\xi) = 2\pi i \widehat{xu}(\xi).$$

Smoothness is proved inductively using this idea.
For b) we write

$$2\pi i \xi \hat{u}(\xi) = 2\pi i \int u(x)\xi e^{-2\pi i x\xi} dx = -\int u(x)(e^{-2\pi i x\xi})' \, dx$$

$$= \int u'(x)e^{-2\pi i x\xi} dx$$

by integration by parts.
Thus

(1.50) $$\widehat{u'}(\xi) = 2\pi i \xi \hat{u}(\xi).$$

So b) is proved by an obvious inductive argument.

Exercise 1.9.6 Let $h(x) = e^{-\pi a x^2}$ where $a > 0$. Then

$$\hat{h}(\xi) = \frac{1}{\sqrt{a}} e^{-\frac{\pi \xi^2}{a}}.$$

(*Hint.* Differentiate under the integral sign in $\hat{h}(\xi) := \int e^{-\pi a x^2} e^{-2\pi i x\xi} d\xi$ and use integration by parts to get $\hat{h}'(\xi) = \frac{-2\pi \xi}{a} \xi \cdot \hat{h}(\xi)$. Solve this differential equation, using that $\int e^{-\pi a x^2} dx = \frac{1}{\sqrt{a}}$, to fix the constant.)

Because Fourier transforms exchange differentiation and multiplication, a natural domain for it is the *Schwarz algebra* $S(\mathbb{R})$ of the real line, defined as follows.

Definition 1.9.7 The Schwarz algebra $S(\mathbb{R})$ is

$$S(\mathbb{R}) := \{f \in C^\infty(\mathbb{R}) \mid \forall m, n \geq 0, \ \sup_{x \in \mathbb{R}} |(1 + |x|)^m \, \partial^n f(x)| < \infty\},$$

where $\partial^n f$ is the n-th derivative of f.

The Schwarz space S has a natural structure of topological vector space with seminorms $p_{n,m}(f) := \sup_{x \in \mathbb{R}, 0 \leq k \leq m} (1 + |x|)^n |\partial^k f(x)|$.

Exercise 1.9.8 Prove that the convolution $f * g$ of two Schwarz functions $f, g \in S$ is again in S.
(*Hint.* The inequality $1 + |x| \leq (1 + |x - y|) \cdot (1 + |y|)$ is helpful for this.)

By the Exercise, the convolution of two Schwarz functions is again Schwarz, and with $F_{\mathbb{R}}$ the Fourier transform, $F_{\mathbb{R}}u \in S$ if $u \in S$. So Fourier transform restricts to a (continuous) map $F_{\mathbb{R}}: S(\mathbb{R}) \to S(\mathbb{R})$.

Theorem 1.9.9 *Let $F_{\mathbb{R}}: S(\mathbb{R}) \to S(\mathbb{R})$ be the Fourier transform:*

a) *If $u, v \in S(\mathbb{R})$, then $\int \hat{u}v = \int u\hat{v}$.*
b) *If $u \in S(\mathbb{R})$, then $\|\hat{u}\|_{L^2(\mathbb{R})} = \|u\|_{L^2(\mathbb{R})}$.*
c) *If $u, v \in S(\mathbb{R})$, then $\widehat{u * v} = \hat{u}\hat{v}$.*
d) *If $u \in S(\mathbb{R})$, then $F_{\mathbb{R}}\hat{u}(x) = u(-x)$. Equivalently (the Fourier inversion formula)*

$$(1.51) \qquad u(x) = \int \hat{u}(\xi)e^{2\pi i x\xi}d\xi$$

holds for all Schwarz functions u.

The *inverse* Fourier transform is given thus by

$$F_{\mathbb{R}}^{-1}u(x) := \hat{u}(-x).$$

We sometimes denote $\check{u}(x) := \hat{u}(-x)$.

We will need two Lemmas.

Lemma 1.9.10 (Minkowski's Inequality for Integrals) *Suppose that (X, μ) and (Y, v) are σ-finite measure spaces and k measurable on $X \times Y$. Let $1 \le p < \infty$. If $k \ge 0$, then*

$$\left[\int\left(\int k(x, y)dv(y)\right)^p\right]^{\frac{1}{p}} \le \int\left[\int k(x, y)^p d\mu(x)\right]^{\frac{1}{p}}.$$

And if k is measurable on $X \times Y$, $k(\cdot, y) \in L^p(X)$ for a.e. y, and $\int \|k(\cdot, y)\|dv(y) < \infty$, then $k(x, \cdot) \in L^1(v)$ for a.e. x, the function $k(\cdot, y): x \mapsto \int k(x, y)dv(y)$ is in $L^p(X)$, and

$$\left\|\int k(\cdot, y)dv(y)\right\|_p \le \int \|k(\cdot, y)\|_p dv(y).$$

Exercise 1.9.11 Use Minkowski's inequality to give another proof of Exercise 1.9.1: If $f \in L^1(\mathbb{R})$ and $g \in L^p(\mathbb{R})$, then $(f*g)(x)$ exists a.e. x, $f*g \in L^p(\mathbb{R})$, and $\|f * g\|_p \le \|f\|_1 \cdot \|g\|_p$. (*Hint.* Let $k(x, y) = f(x - y)$.)

Lemma 1.9.12 *Let $1 \le p < \infty$ and $g \in L^1(\mathbb{R})$ with $\int g(x)dx = a$. Let*

$$g_t(x) = \frac{1}{t}g(x/t).$$

Then if $f \in L^p(\mathbb{R})$, then $g_t * f \in L^p(\mathbb{R})$, $\|g_t * f\|_p \leq \|g\|_1 \cdot \|f\|_p$, and $g_t * f \to af$ in $L^p(\mathbb{R})$ as $t \to 0$. If f is bounded and continuous, then $g_t * f \to af$ uniformly on compact sets as $t \to 0$.

Proof The last statement is Exercise 1.9.4. That if $f \in L^p(\mathbb{R})$, then $g_t * f \in L^p(\mathbb{R})$ and $\|g_t * f\|_p \leq \|g\|_1 \cdot \|f\|_p$ is Exercise 1.9.1 or Exercise 1.9.11. We show that $g_t * f \to f$ in L^p.

Let $\tau_s(f)(x) := f(x - s)$. Compute

$$(g_t * f)(x) - af(x) = \int (\tau_{tz}(f)(x) - f(x)) \cdot g(z)\, dz.$$

Then apply Minkowski's inequality to get $\|g_t * f - af\|_p \leq \int \|\tau_{tz}(f) - f\|_p \cdot |g(z)|\, dz$. Now $\|\tau_{tz}(f) - f\|_p \cdot |g(z)|$ converges to zero for all z as $t \to 0$, and $|\tau_{tz}(f) - f|_p \cdot |g(z)| \leq 2\|f\|_p \cdot |g(z)|$ and $g \in L^1$ so the functions $z \mapsto \|\tau_{tz}(f) - f\|_p \cdot |g(z)|$ are dominated by an L^1-function. So we may apply Dominated Convergence to deduce that $\lim_{t \to 0} \int \|\tau_{tz}(f) - f\|_p \cdot |g(z)|\, dz = 0$ as required.   □

Proof of Theorem 1.9.9 Parts a), c) are routine.

We establish the Fourier inversion formula (1.51) in d):

$$(1.52) \qquad\qquad f(x) = \int \int f(y) e^{2\pi i(x-y)\xi}\, dy\, d\xi,$$

where $f \in S, x \in \mathbb{R}$.

By Lemma 1.9.12, if $g(x) = e^{-\pi x^2}$, so that $\int g(x) dx = 1$, then $(g_t * f) \to f(x)$ uniformly on compact sets, since f is continuous and bounded.

Now Exercise 1.9.6 shows that $g_t(x - y) = \hat{h}(y)$, where $h(\xi) := e^{2\pi i x\xi - \pi t^2\xi^2}$. Hence

$$(f * g_t)(x) = \int f(y) g_t(x-y) dy = \int f(y)\hat{h}(y) dy = \int \hat{f}(\xi) h(\xi) d\xi$$

$$= \int \hat{f}(\xi) e^{2\pi i x\xi - \pi t^2\xi^2} d\xi,$$

and the Dominated Convergence Theorem implies that the right hand side converges as $t \to 0$ to $\int \hat{f}(\xi) e^{2\pi i x\xi} d\xi$.

From a), we have

$$\langle \hat{u}, v \rangle = \int \hat{u}(\xi) v(\bar{\xi}) d\xi = \int u(\xi) \hat{\bar{v}}(\xi) d\xi.$$

On the other hand, $\hat{\bar{v}}(\xi) = \overline{\hat{v}(-\xi)}$, so

$$\langle \hat{u}, v \rangle = \int u(\xi) \overline{\hat{v}(-\xi)} d\xi = \langle u, \check{v} \rangle,$$

with $\check{v}(x) := \hat{v}(-x)$. Replacing v by $\hat{v}$ and using the Fourier inversion formula give $\langle \hat{u}, \hat{v} \rangle = \langle u, v \rangle$. In particular b) holds.   □

Remark 1.9.13 The statement of the Fourier inversion formula we have given can be strengthened as follows: If $f \in L^1$ and $\hat{f} \in L^1$, then f agrees a.e. with a continuous function f_0 and $\check{\hat{f}} = f_0$. Indeed, in the notation of the proof, $f * g_t \to f$ in L^1 by Lemma 1.9.12. On the other hand $(f * g_t)(x) = \int \hat{f}(\xi) e^{2\pi i x \xi - \pi t^2 \xi^2} d\xi$. The latter expression converges for every $x \in \mathbb{R}$, since $\hat{f} \in L^1(\mathbb{R})$, and Dominated Convergence, as $t \to 0$. If we set $f_0(x) := \int \hat{f}(\xi) e^{2\pi i x \xi} d\xi$, then f_0 is the Fourier transform of an L^1-function. Hence f_0 is continuous, by the Riemann–Lebesgue Lemma, and $f = f_0$ a.e.

Exercise 1.9.14 If g, g_t are as in Lemma 1.9.12, with $\int g = 1$, then $g_t * u \to u$ for all $u \in L^2(\mathbb{R})$. Deduce that the elements $g_t \in L^1(\mathbb{R}) \subset C^*(\mathbb{R})$ define an "approximate unit" for $C^*(\mathbb{R})$ in the sense that $g_t * f \to f$ in the norm of $C^*(\mathbb{R})$, for all $f \in C^*(\mathbb{R})$.

Corollary 1.9.15 (Plancherel Theorem) *The Fourier transform* $F_\mathbb{R} \colon S(\mathbb{R}) \to S(\mathbb{R})$ *extends to a unitary isomorphism* $U_\mathbb{R} \colon L^2(\mathbb{R}) \to L^2(\mathbb{R})$ *satisfying*

$$U_\mathbb{R} \lambda(f) U_\mathbb{R}^* = M_{\hat{f}},$$

for any $f \in S$.
    Therefore,

$$C^*(\mathbb{R}) \cong C_0(\mathbb{R})$$

by a C-algebra isomorphism sending the operator* $\lambda(f)$ *to the function* $\hat{f} \in C_0(\mathbb{R})$, *for* $f \in S(\mathbb{R}) \subset C^*(\mathbb{R})$.

Example 1.9.16 Let $f(x) = \frac{1}{1+x^2}$. Then $f \in L^1(\mathbb{R}) \subset C_0(\mathbb{R})$ and its Fourier transform converges absolutely to $\hat{f}(\xi) = \pi e^{-2|\xi|}$.
    The proof uses some standard methods from basic complex analysis.
    Let C_R' be the upper half of the circle of radius R in the complex plane, oriented counter-clockwise, and let C_R be the closed contour in $\mathbb{C}$ consisting of the segment $[-R, R]$ joined to C_R'. Suppose $\xi < 0$. Observe that $\int_{C_R'} \frac{e^{-i\xi z} dz}{1+z^2} \to 0$ as $R \to \infty$. On the other hand, $\int_{C_R} \frac{e^{-i\xi z}}{1+z^2} dz = \int_{C_R} \frac{g(z)}{z-i} dz$ where $g(z) = \frac{e^{-i\xi z}}{z+i}$, an analytic function on the upper half plane, so by the Cauchy Integral formula $\int_{C_R} \frac{e^{-i\xi z}}{1+z^2} dz = 2\pi i g(i) = \pi e^{\xi}$.
    It follows that

$$\hat{f}(\xi) = \pi e^{2\pi \xi}$$

when $\xi < 0$. On the other hand, since $f(x) = \frac{1}{1+x^2}$ is real-valued, $\hat{f}(-\xi) = \overline{\hat{f}(\xi)}$, so that if $\xi > 0$, by the computation above, $\hat{f}(-\xi) = \pi e^{-\xi}$ and hence $\hat{f}(\xi) = \overline{\pi e^{-\xi}} = \pi e^{-\xi}$.

Therefore,

$$f(x) = \frac{1}{1+x^2} \Rightarrow \hat{f}(\xi) = \pi e^{-|\xi|}, \quad \xi \in \mathbb{R}$$

as claimed.

Exercise 1.9.17 In this exercise $\|\cdot\|$ denotes the norm in $L^2(\mathbb{R})$:

a) Prove that $\|x^n e^{-\frac{x^2}{2}}\| \le 2^{-1/2}\sqrt{n!}$ for all nonnegative integers n.

b) Prove that the series $\sum_{n=0}^{\infty} \frac{(-2\pi i\xi x)^n}{n!} e^{-\frac{x^2}{2}}$ of vectors in $L^2(\mathbb{R})$ converges in $L^2(\mathbb{R})$ to $e^{-2\pi i x\xi - \frac{x^2}{2}}$.

c) Let $f \in L^2(\mathbb{R})$. Show that if f is orthogonal to $x^n e^{-\frac{x^2}{2}}$ for $n = 0, 1, 2, \ldots$ then $\int f(x) e^{-\frac{x^2}{2}} e^{-2\pi i x\xi} dx = 0$ for all ξ and deduce from the Plancherel Theorem that $f = 0$.

Exercise 1.9.18 Verify the Plancherel Theorem $\|\chi_R\|_2 = \|\widehat{\chi_R}\|_2$ by direct computation, where $\chi_R = \chi_{[-R,R]} \in L^2(\mathbb{R})$ is the characteristic function of an interval (use the fact that $\int \frac{\sin^2 x}{x^2} dx = \pi$).

Exercise 1.9.19 Show that if $f \in L^2(\mathbb{R})$, then $\hat{f}(\xi) = \lim_{R\to\infty} \int_{-R}^{R} \hat{f}(\xi) e^{2\pi i x\xi} dx$ holds in the L^2. That is, if $\hat{f}_R(\xi) := \int_{-R}^{R} \hat{f}(\xi) e^{2\pi i x\xi} dx$, then $\hat{f}_R \to \hat{f}$ in L^2.

(*Hint.* We can *define* an operator on L^2 by $f \mapsto \lim_{R\to\infty} \int_{-R}^{R} \hat{f}(\xi) e^{2\pi i (\cdot)\xi} dx$ by defining it initially on simple functions, verifying that the map is isometric in the Hilbert space norm, and then extending it by continuity. It agrees with Fourier transform on a dense subset of $L^2(\mathbb{R})$.)

Exercise 1.9.20 The proof of the Fourier Inversion Theorem involves a regularization of divergent integrals of general interest. Suppose that $\Phi \in C_0(\mathbb{R}) \cap L^1(\mathbb{R})$ such that $\Phi(0) = 1$ and $\phi := \check{\Phi} \in L^1$. For $f \in \mathcal{S}$ let

$$(1.53) \qquad f^t(x) := \int \hat{f}(\xi)\Phi(t\xi)e^{2\pi i x\xi} d\xi.$$

a) The integral (1.53) converges absolutely.

b) Let $\phi_t(x) := \frac{1}{t}\phi(x/t)$. Show that $f^t = f * \phi_t$.

c) Use the Fourier Inversion formula (use the strengthened version in Exercise 1.9.13) to deduce that $\int \phi = 1$ and deduce that $f^t \to f$ in L^p for $1 \le p < \infty$ and $t \to 0$, and uniformly on compact sets (see Lemma 1.9.12).

Another good example where the technique of Example 1.9.20 can be applied is to harmonic function theory and the Poisson transform.

Example 1.9.21 Example 1.9.16 shows that if $\phi(x) = \frac{1}{\pi} \cdot \frac{1}{1+x^2} \in L^1(\mathbb{R})$, then $\Phi(\xi) := \check{\phi}(\xi) = e^{-2|\xi|}$ and

$$\phi_t(x) = \frac{1}{t}\phi(x/t) = \frac{1}{\pi} \cdot \left(\frac{t}{t^2 + x^2}\right),$$

which is called the *Poisson kernel*.

It follows from the discussion above that if $f \in L^2(\mathbb{R})$ then the integral defining $(f * \phi_t)(x)$ exists a.e., $f * \phi_t$ defines an element of $L^2(\mathbb{R})$, and

$$(f * \phi_t)(x) = \int f(s)\phi_t(x - s)ds = \frac{1}{\pi}\int f(x - s) \cdot \left(\frac{t}{t^2 + s^2}\right) ds$$

holds a.e. By our results, $f * \phi_t \to f$ in L^2, and the convergence is uniform on compact sets if f is bounded and continuous.

We may interpret $f * \phi_t$ geometrically with a (harmonic) function on the upper half plane in the following way. Writing $y > 0$ in place of t, we let Pf be defined on the upper half plane $z = x + iy$, $y > 0$, by

$$(Pf)(x + iy) := (f * \phi_y)(x) = \int f(x - s) \cdot \left(\frac{y}{s^2 + y^2}\right) ds$$

$$= \int f(s) \cdot \left(\frac{y}{(x - s)^2 + y^2}\right) ds.$$

Note that $\frac{y}{(x-s)^2+y^2} = -\text{Im}(\frac{1}{z-s})$. So if f is real-valued, then noting that $f \in L^2$ by assumption and $s \mapsto \frac{1}{z-s}$ is in L^2, since $z = x + iy$, $y > 0$, the integral $\int \frac{f(s)}{z-s} ds$ converges absolutely, and

$$(Pf)(z) = -\text{Im}\left(\int \frac{f(s)}{z - s} ds\right),$$

which shows that Pf is the imaginary part of an analytic function, if f is real-valued, and in particular Pf is harmonic in the upper half plane. Since a complex combination of harmonic functions is harmonic, it follows that if $f \in L^2$, then Pf a harmonic function in the upper half plane $\text{Im}(z) > 0$, and

$$\lim_{y \to 0} Pf(x + iy) = f(x)$$

in L^2, and uniformly on compact subsets of $\mathbb{R}$ if f is bounded and continuous.

We close this section with some applications of the technique explained in Exercise 1.9.20 to Fourier analysis on $\mathbb{T}$.

Theorem 1.9.22 *Suppose $f \in C(\mathbb{R})$ and suppose there exists $\varepsilon > 0$ such that $|f(x)| \le C(1 + |x|)^{-1-\varepsilon}$ for all x and $|\hat{f}(\xi)| \le C(1 + |\xi|)^{-1-\varepsilon}$ for all ξ:*

a) *Let $\tau_k(f)(x) = f(x + k)$. Then $\sum_{k \in \mathbb{Z}} \tau_k(f)$ converges pointwise and a.e. to a periodic function Pf on $\mathbb{R}$ such that $\|Pf\|_{L^1(\mathbb{T})} \le \|f\|_1$. Moreover,*

$$F_{\mathbb{T}}(Pf)(k) = \hat{f}(k), \quad \forall k \in \mathbb{Z}.$$

b) *(Poisson Summation) In the same notation, $\sum_{k \in \mathbb{Z}} f(x+k) = \sum_{k \in \mathbb{Z}} \hat{f}(k)e^{2\pi ixk}$, where both series converge absolutely and uniformly on $\mathbb{T}$.*

Proof

a) We have

$$\int_{\mathbb{R}} |f(x)| dx = \sum_{k \in \mathbb{Z}} \int_k^{k+1} |f(x)| dx = \sum_{k \in \mathbb{Z}} \int_0^1 |\tau_k(f)(x)| dx = \sum_{k \in \mathbb{Z}} \|\tau_k(f)\|_{L^1([0,1])}.$$

Consequently, the series $\sum \tau_k(f)$ in $L^1([0, 1])$ is absolutely summable, and hence summable, so $\sum \tau_k(f)$ converges norm absolutely in $L^1([0, 1]) \cong L^1(\mathbb{T})$. Now if $(g_k) \subset L^1(\mathbb{T})$ is any sequence such that $\sum \|g_k\|_1 < \infty$, then $\sum g_n$ converges a.e. Indeed, by the Monotone Convergence Theorem $\int_{\mathbb{T}} \sum |g_k| = \sum \int_{\mathbb{T}} |g_k| = \sum \|g_k\|_{L^1(\mathbb{T})} < \infty$, and so $h(x) := \sum |g_k(x)|$ is in $L^1(\mathbb{T})$. Hence $0 \le h(x) < \infty$ a.e so $\sum |g_k(x)| < \infty$ a.e. and for each x for which the sum is finite, the sum $\sum g_k(x)$ converges, by completeness of $\mathbb{C}$.
Applying this to $g_k = \tau_k(f)$ gives that $\sum_k \tau_k(f)$ converges a.e. and in $L^1(\mathbb{T})$. Let Pf be the limit function. Then $\|Pf\|_{L^1(\mathbb{T})} \le \|f\|_1$, and $F_{\mathbb{T}}(Pf)(k) = \hat{f}(k)$ is easily verified by direct computation.

b) The decay condition on f implies immediately that the series $\sum f(x + k)$ converges absolutely and uniformly. Hence Pf is continuous on $\mathbb{T}$. Hence $Pf \in L^2(\mathbb{T})$ and so the series $\sum_{k \in \mathbb{Z}} (\hat{P}f)(k)e^{2\pi ixk}$ converges in $L^2(\mathbb{T})$ to Pf. By a) this is the same as the series $\sum_{k \in \mathbb{Z}} \hat{f}(k)e^{2\pi ixk}$. So the latter series converges in $L^2(\mathbb{T})$ to Pf. The latter series also converges absolutely uniformly, due to the decay assumption on $\hat{f}$. This proves b). $\square$

Corollary 1.9.23 *Suppose that $C \ge 0, \varepsilon > 0$ and $\Phi \in C(\mathbb{R}^n)$ with $|\Phi(\xi)| \le C(1 + |\xi|)^{-1-\varepsilon}$ and that $|\check{\Phi}(x)| \le C(1 + |x|)^{-1-\varepsilon}$. Assume $\Phi(0) = 1$.*
Let $f \in L^1(\mathbb{T})$, and f^t be defined by the absolutely convergent integral $f^t(x) := \sum_{k \in \mathbb{Z}} (F_{\mathbb{T}}f)(k)\Phi(tk)e^{2\pi ikx}$. Then if $f \in L^p(\mathbb{T}), 1 \le p < \infty$, then $\|f^t - f\|_{L^p(\mathbb{T})} \to 0$ as $t \to 0$, and if $f \in C(\mathbb{T})$, then $f^t \to f$ uniformly on $\mathbb{T}$ as $t \to 0$.

Proof Let $\phi(x) = \check{\Phi}(x)$ and $\phi_t(x) = \frac{1}{t}\phi(x/t)$. By Theorem 1.9.22

(1.54)
$$\psi_t(x) := \sum_{k\in\mathbb{Z}} \phi_t(x-k) = \sum_{k\in\mathbb{Z}} \Phi(tk)e^{2\pi ixk}.$$

The function ψ_t is periodic, and if $*$ denotes convolution in $\mathbb{T}$, then $F_\mathbb{T}(f*\psi_t)(k) = (F_\mathbb{T}f)(k) \cdot (F_\mathbb{T}\psi_t)(k)$. And $(F_\mathbb{T}\psi_t)(k) = \Phi(tk)$ by (1.54). Hence $F_\mathbb{T}(f*\psi_t)(k) = (F_\mathbb{T}f)(k)\Phi(tk) = (F_\mathbb{T}f^t)(k)$. It follows that $f^t = f*\psi_t$. By Young's inequality, $\|f^t\|_{L^p(\mathbb{T})} = \|f*\psi_t\|_{L^p(\mathbb{T})} \le \|\psi_t\|_1 \cdot \|f\|_{L^1(\mathbb{T})} = \|f\|_{L^1(\mathbb{T})} \cdot \|\phi\|_{L^1(\mathbb{T})}$. Hence the operators $f \mapsto f^t$ are uniformly bounded on $L^p(\mathbb{T})$, $1 \le p < \infty$. Since $f^t(x) := \sum_{k\in\mathbb{Z}}(F_\mathbb{T}f)(k)\Phi(tk)e^{2\pi ikx}$ it is clear that if $f \in \mathbb{C}[z,\bar{z}]$, so that $F_\mathbb{T}f$ vanishes except on a finite set of integers, then $f_t \to f$ uniformly as $t \to 0$. Since $\mathbb{C}[z,\bar{z}]$ is dense in $C(\mathbb{T})$ and in $L^p(\mathbb{T})$, $1 \le p < \infty$, we deduce that $f^t \to f$ uniformly for $f \in C(\mathbb{T})$, and $f^t \to f$ in $L^p(\mathbb{T})$ for $f \in L^p(\mathbb{T})$, as claimed.   □

Example 1.9.24 With notation as in Corollary 1.9.23 let $\Phi(\xi) = e^{-2\pi|\xi|}$, so that $\phi(x) = \frac{1}{\pi} \cdot \frac{1}{1+x^2}$, by Example 1.9.16, the *Poisson kernel*. The methods above give the following. If $f \in L^1(\mathbb{T})$, let

(1.55)
$$f^t(z) = \sum_{n\in\mathbb{Z}}(F_\mathbb{T}f)(k)\Phi(tk)e^{2\pi ikx}.$$

The integral converges absolutely for all $t > 0$. By Corollary 1.9.23, $f^t \to f$ as $t \to 0$, where the convergence is uniform if f is continuous and in L^2 if f is in L^2. If $r := e^{-2\pi t} < 1$, then with $\hat{f} := F_\mathbb{T}f$, (1.55) can be written

(1.56)
$$f(rz) = \sum_{n\in\mathbb{Z}} r^{|k|} f_n z^n, \quad z \in \mathbb{T}.$$

Equivalently, for rz considered as a point in the unit disk,

(1.57) $$\tilde{f}(z) := \sum_{n\in\mathbb{Z}}|z|^{-|k|} f_n z^n = \sum_{n=0}^{\infty}|z|^{-n} f_n z^n + \sum_{n=1}^{\infty}|z|^{-n} f_{-n}\bar{z}^n, \quad z \in \mathbb{D}.$$

The sums are all absolutely convergent, and the results above show that $\tilde{f}(z) \to f$ radially, as $z \to \partial\mathbb{D}$, in the sense(s) discussed, and in particular uniformly. The decomposition of $\tilde{f}$ into a sum of a holomorphic function and an anti-holomorphic function in the disk implies that $\tilde{f}$ is harmonic. The discussion above shows that $f^t = f*\psi_t$ where $\psi_t(z) = \sum_{z\in\mathbb{Z}} e^{-2\pi t|k|}z^k$ and a computation shows, substituting $r = 2^{-2\pi t}$ that

$$\psi_r(z) = \frac{1-r^2}{|1-rz|^2}, \quad z \in \mathbb{T}.$$

The formula for f^t translates into the familiar Poisson transform $f \mapsto \tilde{f}$ where

$$\tilde{f}(z) = \int_{\mathbb{T}} f(w) \cdot \left(\frac{1 - |z|^2}{|1 - \bar{w}z|^2} \right) d\mu(w), \quad z \in \mathbb{D},$$

with μ normalized Lebesgue measure on $\mathbb{T}$.

The Szegö or Toeplitz projection $P_+ \in \mathbb{B}(L^2(\mathbb{T}))$ discussed in connection with Toeplitz operators in Section 2.3 maps an element $f \in L^2(\mathbb{T})$ to its truncated Fourier series $\sum_{n=0}^{\infty} \hat{f}(n)z^n$. As discussed earlier, this suggests a convolution formula

$$(P_+ f)(z) = f * \chi, \quad \chi(z) = \frac{1}{1 - z}.$$

If one attempts to regularize the singular operator of convolution with χ, we can define

$$f_r(z) := \int_{\mathbb{T}} \frac{f(w)}{1 - r\bar{z}w} d\mu(w), \quad z \in \mathbb{T}$$

and attempt to take a limit as $r \to 1$. The limit certainly exists for trigonometric polynomials $f(z) = \sum_n a_n z^n$. Furthermore, $\|f_r\|_{L^2(\mathbb{T})}^2 = \sum_{n \geq 0} r^{2n} |a_n|^2 \leq \sum_n |a_n|^2 = \|f\|_{L^2(\mathbb{T})}^2$. Since $f_r \to f$ for $f \in \mathbb{C}[z, \bar{z}]$ as $r \to 1$, by density of $\mathbb{C}[z, \bar{z}]$ in $L^2(\mathbb{T})$, and uniform boundedness of the operators $f \mapsto f_r$, it follows that $f_r \to f$ in $L^2(\mathbb{T})$. But $f_r \to f$ in $C(\mathbb{T})$ fails. Equivalently, $\tilde{f}$ is not necessarily in the disk algebra if f is merely continuous on $\mathbb{T}$, even though it is the case if $f \in \mathbb{C}[z, \bar{z}]$.

To see an example, use the Riemann Mapping Theorem to find $f : \mathbb{D} \to \mathbb{C}$ an analytic bijection from $\mathbb{D}$ onto the rectangle $-1 < \text{Re}(z) < 1, -R < \text{Im}(z) < R$ with $f(0) = 0$. Then f extends continuously to $\partial \mathbb{D}$ and defines an element of the disk algebra, and since $f(0) = 0$ it is clear from the Fourier series of f that $P_+ \bar{f} = 0$. Hence if $u = f + \bar{f} = 2\text{Re}(f) \in C(\mathbb{T})$, then $P_+ u = P_+(f + \bar{f}) = f + 0 = f$. But $\|u\|_{C(\mathbb{T})} = 2$, but $P_+ u = f$, and $\|f\|_{C(\mathbb{T})} = R$.

Exercise 1.9.25 (Heisenberg's Inequality (See [83])) Let $f \in L^2(\mathbb{R})$:

a) $\left(\int x^2 |f(x)|^2 \, dx \right) \cdot \left(\int \xi^2 |\hat{f}(\xi)|^2 \, d\xi \right) \geq \frac{\|f\|_2^4}{16\pi^2}$.

b) For any $a, \eta \in \mathbb{R}$,

$$\left(\int (x - a)^2 |f(x)|^2 \, dx \right) \cdot \left(\int (\xi - \eta)^2 |\hat{f}(\xi)|^2 \, d\xi \right) \geq \frac{\|f\|_2^4}{16\pi^2}.$$

c) Part b) shows that it is not possible for both f and $\hat{f}$ be sharply localized at single points. Why?

d) Verify Heisenberg's inequality for the Gaussians $e^{-\pi a x^2}$.

Chapter 2
An Introduction to Index Theory and Noncommutative Geometry

In this chapter we introduce some of the basic ideas of Index Theory and of the Noncommutative Geometry of Alain Connes. Our main specific objective is to state and prove the Toeplitz Index Theorem, a nice relationship between a certain analytic invariant of a non-vanishing continuous, complex-valued function on the circle, and the topology of the function (actually the winding number of the function around the origin.)

If $T\colon \mathbb{C}^n \to \mathbb{C}^n$ is a linear map, represented by a matrix, then one learns in linear algebra to find a certain number of linear functions in $b_1, \ldots, b_k$, which are constraints for a vector b to lie in the range of T. On the other hand, for a given, fixed b, the number of parameters which appear in the general solution to $Tx = b$, is another positive integer l. The *difference* $l - k$ is the *index* of T. Actually, a simple exercise shows that the index does not depend on T (in finite dimensions). On the other hand, if one considers more generally, the class (roughly) of operators for which both the parameters above are finite, but where H_i are not required to be finite-dimensional, then one obtains a very interesting and nontrivial theory. This is mainly due to the fact that elliptic operators on compact manifolds are Fredholm in this sense. The Fredholm indices of such operators are computed by the Atiyah–Singer Theorems, as we discuss later in this book. The most basic of these Index Theorems is the Toeplitz Index Theorem.

2.1 Trace-Class and Schatten Classes of Compact Operators

Certain algebraic ideals (they are not closed in the norm topology) of the C*-algebra of compact operators play an essential role in Noncommutative Geometry. These are called *Schatten* ideals, and they are the noncommutative analogues of the l^p-spaces of basic analysis.

© Springer Nature Switzerland AG 2024
H. Emerson, *An Introduction to C*-Algebras and Noncommutative Geometry*,
Birkhäuser Advanced Texts Basler Lehrbücher,
https://doi.org/10.1007/978-3-031-59850-0_2

If T is a diagonal operator with respect to an orthonormal basis of a Hilbert space, with, say, nonnegative diagonal entries $a_0, a_1, \ldots$, then T is compact if and only if $a_n \to 0$ as $n \to \infty$. Such T resembles an "infinitesimal" in the sense that, for all $\varepsilon > 0$, the restriction of T to the orthogonal complement of a sufficiently large finite-dimensional subspace has norm $< \varepsilon$.

If S is a general nonzero compact operator, then S^*S is self-adjoint and hence diagonalizable by Theorem 1.6.23, and has spectrum consisting of 0 and a countable collection of eigenvalues which we may list in decreasing order $\lambda_0 \geq \lambda_1 \geq \lambda_2 \geq \cdots$ where we list including multiplicity.

The *singular values* of S are given by the numbers $\mu_n := \sqrt{\lambda_n}$. They are thus the nonzero eigenvalues of $T := |S| := \sqrt{S^*S}$. Note that $\mu_0 = \|T\| = \|S\|$. Since for any $S \in \mathbb{B}(H)$, SS^* and S^*S have the same nonzero eigenvalues, $\mu_k(S) = \mu_k(S^*)$ for any S.

A similar discussion as the one above for diagonal operators applies to $T := |S|$: working with the orthonormal basis diagonalizing T, we see that for all $\varepsilon > 0$, there exists a finite-dimensional subspace $F \subset H$ such that $\|T_{F^\perp}\| < \varepsilon$. Indeed, one can take F to be the direct sum of all the eigenspaces of T with eigenvalue $\geq \varepsilon$. In fact, this argumentation leads to a geometric formula for the singular values.

Lemma 2.1.1 *If S is a compact operator with singular values $\mu_0 \geq \mu_1 \geq \cdots$ then*

$$\mu_k = \inf\{\|S_{|F^\perp}\| \mid F \subset H, \ \dim F \leq k\}.$$

Proof Let $T = |S|$. The previous discussion shows that

$$\mu_k \geq \inf\{\|T_{|F^\perp}\| \mid F \subset H, \ \dim F \leq k\}.$$

Conversely, suppose that $F \subset H$ with $\dim F \leq k$. Let $v_0, \ldots, v_k$ be unit pairwise orthogonal eigenvectors for T with eigenvalues $\mu_0, \ldots, \mu_k$. Since $\dim(H/F^\perp) = \dim(F) \leq k$, the quotient map

$$\mathrm{span}\{v_0, \ldots, v_k\} \to H/F^\perp$$

cannot be injective since the domain has dimension $k + 1$. Hence there exists a unit vector $v \in F^\perp \cap \mathrm{span}\{v_0, \ldots, v_k\} \to H/F^\perp$. Since $v \in H_0 \oplus \cdots \oplus H_k$, then $\|Tv\| \geq \mu_k\|v\|$. Hence $\|T|_{F^\perp}\| \geq \mu_k$.

Finally, we note that $\|Tv\| = \|Sv\|$ for any v, for $\|Tv\|^2 = \langle Tv, Tv \rangle = \langle S^*Sv, v \rangle = \langle Sv, Sv \rangle = \|Sv\|^2$. So the corresponding result holds with S in place of T. $\qquad\square$

Exercise 2.1.2 If T is compact with singular value sequence $\mu_0 \geq \mu_1 \geq \cdots$ then

$$\mu_n = \inf\{\|T - S\| \mid \mathrm{rank}(S) \leq n\}.$$

Exercise 2.1.3 Let $A \in \mathbb{B}(H)$ be any bounded operator on a Hilbert space and let $B = \begin{bmatrix} 0 & A^* \\ A & 0 \end{bmatrix}$, acting on $H \oplus H$. Then B is self-adjoint and is compact if A is compact. Show that the eigenvalues of B are of the form $\pm\mu$ where μ is a singular value of A.

Definition 2.1.4 If $1 \leq p \leq \infty$, the *Schatten p-class* $\mathcal{L}^p(H)$ is the collection of compact operators $T \in \mathcal{K}(H)$ such that $\sum_{n=0}^{\infty} \mu_n(T)^p < \infty$, where $\mu_0(T) \geq \mu_1(T) \geq \cdots$ is the list of singular values of T.

The Schatten p-norm of such an operator is defined

$$\|T\|_{\mathcal{L}^p} := \left(\sum_{n=0}^{\infty} \mu_n^p(T) \right)^{\frac{1}{p}}.$$

Exercise 2.1.5 Show that $\mathcal{L}^p(H)$ is a linear subspace of $\mathbb{B}(H)$ closed under unitary conjugation: that is, $T \in \mathcal{L}^p$ if and only if $uTu^* \in \mathcal{L}^p$ for any unitary u and adjoint: $T \in \mathcal{L}^p(H)$ if and only if $T^* \in \mathcal{L}^p(H)$, and $\|T\|_{\mathcal{L}^p} = \|T^*\|_{\mathcal{L}^p}$. (*Hint*. For the first statement use Lemma 2.1.1)

If $T \in \mathcal{L}^p(H)$ and $W \in \mathbb{B}(H)$ then Lemma 2.1.1 shows that $WT \in \mathcal{L}^p(H)$ and $\|WT\|_{\mathcal{L}^p} \leq \|W\| \cdot \|T\|_{\mathcal{L}^p}$. Using closure of $\mathcal{L}^p$ under adjoint, one obtains as well that $TW \in \mathcal{L}^p$ and $\|WT\|_{\mathcal{L}^p} \leq \|W\| \cdot \|T\|_{\mathcal{L}^p}$. Lemma 2.1.1 implies that $\mathcal{L}^p(H)$ is a linear subspace of $\mathcal{K}(H)$ and that $\|S + T\|_{\mathcal{L}^p} \leq \|S\|_{\mathcal{L}^p} + \|T\|_{\mathcal{L}^p}$. Thus, the $\mathcal{L}^p(H)$ are ideals in $\mathbb{B}(H)$, and normed vector spaces in their own right. One can also show they are complete in the Schatten norm, and so are Banach spaces. However, they are clearly not closed in the operator norm. So they are non-closed, dense ideals in $\mathcal{K}(H)$.

We will be mainly interested in the cases $p = 1$ and $p = 2$.

Definition 2.1.6 A compact operator $T \in \mathcal{K}(H)$ is *trace-class* if it is in $\mathcal{L}^1(H)$. T is *Hilbert-Schmidt* if $T \in \mathcal{L}^2(H)$.

Lemma 2.1.7 *If $(e_i)_{i \in I}$ and $(\varepsilon_j)_{j \in J}$ are two orthonormal bases for H, and $T \in \mathcal{K}(H)$, then*

$$\sum_{i \in I} \|Te_i\|^2 = \sum_{j \in J} \|T\varepsilon_j\|^2 \in [0, \infty].$$

Proof First observe that from Parseval's Identity,

$$\sum_{i \in I} \|Te_i\|^2 = \sum_{i,j \in I} |\langle Te_i, e_j \rangle|^2 = \sum_{i,j \in I} |\langle e_i, T^*e_j \rangle|^2$$

$$= \sum_{i,j \in I} |\langle T^*e_j, e_i \rangle|^2 = \sum_{i \in I} \|T^*e_i\|^2.$$

Next,

$$\sum_{i \in I} \|T e_i\|^2 = \sum_{i \in I, j \in J} |\langle T e_i, \varepsilon_j \rangle|^2 = \sum_{i,j} |\langle e_i, T^* \varepsilon_j \rangle|^2 = \sum_{j \in J} \|T^* \varepsilon_j\|^2 = \sum_{j \in J} \|T \varepsilon_j\|^2$$

by Parseval again, and the first observation.   □

Corollary 2.1.8 *If $S \in \mathcal{K}(H)$ is a compact operator on H and $(e_i)_{i \in I}$ is any orthonormal basis for H, then S is Hilbert-Schmidt if and only if $\sum_{i \in I} \|S e_i\|^2 < \infty$, and in this case*

$$\|S\|_{\mathcal{L}^2}^2 = \sum_{i \in I} \|S e_i\|^2.$$

Proof Let S be compact and $T = |S|$. The previous Lemma implies that $\sum_{i \in I} \|T e_i\|^2$ is independent of the basis. In particular, we may assume $(e_i)_{i \in I}$ consists of eigenvectors for T. This yields that

$$\sum_{i \in I} \|T e_i\|^2 = \sum_{i \in I} \mu_i^2,$$

where μ_i is the eigenvalue corresponding to e_i. These are the singular values of S. Therefore $\sum_{i \in I} \|T e_i\|^2 = \|S\|_{\mathcal{L}^2}^2$. Finally,

$$\|T e_i\|^2 = \langle \sqrt{S^* S} e_i \sqrt{S^* S} e_i \rangle = \langle S^* S e_i e_i \rangle = \|S e_i\|^2.$$

This proves the result.   □

Hilbert-Schmidt operators appear in geometric contexts, due to the following.

Proposition 2.1.9 *Let $k \in L^2(X \times X)$, where (X, μ) is a measure space. Then the integral operator I_k is Hilbert-Schmidt and*

$$\|I_k\|_{\mathcal{L}^2} = \|k\|_{L^2(X \times X)}.$$

Proof Let $(e_i)_{i \in I}$ be an orthonormal basis for $L^2(X)$. Then the functions $e_{i,j}(x, y) := e_i(x)\overline{e_j(y)}$ form an orthonormal basis for $L^2(X \times X)$. From a quick calculation, we have

$$\langle e_{ij}, k \rangle = \langle e_i, I_k e_j \rangle.$$

Hence

$$\|k\|_{L^2(X \times X)}^2 = \sum_{i,j} |\langle e_{i,j}, k \rangle|^2 = \sum_{i,j} |\langle e_i, I_k e_j \rangle|^2 = \sum_i \|I_k e_i\|^2$$

from which the result follows from Corollary 2.1.8. $\square$

We now discuss trace-class operators.

Lemma 2.1.10 *Let $S \in \mathcal{K}(H)$ and $(e_i)_{i \in I}$ an orthonormal basis for H. The sum*

$$\sum_i \langle |S|e_i, e_i \rangle$$

is independent of the orthonormal basis, and S is trace-class if and only if $\sum_i \langle |S|e_i, e_i \rangle < \infty$, and in this case

$$\|S\|_{\mathcal{L}^1} = \sum_i \langle |S|e_i, e_i \rangle.$$

We leave the proof to the reader, as it is similar to the corresponding statement for Hilbert-Schmidt operators. Note that

$$\sum_i \langle |S|e_i, e_i \rangle = \sum_i \langle \sqrt{S}e_i, \sqrt{S}e_i \rangle = \sum_i \|\sqrt{|S|}e_i\|^2$$

so independence of this expression on the basis follows from the discussion of Hilbert-Schmidt operators.

Lemma 2.1.11 *If $S \in \mathcal{K}(H)$ then S is trace-class if and only if S is a product of two Hilbert-Schmidt operators.*

Proof From the remarks before the Lemma, $|S|$ is trace-class if and only if $\sqrt{|S|}$ is Hilbert-Schmidt. In particular, if $|S|$ is trace-class, then $|S|$ is a product of two Hilbert-Schmidt operators. Now write S trace-class in polar decomposition, then $S = (U\sqrt{|S|}) \cdot \sqrt{|S|}$ expresses S as a product of two Hilbert-Schmidt operators.

Conversely, suppose first that $S \in \mathcal{K}(H)$ and $|S| = QT$ is a product of two Hilbert-Schmidt operators. If $(e_i)_{i \in I}$ is an orthonormal basis then

$$(2.1) \quad \sum_i \langle |S|e_i, e_i \rangle = \sum_i \langle QTe_i, e_i \rangle = \sum_i \langle Te_i, Q^*e_i \rangle$$

$$\leq \sum_i \|Te_i\| \cdot \|Q^*e_i\| \leq \left(\sum_i \|Te_i\|^2 \right)^{\frac{1}{2}} \cdot \left(\sum_i \|Q^*e_i\|^2 \right)^{\frac{1}{2}}$$

by the Cauchy-Schwarz inequality (twice). Hence $|S|$, and S, are trace-class.

If we know that $S = BC$ is a product of two Hilbert-Schmidt operators, write $S = U|S|$ in polar decomposition. Since $UU^*|S| = |S|$, it follows that $|S| = U^*S = (U^*B)C$. This expresses $|S|$ as a product of two Hilbert-Schmidts. $\square$

The following shows that the familiar recipe of summing the diagonal entries of a matrix (taking the trace) extends to a well-defined map on trace-class operators.

Theorem 2.1.12 *If $S \in \mathcal{L}^1(H)$ and $(e_i)_{i \in I}$ is an orthonormal basis of H then*

$$\sum_{i \in I} |\langle Se_i, e_i \rangle| < \infty.$$

The convergent sum

$$\mathrm{Trace}(S) := \sum_{i \in I} \langle Se_i, e_i \rangle$$

is independent of the choice of orthonormal basis.

Proof We can write $S = Q^*T$ as a product of two Hilbert-Schmidt operators. From definitions and the Cauchy-Schwarz inequality

$$|\langle Se_i, e_i \rangle| = |\langle Te_i, Qe_i \rangle| \leq \|Te_i\| \cdot \|Qe_i\| \leq \frac{\|Te_i\|^2 + \|Qe_i\|^2}{2}.$$

We get

$$\sum_i |\langle Se_i, e_i \rangle| \leq \frac{1}{2} \cdot \sum_i \|Te_i\|^2 + \|Qe_i\|^2 = \frac{1}{2} \cdot (\|T\|_{\mathcal{L}^2}^2 + \|Q\|_{\mathcal{L}^2}^2)$$

proving absolute convergence of the series $\sum_i \langle Se_i, e_i \rangle$. $\square$

From this discussion we see that if $Q, T \in \mathcal{L}^2(H)$ then $Q^*T \in \mathcal{L}^1(H)$ and $\mathrm{Trace}(Q^*T)$ is well defined. The functional $\mathrm{Trace}(Q^*, T)$ is sesquilinear and $\mathrm{Trace}(Q^*Q) = \|Q\|_{\mathcal{L}^2}^2$. Thus, we obtain the following

Corollary 2.1.13 *The complex vector space $\mathcal{L}^2(H)$ is a Hilbert space with the inner product*

$$\langle Q, T \rangle := \mathrm{Trace}(Q^*T).$$

By Proposition 2.1.9 and polarization we get:

Proposition 2.1.14 *Let (X, μ) be a measure space, $H = L^2(X)$. Then the map $L^2(X \times X) \to \mathcal{L}^2(H)$, $k \mapsto I_k$, is a unitary isomorphism of Hilbert spaces.*

Proof By Proposition 2.1.9, $I_k \in \mathcal{L}^2(H)$ if $k \in L^2(X \times X)$, and $\mathrm{Trace}(I_k^* I_k) = \|I_k\|_{\mathcal{L}^2(H)}^2 = \|k\|_{L^2(X \times X)}^2$. By polarization,

$$\mathrm{Trace}(I_k^* I_l) = \langle f, g \rangle_{L^2(X \times X)}$$

for any $k, l \in L^2(X \times X)$. So the map $k \mapsto I_k$ is a Hilbert space isometry. We leave it as an exercise to verify it has dense range. Hence it is a unitary isomorphism.   □

We conclude this discussion with an important result in connection with index theory. We sketch the proof, as it requires a bit of familiarity with manifolds; the case where the manifold is the circle can be handled directly and is left as an exercise below.

Theorem 2.1.15 *Suppose that X is a compact Riemannian manifold and μ a Borel probability measure on X. Suppose that $k \in C^\infty(X \times X)$ is a smooth kernel. Then I_k is trace-class, and*

$$\mathrm{Trace}(I_k) = \int_X k(x, x) d\mu(x).$$

Proof We start with some observations about integral operators with continuous kernels. Let $H = L^2(X)$. If k_1 and k_2 are continuous kernels, then their convolution

$$(k_1 * k_2)(x, y) = \int_X k_1(x, z) k_2(z, x) \, d\mu(z)$$

is a continuous kernel as well, and $I_{k_1} I_{k_2} = I_{k_1 * k_2}$ as Hilbert-Schmidt operators on H. Similarly, a simple computation shows that $k^*(x, y) = \overline{k(y, x)}$ is a continuous kernel and $I_{k^*} = I_k^*$.

Now the operator $I_{k_1}^* I_{k_2} = I_{k_1^* k_2}$, for a pair of continuous kernels k_i, is trace-class, since it is a product of Hilbert-Schmidts. We have, using Proposition 2.1.14

$$\mathrm{Trace}(I_{k_1}^* I_{k_2}) = \langle I_{k_1}, I_{k_2} \rangle_{\mathcal{L}^2(H)} = \langle k_1, k_2 \rangle_{L^2(X \times X)}$$

$$= \int_{X \times X} \overline{k_1(y, x)} \cdot k_2(x, y) d\mu(x) d\mu(y)$$

$$= \int_X \left(\int_X k_1^*(x, y) k_2(y, x) d\mu(y) \right) d\mu(x)$$

(2.2) $$= \int_X (k_1^* * k_2)(x, x) d\mu(x).$$

Now choose $k \in C^\infty(X \times X)$. Let Δ be the Laplacian on X, which we may consider as a densely defined unbounded operator on $L^2(X)$ with domain $C^\infty(X)$. From Sobolev theory, $(1 + \Delta)^{-d} = I_k$ for a suitable *continuous* kernel k, for d sufficiently large relative to $\dim X$. On the other hand, the composite densely defined operator on $C^\infty(X)$ given by $(1 + \Delta)^d I_k$ is an integral operator with smooth kernel $k_2(x, y) = ((1 + \Delta)_x^d k)(x, y)$. We thus get that $k_1^* * k_2 = k$. We can factorize

$$I_k = (1 + \Delta)^{-d} \cdot (1 + \Delta)^d I_k$$

into a product of two integrals operators with continuous kernels and from the above discussion.

$$\text{Trace}(I_k) = \int_X (k_1^* * k_2)(x, x)d\mu(x) = \int_X k(x, x)d\mu(x)$$

proving the result.   □

The formula

$$\text{Trace}(I_k) = \int_{\mathbb{T}} k(\theta, \theta)\, d\theta$$

for suitable $k \in C^\infty(\mathbb{T} \times \mathbb{T})$, will play a key role in connection with the Toeplitz Index Theorem discussed below.

Exercise 2.1.16 Let G be a compact Lie group (such as $G = \mathbb{T}$) equipped with normalized Haar measure, and $f \in C^\infty(G)$ be a smooth function, $\lambda(f)$ convolution by f, $\lambda(f) \in \mathcal{K}(L^2(G))$. Deduce from Theorem 2.1.15 that $\lambda(f)$ is trace-class and

$$\text{Trace}(\lambda(f)) = f(e),$$

where $e \in G$ is the identity element.

See Corollary 3.2.8 for more information about the spectrum of a compact operator, deduced from the (holomorphic) functional calculus developed in Chapter 3.

2.2 The Toeplitz Algebra

The Toeplitz algebra is an important C*-algebra connected with analytic function theory in the disk. It eventually will be shown to play a big role in K-theory and the Index Theorem.

Let $H = L^2(\mathbb{T})$, with its standard orthonormal basis $\{z^n\}_{n \in \mathbb{Z}}$ of characters, μ normalized Lebesgue measure on $\mathbb{T}$.

The *Szegö projection* (or *Toeplitz projection*) $P_+ \in \mathbb{B}(H)$ is the orthogonal projection onto the closed subspace $\mathbf{H}^2 := \text{span}\{z^n \mid n \geq 0\} \cong l^2(\mathbb{N})$ of H. Explicitly:

$$(2.3) \qquad (P_+ f)(z) = \sum_{n=0}^{\infty} \left(\int_{\mathbb{T}} f(w)\overline{w}^n d\mu(w) \right) \cdot z^n.$$

The sum is a convergent series of vectors in $L^2(\mathbb{T})$ for every $f \in L^2(\mathbb{T})$.

Definition 2.2.1 If $f \in C(\mathbb{T})$, the *Toeplitz operator with symbol* f is by definition the operator $T_f := P_+ M_f$ acting on the subspace $\mathbf{H}^2 \subset L^2(\mathbb{T})$.

The C*-algebra generated by the Toeplitz operators on $\mathbf{H}^2$ is called the *Toeplitz algebra* and will be denoted $\mathcal{T}$.

Operators in $\mathcal{T}$ will be called *pseudo-Toeplitz operators*.

Using the standard orthonormal basis $1, z, z^2, \ldots$ for $\mathbf{H}^2$, we can expand a Toeplitz operator T_f into an infinite matrix, its Fourier transform as an operator. The m, nth entry is by definition

$$\langle T_f e_n, e_m \rangle = \langle P_+(f z^n), z^m \rangle = \sum_{k \geq -m} \hat{f}(k) \langle z^{m+k}, z^n \rangle = \hat{f}(n - m).$$

Thus,

(2.4)
$$\widehat{T_f} = \begin{bmatrix} \hat{f}(0) & \hat{f}(-1) & \hat{f}(-2) & \cdot \cdot & \cdot \cdot \\ \hat{f}(1) & \hat{f}(0) & \hat{f}(-1) & \hat{f}(-2) & \cdot \cdot \\ \hat{f}(2) & \hat{f}(1) & \hat{f}(0) & \hat{f}(-1) & \cdot \cdot \\ \cdot \cdot & \hat{f}(2) & \hat{f}(1) & \hat{f}(0) & \cdot \cdot \\ \cdot \cdot & \cdot \cdot & \cdot \cdot & \cdot \cdot & \cdot \cdot \end{bmatrix}$$

so that along the various diagonals of the matrix, we see the values of $\hat{f}$; the negative values appear above the diagonal and the positive ones below.

Note also that if $f \in \mathbb{C}[z, \bar{z}]$ then $\widehat{T_f}$ has only finitely many diagonals, that is, the support of $\widehat{T_f}$ is contained in a neighborhood of the diagonal: there exists a constant $C \geq 0$ such that

$$(\widehat{T_f})_{n,m} \neq 0 \Rightarrow |n - m| \leq C.$$

(Such matrices are said to have *finite propagation*.)

For example, $\widehat{T_z} = S$ the shift, $S(e_n) = e_{n+1}$, $n = 0, 1, 2, \ldots$

$$S = \begin{bmatrix} 0 & 0 & 0 & \cdot \cdot & \cdot \cdot \\ 1 & 0 & 0 & 0 & \cdot \cdot \\ 0 & 1 & 0 & 0 & \cdot \cdot \\ \cdot \cdot & 0 & 1 & 0 & \cdot \cdot \\ \cdot \cdot & \cdot \cdot & \cdot \cdot & 1 & \cdot \cdot \cdot \cdot \end{bmatrix}$$

Exercise 2.2.2 If S is the shift on $l^2 \mathbb{N}$ then prove that

a) S is an isometry $l^2(\mathbb{N}) \to l^2(\mathbb{N})$, that is, $S^* S = 1$. Furthermore, $1 - S S^* = P_0$ where P_0 is orthogonal projection to $\mathbb{C} e_0$.

b) $1 - S^n(S^n)^* = \sum_{k=0}^{n-1} P_k$, with P_k the rank-one projection onto $\mathbb{C}e_k$.
c) $S^i P_0(S^j)^* = E_{i,j}$ for all $i, j \in \mathbb{N}$, where $E_{i,j}$ is the rank-one operator whose matrix has a 1 in the i, jth spot and zeros everywhere else.

Exercise 2.2.2 c) implies that the C*-algebra generated by S, and hence the Toeplitz algebra, contains all operators on $l^2(\mathbb{N})$ whose matrix representations contain only finitely many nonzero entries. Hence:

Proposition 2.2.3 *The Toeplitz algebra $\mathcal{T}$ contains $\mathcal{K}(\mathbf{H}^2)$ as a norm-closed ideal closed under adjoint.*

Proof The *-algebra of operators on $\mathbf{H}^2$ whose matrices with respect to the standard basis have only finitely many entries, are dense in $\mathcal{K}(L^2(\mathbb{T}))$, by Exercise 1.6.6, and Lemma 1.6.9. Since $\mathcal{T}$ is closed in the operator norm topology, $\mathcal{T}$ contains $\mathcal{K}$ as claimed. $\square$

We next observe that the Toeplitz algebra is actually generated by a single operator.

Proposition 2.2.4 *The Toeplitz algebra is generated as a C*-algebra by $S := T_z$.*

Proof First note that $C^*(S)$ contains all Toeplitz operators T_f where $f \in \mathbb{C}[z, \bar{z}]$. For example, if $f(z) = 2\bar{z} + 1 + 3z + 4z^2$ then

$$(2.5) \qquad \widehat{T_f} = \begin{bmatrix} 1 & 2 & 0 & 0 & \\ 3 & 1 & 2 & 0 & \ddots \\ 4 & 3 & 1 & 2 & \\ 0 & 4 & 3 & 1 & \\ & \ddots & & & \ddots \end{bmatrix} = 2S^* + 1 + 3S + 4S^2.$$

Furthermore, since $T_f = P_+ M_f|_{\mathbf{H}^2}$, it is clear that $\|T_f\| \leq \|M_f\| = \|f\|$, so if $f_n \in \mathbb{C}[z, \bar{z}]$ is a uniformly convergent sequence, converging to $f \in C(\mathbb{T})$, then $T_{f_n} \to T_f$.

Consequently $C^*(S)$ contains all T_f's with $f \in C(\mathbb{T})$, and hence it contains the generators of $\mathcal{T}$. So it contains $\mathcal{T}$. Obviously $C^*(S) \subset \mathcal{T}$ since $S \in \mathcal{T}$. Hence $C^*(S) = \mathcal{T}$ as claimed. $\square$

Exercise 2.2.5 Prove the following.

a) If T is a Toeplitz operator then $T_z^* T T_z = T$. What about the converse?
b) Show that T_f Toeplitz implies T_f^* is Toeplitz, and that $T_f^* = T_{f^*}$.
c) Let $\lambda : \mathbb{T} \to \mathbf{U}(L^2(\mathbb{T}))$ be the regular representation. It leaves the Hardy space $\mathbf{H}^2$ invariant so we will consider for the moment λ as a representation of $\mathbb{T}$ on $\mathbf{H}^2$. Show that $\lambda(w)^* T_z \lambda(w) = w T_z$ for any $w \in \mathbb{T}$. What does this say about the spectrum of T_z as a subset of the complex plane?
d) Prove that if $\pi : \mathbb{T} \to \mathbf{U}(H)$ is a *finite-dimensional* representation of $\mathbb{T}$ and $T \in \mathbb{B}(H)$ such that $\pi(z) T \pi(z)^* = zT$ for all $z \in \mathbb{T}$, then the spectrum of T is

$\{0\}$. Then give an example of an operator T on $\mathbb{C}^2$ and a unitary representation of $\mathbb{T}$ on $\mathbb{C}^2$ such that $\pi(z)T\pi(z)^* = zT$ for all $z \in \mathbb{T}$, but T is not the zero operator.

The previous exercise shows that the Toeplitz algebra $\mathcal{T}$ carries an action of the group $\mathbb{T}$ as C*-algebra automorphisms: the automorphism of $\mathcal{T}$ corresponding to $z \in \mathbb{T}$ is given by $\alpha_z(T) := \lambda(z)T\lambda(z)^*$.

Commutators play an important role in Noncommutative Geometry. If P_+ is the Szegö projection, considered as projection to $l^2(\mathbb{N})$ inside $l^2(\mathbb{Z})$, by multiplication by the characteristic function of the natural numbers, λ the regular representation of $\mathbb{Z}$, then the commutator $[P_+, \lambda(n)] = P_+\lambda(n) - \lambda(n)P_+$ for $n \geq 0$ is given by

$$[\chi_{[0,\infty)}, \lambda(n)] = \lambda(n) \cdot \chi_{[-n,0)},$$

where $\chi_{[-n,0)}$ is the operator of multiplication by the corresponding interval. In particular the commutator has finite rank.

Taking Fourier transform we see that $[P_+, M_f] \in \mathbb{B}(L^2\mathbb{T}))$ has finite rank for $f \in \mathbb{C}[z, \bar{z}]$, and in particular $[P_+, M_f]$ is compact for $f \in \mathbb{C}[z, \bar{z}]$, and hence, by density of $\mathbb{C}[z, \bar{z}]$ in $C(\mathbb{T})$ and the norm continuity of the commutator operation that

Lemma 2.2.6 *If $f \in C(\mathbb{T})$, then the commutator $[M_f, P_+]$ is compact.*

Corollary 2.2.7 *If T_f and T_g are two Toeplitz operators, then*

$$T_f T_g - T_{fg}$$

is a compact operator.

Proof Now, let T_f and T_g be two Toeplitz operators. Then as operators on $\mathbf{H}^2$,
(2.6)
$$T_f T_g = P_+ M_f P_+ M_g = P_+ M_f(M_g + [P_+, M_g]) = T_{fg} + P_+ M_f[P_+, M_g],$$

By Lemma 2.2.6, $P_+ M_f[P_+, M_g]$ is a compact operator. $\qquad\square$

In particular, up to "compact operator" error, the Toeplitz algebra is commutative. We will show shortly that the C*-algebra quotient $\mathcal{T}/\mathcal{K}$ is isomorphic to $C(\mathbb{T})$.

Theorem 2.2.8 *The Toeplitz algebra contains $\mathcal{K}(L^2(\mathbb{T}))$ as a closed *-subalgebra and an ideal. Furthermore,*

a) *The equalities*

$$\mathrm{dist}(T_f, \mathcal{K}) := \inf\{\|T_f - S\| \mid S \in \mathcal{K}(L^2(\mathbb{T}))\} = \|T_f\| = \|f\|_{C(\mathbb{T})}$$

hold for any $f \in C(\mathbb{T})$.
b) *Every pseudo-Toeplitz operator $T \in \mathcal{T}$ can be written uniquely in the form $T = T_f + S$ where $f \in C(\mathbb{T})$ and $S \in \mathcal{K}$.*
c) *If $\mathcal{T}/\mathcal{K}$ is given the quotient norm $\|T + \mathcal{K}\| := \mathrm{dist}(T, \mathcal{K})$, and quotient *-algebra structure, then $\mathcal{T}/\mathcal{K}$ is a C*-algebra, and, moreover, it is isomorphic to $C(\mathbb{T})$, by*

the map $q \colon C(\mathbb{T}) \to \mathcal{T}/\mathcal{K}$.

$$q(f) = T_f \bmod \mathcal{K}.$$

If $T = T_f + S$ is a pseudo-Toeplitz operator, then we call f the *symbol* of T. That the symbol is uniquely defined is a consequence of the work done above.

Proof Statement a) is Proposition 2.2.3.

Consider the matrix

$$(2.7) \qquad T_f = \begin{bmatrix} a_0 & a_{-1} & a_{-2} & a_{-3} & \cdots \\ a_1 & a_0 & a_{-1} & a_{-2} & \cdots \\ \hline a_2 & a_1 & a_0 & a_{-1} & a_{-2} \\ \cdots a_3 & a_2 & a_1 & a_0 & a_{-1} \\ \cdots & & a_2 & a_1 & a_0 \\ & & & \cdots & & \cdots \end{bmatrix}$$

of a Toeplitz operator T_f. We have indicated a partition of the matrix. For each n let P_n be projection to the closed span of $e_n, e_{n+1}, \ldots$. Truncating the matrix as shown to the bottom right hand corner amounts to replacing T_f by $P_2 T_f P_2$, which has matrix

$$(2.8) \qquad P_2 T_f P_2 = \begin{bmatrix} 0 & 0 & 0 & 0 & \cdots \\ 0 & 0 & 0 & 0 & \cdots \\ \hline 0 & 0 & a_0 & a_{-1} & a_{-2} \\ 0 & 0 & a_1 & a_0 & a_{-1} \\ \cdots \cdots & a_2 & a_1 & a_0 \\ & & & \cdots & & \cdots \end{bmatrix}$$

It is obvious from looking at the two corresponding matrices that T_f and $P_2 T_f P_2$ have the same operator norm. Thus, we prove inductively the interesting fact that truncation is isometric when applied to Toeplitz operators:

$$(2.9) \qquad \|P_n T_f P_n\| = \|T_f\|, \ n = 0, 1, \ldots.$$

Now let S be a finitely supported matrix. Then its truncations $P_n S P_n$ are zero for n large enough, and hence

$$\|S - T_f\| \ge \|P_n(S - T_f)P_n\| = \|P_n T_f P_n\| = \|T_f\|.$$

This shows that

$$\mathrm{dist}(T_f, F) \ge \|T_f\|,$$

where F is the *-algebra of operators with matrices of finite support. Since F is dense in $\mathcal{K}$ it follows that

$$\text{dist}(T_f, \mathcal{K}) \geq \|T_f\|.$$

On the other hand, $\text{dist}(T_f, \mathcal{K}) \leq \|T_f\|$ is clear. So

$$\text{dist}(T_f, \mathcal{K}) = \|T_f\|.$$

It remains to prove that $\|T_f\| = \|f\|$. Clearly $\|T_f\| \leq \|f\|$. If $\varepsilon > 0$ and $f \in \mathbb{C}[z, \bar{z}]$ then since $\|f\| = \|M_f\|$, there exists $v \in \mathbb{C}[z, \bar{z}] \subset L^2(S^1)$ such that $\|M_f v\| \geq \|f\| - \varepsilon$ and $\|v\| = 1$. Since $M_f v \in \mathbb{C}[z, \bar{z}]$, $M_z^n M_f v \in \mathbb{C}[z]$ for n large enough, and M_f and the unitary M_z commute so we may argue so $\|T_f\| \geq \|T_f M_z^n v\| = \|P_+ M_f M_z^n v\| = \|P_+ M_z^n M_f v\| = \|z^n M_f v\| = \|M_f v\| \geq \|f\| - \varepsilon$. Hence $\|T_f\| \geq \|f\|$ for $f \in \mathbb{C}[z, \bar{z}]$ and so $\|f\| = \|T_f\|$. It follows that the map $q \colon C(\mathbb{T}) \to \mathcal{T}/\mathcal{K}$ is isometric on the dense subset $\mathbb{C}[z, \bar{z}]$ and hence extends to an isometry of normed spaces.

In particular $\mathcal{T}/\mathcal{K}$ is a C*-algebra isomorphic to $C(\mathbb{T})$.   □

Exercise 2.2.9 Let P_n be projection to the closed span of $e_n, e_{n+1}, \ldots$ in $l^2(\mathbb{N})$, as in the proof of Theorem 2.2.8. Prove that $\lim_{n \to \infty} \|P_n S\| = 0$ for any compact operator S.

Exercise 2.2.10 (Cuntz Algebras) Given any two intervals $[a, b]$, $[c, d]$ of $\mathbb{R}$ there is a canonical affine map

$$\phi(x) := \left(\frac{b-a}{d-c}\right)x + \frac{ad-bc}{d-c},$$

from $[c, d]$ to $[a, b]$, with constant positive derivative $\phi'(x) = \frac{b-a}{d-c}$. We define a linear operator

$$s \colon L^2([a, b]) \to L^2([c, d]), \quad (s\xi)(x) = \sqrt{\phi'(x)} \cdot \xi(\phi(x)) = \sqrt{\frac{b-a}{d-c}} \cdot \xi(\phi(x)),$$

a Hilbert space map between the two corresponding L^2-spaces.

a) s is an isometry.
b) Now divide the unit interval $[0, 1]$ into n subintervals $I_1, \ldots I_k$ of equal length, and we regard $L^2(I_k)$ as a closed subspace of $L^2([0, 1])$ in the obvious way by extending functions by zero. Let

$$s_k \colon L^2([0, 1]) \to L^2(I_k) \to L^2([0, 1])$$

be the composition, the first being one of the interval isometries described above, and the second map the inclusion.

Prove that $s_1, \ldots, s_k \in \mathbb{B}(L^2([0, 1])$ are isometries with orthogonal ranges, compute the range projections $s_i s_i^*$, as Hilbert space projections on $L^2([0, 1])$, and verify that

$$\sum_{i=1}^{n} s_i s_i^* = 1.$$

The C*-algebra generated by $s_1, \ldots, s_n$ is the *Cuntz algebra* O_n (see [55].) A more general, but similar class of algebras related to topological Markov chains is discussed in Exercise 1.7.13, see [58].

Remark 2.2.11 The Cuntz algebra O_n turns out to be unique in the following strong sense (see [55]): *any two C*-algebras generated by n isometries $s_1, \ldots, s_n$, and $t_1, \ldots, t_n$, with orthogonal ranges summing to the identity, are isomorphic by a map sending s_i to t_i.*

Exercise 2.2.12 Let $s_1, \ldots, s_n$ be isometries of a Hilbert space H such that $\sum_{i=1}^{n} s_i s_i^* = 1$.

a) Prove that the s_i have orthogonal ranges.
b) Prove that the linear span of the elements $s_i s_j^*$, $i, j = 1, 2, \ldots n$ form a *-algebra isomorphic to $M_n(\mathbb{C})$.
c) What about the *-algebra (or C*-algebra) generated by $s_i s_j s_k^* s_l^*$?

2.3   The Toeplitz Trace Theorem

The connection between the Toeplitz algebra and analytic function theory leads to an interesting identity which we will identify in the next section as an Index Theorem.

Exercise 2.3.1 Use the Cauchy-Schwarz inequality to prove that if $f \in \mathbf{H}^2$ with Fourier series $f = \sum_{n=0}^{\infty} a_n z^n \in \mathbf{H}^2$, then for each $z \in \mathbb{D}$, the power series $\tilde{f}(z) := \sum_{n=0}^{\infty} a_n z^n$ converges absolutely and

$$|\tilde{f}(z)| \leq \frac{\|f\|_{\mathbf{H}^2}}{\sqrt{1 - |z|^2}}.$$

Deduce that $\tilde{f}$ is analytic in $\mathbb{D}$.

Let $f \in \mathbf{H}^2$, with Fourier series $\sum_{n=0}^{\infty} a_n z^n$. By Exercise 2.3.1 the function $\sum_{n=0}^{\infty} a_n z^n$ is a power series defining an analytic function $\tilde{f}$ in $\mathbb{D}$.

Exercise 2.3.2 Show that the map $f \mapsto \tilde{f}(z)$ defines a continuous linear functional $\mathbf{H}^2 \to \mathbb{C}$ for each $z \in \mathbb{D}$ and deduce by the Riesz Representation Theorem that there exist $k_z \in \mathbf{H}^2$ such that

$$\tilde{f}(z) = \langle k_z, f \rangle,$$

the inner product in $\mathbf{H}^2$. Show that

$$k_z(w) = \sum_{n=0}^{\infty} (\bar{z}w)^n = \frac{1}{1 - \bar{z}w}$$

is the required vector.

The function

$$k(z, w) := \overline{k_z(w)} = \frac{1}{1 - \bar{w}z}$$

is called the *Szegö kernel*, and we have

(2.10) $$\tilde{f}(z) = \int_{\mathbb{T}} \frac{f(w)}{1 - \bar{w}z} d\mu(w),$$

with μ normalized Lebesgue measure on $\mathbb{T}$. The integral converges absolutely for $f \in \mathbf{H}^2$ and $|z| < 1$. Notice that we may identify the integral (2.10) with the contour integral

$$\frac{1}{2\pi i} \cdot \oint_{\mathbb{T}} \frac{f(w)}{w - z} dw,$$

over the unit circle oriented counter-clockwise. Therefore, by the Cauchy Theorem, if f extends analytically to the disk then $\int_{\mathbb{T}} \frac{f(w)}{1 - \bar{w}z} d\mu(w) = \tilde{f}(z)$, giving another point of view on (2.10).

Exercise 2.3.3 Use Residue Theorem to prove that

$$\oint_{\mathbb{T}} \frac{w^n}{w - z} dw = 0$$

if $n < 0$. (*Hint.* The Residue Theorem equates the integral with the sum of the residues of the meromorphic function $g(w) = \frac{w^n}{w-z}$. For any $n \in \mathbb{Z}$, one of these poles happens at $w = z$ with residue z^n. If $n < 0$ there is a pole at $w = 0$; check that the residue there is $-z^n$.)

The discussion shows that the Szegö projection has the following alternative geometric meaning. If $f \in L^2(\mathbb{T})$ is in a suitable (dense) class of functions (*e.g.*, trigonometric polynomials), then $P_+ f \in \mathbf{H}^2$ is the boundary value of the analytic function in the disk given by

$$I_k f(z) = \int_{\mathbb{T}} \frac{f(w)}{1 - \bar{w}z} d\mu(w) = \frac{1}{2\pi i} \oint_{\mathbb{T}} \frac{f(w)}{w - z} dw.$$

By "boundary value" we refer to the continuous function on $\mathbb{T}$

$$z \mapsto \lim_{z' \to z, \, |z'| < 1} I_k f(z').$$

Exercise 2.3.4 Let X be a compact Hausdorff space with a probability measure μ on it, let $f \in C(X)$ and $k \in L^2(X \times X)$, I_k the integral operator

$$(I_k v)(x) = \int_X k(x, y)v(y)d\mu(y),$$

with kernel k. Let M_f be the multiplication operator $(M_f v)(x) = f(x)v(x)$. Show that the commutator $[M_f, I_k]$ is the integral operator with kernel $k_f(x, y) = (f(x) - f(y)) \cdot k(x, y)$.

The previous exercise suggests a more geometric way of understanding commutators with the Szegö projection.

We have noted above that for f in a suitable class of functions on the circle, e.g., for $f \in \mathbb{C}[z, \bar{z}]$, then

$$P_+ f(x) = \lim_{z' \to z, \, |z'| < 1} (I_k f)(z'),$$

where for $|z| < 1$,

$$(I_k f)(z) = \int_{\mathbb{T}} \frac{f(w)}{1 - \bar{w}z} d\mu(w).$$

Using this formula, we deduce that the commutator $[M_f, P_+]$ is given by

$$([M_f, P_+]v)(z) = \lim_{z' \to z, \, |z'| < 1} \int_{\mathbb{T}} \frac{f(w) - f(w)}{1 - \bar{w}z'} \cdot v(w) \, d\mu(w),$$

at least for $f, v \in \mathbb{C}[z, \bar{z}]$. This equals

$$\lim_{z' \to z, \, |z'| < 1} \int_{\mathbb{T}} w \cdot \left(\frac{f(w) - f(w)}{w - z'} \right) \cdot v(w) \, d\mu(w),$$

Consider the function

$$k_f(z, w) = w \cdot \left(\frac{f(w) - f(z)}{w - z} \right),$$

defined initially on the complement of the diagonal $z = w$ in $\mathbb{T} \times \mathbb{T}$. If the complex derivative $f'(z)$ exists everywhere on $\mathbb{T}$ (for example if $f \in \mathbb{C}[z, \bar{z}]$) then we can define k_f along the diagonal by

$$k_f(z, z) := z \cdot \frac{\partial f}{\partial z}$$

and this will give a continuous extension of k_f to $\mathbb{T} \times \mathbb{T}$ for which we may write

$$[M_f, P_+] v(z) = \int_{\mathbb{T}} k_f(z, w) v(w) d\mu(w),$$

realizing the operator commutator as an integral operator with continuous (whence L^2) kernel and hence a compact operator.

Actually, $z \cdot \frac{\partial f}{\partial z}$ only depends on the *angular* differentiability of f. To see this, write f as a function of θ with $z = e^{i\theta}$, we have

$$(2.11) \quad \frac{\partial f}{\partial z}(e^{i\theta}) = \lim_{\alpha \to 0} \frac{f(e^{i\theta}) - f(e^{i(\theta+\alpha)})}{e^{i\theta} - e^{i(\theta+\alpha)}}$$

$$= e^{-i\theta} \cdot \lim_{\alpha \to 0} \left(\frac{f(e^{i\theta}) - f(e^{i(\theta+\alpha)})}{\alpha} \right) \cdot \left(\frac{\alpha}{1 - e^{i\alpha}} \right)$$

$$= i e^{-i\theta} \cdot \frac{\partial f}{\partial \theta}(e^{i\theta}),$$

since $\lim_{\alpha \to 0} \frac{\alpha}{1 - e^{i\alpha}} = -i$. Thus,

$$k_f(e^{i\theta}, e^{i\theta}) = e^{i\theta} \cdot \frac{\partial f}{\partial z}(e^{i\theta}) = i \cdot \frac{\partial f}{\partial \theta}(e^{i\theta})$$

In particular, k_f is continuous on $\mathbb{T} \times \mathbb{T}$ if f is C^1 on $\mathbb{T}$, and the commutator $[P_+, f]$ is given by the integral operator I_{k_f}.

Proposition 2.3.5 *If* $f \in C^\infty(\mathbb{T})$, *define, in, polar coordinates* θ, θ' *on* $\mathbb{T}$,

$$(2.12) \qquad k_f(\theta, \theta') = \begin{cases} \left(\frac{f(e^{i\theta}) - f(e^{i\theta'})}{e^{i\theta} - e^{i\theta'}} \right) & \text{if } \theta \neq \theta' \\ -i \cdot \frac{\partial f}{\partial \theta}(e^{i\theta}) & \text{if } \theta = \theta' \end{cases}$$

Then k_f *is smooth on* $\mathbb{T} \times \mathbb{T}$, *and hence the integral operator* I_{k_f} *with kernel* k *is a trace-class operator.*

For such f, *the commutator* $[P_+, M_f]$ *equals* I_{k_f}.

Proof We have proved above that $[P_+, M_f] = I_{k_f}$ for $f \in \mathbb{C}[z, \bar{z}]$. Under Fourier transform, $C^\infty(\mathbb{T})$ is the space of sequences $(a_n)_{n\mathbb{Z}}$ of rapid decay: that is $\sup_{n\in\mathbb{Z}} |p(n) \cdot a_n|. < \infty$ for any polynomial p. Using this, if f smooth has Fourier series $\sum a_n z^n$ then the partial sums $f_n(z) = \sum_{|k|\leq n} a_k z^k$ along with all of their derivatives $f_n^{(k)}$, converge absolutely and uniformly to f on $\mathbb{T}$. It follows that $k_{f_n} \to k_f$ uniformly on $\mathbb{T} \times \mathbb{T}$, and hence $k_{f_n} \to k_f$ in $L^2(\mathbb{T} \times \mathbb{T})$ and hence $I_{k_{f_n}} \to I_{k_f}$ in operator norm. We get

$$[P_+, M_f] = \lim_{n\to\infty} [P_+, M_{f_n}] = \lim_{n\to\infty} I_{k_{f_n}} = I_{k_f}$$

as claimed, for any $f \in C^\infty(\mathbb{T})$.   □

We now deduce a rather remarkable identity.

It is a basic but nontrivial result of topology that the fundamental group of the circle is isomorphic to the group of integers. If $\varphi \colon \mathbb{T} \to \mathbb{T}$ is a loop in the compact space $\mathbb{T}$, with $\varphi(1) = 1$, then the integer corresponding to the class $[\gamma] \in \pi_1(\mathbb{T})$ is called the *winding number* of γ. It is an invariant of the homotopy-class of φ amongst maps $\mathbb{T} \to \mathbb{T}$.

Intuitively, it is the number of times (possibly negative) it wraps the circle around itself.

More generally, since $\mathbb{T} \subset \mathbb{C}^*$ is a deformation retract, for any non-vanishing continuous function $f \colon \mathbb{T} \to \mathbb{C}^*$ the composition

$$\mathbb{T} \xrightarrow{f} \mathbb{C}^* \xrightarrow{r} \mathbb{T}$$

defines an element of $\pi_1(\mathbb{T})$ and we denote the corresponding integer by $\mathrm{wind}_f(0)$.

Remark 2.3.6 If $f \in C^1(\mathbb{T})$ then the winding number can be expressed as a contour integral

$$(2.13) \qquad \mathrm{wind}_f(0) = \frac{1}{2\pi i} \oint_f \frac{dz}{z},$$

where in the formula, f is understood to be the closed curve, i.e., closed contour, $[0, 2\pi] \xrightarrow{\exp} \mathbb{T} \xrightarrow{f} \mathbb{C}^*$ in the complex plane. To be slightly more explicit, if we view f as a function of the argument parameter $\theta \in [0, 2\pi]$, then

$$(2.14) \qquad \mathrm{wind}_f(0) = \frac{1}{2\pi i} \int_0^{2\pi} \frac{f'(\theta)}{f(\theta)} d\theta.$$

Consider the bilinear functional $C(\mathbb{T}) \times C(\mathbb{T}) \to \mathbb{C}$,

$$\varphi(f, g) := \mathrm{Trace}(f[P_+, g]).$$

where for brevity of notation we have written simply f in place of M_f.

A computation show that it has the anti-symmetry property $\varphi(f, g) = -\varphi(g, f)$:

$$\varphi(f, g) = \text{Trace}(f[P_+, g]) = \text{Trace}([P_+, g]f) = \text{Trace}(P_+ fg - gP_+ f)$$
$$= \text{Trace}\,(P_+ fg - g(fP_+ + [P_+, f]))$$
$$= \text{Trace}([P_+, fg]) - \text{Trace}(g[P_+, f])$$
$$(2.15) \qquad = -\text{Trace}(g[P_+, f]) = -\varphi(g, f).$$

The functional φ is an example of a *cyclic cocycle*. What is remarkable, is that $\varphi(f, g)$ is an *integer* when $fg = 1$:

Theorem 2.3.7 *Let* $f \in C^\infty(\mathbb{T})$ *be a non-vanishing smooth function on* $\mathbb{T}$. *Then the operator*

$$M_{1/f} \cdot [P_+, M_f]$$

on $L^2(\mathbb{T})$ *is trace-class, and*

$$(2.16) \qquad\qquad \text{Trace}(M_{1/f} \cdot [P_+, M_f]) = \text{wind}_f(0).$$

Proof By Proposition 2.3.5, $[P_+, M_f]$ is an integral operator with smooth kernel. Hence $g[P_+, M_f]$ is also an integral operator with smooth kernel, for any smooth g. By Theorem 2.1.15, the trace of an integral operator on $L^2(\mathbb{T})$ with smooth kernel is the integral of the kernel along the diagonal, given in this case by $-ig(\theta)f'(\theta)$. We obtain therefore a geometric formula for the cyclic cocycle φ:

$$(2.17) \qquad \varphi(g, f) := \text{Trace}(M_g[P_+, M_f]) = \frac{1}{2\pi i} \int_{\mathbb{T}} g(\theta)f'(\theta)d\theta.$$

$\square$

We are going to interpret the invariant appearing in the Theorem in the next section as a Fredholm index.

2.4   The Calkin Algebra

In our discussion of the Toeplitz algebra in the previous section, several important points emerged. One of these was that a (pseudo-)Toeplitz operator T, *when considered up to compact perturbation*, is exactly equivalent to a continuous function on the circle, its 'symbol.'

More precisely, the quotient Banach *-algebra $\mathcal{T}/\mathcal{K}$, with the quotient norm, is a C*-algebra isomorphic to $C(\mathbb{T})$ by the quotient map $q : C(\mathbb{T}) \to \mathcal{T}/\mathcal{K}$, $f \mapsto T_f + \mathcal{K}$.

The Toeplitz Index Theorem, a (very) special case of the Atiyah–Singer Index Theorem, asserts the equality of two integer invariants of a pseudo-Toeplitz operator T with non-vanishing symbol.

The first is the winding number of the symbol, discussed at the end of the previous section. The second, called the Fredholm index, is analytic in nature: it is by definition the difference in dimensions of the solutions spaces to the equations $T\xi = 0$ and to $T^*\xi = 0$.

It is not obvious that our assumption that the symbol is non-vanishing, implies that these dimensions are even finite.

The Toeplitz Index Theorem represents, therefore, an interesting bridge between analysis, on the one hand, and topology, on the other.

The general framework of index theory is based on another C*-algebra quotient, called the Calkin algebra.

By Corollary 1.6.10, for any Hilbert space H, the C*-algebra $\mathcal{K}(H)$ of compact operators on H is a closed *-ideal in $\mathbb{B}(H)$. By standard algebra, the quotient (vector space) $\mathbb{B}(H)/\mathcal{K}(H)$ is an algebra, under multiplication of cosets. Since $\mathcal{K}(H)$ is invariant under adjoint, $\mathbb{B}(H)/\mathcal{K}(H)$ inherits an adjoint operation as well.

Also, if we give the quotient $\mathbb{B}(H)/\mathcal{K}(H)$ the quotient norm

$$\|T + \mathcal{K}\| := \operatorname{dist}(T, \mathcal{K}) := \inf\{\|T + S\|; \mid S \in \mathcal{K}\},$$

then we obtain a Banach *-algebra; this is a rather general fact, as the following Exercise shows.

Exercise 2.4.1 Let A be a Banach algebra, and let $J \subset A$ be an ideal which is also a closed subspace of A) in the norm topology. Prove that the quotient vector space A/J with its operation of multiplication of cosets, and the norm

$$\|a + J\| : \operatorname{dist}(a, J)$$

is a Banach algebra. (*Hint.* See Section 4.4 Lemma 4.4.10 for the proof.)

When we consider the smaller quotient $\mathcal{T}/\mathcal{K}$ in the previous section, we verified the C*-identity for the quotient norm. Similar arguments prove that actually the C*-identity holds for $Q(H) := \mathbb{B}(H)/\mathcal{K}(H)$ as well, as we show.

Let $e_1, e_2, \ldots$ be an orthonormal basis for H and P_n be projection to the span of $e_n, e_{n+1}, \ldots$.

Lemma 2.4.2 *If $T \in \mathbb{B}(H)$ then*

$$\operatorname{dist}(T, \mathcal{K}) = \lim_{n \to \infty} \|T P_n\| = \lim_{n \to \infty} \|P_n T\|.$$

The reader should compare to (2.9); that is, we have already verified the Lemma for Toeplitz operators T_f (indeed, it is obvious for them, since the sequence $\|P_1 T_f\|, \|P_2 T_f\|, \ldots$ was observed there to be *constant*.)

Proof It is clear that $\mathrm{dist}(T, \mathcal{K}) \leq \|T - T(1 - P_n)\| = \|TP_n\|$, since $1 - P_n$ is compact. Since this is true for all n, we get

$$\mathrm{dist}(T, \mathcal{K}) \leq \liminf_{n\to\infty} \|TP_n\|.$$

On the other hand, if $S \in \mathcal{K}$ then

$$\|TP_n\| = \|TP_n + SP_n - SP_n\| = \|(T + S)P_n - SP_n\| \leq \|T + S\| + \|SP_n\|$$

by the triangle equality and the fact that $\|P_n\| = 1$ for all n. Hence, since $\lim_{n\to\infty} \|SP_n\| = 0$ by Exercise 2.2.9, we get

$$\limsup_{n\to\infty} \|TP_n\| \leq \mathrm{dist}(T, \mathcal{K}).$$

Putting these two results together gives that $\lim_{n\to\infty} \|P_n T\| = \mathrm{dist}(T, \mathcal{K})$. Since this is true for all $T \in \mathbb{B}(H)$ and since $\|T\| = \|T^*\|$ for any bounded operator, it follows that $\lim_{n\to\infty} \|TP_n\| = \mathrm{dist}(T, \mathcal{K})$ as well. $\qquad\square$

Corollary 2.4.3 *The Banach algebra* $Q(H) := \mathbb{B}(H)/\mathcal{K}(H)$ *endowed with the quotient norm* $\|T + \mathcal{K}\| := \mathrm{dist}(T, \mathcal{K})$ *and adjoint* $(T + \mathcal{K})^* := T^* + \mathcal{K}$ *is a C*-algebra.*

The C*-algebra Q is called the *Calkin algebra*.

Proof The proposed norm satisfies the C*-identity because

$$\mathrm{dist}(T, \mathcal{K})^2 = \mathrm{dist}(T, \mathcal{K}) \cdot \mathrm{dist}(T^*, \mathcal{K}) \geq \mathrm{dist}(T^*T, \mathcal{K}) = \lim_{n\to\infty} \|T^*TP_n\|$$

$$(2.18) \qquad \geq \lim_{n\to\infty} \|P_n TT^* P_n\| = \lim_{n\to\infty} \|TP_n\|^2 = \mathrm{dist}(T, \mathcal{K})^2.$$

where the first step uses the fact that Q is a Banach algebra with the distance norm. $\qquad\square$

The Calkin algebra comes with a quotient map $\rho \colon (\mathbb{B}(H)$ to $Q(H))$ It is a surjective *-homomorphism with kernel $\mathcal{K}$.

The construction of Toeplitz operators from $f \in C(\mathbb{T})$ determines a C*-algebra injection $C(\mathbb{T}) \to Q$.

Proposition 2.4.4 *The map*

$$\tau \colon C(\mathbb{T}) \to Q(H^2), \quad \tau(f) := \pi(T_f),$$

*is an injective, unital *-homomorphism.*

In particular, if $f \in C(\mathbb{T})$ does not vanish anywhere on the circle, then $T_f + \mathcal{K}$ is an invertible in the C*-algebra $Q(\mathbf{H}^2)$.

Proof By the definitions, $\mathcal{T}$ is a C*-subalgebra of $\mathbb{B}(\mathbf{H}^2)$, that is, there is an injective *-homomorphism $\mathcal{T} \to \mathbb{B}(\mathbf{H}^2)$. This *-homomorphism maps the ideal $\mathcal{K}(\mathbf{H}^2)$ to itself, and hence induces a C*-algebra homomorphism $\mathcal{T}/\mathcal{K} \to \mathbb{B}/\mathcal{K}$, which is injective by the definitions.

Now since $C(\mathbb{T}) \to \mathcal{T}/\mathcal{K}$ has already been shown to be a C*-algebra isomorphism in Theorem 2.2.8 c), the result follows. $\square$

We have already noted that a Toeplitz operator T_f with $f \in C(\mathbb{T})$ not vanishing anywhere, is invertible when considered as an element of the Calkin C*-algebra $Q(\mathbf{H}^2)$. That is, T_f is "invertible mod compacts," when f is non-vanishing. Operators which are invertible mod compacts are called *Fredholm operators*.

Lemma 2.4.5 *The following conditions are equivalent for a bounded operator $T : H \to K$ between two Hilbert spaces, and such T is called* Fredholm *if it satisfies them.*

1) *There exist bounded operators $Q, Q' : K \to H$ such that $QT - \mathrm{id}_H$ and $TQ' - \mathrm{id}_K$ are finite-rank operators.*
2) *There exist bounded operators $Q, Q' : K \to H$ such that $QT - \mathrm{id}_H$ and $TQ' - \mathrm{id}_K$ are compact operators.*
3) *The range of T is closed and $\ker(T)$ and $\mathrm{coker}(T) := K/\mathrm{ran}(T)$ are each finite-dimensional vector spaces.*

If T is Fredholm, the index *of T is defined to be*

$$(2.19) \qquad\qquad \mathrm{Index}(T) := \dim \ker(T) - \dim \mathrm{coker}(T).$$

Remark 2.4.6 Note that if T is Fredholm, then as T has closed range, $\mathrm{coker}(T) := H/\mathrm{ran}(T) \cong \mathrm{ran}(T)^{\perp} = \ker(T^*)$, so

$$(2.20) \qquad\qquad \mathrm{Index}(T) = \dim \ker(T) - \dim \ker(T^*).$$

In particular, $\mathrm{Index}(T) = 0$ for any self-adjoint Fredholm operator.

Exercise 2.4.7 If H is a *finite*-dimensional Hilbert space and $T : H \to H$ is a linear map then $\mathrm{Index}(T) = 0$. What about an operator between two finite-dimensional Hilbert spaces of possibly different dimensions?

Proof Suppose T as in 2).

Suppose Q, Q' are bounded such that $QT - 1$ and $TQ' - 1$ are compact operators. Let $A := QT - 1$, $B = TQ' - 1$. Then the kernel of T is contained in the kernel of $QT = 1 + A$, which is the -1-eigenspace of A. But all eigenspaces of a compact operator corresponding to nonzero eigenvalues are finite-dimensional (Exercise 1.6.14 a)). So the kernel of T is finite-dimensional.

We show next that $\operatorname{ran}(T)$ is closed if it satisfies 2). Note that $\operatorname{ran}(QT) = \operatorname{ran}(1 + A)$ is closed for A compact. The restriction of QT is a bijective bounded operator $\ker(QT)^{\perp} \to \operatorname{ran}(QT)$ between Hilbert spaces, and hence (by the Open Mapping Theorem) there exists $C > 0$ such that $\|QTv\| \geq C\|v\|$ for all $v \in \ker(QT)^{\perp}$. We deduce

$$\|Q\| \cdot \|Tv\| \geq C\|v\|, \quad \forall v \in \ker(QT)^{\perp},$$

and it follows immediately that $T\left(\ker(QT)^{\perp}\right)$ is closed. We get that

$$\operatorname{ran}(T) = T\left(\ker(QT)^{\perp}\right) + T\left(\ker(QT)\right)$$

is the sum of a closed subspace and a finite-dimensional subspace. Hence it is closed.

Finally, since $\operatorname{ran}(T)$ is closed, $H/\operatorname{ran}(T) \cong \operatorname{ran}(T)^{\perp} = \ker(T^*)$. Applying the above argument using $TQ = 1 + B$ shows that this is finite-dimensional.

Assume 3). We prove 1). T restricts to a bounded bijective linear map $\ker(T)^{\perp} \to \operatorname{ran}(T)$ between two Hilbert spaces, so there is a (unique) bounded linear map $S\colon \operatorname{ran}(T) \to \ker(T)^{\perp}$ such that $ST = \operatorname{id}_{\ker(T)^{\perp}}$ and $TS = \operatorname{id}_{\operatorname{ran}(T)}$. We can extend S to K by setting it equal to zero on $\operatorname{ran}(T)^{\perp}$. The extension is now a bounded linear map $Q\colon K \to H$ such that QT is the identity operator on $\ker(T)^{\perp}$, and is zero on $\ker(T)$. Thus, $1 - QT$ is the orthogonal projection operator onto $\ker(T)$, a finite-rank, operator. Similarly, $1 - TQ$ is zero on $\operatorname{ran}(T)$ and the identity on $\operatorname{ran}(T)^{\perp}$ so is orthogonal projection to $\ker(T^*)$. $\square$

2.5   General Properties of the Fredholm Index

Corollary 2.5.1 *A bounded operator $T \in \mathbb{B}(H)$ is a Fredholm operator if and only if its image $\pi(T) \in Q(H) := \mathbb{B}(H)/\mathcal{K}(H)$ is invertible, where $\pi\colon \mathbb{B}(H) \to Q(H)$ is the quotient map to the Calkin algebra.*

Proof Let T be Fredholm. By Lemma 2.4.5, $\pi(T) \in Q(H)$ is both left and right-invertible. Hence it is invertible.

Conversely, if T has invertible image in the Calkin algebra, let $S \in \mathbb{B}(H)$ such that $\pi(S) = \pi(T)^{-1}$, i.e., let S be a pre-image of its inverse. Then $ST - 1$ and $TS - 1$ are both compact operators. So T is Fredholm. $\square$

It is immediate from Corollary 2.5.1 that the product of two Fredholm operators is Fredholm and the adjoint of a Fredholm operator is Fredholm.

Corollary 2.5.2 *If $f \in C(\mathbb{T})$, then the Toeplitz operator $T_f \in \mathbb{B}(\mathbf{H}^2)$ is Fredholm. In particular, T_f has closed range and $\dim\ker T_f$ and $\dim\ker T_f^*$ are finite-dimensional.*

This is immediate from Lemma 2.4.5

One can be a bit more precise: if $f \in C(\mathbb{T})$ is non-vanishing on the circle, then $T_{1/f}$ is a parametrix for T_f, by (2.6).

We now establish further properties of the Fredholm index defined in the previous section: recall that if T is Fredholm, the *index* of T is defined to be $\mathrm{Index}(T) := \dim \ker(T) - \dim \mathrm{coker}(T)$.

Theorem 2.5.3 *Let H be a Hilbert space and* $\mathrm{Fred}(H)$ *denote the set of Fredholm operators on H.*

a) *If S and T are Fredholm on H then so is ST and T^*, and* $\mathrm{Index}(ST) = \mathrm{Index}(S) + \mathrm{Index}(T)$, *while* $\mathrm{Index}(T^*) = -\mathrm{Index}(T)$.
b) *If T is Fredholm and S is a compact operator then* $\mathrm{Index}(T + S) = \mathrm{Index}(T)$.
c) *The subspace* $\mathrm{Fred}(H) \subset \mathbb{B}(H)$ *is a open.*
d) *The function* $\mathrm{Index}: \mathrm{Fred}(H) \to \mathbb{Z}$ *is continuous.*

Remark 2.5.4 The theorem may be summarized by saying that the Fredholm index induces a continuous group homomorphism

$$\mathrm{Index}: \mathbf{GL}(Q) \to \mathbb{Z},$$

from the topological group of invertibles in the Calkin algebra, under multiplication, to the group of integers under addition.

For the proof we will need several lemmas, the first of which is a matter of elementary linear algebra. A sequence of vector spaces and vector space maps

$$0 \to V_1 \xrightarrow{f_1} V_2 \xrightarrow{f_2} \cdots \xrightarrow{f_{n-1}} V_n \to 0$$

is said to be *exact* if $\mathrm{ran}(f_i) = \ker(f_{i+1})$ for $i = 0, 2, \ldots, n$ (where f_0 is understood to be the inclusion of the zero subspace, and f_n the map *to* the zero subspace.)

In particular, f_1 is injective and f_{n-1} is surjective.

Lemma 2.5.5 *Let*

$$0 \to V_1 \to V_2 \to \cdots \to V_n \to 0$$

be an exact sequence of vector spaces. Then $\sum_{k=1}^{n}(-1)^k \dim(V_k) = 0$.

Proof By induction. If $n = 3$ the result follows from $V_2/V_1 \cong V_3$.

If the result holds for all sequences of length $n \geq 2$ and if $n > 2$ and we are given an exact sequence

$$0 \to V_1 \xrightarrow{f_1} V_2 \xrightarrow{f_1} \cdots \xrightarrow{f_{n-1}} V_n \xrightarrow{f_n} V_{n+1} \to 0,$$

then observe that the sequences

$$0 \to V_1 \xrightarrow{f_1} V_2 \xrightarrow{f_1} \cdots \xrightarrow{f_{n-1}} \mathrm{ran}(f_{n-1}) \to 0$$

and

$$0 \to \operatorname{ran}(f_{n-1}) \to V_n \xrightarrow{f_n} V_{n+1} \to 0$$

are exact; now use the inductive hypothesis, the case $n = 2$, and a small amount of algebra. ∎

Lemma 2.5.6 *Let* $T: H_1 \to H_2$ *be a bounded linear operator with closed range. Then there exists* $\varepsilon > 0$ *such that if* $\|S - T\| < \varepsilon$ *then* $\ker(S)$ *injects in* $\ker(T)$. *In particular,* $\dim \ker S \le \dim \ker(T)$.

Proof Since T has closed range, its restriction to $\ker(T)^{\perp}$ is bijective onto a closed subspace of a Hilbert space. So, by the Open Mapping Theorem (see [54]) there exists $C > 0$ such that $\|T\xi\| \ge C$ for all unit vectors ξ in $\ker(T)^{\perp}$.

Choose $\varepsilon = \frac{C}{2}$. Let $P \in \mathbb{B}(H)$ be projection onto $\ker(T)^{\perp}$. Of course then $1 - P$ is projection onto the kernel of T. Let ξ be a unit vector in $\ker(S)$. Then

$$\varepsilon > \|(S - T)\xi\| = \|T\xi\| = \|T(P\xi + (1 - P)\xi)\| = \|TP\xi\| \ge C\|P\xi\|,$$

so $\|P\xi\| < \frac{1}{2}$. Hence $\|(1 - P)\xi\|^2 \ge \|\xi\|^2 - \|P\xi\|^2 = 1 - \|P\xi\|^2 > \frac{3}{4}$ for every unit vector ξ in $\ker(S)$, which shows that the restriction of $1 - P$ to $\ker(S)$ is injective. ∎

Proof of Theorem 2.5.3 From Corollary 2.5.1 it is clear that the product of two Fredholm operators is Fredholm, and that the topological subspace $\operatorname{Fred}(H) \subset \mathbb{B}(H)$ of Fredholm operators on H, is open, since $\operatorname{Fred}(H) = \pi^{-1}(\mathbf{GL}(Q))$, π is continuous, and the group $\mathbf{GL}(Q)$ of invertibles in Q is open, because the invertibles in any C*-algebra are open—see Corollary 3.1.9 of Chapter 3 where we prove it in connection with our development of basic spectral theory.

Thus, c) is proved, and part of a).

Now let T_1 and T_2 be Fredholm. The sequence of finite-dimensional vector spaces

$$0 \to \ker(T_2) \to \ker(T_1 T_2) \xrightarrow{T_2} \ker(T_1) \to H/\operatorname{ran}(T_2)$$

(2.21) $$\xrightarrow{T_1} H/\operatorname{ran}(T_1 T_2) \to H/\operatorname{ran}(T_1) \to 0$$

is routinely checked to be exact. An application of Lemma 2.5.5 to this sequence yields the result. This proves the remainder of a): the additivity of the index.

Next, we prove that the index is invariant under perturbation by *finite-rank* operators. By the Rank-Nullity Theorem of basic linear algebra, $\dim \ker(\lambda - F) = \dim \ker(\bar{\lambda} - F^*)$ for every finite-rank operator F and every $\lambda \in \mathbb{C}$. Applying this to the Fredholm operator $\lambda - F$ gives

$$\operatorname{Index}(\lambda - F) = \dim \ker(\lambda - F) - \dim \ker(\bar{\lambda} - F^*) = 0.$$

Hence, any finite-rank perturbation of a multiple of the identity operator has zero index.

Now if T is Fredholm and F has finite-rank, let Q be a parameterix for T with $QT = 1 + F'$, F' finite-rank. We get that $\text{Index}(Q) + \text{Index}(T) = \text{Index}(QT) = \text{Index}(1 + F) = 0$ by the result just proved. So $\text{Index}(Q) = -\text{Index}(T)$. Furthermore, as $F' + QF$ also has finite-rank, $\text{Index}(1 + F' + QF) = 0$. As $Q(T + F) = QT + F = 1 + F' + QF$, we deduce $\text{Index}(Q(T + F)) = 0$. Since this equals $\text{Index}(Q) + \text{Index}(T + F) = -\text{Index}(T) + \text{Index}(T + F)$, we get $\text{Index}(T + F) = \text{Index}(T)$ as claimed, proving b).

Next, we show continuity of the index. Coupled with its invariance under finite-rank perturbation, this will imply invariance under *compact* perturbation, and conclude the proof of the Theorem.

We will show:

Claim If T is Fredholm, then there exists $\varepsilon > 0$ such that $\|S - T\| < \varepsilon$ then S is Fredholm and $\text{Index}(S) \geq \text{Index}(T)$.

Once the claim is proved, the equality $\text{Index}(S) = \text{Index}(T)$ follows, since we can replace S by S^* and T by T^* in the claim without changing their distance apart, and the index changes sign when we take an adjoint.

To clarify things, we will use the decomposition $H = \ker(T)^{\perp} \oplus \ker(T)$. With respect to this decomposition we can write $T = \begin{bmatrix} T_0 & 0 \\ T_1 & 0 \end{bmatrix}$, with $T_0 \colon \ker(T)^{\perp} \to \ker(T)^{\perp}$, $T_1 \colon \ker(T)^{\perp} \to \ker(T)$. Since T_1 has finite-rank, $\text{Index}(T) = \text{Index}(T_0)$. Furthermore, if S as an operator on H, then $S - \begin{bmatrix} 0 & 0 \\ T_1 & 0 \end{bmatrix}$ is a finite-rank perturbation of S at the same distance from $T = \begin{bmatrix} T_0 & 0 \\ 0 & 0 \end{bmatrix}$ as the distance from T to S, and the index of S is the same as the index of $S - \begin{bmatrix} 0 & 0 \\ T_1 & 0 \end{bmatrix}$. These observations show that if we can prove the result for Fredholm operators T in whose matrix form $T = \begin{bmatrix} T_0 & 0 \\ T_1 & 0 \end{bmatrix}$, the term T_1 is zero, then we will be done in general.

So assume $T = \begin{bmatrix} T_0 & 0 \\ 0 & 0 \end{bmatrix}$, that is, that T maps $\ker(T)^{\perp}$ into itself. Note that $T_0 \colon \ker(T)^{\perp} \to \ker(T)^{\perp}$ has trivial kernel closed range (since it is Fredholm.) Also, T_0^* is also Fredholm so has closed range as well.

By Lemma 2.5.6 we can choose

- $\varepsilon_1 > 0$ such that if $A \in \mathbb{B}(\ker(T)^{\perp})$ and $\|A - T_0\| < \varepsilon_1$, then $\dim \ker(A) \leq \dim \ker(T_0) = 0$ (making A injective).
- $\varepsilon_2 > 0$ that if $A' \in \mathbb{B}(\ker(T)^{\perp})$ and $\|A' - T_0^*\| < \varepsilon_2$ then $\dim \ker(S) \leq \dim \ker(T_0^*)$.

Now let S be a bounded operator on H at distance $< \varepsilon$ to T.

Write $S = \begin{bmatrix} A & B \\ C & D \end{bmatrix}$. We have

$$\|A - T_0\| = \|PSP - PTP\| = \|P(S - T)P\| \leq \|S - T\| < \varepsilon.$$

Therefore, A is injective by choice of $\varepsilon < \varepsilon_1$.

Since $\begin{bmatrix} A & 0 \\ 0 & 0 \end{bmatrix}$ is a finite-rank perturbation of S (exercise), $\text{Index}(S) = \text{Index}(\begin{bmatrix} A & 0 \\ 0 & 0 \end{bmatrix})$, and the latter is easily checked to equal $\text{Index}(A)$, which, since A is injective, equals $-\dim\text{coker}(A)$.

Thus, we have shown that $\text{Index}(T) = \text{Index}(T_0) = -\dim\ker(T_0^*)$ and that $\text{Index}(S) = \text{Index}(A) = -\dim\ker(A^*)$. On the other hand, A^* and T_0^* are at distance at most ε as well, and hence by our choice of $\varepsilon < \varepsilon_2$, $\dim\ker(A^*) \leq \dim\ker(T_0^*)$. Now putting everything together gives

$$\text{Index}(S) = -\dim\ker(A^*) \geq \dim\ker(T_0^*) = \text{Index}(T)$$

for all S with $\|S - T\| < \varepsilon$.

Finally, continuity of the index coupled with density of finite-rank operators in the compact operators implies the invariance of $\text{Index}(T)$ under compact, not just finite-rank, perturbations, and completes the proof. $\square$

2.6 The Toeplitz Index Theorem

We now prove the Toeplitz index theorem, giving a Fredholm Index interpretation to the invariant (2.16). The conclusion will be that the Fredholm index of a Toeplitz operator T_f is minus the number of times the map f wraps the circle around itself.

Lemma 2.6.1 *Let $T \colon H \to K$ be a Fredholm operator between two (possibly different) Hilbert spaces. If Q is a bounded operator $K \to H$ such that $1 - TQ$ and $1 - QT$ are each* trace-class *operators (such Q exists for any Fredholm operator, by Lemma 2.4.5), then*

$$(2.22) \qquad \text{Index}(T) = \text{Trace}(1 - QT) - \text{Trace}(1 - TQ).$$

Proof We have already noted in the proof of Lemma 2.4.5 that if $T \colon H \to K$ is a Fredholm operator between two (possibly different) Hilbert spaces then there exists a Fredholm operator $G \colon K \to H$ such that

$$(2.23) \qquad 1 - GT = \text{pr}_{\ker T}, \qquad 1 - TG = \text{pr}_{\ker(T^*)}.$$

where $\text{pr}_{\ker T}$ is the projection to the kernel of T, $\text{pr}_{\ker(T*)}$ projection to the cokernel of T. If $Q = G$ then the statement regarding traces is obvious.

Now suppose that $Q \in \mathbb{B}(H)$ and $1 - QT$ and $Q - TQ$ are finite rank. It follows that $T(Q - G)$ and $(Q - G)T$ are each trace-class. We have, for $\Delta = Q - G$,

$$\text{Trace}(1 - QT) - \text{Trace}(1 - TQ)$$
$$= \text{Trace}\,(1 - (G + \Delta)T) - \text{Trace}\,(1 - T(G + \Delta))$$
$$= \text{Trace}(1 - GT) - \text{Trace}(\Delta T) - \text{Trace}(1 - TG) + \text{Trace}(T\Delta)$$

$$(2.24) \qquad = \text{Trace}(1 - GT) - \text{Trace}(1 - TG) = \text{Index}(T)$$

completing the proof.   □

The computation of the index using the trace given in the following is a special case of Connes' Chern character formula, discussed in Chapter 10.

Theorem 2.6.2 *Let A be a unital C*-algebra and $\pi: A \rightarrow \mathbb{B}(H)$ a unital representation of A on a Hilbert space H. Let $P \in \mathbb{B}(H)$ be a projection such that*

$$[\pi(a), P] \in \mathcal{K}(H)$$

for all $a \in A$.

Then if $u \in A$ is invertible, then $T_u := P\pi(u)|_{PH}$ is a Fredholm operator on PH, and so has an index $\text{Index}(T_u) \in \mathbb{Z}$.

a) *The map $u \mapsto \text{Index}(T_u)$ is constant on connected components of $\mathbf{GL}(A) := \{u \in A \mid u \text{ invertible}\}$, an open subset of A with the subspace topology.*
b) *If $A^\infty \subset A$ is a dense *-subalgebra with the property that*

$$[\pi(a), P] \in \mathcal{L}^1(H)$$

for all $a \in A^\infty$, then every connected component of $\mathbf{GL}(A)$ intersects A^∞, and

$$(2.25) \qquad\qquad \text{Index}(T_u) = \text{Trace}(u^{-1}[u, P])$$

holds for all invertibles $u \in A^\infty$.

Proof We suppress the representation π to simplify notation, thus, writing a instead of $\pi(a)$.

That T_u is a Fredholm operator on PH for u invertible in A follows immediately from the assumptions. We verify (2.25).

The bounded operator $T_{u^{-1}} := Pu^{-1}P$ on PH provides an essential inverse. Hence

$$\text{Index}(T_u) = \text{Trace}(1 - T_{u^{-1}}T_u) - \text{Trace}(1 - T_u T_{u^{-1}}),$$

by Lemma 2.6.1. This identity involves traces of operators on PH. In terms of operators on H, this can be written

$$\text{Index}(T_u) = \text{Trace}(P - Pu^{-1}PuP) - \text{Trace}(P - PuPu^{-1}).$$

Using the tracial property and the fact that $P^2 = P$ we get

$$(2.26) \qquad \text{Index}(T_u) = \text{Trace}\left(P(1 - u^{-1}Pu)\right) - \text{Trace}\left(P(1 - uP^{-1})\right).$$

We have

$$P - u^{-1}Pu = u^{-1}(PuP - Pu) + P - u^{-1}PuP$$

and this is a sum of two trace-class operators since $uP = Pu$ modulo trace-class operators, by hypothesis. Taking traces and using the tracial property gives

$$\text{Trace}(P - u^{-1}Pu) = \text{Trace}(PuPu^{-1} - P) + \text{Trace}\left(P \cdot (1 - u^{-1}Pu)\right)$$

which equals

$$\text{Trace}(P \cdot (1 - u^{-1}Pu) - \text{Trace}(P(1 - uPu^{-1}))$$

which is (2.26). Hence $\text{Index}(T_u) = \text{Trace}(P - u^{-1}Pu)$. Finally, note that

$$\text{Trace}(P - u^{-1}Pu) = \text{Trace}\left(u^{-1}(uP - Pu)\right) = \text{Trace}(u^{-1}[u, P])$$

giving (2.25). □

Theorem 2.6.3 (The Toeplitz Index Theorem) *Let* $f: \mathbb{T} \to \mathbb{C}^*$ *be a non-vanishing smooth function on the circle,* T_f *the corresponding Toeplitz operator. Then the Fredholm index of* T_f *agrees with minus the topological winding number of* f:

$$(2.27) \qquad \text{Index}(T_f) = -\text{wind}_f(0).$$

If f *is smooth,*

$$\text{Index}(T_f) = \frac{-1}{2\pi i} \int_0^{2\pi} \frac{f'(\theta)}{f(\theta)} d\theta.$$

Proof In Theorem 2.6.2, put $A = C(\mathbb{T})$ and $A^\infty := C^\infty(\mathbb{T})$, $P = P_+$ and $\pi(f) = M_f$. The commutators $[M_f, P_+]$ for $f \in C^\infty(\mathbb{T})$ are integral operators with smooth kernels by Lemma 2.3.5 and hence are trace-class by Theorem 2.1.15. Finally, we apply Theorem 2.3.7. □

Example 2.6.4 Let $f(z) = z^{-3}$, then as a function of $\theta \in [0, 2\pi]$, $f(\theta) = e^{-3i\theta}$, $\frac{f'(\theta)}{f(\theta)} = -3i$ and $\text{wind}_{z^{-3}}(0) = -3$ by the formula (2.14).

On the other hand, in terms of the standard orthonormal basis $1, z, z^2, \ldots$ for $\mathbf{H}^2$, $T_{z^{-3}}$ shifts sequences by 3 units to the left: $T_{z^{-3}}(z^k) = z^{k-3}$ if $k \geq 3$, and $T_{z^{-3}}(z^k) = 0$ if $k < 3$. Hence $\ker(T_{z^{-3}})$ is the span of $1, z, z^2$, it is 3-dimensional. And $\text{coker}(T_{z^{-3}}) = 0$ so $\text{Index}(T_{z^{-3}}) = 3 = -\text{wind}_{z^{-3}}(0)$.

Exercise 2.6.5 Prove that if $f: \mathbb{T} \rightarrow \mathbb{T}$ is not surjective, then $\text{wind}_f(0) = 0$. Hence, for a non-surjective map $\mathbb{T} \rightarrow \mathbb{T}$, the corresponding Toeplitz operator T_f has zero index.

To conclude this section, the Toeplitz Index theorem expresses the Fredholm index of a Toeplitz operator $T = T_f \in \mathbb{B}(\mathbf{H}^2)$ in terms of a topological (homotopy) invariant of its symbol $f: \mathbb{T} \rightarrow \mathbb{C}^*$, namely, its winding number.

In a similar way, the Atiyah–Singer Index Theorem, one of the main theorems described in this book, computes the Fredholm index $\text{Index}(D)$ of an *elliptic differential operator*

$$D: C^\infty(X, E) \subset L^2(X, E) \rightarrow L^2(X, F),$$

with symbol σ_D, between spaces of smooth vector bundle sections, over a smooth, compact manifold X, in terms of an appropriate topological invariant of its symbol—a kind of generalized winding number. In fact, this invariant, which is of course is more complicated than a winding number, since general manifolds have more complicated topology than the circle, turns out to be be described extremely conveniently using K-theory.

Chapter 3
Spectral Theory and Representations

The mathematical concept of spectrum, and the term, arose out of the work of Hilbert and his students; the term is said to originate in an 1897 paper of W. Wirtinger.

With the later development of quantum mechanics, the a connection between the mathematical concept of spectrum and atomic spectra in physics emerged: when a chemical element or compound makes a transition from a higher to a lower energy state, a photon is emitted, resulting in the production of light, which has a frequency. These frequencies, or wavelengths are the spectrum of the element. Spectrum in physics is more generally is concerned with frequencies of vibrations, and these frequencies appear as points in the spectrum of an appropriate operator. For instance, the frequencies at which a drum vibrates consists of the mathematical spectrum of the Laplacian; the aptly named paper of Kac [106] discusses the extent to which the shape of the drum can be reconstituted from this spectrum.

The spectrum of an operator T on a finite-dimensional complex vector space is its set of eigenvalues, but equivalently, the spectrum parameterizes the maximal ideals in the (commutative) ring $\mathbb{C}[T]$ generated by T, and this concept extends to commutative Banach algebras and C*-algebras. In the case of C*-algebras, the C*-algebra generated by a bounded operator T on a Hilbert space is commutative if and only if the operator T is *normal*, and in this case the spectrum of T is in natural correspondence with the space of characters of $C^*(T)$, that is, *-homomorphisms $\chi : C^*(T) \to \mathbb{C}$. Motivated by these examples one defines the spectrum of any commutative C*-algebra to be the the space of characters, and Gelfand's celebrated theorem sets up a canonical isomorphism between any commutative C*-algebra, and the C*-algebra of continuous functions on its spectrum, so that in particular, commutative unital C*-algebras are all of the form $C(X)$ for some X. The precise version of Gelfand's theorem is a stronger statement than this: there is a canonical (contravariant) isomorphism of categories between the category of commutative unital C*-algebras, and the category of compact Hausdorff spaces. In this sense, C*-algebra theory may be thought of as 'noncommutative topology.'

© Springer Nature Switzerland AG 2024
H. Emerson, *An Introduction to C*-Algebras and Noncommutative Geometry*,
Birkhäuser Advanced Texts Basler Lehrbücher,
https://doi.org/10.1007/978-3-031-59850-0_3

3.1   Spectrum in a Banach Algebra

An element a in a unital algebra A is invertible if there exists $b \in A$ such that $ab = ba = 1$.

Definition 3.1.1 Let A be a unital Banach algebra and $a \in A$. The *spectrum* $\mathrm{Spec}_A(a)$ of a is the set $\lambda \in \mathbb{C}$ such that $\lambda - a$ is not invertible.

Here λ really means $\lambda \cdot 1$, where 1 is the unit, but we generally just write λ. Invertibility of course makes no sense unless the algebra is unital.

Exercise 3.1.2 If A is a Banach algebra and u is an invertible in A then $\mathrm{Spec}_A(a) = \mathrm{Spec}_A(uau^{-1})$.

If $\alpha \colon A \to B$ is a unital homomorphism of unital Banach algebras, continuous or not, then $\mathrm{Spec}_B(\alpha(a)) \subset \mathrm{Spec}_A(a)$ for every $a \in A$.

Remark 3.1.3 Invertibility sometime depends on in which algebra one allows the inverse to be in, as in, for instance, if $A \subset B$ is a unital subalgebra of a unital Banach algebra B then by Exercise 3.1.2 $\mathrm{Spec}_B(a) \subset \mathrm{Spec}_A(a)$ but the containment may be strict.

For example, consider the Banach subalgebra $\mathcal{A}(\mathbb{D})$ of $C(\mathbb{T})$ (Example 1.1.12 of Chapter 2).

The function $f(z) = z$ in $C(\mathbb{T})$ is invertible in $C(\mathbb{T})$ but not in $\mathcal{A}(\mathbb{D})$, since its inverse would have to be $\frac{1}{z}$, which has a singularity at the origin. Hence $0 \in \mathrm{Spec}_{\mathcal{A}(\mathbb{D})}(z) - \mathrm{Spec}_{C(\mathbb{T})}(z)$.

Example 3.1.4 If $A = M_n(\mathbb{C})$ the spectrum reduces to the usual notion of eigenvalue of a matrix, since $T \in M_n(\mathbb{C})$ is invertible exactly when $\det(\lambda - T) \neq 0$. So if $\lambda \in \mathrm{Spec}(A)$ then $\lambda - A$ is not invertible and hence has a nonzero kernel, spanned by the λ-eigenvectors of A.

For infinite dimensional Hilbert spaces H, the spectrum of $T \in \mathbb{B}(H)$ is the set of $\lambda \in \mathbb{C}$ such that $\lambda - T$ is not bijective, since bijectivity of an operator is equivalent to its invertibility in $\mathbb{B}(H)$, by the Open Mapping Theorem, see [54]. But bijectivity may fail by failure of surjectivity without injectivity failing, in infinite dimensions, so spectral elements need not be eigenvalues in general.

Exercise 3.1.5 If X is a compact Hausdorff space and $f \in C(X)$, then f is invertible if and only if f does not vanish anywhere, so *the spectrum of f in $C(X)$ is the range of f.*

Exercise 3.1.6 Let A be a unital C*-algebra. Show that $a \in A$ is invertible if and only if a^* is invertible and in this case $(a^*)^{-1} = (a^{-1})^*$. Deduce that if A is a C*-algebra and $a \in A$ then $\mathrm{Spec}_A(a^*) = \{\bar{\lambda} \mid \lambda \in \mathrm{Spec}_A(a)\}$.

Exercise 3.1.7 If T_z is the Toeplitz operator with symbol z, then $T_z^* T_z = 1$ is invertible, but $T_z T_z^*$ is not; however, it is true in general that for a, b in a unital Banach algebra, $\mathrm{Spec}(ab) - \{0\} = \mathrm{Spec}(ba) - \{0\}$ (see Exercise 1.1.4 of Chapter 2.)

The following lemma shows that the open disk in A centered at 1 consists entirely of invertibles.

Lemma 3.1.8 *If A is unital, $a \in A$ and $\|a - 1\| < 1$, then a is invertible and the series $\sum_{n=0}^{\infty}(1 - a)^n$ converges norm absolutely in A to a^{-1}.*

Proof In a complete normed linear space, e.g., in a C*-algebra, or Hilbert space, if a series $\sum_n b_n$ converges absolutely, that is, if $\sum \|b_n\|$ converges, then the series converges. This is because of the triangle inequality implies that the sequence of partial sums of such a series is a Cauchy sequence.

Now since $\|a - 1\| < 1$, the series $\sum_{n=0}^{\infty}\|1 - a\|^n$ converges. Since $\|(1 - a)^n\| \leq \|1 - a\|^n$, the series $\sum_{n=0}^{\infty}\|(1 - a)^n\|$ converges, that is, $\sum_{n=0}^{\infty}(1 - a)^n$ is an absolutely convergent series in A. Hence it converges, say to b. By considering the partial sums of $(1 - a) \cdot \sum_{n=0}^{\infty}(1 - a)^n$ one sees easily that $(1 - a)b = b - 1$. Hence $b - ab = b - 1$, so $ab = 1$. Similarly, $ba = 1$. □

Corollary 3.1.9 *Let A be a unital Banach algebra.*

a) *Let $a \in A$ be invertible. If $b \in A$ and $\|a - b\| < \frac{1}{\|a^{-1}\|}$, then b is invertible, and the series $\sum_{n=0}^{\infty}(1 - ba^{-1})^n a$ converges to b^{-1}. In particular, the invertibles in A form an open subset of A in the norm topology.*

b) *If $|\lambda| > \|a\|$ then $\lambda - a$ is invertible, and the norm absolutely convergent series*

$$\frac{1}{\lambda} \sum_{n=0}^{\infty} (\frac{a}{\lambda})^n$$

converges to $(\lambda - a)^{-1}$. In particular, $\mathrm{Spec}_A(a) \subset \{\lambda \in \mathbb{C} \mid |\lambda| \leq \|a\|\}$.

To proceed further with spectral theory, we need to develop a some calculus for Banach algebra valued functions.

Fix a Banach space A, which later will be a Banach algebra, or C*-algebra. A function $f : (a, b) \to A$ from an open interval in $\mathbb{R}$ to A, is differentiable at $t_0 \in (a, b)$ if $\lim_{t \to t_0} \frac{f(t) - f(t_0)}{t - t_0}$ exists in A, in this case we denote the limit $f'(t_0)$, and is differentiable on the interval if it is at every point. The standard properties of the derivative, like its linearity, the Leibnitz rule $(f_1 f_2)'(t) = f_1'(t)f_2(t) + f_1(t)f_2'(t)$ go through for A-valued functions, we may speak of $C^1, C^2, \ldots, C^k$ or C^{∞}-functions in the evident way, and so on.

Exercise 3.1.10 Prove that if $f, g : (a, b) \to A$ are differentiable then so is fg and $(fg)'(t) = f(t)g'(t) + f(t)g'(t)$.

Similarly, the Riemann integral is defined for continuous functions $f : [a, b] \to A$ is defined using nets. If $\mathcal{P}$ is the set of all partitions of $[a, b]$, and $P \in \mathcal{P}$ is one of them with points $a = x_0 < x_1 < \cdots < x_n = b$, we associate to it the element

$$(3.1) \qquad \langle f, P \rangle = \sum_{i=1}^{n} f(t_i)(t_i - t_{i-1}) \in A.$$

The net $(\langle f, P \rangle)_{P \in \mathcal{P}}$ is Cauchy in A and hence converges since A is complete (a Banach space). We define

$$\int_{a}^{b} f(t)dt := \lim_{P \in \mathcal{P}} \langle f, P \rangle.$$

The linearity and other expected basic properties of the integral are easily checked. In fact, the space $C([a, b], A)$ is itself a Banach space, and integration defines a linear functional $C([a, b], A) \to A$ which is continuous, since

$$\left\| \int_{a}^{b} f(t)dt \right\| \leq \sup_{t \in [a,b]} \|f(t)\|(b - a) = \|f\| \cdot (b - a)$$

is easily checked from the definition.

Let $f : [a, b] \to A$ be a C^1-function on some open neighborhood of $[a, b]$. The derivative of f being continuous implies that the function

$$\tilde{f} : [a, b] \times [a, b] \to A, \quad \tilde{f}(s, t) := \frac{f(s) - f(t)}{s - t} \text{ if } t = s, \text{ else } \tilde{f}(t, t) := f'(t)$$

is continuous. Since the square is compact, it is uniformly continuous, and it follows that for all $\varepsilon > 0$ there exists $\delta > 0$ such that if $|s - t| < \delta$ then $\left\| \frac{f(s) - f(t)}{s - t} - f'(s) \right\| < \frac{\varepsilon}{b-a}$. For a sufficiently fine partition P with points $a = x_0 < x_1 < \cdots < x_n = b$, we have thus

$$\frac{f(t_{i+1}) - f(t_i)}{t_{i+1} - t_i} = f'(t_i) + a_i,$$

where $a_i \in A$ has norm $< \frac{\varepsilon}{b-a}$. Now pairing f' with P yields

$$\langle f', P \rangle = \sum_{i=1}^{n} f'(t_i)(t_i - t_{i-1}) = \sum_{i=1}^{n} \big(f(t_{i+1}) - f(t_i) \big) - a_i(t_i - t_{i-1})$$

and $\sum_{i=1}^{n} \big(f(t_{i+1}) - f(t_i) \big) - a_i(t_i - t_{i-1})$ is within ε of $f(b) - f(a)$.

Using approximation of domains in the plane by rectangles, one similarly defines the integral $\int \int_D f$ of a continuous function $f : D \to A$, on a suitable class of regions $D \subset \mathbb{C}$ of the plane. Fubini's theorem holds, so such integrals can be computed by the method of iterated integrals. The class of regions for which all this can be checked includes those enclosed by piecewise smooth, simple closed curves in $\mathbb{C}$.

We now discuss line integrals. Let $\gamma : [0, 1] \to W \subset \mathbb{C}$ be a smooth curve with $\gamma(t) = x(t) + iy(t)$, W the domain of a continuous function $f : W \to A$. Set

$$(3.2) \qquad \int_\gamma f\,dx := \int_0^1 f\big(\gamma(t)\big) x'(t)\,dt, \quad \int_\gamma f\,dy := \int_0^1 f\big(\gamma(t)\big) y'(t)\,dt,$$

these "line integrals" and any complex linear combination of them define continuous linear functionals $C_b(D, A) \to A$. An important such linear combination is the contour integral

$$(3.3) \qquad \int_\gamma f\,dz := \int_\gamma f\,dx + i \int_\gamma f\,dy.$$

If γ is merely a piecewise smooth curve, it is the union of finitely many smooth segments, and by adding up the relevant integrals, one extends the all the definitions (3.2) above to work for piecewise smooth curves as well.

Suppose $D = [a, b] \times [c, d]$ is a rectangle in the complex plane. A direct calculation using Fubini's theorem yields Green's Theorem for the rectangle:

$$\int_{\partial D} f\,dx + g\,dy = \int \int_D \left(\frac{\partial g}{\partial x} - \frac{\partial f}{\partial y} \right) dx\,dy,$$

where the boundary is oriented positively in the usual way. Green's theorem can then be extended to all D which are interiors of piecewise smooth simple closed curves.

For purposes of spectral theory, we are most interested in holomorphic functions. A function $f : W \to A$ is *holomorphic* at a point $z_0 \in W$ if

$$\frac{\partial f}{\partial z}(z_0) := \lim_{z \in W z \to z_0} \frac{f(z) - f(z_0)}{z - z_0} \quad \text{exists.}$$

f is holomorphic in W if it is holomorphic at every point of W.

Exercise 3.1.11 Let A be a unital Banach algebra, and a be an element. Then

$$f(z) := (z - a)^{-1}$$

is a holomorphic A-valued function defined on the open subset $U := \mathbb{C} \setminus \mathrm{Spec}_A(a)$.

If f is holomorphic at $z_0 = x_0 + iy_0$, then in particular the limits

$$\lim_{x \to x_0} \frac{f(x + iy_0) - f(z_0)}{x - x_0}, \quad \lim_{x \to x_0} \frac{f(x_0 + iy) - f(z_0)}{iy - iy_0}$$

exist in A, i.e., $\frac{\partial f}{\partial x}$ and $-i\frac{\partial f}{\partial y}$ exist at z_0, where $\frac{\partial}{\partial x}$ and $\frac{\partial}{\partial y}$ are the standard vector fields on $\mathbb{C}$, and are each equal to $\frac{\partial f}{\partial z}$ whence to each other; we get the Cauchy-Riemann equation

$$f_x = -if_y \in A.$$

Now let D be a region whose boundary is a piecewise smooth curve γ. Let f be an A-valued function which is holomorphic on a neighborhood of D. With $dz :=$ $dx + i\,dy$ as in (3.3) above, we obtain the analogue of the Cauchy-Goursat Theorem as an immediate consequence of Green's Theorem

$$\oint_\gamma f\,dz = 0.$$

The existence of anti-derivatives of analytic functions in simply connected open subsets $U \subset \mathbb{C}$ is a consequence, From existence of anti-derivatives one obtains that $\oint_\gamma f(z)dz = 0$ for arbitrary piecewise smooth closed curves in U, for f analytic in U, and U simply connected.

If γ is a simple closed piecewise smooth curve then it splits $\mathbb{C}$ into two components, one bounded and one unbounded. If w is in one of these components, then by standard complex analysis

$$\oint_\gamma \frac{dz}{z-w} = 2\pi i \cdot \mathrm{wind}_w(\gamma),$$

the winding number of γ around α. The winding number is zero if w is in the unbounded component and is $+1$ if w is inside and the curve is oriented positively with respect to its interior. More generally, we say a system Γ of pairwise disjoint closed curves γ_j is *positively oriented* if $\mathrm{wind}_w(\Gamma) := \sum_j \mathrm{wind}_w(\gamma_j)$ is either 0 or $+1$ for all $w \notin \Gamma$.

Theorem 3.1.12 *Suppose that $\Gamma \subset U \subset \mathbb{C}$ is a positively oriented system of closed curves such that* $\mathrm{wind}_w(\Gamma) = 0$ *for all* $w \notin U$. *Let* $f : U \to A$ *be analytic, where A is a Banach space. Then*

$$f(z) \cdot \mathrm{wind}_z(\Gamma) = \frac{1}{2\pi i} \oint_\Gamma \frac{f(w)}{w-z}\,dw, \quad \forall z \notin |\gamma|.$$

Furthermore,

$$\oint_\Gamma f(w)dw = 0.$$

Proof The set

$$\text{out}(\Gamma) := \{w \notin \Gamma \mid \text{wind}_w(\Gamma) = 0\}$$

is open and contains $\mathbb{C}\backslash U$. Now for f analytic on U consider the function $\tilde{f}(z, w) = \frac{f(z)-f(w)}{z-w}$ if $z \neq w$ and else $= f'(z)$, for $z, w \in U$. It is routine to check that $\tilde{f}$ is continuous on $U \times U$. Set

$$f_1(w) = \frac{1}{2\pi i} \oint_\Gamma \tilde{f}(z, w) dz.$$

The proof will be clearly be finished if we can show that $f_1(w) = 0$ for all $w \notin |\Gamma|$. However, the function

$$f_2(w) = \frac{1}{2\pi i} \oint \frac{f(z)}{z-w} dz, \quad w \notin |\Gamma|$$

is analytic on $U \setminus |\Gamma|$ and agrees with f_1 on $\text{out}(\Gamma)$ by the hypothesis that $\text{wind}_w(\Gamma) = 0$. As $\text{out}(\Gamma)$ is open, f_1, f_2 piece together to give an analytic function h on $\text{out}(\Gamma) \cup U$ and by hypothesis this union is $\mathbb{C}$. So h extends to an entire function. But clearly f_2 vanishes at ∞, whence h is bounded and so by Liouville's Theorem $h = 0$ everywhere, whence $f_1 = 0$ as well.   □

By differentiation under the integral sign we get the generalized Cauchy Integral formula

$$f^{(n)}(z) \cdot \text{wind}_z(\Gamma) = \frac{n!}{2\pi i} \int_\gamma \frac{f(z)}{(w-z)^{n+1}} \, dw.$$

Theorem 3.1.13 (Liouville's Theorem) *If $f : \mathbb{C} \to A$ is bounded and holomorphic everywhere, then f is constant.*

Indeed, we show that the complex derivative f' vanishes everywhere. For if $z_0 \in \mathbb{C}$ and if $|f(z)| \leq C$ for all z then the Cauchy integral formula applied to the circle of radius n around z_0, gives

$$\|f'(z_0)\| = \|\int_{\gamma_n} \frac{f(z)}{(z-z_0)^2} dz\| \leq \frac{2\pi C}{n},$$

which implies the result by letting $n \to \infty$.

Corollary 3.1.14 *The spectrum $\text{Spec}(a)$ of any element of a unital Banach algebra A is a nonempty compact subset of the complex plane.*

Proof The first statement is immediate from Corollary 3.1.9, the first part of which implies that $\mathbb{C} - \text{Spec}(a)$ is open, and the second part that it is bounded.

To prove that the spectrum is nonempty, assume the contrary. We can then apply Liouville's theorem for C*-algebra-valued functions to the function

$$f : \mathbb{C} \to A, \quad f(\lambda) := (\lambda - a)^{-1},$$

which is entire. Clearly $\lim_{|\lambda| \to \infty} \| f(\lambda) \| = 0$. In particular, f is bounded, whence is constant, and hence is zero, which is ridiculous.

This contradiction implies that $\mathrm{Spec}(a) \neq \emptyset$. ◻

We omit the proof of the following Theorem, which also is essentially the same as the case $A = \mathbb{C}$.

Theorem 3.1.15 *Let A be a Banach space and $f : W \to A$ be a holomorphic function defined on an open set W. Then at any point $z_0 \in W$, f has a power series expansion*

$$f(z) = \sum_{n=0}^{\infty} a_n (z - z_0)^n$$

which converges absolutely and uniformly on compact subsets of the open disk $|z - z_0| < R$ to f, where R its the distance from z_0 to $\mathbb{C} \setminus W$. Moreover,

$$\frac{1}{R} = \lim_{n \to \infty} \| a_n \|^{\frac{1}{n}}$$

gives the radius of convergence in terms of the coefficients.

Let A be a unital Banach algebra. The *spectral radius* of an element $a \in A$ is defined

$$r(a) := \sup_{\lambda \in \mathrm{Spec}(a)} |\lambda|.$$

If $|\lambda| > \|a\|$, then $\lambda - a$ is invertible, as discussed above. Hence $r(a) \leq \|a\|$.

Theorem 3.1.16 (Spectral Radius Formula) *If A is a unital Banach algebra and $a \in A$ then*

$$r(a) = \lim_{n \to \infty} \| a^n \|^{\frac{1}{n}}.$$

Proof Let $f(\lambda) = (\lambda - a)^{-1}$ for $\lambda \notin \mathrm{Spec}_A(a)$, then f is analytic on $\mathbb{C} \setminus \mathrm{Spec}_A(a)$, and it has a power series expansion

$$f(\lambda) = \frac{1}{\lambda} \sum_{n=0}^{\infty} a^n \lambda^{-n}, \quad |\lambda| > \|a\|$$

Set $g(\lambda) = f(\frac{1}{\lambda})$ if $\lambda \neq 0$, and set $g(0) = 0$. The power series expansion of g at 0 is obtained by substituting $\frac{1}{\lambda}$ into λ in the power series expansion of f, thus

$$(\lambda^{-1} - a)^{-1} = g(\lambda) = \lambda \cdot \sum_{n=0}^{\infty} a^n \lambda^n.$$

Now if $|\lambda| < \frac{1}{r(a)}$ then $\lambda^{-1} \notin \text{Spec}(a)$ and so g will be holomorphic at λ. Thus g is holomorphic in the ball of radius $\frac{1}{r(a)}$ centered at zero. Hence, by the machinery of power series, the series converges absolutely on compact subsets of $\{\lambda \in \mathbb{C} \mid |\lambda| < \frac{1}{r(a)}\}$ and the radius of convergence of the series is $\frac{1}{r(a)}$. Since the nth coefficient of our series is now a^n, the result now follows from the radius of convergence formula of Theorem 3.1.15. □

Theorem 3.1.17 *Let A be a unital C^*-algebra and $a \in A$ a self-adjoint. Then* $\|a\| = r(a)$.

Proof By Theorem 3.1.16, $\lim_{n \to \infty} \|a^{2^n}\|^{\frac{1}{2^n}} = r(a)$ holds for any a even in a Banach algebra. . On the other hand, $\|a^2\| = \|a\|^2$ for self-adjoint elements in a C^*-algebra, by the C^*-identity, and inductively, $\|a^{2^n}\| = \|a\|^{2^n}$. The result follows. □

From Theorem 3.1.17, $r(a^*a) = \|a^*a\|$ for any $a \in A$, since a^*a is self-adjoint. So combining this with the C^*-identity gives the following purely *algebraic* description of the norm on a C^*-algebra:

$$\|a\|^2 = \sup\{|\lambda| \mid \lambda - a^*a \text{ is not invertible}\}.$$

In particular, C^*-algebras are *rigid* in the following sense.

Corollary 3.1.18 *Any unital *-homomorphism $\varphi\colon A \to B$ between unital C^*-algebras is contractive:*

$$\|\varphi(a)\| \leq \|a\|, \quad \forall a \in A.$$

Proof What is obvious is that if $\lambda - a$ is invertible, then so is $\varphi(\lambda - a) = \lambda - \varphi(a)$. Thus $\text{Spec}(\varphi(a)) \subset \text{Spec}(a)$, so $r(\varphi(a)) \leq r(a)$ holds for any $a \in A$.

Now, since $\|a\| = r(a)$ for self-adjoint elements, and since $\varphi(a)$ is self-adjoint if a is, we see that $\|\varphi(a)\| = r(\varphi(a)) \leq r(a) = \|a\|$ for self-adjoints, using Theorem 3.1.17. Now in general, a^*a is self-adjoint, and using the C^*-identity we get the result. □

Corollary 3.1.19 *A C^*-algebra isomorphism $\varphi\colon A \to B$ is isometric:*

$$\|\varphi(a)\| = \|a\|, \quad a \in A.$$

Proof $\varphi(r(a)) = r(\varphi(a))$ is clear for isomorphisms. Now proceed as in the previous proof to show that the norm is also preserved. □

It will follow from the Spectral Permanence Theorem 3.5.1 that the hypothesis can be weakened from *isomorphism* to *injective*.

Essential spectrum of a bounded operator

An interesting example of a spectrum is to take a bounded operator $T \in \mathbb{B}(H)$ and look at its image in the C*-algebra $Q(H) := \mathbb{B}(H)/\mathcal{K}(H)$. (it was shown to be a C*-algebra under coset multiplication in the previous chapter.)

Definition 3.1.20 The *essential spectrum* $\mathrm{Spec}_{\mathrm{ess}}(T)$ of a bounded operator T, is the spectrum of T in $Q(H)$, the set of $\lambda \in \mathbb{C}$ such that $\lambda - \pi(T)$ is not invertible in Q.

By Exercise 3.1.2, $\mathrm{Spec}_{\mathrm{ess}}(T) \subset \mathrm{Spec}(T)$. The essential spectrum is the part of the spectrum which remains unchanged when T is replaced by a compact perturbation of T. By the definitions, T is *Fredholm* if and only if $0 \notin \mathrm{Spec}_{\mathrm{ess}}(T)$.

Exercise 3.1.21 Let $T \in \mathbb{B}\big(l^2(\mathbb{N})\big)$ be the multiplication operator M_f where $f \in l^{\infty}(\mathbb{N})$. Say $\|f\| \leq 1$ for simplicity. Prove that $\mathrm{Spec}(T) = \overline{\mathrm{ran}(f)}$, a closed subset of the unit disk, while

$$\mathrm{Spec}_{\mathrm{ess}}(T) = \mathrm{ran}(f)' \cup \{\lambda \in \mathbb{C} \mid f^{-1}(\lambda) \subset \mathbb{N} \text{ is infinite}\},$$

where $\mathrm{ran}(f)'$ means the set of *limit points*.

Exercise 3.1.22 Prove that if T is essentially unitary and $\mathrm{Spec}_{\mathrm{ess}}(T) \subset \mathbb{T}$ is a *proper* subset of the circle, then $\mathrm{Index}(T) = 0$.

Exercise 3.1.23 Let T be any bounded operator. If $\lambda \notin \mathrm{Spec}_{\mathrm{ess}}(T)$ then $\lambda - T$ is Fredholm. Show that $\mathrm{Index}(\lambda - T)$ is constant on connected components of $\mathbb{C} \setminus \mathrm{Spec}_{\mathrm{ess}}(T)$ and vanishes on unbounded components.

Exercise 3.1.24 Prove that $\mathrm{Spec}_{\mathrm{ess}}(T_f) = \mathrm{ran}(f)$, if $f \in C(\mathbb{T})$ and T_f the corresponding Toeplitz operator.

3.2 The Holomorphic Functional Calculus

Let A be a commutative, unital Banach algebra and $a \in A$, with spectrum $\mathrm{Spec}_A(a) \subset \mathbb{C}$.

Let U be an open neighborhood of $\mathrm{Spec}(a)$ and f an analytic (holomorphic) function on U. Let γ be a simple, closed, positively oriented piecewise smooth curve in U so that $\mathrm{Spec}(a) \subset \mathrm{ins}(\gamma) := \{w \notin |\gamma| \mid \mathrm{wind}_w(\gamma) = +1\}$ and such that $\mathbb{C} \setminus U \subset \mathrm{out}(\gamma) := \{w \notin |\gamma| \mid \mathrm{wind}_w(\gamma) = 0\}$. Recall that the winding number is defined $\mathrm{wind}_w(\gamma) = \frac{1}{2\pi i} \oint_{\gamma} \frac{1}{z-w} dz$.

We are going to define a quantity

$$(3.4) \qquad f(a) := \frac{1}{2\pi i} \oint_\gamma f(w)(w-a)^{-1} dw.$$

The first observation is that the formula does not depend on γ. For if γ' were another such positively oriented closed curve then $\text{wind}_w(\gamma - \gamma') = \text{wind}_w(\gamma) - \text{wind}_w(\gamma') = 0 - 0 = 0$ if $w \notin U$, and $\text{wind}_w(\gamma - \gamma') = 1 - 1 = 0$ if $w \in \text{Spec}(a)$, so that the cycle $\Gamma := \gamma - \gamma'$ is a system of closed curves in $V := U \setminus \text{Spec}(a)$ such that $\text{wind}_w(\Gamma) = 0$ for all $w \notin V$. Since $f(w)(w-a)^{-1}$ is analytic on V,

$$\oint_\Gamma f(w)(w-a)^{-1} dw = 0$$

by Cauchy's Theorem 3.1.12. That is,

$$\int_\gamma f(w)(w-a)^{-1} \, dw = \int_{\gamma'} f(w)(w-a)^{-1} \, dw.$$

The same argument extends to systems Γ of closed curves; we refer to [54] for the details.

Theorem 3.2.1 *Assume that* $\text{Spec}(a) \subset U \subset \mathbb{C}$ *where* U *is open. Let* Γ *and* Γ' *be systems of positively oriented piecewise smooth curves in* U *such that* $\text{Spec}(a) \subset \text{ins}(\Gamma) \subset U$ *and* $\text{Spec}(a) \subset \text{ins}(\Gamma') \subset U$. *Then for any* f *analytic on* U,

$$\oint_\Gamma f(z) \cdot (z-a)^{-1} dz = \oint_{\Gamma'} f(z) \cdot (z-a)^{-1} dz.$$

Definition 3.2.2 Suppose that f is analytic on a neighborhood U of $\text{Spec}(a)$. Let Γ be a system of positively oriented simple closed curves in U such that $\text{Spec}(a) \subset \text{ins}(\Gamma)$ and $\mathbb{C} \setminus U \subset \text{out}(\Gamma)$. We define

$$f(a) := \frac{1}{2\pi i} \oint_\Gamma f(z)(z-a)^{-1} dz.$$

Example 3.2.3 Let A be the Banach algebra (the C*-algebra) $C(X)$, with $X \subset \mathbb{C}$ any compact subset. Let $a \in A$ be the restriction of $f(w) = w$ to X. Then $\text{Spec}_A(a) = X$. Let U be an open neighborhood of X. Let Γ be a system of positively oriented closed curves as in Definition 3.2.2. If f is analytic on U then by the usual Cauchy Integral formula

$$f(w) = \frac{1}{2\pi i} \oint_\Gamma f(w)(z-w)^{-1} dz$$

holds for all $w \in X$.

The right hand side is the integral in (3.4). In other words, the map $f \mapsto f(a)$ defined in (3.4), agrees with the natural inclusion of algebras

$$\mathrm{Hol}(U) \to C(X), \quad f \mapsto f|_X.$$

Proposition 3.2.4 *Suppose* $\mathrm{Spec}(a) \subset U \subset\subset \mathbb{C}$ *and let* Γ *be a system of piecewise smooth curves in* U *as in Definition 3.2.2. Suppose that* f *is analytic on* $|z| < R$ *for some* $R > r(a)$. *Then* $\sum_{n=0}^{\infty} c_n a^n$ *converges absolutely in A and*

(3.5)
$$f(a) = \sum_{n=0}^{\infty} c_n a^n.$$

The proof is left as an exercise.

Proposition 3.2.5 *In the above notation,*

$$f(a) \cdot g(a) = (f \cdot g)(a).$$

That is, the map $f \mapsto f(a)$ is an algebra homomorphism $\mathrm{Hol}\,(\mathrm{Spec}(a)) \to A$.

Proof Let α and β be two curves with $\mathrm{Spec}_A(a)$ on the inside of both of them and α inside β. Using α to define $f(a)$ and β to define $g(a)$, we write

$$f(a) \cdot g(a) = \frac{1}{(2\pi i)^2} \int_\alpha \int_\beta f(z)g(w)(z-a)^{-1}(w-a)^{-1}dwdz$$

$$= \frac{1}{(2\pi i)^2} \oint_\alpha \int_\beta f(z)g(w)$$

$$\cdot \left(\frac{(z-a)^{-1} - (w-a)^{-1}}{w-z} \right) dwdz \quad \text{by algebra}$$

$$= \frac{1}{(2\pi i)^2} \oint_\alpha f(z) \cdot \oint_\beta \frac{g(w)}{w-z}dw \cdot (z-a)^{-1}dz$$

(3.6)
$$+ \frac{1}{(2\pi i)^2} \oint_\beta g(w)(w-a)^{-1} \oint_\alpha \frac{f(w)}{z-w}dzdw$$

but in the second term, for each w we see the integral

$$\oint_\alpha \frac{f(w)}{z-w}dz$$

and w is a point on β, and hence is outside the loop α, so the integral is zero by Theorem 3.1.12. Thus, the second term vanishes, while in the first integral, we see the integral

$$\oint_\beta \frac{g(w)}{w-z}dw,$$

where z is on α, and hence is inside β. From the Cauchy Integral Formula

$$\oint_\beta \frac{g(w)}{w - z} dw = 2\pi i \cdot g(z).$$

Going back to (3.6) we see that we have showed that

$$f(a) \cdot g(a) = \frac{1}{2\pi i} \oint_\alpha f(z)g(z)(z - a)^{-1} dz,$$

which is $(f \cdot g)(a)$, defined using the curve α.

The argument is only computationally more complicated for systems of curves, we refer to [54] for the details. $\square$

Exercise 3.2.6 If (f_n) is a sequence of analytic functions in U and $f_n \to f$ uniformly on compact subsets of U (so that f is therefore analytic on U as well), then

$$f_n(a) \to f(a)$$

in the Banach algebra A.

An important result concerning the holomorphic calculus is the

Theorem 3.2.7 (Spectral Mapping Theorem) *If* $f \in \mathrm{Hol}\,(\mathrm{Spec}(a))$ *then*

$$\mathrm{Spec}\,(f(a)) = f\,(\mathrm{Spec}(a))\,.$$

Proof Let U, Γ be as in Definition 3.2.2. Let $\lambda \in \mathrm{Spec}(a)$ and f be holomorphic on U. Let $g(z) = \frac{f(z)-f(\lambda)}{z-\lambda}$. Then g is also holomorphic on U. Since $(z - \lambda)g(z) = f(z) - f(\lambda)$, $(a - \lambda)g(a) = f(a) - f(\lambda)$. If $f(\lambda) \notin \mathrm{Spec}\,(f(a))$ then $f(a) - f(\lambda)$ would be invertible, which would imply that $a - \lambda$ is invertible, a contradiction. Hence $f(\lambda) \in \mathrm{Spec}\,(f(a))$ as claimed.

Conversely, suppose $\mu \notin f\,(\mathrm{Spec}(a))$. Then $g(z) := (f(z) - \mu)^{-1}$ is holomorphic on a (possibly smaller) neighborhood of $\mathrm{Spec}(a)$ and so $g(a)(f(a)-\mu)^{-1} = 1$. Hence $\mu \notin \mathrm{Spec}\,(f(a))$. $\square$

Corollary 3.2.8 *If* $\lambda \neq 0$ *is an isolated point of* $\mathrm{Spec}(T)$, *where* T *is a compact operator, then* λ *is an eigenvalue of* T.

Proof Let U be a neighborhood of λ, V an open set containing $\mathrm{Spec}(a) - \{\lambda\}$, and $U \cap V = \emptyset$. Let γ be a positively oriented loop around λ in U, and Γ' be a positively oriented closed path or system of closed paths containing $\mathrm{Spec}(a) - \{\lambda\}$ in its inside. Let $\Gamma = \gamma + \Gamma'$. Set

$$E_\lambda := \frac{1}{2\pi i} \oint_\gamma (z - T)^{-1} dz.$$

I claim that E_λ is an idempotent. Indeed, if f is the function on $U \cup V$ equal to 1 on U and 0 on V, then f is clearly holomorphic on $U \cup V$, and hence we may apply the holomorphic functional calculus to form

$$f(T) = \frac{1}{2\pi i} \oint_\Gamma f(z)(z - T)^{-1} dz.$$

This reproduces the formula for E_λ above. From the fact that $f^2 = f$ as holomorphic functions on $U \cup V$, and that the functional calculus is an algebra homomorphism, we deduce that $f(T) = E_\lambda$ is an idempotent. Since T commutes with E_λ, $T = \begin{bmatrix} T' & 0 \\ 0 & T'' \end{bmatrix}$ where $T'' = T(1 - E_\lambda)$, and the decomposition is into the direct (non-orthogonal) sum of $E_\lambda H$ and $(1 - E_\lambda)H$.

Provided that $\lambda \neq 0$, it is easy to see that $T' = E_\lambda T$ is invertible as an operator $E_\lambda H \to E_\lambda H$ (construct an inverse). Hence $0 \notin \mathrm{Spec}(T')$. Since $\mathrm{Spec}(f(T)) \subset \{0, \lambda\}$, it follows that $\mathrm{Spec}(T')$ consists of λ alone.

Now if T is compact, so is T', and since $T' : E_\lambda H \to E_\lambda H$ is invertible, $E_\lambda H$ is finite-dimensional. Since in finite-dimensions, every element of the spectrum of an operator is an eigenvalue, λ is an eigenvalue of T', whence of T. $\square$

3.3 Characters and Gelfand's Theorem

Let A be a unital commutative Banach algebra. The set of algebraic, i.e., not necessarily closed, proper ideals in A is a poset to which Zorn's lemma can be applied. We deduce the existence of *maximal proper* ideals $\mathcal{M}$, i.e., proper ideals of A which are contained in no larger proper ideal.

Maximality of a proper ideal $\mathcal{M}$ implies that it is closed. For otherwise, the closure of $\mathcal{M}$ would be a larger ideal. If this larger ideal is A itself, then $\mathcal{M}$ would be dense in A, and hence $\mathcal{M}$ itself would nontrivially intersect the open subset $\{a \in A \mid \|a - 1\| < 1\}$, which consists entirely of invertibles. An ideal containing an invertible can only of course be A itself. This contradicts properness of $\mathcal{M}$.

Exercise 3.3.1 Let A be a Banach algebra and $J \subset A$ be any closed ideal. Show that A/J with the quotient norm

$$\|a + J\| := \inf_{x \in J} \|a + x\|$$

and quotient vector space and algebra structure, is a Banach algebra, and that the quotient map $\pi : A \to A/J$ is a contractive homomorphism of Banach algebras.

By the Exercise, $A/\mathcal{M}$ with the quotient norm is a Banach algebra for any maximal ideal, so for any $a \in A$ we can speak of the spectrum of the coset $a + \mathcal{M}$ in the Banach algebra $A/\mathcal{M}$.

Lemma 3.3.2 *If $\mathcal{M}$ is any maximal ideal in a commutative, unital Banach algebra A, then the spectrum of any element $a + \mathcal{M}$ in the Banach algebra $A/\mathcal{M}$ consists of a single point in the spectrum of a in A. The mapping sending $a + \mathcal{M}$ to λ if $\mathrm{Spec}_{A/\mathcal{M}}(a+\mathcal{M}) = \{\lambda\}$ is an isometric isomorphism $A/\mathcal{M} \cong \mathbb{C}$ of Banach algebras.*

Proof If $a \notin \mathcal{M}$, i.e., if $a + \mathcal{M}$ is a nonzero element of $A/\mathcal{M}$, then $a + \mathcal{M}$ generates a nonzero principal ideal $\langle a + \mathcal{M}\rangle := \{ab + \mathcal{M} \mid b \in A\} \subset A/\mathcal{M}$ which, clearly, is proper in $A/\mathcal{M}$ if and only if $a + \mathcal{M}$ is not invertible in $A/\mathcal{M}$. The inverse image $\pi^{-1}(\langle a + \mathcal{M}\rangle) \subset A$ of this ideal is an ideal of A containing $\mathcal{M}$. Since $\mathcal{M}$ is maximal, this inverse image must be all of A. Thus the ideal $\langle a+\mathcal{M}\rangle$ we started with is actually $A/\mathcal{M}$. In particular, $a + \mathcal{M}$ must be invertible in $A/\mathcal{M}$. This shows that any nonzero element of $A/\mathcal{M}$ is invertible in $A/\mathcal{M}$.

From this, we deduce that the the spectrum

$$\mathrm{Spec}(a + \mathcal{M}) := \{\lambda \in \mathbb{C} \mid \lambda - a + \mathcal{M} \text{ invertible in } A/\mathcal{M}\}$$

of an element $a + \mathcal{M}$ of $A/\mathcal{M}$ is the same as $\{\lambda \in \mathbb{C} \mid a + \mathcal{M} = \lambda + \mathcal{M}\}$, or, equivalently, the same as $\{\lambda \in \mathbb{C} \mid \lambda - a \in \mathcal{M}\}$. Note that there can be at most one scalar λ such that $\lambda - a \in \mathcal{M}$, since if there were two, say λ_1 and λ_2, then we would get $\lambda_1 - \lambda_2 \in \mathcal{M}$, but $\mathcal{M}$, being proper, can contain no nonzero scalar. Furthermore, by Liouville's theorem, $\mathrm{Spec}(a + \mathcal{M})$ is nonempty.

This all shows that the Banach algebra $A/\mathcal{M}$ consists exactly of multiples $\lambda + \mathcal{M}$ of the unit, with $a + \mathcal{M}$ corresponding to the multiple $\lambda + \mathcal{M}$ if and only if $\{\lambda\} = \mathrm{Spec}_{A/\mathcal{M}}(a + \mathcal{M})$.   □

If A is a commutative unital Banach algebra we call a nonzero homomorphism $\chi : A \to \mathbb{C}$ of Banach algebras a *character* of A. If $\mathcal{M} := \ker(\chi)$ then χ determines an isomorphism $A/\mathcal{M} \cong \mathbb{C}$ and since $\mathbb{C}$ has no nonzero proper ideals, neither does $A/\mathcal{M}$, and hence $\mathcal{M}$ is a maximal ideal in A.

Hence the proof of Lemma 3.3.2 provides an isomorphism $A/\mathcal{M} \cong \mathbb{C}$ mapping $a + \mathcal{M}$ to the unique point $\lambda \in \mathrm{Spec}_A(a) \subset \mathbb{C}$ such that $\mathrm{Spec}_{A/\mathcal{M}}(a + \mathcal{M}) = \{\lambda\}$. Equivalently, λ is determined by the property that $\lambda - a \in \mathcal{M}$. Since $\chi(a)$ satisfies this condition, $\chi(a) = \lambda$.

In particular, $\chi(a) \in \mathrm{Spec}_A(a)$ for any character and any $a \in A$. Since $|\lambda| \leq r(a) \leq \|a\|$, we obtain the following facts about characters.

Lemma 3.3.3 *If A is a commutative unital Banach algebra and $\chi : A \to \mathbb{C}$ is a character of A then $\chi(a) \in \mathrm{Spec}_A(a)$ for all $a \in A$, and (hence) $|\chi(a)| \leq \|a\|$ for any a, that is, χ is automatically contractive.*

In particular, characters of Banach algebras are automatically continuous.

Lemma 3.3.4 *If A is a commutative unital Banach algebra and $a \in A$, then for every $\lambda \in \mathrm{Spec}_A(a)$, there is a character $\chi : A \to \mathbb{C}$ such that $\chi(a) = \lambda$.*

In particular, for any commutative unital Banach algebra and any $a \in A$, the spectrum $\mathrm{Spec}_A(a)$ consists precisely of the values $\chi(a)$ of characters of A at a.

Proof Since $\lambda - a$ is not invertible, it generates a proper ideal $\langle \lambda - a \rangle$ in A. This is contained in a maximal ideal $\mathcal{M}$, by Zorn's lemma, which is the kernel of some character $\chi : A \to \mathbb{C}$. Moreover, $\chi(a) = \lambda$ if and only if $\lambda - a \in \mathcal{M}$, and since $\lambda := \chi(a)$ satisfies this condition, $\chi(a) = \lambda$.   □

Exercise 3.3.5 Prove that point evaluation at a point of $\mathbb{D}$ determines a character of the disk algebra $\mathcal{A}(\mathbb{D})$.

Exercise 3.3.6 If G is a locally compact abelian group and χ is a character of G (a continuous homomorphism $G \to \mathbb{T}$) then

$$\alpha(f) := \int_G f(g)\overline{\chi(g)} \, d\mu(g)$$

defines a character the Banach algebra $L^1(G)$, and the C*-algebra $C^*(G)$ of G.

In the case of C*-algebras, we have the special general result about characters.

Lemma 3.3.7 *If A is a commutative C*-algebra and $\chi : A \to \mathbb{C}$ is a character, then $\chi(a^*) = \overline{\chi(a)}$, that is, χ is automatically a *-homomorphism.*

This implies that $A/\mathcal{M}$ is a C*-algebra, since it is isometrically *-isomorphic to $\mathbb{C}$. It also implies that a maximal ideal in a C*-algebra is automatically both closed, and a *-ideal, i.e., closed under adjoint.

Proof It clearly suffices to show that $\chi(a) \in \mathbb{R}$ for a self-adjoint. Write $\chi(a) = \alpha + i\beta$. Then $|\chi(a + it)|^2 = \alpha^2 + \beta^2 + 2\beta t + t^2$ for any $t \in \mathbb{R}$. On the other hand, χ is contractive, so $|\chi(a + it)|^2 \le \|a + it\|^2 = \|(a + it)^*(a + it)\| = \|a^2 + t^2\| \le \|a\|^2 + t^2$. The resulting inequality

$$\alpha^2 + \beta^2 + 2\beta t + t^2 \le \|a\|^2 + t^2$$

valid for all $t \in \mathbb{R}$ implies immediately that $\beta = 0$.   □

Corollary 3.3.8 *If A is a C*-algebra and $a \in A$ is any self-adjoint element, then $\mathrm{Spec}(a) \in \mathbb{R}$.*

Proof Indeed, the elements of $\mathrm{Spec}_A(a)$ are precisely the values of characters $\chi : A \to \mathbb{C}$ by Lemma 3.3.4, and such a character maps self-adjoints to $\mathbb{R}$ by Lemma 3.3.7.   □

3.4   Gelfand's Theorem

Definition 3.4.1 If A is a commutative unital Banach algebra, $\widehat{A}$ denotes the space of characters $\chi : A \to \mathbb{C}$ endowed with the topology of pointwise convergence on A.

We remind the reader that in the case where A is a C*-algebra we do not need to require additionally that characters are *-homomorphisms, since this is automatic.

We sometimes refer to $\widehat{A}$ as the *spectrum* of A, or *Gelfand spectrum*.

Proposition 3.4.2 *If $A = C(X)$ for a compact Hausdorff space X, then the characters of A are in natural 1–1 correspondence with the points of X, with a point $x \in X$ corresponding to the *-homomorphism $C(X) \to \mathbb{C}$ of evaluation $f \mapsto f(x)$ of functions at x.*

Thus, the spectrum of $C(X)$ is X.

Proof Suppose $\alpha \colon C(X) \to \mathbb{C}$ is a character of $C(X)$. If f and g have disjoint supports, then $fg = 0$ so it follows $\alpha(f)\alpha(g) = 0$, whence either $\alpha(f) = 0$ or $\alpha(g) = 0$. Now say $x \in \operatorname{supp}(\alpha)$ if for all neighborhoods U of x, there exists $f \in C_c(U)^+$ such that $\alpha(f) \neq 0$. The complement of $\operatorname{supp}(\alpha)$ is open by definition, so $\operatorname{supp}(f)$ is closed. Suppose that $\operatorname{supp}(\alpha) = \emptyset$. By definition $x \in X - \operatorname{supp}(\alpha)$ implies there exists a neighborhood U of x such that $\alpha(C_c(U)) = 0$. If $\operatorname{supp}(\alpha) = \emptyset$ every point satisfies this condition. Since X is compact, there exists a finite collection $\{U_i\}$ of such open sets. Let (ρ_i) be a subordinate partition of unity, so $0 \leq \rho_i \leq 1$, $\operatorname{supp}(\rho_i) \subset U_i$, and $\sum_i \rho_i(x) = 1$ for all $x \in X$. We get

$$\alpha(f) = \alpha\Big(\sum_i \rho_i f\Big) = \sum \alpha(\rho_i f) = 0$$

for all $f \in C(X)$, a contradiction to $\alpha \neq 0$. Hence $\operatorname{supp}(\alpha) \neq \emptyset$. Next, $\operatorname{supp}(\alpha)$ contains at most one point, for if $x, y \in \operatorname{supp}(\alpha)$ then there exist U and V disjoint neighborhoods of x, y, and $f \in C_c(U)$ and $g \in C_c(V)$ such that $\alpha(f) \neq 0$ and $\alpha(g) \neq 0$, which contradicts $\alpha(f)\alpha(g) = \alpha(fg) = \alpha(0) = 0$. So $\operatorname{supp}(\alpha)$ contains exactly one point point, say $\{x_0\}$.

Next, I claim that if $f(x_0) = 0$ then $\alpha(f) = 0$. Suppose $f(x_0) = 0$. Choose $\varepsilon > 0$ and any neighborhood U_0 of x_0 such that $|f| < \varepsilon$ on U_0. If $x \neq x_0$ there exists a neighborhood U of x such that α is zero on $C_c(U)$. Find a finite collection $\{U_i\}$ of such U so that $U_0 \cup \bigcup_i U_i$ covers X, and a partition of unity $\{\rho_i\}$ subordinate to the cover by U_0 and the U_i. Then

$$\alpha(f) = \alpha\Big(\sum \rho_i f\Big) = \alpha(\rho_0 f).$$

Since $|\rho_0 f| < \varepsilon$ and characters are contractive, $|\alpha(\rho_0 f)| < \varepsilon$. This shows that $|\alpha(f)| < \varepsilon$, and since ε is arbitrary, $\alpha(f) = 0$.

Finally, $f - f(x_0) \cdot 1$ vanishes at x_0 whence $\alpha(f) = f(x_0)$ follows. $\square$

Lemma 3.4.3 *For any commutative unital Banach algebra A, $\widehat{A}$ is a compact Hausdorff space.*

Proof Since any characters is contractive, we may identity $\widehat{A}$ with a collection of functions $A_1 := \{a \in A \mid \|a\| \leq 1\}$ to $\overline{\mathbb{D}}$, that is, as an element of the Cartesian product $X := \prod_{A_1} \overline{\mathbb{D}}$. Endowing this product with the product topology results in a compact Hausdorff space since $\overline{\mathbb{D}}$ is compact Hausdorff (and the Tychonoff theorem). The resulting map $\widehat{A} \to X$ is clearly injective since a character is determined by its values on A_1, and is continuous, by the definitions. Moreover, its range is closed. Indeed, suppose that $\chi : A_1 \to \mathbb{C}$ is a map which is a limit point of the image of $\widehat{A}$. So there is a sequence of characters (χ_n) such that $\chi_n(a) \to \chi(a)$ for all $a \in A_1$. First we extend χ to A by setting $\tilde{\chi}(a) := \|a\| \cdot \chi(\frac{a}{\|a\|})$. If $a \in A$ is any nonzero element, $\chi_n(a) = \|a\| \cdot \chi_n(\frac{a}{\|a\|}) \to \|a\| \cdot \chi(\frac{a}{\|a\|}) = \tilde{\chi}(a)$. Hence $\chi_n \to \tilde{\chi}$ pointwise on all of A. Now the fact that $\tilde{\chi}$ is a character, and hence that χ is the restriction to A_1 of a character, follows immediately using limits. We leave the details to the reader. $\square$

If A is a commutative, unital Banach algebra, $C(\widehat{A})$ is a C*-algebra, and in particular a Banach algebra. If $a \in A$ is any element, we let $\hat{a}$ denote the function on $\widehat{A}$ defined by $\hat{a}(\chi) := \chi(a)$. Clearly $\hat{a}$ is continuous on $\widehat{A}$. The map $a \mapsto \hat{a}$ is the *Gelfand transform*

$$A \to C(\widehat{A}).$$

Theorem 3.4.4 (Gelfand) *For any commutative unital Banach algebra A, the Gelfand transform $A \to C(\widehat{A})$ is a contractive, injective, Banach algebra homomorphism. If A is a C*-algebra, it is a C*-algebra isomorphism.*

Proof The Gelfand transform is easily checked to be an algebra homomorphism. Using Lemma 3.3.4, we get, for any $a \in A$,

$$\|\hat{a}\| = \sup_{\chi \in \widehat{A}} |\hat{a}(\chi)| = \sup_{\chi \in \widehat{A}} |\chi(a)| = \sup_{\lambda \in \mathrm{Spec}_A(a)} |\lambda| = r(a) \leq \|a\|,$$

so the Gelfand transform is a contractive homomorphism of Banach algebras. If A is a C*-algebra, $\chi(a^*) = \overline{\chi(a)}$ for any character, from Lemma 3.3.7. It follows that $\widehat{a^*} = \hat{a}^*$ if A is a C*-algebra, so in this case the Gelfand transform is a *-homomorphism. Moreover, if a is self-adjoint, then since $r(a) = \|a\|$ by Theorem 3.1.17, we get that $\|\hat{a}\| = \|a\|$ and the Gelfand transform is isometric, whence injective. Clearly $\hat{1}$ is the constant function 1 on $\widehat{A}$. Moreover, if $\chi_1 \neq \chi_2$ are different characters, then by definition, there is some $a \in A$ such that $\chi_1(a) \neq \chi_2(a)$, so the image of A in $C(\widehat{A})$ is a *-subalgebra of $C(\widehat{A})$ which separates points of $\widehat{A}$, contains the constant functions, and is closed under conjugation, so by the Stone-Weierstrass theorem it is dense in $C(\widehat{A})$. Since the Gelfand transform is isometric, however, and $C(\widehat{A})$ is complete, the range of the Gelfand transform is closed by a standard argument. Hence its image is $C(\widehat{A})$. $\square$

From the above discussion we obtain the following critically important theorem.

Corollary 3.4.5 *Every commutative unital C*-algebra is isomorphic to $C(X)$ where X is a compact Hausdorff space.*

This result is why general C*-algebras are sometimes considered philosophically as generalized spaces, or "noncommutative" spaces.

One can formulate Corollary 3.4.5 more precisely as follows. Let **Top** be the category with objects compact Hausdorff spaces and morphisms continuous maps and $\mathbf{C}^*_{\mathbf{Ab}}$ the category of commutative unital C*-algebras and unital C*-algebra homomorphisms. Define a (contravariant) functor **Top** $\to \mathbf{C}^*_{\mathbf{Ab}}$ by sending an object X of **Top** to the object $C(X)$ of $\mathbf{C}^*_{\mathbf{Ab}}$ and a morphism $\phi: X \to Y$ to the induced *-homomorphism $C(Y) \to C(X)$, $f \mapsto f \circ \phi$. This functor is a (contravariant) equivalence of categories. Equivalently, it is an equivalence between **Top** and the opposite category $\mathbf{C}^*_{\mathbf{Ab}}{}^{op}$.

Exercise 3.4.6 Let X and Y be compact Hausdorff spaces.

a) As discussed in Remark 3.5.4, any *-homomorphism $\alpha: C(X) \to C(Y)$ has the form $\alpha(f) = f \circ \phi$ for a unique continuous map $\phi: Y \to X$. Show that if α is injective then ϕ is surjective, and if α is surjective, then ϕ is injective.

b) Deduce from a) that *any* injective *-homomorphism between C*-algebras—commutative or not—is isometric. (*Hint.* Note that to show that a *-homomorphism $\alpha: A \to B$ between C*-algebras is isometric it suffices to show that $\|\alpha(a)\| = \|a\|$ for all a self-adjoint (by the C*-identity.) Now if $a \in A$ is self-adjoint, then $C^*(a)$ is a commutative unital C*-algebra, as is $C^*(\alpha(a))$. Use Gelfand's Theorem and part a).)

Exercise 3.4.7 A C*-algebra is *separable* if it is as a topological space, that is, if it contains a countable dense set. Show that if X is locally compact Hausdorff, then $C_0(X)$ is separable if and only if X is second countable. Thus, a general commutative C*-algebra A is separable if and only if its spectrum $\widehat{A}$ is second countable.

Exercise 3.4.8 Viewing the circle $\mathbb{T}$ as $\mathbb{R}/\mathbb{Z}$, we have a copy of $\mathbb{Q} \cap [0, 1]$ inside of it (with 0 and 1 identified). Let A be the C*-algebra generated by $C(\mathbb{T})$ and the characteristic functions $\chi_{[p,q]}$ of the characteristic functions of closed intervals $[p, q]$, $p < q$, $p, q \in \mathbb{Q}$, with rational endpoints. Clearly A is commutative, and $C(\mathbb{T})$ is a subalgebra. Therefore $\widehat{A}$ is a compact space which maps to $\mathbb{T}$. Describe $\widehat{A}$ and the map.

Exercise 3.4.9 It is clear that an inductive limit of commutative C*-algebras $A_n \cong C_0(X_n)$ results in a commutative C*-algebra $\varinjlim C_0(X_n)$—what is its spectrum?

Let I be a directed system. An *inverse system of compact Hausdorff spaces* over I is a family $\{X_i \mid i \in I\}$ of compact Hausdorff spaces and a family $\{\phi_{ij}: X_j \to X_i \mid$ for $i \le j\}$ of continuous maps such that $\phi_{ij} \circ \phi_{jk} = \phi_{ik}$ for all $i \le j \le k$. The corresponding *inverse limit* space is defined to be the set

$$\varprojlim X_j := \{(x_i)_{i \in I} \in \prod_i X_i \mid x_i = \phi_{ij}(x_j) \ \forall \ i \le j\},$$

topologized as a subspace of $\prod_i X_i$.

It is a closed subspace of a compact Hausdorff space, whence is itself compact Hausdorff.

The inverse limit satisfies the following universal property: if Y is a compact Hausdorff space and $\{\alpha_i : Y \to X_i; \mid i \in I\}$ is a family of maps such that if $i \geq j$ then $\phi_{ji} \circ \alpha_i = \alpha_j$, then there is a unique map $\alpha : Y \to \varprojlim X_i$ such that $\pi_i \circ \alpha = \alpha_i$ for all $i \in I$.

The maps ϕ_{ij} of an inverse system give rise to *-homomorphisms $\hat{\phi}_{ij} : C(X_j) \to C(X_i)$, for $i \leq j$, by $\hat{\phi}_{ij}(f) = f \circ \phi_{ji}$ and these make up a directed system of C*-algebras.

Prove that

$$C(\varprojlim X_i) \cong \varinjlim C(X_i).$$

The following exercise introduces some interesting compactifications of $\mathbb{R}$, and of more general metric spaces, which in a number of cases are in a sense more naturally defined in terms of their C*-algebras of continuous functions, than as spaces described in terms of a locus of points.

The simplest example of such a space is the Stone-Cech compactification βX say of the discrete space $X := \mathbb{N}$ of natural numbers. Then $\beta \mathbb{N}$ is by definition the Gelfand spectrum of the C*-algebra $C_b(\mathbb{N}) = l^\infty(\mathbb{N})$. To describe the actual *points* of $\beta \mathbb{N}$ is not so easy.

Exercise 3.4.10 Let (X, d) be a noncompact metric space. Define

$$C(\eta X) := \{f \in C(X) \mid \lim_{x \to \infty} \sup_{d(x,y) \leq R} |f(x) - f(y)| = 0, \; \forall R > 0\}.$$

Show that $C(\eta X)$ is a commutative C*-algebra. It is Gelfand dual ηX is, by definition, called the *Higson corona of X*, and plays a role in Index Theory. The Higson corona construction is part of the more general field of *coarse geometry*, information about which can be found in [98].

a) With $\mathbb{R}$ given its standard metric, prove that smooth functions f on $\mathbb{R}$ with $f' \in C_0(\mathbb{R})$ are dense in $C(\eta \mathbb{R})$.
b) Prove that if $\phi : X \to Y$ is a proper Lipschitz map between metric spaces then

$$f \mapsto f \circ \phi$$

defines a *-homomorphism $C(\eta Y) \to C(\eta X)$, whence that a proper Lipschitz map $X \to Y$ determines a map $\eta Y \to \eta X$.
c) Prove that any group G of isometries of (X, d) acts naturally by homeomorphisms of the compact space ηX.
d) Suppose (X, d) is a discrete metric space and $X \subset \overline{X}$ is a compactification of X satisfying the following condition, making it a *coarse compactification*. If (x_n) and (y_n) are sequences in X such that $\sup_n d(x_n, y_n) < \infty$, then (x_n) converges

to a boundary point $\xi \in \bar{X} - X$ if and only if (y_n) does, and they converge to the same boundary point.

Prove that there is a surjection continuous map $\eta X \to \overline{X}$.

e) Let X be the metric space $\mathbb{D}$ with the Poincaré metric (defining the hyperbolic plane.) Show that the usual compactification of $\mathbb{D}$ by $\partial \mathbb{D} = \mathbb{T}$ is a coarse compactification of X in the sense of d).

Exercise 3.4.11 Let (X, d) be a noncompact metric space. Recall that $f \in C(X)$ is *uniformly continuous* on X if for all $\varepsilon > 0$ there exists $\delta > 0$ such that $d(x, y) < \delta$ implies $|f(x) - f(y)| < \varepsilon$.

Define

$$C_u(X) := \{f \in C_b(X) \mid f \text{ is uniformly continuous on } X\}.$$

Show that $C_u(X)$ is a unital commutative C*-algebra, determining, therefore, by Gelfand duality a compact space $\overline{X}^u$, the *uniform compactification* of X.

a) Prove that smooth functions f on $\mathbb{R}$ with $f' \in C_b(\mathbb{R})$ are dense in $C_u(\mathbb{R})$.
b) Prove that any group G of isometries of (X, d) acts naturally by homeomorphisms of the compact space $\overline{X}^u$.
c) Show that for $\mathbb{T}$ the circle, lifting functions to periodic functions determines an injection $C(\mathbb{T}) \to C_u(\mathbb{R})$, and an induced surjection $\overline{\mathbb{R}}^u \to \mathbb{T}$.

We close this section on some remarks on Gelfand's theorem for (potentially) non-unital, commutative C*-algebras.

If A is a commutative C*-algebra, A^+ the unitization of A, then Gelfand's theorem provides an isomorphism $A^+ \cong C(\widehat{A^+})$. The unitization A^+ comes equipped with an augmentation $\varepsilon \colon A^+ \to \mathbb{C}$ and ε is a point of $\widehat{A^+}$, hence determines an ideal: the ideal of continuous functions on $\widehat{A^+}$ which vanish at ε. This ideal identifies under Gelfand's isomorphism to the ideal $A \subset A^+$.

Theorem 3.4.12 *If A is a commutative C*-algebra, and A^+ the unitization of A, then A is isomorphic to the ideal $J \subset C(\widehat{A^+})$ of continuous functions on $\widehat{A^+}$ which vanish at ε.*

Since the complement of a point in a compact Hausdorff space is locally compact, it follows that any commutative C*-algebra is isomorphic to $C_0(X)$, for some locally compact Hausdorff space X.

3.5 Functional Calculus, Isospectral Subalgebras

If A is a unital subalgebra (or another type of subalgebra) of a C*-algebra B, and if $a \in A$, is the spectrum of a the same in A and in B? This question has important implications for functional calculus constructions, and is important in Noncommutative Geometry.

Firstly, if A is a C*-subalgebra of B, then the answer is 'yes.'

Theorem 3.5.1 (Spectral Permanence) *Let A be a unital C*-subalgebra of the unital C*-algebra B. Then* $\mathrm{Spec}_A(a) = \mathrm{Spec}_B(a)$ *for all $a \in A$.*

Proof Clearly $\mathrm{Spec}_B(a) \subset \mathrm{Spec}_A(a)$ for any $a \in A$. We need to prove that if $a \in A$ is invertible in B, then it is also invertible in A. Notice first that it suffices to prove the statement for self-adjoints. For if a is invertible in B so is a^* (in B), and hence so are a^*a and aa^*. If we can show these are both invertible in A, invertibility of a in A will follow (if a^*a is invertible then a has a left inverse, etc.).

If a is any-self-adjoint, it is spectrum is real, by Corollary 3.3.8. Since a is self-adjoint in both A and B, both $\mathrm{Spec}_A(a)$ and $\mathrm{Spec}_B(a)$ are subsets of $\mathbb{R}$. Let $\lambda \notin \mathrm{Spec}_B(a)$. We show that $\lambda \notin \mathrm{Spec}_A(a)$. Assume otherwise. Choose a sequence of complex numbers λ_n lying off the real axis and converging to λ. Since $\lambda_n \notin \mathbb{R}$ for all n, $\lambda_n - a$ is invertible in both A and B for all n. As inversion is continuous in any C*-algebra, $(\lambda_n - a)^{-1} \to (\lambda - a)^{-1}$ in B. But since the inclusion $A \to B$ is isometric, it follows that $(\lambda_n - a)^{-1}$ also converges in A (it is a Cauchy sequence and hence converges to something). A routine argument shows that it converges to the inverse of $\lambda - a$, i.e., we obtain a contradiction to the assumption that $\lambda \in \mathrm{Spec}_A(a)$. $\qquad\square$

Corollary 3.5.2 *A unital *-homomorphism $A \to B$ is injective if and only if*

$$(3.7) \qquad\qquad \mathrm{Spec}(a) = \mathrm{Spec}(\varphi(a)) \ \forall a \in A.$$

Proof The Principal of Spectral Permanence Theorem 3.5.1 asserts that injective unital *-homomorphisms are isospectral in the sense of (3.7).

For the converse, $\|\varphi(a)\|^2 = \|\varphi(a)^*\varphi(a)\| \|\varphi(a^*a)\| = r(\varphi(a^*a))$ by Theorem 3.1.17, for any $a \in A$, but since φ is assumed isospectral this equals $r(a^*a) = \|a\|^2$. $\qquad\square$

Accordingly, from now on, when referring to the spectrum of an element a of a C*-algebra, we will just write $\mathrm{Spec}(a)$ rather than $\mathrm{Spec}_A(a)$, since the spectrum does not depend on the containing C*-algebra.

We now develop the *functional calculus*, for normal elements of a C*-algebra.

An element a in a C*-algebra A is *normal* if $a^*a = aa^*$. The set of all elements of A of the form $\sum_{n,m} \lambda_{n,m} a^n (a^*)^m$ is then a commutative *-subalgebra of A, whose closure is the C*-algebra $C^*(a)$ generated by a, that is, the smallest unital C*-subalgebra of A containing a. Clearly then $C^*(a)$ is commutative. From spectral permanence, the spectrum of a is the same if we regard a as an element of $C^*(a)$, or of the A we started with. So for brevity of notation, we replace A by $C^*(a)$ in the following. That is, we will assume that a generates A.

Lemma 3.5.3 *If $A = C^*(a)$ and a is normal then the compact Hausdorff spaces $\widehat{A}$ and $\mathrm{Spec}(a)$ are homeomorphic by the map $\widehat{A} \to \mathrm{Spec}(a)$ of evaluation of characters at a.*

Proof Note that any character of $A = C^*(a)$ is determined by its value at a. This shows injectivity of the evaluation map. Surjectivity is immediate from the fact that $\mathrm{Spec}(a)$ consists exactly of values of characters at a (Lemma 3.3.4). $\square$

By Gelfand's theorem, $A \cong C(\widehat{A})$. By the Lemma, $\widehat{A} \cong \mathrm{Spec}(a)$, whence $C(\widehat{A}) \cong C(\mathrm{Spec}(a))$ as C*-algebras. Thus

(3.8) $$C(\mathrm{Spec}(a)) \to A.$$

This map might reasonably be called *Gelfand functional calculus*, or "continuous" functional calculus. It extends holomorphic functional calculus (as we show below.)

Definition 3.5.4 (Functional Calculus for Normal Elements) Let a be a normal element generating a unital C*-algebra A. Let $f: \mathrm{Spec}(a) \to \mathbb{C}$ be a continuous function on $\mathrm{Spec}(a)$. Then $f(a)$ denotes the element of A corresponding to f under the isomorphism (3.8).

Proposition 3.5.5 *If $a \in A$ is normal and $f(z) = \sum a_{n,m} z^n \bar{z}^m$ is a polynomial in $z, \bar{z}$, then*

$$f(a) = \sum c_{n,m} a^n (a^*)^m,$$

where $f(a)$ is defined by the Gelfand functional calculus.

If f is holomorphic on a neighborhood U of $\mathrm{Spec}(a)$ and if Γ is a positively oriented system of closed curves in U as in Definition 3.2.2, then

$$f(a) = \frac{1}{2\pi i} \oint_\Gamma f(z)(z - a)^{-1} \, dz.$$

That is, the holomorphic functional calculus and the Gelfand functional calculus agree.

Proof Since the functional calculus is a C*-algebra homomorphism it suffices to check that it maps $z|_{\mathrm{Spec}(a)}$ to a, since $z|_{\mathrm{Spec}(a)}$ generates $C(\mathrm{Spec}(a))$ as a C*-algebra. To check that $f(a) = a$ when $f(z) = z$, note that f corresponds under $\mathrm{Spec}(a) \cong \widehat{A}$ to the continuous function $\hat{f}(\chi) = f(\chi(a)) = \chi(a)$. This continuous function on $\widehat{A}$ corresponds under $C(\widehat{A}) \cong A$ to a itself. This proves the first statement.

Suppose now U, f, Γ is as in the statement. We argue that

$$\frac{1}{2\pi i} \oint_\Gamma f(z)(z - a)^{-1} \, dz$$

agrees with Gelfand's $f(a)$. If $\gamma: A \to C(\mathrm{Spec}(a))$ is Gelfand's isomorphism, then we may apply γ to the A-valued contour integral above. One has $\gamma(a) = z|_{\mathrm{Spec}(a)}$ and it follows that $\gamma((z - a)^{-1})$ is the continuous function on $\mathrm{Spec}(a)$, $\lambda \mapsto \frac{1}{z - \lambda}$.

Since γ is continuous, application of γ, for each λ, to the contour integral produces the function

$$\lambda \mapsto \frac{1}{2\pi i} \oint_\Gamma \frac{f(z)}{z - \lambda} \, dz$$

on $\mathrm{Spec}(a)$, and this equals f by Cauchy's Theorem.   □

Since $z|_{\mathrm{Spec}(a)}$ generates $C(\mathrm{Spec}(a))$ as a C*-algebra, any *-homomorphism from $C(\mathrm{Spec}(a))$ to another C*-algebra, is completely determined by the image of z ($\bar{z}$ is then sent to the adjoint of this element), and then polynomials in z and $\bar{z}$ are sent to the corresponding combinations in the C*-algebra. This leads to a compact formulation of the above discussion as a uniqueness result about functional calculus:

Lemma 3.5.6 *Let a be a normal element of a C*-algebra. Then the functional calculus is the unique unital *-homomorphism*

$$C(\mathrm{Spec}(a)) \to C^*(a)$$

which maps the inclusion function $z\colon \mathrm{Spec}(a) \to \mathbb{C}$, to a.

Lemma 3.5.6 implies the following result, the *Spectral mapping theorem*.

Proposition 3.5.7 *If a is a normal element of a unital C*-algebra and $f \in C(\mathrm{Spec}(a))$ then $\mathrm{Spec}(f(a)) = f(\mathrm{Spec}(a))$.*

Proof The functional calculus is a C*-algebra isomorphism so $\mathrm{Spec}(f(a)) = \mathrm{Spec}(f)$, where $\mathrm{Spec}(f)$ means of course the spectrum of f as an element of the C*-algebra $C(\mathrm{Spec}(a))$. Since the spectrum of a continuous function is its range, the result follows.   □

Corollary 3.5.8 *Let a be normal, $g \in C(\mathrm{Spec}(a))$ and f be a continuous function on $g(\mathrm{Spec}(a))$. Then $(f \circ g)(a) = f(g(a))$.*

In the statement of the Corollary, $(f \circ g)(a)$ refers to the functional calculus for a applied to $f \circ g$ and $f(g(a))$ refers to applying first the functional calculus for a using g, then the functional calculus for $g(a)$ using f.

Proof We apply the uniqueness result of functional calculus to $b := g(a)$. The map $f \mapsto f \circ g$ is a *-homomorphism $C\left[g(\mathrm{Spec}(a))\right] \to C(\mathrm{Spec}(a))$. By the Spectral Mapping Theorem $g(\mathrm{Spec}(a)) = \mathrm{Spec}(g(a))$. Hence this is a *-homomorphism $C(\mathrm{Spec}(g(a)) \to C(\mathrm{Spec}(a))$. Composing with the functional calculus for a then gives a *-homomorphism $C(\mathrm{Spec}(g(a)) \to C(\mathrm{Spec}(a)) \to C^*(a)$. This maps the inclusion function $z\colon \mathrm{Spec}(g(a)) \to \mathbb{C}$ to the function $g(a)$ by the definitions. By uniqueness, it must agree with functional calculus for $g(a)$. In other words, $f(g(a)) = (f \circ g)(a)$.   □

Another result that is useful and follows immediately from density in $C(\mathrm{Spec}(a))$ of polynomials in $z, \bar{z}$, is the following.

Corollary 3.5.9 *If a is a normal element of a C*-algebra A and $b \in A$ commutes with a and a^* then b commutes with $f(a)$ for every $f \in C(\mathrm{Spec}(a))$.*

Corollary 3.5.10 *if a is a self-adjoint in a unital C*-algebra A then there exist unique positive elements a_1 and a_2 in A such that $a = a_1 - a_2$ and $a_1 a_2 = a_2 a_1 = 0$.*

Proof For existence set $f_1(t) = \max(t, 0)$ and $f_2(t) = -\min(t, 0)$, then $a_1 := f_1(a)$ and $a_2 := f_2(a)$ are the required elements. If A happens to (isomorphic to) $C(X)$, for X compact Hausdorff, the uniqueness part of the statement is easily checked by hand. We reduce the general case to this one by showing that if $a = b_1 - b_2$ has a second decomposition in A of the same kind, then the b_i's commute with the a_j's, whence $C^*(a_1, a_2, b_1, b_2)$ will be commutative, whence, by Gelfand's theorem, isomorphic to $C(X)$ for X compact Hausdorff, the special case just discussed. But given that $a = b_1 - b_2$ with $b_1 b_2 = b_2 b_1 = 0$, it follows immediately that $b_1 a = b_1^2 = a b_1$, so b_1 commutes with a, and similarly b_2 commutes with a. Now by Corollary 3.5.9, each b_i then commutes with $f(a)$ for any continuous function f on $\mathrm{Spec}(a)$, so in particular each b_i commutes with both $a_1 = f_1(a)$ and $a_2 = f_2(a)$.   □

Exercise 3.5.11 The function $f(t) = \sqrt{1-t}$ is the uniform limit on any closed interval $|t| \le 1 - \varepsilon$, of its Taylor series expansion at $t = 0$: thus, for appropriate constants $c_0, c_1, \ldots,$

$$\sqrt{1-t} = \sum_{n=0}^{\infty} c_n t^n,$$

and the sum is norm convergent in the C*-algebra $C([0, 1 - \varepsilon])$. Use the fact that

$$a(a^*a)^n = (aa^*)^n a$$

in combination with the Spectral Theorem to show that for any unital C*-algebra A and any $a \in A$ with $\|a\| \le 1$,

$$a(1 - a^*a)^{\frac{1}{2}} = (1 - aa^*)^{\frac{1}{2}} a.$$

Isospectral subalgebras

We now discuss some examples of a geometric kind in which spectral permanence holds for subalgebras $B \subset A$ of a C*-algebra which are not C*-algebras, but are *dense* subalgebras.

Let A be a C*-algebra. Assume that A comes equipped with a *time evolution*: a 1-parameter group $(\sigma_t)_{t \in \mathbb{R}}$ of automorphisms of A such that $t \mapsto \sigma_t(a)$ is continuous for all $a \in A$.

An element $a \in A$ is *smooth* if the map $\mathbb{R} \to A, t \mapsto \sigma_t(a)$, is smooth. Let A^∞ be the subalgebra of smooth elements of A. The map $\delta: A^\infty \to A$,

$$(3.9) \qquad\qquad\qquad \delta(a) := \frac{d}{dt}\sigma_t(a)$$

is a *derivation*: $\delta(ab) = a\delta(b) + \delta(a)b$.

Exercise 3.5.12 Suppose that a time evolution with derivation δ is inner in the sense that $\sigma_t(a) = e^{-itH}ae^{-itH}$ for some $H \in A$ self-adjoint. Show that $\delta(a) = i[H, a]$.

Exercise 3.5.13 Show that $A^\infty \subset A$ is dense in A and that if $a \in A^\infty$ is invertible in A then a is invertible in A^∞. Deduce that the spectrum of $a \in A^\infty$ is the same as the spectrum of a in A.

Example 3.5.14 Let M be a compact manifold M admitting a smooth flow $(\tau_t)_{t\in\mathbb{R}}$ then $C(M)$ gets a time evolution $\sigma_t(f)(p) := f(\tau_{-t}p)$ and derivation $\delta(f) := \mathcal{L}_X(f)$, where $\mathcal{L}_X$ denotes the Lie derivative with respect to the tangent vector field of the flow

$$(\mathcal{L}_X f)(p) := \frac{d}{dt}f(\tau_t p), \quad p \in M.$$

The smooth elements of $C(M)$ are the continuous functions which are smooth along the orbits of the flow.

Let A have a time evolution $(\sigma_t)_{t\in\mathbb{R}}$ as above. We show that the *-subalgebra A^∞ of smooth elements is stable under *smooth* functional calculus. Let $f \in C_c^\infty(\mathbb{R})$ be a smooth, compactly supported function. Since f is smooth, its Fourier transform

$$\hat{f}(\xi) = \int f(x)e^{-ix\xi}dx$$

is defined and is a smooth function of rapid decay on $\mathbb{R}$: for every n the function $|\hat{f}(\xi)| \cdot |\xi|^n$ is bounded. The Fourier inversion formula then expresses f as an integral

$$f(x) = \frac{1}{2\pi}\int \hat{f}(\xi)e^{ix\xi}d\xi.$$

Exercise 3.5.15 Let $a \in A$ be self-adjoint.

a) Prove that the integral $\int \hat{f}(\xi)e^{i\xi a}d\xi$ converges absolutely in A and that

$$(3.10) \qquad\qquad\qquad f(a) = \frac{1}{2\pi}\int \hat{f}(\xi)e^{i\xi a}d\xi$$

holds in A.

b) By the usual technique of differentiating under the integral sign using (3.10), prove that $f(a)$ is smooth if a is smooth. Hence A^∞ is stable under smooth functional calculus. It is therefore also stable under holomorphic functional calculus.

c) Use (3.10) to deduce the 'chain rule'

$$\delta\big(f(a)\big) = \delta(a)\, f'(a),$$

where $f(a)$ and $f'(a)$ refer to continuous functional calculus for a.

The smooth elements of a flow are an example of an "isospectral subalgebra" in the following sense.

Definition 3.5.16 A unital, dense subalgebra $B \subset A$ of a unital C*-algebra is *isospectral* in A if $\mathrm{Spec}_{\mathcal{A}}(a) = \mathrm{Spec}_A(a)$ for all $a \in \mathcal{A}$.

Such subalgebras are important in Noncommutative Geometry in its applications to K-theory.

Example 3.5.17 $C^\infty(M) \subset C(M)$ is isospectral, since f is invertible in $C(M)$ (resp. $C^\infty(M)$) if and only if $f(x) \neq 0$ for all $x \in X$.

Exercise 3.5.18 Let $B \subset A$ be a dense subalgebra.

a) Show that $B \subset A$ is isospectral if and only if $b \in B$ invertible in A implies $b^{-1} \in B$.

b) Show that if $B \subset A$ is isospectral, and if there is a norm on B making B a Banach algebra, then B is closed under holomorphic functional calculus. That is, if $b \in B$ and f is holomorphic on a neighborhood of $\mathrm{Spec}_A(b) = \mathrm{Spec}_B(b)$ then $f(b) \in B$.

c) Show that if $p \in A$ is a projection, then there exists an idempotent $q \in B$ such that $\|p - q\| < 1$. (*Hint.* By density find $b \in B$ arbitrarily close to p, and for such b, $b^2 - b$ is small, and if it small enough $\frac{1}{2} \notin \mathrm{Spec}(b)$ and hence there is a holomorphic function h in a neighborhood of $\mathrm{Spec}(b)$ such that $h(b)$ is idempotent.)

Part c) implies that the inclusion $B \to A$ induces an isomorphism $K_0(B) \to K_0(A)$. This is one of the essential points about isospectral subalgebras.

Other important examples of isospectral subalgebras come from densely defined derivations. We have already discussed one instance above: that of the derivation associated to a flow.

More generally, let $\mathcal{A} \subset A$ be a dense subalgebra of a C*-algebra. Let $\mathcal{D}$ be an A- bimodule, equipped with a Banach space norm such that $\|avb\| \leq \|a\|\,\|v\|\,\|b\|$ for all $a \in A$, $b \in A$, $v \in \mathcal{D}$.

In this notation, a *derivation* $\delta\colon \mathcal{A} \to \mathcal{D}$ is a linear map satisfying $\delta(ab) = \delta(a)b + a\delta(b)$ for all $a, b \in \mathcal{A}$.

Say δ is *closable* if whenever $a_n, a, b \in \mathcal{A}$ such that $a_n \to a$ and $\delta(a_n) \to b$, then $\delta(a) = b$.

The following exercise shows that δ extends to a derivation of a dense Banach subalgebra dom(δ) of A such that dom$(\delta) \subset A$ is isospectral.

Exercise 3.5.19 Let $\delta \colon \mathcal{A} \to \mathcal{D}$ be a densely defined closable derivation as in the discussion above.

a) Define for $a \in \mathcal{A}$, $\|a\|_1 := \max\{\|a\|, \|\delta(a)\|\}$. Prove that

$$\|ab\|_1 \leq 2\|a\|_1 \|b\|_1.$$

Let now $\|a\|_\delta := \sup_{\|b\|_1 \leq 1} \|ab\|_1$. Prove that $\|\cdot\|_\delta$ is an equivalent norm to $\|\cdot\|_1$ but now is a Banach algebra norm $\|ab\|_\delta \leq \|a\|_\delta \cdot \|b\|_\delta$.

b) Let dom(δ) of $\mathcal{A}$ be the completion of $\mathcal{A}$ with respect to $\|\cdot\|_\delta$, equivalently, with respect to $\|\cdot\|_1$. Prove that δ extends to a derivation of dom(δ) into $\mathcal{D}$. The new derivation is *closed* (the graph of δ is closed in $A \oplus \mathcal{D}$.) (*Hint.* An element of dom(δ) is an element $a \in A$ for which there exist $a_n \in \mathcal{A}$ such that $a_n \to a$ and $\delta(a_n)$ converges in B.)

c) Suppose that $a \in$ dom(δ) is invertible in A. Prove that $a^{-1} \in$ dom(δ). (*Hint.* Choose a sequence $(b_n) \subset$ dom(δ) such that $b_n \to a^{-1}$. Apply the derivation rule to the sequence (ab_n) and deduce that $\delta(b_n) \to a^{-1}\delta(a)a^{-1}$. Deduce the result).

d) Prove that dom$(\delta) \subset A$ is isospectral.

Exercise 3.5.20 Let $(f_n) \subset C_b(\mathbb{R})$ be a bounded sequence of differentiable functions on $\mathbb{R}$ such that $f_n \to 0$ uniformly and $f_n' \to h$ uniformly, for some continuous function h on $\mathbb{R}$. Prove that $h = 0$. (*Hint.* Show $\int_I h = 0$ for any interval I.)

Deduce that if $\delta = \mathcal{L}_X$ is the densely defined derivation of $C^\infty(M)$ determined by Lie derivative with respect to a vector field X, as in Example 3.5.14, then $\delta \colon C^\infty(M) \to C(M)$ is closable.

Exercise 3.5.21 Let G be a countable group and $\phi \colon G \to \mathbb{R}$ be a group homomorphism. On $\mathbb{C}[G]$ define $\delta(\sum_{g \in G} a_g[g]) := \sum_{g \in G} \phi(g)a_g[g]$. Prove that δ is a densely defined derivation $\mathbb{C}[G] \to C^*(G)$ and that δ is closable. Hence dom(δ) is a dense subalgebra of $C^*(G)$ which is closed under holomorphic functional calculus.

Exercise 3.5.22 Let A be a C*-algebra and $\pi \colon A \to \mathbb{B}(H)$ be a representation. Let $F \in \mathbb{B}(H)$ be a self-adjoint operator such that the commutators $[\pi(a), F]$ are in the Schatten trace-class ideal $\mathcal{L}^1(H)$ for $a \in \mathcal{A}$, where $\mathcal{A} \subset A$ is a dense *-subalgebra.

Prove that $\mathcal{L}^1(H)$ has a natural A-bimodule structure and that $\delta \colon A \to \mathcal{L}^1(H)$, $\delta(a) := [\pi(a), F]$ is a closable derivation. Hence dom(δ) is a dense Banach subalgebra of A which is isospectral in A and for which $[\pi(a), F] \in \mathcal{L}^1(H)$ for all $a \in$ dom(δ).

Chapter 4
Positivity, Representations, Tensor Products, and Ideals in C*-Algebras

In classical mechanics an observable is a continuous function on a space X and an element of the commutative C*-algebra $C(X)$. In quantum mechanics an observable is a bounded operator on a Hilbert space, or an element of a C*-subalgebra $A \subset \mathbb{B}(H)$, such as $M_n(\mathbb{C})$, if the system is very simple, like that of an atom which can be in n quantum states, corresponding to energy levels of the Hamiltonian. A "microscopic" state of the *classical* system is a point of the space and corresponds to a C*-algebra character $C(X) \to \mathbb{C}$. In quantum mechanics one is forced to work with general states in the C*-algebraic sense: *positive* linear functionals $\varphi \colon A \to \mathbb{C}$, i.e., linear functionals taking positive values at self-adjoint elements of A with positive spectrum. For example, the the energy states of the Hamiltonian are the vector states corresponding to the eigenfunctions of the Hamiltonian H.

States are the quantum analogues of probability measures on a compact space. This requirement of working with states, due to failure of various operators to commute, is what given quantum mechanics its probabilistic nature.

Vector states associated to a representation $\pi \colon A \to \mathbb{B}(H)$ of a C*-algebra on a Hilbert space are obtained by the formula $\varphi(a) = \langle \pi(a)v, v \rangle$, where v is a unit vector. Thus, a representation gives rise to a family of states. The GNS construction gives something like a converse: A state on a C*-algebra gives rise to a representation of the C*-algebra, from which the state can be recovered as a vector state. A consequence is the famous theorem of Gelfand, Naimark and Segal [86] published in 1943, stating that every abstract C*-algebra is isomorphic to a C*-subalgebra of $\mathbb{B}(H)$ for some H.

In this chapter we discuss positivity, states, and representations of C*-algebras, establish the GNS theorem, define tensor products, and discuss the general procedure of completions of *-algebras, as well as prove some basic results on ideals deferred up to now, i.e., that the quotient of a C*-algebra is a closed ideal is a C*-algebra. We close by studying an example and analyze the space of irreducible representations of the examples $C_0(X) \rtimes G$ of noncommutative, orbifold examples, where G is discrete and acts properly, co-compactly on X. The computation comes

© Springer Nature Switzerland AG 2024
H. Emerson, *An Introduction to C*-Algebras and Noncommutative Geometry*,
Birkhäuser Advanced Texts Basler Lehrbücher,
https://doi.org/10.1007/978-3-031-59850-0_4

from the paper [67], where the description of the topology on the spectrum is given directly in terms of the space X and the action.

The books [7] and [126, 127], are good sources for representation theory and ideal theory of C*-algebras.

4.1 Positivity and States

Definition 4.1.1 Let A be a unital C*-algebra. An element $a \in A$ is *positive* if a is self-adjoint and $\mathrm{Spec}(a) \subseteq [0, \infty)$.

We write $a \geq 0$ if a is positive. More generally, if a and b are self-adjoint, we write $a \leq b$ if $b - a \geq 0$.

Exercise 4.1.2 Let p and q be two projections in $\mathbb{B}(H)$. Prove that $p \leq q$ if and only if p is a subprojection of q.

Remark 4.1.3 The square a^2 of any self-adjoint element a in a unital C*-algebra is positive, for by the Spectral Mapping Theorem (Proposition 3.5.7), $\mathrm{Spec}(a^2) = \{\lambda^2 \mid \lambda \in \mathrm{Spec}(a)\}$, and since $a = a^*$, $\mathrm{Spec}(a) \subset \mathbb{R}$, so $\mathrm{Spec}(a^2) \subset [0, \infty)$. Conversely, if $a \geq 0$, then $f(t) := \sqrt{t}$ is a continuous function on $\mathrm{Spec}(a)$, so $\sqrt{a}$ is defined and has square equal to a. Thus positivity for elements of a C*-algebra A is equivalent to being the square of a self-adjoint element of A.

It is clear that if $f \in C(X)$ for a compact Hausdorff space X, then $f \geq 0$ in the C*-algebra sense if and only if $f(x) \geq 0 \ \forall x \in X$.

A very useful criterion for a self-adjoint operator on a Hilbert space to be positive is given by the following.

Proposition 4.1.4 If $T \in \mathbb{B}(H)$ is a self-adjoint, then $T \geq 0$ if and only if $\langle T\xi, \xi \rangle \geq 0$ for all $\xi \in H$.

Proof If $T \geq 0$, then $T = S^2$ for a self-adjoint S by Remark 4.1.3, and then $\langle T\xi, \xi \rangle = \langle S^2\xi, \xi \rangle = \langle S\xi, S\xi \rangle \geq 0$ holds evidently for all $\xi \in H$.

Conversely, suppose that $\langle T\xi, \xi \rangle \geq 0$ for all ξ. Then T is self-adjoint. By Proposition 3.5.10 of the previous chapter, we can write $T = T_1 - T_2$ for positive elements $T_i \in \mathbb{B}(H)$ such that $T_1 T_2 = 0$. Now

$$\|T_2^{\frac{3}{2}}\xi\|^2 = \langle T_2^{\frac{3}{2}}\xi, T_2^{\frac{3}{2}}\xi \rangle = \langle T_2^2\xi, T_2\xi \rangle.$$

But since $T_1 T_2 = 0$, $T T_2 = -T_2^2$. So $\langle T_2^2\xi, T_2\xi \rangle = -\langle T T_2\xi, T_2\xi \rangle \leq 0$ follows. Therefore $\|T_2^{\frac{3}{2}}\xi\|^2 = 0$ so $T_2^{\frac{3}{2}}\xi = 0$. Since ξ was arbitrary, this shows that $T_2^{\frac{3}{2}} = 0$. Hence $T_2 = 0$. So $T = T_1$ and T is positive. □

Note that Proposition 4.1.4 implies immediately that any operator of the form T^*T is positive. One of the trickier results in elementary C*-algebra theory is that

a^*a is positive for any element a of any C*-algebra. Note that the result is clear for elements of abelian C*-algebras, since these are isomorphic to $C(X)$ for compact Hausdorff spaces X, and here the result is obvious, since a function of the form $f^*f = |f|^2$ obviously takes positive values only.

The following criterion for positivity will be useful for several arguments.

Lemma 4.1.5 *If a is a self-adjoint in a C*-algebra with $\|a\| \leq 1$, then $a \geq 0$ if and only if $\|1 - a\| \leq 1$.*

Proof The statement is clearly true for complex numbers, then follows for $f \in C(X)$ by considering the values of f. Now the result for general a self-adjoint and contractive follows from the Spectral Theorem, giving an isomorphism $C^*(a) \cong C(\mathrm{Spec}(a))$. □

Corollary 4.1.6 *The sum of two positive elements is positive.*

Proof Let a and b be positive; assume first that $\|a\| \leq 1$ and $\|b\| \leq 1$. Then $\|\frac{a+b}{2}\| \leq 1$ as well, and

$$\|1 - (\frac{a+b}{2})\| = \frac{1}{2}\|(1-a) + (1-b)\| \leq \frac{1}{2}(\|1-a\| + \|1-b\|) \leq 1$$

since a and b are positive and have norm ≤ 1. This shows that $\frac{a+b}{2}$ is positive, and hence that $a + b$ is positive.

Now for the general case, replace a and b by ta and tb for suitable $t \geq 0$ making both ta and tb have norm ≤ 1. We deduce that $ta + tb = t(a + b)$ is positive and hence that $a + b$ is. □

Theorem 4.1.7 *If A is a C*-algebra and $a \in A$, then a^*a is positive.*

Hence an element of a C*-algebra is positive if and only if it is of the form a^*a for some a.

Proof Write $a^*a = u - v$ as a difference of positive elements as in Proposition 3.5.10. Let $e = av$. We are going to show $v = 0$.

First of all,

$$e^*e = va^*av = v(u - v)v = -v^3$$

since $uv = vu = 0$ and $v = v^*$. On the other hand, if we write $e = x + iy$ with x, y self-adjoint, then we see that $e^*e = x^2 + y^2 + i(xy - yx)$, while $ee^* = x^2 + y^2 - i(xy - yx)$. Hence $e^*e + ee^* = x^2 + y^2$. Since x^2 and y^2 are certainly positive, their sum is positive. Hence $e^*e + ee^* \geq 0$. On the other hand, we have just observed that $e^*e = -v^3$. Thus $ee^* = -e^*e + x^2 + y^2 = v^3 + x^2 + y^2$, which is another sum of positive elements, whence positive. Thus, ee^* is positive. On the other hand since $\mathrm{Spec}(ee^*) \setminus \{0\} = \mathrm{Spec}(e^*e) \setminus \{0\}$, by Exercise 3.1.7 of Chapter 2, and since ee^* has been shown to be positive, it must be that $e^*e \geq 0$ as well. But again, $e^*e = -v^3 \leq 0$. So the spectrum of e^*e can consist only of 0. Since e^*e is self-adjoint, this implies that $e^*e = 0$ and hence that $-v^3 = 0$, whence, since v is self-adjoint, $v = 0$.   □

Exercise 4.1.8 Prove that if A is any unital *-algebra, then $1 + a^*a$ is invertible for all $a \in A$, holds if and only if $\mathrm{Spec}(a^*a) \subset [0, \infty)$ for all $a \in A$. (*Hint.* For a fixed, and any $\lambda > 0$, the hypothesis implies that $0 \notin \mathrm{Spec}(1 + \lambda a^*a)$ and hence that $-\frac{1}{\lambda} \notin \mathrm{Spec}(a^*a)$.)

Since the converse is trivially true, we see that the property of C*-algebras that $a^*a \geq 0$ for all $a \in A$ is equivalent to the property that $1 + a^*a$ is invertible, for all $a \in A$.

Definition 4.1.9 Let A be a unital C*-algebra. A *state* on A is a linear functional $\varphi \colon A \to \mathbb{C}$ such that $\varphi(1) = 1$, and if $a \geq 0$, then $\varphi(a) \geq 0$.

Example 4.1.10 States of $C(X)$ are in 1–1 correspondence with Borel probability measures on X (the Riesz Representation Theorem).

Lemma 4.1.11 *Let* $\varphi : A \to \mathbb{C}$ *be a linear functional. Then* φ *is a state if and only if* $\varphi(1) = 1$, φ *is contractive, and* $\varphi(a) \in \mathbb{R}$ *if* $a = a^*$.

In particular, states are automatically continuous.

Proof Suppose φ is a state. If a is any self-adjoint, then $a = u - v$ where u, v are positive, whence $\varphi(a) = \varphi(u) - \varphi(v)$ is a difference of positive real numbers so $\varphi(a) \in \mathbb{R}$ for $a = a^*$.

Since $a \leq \|a\|$ holds for any $a \in A$ self-adjoint, $\varphi(a) \leq \|a\| \cdot \varphi(1) = \|a\|$.

Conversely, let $\varphi \colon A \to \mathbb{C}$ be linear and satisfy the stated conditions. If $a \in A$ with $\|a\| \leq 1$, then since a^*a is positive and $\|a^*a\| = \|a\|^2 \leq 1$, an application of Lemma 4.1.5 gives that $\|1 - a^*a\| \leq 1$. Since φ is a contraction, $|1 - \varphi(a^*a)| = |\varphi(1 - a^*a)| \leq 1$, and furthermore, $|\varphi(a^*a)| \leq 1$, and these two conditions together give that $\varphi(a^*a) \geq 0$ for any a. Since positive elements are exactly those of the form a^*a, this shows that φ maps positive elements to nonnegative real numbers as required.   □

Theorem 4.1.12 *If* A *is any unital C*-algebra and* $S(A)$ *is the state space of* A, *then*

$$\|a\| = \sup_{\varphi \in S(A)} |\varphi(a)|$$

for any self-adjoint element $a \in A$.

Proof If $a \in A$ is self-adjoint, then $C^*(a) \cong C\big(\mathrm{Spec}(a)\big)$ with a corresponding to the inclusion $z : \mathrm{Spec}(a) \to \mathbb{C}$. Since $\mathrm{Spec}(a)$ is a compact subset of $\mathbb{R}$, the function $|z|$ obtains its maximum on $\mathrm{Spec}(a)$, say at λ. Then the map $\psi_0 \colon C^*(a) \cong C\big(\mathrm{Spec}(a)\big) \to \mathbb{C}$ corresponding to evaluation $f \mapsto f(\lambda)$ of functions at λ is a character of $C^*(a)$ satisfying $|\psi_0(a)| = \|a\|$. Being a character, it is contractive on $C^*(a)$. By the Hahn–Banach Theorem it can be extended to a contractive linear functional $\psi \colon A \to \mathbb{C}$. Then $\varphi(a) := \frac{1}{2}(\psi(a) + \overline{\psi(a^*)})$ is still contractive, extends

ψ_0, maps self-adjoints of A to real numbers, and has $\varphi(1) = 1$, so it is a state by Lemma 4.1.11 and by design satisfies $|\varphi(a)| = \|a\|$. □

Remark 4.1.13 Note that the proof shows that for each self-adjoint $a \in A$ there exists a state $\varphi \in S(A)$ such that $\|a\| = |\varphi(a)|$.

Exercise 4.1.14 Let A be a unital C*-algebra. Answer the following questions about positivity:

a) $a \leq \|a\| \cdot 1$ for all $a \in A$ self-adjoint.
b) If $x \leq y$ are self-adjoints, then $a^*xa \leq a^*ya$ for all $a \in A$.
c) If $a, x \in A$ with $x \geq 0$, then $a^*xa \leq \|x\|a^*a$.
d) If $x \leq y$ and both are invertible, then $y^{-1} \leq x^{-1}$. (*Hint* Observe first that $y^{-\frac{1}{2}}xy^{-\frac{1}{2}} \leq 1$. Deduce that $\|x^{\frac{1}{2}}y^{-\frac{1}{2}}\| \leq 1$ and then that $x^{\frac{1}{2}}y^{-1}x^{\frac{1}{2}} \leq 1$.)

Remark 4.1.15 There is a special class of states appearing in the C*-algebraic formulation of quantum statistical mechanics, called KMS, or equilibrium states, which physically measure the properties of a system in thermal equilibrium. Given a 1-parameter family $\{\sigma_t\}_{t \in \mathbb{R}}$ of automorphisms of A, a KMS$_\beta$ state is a state φ of A with a certain twisted tracial property with respect to this evolution. This set-up appears in connection with a number of examples related to number theory, see [28, 53], corresponding physically to systems with interaction in which at high temperature (small β) disorder is predominant, resulting in uniqueness of KMS$_\beta$ states, while at low temperatures order sets in and allows a multitude of possible KMS$_\beta$ states, i.e., thermodynamical phases (called spontaneous symmetry breaking). In a number of examples, these state spaces admit interesting descriptions (see [57, 115]).

4.2 The GNS Theorem

Definition 4.2.1 A *representation* of a C*-algebra A is *-homomorphism $\pi : A \rightarrow \mathbb{B}(H)$ for some Hilbert space H. The representation is *faithful* if π is an injective *-homomorphism, and is *nondegenerate* if $\pi(A)H = H$. It is *irreducible* if there is no proper nonzero closed subspace $H' \subset H$ of H such that $\pi(A)H' \subset H'$. It is *nondegenerate* if $\pi(A)H$ is dense in H, and is *cyclic* if there is a vector $\xi \in H$ such that $\pi(A)\xi$ is dense in H.

Exercise 4.2.2 Let A be a C*-subalgebra and π a representation of A on H. If $H' \subset H$ is a closed subspace invariant under a, then $H^\perp$ is also invariant under $\pi(A)$.

With respect to this decomposition,

$$\pi(a) = \begin{bmatrix} \pi'(a) & 0 \\ 0 & \pi''(a) \end{bmatrix}$$

for a pair of representations of A on H, $H^\perp$.

Two representations $\pi: A \to \mathbb{B}(H)$ and $\rho: A \to \mathbb{B}(K)$ are *equivalent* if there is a unitary $u: H \to K$ such that $\pi(a) = \pi(u)^*\rho(a)\pi(u)$ for all $a \in A$.

The complex numbers, and more generally, matrix algebras of any size, have unique irreducible representations, up to unitary equivalence.

Proposition 4.2.3 *Up to unitary equivalence, the C*-algebra $M_n(\mathbb{C})$ has a unique irreducible representation—the standard one, given by the action of $M_n(\mathbb{C})$ on $\mathbb{C}^n$ by matrix multiplication.*

We need the following:

Lemma 4.2.4 *If $\varphi: M_n(\mathbb{C}) \to M_m(\mathbb{C})$ is a unital *-homomorphism, then $n|m$, and there is a unitary $u \in M_m(\mathbb{C})$ such that*

$$\varphi(A) = u \begin{bmatrix} A & 0 & \cdots \\ 0 & A & 0 \\ \cdots\cdots\cdots \\ 0 & \cdots & A \end{bmatrix} u^*$$

for all $A \in M_n(\mathbb{C})$).

Of course the number of diagonal summands in the matrix is $\frac{m}{n}$.

Proof Let us start by recalling that since $M_n(\mathbb{C})$ is always simple (Exercise 4.4.6), every nonzero *-homomorphism $M_n(\mathbb{C}) \to D$ to another C*-algebra is automatically injective, whence isometric.

The idea of the proof is to think of the matrix units $e_{ij} \in M_n(\mathbb{C})$ geometrically in terms of the Hilbert space $\mathbb{C}^m$ on which $M_n(\mathbb{C})$ acts via the representation $\varphi: M_n(\mathbb{C}) \to M_m(\mathbb{C}) \cong \mathbb{B}(\mathbb{C}^m)$. Each e_{ii} acts as a projection ε_{ii} onto a subspace E_i of $\mathbb{C}^m$. The subspaces E_i are pairwise orthogonal and $\oplus_{i=1}^n E_i = \mathbb{C}^m$. (In particular $n|m$.)

The matrix units e_{ij} for $i \neq j$ act as partial isometries ε_{ij}, the range projection for ε_{ij} is $\varepsilon_{ij}\varepsilon_{ij}^* = \varepsilon_{ij}\varepsilon_{ji} = \varepsilon_{ii}$, and the source projection is ε_{jj}. Thus ε_{ij} restricts to a unitary isomorphism $E_j: \to E_i$, and it is the zero operator on the orthogonal complement $E_j^\perp = \oplus_{k \neq j} E_k$.

Consider E_1. It has an orthonormal basis. Using $\varepsilon_{j1}: E_1 \to E_j$, we construct the corresponding image orthonormal bases of $E_2, E_3, \ldots, E_n$. Putting all of these orthonormal bases together gives an orthonormal basis of $\mathbb{C}^m$. We leave it to the reader to check that with respect to this basis, $\varphi(A)$ has the form

$$\begin{bmatrix} A & 0 & \cdots \\ 0 & A & 0 \\ \cdots\cdots\cdots \\ 0 & \cdots & A \end{bmatrix}$$

as claimed. $\square$

Exercise 4.2.5 If A is unital and $\pi: A \to \mathbb{B}(H)$ is a representation, then there exists a subspace H' of H so that with respect to the induced decomposition of H,

$$\pi(a) = \begin{bmatrix} \pi'(a) & 0 \\ 0 & 0 \end{bmatrix}$$

where $\pi': A \to \mathbb{B}(H')$ is nondegenerate.

Exercise 4.2.6 Prove that any representation $\pi: A \to \mathbb{B}(H)$ is a direct sum of (countably many) cyclic representations. That is, there is a decomposition $H = H_1 \oplus H_2 \oplus \cdots$ and cyclic representations $\pi_n: A \to \mathbb{B}(H_n)$ such that $\pi(a) = \bigoplus_{n=1}^{\infty} \pi_n(a)$.

Let $\pi: A \to \mathbb{B}(H)$ be a representation. If ξ is a unit vector in H, let $\varphi: A \to \mathbb{C}$ be the linear functional $\varphi(a) = \langle \pi(a)\xi, \xi \rangle$. Then clearly $\varphi(1) = 1$ and $\varphi(a^*a) = \|\pi(a)\|^2 \geq 0$ so φ is actually a state, usually called a *vector state*.

The GNS construction, explained below, reverses this and produces a representation from a state. Let $\varphi: A \to \mathbb{C}$ be a state. Define a sesquilinear form on A by $\langle a, b \rangle := \varphi(b^*a)$. By the generalized Cauchy–Schwarz inequality (Theorem 5.4.15),

$$(4.1) \qquad\qquad |\varphi(ab^*)|^2 \leq \varphi(a^*a)\varphi(b^*b)$$

for all $a, b \in A$, a fact used in the proof below.

Proposition 4.2.7 *Let φ be a state of A. Then there is a cyclic representation $\pi: A \to \mathbb{B}(H)$ and a cyclic vector ξ such that $\varphi(a) = \langle \pi(a)\xi, \xi \rangle$ for all $a \in A$.*

Furthermore, if $\rho: A \to \mathbb{B}(K)$ is another cyclic representation, and $\eta \in K$ is a cyclic vector giving the same state, i.e., for which $\varphi(a) = \langle \rho(a)\eta, \eta \rangle$, then π and ρ are unitarily equivalent by a unitary isomorphism sending ξ to η.

Proof In the notation above set $\langle a, b \rangle := \varphi(b^*a)$. Let $N \subset A$ be the subset of A consisting of a such that $\langle a, a \rangle = 0$. By Equation (4.1) N is a linear subspace of A. Form the quotient vector space $H_0 := A/N$. Then $\langle \cdot, \cdot \rangle$ descends to an inner product on H_0. Let H be its completion, a Hilbert space. Write $\|a\|_H := \langle a, a \rangle := \varphi(a^*a)$ to distinguish from the C*-algebra norm.

If $a, b \in A$, then since $b^*a^*ab \leq \|a\|^2 b^*b$, and since φ is a state,

$$(4.2) \qquad \|ab\|_H^2 = \langle ab, ab \rangle = \varphi(b^*a^*ab) \leq \|a\|^2 \varphi(b^*b) = \|a\|_H^2 \|b\|_H^2.$$

So if $b \in N$ then $ab \in N$. Hence N is actually a left ideal in A, and so there is a well-defined bilinear multiplication, $A \times H_0 \to H_0$, $(a, b + N) \mapsto ab + N$, where $b + N$ etc denotes the coset of b in $H_0 = A/N$. Equation (4.2) also shows that the linear map $H_0 \to H_0$ of left multiplication by $a \in A$ is bounded by $\|a\|$. Consequently, left multiplication by a extends to a linear map $H \to H$ which is a bounded operator with norm $\leq \|a\|$. Denote this bounded linear map by $\pi(a)$. It is easily verified that $\pi: A \to \mathbb{B}(H)$ is a representation of A and that if $\xi \in H$ is the coset of 1 in $H_0 \subset H$, then $\langle \pi(a)\xi, \xi \rangle = \varphi(a)$.

For the uniqueness statement, suppose $\rho: A \to \mathbb{B}(K)$ is another representation of A and $\eta \in H$ is a vector such that $\varphi(a) = \langle \rho(a)\eta, \eta \rangle$. Notice that from the definitions $\varphi(a^*a) = \|\pi(a)\|^2$ and for the same reason $= \|\rho(a)\|^2$. In particular if $a \in A$, then $\pi(a)\xi = 0$ if and only if $\rho(a)\eta = 0$, and the map $U: H \to K$, $U(\pi(a)\xi) := \pi(a)\eta$ is then well defined . It is clearly linear and isometric on the dense subspace $\pi(A)\xi \subset H$ and hence extends uniquely to a unitary map $H \to K$. Since it was also assumed that $\rho(A)\eta$ is dense in K, it is a unitary isomorphism. From the definitions, if $\xi_1 := \pi(a)\xi \in H$, then $U\xi_1 = \rho(a)\eta \in K$ so

$$(U^*\rho(b)U)\xi_1 = U^*\rho(b)\rho(ba)\eta = U^*\rho(ba)\eta = \pi(ba)\xi = \pi(b)\pi(a)\xi = \pi(b)\xi_1.$$

This implies that π and ρ are unitarily equivalent by a unitary mapping ξ to η. □

Example 4.2.8 Let G be a discrete group, and $\tau: C_r^*(G) \to \mathbb{C}$ the tracial state which on elements of $\mathbb{C}[G]$ is given by $\tau(\sum \lambda_g[g]) = \lambda_e$ (see Exercise 1.3.10).

Then the cyclic representation associated to τ is the *left regular representation*

$$\lambda: C^*(G) \to \mathbb{B}(l^2(G)),$$

with cyclic vector the point mass e_1 at the identity element of G.

Thus

$$\tau(T) = \langle \lambda(T)e_1, e_1 \rangle$$

represents τ as a vector state affiliated with the regular representation.

Example 4.2.9 Let X be a compact Hausdorff space and μ a Borel probability measure on X, φ_μ the corresponding state of $C(X)$. The associated GNS construction forms the Hilbert space obtained by completing $C(X)$ with the norm $\|f\|^2 = \int_X |f(x)|^2 d\mu(x)$, and represents $C(X)$ on $L^2(X, \mu)$ as multiplication operators. The cyclic vector is the constant function $1 \in L^2(X)$.

Corollary 4.2.10 *Every C*-algebra is isomorphic to a C*-subalgebra of $\mathbb{B}(H)$ for some Hilbert space H.*

Proof From each unitary equivalence class of representation of A, pick a representative $\pi: A \to \mathbb{B}(H_\pi)$. Let X be the set of such representations. Let $H = \oplus_{\pi \in X} H_\pi$ and represent A on H diagonally by $\rho(a)(\xi_\pi)_{\pi \in X} := (\pi(a)\xi_\pi)_{\pi \in X}$. We need to show that ρ is injective. But if $\rho(a) = 0$ for some $a \in A$, then $\pi(a)\xi = 0$ for every $\pi \in X$ and every vector $\xi \in H_\pi$. Since every state of A can be represented as a vector state for some $\pi \in X$ and some $\xi \in H_\pi$, we get that $\varphi(a^*a) = 0$ for every state of A, which contradicts Theorem 4.1.12.

As an alternative proof, by Remark 4.1.13 for each $a \in A$ there exists $\varphi_a \in S(A)$ such that $\varphi(a^*a) = \|a\|^2$. Taking the direct sum of the corresponding cyclic representations gives another faithful representation of A. □

Exercise 4.2.11 Prove that the C*-algebra $M_2(\mathbb{C})$ has a unique irreducible representation, up to unitary equivalence, which is the standard representation (on $\mathbb{C}^2$).

Exercise 4.2.12 The following exercise addresses the question of whether or not a representation can be *extended* from a C*-subalgebra to a larger one.

Let A be a unital C*-algebra, $B \subset A$ a C*-subalgebra containing the unit of A, and $\pi : B \to \mathbb{B}(H)$ be a representation of B. Follow the steps below to prove the following:

Proposition 4.2.13 *In the above notation, there is a Hilbert space K and a representation $\rho : A \to K$ whose restriction $\rho|_B : B \to \mathbb{B}(K)$ splits into a direct sum of representations*

$$\rho(b) = \begin{bmatrix} \rho'(b) & 0 \\ 0 & \rho''(b) \end{bmatrix}$$

of B, where ρ' is unitarily equivalent to π.

a) Let ψ be a state of B, and $\pi_\psi : B \to \mathbb{B}(H_\psi)$ the associated GNS representation. Let φ be a state of A extending ψ, and $\pi_\varphi : A \to \mathbb{B}(H_\varphi)$ the corresponding GNS representation. Prove that the inclusion $B \to A$ induces a Hilbert space isometry $U : H_\psi \to H_\varphi$, that its image is invariant under $\pi_\varphi(B)$, and that $U\pi_\psi(b)U^*|_{U(H_\psi)} = \pi_\varphi(b)|_{U(H_\psi)}$ for all $b \in B$.

b) Prove that if $\pi : B \to \mathbb{B}(H)$ is a cyclic representation, then the conclusion of Proposition 4.2.13 holds.

c) Using Exercise 4.2.6 to deduce the Proposition for general representations π.

Exercise 4.2.14 Let A be a unital C*-algebra. Equip $M_2(A)$ with the obvious vector space structure, adjoint

$$\begin{bmatrix} a & b \\ c & d \end{bmatrix}^* := \begin{bmatrix} a^* & c^* \\ b^* & d^* \end{bmatrix},$$

and multiplication

$$\begin{bmatrix} a & b \\ c & d \end{bmatrix} \cdot \begin{bmatrix} a' & b' \\ c' & d' \end{bmatrix} := \begin{bmatrix} aa' + bc' & ab' + bd' \\ ca' + dc' & cb' + dd' \end{bmatrix}.$$

The result is a *-algebra.

Let $\pi : A \to \mathbb{B}(H)$ be a faithful representation of A on a Hilbert space H. Construct from this a faithful *-homomorphism $\bar{\pi} : M_2(A) \to \mathbb{B}(H \oplus H)$. We may define then

$$\left\| \begin{bmatrix} a & b \\ c & d \end{bmatrix} \right\| := \left\| \bar{\pi}\left(\begin{bmatrix} a & b \\ c & d \end{bmatrix} \right) \right\|,$$

where the norm on the right hand side is the norm in $\mathbb{B}(H \oplus H)$. Show that no nonzero element of $M_2(A)$ has zero norm, that $M_2(A)$ is already complete with respect to this norm, and that $M_2(A)$ (with the given norm) is a C*-algebra.

Remark 4.2.15 Since a C*-algebra can have only one C*-norm, the norm defined on $M_2(A)$ does not, in fact, depend on π.

Example 4.2.16 The following construction produces an injective representation of the inductive limit C*-algebra $U(2^\infty)$ of Example 1.7.9.

We will take $H = L^2([0, 1])$.

A standard argument used to prove the Bolzano–Weierstrass theorem proceeds as follows. Divide the interval $[0, 1]$, which we denote just by I, into $I_1 := [0, \frac{1}{2}]$ and $I_2 := L^2([\frac{1}{2}, 1])$. We can further subdivide I_1 into equal-length subintervals I_{11}, I_{12} and similarly subdivide I_2 into I_{21}, I_{22}. Continue in this way, for example dividing I_{21} (the interval $[\frac{1}{2}, \frac{3}{4}]$) into I_{211} (the interval $[\frac{1}{2}, \frac{5}{8}]$) and I_{212} (the interval $[\frac{5}{8}, \frac{3}{4}]$).

We obtain a family of intervals I_μ, where μ ranges over all finite sequences of 1's and 2's.

Now let μ and ν be distinct sequences of equal length. Then the intervals I_μ and I_ν are disjoint, and of the same length, and hence same measure. There is then a canonical Hilbert space isometry (c.f. Exercise 2.2.10) $s_{\nu,\mu} \colon L^2(I_\nu) \to L^2(I_\mu)$ induced by translating the one interval onto the other. Viewing $L^2(I_\nu)$ and so on to be closed subspaces of $L^2([0, 1])$, we can then extend $s_{\nu,\mu}$ to be zero on the orthogonal complement of the closed subspace $L^2(I_\nu)$ to get a partial isometry $s_{\mu,\nu} \colon L^2([0, 1]) \to L^2([0, 1])$.

Now we may parameterize a basis for $\mathbb{C}^{2^n}$ by finite sequences of 1's and 2's of length n, since the number of such sequences is 2^n. If μ and ν are such sequences (of the same length), let $E_{\mu,\nu} \in M_{2^n}(\mathbb{C})$ be the corresponding matrix unit, and let $E_{\mu,\nu}$ act on $L^2([0, 1])$ by $s_{\mu,\nu}$.

Exercise 4.2.17 Prove that this defines an injective representation of $U(2^\infty)$ as bounded operators on the Hilbert space $L^2([0, 1])$.

In other words, $U(2^\infty)$ is isomorphic to the C*-subalgebra of $\mathbb{B}(L^2([0, 1]))$ generated by the partial isometries $s_{\mu,\nu} \colon L^2([0, 1]) \to L^2([0, 1])$, where μ and ν range over all pairs of finite sequences of the same length.

The same idea produces a representation of $U(n^\infty)$ for any n.

Exercise 4.2.18 Prove that the universal UHF algebra $\mathcal{N}$ of Example 1.7.10 can be realized as a C*-subalgebra of bounded operators on $L^2([0, 1])$ in the following way. If n, k, l are natural numbers with $0 \le k, l < n$, let

$$s_{n;k,l} \colon L^2([0, 1]) \to L^2([0, 1])$$

be the partial isometry which is zero on the orthogonal complement of the closed subspace $L^2([\frac{l}{n}, \frac{l+1}{n}])$, and which maps $L^2([\frac{l}{n}, \frac{l+1}{n}])$ isomorphically onto

$L^2([\frac{k}{n}, \frac{k+1}{n}]$ by the obvious map, induced by translating the one interval onto the other.

Prove that $\mathcal{N}$ is isomorphic to the C*-algebra generated by the $s_{n;k,l}$.

We close with some general remarks on containment and weak containment of representations.

In general, a representation of a C*-algebra need not have any nontrivial representations contained in it as summands. So a general representation does not decompose into a direct sum of irreducible representations, as with representations of compact groups.

Indeed, if X is compact and μ a Borel measure with full support, then the representation $M : f \mapsto M_f$ of $C(X)$ by multiplication operators on $L^2(X, \mu)$ contains the representation π_x of point evaluation $f \mapsto f(x)$ for some $x \in X$, if and only if x is an atom for the measure.

How are the representations on $L^2(X, \mu)$ and on the Hilbert space direct sum $\oplus_{x \in X} \mathbb{C}$ of the point evaluation representations π_x related? Notably, they give the same norm on $C(X)$: since $\|M_f\| = \sup_{x \in X} |f(x)|$ if μ is a Borel measure with full support.

The right notion is that of *weak containment* of representations.

Definition 4.2.19 If π, ρ are representations of a C*-algebra A, we say $\pi \leq \rho$ if $\|\pi(a)\| \leq \|\rho(a)\|$ for all $a \in A$.

If $x \in X$, then $\pi_x \leq M$, in the notation above.

In this book the author periodically uses the term *spectrum* $\widehat{A}$ of a C*-algebra A to refer to the space of unitary equivalence classes of irreducible representations of A. We do not discuss it in any kind of systematic way. In this topology, a representation π determines the closed set of (classes of) irreducible representations ρ which are weakly contained in π.

Exercise 4.2.20 Show that if μ is a Borel measure on X, $M: C(X) \to \mathbb{B}(L^2(X, \mu))$ is the representation by multiplication operators, and $\pi_x: C(X) \to \mathbb{C}$ is the representation of $C(X)$ by point evaluation at x, then $\pi_x \leq M$ if and only if $x \in \text{supp}(\mu)$.

Exercise 4.2.21 Let $A_\hbar = C(\mathbb{T}) \rtimes_\hbar \mathbb{Z}$ be the irrational rotation algebra. If $x \in \mathbb{T}$, $C(\mathbb{T})$ is naturally represented on $l^2(\mathbb{Z}x)$, with $\mathbb{Z}_x$ the orbit of x. The group $\mathbb{Z}$ acts by shifting:

a) Verify that these constitute a covariant pair and give an *irreducible* representation π_x of $A_\hbar$.

b) Show that π_x and π_y are unitarily equivalent irreducible representations if x and y are in the same orbit of the $\mathbb{Z}$-action. Thus, one gets a map $\mathbb{Z}\backslash\mathbb{T} \to \widehat{A_\hbar}$ from the space of orbits into the "spectrum" of $A_\hbar$.

The issue of comparing norms from different representations of *-algebras and C*-algebras is discussed further in the next section.

4.3   Generalities About Completions of *-Algebras

Many C*-algebras are defined by an algebraic construction followed by a "completion" procedure. The algebraic construction makes a *-algebra. The completion procedure completes it to a C*-algebra. C*-algebras are themselves "rigid"—a C*-algebra has a unique norm with respect to which it is a C*-algebra. But a given *-algebra may be endowed with many different submultiplicative norms satisfying the C*-norm axioms (except completeness, of course). Each gives rise to a potentially different C*-algebra, by completion.

Definition 4.3.1 A *pre-C*-algebra* will mean a *-algebra $\mathcal{A}$ which is equipped with a semi-norm $\|\cdot\| \colon \mathcal{A} \to [0, \infty)$ satisfying $\|ab\| \le \|a\| \cdot \|b\|$ for all $a, b \in A$, and $\|a^*a\| = \|a\|^2$ for all $a \in \mathcal{A}$.

Exercise 4.3.2 Show that if $(\mathcal{A}, \|\cdot\|)$ is a pre-C*-algebra, then the elements of norm zero in $\mathcal{A}$ form a *-ideal $\mathrm{Null}(\|\cdot\|)$. Furthermore, if we define

$$\|a + \mathrm{Null}(\|\cdot\|)\| := \inf\{\|a + b\| \mid \|b\| = 0\},$$

then this defines a *norm* (not just a semi-norm) on the *-algebra $\mathcal{A}/\mathrm{Null}(\|\cdot\|)$. Show that the quotient map $\mathcal{A} \to \mathcal{A}/\mathrm{Null}(\|\cdot\|)$ is isometric.

Remark 4.3.3 A typical (and in fact the general) way of getting a pre-C*-algebra structure on a *-algebra $\mathcal{A}$ would be to find a *-representation $\pi \colon \mathcal{A} \to \mathbb{B}(H)$ of $\mathcal{A}$ as bounded operators on a Hilbert space. We could then define a semi-norm $\|\cdot\|_\pi$ satisfying the C*-identity by setting

$$\|a\|_\pi := \|\pi(a)\|,$$

where the norm on the right hand side is the operator norm in $\mathbb{B}(H)$.

The Gelfand–Naimark–Segal Theorem combined with Proposition 4.3.7 below implies that every pre-C*-algebra structure on a given *-algebra arises in this way.

Definition 4.3.4 A *completion* of a pre-C*-algebra $(\mathcal{A}, \|\cdot\|)$ is a pair consisting of a C*-algebra A and an isometric *-homomorphism $\pi \colon \mathcal{A} \to A$ (with respect to the given semi-norm on $\mathcal{A}$ and the C*-algebra norm on A) which satisfies the following universal property. For any C*-algebra B and any contractive *-homomorphism $\varphi \colon \mathcal{A} \to B$, there is a unique *-homomorphism $\overline{\varphi} \colon A \to B$ such that $\overline{\varphi} \circ \pi = \varphi$.

Exercise 4.3.5 Assume the existence of a completion A of a pre-C*-algebra $(\mathcal{A}, \|\cdot\|)$. Suppose that B is a C*-algebra and $\varphi \colon \mathcal{A} \to B$ is an *isometric* *-homomorphism. Show that the induced map $\overline{\varphi} \colon A \to B$ is also isometric.

Exercise 4.3.6 Prove that if $\pi \colon \mathcal{A} \to A$ is a completion, then π has dense range.

Completions exist:

Proposition 4.3.7 *Every pre-C*-algebra has a completion; and if $\pi : \mathcal{A} \to A$ and $\pi' : \mathcal{A} \to A'$ are any two completions of $(\mathcal{A}, \|\cdot\|)$, then there is a unique C*-algebra isomorphism $\alpha : \mathcal{A} \to \mathcal{A}'$ such that $\pi' = \alpha \circ \pi$.*

Proof Let $(\mathcal{A}, \|\cdot\|)$ be as in the statement. The set of elements of $\mathcal{A}$ of norm zero form a *-ideal in $\mathcal{A}$. The quotient of $\mathcal{A}$ by this *-ideal then an algebra to which the semi-norm descends, giving a a norm satisfying the C*-identity. To avoid introducing new notation, we just assume that $\|\cdot\|$ was a norm to begin with.

As any metric space has a completion, we may apply this to the current situation, obtaining a metric space A into which $\mathcal{A}$ is embedded isometrically and densely. Any Lipschitz map from $\mathcal{A}$ into a complete metric space extends uniquely to A. In particular, the norm on $\mathcal{A}$ extends to a norm on A. The algebra operations on $\mathcal{A}$ similarly extend to A: For example addition by a fixed $a \in \mathcal{A}$ regarded as a map $\mathcal{A} \to A$ is isometric, hence Lipschitz, and so extends continuously to a map $A \to A$. The adjoint $*$ also extends, by the same arguments. Thus A is actually a C*-algebra.

If $\varphi : \mathcal{A} \to B$ is a contractive *-homomorphism to a C*-algebra B, then φ maps null vectors in $\mathcal{A}$ to zero and induces a (contractive) *-homomorphism $\mathcal{A}/\mathrm{Null}(\|\cdot\|) \to B$. Since a contractive map from a metric space into a complete metric space extends uniquely to the completion, φ extends uniquely to a contractive map $A \to B$. This extension is easily seen to be a *-homomorphism. $\square$

Exercise 4.3.8 Show, using the universal property of completions, that if $\pi : \mathcal{A} \to A$ is a C*-algebra completion of a pre-C*-algebra $(\mathcal{A}, \|\cdot\|)$, then $\ker(\pi) = \{a \in \mathcal{A} \mid \|a\| = 0\}$. *Hint.* Compare the completion of $(\mathcal{A}, \|\cdot\|)$ to the completion of the pre-C*-algebra $\mathcal{A}/\mathrm{Null}(\|\cdot\|)$.

Exercise 4.3.9 Let $\mathcal{A}$ be any dense *-subalgebra of $C(X)$, where X is a compact Hausdorff space. Prove that if $\pi : \mathcal{A} \to A$ is a completion of $\mathcal{A}$, then $\|\pi(f)\| = \||f|_F\|$ for some closed subset $F \subset X$. In particular, A is C*-isomorphic to $C(F)$.

With the language of completions, we revisit group C*-algebras in the following example.

Example 4.3.10 As we have seen, a unitary representation $\pi : G \to \mathrm{U}(H)$ of a discrete group G on a Hilbert space H induces a *-algebra homomorphism $\pi : \mathbb{C}[G] \to \mathrm{U}(H)$,

$$\pi\left(\sum_{g \in G} \lambda_g [g]\right) := \sum_{g \in G} \lambda_g \pi(g) \in \mathbb{B}(H),$$

and hence every unitary representation of G gives a a pre-C*-algebra $(\mathbb{C}G, \|\cdot\|_\pi)$ where $\|f\|_\pi := \|\pi(f)\|$, for $f \in \mathbb{C}[G]$, and a corresponding C*-algebra completion $C_\pi^* G$.

In this notation, $C_\lambda^*(G) = C^*(G)$ where $\lambda : G \to \mathrm{U}(l^2(G))$ is the left regular representation.

Exercise 4.3.11 Let $G = \mathbb{Z}$. Compute $C^*_\varepsilon(\mathbb{Z})$ where $\varepsilon\colon \mathbb{Z} \to \mathbb{T}$ is the trivial representation.

Exercise 4.3.12 Let $F \subset \mathbb{T}$ be a closed subset of the circle. Restriction of Laurent polynomials $f \in \mathbb{C}[z, z^{-1}]$ to F determines a completion $\mathbb{C}[z, z^{-1}] \to C(F)$ of the *-algebra of Laurent polynomials.

Using Fourier transform, we may regard this as giving a completion $\pi_F\colon \mathbb{C}[\mathbb{Z}] \to C(F)$ of the group algebra $\mathbb{C}[\mathbb{Z}]$ of the integers.

Prove that if F has an accumulation point, then π_F is injective on $\mathbb{C}[\mathbb{Z}]$.

The case of the group of integers has a special property (amenability), with regard to completions, not possessed by all discrete groups. If $F \subset F' \subset \mathbb{T}$ are two nested closed subsets of the circle, then $\|\pi(f)\|_F \leq \|f\|_{F'}$ for all $f \in \mathbb{C}[G]$. In particular, $\|f\|_F \leq \|f\|_{\mathbb{T}}$ for all $f \in \mathbb{C}[z, z^{-1}]$ (and it follows that the *-homomorphism $\pi_F\colon \mathbb{C}[z, z^{-1}] \to C(F)$ extends continuously to a C*-algebra homomorphism $\overline{\pi}_F\colon C(\mathbb{T}) \to C(F)$).

If $G = \mathbb{F}_n$ is a free group on $n \geq 2$ generators, then the *-algebra homomorphism $\varepsilon\colon \mathbb{C}[G] \to \mathbb{C}$ induced by the trivial representation does *not* extend continuously to $C^*_r(G)$ ($\mathbb{F}_n$ is not amenable, $n \geq 2$).

Example 4.3.13 Another important example where a completion procedure is used is in connection with direct limits. We have already discussed direct limits, especially limits in which the structure maps of the system are injective. We now discuss the general case in the language of pre-C*-algebras.

Assume that $\{\varphi_{ij}\colon A_j \to A_i\}_{i \geq j}$ is a direct system of C*-algebras and *-homomorphisms. We start with forming the algebraic limit $\mathcal{A} := \sqcup A_i / \sim$ where $\sim$ is the equivalence relation generated by identifying $a \in A_i$ with $\varphi_{ji}(a_i) \in A_j$ for any $i \leq j$. Let $\varphi_i\colon A_i \to \mathcal{A}$ be the evident maps of A_i into $\mathcal{A}$. Note that if one of the structure maps $\varphi_{ij}\colon A_j \to A_i$ is not injective, and maps an element $a \in A_j$ to $0 \in A_i$, then $\varphi_j(a) = 0 \in \mathcal{A}$.

As discussed previously, $\mathcal{A}$ has a natural *-algebra structure. As before, if $a \in A_i$, we set $\|\varphi_i(a)\| := \lim_{j \to \infty} \|\varphi_{ji}(a)\|$. The limit exists because it is a decreasing net of positive real numbers, because *-homomorphisms are automatically contractive, by Lemma 1.7.6. We therefore have a pre-C*-algebra $\mathcal{A}, \|\cdot\|$. The direct limit is defined to be the completion of this pre-C*-algebra.

Exercise 4.3.14 If $\varphi_i\colon A_i \to A := \varinjlim A_i$ is a directed system, prove that

$$\ker(\varphi_i) = \{a \in A_i \mid \lim_{j \to \infty} \|\varphi_{ji}(a)\| = 0\},$$

for all $i \in I$.

Exercise 4.3.15 Let $A_i := C([0, 1])$ for $i = 1, 2, \ldots$, and let $\varphi_{ij}(f)(x) = f(2^{j-i}x)$, for $i \geq j$ and $f \in A_j = C([0, 1])$. The *-homomorphisms in the system are not injective, since they are induced by non-surjective self-maps of the interval. Show that $\varinjlim A_i \cong \mathbb{C}$ by evaluation of functions at 0.

4.4 Ideals and Quotients of C*-Algebras

We have already discussed several instances where the Banach algebra structure on the quotient of a C*-algebra by a closed *-ideal makes the quotient into a C*-algebra (the Calkin algebra, the quotient of the Toeplitz algebra by the compact operator ideal in it). In this section we establish the general result: The quotient of a C*-algebra by a closed ideal is always a C*-algebra with the distance norm.

This relies on a technical result concerning existence of approximate units.

An *approximate* unit in a C*-algebra A is a net $(u_\lambda)_{\lambda \in \Lambda}$ in A such that $u_\lambda a \to a$ as $\lambda \to \infty$ for all $a \in A$. We are going to show that an approximate unit exists in any C*-algebra. In fact, one can put the index set for the net to be simply $\Lambda := \{a \in A_+ \mid \|a\| < 1\}$ and the net the "identity net," whose value at $a \in \Lambda$ is a itself, but the important point is that the ordering on the index set Λ needed to make a net is the usual ordering of positive elements $a \le b$ if and only if $b - a \ge 0$.

It is clear that this relation is reflexive, transitivity is an extremely easy exercise, but the upper bound requirement, making this a directed set, needs to be verified. Clearly since the sum of two positive elements is positive (Corollary 4.1.6), if $a, b \in A_+$, then $a+b$ is an upper bound for a, b, so all of A_+ is a directed set, but if a, b are contractions, it is of course not true that $a + b$ needs to be a contraction. However, the map

$$\alpha \colon A_+ \to \Lambda, \quad \alpha(a) := a(1+a)^{-1}$$

is an order isomorphism $A_+ \cong \Lambda$ with inverse $\beta \colon \Lambda \to A_+$, $\beta(a) := a(1-a)^{-1}$, and since the upper bound condition holds for the ordered set A_+, it does for Λ as well. To check these facts it helps to note that $a(1+a)^{-1} = 1 - (1+a)^{-1}$.

Exercise 4.4.1 If a and b are in Λ find an explicit formula for an upper bound of a, b in Λ.

Proposition 4.4.2 Λ *is an approximate unit for* A. *That is, if* $a \in A$ *and* $\varepsilon > 0$, *then there exists* $u_0 \in \Lambda$ *such that if* $u \in \Lambda$ *and* $u \ge u_0$, *then* $\|ua - a\| < \varepsilon$.

In the commutative case, the content of the proposition is as follows. Given $f \in C_0(X)$, and $\varepsilon > 0$, there exists $\rho_0 \in C_0(X)$ such that $0 \le \rho < 1$, and for all $\rho_0 \le \rho < 1$, $\|\rho f - f\| < \varepsilon$. It is clear, at any rate, if $f \in C_c(X)$, then one should take ρ_0 strictly less than, but very close to 1 on the support of f, which would certainly do the trick. An approximation argument extends this to where $f \in C_0(X)$, in which case the support need not be compact necessarily. Alternatively, one can argue that one can simply choose, depending on $\varepsilon > 0$, ρ_0 to be of the form $\mu f (1 + \mu f)^{-1}$, where μ is large. The relevant estimates are done in the proof below.

Proof If we can show that $ua \to a$ as $u \to \infty$ in Λ, for a positive and of norm < 1, then it follows that $ua \to a$ for any positive element. Since any self-adjoint can be written as a difference of two positive elements, $ua \to a$ for a self-adjoint.

The result for arbitrary a follows from writing a as a linear combination of two self-adjoints.

So assume that $a \in A_+$, $\|a\| < 1$ (that is, that $a \in \Lambda$). Choose $\varepsilon > 0$. Let $u_0 = \varepsilon^{-2}a(1 + \varepsilon^{-2}a)^{-1}$. Let $u \geq u_0$. Since $1 - u \leq 1$, $(1 - u)^2 \leq 1 - u$. Since $u \geq u_0$, $1 - u \leq 1 - u_0 = (1 + \varepsilon^{-2}a)^{-1}$. Putting these facts together gives

$$(1 - u)^2 \leq (1 + \varepsilon^{-2}a)^{-1}.$$

So we deduce

$$a(1 - u)^2 a \leq a(1 + \varepsilon^{-2}a)^{-1}a.$$

Since $a(1 + \varepsilon^{-2}a)^{-1}a \leq \varepsilon^2 a$, we get $a(1 - u)^2 a \leq \varepsilon^2 a$. Hence $\|a(1 - u)^2 a\| \leq \varepsilon^2\|a\| < \varepsilon^2$. By the C*-identity, on the other hand, $\|a(1 - u)^2 a\| = \|a(1 - u)\|^2$. Thus $\|a - au\| < \varepsilon$ as required. $\qquad\square$

Exercise 4.4.3 If H is any Hilbert space, it has an orthonormal basis $\mathcal{E}$ and the dimension of H is by definition the cardinality of $\mathcal{E}$. Make a directed set Λ consisting of all finite subsets of $\mathcal{E}$ under the subset relation, and let $(P_F)_{F \in \Lambda}$ be the net in which P_F is the projection to $\mathrm{span}(F)$ for any finite subset $F \subset \mathcal{E}$. Show that $(P_F)_{F \in \Lambda}$ is an approximate unit for $\mathcal{K}(H)$.

Exercise 4.4.4 Show that if A is separable, then A has an approximate unit which is an increasing sequence $u_1 \leq u_2 \leq \cdots$ of positive contractions in A. *Hint.* Let $a_1, a_2, \ldots$ be a countable dense set in A. Using Proposition 4.4.2, inductively construct (u_n) such that $0 \leq u_1 \leq u_2 \leq \cdots$ and $\|u_n a_i - a_i\| < \frac{1}{n}$ for $1 \leq i \leq n$.

A *closed ideal* in a C*-algebra is an algebraic ideal $J \subset A$ which is also closed in the norm topology. This implies by the following Lemma that J is also closed under adjoints.

Lemma 4.4.5 *If J is a closed ideal in a C*-algebra A, then J is automatically closed under adjoint (and hence is, in particular, a C*-subalgebra of A).*

Proof If $a \in J$, then a^*a, and hence the (non-unital) C*-algebra $C^*(a^*a)$ generated by a^*a is contained in J. Indeed, $C^*(a^*a)$ is the completion of the *-algebra of polynomials $\sum_{k=0}^n \lambda_k(a^*a)^k$ in J.

In particular, $C^*(a^*a)$ has an approximate unit (u_λ) consisting of positive contractions. If we can show that $u_\lambda a \to a$, it will follow, since the adjoint is continuous, that $u_\lambda a^* \to a^*$, as well. Since $u_\lambda a^* \in J$ for all λ, it will then follow that $a^* \in J$ since J is closed.

But using the C*-identity and the facts that u_λ is self-adjoint and a contraction, $\|au_\lambda - a\|^2 = \|a(u_\lambda - 1)\|^2 = \|(u_\lambda - 1)a^*a(1 - u_\lambda)\| \leq \|(u_\lambda - 1)a^*a\| \cdot \|u_\lambda - 1\| \leq \|(u_\lambda - 1)a^*a\| = \|u_\lambda a^*a - a^*a\|$. Since $u_\lambda a^*a \to a^*a$, we see that $u_\lambda a \to a$ as well, as required. $\qquad\square$

Exercise 4.4.6 A C*-algebra is *simple* if it has no nonzero, proper closed ideals. Prove that $M_n(\mathbb{C})$ is simple.

Exercise 4.4.7 If X is a locally compact Hausdorff space and $F \subset X$ is a closed subset, then show that $J_F := \{f \in C_0(X) \mid f_{|F} = 0\}$ is a closed ideal in $C_0(X)$. It is a standard result of basic analysis that all closed ideals in $C_0(X)$ arise as J_F for some F. That is, closed ideals in $C_0(X)$ are in 1–1 correspondence with closed subsets of X.

Example 4.4.8 The results of Section 1.6 show that the C*-subalgebra $\mathcal{K}(H)$ of $\mathbb{B}(H)$ is a closed ideal.

Exercise 4.4.9 Prove that $\mathcal{K}(H)$ is the *unique* closed ideal in $\mathbb{B}(H)$.

Let $J \subset A$ be a closed ideal. The quotient vector space A/J has an evident structure of a *-algebra. We equip it with the quotient norm

$$\|a + J\| := \inf\{\|a - b\| \mid b \in J\}.$$

Since J is assumed closed, A/J is complete as a normed vector space, by an easy exercise.

Lemma 4.4.10 *The cosets in A/J with the quotient norm form a Banach algebra.*

Proof Indeed,

$$\|ab + J\| := \inf\{\|ab - c\| \mid c \in J\} \le \inf\{\|ab - (c_1 b - c_1 c_2 + ac_2)\| \mid c_1, c_2 \in J\}$$

$$(4.3) \qquad = \inf\{\|(a - c_1)(b - c_2)\| \mid c_i \in J\} \le \inf\{\|a - c_1\| \cdot \|b - c_2\| \mid c_i \in J\}$$

$$\le \inf\{\|a - c_1\| \mid c_1 \in J\} \cdot \inf\{\|b - c_2\| \mid c_2 \in J\}.$$

$\square$

We have already met one example: that of the Calkin algebra $Q(H) := \mathbb{B}(H)/\mathcal{K}(H)$. We proved in Chapter 1 that the C*-identity holds for the quotient norm; we now prove this result in general.

Lemma 4.4.11 *If $J \subset A$ is a closed ideal and (u_λ) is an approximate unit of positive contractions in J, then $\|a + J\| = \lim_{\lambda \to \infty} \|a u_\lambda - a\|$ for all $a \in A$.*

Proof If $b \in J$, then

$$\|a - a u_\lambda\| = \|a + b - b + b u_\lambda - b u_\lambda - a u_\lambda\|$$

$$= \|a + b - (a + b)u_\lambda + b u_\lambda - b\|$$

$$\le \|a + b - (a + b)u_\lambda\| + \|b u_\lambda - b\|$$

$$= \|(a + b)(1 - u_\lambda)\| + \|b u_\lambda - b\|$$

$$\le \|a + b\| + \|b u_\lambda - b\|.$$

Therefore $\limsup_{\lambda \to \infty} \|a - au_\lambda\| \leq \|a + b\| \leq \|a + J\|$. On the other hand $\|a + J\| \leq \|a - au_\lambda\|$ for all λ so $\|a + J\| \leq \liminf_{\lambda \to \infty} \|a - au_\lambda\|$. The result follows immediately.   □

Corollary 4.4.12 *The Banach *-algebra A/J with the quotient norm satisfies the C*-identity, and hence is a C*-algebra.*

Proof If $a \in A$, then by the lemma,

$$\|(a + J)^*(a + J)\| = \|a^*a + J\| = \lim_{\lambda \to \infty} \|a^*au_\lambda - a^*a\| = \lim_{\lambda \to \infty} \|a^*a(u_\lambda - 1)\|$$

$$\geq \lim_{\lambda \to \infty} \|(u_\lambda - 1)a^*a(u_\lambda - 1)\| = \lim_{\lambda \to \infty} \|a(u_\lambda - 1)\|^2 = \lim_{\lambda \to \infty} \|au_\lambda - a\|^2 = \|a + J\|^2.$$

The other inequality $\|(a + J)^*(a + J)\| \leq \|a^* + J\|\|a + J\| = \|a + J\|^2$ follows from the fact that A/J is a Banach *-algebra with the quotient norm.   □

Example 4.4.13 If $F \subset X$, F closed, X locally compact Hausdorff,

$$J_F = \{f \in C_0(X) \mid f|_F = 0\}$$

the closed ideal of $C_0(X)$ of Exercise 4.4.7, then the C*-algebra quotient $C_0(X)/J_F$ is naturally isomorphic to $C_0(F)$, by the homomorphism $f + J_F \mapsto f|_F$, that is, by the map of restriction of functions to F.

Corollary 4.4.14 *If $\alpha\colon A \to B$ is a *-homomorphism between C*-algebras, then the image of α is closed and hence is a C*-subalgebra of B.*

Proof The kernel $\ker(\alpha)$ is a closed ideal in A. By standard algebra, α induces an injective *-homomorphism $\dot{\alpha}\colon A/\ker(\alpha) \to B$. Since it is injective, it is isometric, and hence its range is closed in B.   □

Exercise 4.4.15 Show that if a unital C*-algebra A has no nontrivial *algebraic* ideals, then it has no *closed* ideals either.

Exercise 4.4.16 Prove that $C_0(\mathbb{R})$ is a closed ideal in the C*-algebras $C_b(\mathbb{R})$ of bounded continuous functions on $\mathbb{R}$, $C(\eta\mathbb{R})$ the C*-algebra of functions of vanishing variation on $\mathbb{R}$ (see Exercise 3.4.10), and $C_u(\mathbb{R})$, bounded uniformly continuous functions on $\mathbb{R}$.

Deduce that $\mathbb{R}$ embeds as an open, dense subset of each of $\beta\mathbb{R}$, $\eta\mathbb{R}$ and $\overline{\mathbb{R}}^u$. The term "compactification" is generally used for a compact space containing $\mathbb{R}$ as an open, dense subset.

4.5 Tensor Products of C*-Algebras

If A is any C*-algebra, and n any positive integer, then the *-algebra $M_n(A)$ of n-by-n matrices with entries in A is a C*-algebra. It is an example of a tensor product: In this case, it is the tensor product $A \otimes M_n(\mathbb{C})$. Another example is the C*-algebra $C(X \times Y)$, with X, Y compact Hausdorff: This C*-algebra turns out to agree with the tensor product $C(X) \otimes C(Y)$—tensor product is Gelfand dual to product of topological spaces.

We start with tensor products of vector spaces.

Let V_1 and V_2 be vector spaces over $\mathbb{C}$. Their *tensor product* is a vector space denoted $V_1 \otimes V_2$ equipped with a bilinear map $V_1 \times V_2 \to V_1 \otimes V_2$, usually denoted simply $(v_1, v_2) \mapsto v_1 \otimes v_2$, and satisfying the following universal property: If $f : V_1 \times V_2 \to W$ is any bilinear map to a vector space W, then there is a unique *linear* map $\bar{f} : V_1 \otimes V_2 \to W$ such that $\bar{f}(v_1 \otimes v_2) = f(v_1, v_2)$.

In other words, the diagram

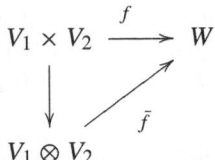

can be made to commute, by a unique *linear* $\bar{f}$ and initial *bilinear* f.

Assuming the existence of such an object, notice that the assumption that $(v_1, v_2) \mapsto v_1 \otimes v_2$ is bilinear means that

$$(4.4) \qquad (\lambda v_1) \otimes v_2 = v_1 \otimes (\lambda v_2) = \lambda (v_1 \otimes v_2),$$

for all scalars λ, vectors v_1, v_2, and that

$$(4.5) \quad (v_1 + v_1') \otimes v_2 = v_1 \otimes v_2 + v_1' \otimes v_2 \text{ and } v_1 \otimes (v_2 + v_2') = v_1 \otimes v_2 + v_1 \otimes v_2'.$$

To prove existence, one can construct $V_1 \otimes V_2$ in the following way. Consider initially the free vector space with basis the elements of $V_1 \times V_2$. This, by definition, is the vector space consisting of all formal, finite linear combinations of elements of $V_1 \times V_2$. Then let $V_1 \otimes V_2$ denote the quotient, in the category of vector spaces, of this vector space, by the subspace spanned by the vectors

$$(\lambda v_1, v_2) - \lambda(v_1, v_2), \quad (v_1, \lambda v_2) - \lambda(v_1, v_2)$$

and the vectors

$$(v_1 + v_1', v_2) - (v_1, v_2) - (v_1', v_2), \quad (v_1, v_2 + v_2') - (v_1, v_2) - (v_1, v_2').$$

If one denotes the equivalence class of the basis vector (v_1, v_2) by $v_1 \otimes v_2$, then the symbols $v_1 \otimes v_2$ clearly span $V_1 \otimes V_2$, satisfy the bilinearity relations (4.4) and (4.5), and it can be easily checked that $V_1 \otimes V_2$, together with the (bilinear) quotient map $(v_1, v_2) \mapsto v_1 \otimes v_2$, satisfies the required universal property.

We record a basic consequence of the universal property of a tensor product.

Lemma 4.5.1 *If $T: V_1 \to W_1$ and $T_2: V_2 \to W_2$ are two linear maps, then there is a unique linear map*

$$T_1 \otimes T_2: V_1 \otimes V_2 \to W_1 \otimes W_2$$

such that $(T_1 \otimes T_2)(v_1 \otimes v_2) = T_1 v_1 \otimes T_2 v_2$.

Exercise 4.5.2 Show that if $w_1, \ldots, w_n$ is a basis for W, then $V \otimes W \cong W \oplus \cdots \oplus W$ under the map

$$v \otimes w \mapsto (\lambda_1(w)v, \ldots, \lambda_n(w)v),$$

where $\lambda_1(w), \ldots, \lambda_n(w)$ are the coefficients of w with respect to the basis.

Show that in the notation of Lemma 4.5.1, $T \otimes 1_W \cong T \oplus T \cdots \oplus T$ with respect to this decomposition, for any $T \in \mathrm{End}(V)$, where here 1_W denotes the identity operator on W.

Exercise 4.5.3 Although by definition every element of a tensor product $V_1 \otimes V_2$ of vector spaces can be written in the form $\sum_i v_i \otimes w_i$ for some $v_1, \ldots, v_n \in V_1$ and $w_1, \ldots, w_n \in V_2$, it is less clear when two such expressions are equal in $V_1 \otimes V_2$. The following exercise helps with this:

a) If V_1, V_2 are vector spaces, show that any element of $V_1 \otimes V_2$ can be written in the form $\sum_i v_i \otimes w_i$ where the w_i are *linearly independent*. (*Hint*. Prove it by induction on n where $v = \sum_{i=1}^n v_i \otimes w_i$.)

b) Prove that if $w_1, \ldots, w_n$ are linearly independent vectors in V_1 and $v_1, \ldots, v_n$ and $v_1', \ldots, v_n'$ are arbitrary vectors in V_2, then $\sum_i v_i \otimes w_i = \sum_i v_i' \otimes w_i$ implies $v_i = v_i'$ for all i. (*Hint*. Think about applying maps $f \otimes 1: V_1 \otimes V_2 \to V_2$, where $f \in V_1^*$ is appropriately chosen.)

c) Prove that if $e_1, \ldots, e_n$ is a basis for V_1 and $f_1, \ldots, f_m$ a basis for V_2, then $\{e_i \otimes f_j \mid i = 1, \ldots m, j = 1, \ldots, m\}$ is a basis for $V_1 \otimes V_2$. Deduce that $\mathbb{C}^n \otimes \mathbb{C}^m \cong \mathbb{C}^{nm}$. Thus if V_1 and V_2 are finite-dimensional vector spaces, then $\dim(V_1 \otimes V_2) = \dim(V_1) \dim(V_2)$.

Now let A and B be a pair of algebras; their tensor product $A \otimes B$ in the category of vector spaces has a natural structure of an algebra with

$$\left(\sum_i a_i \otimes b_i \right) \cdot \left(\sum_i c_i \otimes d_i \right) := \sum_{i,j} a_i c_i \otimes b_i d_j.$$

In particular, if $\text{End}(V_i)$ denotes as usual the linear maps $V_i \to V_i$, $i = 1, 2$, then $\text{End}(V_i)$ are algebras, and so $\text{End}(V_1) \otimes \text{End}(V_2)$ is an algebra.

Now, according to Lemma 4.5.1, for any $T_1 \in \text{End}(V_1)$ and $T_2 \in \text{End}(V_2)$ there is a unique linear map $T_1 \otimes T_2 \in \text{End}(V_1 \otimes V_2)$ such that $(T_1 \otimes T_2)(v_1 \otimes v_2) = T_1 v_1 \otimes T_2 v_2$. This produces a map

$$\text{End}(V_1) \times \text{End}(V_2) \to \text{End}(V_1 \otimes V_2)$$

mapping a pair (T_1, T_2) to $T_1 \otimes T_2$. It is easy to check that this map is bilinear. Hence by the universal property it determines a linear map

(4.6) $$\text{End}(V_1) \otimes \text{End}(V_2) \to \text{End}(V_1 \otimes V_2),$$

mapping the tensor $T \otimes S$ to the corresponding endomorphism. This map is easily checked to be an algebra homomorphism.

Lemma 4.5.4 *The algebra homomorphism* (4.6) *is injective; it is surjective if V_1 and V_2 are both finite-dimensional.*

Proof Let $\sum T_i \otimes S_i \in \text{End}(V_1) \otimes \text{End}(V_2)$, and suppose that its action on $V_1 \otimes V_2$ is the zero operator. By Exercise 4.5.3a), we may assume without loss of generality that the T_i are linearly independent in $\text{End}(V_1)$. Now the hypothesis implies that

(4.7) $$\sum_i \langle T_i v, v' \rangle \langle S_i w, w' \rangle = 0$$

for all $v, v' \in V_1$, $w, w' \in V_2$, from which it is immediate that the operator $L := \sum \langle S_i w, w' \rangle T_i$ is the zero operator, since $\langle Lv, v' \rangle$ is equal to the expression (4.7) and so is zero, for all v, v'. The linear independence of the T_i now implies that each $\langle S_i w, w' \rangle = 0$. As w, w' were arbitrary vectors in V_2, each $S_i = 0$.

Surjectivity of (4.6) holds for dimension reasons when V_1 and V_2 are finite-dimensional, since

$$\dim \text{End}(V_1 \otimes V_2) = \big(\dim(V_1) \dim(V_2)\big)^2 = \dim(V_1)^2 \dim(V_2)^2$$
$$= \dim \text{End}(V_1) \dim \text{End}(V_2)$$
$$= \dim \big(\text{End}(V_1) \otimes \text{End}(V_2)\big).$$

We leave the converse to the reader. $\qquad\square$

Now let A and B be a pair of algebras; their tensor product $A \otimes B$ in the category of vector spaces has a natural structure of an algebra as well, with

$$\big(\sum_i a_i \otimes b_i\big) \cdot \big(\sum_i c_i \otimes d_i\big) := \sum_{i,j} a_i c_i \otimes b_i d_j.$$

If A, B are *-algebras, then so is $A \otimes B$, using

$$\left(\sum a_i \otimes b_i\right)^* := \sum a_i^* \otimes b_i^*.$$

We call $A \otimes B$ with this (*-)algebra structure the tensor product in the category of (*)-algebras.

Example 4.5.5 Let A be a C*-algebra. Then the tensor product $A \otimes M_n(\mathbb{C})$ in the category of *-algebras is isomorphic to the *-algebra $M_n(A)$ of n-by-n matrices with entries in A (c.f. Exercise 4.2.14). The map sends $a \otimes T \in A \otimes M_2(\mathbb{C})$ to the matrix $(aT)_{ij} := aT_{ij}$.

Exercise 4.5.6 Prove that the map just described is an isomorphism of *-algebras $A \otimes M_n(\mathbb{C}) \cong M_n(A)$.

Exercise 4.5.7 Prove that $M_n(\mathbb{C}) \otimes M_m(\mathbb{C}) \cong M_{nm}(\mathbb{C})$ for any positive integers n, m.

Exercise 4.5.8 Prove that if A, A', B, B' are algebras (or *-algebras), $\alpha : A \to A'$ an algebra (or *-algebra) homomorphism, $\beta : B \to B'$ another, then there is a unique (*-)algebra homomorphism $\alpha \otimes \beta : A \otimes B \to A' \otimes B'$ mapping $a \otimes b$ to $\alpha(a) \otimes \beta(b)$. Prove furthermore that $\alpha \otimes \beta$ is injective if α and β are each injective. (*Hint*. To prove injectivity, adapt the proof of Lemma 4.5.4, using more general linear functionals than the inner product functions $S \mapsto \langle Sv, v' \rangle$ used there.)

Exercise 4.5.9 Let A, B be two *-algebras, and let $\pi : A \to \text{End}(H)$, $\rho : B \to \text{End}(K)$ be two representations of A, B as linear operators on vector spaces H, K. Prove that there is a unique representation

$$\pi \otimes \rho : A \otimes B \to \text{End}(H \otimes K),$$

such that

$$(\pi \otimes \rho)(a \otimes b) = \pi(a) \otimes \rho(b) \in \text{End}(H) \otimes \text{End}(K) \subset \text{End}(H \otimes K)$$

for all $a \in A, b \in B$. Prove that $\pi \otimes \rho$ is injective if π and ρ are each injective.

Tensor products in the category of Hilbert spaces

If V_1 and V_2 are Hilbert spaces, their tensor product $V_1 \otimes V_2$ can be made into a pre-Hilbert space by setting

$$(4.8) \qquad \langle v_1 \otimes v_2, v_1' \otimes v_2' \rangle := \langle v_1, v_1' \rangle \langle v_2, v_2' \rangle.$$

By the universal property of tensor products, it determines a corresponding sesquilinear form $V_1 \otimes V_2 \times V_1 \otimes V_2 \to \mathbb{C}$.

Lemma 4.5.10 *The sesquilinear form on $V_1 \otimes V_2$ determined by (4.8) is nondegenerate.*

Proof Indeed, suppose otherwise that there is an element x of $V_1 \otimes V_2$ such that $\langle x, y \rangle = 0$ for all y. We may write $x = \sum v_i \otimes w_i$ where the w_i's are linearly independent. The assumption then implies that

$$\langle v \otimes w, \sum_i v_i \otimes w_i \rangle = 0$$

for all vectors $v \in V_1$, $w \in V_2$. We may rewrite this in the form

$$\sum \langle w, \langle v, v_i \rangle w_i \rangle = 0, \ \forall v \in V_1, \ w \in V_2.$$

In particular, for every $v \in V_1$, the linear functional

$$w \mapsto \langle w, \sum \langle v, v_i \rangle w_i \rangle$$

is the zero linear functional on V_2. Since it is inner product with the vector $\sum \langle v, v_i \rangle w_i$, we conclude that this latter vector is zero in V_2. Since the w_i's were assumed linearly independent, $\langle v, v_i \rangle = 0$ for all i. Finally, since v was arbitrary, we get that $v_i = 0$ for all i. Hence $x = 0$. $\qquad\square$

Definition 4.5.11 The t*ensor product* $V_1 \otimes V_2$ of two Hilbert spaces V_1 and V_2 is defined to be the completion of the vector space tensor product of V_1 and V_2 with respect to the norm

$$\left\| \sum v_i \otimes w_i \right\|^2 := \sum_{i,j} \langle v_i, v_j \rangle \langle w_i, w_j \rangle.$$

In particular, $\|v \otimes w\| = \|v\| \cdot \|w\|$ for all vectors $v \in H$, $w \in K$.

The inner product $\langle \cdot, \cdot \rangle$ extends to the completion, so that the completion (still denoted $V_1 \otimes V_2$) is a Hilbert space.

Of course the tensor product *in the category of Hilbert spaces* involves a completion and is *not* the same as the tensor product in the category of vector spaces. However, we will use the same notation, assuming the context makes it clear which we are using.

If H and K are Hilbert spaces, $H \otimes K$, unless otherwise specified, refers to the *Hilbert space* completion of H and K.

Exercise 4.5.12 Prove that if $\{e_i\}_{i \in I}$ is an orthonormal basis for H and if $\{e'_j\}_{j \in I}$ is an orthonormal basis for K, then $\{e_i \otimes e'_j\}_{i \in I, j \in J}$ is an orthonormal basis for $H \otimes K$.

Lemma 4.5.13 *Let H and K be Hilbert spaces, $T \in \mathbb{B}(H)$, $S \in \mathbb{B}(K)$ bounded linear operators. Then there is a unique bounded linear operator $T \otimes S \colon H \otimes K \to H \otimes K$ such that*

$$(T \otimes S)(v \otimes w) = Tv \otimes Sw.$$

Furthermore, $\|T \otimes S\| = \|T\| \cdot \|S\|$, *and* $(T \otimes S)^* = T^* \otimes S^*$.

Proof Start off with the tensor product of H and K in the category of *vector spaces*. By Lemma 4.5.1, there is a unique linear endomorphism $T \otimes S$ on this vector space, such that $(T \otimes S)(v \otimes w) = T(v) \otimes S(w)$. In order to show that $T \otimes S$ extends to a bounded linear operator on the completion, i.e., the Hilbert space tensor product, we show that

$$\left\| \sum_{i=1}^{n} Tv_i \otimes Sw_i \right\| \leq \|T\| \cdot \|S\| \cdot \left\| \sum v_i \otimes w_i \right\|$$

for each finite collection of vectors $v_1, \ldots, v_n \in H$, $w_1, \ldots, w_n \in K$.

We show this for $S = \mathrm{id}_K$. The case $T = \mathrm{id}_H$ is similar, and together, these two partial results imply the result desired.

We can find a finite orthonormal basis $e_1, \ldots, e_m$ for the span of $w_1, \ldots, w_n$, and in this way rewrite the vector $\sum v_i \otimes w_i$ in the form $\sum u_i \otimes e_i$ for some collection of vectors $u_1, \ldots, u_m$. We compute

$$\|(T \otimes 1)(\sum u_i \otimes e_i)\|^2 = \left\| \sum Tu_i \otimes e_i \right\|^2 = \sum_{i,j} \langle Tu_i \otimes e_i, Tu_j \otimes e_j \rangle$$

$$= \sum_{i,j} \langle Tu_i, Tu_j \rangle \langle e_i, e_j \rangle$$

$$= \sum_{i} \|Tu_i\|^2 \leq \|T\|^2 \cdot \sum \|u_i\|^2$$

$$\text{(4.9)} \qquad = \|T\|^2 \cdot \left\| \sum u_i \otimes e_i \right\|^2 = \|T\|^2 \cdot \left\| \sum v_i \otimes w_i \right\|^2.$$

This proves the claim and also proves, as sketched above, that $T \otimes S$ is bounded and hence extends continuously to the Hilbert space tensor product $H \otimes K$ to itself, and, furthermore, that $\|T \otimes S\| \leq \|T\| \cdot \|S\|$.

If $\varepsilon > 0$, then we can find a unit vector $\xi \in H$ and a unit vector $\eta \in K$ so that $\|T\xi\| \geq \|T\| - \varepsilon$, $\|S\eta\| \geq \|S\| - \varepsilon$. Then the unit vector $\xi \otimes \eta$ satisfies

$$\|(T \otimes S)(\xi \otimes \eta)\| = \|T\xi \otimes S\eta\| = \|T\xi\|\|S\eta\| \geq (\|T\| - \varepsilon)(\|S\| - \varepsilon),$$

whence $\|T \otimes S\| \geq \|T\|\|S\|$ follows by letting $\varepsilon \to 0$.

The statement about the adjoints is left to the reader to prove. $\square$

Exercise 4.5.14 Prove directly using the definitions that if $T \geq 0$ is a positive operator on H such that

$$\langle Tv, v \rangle \geq \lambda \|v\|^2$$

for all nonzero vectors $v \in H$, then

$$\langle (T \otimes 1)x, x \rangle \geq \lambda \|x\|^2$$

for all nonzero vectors $x \in H \otimes K$. (*Hint.* Start by proving it for x in the algebraic tensor product, and write such an x in the form $x = \sum v_i \otimes w_i$, where the w_i's are orthonormal vectors.)

Exercise 4.5.15 Let A and B be self-adjoint operators on Hilbert spaces H, K. Prove that

$$\text{Spec}(A \otimes 1 + 1 \otimes B) = \text{Spec}(A) + \text{Spec}(B)$$

and that

$$\text{Spec}(A \otimes B) = \text{Spec}(A) \cdot \text{Spec}(B).$$

The first set $\text{Spec}(A) + \text{Spec}(B)$ refers to all sums $\lambda + \mu$, with $\lambda \in \text{Spec}(A)$, $\mu \in \text{Spec}(B)$.

Tensor products in the category of C-algebras*

If A and B are C*-algebras, we may complete their algebraic tensor product, that is, the tensor product in the category of *-algebras, which we, for the moment, denote by $A \otimes_{\text{alg}} B$, to a C*-algebra $A \otimes B$, using the following method.

Let $\pi \colon A \to \mathbb{B}(H)$ be a representation of A, $\rho \colon B \to \mathbb{B}(K)$ a representation of B. By Exercise 4.5.8 they combine to give a *-algebra homomorphism $\pi \otimes \rho \colon A \otimes_{\text{alg}} B \to \mathbb{B}(H) \otimes_{\text{alg}} \mathbb{B}(K)$. By Lemma 4.5.13, the algebraic tensor product $\mathbb{B}(H) \otimes_{\text{alg}} \mathbb{B}(K)$ can be viewed as a *-subalgebra of $\mathbb{B}(H \otimes K)$. Therefore we obtain a representation $\pi \otimes \rho$ of $A \otimes_{\text{alg}} B$ on $H \otimes K$ mapping $a \otimes b$ to the bounded operator $\pi(a) \otimes \rho(b)$.

Definition 4.5.16 The *minimal (or spatial) tensor product* $A \otimes B$ of two C*-algebras A and B is the completion of their tensor product $A \otimes_{\text{alg}} B$ in the category of *-algebras, with respect to the norm

$$(4.10) \qquad \left\| \sum a_i \otimes b_i \right\| := \sup_{\pi, \rho} \left\| \sum a_i \otimes b_i \right\|_{\pi, \rho} := \left\| \sum \pi(a_i) \otimes \rho(b_i) \right\|,$$

where the supremum is taken over all representations π, ρ of A, B.

The minimal tensor product, although it may feel somewhat inexplicit, has the advantage of satisfying the following important universal property.

Proposition 4.5.17 *If A_1, A_2, B_1, B_2 are C*-algebras, and if $\alpha\colon A_1 \to B_1$ and $\alpha_2\colon A_2 \to B_2$ are *-homomorphisms, then there is a unique *-homomorphism*

$$\alpha_1 \otimes \alpha_2 \colon A_1 \otimes A_2 \to B_1 \otimes B_2$$

such that $(\alpha_1 \otimes \alpha_2)(a_1 \otimes a_2) = \alpha_1(a_1) \otimes \alpha_2(a_2)$.

Proof The maps α_i combine to a *-homomorphism $\alpha_1 \otimes \alpha_2 \colon A_1 \otimes_{\text{alg}} A_2 \to B_1 \otimes_{\text{alg}} B_2$, and we need to show that it extends continuously to the completions, i.e., that it is contractive with respect to the spatial tensor product norms, defined in (4.10). Let π_i be representations of B_i. Then $\pi_i \circ \alpha_i$ are representations of A_i. It is immediate then that if $x \in A_1 \otimes_{\text{alg}} A_2$ then

$$\|(\pi_1 \otimes \pi_2)\big((\alpha_1 \otimes \alpha_2)(x)\big)\| = \|(\pi \circ \alpha_1 \otimes \pi_2 \circ \alpha_2)(x)\| \leq \|x\|.$$

Taking sups over all π_1, π_2 gives that

$$\|(\alpha_1 \otimes \alpha_2)(x)\| \leq \|x\|$$

as required. $\qquad\square$

Exercise 4.5.18 Suppose that $\pi \sim_u \pi'$ are unitarily equivalent representations of A and that $\rho \sim_v \rho'$ are unitarily equivalent representations of B. Prove that the representations $\pi \otimes \rho$ and $\pi' \otimes \rho'$ of $A \otimes_{\text{alg}} B$ are also unitarily equivalent. Deduce that $\|\cdot\|_{\pi,\rho} = \|\cdot\|_{\pi',\rho'}$.

Exercise 4.5.19 If A is a *-algebra and π, ρ are two representations of A as bounded operators on Hilbert spaces H, K, write $\pi \leq \rho$ if π is unitarily equivalent to a subrepresentation of ρ. That is, there is an orthogonal decomposition of the Hilbert space $K = K' \oplus K''$ such that with respect to this decomposition

$$\rho(a) = \begin{bmatrix} \rho'(a) & 0 \\ 0 & \rho''(a) \end{bmatrix}$$

for a pair of representations ρ', ρ'' of A on K', K'', such that ρ' unitarily equivalent to π. Start by checking the easy fact that if $\pi \leq \rho$ then $\|\pi(a)\| \leq \|\rho(a)\|$ for all $a \in A$:

a) If A and B are *-algebras, $\pi \leq \pi'$ are representations of A, ρ a representation of B, prove that $\pi \otimes \rho \leq \pi' \otimes \rho$, and hence that $\|\cdot\|_{\pi,\rho} \leq \|\cdot\|_{\pi',\rho}$.
b) In the definition (4.10), prove that it suffices to take the sup over only *injective* representations π, ρ. (*Hint.* By the GNS theorem A has at least one injective representation ρ_0. Then $\rho := \pi \oplus \rho_0$ is still injective, and $\pi \leq \rho$.)
c) Suppose B has a representation ρ_0 with the property that any representation of B is unitarily equivalent to a subrepresentation of some number (possibly infinite)

of copies of ρ_0. That is, suppose that for all ρ, there exists an index set Λ such that

$$\rho \leq \oplus_{\lambda \in \Lambda} \rho_0.$$

Prove that in the definition of the norm (4.10) on $A \otimes B$, it suffices to use the single representation ρ_0 for B.

d) Prove that $B := M_n(\mathbb{C})$ satisfies the property of c) for $\rho_0 \colon M_n(\mathbb{C}) \to \mathbb{B}(\mathbb{C}^n)$ the standard representation.

Example 4.5.20 Let A be a C*-algebra. Then the tensor product of *-algebras $A \otimes M_n(\mathbb{C})$ (in the category of *-algebras) is isomorphic to $M_n(A)$, as discussed in Example 4.5.5. Under this identification, if $\pi \colon A \to \mathbb{B}(H)$ is an injective representation of A, and if $\rho_0 \colon M_n(\mathbb{C}) \to \mathbb{B}(\mathbb{C}^n)$ is the standard representation of $M_n(\mathbb{C})$, then the tensor product representation $\pi \otimes \rho_0$ can be identified with the representation

$$\bar{\pi} \colon M_n(A) \to \mathbb{B}(H \oplus \cdots \oplus H), \quad \bar{\pi}(T) = \begin{bmatrix} \pi(T_{11}) & \cdots & \pi(T_{1n}) \\ \cdots & \cdots & \cdots \\ \pi(T_{n1}) & \cdots & \pi(T_{nn}) \end{bmatrix}$$

of $M_n(A)$ on $H^n = H \oplus \cdots \oplus H$. Note that it is obviously injective. Furthermore,

$$\sup_{i,j} \|\pi(T_{ij})\| \leq \|\bar{\pi}(T)\| \leq n^2 \sup_{i,j} \|\pi(T_{ij})\|,$$

holds for any T, so that $M_n(A)$, or, equivalently, the algebraic tensor product $A \otimes M_n(\mathbb{C})$ is already complete with respect to $\|\cdot\|_{\bar{\pi}}$. Thus, it is a C*-algebra.

Note that a C*-algebra can have only one norm, so norm produced by π in the previous paragraph is actually independent of π amongst injective representations of A.

Example 4.5.21 Another extremely important example of a tensor product is the tensor product $A \otimes \mathcal{K}$ of any C*-algebra, with the compact operators (say, on a separable Hilbert space). The norm on the tensor product in this case is rather easy to understand. The compact operators $\mathcal{K}(H)$ have a unique irreducible representation, the obvious one, on H. Choosing any injective representation of A on a Hilbert space L gives a tensor product representation

$$A \otimes \mathcal{K} \to \mathbb{B}(L \otimes H).$$

Another way of describing this is as follows. Fix an orthonormal basis $e_1, e_2, \ldots$ for H. The tensor product Hilbert space $L \otimes H$ can then be identified with $L \oplus L \oplus \cdots$, and operators on $L \otimes H$ can be represented as $\mathbb{N}$-by-$\mathbb{N}$ matrices A with entries A_{ij} in $\mathbb{B}(L)$. Elements of the algebraic tensor product of A and finitely supported elements of $\mathcal{K}(H)$ correspond to matrices with only finitely many nonzero entries. The entries can be considered elements of A, or as operators on L by the given representation.

The tensor product $A \otimes \mathcal{K}$ is thus the closure of this algebra $M_\infty(A)$ of infinite matrices (with only finitely nonzero entries) with entries in A.

C*-algebras A and B such that $A \otimes \mathcal{K}$ is isomorphic to $B \otimes \mathcal{K}$ are said to be *Morita equivalent*. Since $\mathcal{K} \otimes \mathcal{K} \cong \mathcal{K}$, A is Morita equivalent to $A \otimes \mathcal{K}$, for any A.

Proposition 4.5.22 *For any pair of injective representations* $\pi \colon A \to \mathbb{B}(H)$ *and* $\rho \colon B \to \mathbb{B}(K)$, *the norms* $\|\cdot\|_{\pi,\rho}$ *are equal.*

That is, in computing the norm on $A \otimes B$, we can do so with any, *fixed* pair of injective representations. This is obviously quite helpful in thinking about specific examples, where typically, there is an "obvious" such pair.

Proof Suppose that π and ρ are injective representations on H, K. Let ρ' be another injective representation. Let $x = \sum a_i \otimes b_i \in A \otimes_{\mathrm{alg}} B$. Let (P_n) be a sequence of finite-rank projections in $\mathbb{B}(H)$ with P_n of rank n, and $P_n\xi \to \xi$ for all $\xi \in H$. By Exercise 1.2.24,

$$\|T\| = \sup_n \|(P_n \otimes \mathrm{id}_K)T(P_n \otimes \mathrm{id}_K)\|, \quad \forall T \in \mathbb{B}(H \otimes K).$$

In particular, this applies to the operator $T = \sum \pi(a_i) \otimes \rho(b_i)$ and implies that

$$\Big\| \sum a_i \otimes b_i \Big\|_{\pi,\rho} = \sup_n \|(P_n\pi(a_i)P_n \otimes \rho(b_i))\|.$$

Therefore, to prove that $\|\cdot\|_{\pi,\rho} = \|\cdot\|_{\pi,\rho'}$ it suffices to fix n and prove that

$$(4.11) \qquad \Big\| \sum_i P_n\pi(a_i)P_n \otimes \rho(b_i) \Big\| = \Big\| \sum_i P_n\pi(a_i)P_n \otimes \rho'(b_i) \Big\|.$$

By the definitions, the left hand side of this equation is the operator norm on $\mathbb{B}(P_nH \otimes K)$, and the right hand side is the operator norm on $\mathbb{B}(P_nH \otimes K')$. On the other hand, $M_n(\mathbb{C}) \otimes B$ (equivalently, $M_n(\mathbb{C}) \otimes_{\mathrm{alg}} B$) is a C*-algebra and has a unique norm. Now let $i \colon M_n(\mathbb{C}) \to \mathbb{B}(P_nH)$ be the inclusion; then the representation $i \otimes \rho \colon M_n(\mathbb{C}) \otimes B \to \mathbb{B}(P_nH \otimes K)$ is injective and so results in the same norm on $M_n(\mathbb{C}) \otimes B$ as does the one $i \otimes \rho'$ using ρ' instead (as ρ' is also assumed injective) because $M_n(\mathbb{C}) \otimes B$ is already a C*-algebra. This proves the equality (4.11). Now, reversing the roles of π and ρ in the obvious way gives that

$$\|\cdot\|_{\pi,\rho} = \|\cdot\|_{\pi',\rho'}$$

as required. $\square$

The above proof is due to N. Brown and N. Ozawa.

Exercise 4.5.23 Prove that if X is any locally compact Hausdorff space and A is any C*-algebra, then $C_0(X) \otimes A \cong C_0(X, A)$, where $C_0(X, A)$ has the C*-norm explained in Exercise 1.1.11.

We conclude this chapter with two basic results about the spatial tensor product.

Proposition 4.5.24 *Let* $\{A_i, \phi_{ij} \mid i < j\}$ *be an inductive system of C*-algebras with injective structure maps. Then* $\{A_i \otimes B, \phi_{ij} \otimes \mathrm{id}_B \mid i < j\}$ *is another inductive system (with injective structure maps), and*

$$(\varinjlim_i A_i) \otimes B \cong \varinjlim_i A_i \otimes B$$

for any C-algebra B.*

Proof Choose an injective representation $\pi \colon \varinjlim_i A_i$. If $\phi_i \colon A_i \to \varinjlim_i A_i$ are the canonical inclusions, then the representations $\pi \circ \phi_i \colon A_i \to \mathbb{B}(H)$ are also injective. Fixing an injective representation of B, we can compute the norm on $A_i \otimes B$ using $\|\cdot\|_{\pi \circ \phi_i, \rho}$. The norm of an element of the algebraic inductive limit $\varinjlim_i A_i \otimes B$ is its norm in $A_i \otimes B$, for i sufficiently large, and hence its $\|\cdot\|_{\pi \circ \phi_i, \rho}$ norm. On the other hand, the norm on $(\varinjlim_i A_i) \otimes B$ is the $\|\cdot\|_{\pi, \rho}$ norm. Since this reduces to the $\|\cdot\|_{\pi \circ \phi_i, \rho}$ norm on elements of $\phi_i(A_i) \subset \varinjlim_i A_i$, for sufficiently large i, the two ways of completing are the same, proving the result.   □

Finally, we prove the C*-algebraic version of the injectivity part of Exercise 4.5.8.

Proposition 4.5.25 *If* A_1, A_2, B_1, B_2 *are C*-algebras, and if* $\alpha \colon A_1 \to B_1$ *and* $\alpha_2 \colon A_2 \to B_2$ *are injective *-homomorphisms, then*

$$\alpha_1 \otimes \alpha_2 \colon A_1 \otimes A_2 \to B_1 \otimes B_2$$

is also injective.

This follows from the following.

Lemma 4.5.26 *Suppose that* A, B *are unital C*-algebras and* $i \colon B \to A$ *an injective *-homomorphism. Then* $i \otimes \mathrm{id}_D \colon B \otimes D \to A \otimes D$ *is an injective *-homomorphism, for any C*-algebra D.*

Proof From Exercise, 4.5.8 the map $i \otimes \mathrm{id}_D \colon B \otimes_{\mathrm{alg}} D \to A \otimes_{\mathrm{alg}} D$ is injective on the algebraic tensor product and extends to a C*-algebra homomorphism $i \otimes \mathrm{id}_D \colon B \otimes D \to A \otimes D$, which we need to show is isometric. Thus, it suffices to show that if $x \in B \otimes_{\mathrm{alg}} D \subset B \otimes D$, then $\|(i \otimes \mathrm{id}_D)(x)\| \geq \|x\|$. But by Proposition 4.2.13, given any representation π of B, there is a representation π' of A such that $\pi \leq \pi'|_B$. It follows that $\pi \otimes \rho \leq \pi'|_B \otimes \rho$ for any representation ρ of D, and hence that

$$\|\cdot\|_{\pi, \rho} \leq \|\cdot\|_{\pi'|_B, \rho}.$$

Taking sups over all π and ρ gives $\|x\| \leq \|i(x)\|$. That is, $\|i(x)\| \geq \|x\|$ for all $x \in B \otimes_{\text{alg}} D$, as required. $\square$

Exercise 4.5.27 This exercise does Example 1.7.9 again, using tensor products. Let $A_n := M_2(\mathbb{C}) \otimes \cdots M_2(\mathbb{C})$ (n times). Let $\varphi_n : A_n \to A_{n+1}$ be the map $\varphi_n(T) := T \otimes 1_{M_2(\mathbb{C})}$:

a) Prove that

$$A_1 \xrightarrow{\varphi_1} A_2 \xrightarrow{\varphi_2} A_3 \to \cdots$$

is an inductive system of C*-algebras and that

$$\varinjlim_n A_n \cong U(2^\infty),$$

with $U(2^\infty)$ the UHF algebra of Type 2^∞.

b) Let $v = \begin{bmatrix} 1 & 0 \\ 0 & -1 \end{bmatrix}$ and let $\alpha_n : A_n \to A_n$ be the C*-algebra automorphism of conjugation by the unitary element $v \otimes \cdots \otimes v \in M_2(\mathbb{C}) \otimes \cdots \otimes M_2(\mathbb{C}) = A_n$. Prove that the family $\{\alpha_n : A_n \to A_n\}_{n=1}^\infty$ assembles to give an automorphism of $U(2^\infty)$ of order 2.

Exercise 4.5.28 Let $\pi : A \to B$ be a surjective *-homomorphism of unital C*-algebras:

a) Prove that π has the "path-lifting property": if $\gamma : [0, 1] \to B$ is a continuous path in B, and if $a \in A$ with $\pi(a) = \gamma(0)$, then there exists a continuous path $\tilde{\gamma} : [0, 1] \to A$ such that $\tilde{\gamma}(0) = a$ and $\pi \circ \tilde{\gamma} = \gamma$. (*Hint.* Paths in A correspond to elements of the C*-algebra $C([0, 1], A)$, and π determines a surjective *-homomorphism $C([0, 1], A) \to C([0, 1], B)$.)

b) Show that the restriction of π to a map $A_{sa} \to B_{sa}$ to the space of self-adjoints in A, to the space of self-adjoints in B, has the path-lifting property.

c) Show that the restriction of π to a map $\mathbf{U}(A) \to \mathbf{U}(B)$ also has the path-lifting property. (*Hint.* Start by showing that if $(u_t)_{t \in [0,1]}$ is a path of unitaries with $u_0 = 1 \in A$, then for some $\varepsilon > 0$, then the portion $(u_t)_{t \in [0,\varepsilon]}$ of the given path with $0 \leq t \leq \varepsilon$ lifts to a path $(\tilde{u}_t)_{t \in [0,\varepsilon]}$ by using a logarithm to write $u_t = e^{ix_t}$ for a path $(x_t)_{t \in [0,\varepsilon]}$ of self-adjoints.)

4.6   Structure of Crossed Products by Proper Actions of Discrete Groups

A condition on a group action $G \times X \to X$ that ensures that the space of orbits $G \backslash X$, with the quotient topology, is Hausdorff, is that the action is *proper* (see Proposition 4.6.4). The space of orbits can thus be studied purely topologically, since the quotient $G\backslash X$ is a reasonable topological space (unlike the space of orbits of irrational rotation on the circle).

Before proceeding to a study of crossed products from proper actions, we make a remark about "noncommutative spaces." In this book, I define a *noncommutative space* as a Morita equivalence class of C*-algebras. Many of the key invariants of C*-algebras do not distinguish between two Morita equivalent C*-algebras, one of the most important being the space of unitary equivalence classes of irreducible representations, which, at the risk of overusing the term, we sometimes refer to as the *spectrum* $\hat{A}$ of A. As a set, the meaning of $\hat{A}$ is clear, but it also has a topology which we do not discuss in this book. But if A is commutative, this reduces to the Gelfand spectrum of characters of A. In general, $\hat{A}$ is non-Hausdorff. However, it is sometimes possible to get quite a satisfactory parameterization of $\hat{A}$, as a set. This is the case with crossed products from proper actions.

Definition 4.6.1 An action $G \times X \to X$ of a discrete group on a locally compact space X is *proper* if for every pair K, K' of compact subsets of X, the subset $\{g \in G \mid g(K) \cap K' \neq \emptyset\}$ of G is finite.

Example 4.6.2 Some examples of proper actions are:

a) Any action of a finite group is proper.
b) The action of $\mathbb{Z}$ on $\mathbb{R}$ by translation is proper, or of $\mathbb{Z}^n$ on $\mathbb{R}^n$.
c) The translation action of a discrete subgroup $G \subset G'$ of a locally compact group G', on G', is proper.

The action of the integers by irrational rotation is definitely not proper. Indeed, no infinite group can act properly on a compact space; proper actions of infinite groups are always on noncompact spaces.

Exercise 4.6.3 Prove that if G is a discrete group and H is a subgroup, the left multiplication action of G on G/H is proper if and only if H is finite.

The following is a standard result from basic topology, and we omit the proof.

Proposition 4.6.4 *If G is a discrete group acting properly on a locally compact Hausdorff space X, then $G\backslash X$ with the quotient topology is locally compact and Hausdorff.*

This means that the quotient space $G\backslash X$ is a perfectly reasonable topological space, whose homology, cohomology, *etc* can be computed—in the case of a *proper* action. So there is in a sense no need for noncommutative C*-algebras at this point, if one is interested only in the quotient space. However, the quotient space contains no information about the isotropy groups of points of X. The noncommutative C*-algebra crossed product $C_0(X) \rtimes G$ contains this additional information, as we show.

Before going on, we describe some interesting geometric examples of finite group actions and proper actions.

Example 4.6.5 Let D_4 be the dihedral group of symmetries of a square. We can realize D_4 as generated by a counter-clockwise rotation R of the plane through

$\frac{\pi}{2}$ radians, and the reflection of the plane S across the x-axis. These two group elements, and the group of order 8 they generate, commute with the translation action of $\mathbb{Z}^2$ on the plane $\mathbb{R}^2$ and hence descend to homeomorphisms of the 2-torus $\mathbb{T}^2$. Thus D_4 acts on $\mathbb{T}^2$. The infinite group of maps of the plane generated by D_4 and $\mathbb{Z}^2$, which can be easily checked to be the semi-direct product $\mathbb{Z}^2 \rtimes D_4$, is an infinite group which acts properly on $\mathbb{R}^2$.

Exercise 4.6.6 Let $F \subset \mathbb{T}^2$ be the projection to the torus of the triangle $\{(s, t) \in \mathbb{R}^2 \mid 0 \le s \le t, \ \ 0 \le t \le \frac{1}{2}$ in the plane. Show that the restriction of the quotient map $\pi \colon \mathbb{T}^2 \to D_4 \backslash \mathbb{T}^2$ to F is a homeomorphism. That is, $G \backslash \mathbb{T}^2 \cong F$. Compute the isotropy groups $\mathrm{Stab}_{D_4}(x)$ for all points $x \in F$.

Example 4.6.7 The infinite dihedral group D_∞ is the group of homeomorphisms of $\mathbb{R}$ generated by the translation $T(s) = s + 1$ and reflection $S(x) = -x$. By construction, D_∞ acts on $\mathbb{R}$, and it is rather clear that the action is proper.

Exercise 4.6.8 Prove that the quotient space $D_\infty \backslash \mathbb{R}$ is the interval $[0, \frac{1}{2}]$. Compute the isotropy groups at points of $[0, \frac{1}{2}]$.

The following construction is a very general way of producing (free and) proper actions.

Let M be a finite complex and $G = \pi_1(M)$, the fundamental group of M. Let $X = \tilde{M}$, the universal cover of M. Then G acts on X, by "deck-transformations," that is, so that

$$\pi(gx) = \pi(x), \quad \forall x \in X, \ g \in G,$$

with $\pi \colon X \to M$ the covering map.

This G-action is always proper and free, and the quotient space $G \backslash X$ is homeomorphic to M.

Exercise 4.6.9 Let X be the geometric realization of the Cayley graph of the group $\mathbb{F}_2$. Prove that the left translation action of $\mathbb{F}_2$ on itself induces an action on X. Prove that this action is proper.

Example 4.6.10 Let G be the group $\mathbf{PSL}_2(\mathbb{Z})$. It acts by Möbius transformations on the upper half plane $X := \{z \in \mathbb{C} \mid \mathrm{Im}(z) > 0\}$. A matrix $g = \begin{bmatrix} a & b \\ c & d \end{bmatrix}$ acts by the transformation

$$g(z) = \frac{az + b}{cz + d}.$$

Exercise 4.6.11 Prove that the action described above is proper. Is it free? Find all (conjugacy classes) of nontrivial isotropy.

More generally, any *Fuchsian group* acts properly on the hyperoblic plane; such groups may have torsion and fixed points.

Exercise 4.6.12 Find an example of a noncommutative discrete group G which acts properly by affine isometries of the plane $\mathbb{R}^2$.

We now determine the spectrum of a crossed product by a simple proper action with isotropy.

Let the group $\mathbb{Z}/2$ act on $I := [-1, 1]$ by the homeomorphism $\sigma(x) = -x$. The crossed product $A = C(I) \rtimes \mathbb{Z}/2$ is the same as the corresponding twisted group algebra, i.e., there is no completion involved. Elements of $A = C(I) \rtimes G$ can be written as sums $f + g[\sigma]$; here $[\sigma]$ denotes a unitary in A such that $[\sigma]f[\sigma]^* = f \circ \sigma$. The algebra multiplication in the crossed product is determined by this rule and that ff' is the usual product of functions, and that $[\sigma] = [\sigma]^*$, i.e. $[\sigma]^2 = 1$, the unit in A.

Choose any $x \in I$. We define a C*-algebra representation

$$\pi_x : C(I) \rtimes \mathbb{Z}/2 \to \mathbb{B}(\mathbb{C}^2) \cong M_2(\mathbb{C})$$

by the covariant pair

$$f \mapsto \begin{bmatrix} f(x) & 0 \\ 0 & f(-x) \end{bmatrix}, \quad [\sigma] \mapsto \begin{bmatrix} 0 & 1 \\ 1 & 0 \end{bmatrix}.$$

The induced *-homomorphism $C(I) \rtimes \mathbb{Z}/2 \to M_2(\mathbb{C})$ is given by

$$(4.12) \qquad \pi_x(f + g[\sigma]) = \begin{bmatrix} f(x) & g(x) \\ g(-x) & f(-x) \end{bmatrix}.$$

If $x \neq 0$, then $x \neq -x$, and hence we can find a function g such that $g(-x) = 0$ but $g(x) = 1$, or the other way around. For the first choice, we have $\pi_x(g[\sigma]) = \begin{bmatrix} 0 & 1 \\ 0 & 0 \end{bmatrix}$ and $\pi_x(g) = \begin{bmatrix} 1 & 0 \\ 0 & 0 \end{bmatrix}$ and for the second choice of g, $\pi_x(g[\sigma]) = \begin{bmatrix} 0 & 0 \\ 1 & 0 \end{bmatrix}$ and $\pi_x(g) = \begin{bmatrix} 0 & 0 \\ 0 & 1 \end{bmatrix}$. Hence, for $x \neq 0$, the range of

$$\pi_x : C(I) \rtimes \mathbb{Z}/2 \to M_2(\mathbb{C})$$

contains all matrices. Thus π_x is a surjection for all $x \neq 0$. It is also clearly an irreducible representation, since its range contains all $M_2(\mathbb{C})$. Finally, π_{-x} is unitarily equivalent to π_x for all $x \neq 0$.

Remark 4.6.13 The unitary $\begin{bmatrix} 0 & 1 \\ 1 & 0 \end{bmatrix}$ gives a unitary conjugacy between the representations π_x and π_{-x}.

Now consider what happens when $x = 0$. The representation $\pi_0 : C(I) \rtimes \mathbb{Z}/2 \to M_2(\mathbb{C})$ is given by

$$\pi_0(f + g[\sigma]) = \begin{bmatrix} f(0) & g(0) \\ g(0) & f(0) \end{bmatrix}.$$

The collection of matrices of the form

$$\begin{bmatrix} a & b \\ b & a \end{bmatrix}$$

with $a, b \in \mathbb{C}$ forms a C*-algebra naturally isomorphic to $C^*(\mathbb{Z}/2)$: It is the matrix picture of $C^*(\mathbb{Z}/2)$ acting by the regular representation on $l^2(\mathbb{Z}/2) \cong \mathbb{C}^2$. These matrices are simultaneously diagonalizable with unit eigenvectors $\frac{1}{\sqrt{2}} \begin{bmatrix} 1 \\ 1 \end{bmatrix}$, eigenvector of $\begin{bmatrix} a & b \\ b & a \end{bmatrix}$ with eigenvalue $a + b$, and $\frac{1}{\sqrt{2}} \begin{bmatrix} 1 \\ -1 \end{bmatrix}$, an eigenvector of $\begin{bmatrix} a & b \\ b & a \end{bmatrix}$ with eigenvalue $a - b$.

The conclusion is that the representation π_0 splits as a direct sum of two 1-dimensional representations: namely the spans of the two given eigenvectors. Thus

$$\pi_0 \cong \varepsilon \oplus \chi,$$

with

$$\varepsilon \colon C(I) \rtimes \mathbb{Z}/2 \to C^*(\mathbb{Z}/2) \to \mathbb{C}, \quad \varepsilon(f + g[\sigma]) = f(0) + g(0),$$

and

$$\chi \colon C(I) \rtimes \mathbb{Z}/2 \to C^*(\mathbb{Z}/2) \to \mathbb{C}, \quad \varepsilon(f + g[\sigma]) = f(0) - g(0).$$

Remark 4.6.14 In a suitable topology on the spectrum $\widehat{A}$, with $A = C(I) \rtimes \mathbb{Z}/2$, the two characters ε and χ form a "double point." Actually, as a topological space, $\widehat{A}$ admits the following description. Take the intervals $[-1, 0]$ and $[0, 1]$, form their disjoint union, and identify any *nonzero* x in $[-1, 0]$ with $-x$ in $[0, 1]$. The resulting identification space Z

$$[-1, 0] \sqcup [0, 1] \, / \, \sim$$

carries a quotient topology.

Exercise 4.6.15 The quotient topology on Z is not Hausdorff, but it is T_0.

The space just described parameterizes the spectrum of $C(I) \rtimes \mathbb{Z}/2$, by the map assigning to the equivalence class of nonzero x in the disjoint union, to the (class of the) irreducible representation $[\pi_x]$, and assigns to the two 0's in Z, the two characters ε and χ of $C(I) \rtimes \mathbb{Z}/2$ into which π_0 splits.

Proposition 4.6.16 *The C*-algebra $C(I) \rtimes \mathbb{Z}/2$ is isomorphic to the C*-algebra*

$$C\big(I \times_{\mathbb{Z}/2} M_2(\mathbb{C})\big) := \{T : I \to M_2(\mathbb{C}) \mid T(-x) = \begin{bmatrix} 0 & 1 \\ 1 & 0 \end{bmatrix} \cdot T(x) \cdot \begin{bmatrix} 0 & 1 \\ 1 & 0 \end{bmatrix}, \quad \forall x \in I\}.$$

The isomorphism is given by considering the formula (4.12) as specifying a matrix-valued function, which is easily checked to transform as stated.

In this particular example we can go further. A matrix-valued function T on I such that

$$T(-x) = \begin{bmatrix} 0 & 1 \\ 1 & 0 \end{bmatrix} \cdot T(x) \cdot \begin{bmatrix} 0 & 1 \\ 1 & 0 \end{bmatrix}, \quad \forall x \in I,$$

is completely determined by its restriction to the "fundamental domain" $[0, 1] \subset I$. Moreover, the condition implies that $T(0)$ *commutes* with $\begin{bmatrix} 0 & 1 \\ 1 & 0 \end{bmatrix}$.

Corollary 4.6.17 *The crossed product* $A := C(I) \rtimes \mathbb{Z}/2$ *is isomorphic to the C*-algebra*

$$\left\{ f : [0, 1] \to M_2(\mathbb{C}) \mid f(0) \text{ commutes with } \begin{bmatrix} 0 & 1 \\ 1 & 0 \end{bmatrix} \right\}.$$

Its space of irreducible representations is naturally parameterized by the closed interval $[0, 1]$ *with a double point at* 0 *added, by the parameterization described above.*

We now extend these ideas to general proper actions.

Definition 4.6.18 Let G be a discrete group acting properly on X. Let $\rho : G \to U(L^2 G)$ be the right regular representation. Let $\mathcal{K} := \mathcal{K}(l^2(G))$, and denote by $C(X \times_G \mathcal{K})$ the C*-algebra of bounded, continuous functions

$$f : X \to \mathcal{K}$$

such that

(4.13) $$f(gx) = \rho(g)f(x)\rho(g)^{-1},$$

as operators on $l^2(G)$.

Theorem 4.6.19 *The C*-algebras* $C_0(X) \rtimes G$ *and* $C(X \times_G \mathcal{K})$ *are canonically isomorphic.*

Proof We will construct an isomorphism $\pi : C_0(X) \rtimes G \to C(X \times_G \mathcal{K})$ by specifying a covariant pair, as follows.

Set

$$\pi(f)(x)(e_h) := f(hx), \quad \pi(g)(x)(e_h) := e_{gh}, \quad f \in C_0(X), \ x \in X, \ g, h \in G,$$

where in each case we have given the action of the operator $\pi(\cdot)(x)$ on the standard basis $\{e_h\}_{h \in G}$ of $l^2(G)$. If $g \in G$, we are regarding $\pi(g) := \lambda(g)$, to be the constant, operator-valued function on X.

If $f \in C_0(X)$ and $g \in G$, then

$$(4.14) \quad \left[\pi(g)\pi(f)\pi(g)^{-1}\right](x)(e_h) = \pi(g)\pi(f)(x)\pi(g)^{-1}(e_h)$$

$$= \pi(g)\pi(f)(x)(e_{g^{-1}h}) = \pi(g)f(g^{-1}hx)\,e_{g^{-1}h} = f(g^{-1}hx)e_h$$

whence

$$\pi(g)\pi(f)\pi(g)^{-1} = \pi(f \circ g^{-1}),$$

so that we have defined a covariant pair.

For fixed $x \in X$, and $f \in C_0(X)$, the operator $\pi(f)(x)$ on $l^2(G)$ is multiplication by the function on G with value $f(g^{-1}x)$ at $g \in G$. This function on G vanishes at infinity since the G-action is proper, since, as a *-homomorphism $C_0(X) \to C_b(G)$, it is Gelfand dual to the orbit map $G \to X$, $g \mapsto gx$, which is a proper map, since the G-action is assumed proper.

Hence $\pi(f)(x)$, and more generally, finite combinations $\pi(\sum_g f_g[g])(x) \in C_0(X)[G]$ are compact operators on $l^2(G)$, for any $x \in X$.

Next, if $g \in G$, and $f \in C_0(X)$, then

$$\left[\rho(g)\pi(f)(x)\rho(g)^{-1}\right](x)(e_h) = \rho(g)f(x)(e_{hg}) = \rho(g)f(hgx)e_{hg}$$

$$(4.15) \qquad\qquad\qquad\qquad = \pi(f)(x)(gx).$$

This shows that $\pi(f)$ satisfies (4.13), so is an element of $C(X \times_G \mathcal{K})$. If $g \in G$, then $\pi(g) = \lambda(g)$ is a constant, operator-valued function and also satisfies (4.13), since $\lambda(g)$ commutes with $\rho(g)$.

This shows that our homomorphism maps $C_0(X) \rtimes G$ to $C(X \times_G \mathcal{K})$. The fact that it is an isomorphism is not difficult and is left to the reader.   □

Choose $x \in X$. Then the fiber of $X \times_G \mathcal{K} \to G \backslash X$ over the orbit $Gx \in G$ identifies with $\mathcal{K}(l^2(G))$, by choosing a representative point $x \in X$ in the orbit. We obtain a *-homomorphism

$$\pi_x \colon C(X \times_G \mathcal{K}) \to \mathcal{K}(l^2(G)).$$

The kernel $\ker(\pi_x)$ is an ideal, and π_x itself is a representation of $C(X \times_G \mathcal{K})$. The condition (4.13) implies the following:

Lemma 4.6.20 *The representation* $\pi_x \colon C(X \times_G \mathcal{K}) \to \mathcal{K}(l^2(G))$ *of evaluation of a section at a point of Gx maps the C*-algebra $C(X \times_G \mathcal{K})$ isomorphically into the C*-algebra*

$$\mathcal{K}\big(l^2(G)\big)^{\mathrm{Stab}_G(x)},$$

where $\mathrm{Stab}_G(x) := \{h \in G \mid gx = x\}$ is the stabilizer of x, and for $H \subset G$ a subgroup of G,

$$\mathcal{K}\big(l^2(G)\big)^H := \{T \in \mathcal{K}\big(l^2(G)\big) \mid \rho(h)T\rho(h)^{-1} = T \ \forall g \in H\}.$$

Example 4.6.21 In the example of $\mathbb{Z}/2$ acting on I by reflection, the space $X \times_G \mathcal{K}\big(l^2(G)\big)$ and associated C*-algebra amount to the following. Here $X = I = [-1, 1]$, the generator of $\mathbb{Z}/2$ is $\sigma(x) = -x$, and, evidently, we can use the fundamental domain to identify $G\backslash X$ with $[0, 1]$.

We have already described (Corollary 4.6.17 and discussion) a family of *-homomorphisms

$$\pi_x \colon C(I) \rtimes G \to M_2(\mathbb{C}) = \mathcal{K}\big(l^2(\mathbb{Z}/2)\big), \quad x \in [0, 1],$$

and we noted several facts:

- The range of π_x for $x > 0$ is $M_2(\mathbb{C})$, and the range of π_0 consists of matrices which commute with $\begin{bmatrix} 0 & 1 \\ 1 & 0 \end{bmatrix}$ —and thus is a copy of $C^*(\mathbb{Z}/2) = C^*\big(\mathrm{Stab}_{\mathbb{Z}/2}(0)\big)$.
- For $x = 0$, π_x splits into a direct sum of two 1-dimensional representations, i.e., characters,

$$\varepsilon, \chi \colon C_0(X) \rtimes G \to \mathbb{C}.$$

These factor through the restriction map $C(I) \rtimes G \to C^*(G)$ induced by the evaluation of functions in $C(I)$ at 0, and the *-homomorphisms $C^*(\mathbb{Z}/2) \to \mathbb{C}$ induced by the two group characters ε and χ.

All of this data may be thought of as describing a "bundle" of C*-algebras over $[0, 1]$. The fiber at any $x \in (0, 1]$ is $M_2(\mathbb{C})$. The fiber at $x = 0$ is $\mathbb{C} \oplus \mathbb{C}$.

Of course every point $x \in [0, 1]$ determines an ideal, $\ker(\pi_x)$. The origin $x = 0$ determines two ideals, the kernels of the two characters.

Exercise 4.6.22 Describe the two ideals $\ker(\varepsilon)$ and $\ker(\chi)$ explicitly in $C(I) \rtimes \mathbb{Z}/2$.

Finally, we describe the spectrum (the space of irreducible representations) for a general proper action, omitting as always any discussion of the topology. (This is explicitly described in [67].)

Theorem 4.6.23 *Let G be a discrete group acting properly on X. Then the set of equivalence classes of irreducible representations of $C_0(X) \rtimes G$ is naturally parameterized by the set*

$$\bigsqcup_{x \in F} \widehat{\mathrm{Stab}_G(x)}$$

where $F \subset X$ is a set of representatives of the orbits.
The explicit parameterization is discussed in the proof.

We require the following:

Lemma 4.6.24 *Let G be a discrete group, and $H \subset G$ a finite subgroup. Let $\rho: H \to \mathbf{U}(l^2 G)$ be the representation induced by the right translation action of H on G. Then there are isomorphisms*

$$C_0(G/H) \rtimes H \cong \mathcal{K}(l^2 G)^H \cong \mathbb{C}^*(H) \otimes \mathcal{K}(l^2(G/H)),$$

where $\mathcal{K}(l^2 G)^H$ denotes compact operators which commute with $\rho(H)$.

Proof The representation of $C_0(G/H)$ on $l^2(G)$ by multiplication operators lands in $\mathcal{K}(l^2 G)^H$. If we let G act on $l^2 G$ by the left regular representation, we obtain a covariant pair and isomorphism

$$C_0(G/H) \rtimes G \to \mathcal{K}(l^2 G)^H,$$

and we leave it as an exercise to check that this is an isomorphism.

Next, choose a collection of coset representatives $g_i \in G$ for the cosets in G/H. This gives a decomposition of Hilbert spaces $l^2(G) = \oplus_i l^2(g_i H)$ corresponding to the spatial decomposition of G into cosets. The action $\rho(h)$ of $h \in H$ maps each factor to itself, by a conjugate of the right regular representation of H on $l^2(H)$. Since the coset representatives parameterize the points of G/H, our decomposition corresponds to a tensor product decomposition $l^2(G) \cong l^2(H) \otimes l^2(G/H)$. The representation ρ of H corresponds to the tensor product $\rho_H \otimes 1$, where ρ_H is the right regular representation of H on $l^2(H)$. It follows that the compact operators which commute with $\rho(H)$ are exactly the elements of $\mathcal{K}(l^2(G/H)) \otimes C^*(H)$, with $C^*(H)$ acting on $l^2(H)$ by the *left* regular representation. That is,

$$\mathcal{K}(l^2(G/H)) \otimes C^*(H) \cong \mathcal{K}(l^2 G)^H,$$

proving the result.   □

Proof of Theorem 4.6.23 Suppose that π is an irreducible representation of the C*-algebra $C_0(X) \rtimes G$. Then π is strictly continuous and so extends to a representation of $\mathcal{M}(C_0(X) \rtimes G)$, which is clearly also irreducible. The latter C*-algebra contains its center $C_0(G \backslash X) \cong C_0(X)^G$, the bounded, continuous and G-invariant functions on X, and as the center in an irreducible representation must act by scalar multiples of the identity, it follows that π determines a character of $C_0(G \backslash X)$, and the characters are point evaluations at points, i.e., orbits, and there exists an orbit Gx such that $\pi(h) = h(Gx) \cdot 1$ for $h \in C_0(G \backslash X) \subset \mathcal{M}(C_0(X) \rtimes G)$, where 1 is the identity operator. Now if $f \in C_0(X)$ and f vanishes on a neighborhood of the orbit Gx, then by an easy exercise there exists a G-invariant function h such that $h|_{Gx} = 0$ and $hf = f$. Since $\pi(h) = h(Gx) = 0$, it follows that $\pi(f) = 0$. We deduce that π vanishes on the ideal $C_0(X \setminus Gx) \rtimes G$ and hence factors through the *-homomorphism

$$C_0(X) \rtimes G \to C_0(Gx) \rtimes G$$

and an irreducible representation of $C_0(Gx) \rtimes G$. The exercise below shows that $C_0(Gx) \rtimes G$ is stably isomorphic to $C^*(\mathrm{Stab}_G(x))$. Stably isomorphic C*-algebras have the same irreducible representations, and the irreducible representations of $C^*(H)$ for a finite group H are the irreducible representations of H. The result follows. $\square$

Let us summarize the results of this section. Any irreducible representation of $C_0(X) \rtimes G$, with G discrete acting properly, factors through the orbit restriction map

$$C_0(X) \rtimes G \to C_0(Gx) \rtimes G \cong C_0(G/H_x) \rtimes G,$$

for some x, with $H_x = \mathrm{Stab}_G(x)$. Using the isomorphisms

$$C_0(G/H_x) \rtimes G \cong \mathcal{K}(l^2 G)^{H_x} \cong C^*(H_x) \otimes \mathcal{K}(l^2(G/H_x)),$$

we deduce that an irreducible representation χ of H_x determines a representation π_α of $C_0(G/H_x) \rtimes G$, on $H_\alpha \otimes l^2(G/H_x)$. The association $\alpha \mapsto \pi_\alpha$ is called *induction*.

One can picture the spectrum of $C_0(X) \rtimes G$ in the following way. Take the set

$$\hat{X} = \{(x, \alpha) \mid x \in X, \alpha \in \widehat{\mathrm{Stab}_G(x)}\}.$$

Each (x, α) refers thus to an equivalence class of irreducible representation of $\mathrm{Stab}_G(x)$.

The group G acts on $\hat{X}$ diagonally: If α is an irreducible representation of $\mathrm{Stab}_G(x)$ and $g \in G$, then g conjugates the stabilizer subgroups at x and gx and combining with α gives an irreducible representation of $\mathrm{Stab}_G(gx)$. Taking the quotient by this action gives a space $G \backslash \hat{X}$. This maps bijectively to the space of ireducible representations of $C_0(X) \rtimes G$: There is a canonical bijection between the two sets given by

$$G \backslash \hat{X} \to \widehat{C_0(X) \rtimes G}, \quad (x, \alpha) \mapsto \pi_\alpha.$$

The topology on X induces a Hausdorff locally compact topology on $\hat{X}$ and the quotient topology on $G \backslash \hat{X}$ is then Hausdorff, but this topology does not match the official definition of the topology of the spectrum (which we have omitted in these notes). The latter is not Hausdorff, unless the action is free.

Chapter 5
Module Theory of C*-Algebras

When one first encounters algebraic topology, it is usually in connection with contour integration in complex analysis, or line integrals in multivariable calculus, typically over the plane, or the plane with punctures, or holes. Such integrals are insensitive to small deformations of the curves and suggest studying the curves up to homotopy. Since such regions of the plane are locally simply connected, the *local* properties of curves from this point of view are unimportant, they are locally homotopic to line segments, but the *global* properties of the curves can effect values of line integrals, and looking at the ensemble of all curves up to homotopy reveals topological properties of the region.

Vector bundles and their role in topology are analogous. A vector bundle over X compact Hausdorff is a locally trivial family of vector spaces over X. Any two such vector bundles of the same rank are locally isomorphic, by definition, but they may not be globally isomorphic. The tangent bundle to the 2-sphere is not isomorphic to the trivial bundle of the same rank, as one can prove using the Poincaré–Hopf Theorem (see [34]).

The complex linear space of continuous sections of a complex vector bundle is a $C(X)$-module by fiberwise scalar multiplication. Swan's Theorem or the Serre–Swan Theorem (see [152]) asserts that the $C(X)$-module of continuous sections of a vector bundle is *finitely generated projective* as a $C(X)$-module, that is, a direct summand of a free and finitely generated $C(X)$-module. The concept of projective modules over a ring was introduced by Cartan and Eilenberg in 1956. Projective modules are more general than free modules but retain some of the properties. Over the ring $A = \mathbb{Z}$, any finitely generated projective module is free, and by the Quillen–Suslin Theorem [133] a significantly deeper result, the same is true for polynomial rings like $A = \mathbb{C}[x_1, \ldots, x_n]$. But let

$$A := \mathbb{C}[x_1, x_2, x_3]/\langle x_1^2 + x_2^2 + x_3^2 - 1 \rangle$$

be the ring of polynomial functions on the 2-sphere. Then the module

© Springer Nature Switzerland AG 2024
H. Emerson, *An Introduction to C*-Algebras and Noncommutative Geometry*,
Birkhäuser Advanced Texts Basler Lehrbücher,
https://doi.org/10.1007/978-3-031-59850-0_5

$$M := \{(f_1, f_2, f_3) \in A \oplus A \oplus A \mid xf_x + yf_2 + zf_3 = 0\}$$

is the module of (algebraic) sections of the tangent bundle and is finitely generated and projective over A but not free (for the same reason as above). The group $K_0(A)$ of a ring A classifies finitely generated projective modules (which I often abbreviate f.g.p.) over the ring (see [146] for an introduction to algebraic K-theory) and when applied to C*-algebras generates the homology theory which is one of the main topics of this book.

The module $\Gamma(V)$ of sections of a vector bundle V over compact X is a f.g.p. right $C(X)$-module, and for a general C*-algebra A it seems reasonable to consider f.g.p. modules over A to be "noncommutative vector bundles." One can often construct such modules from geometric ideas and the theory of Morita equivalence, one of the central topics of this book, and the subject of the next chapter.

Morita equivalence for C*-algebras is based on the concept of *Hilbert modules*. A Hilbert module is a generalization of a Hilbert space in which the scalar multiplication is by elements of a C*-algebra A and the inner product is A-valued. The basic example is the space $\Gamma(V)$ of section of a vector bundle, since a Hermitian metric on the bundle gives rise to a $C(X)$-valued inner product by applying it pointwise to sections. But Hilbert modules over noncommutative algebras are ubiquitous in C*-algebra theory and noncommutative geometry. We discuss a number of them in the following chapter.

Hilbert modules are discussed in Blackadar's book [26], and the book [116] is an excellent source, also for spectral theory and unbounded operator theory for operators on Hilbert modules. For original source material on Hilbert modules see Kasparov's work [108].

5.1   Vector Bundles

A *section* of a surjective map $\pi: E \to X$ between topological spaces is a continuous map $s: X \to E$ such that $\pi \circ s = \mathrm{id}_X$. If $Z \subset X$, a *section of E on Z* is a map $s: Z \to E$ such that $\pi \circ s = \mathrm{id}_Z$.

Definition 5.1.1 Let X be a locally compact Hausdorff space. A real or complex *vector bundle* over X is a locally compact Hausdorff topological space E together with a continuous surjective map $\pi: E \to X$, satisfying the following additional properties:

a) The fibers $E_x := p^{-1}(x)$, $x \in X$ are all (real or) complex vector spaces.
b) The vector space operations are fiberwise continuous.
c) For each $p \in X$ there exists a neighborhood U of p and continuous sections $s_1, \ldots, s_n$ of $\pi: E \to X$ on U such that the vectors $s_1(x), \ldots, s_n(x)$ are linearly independent in E_x for all $x \in U$.

We will be primarily interested in complex vector bundles in this book. But real vector bundles arise naturally in geometry. Any real vector bundle can be made into a complex vector bundle by complexifying the bundle fiberwise.

A *vector bundle map* $T : E \to E'$ between vector bundles over X is a continuous map such that T restricts to a (real or complex depending on whether the bundle is real or complex) linear map $E_x \to E'_x$ for all $x \in X$. Equivalently, $\pi' \circ T = \pi$.

The identity map $\mathrm{id}_E : E \to E$ is a vector bundle map. We say that a vector bundle map $T : E \to E'$ is an *isomorphism* if there is a vector bundle map $T' : E' \to E$ such that $T \circ T' = \mathrm{id}_{E'}$ and $T' \circ T = \mathrm{id}_E$.

The space of sections of a vector bundle E is the $C(X)$-module of continuous maps $s : X \to E$ such that $\pi \circ s = \mathrm{id}_X$. The space of sections is clearly linear; it is also a $C(X)$-module, using the module multiplication

$$(sf)(x) := s(x) \cdot f(x),$$

for a section s and continuous function $f \in C(X)$. The $C(X)$-module of sections of E is denoted $\Gamma(E)$.

Example 5.1.2 The first projection map $\mathrm{pr}_1 : X \times \mathbb{C}^n \to X$ endows the project space $X \times \mathbb{C}^n$ (with the product topology) with the structure of a complex vector bundle. Condition c) of Definition 5.1.1 is met since we may take $U = X$ and $s_i(x) := (x, e_i)$ where $e_1, \ldots, e_n \in \mathbb{C}^n$ is the standard basis of $\mathbb{C}^n$. Similarly, $X \times \mathbb{R}^n$ is a real vector bundle.

Such bundles are topologically uninteresting. More generally, we call any vector bundle $\pi : E \to X$ *trivial* if it is isomorphic, as a vector bundle, to a product bundle $X \times \mathbb{C}^n$ (or $X \times \mathbb{R}^n$).

Generally, we denote by 1_n the trivial bundle over X of rank n (real or complex, depending on the context).

Note that the space $\Gamma(E)$ of sections of a trivial bundle $E = X \times \mathbb{C}^n$ is $C(X, \mathbb{C}^n) \cong C(X)^n$, a free $C(X)$-module.

If $\pi : E \to X$ is a vector bundle and $Z \subset X$ is a subspace, the restriction $E_{|Z}$ of E to Z is the topological space $\pi^{-1}(Z)$ with projection map $E_Z \to Z$ the restriction of π. It is an easy exercise to check that $E_{|Z}$ is a vector bundle over Z.

Triviality, or local triviality, can also be described in terms of sections.

Proposition 5.1.3 *A vector bundle $\pi : E \to X$ over X is trivial if and only if there is a finite collection $s_1, \ldots, s_n : X \to E$ of sections of $\pi : E \to X$ such that the vectors $s_1(x), \ldots, s_n(x)$ form a basis for E_x, for all $x \in X$.*

In particular, a vector bundle with one-dimensional fibers is trivial if and only if it has a nonvanishing section.

Proof Given n linearly independent sections $s_1, \ldots, s_n : X \to E$, define a vector bundle isomorphism $\varphi : X \times \mathbb{C}^n \to E$ by $\varphi\big((x, (t_1, \ldots, t_n)\big) := t_1 s_1(x) + \cdots + t_n s_n(x)$. Then φ is fiberwise an isomorphism, and is clearly continuous and a bundle map, so is an isomorphism of vector bundles.

Conversely, if $\varphi\colon X \times \mathbb{R}^n \to E$ is a vector bundle isomorphism, define $s_i(x) := \varphi(x, e_i)$, where e_i is the ith standard basis vector of $\mathbb{C}^n$. Then $s_1, \ldots, s_n$ are fiberwise everywhere linearly independent sections as required.   □

In particular, every vector bundle $\pi\colon E \to X$ is *locally trivial* in the sense that every point of X has a neighborhood U such that $E|_U$ is trivial. Typically, the corresponding isomorphisms $\varphi\colon E|_U \to U \times \mathbb{C}^n$ are called *local trivializations* of E.

The lemma suggests that the failure of a vector bundle $\pi\colon E \to X$ to be trivial can only depend on the global topology of X, since it is automatically locally trivial.

The following example (of a real vector bundle) gives a bit of intuition for how a vector bundle can twist around a topologically interesting space (like the circle) in such a way as not to be trivial.

Example 5.1.4 (The Möbius Bundle) Let $E = [0, 1] \times \mathbb{R}/\sim$ where $\sim$ identifies the points $(0, t)$ and $(1, -t)$, for all $t \in \mathbb{R}$. Thus, E is obtained by taking a vertically bi-infinite strip and identifying the sides with a twist. Projecting to the first coordinate determines a map from E to the unit interval with endpoints identified—that is, to the circle S^1.

To show that E is a vector bundle, let $U \subset S^1$ be the image of the open interval $(0, 1) \subset [0, 1]$. We denote points of S^1 by their equivalence classes $[x]$. This notation reflects, of course, an implicit choice of representative x. There is a unique choice on U, however, and we can just define $s([x]) = [(x, 1)] \in E_{[x]}$. This is a nonvanishing section on U.

Now let U' be the image in S^1 of $[0, 1] \setminus \{\frac{1}{2}\}$. We define a section s' on U' by setting $s'([x]) := [(x, 1)]$ for $x < \frac{1}{2}$ and $s'([x]) := [(x, -1)]$ for $x > \frac{1}{2}$. This procedure makes s' well defined at the endpoints and yields a continuous map $s'\colon U' \subset S^1 \to E$ which clearly does not vanish anywhere.

Exercise 5.1.5 Prove that there is a canonical bijective correspondence between the space of sections $\Gamma(E)$ of the Möbius bundle, and continuous maps $f\colon [0, 1] \to \mathbb{R}$ such that $f(0) = -f(1)$. Deduce, using the intermediate value theorem, that E is not trivial.

Example 5.1.6 (The Tangent Bundle to the n-Sphere) Consider the n-sphere S^n, the space of unit vectors in $\mathbb{R}^{n+1}$ with respect to the usual Euclidean metric. The *tangent bundle* TS^n is the vector bundle over S^n given by

$$TS^n := \{(x, v) \in S^n \times \mathbb{R}^{n+1} \mid x \perp v\}.$$

The first coordinate projection $S^n \times \mathbb{R}^{n+1} \to S^n$ restricts to a continuous surjection $\pi\colon TS^n \to S^n$. It is clear that with the usual vector space operations of $\mathbb{R}^n$, each fiber $\pi^{-1}(x)$ is a vector space; it is the orthogonal complement of x and so is a linear subspace of $\mathbb{R}^{n+1}$.

To prove that it is a real vector bundle, if $x \in S^n$, let $p_{x^\perp}\colon \mathbb{R}^{n+1} \to \mathbb{R}^n$ be the orthogonal projection to the linear subspace $\{x\}^\perp \subset \mathbb{R}^{n+1}$ followed by any, fixed, identification $x^\perp \cong \mathbb{R}^n$. Let $p_{x^\perp}(y)$ denote the restriction of $p_{x^\perp}$ to $T_y S^n$, for any y. Then it is easily checked that as long as y lies on the same side in S^n of the hyperplane $x^\perp$ as x, the linear map $p_{x^\perp}(y) \colon T_y S^n \to \mathbb{R}^n$ is a vector space isomorphism. Since these isomorphisms obviously vary continuously, they trivialize $T S^n$ in a neighborhood of x.

Example 5.1.7 (The Hopf Bundle) The following procedure defines a nontrivial complex vector bundle over n-dimensional complex projective space $\mathbb{CP}^n$, the space of one-dimensional complex subspaces of $\mathbb{C}^{n+1}$. To describe the topology on $\mathbb{CP}^n$, we can identify the set of one-dimensional subspaces of $\mathbb{C}^{n+1}$ with the quotient of the space $\mathbb{C}^{n+1} \setminus \{(0, \ldots, 0)\}$ of nonzero vectors in $\mathbb{C}^{n+1}$ by the equivalence relation which identifies two nonzero vectors if they are scalar multiplies of each other. With this identification, we can give $\mathbb{CP}^n$ the corresponding quotient topology.

There is a completely canonical (continuous) family of one-dimensional vector spaces parameterized by the points L of $\mathbb{CP}^n$: Set

$$H := \{(L, v) \mid v \in L\} \subset \mathbb{CP}^n \times \mathbb{C}^{n+1}$$

with the subspace topology of $\mathbb{CP}^n \times \mathbb{C}^{n+1}$. The first projection map $\mathrm{pr}_1 \colon \mathbb{CP}^n \times \mathbb{C}^{n+1} \to \mathbb{CP}^n$ restricts to a surjection $\pi \colon H \to \mathbb{CP}^n$.

I claim that $\pi \colon H \to \mathbb{CP}^n$ is a vector bundle. First, let us describe points of $\mathbb{CP}^n$ by their homogeneous coordinates: If L is a line in $\mathbb{C}^{n+1}$, and $(z_0, \ldots, z_n)$ is a point on the line, denote by $[z_0, \ldots, z_n]$ the equivalence class of the nonzero vector $(z_0, \ldots, z_n)$.

Let

$$U_i := \{[(z_0, \ldots, z_n)] \mid z_i \neq 0\} \subset \mathbb{CP}^n,$$

for $i = 0, 1, \ldots, n$. Then each U_i is open and $\cup_{i=0}^n U_i = \mathbb{CP}^n$. Since we are dealing with a one-dimensional vector bundle, to verify Condition c) of the definition of vector bundle, it is sufficient to produce a nonvanishing section $s_i \colon U_i \to H$ on each U_i. Since on U_i, the coordinate z_i does not vanish, we can set

$$s_i([z_0, \ldots, z_n]) := \left(\frac{z_0}{z_i}, \ldots, \frac{z_n}{z_i} \right).$$

This is well-defined, continuous, and nonvanishing on U_i, since the ith coordinate is 1.

The one-dimensional vector bundle H is usually called the *Hopf bundle*, or sometimes, "canonical line bundle."

A section of H is thus a continuous map $s \colon \mathbb{CP}^n \to \mathbb{C}^{n+1}$ such that $s(L) \in L$ for all L.

Example 5.1.8 (Induced Bundles) Let Γ be a discrete group, acting properly on X. Then there is a natural way of associating a vector bundle over $\Gamma \backslash X$ to any finite-dimensional representation $\pi \colon \Gamma \to \mathrm{GL}(V)$, V a complex vector space. As a space, let $X \times_{\Gamma, \pi} V$ be the quotient of $X \times V$ by the equivalence relation $(gx, \pi(g)v) =$

(x, v), that is, the quotient of $X \times V$ by the given group action. The first coordinate projection $X \times V$ to X induces a well-defined map $\pi : X \times_{\Gamma,\pi} V \to \Gamma \backslash X$.

The fibers of π are clearly copies of V.

Lemma 5.1.9 $\pi : X \times_{\Gamma} V \to \Gamma \backslash X$ *is a complex vector bundle.*

Proof Let $x \in X$. There is a neighborhood U of x such that $g(U) \cap U = \emptyset$ for all nonidentity elements $g \in \Gamma \setminus \{e\}$.   □

The following is an excellent and important exercise.

Exercise 5.1.10 Let $T = \mathbb{R}/\mathbb{Z}$, $T' = \widehat{\mathbb{Z}}$. Both T and T' are circles, clearly, but they have different roles in the following construction. Define an equivalence relation on $\mathbb{R} \times T' \times \mathbb{C}$ by $(x, \chi, z) \sim (x + n, \chi, \chi(n)z)$. Let L be the quotient space.

The coordinate projections define a map $\pi : L \to T \times T'$. Show that L is a rank-one complex vector bundle over $T \times T'$, whose restriction to each slice $T \times \{\chi\}$ is the induced bundle $\mathbb{R} \times_{\mathbb{Z},\chi} \mathbb{C}$ from the one-dimensional representation χ.

Exercise 5.1.11 Let $\pi : E \to X$ be a vector bundle (either real or complex). Prove that the function $x \mapsto \dim(E_x)$ is a locally constant function on X. Deduce that the fibers of a vector bundle over a connected space all have the same dimensions. This common dimension is the *rank* of the vector bundle.

Exercise 5.1.12 Prove that if $\pi : E \to X$ is a real or complex vector bundle over a locally compact space, then π is an open map.

We close this section with an important definition.

Definition 5.1.13 Let $\varphi : X \to Y$ be a map and $\pi : V \to Y$ be a vector bundle over Y. Then, the pulled-back bundle $\varphi^*(V)$ is the vector bundle over X defined as follows. As a space,

$$\varphi^*(V) := \{(x, v) \in X \times V \mid \varphi(x) = \pi(v)\},$$

topologized as a subspace of $X \times V$. The restriction $\mathrm{pr}_1|_{\varphi^*(V)} \to X$ of the first coordinate map $\mathrm{pr}_1 : X \times V \to X$ supplies the vector bundle projection; note that the fiber of $\mathrm{pr}_1|_{\varphi^*(V)}$ over $x \in X$ is $V_{\varphi(x)}$, so the fibers have natural vector space structures.

It takes only a small amount of thought to check that $\varphi^*(V)$ really is locally trivial. Indeed, suppose that $V \subset Y$ is the domain of a chart with local sections $s_1, \ldots, s_n$. Let $U := \varphi^{-1}(V) \subset X$. Then $s_1 \circ \varphi, \ldots, s_n \circ \varphi$ are continuously defined on U and by the definitions are linearly independent sections of $\varphi^*(V)$.

Definition 5.1.14 If $\varphi : X \to Y$ is a map and $\pi : V \to Y$ is a vector bundle over Y, $\varphi^*(V)$ denotes the vector bundle over X described above.

For a simple example, the pull-back of any vector space (in other words, vector bundle over a point) to any compact X under the map from X to a point is a trivial bundle over X.

Exercise 5.1.15 Prove that if $Z \subset X$ is a subspace and $i : Z \to X$ is the inclusion, then $i^*(V) = V_{|Z}$ for any vector bundle V over X.

Exercise 5.1.16 If E is the Möbius bundle over the circle $\mathbb{T}$ and $f : \mathbb{T} \to \mathbb{T}$ is the map $f(z) = z^2$, prove that $f^*(E)$ is trivial.

Direct sums and tensor products of vector bundles

Let V and W be vector bundles over X. Their *direct sum* $V \oplus W$ is defined as a space to be the (closed) subspace of $V_1 \times V_2$ (with the product topology) consisting of all (v_1, v_2) such that $\pi_1(v_1) = \pi_2(v_2)$. By the definitions, there is an obvious (continuous) map $\pi : V_1 \oplus V_2 \to X$. And each fiber $\pi^{-1}(x)$ is just $V_x \times W_x$, which can be endowed with the usual product vector space structure, making it the direct sum vector space $V_x \oplus W_x$.

Exercise 5.1.17 The direct sum $V_1 \oplus V_2$ with projection map defined above is a vector bundle.

Remark 5.1.18 It is easy to check that there are two natural inclusions $i : V \to V \oplus W$ and $j : W \to V \oplus W$, that these are vector bundle maps, and that the direct sum construction is a categorical coproduct: If V_1 and V_2 are vector bundles, W a third vector bundle, and i_1 and i_2 the inclusions $V_i \to V_1 \oplus V_2$, then for any pair of bundle maps $\varphi_1 : V_1 \to W$ and $\varphi_2 : V_2 \to W$, there is a unique vector bundle map $\varphi : V_1 \oplus V_2 \to W$ such that $\varphi \circ i_1 = \varphi_1$ and $\varphi \circ i_2 = \varphi_2$.

The *tensor product* $V \otimes W$ of two vector bundles V, W over X is defined in roughly the same way. It will be the vector bundle whose fiber at x is $V_x \otimes W_x$. Thus, as a set, $V_1 \otimes V_2$ is by definition, $\bigsqcup_{x \in X} V_x \otimes W_x$. There is of course a natural projection from this set to X.

In order to topologize the tensor product, let us cover X by open sets U on which *both* V and W are trivial. Fix such U. Suppose then that $V_{|U} \cong U \times \mathbb{R}^k$ and $W_{|U} \cong U \times \mathbb{R}^m$ by a certain pair of isomorphisms. It follows, by taking the tensor product of these isomorphisms, that we get a canonical set bijection $\bigsqcup_{x \in U} V_x \otimes W_x$ and $U \times \mathbb{R}^k \otimes \mathbb{R}^m$, which, furthermore, maps each $V_x \otimes W_x$ linearly and isomorphically to $\mathbb{R}^k \otimes \mathbb{R}^m$.

We can now specify a collection of subsets of $\bigsqcup_{x \in U} V_x \otimes U_x$ by taking images, under this isomorphism, of open subsets of $U \times \mathbb{R}^k \otimes \mathbb{R}^m$.

As U varies, the collection of all open subsets so obtained forms a basis for a topology on $\bigsqcup_{x \in X} V_x \otimes W_x$, as the reader will easily check, and makes $V \otimes W$ into a vector bundle over X.

Exercise 5.1.19 Let V, W be vector bundles over X. Let $\mathrm{HOM}(V, W)$ be defined as a set to be $\bigsqcup_{x \in X} \mathrm{Hom}(V_x, W_x)$. Topologize this in such a way as to make a vector bundle, and prove that the fiberwise evaluation maps $V_x \otimes \mathrm{Hom}(V_x, W_x) \to W_x$ piece together to give a natural vector bundle map $V \otimes \mathrm{HOM}(V, W) \to W$. Furthermore, the vector space $\mathrm{Hom}(V, W)$ of vector bundle maps from V to W is precisely the space of sections of $\mathrm{HOM}(V, W)$, by the definitions.

Exercise 5.1.20 If V is a vector bundle, the dual V^* of V is the vector bundle $V^* = \text{HOM}(V, X \times 1)$, where 1 denotes the trivial line bundle over X (real if one is working with real bundles, complex else). Thus V^* is the vector bundle whose fiber at $x \in X$ is the dual V_x^* of the vector space V_x.

Prove that if V and W are vector bundles over X, then $\text{HOM}(V, W) \cong V^* \otimes W$ as vector bundles over X.

Exercise 5.1.21 Prove that $V \otimes V^*$ is trivial for any complex line bundle V. (*Hint.* Identify it with $\text{HOM}(V, V)$ and deduce the existence of a nonvanishing section.)

5.2   Finitely Generated Projective (f.g.p.) Modules and Vector Bundles: Swan's Theorem

Gelfand's theorem interprets the class of commutative C*-algebras geometrically, in the sense of identifying them with continuous functions on their spectra. The Serre–Swan Theorem identifies the finitely generated projective module theory of a commutative C*-algebra, with the vector bundle theory of its spectrum. This idea is essential to the development of K-theory.

The first observation is that projection-valued functions on a compact space X determine canonical vector bundles over X.

Lemma 5.2.1 *Let X be a locally compact Hausdorff space and $p : X \to M_n(\mathbb{R})$ (respectively, $M_n(\mathbb{C})$) be a continuous idempotent-valued map. Let $E := \{(x, v) \in X \times \mathbb{R}^n \mid p(x)v = v\}$ (respectively, $\{(x, v) \in X \times \mathbb{C}^n \mid p(x)v = v\}$), equipped with the subspace topology; let $\pi : E \to X$ be the restriction of the first coordinate projection to E.*

Then E is a real (respectively, complex) vector bundle over X.

Thus, the fiber E_x of E at $x \in X$ is the range of $p(x)$, a subspace of $\mathbb{R}^n$.

Before going to the proof, we review one of the standard tools of vector bundle theory. The property of locally compact Hausdorff spaces stated in the Lemma is called *paracompactness*.

Lemma 5.2.2 *If X is a locally compact Hausdorff space, and if $\mathcal{U} = \{U_\alpha\}_{\alpha \in \mathcal{A}}$ is any open cover of X, then there exist:*

- *An open cover $\mathcal{V} = \{V_i\}_{i \in I}$ of X, such that every V_i is contained in some U_α, and such that if $F \subset I$ then $\cap_{i \in F} V_i \neq \emptyset$ only if F is finite.*
- *A family $\{\rho_i\}_{i \in I}$ of continuous functions $\rho_i \in C_c(X)$ of compact support, such that $0 \leq \rho_i \leq 1$ for all $i \in I$, $\text{supp}(\rho_i) \subset V_i$, and such that $\sum_{i \in I} \rho_i(x) = 1$ for all $x \in X$.*

We refer to the data consisting of the locally finite refinement $\mathcal{V} = \{V_i\}_{i \in I}$ of $\mathcal{U}$, in the above Lemma, and a collection of functions $\{\rho_i\}_{i \in I}$, subordinate to $\mathcal{V}$, as a *partition of unity subordinate to the cover $\mathcal{U}$*.

Partitions of unity are useful for proving the following facts about vector bundles. A *Euclidean structure* on a real vector bundle $\pi : E \to X$ is a family $\{\langle \cdot, \cdot \rangle_x \mid x \in X\}$ of inner products on the fibers of E such that for any two continuous sections s_1, s_2 of E, the function $x \mapsto \langle s_1(x), s_2(x) \rangle$ on X is continuous. A *Hermitian structure* on a complex vector bundle $\pi : E \to X$ is a family of Hermitian inner products on the fibers of E, which is continuous in the same sense.

Proposition 5.2.3 *Any real vector bundle over a locally compact space has a Euclidean structure, and any complex bundle has a Hermitian structure.*

Proof If $\{U_i, \varphi_i\}$ is an atlas for the real vector bundle E, and $\{\rho_i \mid i \in I\}$ is a subordinate partition of unity, then we can set, for $e, e' \in E_x$,

$$\langle e, e' \rangle_x = \sum_i \sqrt{\rho_i(x)} \, \langle \varphi_i(e), \varphi_i(e'), \rangle$$

where the right hand side refers to the usual inner product on $\mathbb{R}^n$.

This defines a Euclidean structure on E. The complex case is similar. $\qquad\square$

With partitions of unity in hand, we now prove Lemma 5.2.1.

Proof *(Of Lemma 5.2.1)* We just do the real case; the complex case works exactly the same.

Let $x_0 \in X$. Then $p(x_0)$ is an idempotent matrix in $\mathbf{GL}(n, \mathbb{R})$, with range a subspace $E_0 \subset \mathbb{R}^n$. Let $v_1, \ldots, v_k$ be a basis for E_0 (assuming E_0 is not the zero subspace, otherwise omit this step) and extend it to a basis $v_1, \ldots, v_k, v_{k+1}, \ldots, v_n$ for $\mathbb{R}^n$. Let $f : X \to M_n(\mathbb{R})$ be defined by setting $f(x)$ equal to the n-by-n matrix with columns

$$p(x)v_1, \ p(x)v_2, \ \ldots, \ p(x)v_k, \ v_{k+1}, \ldots v_n.$$

Then f takes an invertible value at x and hence takes invertible values in a neighborhood U of x_0. In particular, the vectors $p(x)v_1, \ldots, p(x)v_k$ must be linearly independent for $x \in U$, and so they form k linearly independent sections of $\pi : E \to X$ on U. This results in a local trivialization of E on U as required. $\qquad\square$

We will denote the vector bundle $\pi : E \to X$ defined by an idempotent-valued function $p : X \to M_n(\mathbb{R})$ by $\mathrm{Im}(p)$. A section of $\mathrm{Im}(p)$ is by definition a continuous map $s : X \to \mathbb{C}^n$ such that $s(x) \in \mathrm{Im}(p(x))$, or, equivalently, such that $p(x)s(x) = s(x)$ for all $x \in X$.

We recall the following general definition.

Definition 5.2.4 If A is a ring, not necessarily unital, then a *finitely generated projective module* (f.g.p. module) over A is a right A-module of the form $p \cdot A^n$, where $p \in M_n(A)$ is an idempotent.

We typically abbreviate finitely generated projective module to *f.g.p. module*.

Exercise 5.2.5 Let L and L' be f.g.p. modules over A. Prove that their direct sum $L \oplus L'$ is an f.g.p. module over A. If $L \cong pA^n$ and $L' \cong qA^m$, then $L \oplus L' \cong (p \oplus q)A^{n+m}$, where $p \oplus q$ is the block matrix $\begin{bmatrix} p & 0 \\ 0 & q \end{bmatrix}$.

Exercise 5.2.6 If $A = \mathcal{K}$ is the non-unital C*-algebra of compact operators, then f.g.p. modules over $\mathcal{K}$ correspond to finite-rank projections in $\mathcal{K}$.

Now set $A = C(X)$, X compact, so that A is unital. If $p: X \to M_n(\mathbb{C})$ is a continuous, idempotent-valued function, as in Lemma 5.2.1, then the $C(X)$-module $\Gamma(\text{Im}(p))$ of sections of $\text{Im}(p)$ is exactly equal to $p\,C(X)^n$. Therefore, it is a finitely generated projective $C(X)$-module.

We prove below that every vector bundle has the form $\text{Im}(p)$ for some p, and hence that $\Gamma(E)$ is finitely generated projective for *every* vector bundle E over a compact space.

Lemma 5.2.7 *Let $\pi: E \to X$ be a real vector bundle over a compact Hausdorff space X. Then E is isomorphic to a sub-bundle of a trivial bundle $X \times \mathbb{R}^n$ for some n. Furthermore, if $p: X \to M_n(\mathbb{R})$ is defined by setting $p(x)$ equal to the orthogonal projection onto $E_x \subset \mathbb{R}^n$, then p is a continuous, projection-valued map and $\text{Im}(p) = E$.*

In particular, the $C(X)$-module $\Gamma(E)$ is a finitely generated projective $C(X)$-module for any vector bundle E over X.

Proof Suppose $U_1, \ldots, U_m$ is a finite cover of X such that $E|_{U_i}$ is trivial for all i. Let $\varphi_i: E|_{U_i} \to \mathbb{R}^n$ the corresponding trivializations. Let ρ_i be a partition of unity subordinate to this cover, which we assume has m elements. Denote elements of V by pairs (x, v), where $\pi(v) = x$. Define a vector bundle map $\Phi: E \to X \times \mathbb{R}^n \oplus \cdots \oplus \mathbb{R}^n$ by

$$(5.1) \qquad \Phi(x, v) = (x, \oplus_i \rho_i(x) \cdot \varphi_i(x, v)) \in \{\pi(e)\} \times \mathbb{R}^n \oplus \cdots \oplus \mathbb{R}^n.$$

Φ is well defined since ρ_i is zero outside of U_i, the domain of φ_i. For $x \in X$, if $\Phi(x, v) = 0$, then $\rho_i(x) \cdot \varphi_i(x, v) = 0$ for each i, whereas $\varphi_i(x, v) \neq 0$ for every i such that $x \in U_i$, and $\rho_i(x) \neq 0$ for at least one i. So Φ is fiberwise injective and so defines fiberwise injective vector bundle map $E \to X \times \mathbb{R}^{nm}$. $\qquad \square$

For example, if $X \times \mathbb{C}^n$ is a trivial bundle, or is isomorphic to one, then its space of sections is $C(X, \mathbb{C}^n) \cong C(X) \oplus \cdots \oplus C(X)$, a free $C(X)$-module. The existence of nontrivial vector bundles over X, in general, is equivalent to the existence of finitely generated projective $C(X)$-modules, which are not free.

Example 5.2.8 (Following Example 5.1.7) Let $X = \mathbb{CP}^1$ and $\pi: H \to \mathbb{CP}^1$ the Hopf bundle. Then by the very definition, $H = \text{Im}(P)$ where $p: \mathbb{CP}^1 \to M_2(\mathbb{C})$ is the following projection-valued function. A point of $\mathbb{CP}^1$ is a line L in $\mathbb{C}^2$, by definition. So we let $P(L)$ be orthogonal projection onto this line.

To find an explicit formula for P is not difficult, using basic linear algebra. In terms of homogeneous coordinates on $\mathbb{CP}^1$,

$$P([z, w]) = \frac{1}{|z|^2 + |w|^2} \begin{bmatrix} |z|^2 & \bar{w}z \\ \bar{z}w & |w|^2 \end{bmatrix}.$$

Note also that if we restrict it to the natural copy of $\mathbb{C} \subset \mathbb{CP}^1$ by $z \mapsto [z, 1]$, we get the projection-valued map

$$p: \mathbb{C} \to M_2(\mathbb{C}), \quad p(z) = \frac{1}{|z|^2 + 1} \begin{bmatrix} |z|^2 & z \\ \bar{z} & 1 \end{bmatrix}$$

on $\mathbb{C}$. It has the property that

$$\lim_{z \to \infty} p(z) = \begin{bmatrix} 1 & 0 \\ 0 & 0 \end{bmatrix}.$$

In particular, it extends continuously to the one-point compactification $\mathbb{C}^+$ of $\mathbb{C}$, which, of course, is the same as $\mathbb{CP}^1$.

Exercise 5.2.9 Find an explicit formula for a projection-valued function $p: S^2 \to M_3(\mathbb{R})$ whose image $\mathrm{Im}(p)$ is the tangent bundle TS^2.

Lemma 5.2.10 *If E and E' are vector bundles over X compact, then $E \cong E'$ as vector bundles if and only if $\Gamma(E) \cong \Gamma(E')$ are isomorphic as $C(X)$-modules.*

Proof One direction is clear; we prove that if $\tau: \Gamma(E) \to \Gamma(E')$ is a module isomorphism, then $E \cong E'$.

First observe that if s is a section of E, then $\mathrm{supp}(s) \subset U$ if and only if $\rho s = 0$ for all $\rho \in C_c(X \setminus U)$, and since $\rho s = 0$ if and only if $\tau(\rho s) = \rho \tau(s) = 0$, we see that $\mathrm{supp}(\tau s) = \mathrm{supp}(s)$ for all sections s of E.

From this it follows that if s_1 and s_2 are two sections of E which agree at a single point, then $\tau(s_1)$ and $\tau(s_2)$ also agree at that same point.

Now let $(x, v) \in E$. Choose any section $s_v: X \to E$ such that $s_v(x) = v$. Set

$$Tv := \tau(s_v)(x) \in E'_x.$$

By the observations above, Tv does not depend on the choice of section s_v taking value v at x, and we leave it to the reader to check that $T: E \to E'$ is a vector bundle isomorphism. $\qquad \square$

Before stating Swan's Theorem, we formalize a definition. As noted above, the direct sum of two f.g.p. modules is again an f.g.p. module, corresponding to taking the direct sum of projections. This sum operation is obviously compatible with isomorphism of modules. Hence it endows the set of isomorphism classes of f.g.p. modules with a semigroup structure with $[L] + [L'] := [L \oplus L']$. This sum operation matches direct sum of vector bundles.

With these remarks, we have the following fundamental result:

Definition 5.2.11 If A is a C*-algebra, we let $\mathcal{P}(A)$ denote the semigroup of isomorphism classes of f.g.p. modules over A.

If X is a locally compact space, $\mathrm{Vect}(X)$ denotes isomorphism classes of complex vector bundles over X, also a semigroup under direct sum of vector bundles.

Theorem 5.2.12 (Swan's Theorem) *Let X be a compact metrizable space. Then the assignment $E \to \Gamma(E)$ descends to an isomorphism of semigroups $\mathrm{Vect}(X) \cong \mathcal{P}(C(X))$.*

In this correspondence, trivial vector bundles correspond to finitely generated free $C(X)$-modules.

Exercise 5.2.13 (The Poincaré Line Bundle) Let $\mathbb{T} = \mathbb{R}/\mathbb{Z}$ and $\Gamma(\mathcal{P})$ be the linear space of continuous functions f on $\mathbb{R} \times \mathbb{T}$ with the property that

$$f(s + n, t) = e^{-2\pi i n t} f(s, t)$$

for all $(s, t) \in \mathbb{R} \times \mathbb{T}, n \in \mathbb{Z}$. Give $\Gamma(\mathcal{P})$ the structure of a (right) $C(\mathbb{T}^2) = C(\mathbb{R}^2/\mathbb{Z}^2)$-module by

$$(f \cdot \varphi)(s, t) := f(s, t)\varphi(s, t) :$$

a) Check that the module structure maps $\Gamma(\mathcal{P})$ to itself.
b) Find a $C(\mathbb{T} \times \mathbb{T})$-valued inner product making $\Gamma(\mathcal{P})$ into a right Hilbert $C(\mathbb{T} \times \mathbb{T})$-module. Prove that it is finitely generated and projective.
c) Prove that $\Gamma(\mathcal{P})$ is the section module of a rank-one complex vector bundle L over $\mathbb{T} \times \mathbb{T}$; describe the bundle concretely. (See Exercise 5.1.10 and Exercise 5.4.11.)

Remarks on projective modules vs. idempotents vs. projections

We finish this section with a brief discussion of the exact relationship between finitely generated projective modules, and the idempotents which go along with them.

Assume A is a ring, unital or not.

If pA^n and qA^m are isomorphic projective modules over A, with $\alpha' : pA^n \to qA^m$ the isomorphism, $\beta' : qA^m \to pA^n$ its inverse, then we can extend α' to an A-module map $A^n \to qA^m$ which is zero on $(1 - p)A^n$, which we denote by α. Note that α is given by the left multiplication action of an m-by-n matrix with entries in A; similarly, $\beta \in M_{nm}(A)$. Multiplying these matrices one way gives $\alpha\beta = p$, and multiplying them the other way gives $\beta\alpha = q$.

Conversely, if $\alpha \in M_{mn}(A)$ and $\beta \in M_{nm}(A)$ with $\alpha\beta = p$ an idempotent in $M_n(A)$ and $\beta\alpha = q$ an idempotent in $M_m(A)$, then the projective A-modules pA^n and qA^m are isomorphic; the isomorphism is multiplication by the matrix α and its inverse is multiplication by the matrix β.

Proposition 5.2.14 *Let A be a ring. Then two finitely generated projective modules* pA^m *and* qA^n *are isomorphic if and only if there exists* $\alpha \in M_{mn}(A)$, $\beta \in M_{nm}(A)$ *such that* $\alpha\beta = p$, $\beta\alpha = q$.

We call this *algebraic equivalence* of idempotents. If A is unital, then a special case is *similarity*, as in the following easy.

Exercise 5.2.15 Prove that if A is unital, and $p, q \in M_n(A)$ are *similar*, i.e., if there is an invertible $u \in M_n(A)$ such that $upu^{-1} = q$, then p and q are algebraically equivalent.

Proposition 5.2.16 *If A is a unital C*-algebra and p and q are similar idempotents, then they are unitarily equivalent.*

Proof Suppose that $apa^{-1} = q$. Then $ap = qa$, and taking adjoints, $pa^* = a^*q$. It follows that $pa^*a = a^*qa = a^*ap$ so that p commutes with a^*a. Hence it commutes with $|a| = (a^*a)^{\frac{1}{2}}$. Write $a = u|a|$ in polar decomposition. We get

$$up = a|a|^{-1}p = ap|a|^{-1} = qa|a|^{-1} = qu$$

and hence $upu^* = q$.   □

Lemma 5.2.17 *If A is a unital C*-algebra, then every idempotent in A is similar to a projection.*

   Furthermore, if p and q are algebraically equivalent projections, with $\alpha\beta = p$ *and* $\beta\alpha = q$, *then there is a partial isometry* $u \in M_{nm}(A)$ *such that* $uu^* = \alpha$ *and* $u^*u = \beta$.

Proof The idea of the proof is to think of $A \subset \mathbb{B}(H)$ for a Hilbert space H. The range of e is a closed subspace eH of H, and H decomposes as an orthogonal direct sum $eH \oplus (eH)^\perp$ and e has a block matrix representation

$$e = \begin{bmatrix} 1 & R \\ 0 & 0 \end{bmatrix}$$

for some operator R. Let p be the operator with matrix representation $p = \begin{bmatrix} 1 & 0 \\ 0 & 0 \end{bmatrix}$ and s the operator with matrix representation $s = \begin{bmatrix} 1 & R \\ 0 & 1 \end{bmatrix}$. It is clear that $ses^{-1} = p$. It remains to show that p and s are actually in the C*-algebra A. To see this, compute with the matrices that

(5.2) $p\big(1 + ee^* + e^*e - (e + e^*)\big) = ee^*$.

Since $1 + ee^* + e^*e - (e + e^*) = 1 + (e - e^*)(e - e^*)^*$, it is a strictly positive and in particular invertible element of A. Hence by (5.2) we get

(5.3) $$p = ee^* \big(1 + (e - e^*)(e - e^*)^*\big)^{-1}$$

and so $p \in A$. Since $e - p = \begin{bmatrix} 0 & R \\ 0 & 0 \end{bmatrix}$, we get $1 + e - p = s$, so that $s \in A$ as well.

For the second statement, suppose that $p = \alpha'\beta'$ and $q = \beta'\alpha'$ for some α', $\beta' \in A$. Let $\alpha = p\alpha'q$ and $\beta = q\beta'p$. Then the equations $\alpha\beta = p$ and $\beta\alpha = q$ still hold, but now α and β satisfy $p\alpha q = \alpha$ and $q\beta p = \beta$.

Now

$$p = p^*p = \beta^*\alpha^*\alpha\beta \le \|\alpha\|^2 \cdot \beta^*\beta,$$

and similarly

$$q = q^*q = \alpha^*\beta^*\beta\alpha \le \|\beta\|^2 \cdot \alpha^*\alpha.$$

Hence $\alpha^*\alpha$ is invertible in qAq and $\beta^*\beta$ is invertible in pAp. Set $u = \beta|\beta|^{-1}$, the partial isometry in the polar decomposition of $\beta \in pAp$. Then $u^*u = p$ is immediate. Also, uu^* is a projection, since u is a partial isometry. Since $|\beta| \in pAp$, it follows that $p|\beta|^{-2} = |\beta|^{-2}$. Using this we get

$$quu^* = q\beta|\beta|^{-2}\beta^* = q\beta p|\beta|^{-2}\beta^* = \beta|\beta|^{-2}\beta^* = uu^*$$

since $q\beta p = \beta$. This shows that uu^* is a subprojection of q. On the other hand,

$$q = qq^* = \beta\alpha\alpha^*\beta^* \le \|\alpha\|^2\beta\beta^*$$

and since $\beta\beta^* = u\beta^*\beta u^*$, we get $q \le \|\alpha\|^2 u\beta^*\beta u^* \le \|\alpha\|^2\|\beta\|^2 uu^*$. Putting things together gives

$$uu^* \le q \le \|\alpha\|^2\|\beta\|^2 \cdot uu^*,$$

which implies that $uu^* = q$. $\square$

Definition 5.2.18 Let A be any C*-algebra. Two projections $p \in M_n(A)$ and $q \in M_m(A)$ are said to be *Murray–von Neumann equivalent* if there is a partial isometry $u \in M_{mn}(A)$ such that $uu^* = p$ and $u^*u = q$.

This discussion shows that as far as classifying projective modules over a unital C*-algebra, the problem is equivalent to classifying Murray–von Neumann equivalence classes of projections in $M_\infty(A)$.

The following exercise shows that Murray–von Neumann equivalence is not very far from unitary equivalence.

Exercise 5.2.19 Let p and q be projections in a unital C*-algebra A which are Murray–von Neumann equivalent. Let v be the partial isometry implementing the equivalence, with $v^*v = p$, $vv^* = q$:

a)

$$u := \begin{bmatrix} v & 1 - vv^* \\ v^*v - 1 & v^* \end{bmatrix}$$

is a unitary satisfying

$$u \begin{bmatrix} p & 0 \\ 0 & 0 \end{bmatrix} u^* = \begin{bmatrix} q & 0 \\ 0 & 0 \end{bmatrix}.$$

Thus, $\begin{bmatrix} p & 0 \\ 0 & 0 \end{bmatrix}$ and $\begin{bmatrix} q & 0 \\ 0 & 0 \end{bmatrix}$ are unitarily equivalent.

b) The unitary u is connected by a continuous path of unitaries in $M_2(A)$ to the identity $\begin{bmatrix} 1 & 0 \\ 0 & 1 \end{bmatrix}$. (Consider the path

$$u_t := \begin{bmatrix} \cos t \, v & 1 - (1 - \sin t)vv^*) \\ (1 - \sin t)v^*v - 1 & \cos t v^* \end{bmatrix},$$

which connects u to $\begin{bmatrix} 0 & 1 \\ -1 & 0 \end{bmatrix}$, for $t \in [0, \frac{\pi}{2}]$.

The latter matrix can then be connected with $\begin{bmatrix} 1 & 0 \\ 0 & 1 \end{bmatrix}$ by the same trick.)

Exercise 5.2.20 Let X be compact Hausdorff. Prove that if $p: X \to M_n(\mathbb{C})$ and $q: X \to M_n(\mathbb{C})$ are continuous, projection-valued functions, and if $\|p - q\| < 1$ as elements of the C*-algebra $C(X, M_n(\mathbb{C}))$, then $\mathrm{Im}(p) \cong \mathrm{Im}(q)$ as vector bundles.

Remark 5.2.21 This implies a certain "discreteness" of the space of isomorphism classes of vector bundles over a compact, second countable topological space: Prove that this set is countable, using the fact that the C*-algebra $C(X) \otimes M_n(\mathbb{C})$ is separable, for all n.

5.3 Multiplier Algebras

Let A be a C*-algebra, possibly not unital.

Definition 5.3.1 A *multiplier* of A is a linear map $L: A \to A$ such that:

- $L(ab) = L(a)b$ for all $a, b \in A$.
- L is *adjointable* in the sense that there exists a linear map $L^*: A \to A$ such that $L(a)^*b = a^*L^*(b)$ for all $a, b \in A$.

Example 5.3.2 Suppose that A is an ideal in a larger C*-algebra B and $x \in B$. Then left multiplication by x defines a multiplier $L_x \colon A \to A$ (with adjoint $L_x^* = L_{x^*}$). This contains the following two examples:

a) Any bounded, continuous function f on X locally compact Hausdorff, defined by pointwise multiplication, a multiplier of $C_0(X)$.
b) Any bounded operator $T \in \mathbb{B}(H)$ defines a multiplier of $\mathcal{K}(H)$, by multiplication.

For the first example, $A = C_0(X)$ is an ideal in the C*-algebra $B = C_b(X)$, and in the second, $A = \mathcal{K}(H)$ is an ideal in $B = \mathbb{B}(H)$.

Exercise 5.3.3 The Szego projection P_+ is a self-adjoint multiplier of $C^*(\mathbb{T})$.

Exercise 5.3.4 If A is a unital C*-algebra, then $A \cong \mathcal{M}(A)$ by mapping $x \in A$ to the left multiplication operator $L_x \colon A \to A$ (whose adjoint is L_{x^*}).

The linear map $L^* \colon A \to A$ specified by Definition 5.3.1, provided that it exists, is both right A-linear: $L^*(ab) = L^*(a)b$ for all $a, b \in A$, and unique. We call it the *adjoint* of L.

Lemma 5.3.5 *Multipliers are bounded: There exists $C \geq 0$ such that $\|La\| \leq C\|a\|$ for all $a \in A$.*

Proof This is a standard exercise in the Closed Graph Theorem. Note that A is in particular a Banach space, so the Closed Graph Theorem applies to a multiplier $L \colon A \to A$; to show that it is bounded, it suffices to show that the graph $\{(x, y) \in A \oplus A \mid y = L(x)\}$ is closed.

So let $(a_\lambda) \subset A$, $a_\lambda \to a$, and suppose that $L(a_\lambda) \to b$. We need to show that $L(a) = b$. But if $c \in A$, then

$$(b - L(a))^* c = \lim_{\lambda \to \infty} (L(a_\lambda) - L(a))^* c = \lim_{\lambda \to \infty} L(a_\lambda - a)^* c = \lim_{\lambda \to \infty} (a_\lambda - a) L^*(c) = 0,$$

so the result follows from Exercise 1.1.9. $\square$

Exercise 5.3.6 Prove that the adjoint operation on $\mathcal{M}(A)$ is conjugate A-linear in the sense that $(Ta)^* = a^* T^*$. Also, prove that $L^{**} = L$ for any multiplier L.

It is easy to check that a linear combination or product (that is, composition) of multipliers is again a multiplier, whilst the set $\mathcal{M}(A)$ of multipliers has an obvious adjoint operation as well, so $\mathcal{M}(A)$ is a *-algebra containing A. If L is a multiplier and $a \in A$, L_a the multiplier of left multiplication by a, then the composite multiplier $L \circ L_a$ maps b to $L(ab) = L(a)b$, which shows that $L \circ L_a = L_{L(a)}$, so A is a right ideal inside $\mathcal{M}(A)$. Since it is also closed under adjoint, it is a left ideal as well.

Set

$$\|L\| = \sup_{\|a\| \leq 1} \|L(a)\|,$$

and then $\|L\|$ defines a norm on $\mathcal{M}(A)$, which restricts to the given norm on A.

Exercise 5.3.7 Show that if L_1 and L_2 are multipliers, then $\|L_1 L_2\| \leq \|L_1\| \cdot \|L_2\|$, and if $a \in A$ and L_a the corresponding multiplier, then $\|L_a\| = \|a\|$.

Lemma 5.3.8 *If L is a multiplier, L^* its adjoint, then $\|L\| = \|L^*\|$.*

Proof We argue as follows: If $a \in A$, then

$$\|L^*(a)\| = \sup_{\|b^*\| \leq 1} \|b^* L^*(a)\| = \sup_{\|b^*\| \leq 1} \|L(b)^* a\| \leq \sup_{\|b^*\| \leq 1} \|L(b)^*\|$$

$$= \sup_{\|b\| \leq 1} \{\|L(b)\| = \|L\|.$$

The first step is by Exercise 1.1.9, and the last step is because $\|x\| = \|x^*\|$ for x in a C*-algebra. Therefore $\|L^*(a)\| \leq \|L\|$ and so $\|L^*\| \leq \|L\|$. Replacing L by L^* completes the argument. $\qquad \square$

Lemma 5.3.9 $\|L^* L\| = \|L\|^2$ *for all multipliers L.*

Proof Since $\|L^* L\| \leq \|L^*\| \|L\|$ by Exercise 5.3.7, and $\|L^*\| = \|L\|$ by Lemma 5.3.8, $\|L^* L\| \leq \|L\|^2$. On the other hand

$$\|L^* L\| = \sup_{\|a\| \leq 1} \|L^* L(a)\| = \sup_{\|a\|, \|b\| \leq 1} \|b^* L^* L(a)\| = \sup_{\|a\|, \|b\| \leq 1} \|L(b)^* L(a)\|$$

$$\geq \sup_{\|a\| \leq 1} \|L(a)^* L(a)\| = \|L\|^2.$$

$\qquad \square$

Lemma 5.3.10 *In the multiplier norm, $\mathcal{M}(A)$ is complete.*

Proof Let (L_λ) be a Cauchy net of multipliers; then $L_\lambda \to L$ where L is some bounded linear operator on the Banach space A, since the space of all bounded linear operators on a Banach space is complete in the operator norm. So it suffices to show that L is a multiplier. It follows immediately from the continuity of multiplication in A that L is right A-linear. So we are reduced to showing that L is adjointable. But since $L_\lambda \to L$ in norm, it follows from Lemma 5.3.8 that L_λ^* converges, and it is easy to check that it converges to the adjoint of L. $\qquad \square$

Definition 5.3.11 We say that a net (L_i) in $\mathcal{M}(A)$ converges in the *strict* topology to a multiplier L if $L_i(a) \to L(a)$ and $L_i^*(a) \to L^*(a)$ for all $a \in A$.

For example, by the definitions, if (u_i) is an approximate unit in A, then the net (L_{u_i}) converges strictly in $\mathcal{M}(A)$ to 1, the identity of $\mathcal{M}(A)$.

Similarly, if L is any multiplier of A, then $L(x) = \lim_{i \to \infty} L(u_i x) = \lim_{i \to \infty} L(u_i) x$. The conclusion is that the net $(L(u_i))$ (or, if one prefers, the net (u_i) itself), converges, in the strict topology, to L. That is, the map

$$A \to \mathcal{M}(A)$$

is a strictly continuous map.

Proposition 5.3.12 *If X is locally compact Hausdorff, then the natural map gives an isomorphism $\mathcal{M}(C_0(X)) = C_b(X)$, with $C_b(X)$ the C*-algebra of bounded continuous functions on X. The strict topology corresponds to the topology of uniform convergence on compact subsets.*

Proof Since bounded continuous functions clearly act by multipliers of $C_0(X)$, here is a natural *-homomorphism $C_b(X) \to \mathcal{M}(C_0(X))$. We leave it to the reader to verify that

$$\|f\| = \sup_{g \in C_0(X), \|g\| \le 1} \|fg\|,$$

for any $f \in C_b(X)$, so the inclusion $C_b(X) \to \mathcal{M}(C_0(X))$ is isometric.

Let (u_i) be an approximate unit for $C_0(X)$, and L a multiplier of $C_0(X)$; then since $u_i \to 1$ strictly, the net of functions $L(u_i)$ converges strictly as well, as multipliers of $C_0(X)$. It follows that the functions $L(u_i)$ converge uniformly on compact subsets of X. Let f be the target function. It is clearly bounded as $\|L(u_i)\| \le \|L\|$ for all i, so that the functions $L(u_i)$ are all uniformly bounded by a fixed constant, giving that f is as well, and $L(h) = fh$ for any $h \in C_0(X)$. So $L = L_f$.

Hence $C_b(X) = \mathcal{M}(C_0(X))$. $\square$

A nondegenerate homomorphism $\alpha \colon A \to B$ is one for which $\alpha(A)B$ is dense in B.

Corollary 5.3.13 *If A is any C*-algebra, then $\mathcal{M}(A)$ is a unital C*-algebra containing A as a closed and strictly dense ideal.*

*In particular, if $\varphi \colon A \to B$ is a strictly continuous *-homomorphism, then φ extends uniquely to a *-homomorphism $\mathcal{M}(A) \to \mathcal{M}(B)$.*

*Nondegenerate *-homomorphisms are strictly continuous and hence extend.*

If $I \subset \mathbb{R}$ is a bounded open interval, then the inclusion $i \colon C_0(I) \to C_0(\mathbb{R})$ does not extend to the multiplier algebras, so is not strictly continuous or nondegenerate. Indeed, $\mathcal{M}(C_0(\mathbb{R})) = C_b(I)$ and $\mathcal{M}(C_0(\mathbb{R})) = C_b(\mathbb{R})$, and there is no reasonable way to extend a bounded continuous function on I to a bounded continuous function on $\mathbb{R}$.

Exercise 5.3.14 Let H be a Hilbert space. Prove that $\mathcal{M}(\mathcal{K}(H)) \cong \mathbb{B}(H)$, and prove that the strict topology on $\mathbb{B}(H)$ corresponds to the strong* operator topology on $\mathbb{B}(H)$, in which a net (T_i) of bounded operators converges in the strong* topology to T if and only if $\lim_{i \to \infty} T_i \xi = T\xi$ and $\lim_{i \to \infty} T_i^* \xi = T^* \xi$ for all $\xi \in H$.

Exercise 5.3.15 Let $\alpha\colon A \to B$ be a *-homomorphism. Prove that $\alpha(A)B$ is dense in B if and only if α maps any approximate unit for A to an approximate unit for B. These are thus equivalent ways of defining *nondegenerate* *-homomorphism.

Exercise 5.3.16 If $\alpha\colon A \to B$ is a *-homomorphism, then $\alpha(A)$ is strictly dense in B if and only if α is nondegenerate.

Exercise 5.3.17 Suppose $\pi\colon A \to \mathbb{B}(H)$ is a nondegenerate representation of A on a Hilbert space in the sense that $\pi(A)H$ is dense in H. Prove $\pi\colon A \to \mathbb{B}(H)$ is nondegenerate. Hence it is strictly continuous and extends to a *-homomorphim $\mathcal{M}(A) \to \mathbb{B}(H)$.

Exercise 5.3.18 Let A be faithfully and nondegenerately represented as bounded operators on a Hilbert space H. Prove that

$$\mathcal{M}(A) \cong \{T \in \mathbb{B}(H) \mid T\pi(a) \text{ and } \pi(a)T \in \pi(A), \ \forall a \in A\}$$

as C*-algebras. Describe the strict topology on $\mathcal{M}(A)$ in terms of the Hilbert space.

Exercise 5.3.19 Let X be a locally compact Hausdorff space and A be any C*-algebra. Prove that $\mathcal{M}(C_0(X) \otimes A)$ is the C*-algebra of bounded, strictly continuous maps $X \to \mathcal{M}(A)$.

5.4 Hilbert Modules

Definition 5.4.1 A *Hermitian right A-module* is a complex vector space which is also a right A-module, with $\mathbb{C}$-linear A-multiplication, with the following piece of additional structure. We require a Hermitian A-valued form on E: a map

$$\langle\,\cdot\,,\,\cdot\,\rangle\colon E \times E \to A$$

$\mathbb{C}$-linear in the second variable, conjugate linear in the first, and such that:

- $\langle x, ya \rangle = \langle x, y \rangle a$ for all $x, y \in E, a \in A$.
- $\langle x, y \rangle = \langle y, x \rangle^*$ and $\langle x, x \rangle \geq 0$ for all $x \in E$.
- $\langle x, x \rangle = 0$ only $x = 0$.

The *support* of E is the closed $\mathbb{C}$-linear span of the set $\langle x, y \rangle$ of inner products of elements of E, and we say E is *full* if $\mathrm{supp}(E) = A$.

If E is complete with respect to the norm $\|x\|^2 := \|\langle x, x \rangle\|$, then we say E is a right *Hilbert A-module*. In this case, we also refer to $\langle\cdot, \cdot\rangle$ as an (A-valued) *inner product*.

Exercise 5.4.2 If E is a Hermitian right B-module with form $\langle\cdot, \cdot\rangle$, then

$$\langle xb, y \rangle = b^* \langle x, y \rangle$$

for all $x, y \in E, b \in B$.

The fact that $\|x\| := \|\langle x, x \rangle\|^{\frac{1}{2}}$ is a norm is proved below.

Sometimes we refer to a right *semi-Hermitian* A-module, if the form $\langle \cdot, \cdot \rangle$ is possibly degenerate, i.e., if there exists nonzero $x \in E$ such that $\langle x, x \rangle = 0$.

Occasionally, when the need arises, we will denote an A-valued inner project in the form $\langle \cdot, \rangle_A$. We will mostly only do this when there is more than one inner product under consideration at the same time (when we discuss strong Morita equivalence).

The most basic example of a right Hilbert A-module is A itself, with

$$\langle a, b \rangle := a^* b.$$

Note that $\|\langle a, b \rangle\| = \|a^* a\| = \|a\|^2$ so that the norm defined as above by the inner product agrees with the original norm on A.

Example 5.4.3 Let $A = M_n(\mathbb{C})$ and let E be the linear space $M_{k \times n}(\mathbb{C})$ of k-by-n matrices with complex entries. If $x, y \in M_{k \times n}(\mathbb{C})$, then $x^* y \in M_n(\mathbb{C})$, and if $a \in M_n(\mathbb{C})$, then $xa \in M_{k \times n}(\mathbb{C})$ so that $M_{k \times n}(\mathbb{C})$ is a right Hilbert A-module.

Example 5.4.4 A is a right Hilbert A-module over itself, for any A, with right module structure right algebra multiplication, and inner product $\langle a, b \rangle := a^* b$. It is full.

More generally, if J is a closed *right* ideal in a C*-algebra A, then J has the structure of a right Hilbert A-module with the evident right multiplication, and inner product $\langle a, b \rangle := a^* b \in A$. Its support is J.

For instance, we could put $A = C_0(X)$, $J = C_0(U)$, where $U \subset X$ is an open subset.

More generally, let $\pi : E \to X$ be a Hermitian vector bundle over X. Let E denote sections of E which vanish at infinity. We define a $C_0(X)$-valued form

$$\langle s_1, s_2 \rangle_{C_0(X)}(x) := \langle s_1(x), s_2(x) \rangle.$$

We obtain a right Hilbert $C_0(X)$-module.

Exercise 5.4.5 Suppose X is locally compact Hausdorff and $\mathcal{E}$ is a right Hilbert $C_0(X)$-module. Since the support of $\mathcal{E}$ is an ideal of $C_0(X)$, it has the form $C_0(U)$ for some $U \subset X$ open. In this notation, prove that

$$\overline{U} = \cap_{f \in \mathrm{ann}(\mathcal{E})} f^{-1}(0),$$

where $\mathrm{ann}(\mathcal{E}) := \{ f \in C_0(X) \mid \xi f = 0 \ \forall \xi \in \mathcal{E} \}$.

Exercise 5.4.6 Prove that if $x_0 \in X$ is any point, and if we view $\mathbb{C}$ as a (right) $C(X)$-module by evaluation of functions at x_0, then there is *no* $C(X)$-valued inner product making $\mathbb{C}$ (with this $C(X)$-module structure) into a Hilbert $C(X)$-module, unless x_0 is an isolated point of X.

Definition 5.4.7 The *standard Hilbert A-module of rank n* is $A^n := A \oplus \cdots \oplus A$ with right A-module structure $(x_1, \ldots, x_n)a := (x_1 a, \ldots, x_n a)$ and inner product

$$\langle x, y \rangle := \sum_{i=1}^{n} x_i^* x_i.$$

The *standard Hilbert A-module of rank* $\mathbb{N}$, denoted $A^{\mathbb{N}}$, is the completion of the collection of finitely supported sequences $(x_n)_{n=1}^{\infty}$ of elements of A, with right module structure $(x_n)a := (x_n a)$ and inner product

$$\langle x, y \rangle = \sum_{n=1}^{\infty} x_n^* y_n,$$

the completion taken with respect to norm

$$\|x\| := \sqrt{\|\sum_{n=1}^{\infty} x_n^* x_n\|}.$$

Exercise 5.4.8 Prove that

$$A^{\mathbb{N}} = \{(x_n)_{n=0}^{\infty} \mid x_n \in A \; \forall n, \; \sum_n x_n^* x_n \text{ converges in } A\}.$$

In slightly different terms, if H is a (separable) Hilbert space, A is a C*-algebra, then the algebraic tensor product $A \otimes_{\mathbb{C}} H$ (of vector spaces) has a natural A-valued Hermitian form $\langle a \otimes \xi, b \otimes \eta \rangle := a^* b \cdot \langle \xi, \eta \rangle$. The completion of the algebraic tensor product is then a right Hilbert A-module $A \otimes_{\mathbb{C}} H$.

Fixing an isomorphism $H \cong l^2(\mathbb{N})$ determines an obvious unitary isomorphism of right Hilbert A-modules $A \otimes_{\mathbb{C}} H \cong A^{\mathbb{N}}$.

Example 5.4.9 For $\xi, \eta \in C_c(\mathbb{R})$ let $\langle \cdot, \cdot \rangle$ be the following $C^*(\mathbb{Z})$-valued inner product:

$$(5.4) \qquad \langle \xi, \eta \rangle := \sum_{n \in \mathbb{Z}} \Big(\int_{\mathbb{R}} \overline{\xi(x+n)} \eta(x) dx \Big) \cdot [n].$$

Note that (5.4) is in the group algebra $\mathbb{C}[\mathbb{Z}]$, because ξ, η are compactly supported.

Give $C_c(\mathbb{R})$ the right $C^*(\mathbb{Z})$-module structure

$$(5.5) \qquad (\xi \cdot [n])(x) := \xi(x+n).$$

Then $C_c(\mathbb{R})$ completes under this inner product to a right Hilbert $C^*(\mathbb{Z})$-module $\mathcal{E}_{\mathbb{Z}, \mathbb{R}}$.

Exercise 5.4.10 Exhibit an explicit isomorphism $\mathcal{E}_{\mathbb{Z},\mathbb{R}} \cong L^2(\mathbb{R}) \otimes C^*(\mathbb{Z})$.

Example 5.4.11 Let G be a discrete group acting *properly and cocompactly* on a locally compact Hausdorff space X.

Consider the linear space of continuous functions $f : X \to C^*(G)$ such that $f(gx) = [g] \cdot f(x)$ for all $x \in X$. Here we are using group algebra notation, with $[g] \in \mathbb{C}[G] \subset C^*(G)$ the unitary generator corresponding to $g \in G$.

Define for two such functions f_1, f_2, an element

$$\langle f_1, f_2 \rangle \in C(G \backslash X) \otimes C^*(G) \cong C\big(G \backslash X, C^*(G)\big),$$

by

$$\langle f_1, f_2 \rangle(\dot{x}) := f_1(x)^* f_2(x),$$

where $\dot{x}$ is the orbit of x. The expression on the right is well defined, i.e., does not depend on the choice of x, since by the equivariance condition on the functions,

$$f_1(gx)^* f_2(gx) = \big([g]f_1(x)\big)^* \big([g]f_2(x)\big) = f_1(x)^*[g]^*[g]f_2(x) = f_1(x)^* f_2(x).$$

We define a right $C(G \backslash X) \otimes C^*(G)$-module structure by

$$(f \cdot g)(x) := f(x)[g], \quad f \cdot \varphi(x) := f(x)\varphi(\dot{x}).$$

Exercise 5.4.12 Check that the above determines a Hermitian right $C(G \backslash X) \otimes C^*(G)$-module.

Exercise 5.4.13 Prove that $\mathcal{E}_{G,X}$ is f.g.p. as a right $C(G \backslash X) \otimes C^*(G)$-module.

We now return to the general theory of Hilbert modules.

Lemma 5.4.14 *Let E be a semi-Hermitian right A-module (Definition 5.4.1) with semi-Hermitian form $\langle \cdot, \cdot \rangle : E \times E \to A$. Then*

(5.6) $$L(y, x)L(x, y) \leq \|L(x, x)\| \cdot L(y, y)$$

holds for all x, $y \in E$.

The lemma generalizes the Cauchy–Schwarz inequality.

Proof Fix x, $y \in E$, $a \in A$ and $t \in \mathbb{R}$. Then the positivity condition on L gives

(5.7)
$$0 \leq L(xa - ty, xa - ty) = L(xa, xa) - L(ty, xa) - L(xa, ty) + L(ty, ty)$$
$$= a^* L(x, x)a - t\big(L(y, x)a + a^* L(x, y)\big) + t^2 L(y, y).$$

Now set $a = L(x, y)$, and then we derive that $2ta^*a \leq a^*L(x, x)a + t^2L(y, y)$, that is,

$$(5.8) \qquad 2tL(y, x)L(x, y) \leq L(y, x)L(x, x)L(x, y) + t^2L(y, y).$$

In the case $L(x, x) = 0$, we get $2tL(y, x)L(x, y) \leq t^2L(y, y)$ for all $t \in \mathbb{R}$, whence it is immediate that $L(y, x)L(x, y) = 0$ as well. Otherwise, suppose $L(x, x) \neq 0$. Since $a^*ba \leq \|b\| \cdot a^*a$ holds for any a, b in a C*-algebra, $L(y, x)L(x, x)L(x, y) \leq \|L(x, x)\| \cdot L(y, x)L(x, y)$ holds, and hence from (5.8) we get

$$(5.9) \qquad 2tL(y, x)L(x, y) \leq \|L(x, x)\| \cdot L(y, x)L(x, y) + t^2L(y, y).$$

Set $t = \|L(x, x)\|$ and perform simple algebra to get

$$L(y, x)L(x, y) \leq \|L(x, x)\| L(y, y)$$

as required. $\qquad\qquad\qquad\qquad\qquad\qquad\qquad\qquad\qquad\qquad\qquad\qquad\qquad$ $\square$

Theorem 5.4.15 (The Cauchy–Schwarz Inequality for Semi-Hermitian A-Modules) *If E, with A-valued inner product $\langle \cdot, \cdot \rangle$, is a semi-Hermitian right A-module, then*

$$(5.10) \qquad \|\langle x, y \rangle\|^2 \leq \|\langle x, x \rangle\| \cdot \|\langle y, y \rangle\|$$

for all $x, y \in E$.

Exercise 5.4.16 If $E, \langle \cdot, \cdot \rangle$ is a Hermitian right A-module with, prove that $\|x\| := \|\langle x, x \rangle\|^{\frac{1}{2}}$ satisfies the triangle inequality. Hence it is a normed linear space, and can be completed, if necessary, to a Banach space.

In the semi-Hermitian case, $\|\cdot\|$ is a semi-norm, and we may mod out by zero-length vectors to get a Banach space.

Exercise 5.4.17 Let E be a semi-Hermitian right Hilbert A-module, $\|x\|^2 := \|\langle x, x \rangle\|$ the induced semi-norm on E:

a) Prove that for all $x \in E, a \in A$, $\|xa\| \leq \|x\| \cdot \|a\|$. Show by an example that the inequality may be strict.
b) Prove that

$$\|x\|^2 = \sup_{\|y\| \leq 1} \|\langle x, y \rangle\|$$

for any $x \in E$.

Exercise 5.4.18 Let E be a right Hilbert A-module with support $J \subset A$:

a) Prove that J is a closed ideal in A.

b) Prove that if $a \in J$, then $xa = 0$ for all $x \in E$ implies $a = 0$. (*Hint.* Show that $a^*\langle x, y \rangle = 0$ for all $x, y \in E$ and deduce that $a^*J = 0$.)

c) Prove that if A is unital, then $x \cdot 1 = x$ for all $x \in E$, where $1 \in A$ is the unit.

d) Prove that if (u_λ) is an approximate unit for the support ideal J, then $\lim_{\lambda \to \infty} x u_\lambda = x$ for all $x \in E$.

Exercise 5.4.19 Let $\mathcal{E}$ be a right Hilbert A-module.

Prove that the right multiplication action of A on $\mathcal{E}$ extends to an action of the multiplier algebra $\mathcal{M}(A)$ on $\mathcal{E}$. (*Hint.* Show that if $(a_n) \subset A$ with $a_n \to a$ strictly, then$(\xi \cdot a_n)$ is a Cauchy sequence of vectors in $\mathcal{E}$.)

Exercise 5.4.20 Let p be a projection in a C*-algebra A. Prove that pA is a closed right ideal in A and hence is a right Hilbert A-module, with inner product $\langle pa, pb \rangle := a^*pb$. Check that its support is the closed (2-sided) ideal ApA generated by p.

Remark 5.4.21 The focus on (semi-) Hermiitian *right* modules in the discussion above is by convention; a (semi-) Hermitian *left* A-module is defined analogously to right modules; the (semi-)Hermitian form is then required to be conjugate linear in the second variable, linear in the first, and satisfy $a\langle x, y \rangle = \langle ax, y \rangle$, for all $a \in A, x, y \in E$, but otherwise all axioms and the corresponding results (especially the Cauchy–Schwartz inequality) remain the same.

Exercise 5.4.22 Suppose $\pi \colon Y \to X$ is a smooth submersion between smooth manifolds. Let $V = \ker(D\pi) \subset TY$ be the "vertical tangent bundle." Then the inclusion $V \to TY$ of vector bundles restricts on each fiber $\pi^{-1}(x)$ to an isomorphism

$$V|_{\pi^{-1}(x)} \cong T\left(\pi^{-1}(x)\right).$$

Assume that V admits an orientation and Euclidean structure; it then admits a smoothly varying family of volume forms ω_x on the fibers. Define a Hermitian form

$$\langle \cdot, \cdot \rangle \colon C_c^\infty(Y) \to C^\infty(X), \quad \langle \xi, \eta \rangle(x) := \int_{\pi^{-1}(x)} \overline{\xi(y)}\eta(y)\omega_x.$$

Show that $C_c^\infty(Y)$ completes in this way to a right Hilbert $C_0(X)$-module $\mathcal{E}_\pi$.

Such Hilbert modules play an important role in families index theorems.

5.5   Operators on Hilbert Modules, Tensor Products, and Applications

It turns out that for Hilbert module maps, bounded does *not* imply the existence of an adjoint, so this has to be assumed to obtain a C*-algebraic structure on the Hilbert module maps.

Definition 5.5.1 Let E_1 and E_2 be Hilbert A-modules. An A-module map $T \colon E_1 \to E_2$ is *adjointable* if there exists an A-module map $T^* \colon E_2 \to E_1$ such that

$$(5.11) \qquad \langle Tx, y \rangle = \langle x, T^* y \rangle, \quad \forall x \in E_1, y \in E_2.$$

The collection of adjointable operators $E_1 \to E_2$ is a *-algebra denoted $\mathbb{B}(E_1, E_2)$. When $E_1 = E_2$ we write $\mathbb{B}(E)$, which we will show shortly is a C*-algebra.

Example 5.5.2 A multiplier $L \colon A \to A$ of a C*-algebra A is an adjointable A-module map, with adjoint L^*.

Exercise 5.5.3 Let A be a C*-algebra and T be an n-by-n matrix of multipliers T_{ij} of A. Prove that matrix multiplication by T is an adjointable operator on A^n. What is the matrix of T^*?

Exercise 5.5.4 Let $E = A^n$ the standard rank n Hilbert A-module. Let $M_n(A)$ act on E by matrix multiplication. Prove that this gives an *-isomorphism $M_n(A) \cong \mathbb{B}(E)$.

Returning to the general situation, the uniqueness of the adjoint of $T \colon E_1 \to E_2$, if it exists, follows from a standard argument from Hilbert space theory:

$$\langle x, (T_1^* - T_2^*)y \rangle = \langle x, T_1^* y \rangle - \langle x, T_2^* y \rangle = \langle Tx, y \rangle - \langle Tx, y \rangle = 0,$$

for two A-linear maps satisfying (5.11). In Lemma 5.3.5 we proved that multipliers are bounded; we leave it as an exercise to adapt the proof cosmetically to work for general adjointable operators.

Exercise 5.5.5 Using the Closed Graph Theorem along the lines of the proof of Lemma 5.3.5, prove that an adjointable operator $T \colon E_1 \to E_2$ is bounded in the respective Hilbert module norms: There exists $C \geq 0$ such that $\|Tx\| \leq C\|x\|$, $\forall x \in E_1$.

The operator norm $\|T\|$ for $T \in \mathbb{B}(E_1, E_2)$ an adjointable operator is defined in the usual way by

$$\|T\| = \sup_{\|x\| \leq 1} \|Tx\|.$$

Lemma 5.5.6 *Let A be any C*-algebra and E a right Hilbert A-module. Then $\mathbb{B}(E)$, with the operator norm, is a C*-algebra.*

Proof The proof works exactly the same as it does when $A = \mathbb{C}$, where we are talking about bounded operators on a Hilbert space. If T is adjointable, and $x \in E$ is a unit vector, then by the Cauchy–Schwarz inequality

(5.12)
$$\|Tx\|^2 = \|\langle Tx, Tx\rangle\| = \|\langle T^*Tx, x\rangle\| \le \|T^*Tx\| \le \|T^*T\| \le \|T^*\| \cdot \|T\|.$$

Thus $\|Tx\|^2 \le \|T^*\|\|T\|$ for unit vectors x, and hence taking sups, we get

(5.13) $$\|T\|^2 \le \|T^*T\| \le \|T^*\|\|T\|.$$

Interchanging the roles of T and T^* gives that $\|T\| = \|T^*\|$ and making the corresponding adjustment to (5.13), we deduce without further ado that $\|T\|^2 = \|T^*T\|$.   □

In particular, this gives another proof that $M_n(A)$ is a C*-algebra for every A, since

$$M_n(A) = \mathbb{B}(A^n)$$

by Exercise 5.5.4.

Exercise 5.5.7 Here is an "easy" proof that any element of the form a^*a in a C*-algebra A is positive (i.e., $\mathrm{Spec}(a^*a) \subset [0, \infty)$), the content of Theorem 4.1.7. It follows from our general results above that $M_2(A) = \mathbb{B}(A \oplus A)$ is a C*-algebra. Choose any $a \in A$. Consider the element $\tilde{a} = \begin{bmatrix} 0 & a^* \\ a & 0 \end{bmatrix} \in M_2(A)$. Since $\tilde{a}$ is self-adjoint, it generates a commutative C*-algebra, the basic Gelfand calculus applies, and $\tilde{a}$ has real spectrum and $\tilde{a}^2$ has positive spectrum.

But $\mathrm{Spec}(\tilde{a}^2) = \mathrm{Spec}(a^*a) \cup \mathrm{Spec}(aa^*)$, as the reader will verify from the matrix form of $\tilde{a}^2$.

Why does this in fact not give an easier proof than the one given after Theorem 4.1.7?

(*Hint.* In the proof of the key Lemma 5.4.14 we needed $a^*ba \le \|b\| \cdot a^*a$ for any a, b to prove Cauchy–Schwarz and that $\mathbb{B}(E)$ is a C*-algebra, and in particular that $M_2(A)$ is a C*-algebra. But this statement already implies positivity of any a^*a, what we are trying to prove, by setting $b = aa^*$.)

If E_1, E_2 are right Hilbert A-modules, and $x \in E_2$, $y \in E_1$, let

(5.14) $$\theta_{x,y} \colon E_1 \to E_2, \quad \theta_{x,y}(z) := x\langle y, z\rangle.$$

From the right A-linearity of the inner product, $\theta_{x,y}$ is right A-linear, i.e., is a module map, and clearly has range xA the rank-one submodule of E_2 generated by x.

Exercise 5.5.8 Let E_1, E_2 be right Hilbert A-modules:

a) If $x \in E_2$, $y \in E_1$, then $\theta_{x,y}$ is adjointable and $\theta_{x,y}^* = \theta_{y,x}$.
b) If E_3 is a third right Hilbert A-module, $x \in E_3$, $y, x' \in E_2$, $y', z \in E_1$, then
   $\theta_{x,y} \circ \theta_{x',y'} = \theta_{x\langle y, x'\rangle, y'}$.

c) If E_3 is a third right Hilbert A-module, $T : E_2 \to E_3$ an adjointable operator, $x \in E_2$, $y \in E_1$, then $T \circ \theta_{x,y} = \theta_{Tx,y}$. Similarly, if $T : E_1 \to E_2$, $x \in E_3$, $y \in E_2$, then $\theta_{x,y} \circ T = \theta_{x,T^*y}$.

d) Prove that $\|\theta_{x,y}\| \le \|x\| \cdot \|y\|$.

Exercise 5.5.9 If E is a right Hilbert B-module and $x \in E$, prove that the operator

$$T : E \to B, \quad T(z) := \langle x, z \rangle$$

is an adjointable operator $E \to B$ between Hilbert B-modules, and that $T^*(b) = xb$. Check that $T^*T = \theta_{x,x}$ and that TT^* is multiplication by $\langle x, x \rangle$.

Definition 5.5.10 If E_1 and E_2 are Hilbert modules, a finite linear combination of operators $E_1 \to E_2$ of the form (5.14) is a *finite-rank operator* $E_1 \to E_2$. A norm limit of finite-rank operators $E_1 \to E_2$ is a *compact operator* $E_1 \to E_2$. The collection of compact operators $E_1 \to E_2$ is denoted $\mathcal{K}(E_1, E_2)$.

Note that our definition of "finite-rank" operator T on a Hilbert module requires that T has an adjoint. This is thus a stronger notion, even just in the case of operators on Hilbert spaces, than just requiring that the range of T be finite-dimensional, since there are unbounded linear functionals on any Hilbert space.

As with bounded operators we just write $\mathcal{K}(E)$ for the compact operators $E \to E$.

Exercise 5.5.11 Let $E_1 = E_2 = E$. Prove that the collection of finite-rank operators $\sum \lambda_j \theta_{x_j, y_j} : E \to E$ (finite sum, $\lambda_j \in \mathbb{C}$, $x_j, y_j \in E$), is a *-subalgebra of $\mathbb{B}(E)$, an algebraic ideal in $\mathbb{B}(E)$. Deduce that $\mathcal{K}(E)$ is a closed ideal of $\mathbb{B}(E)$.

The previous exercise shows that $\mathcal{K}(E)$ is a closed ideal in $\mathbb{B}(E)$ and hence is a C*-algebra in its own right.

The quotient C*-algebra $\mathbb{B}(E)/\mathcal{K}(E)$ is a Hilbert module version of the Calkin algebra.

Exercise 5.5.12 Use the result of Exercise 5.5.11 to deduce that if E is a Hilbert module and $T \in \mathbb{B}(E)$, then $T \in \mathcal{K}(E)$ if and only if $T^*T \in \mathcal{K}(E)$. (*Hint.* Consider the question in the Calkin algebra $\mathbb{B}(E)/\mathcal{K}(E)$.)

Proposition 5.5.13 *If A is a C*-algebra, regarded as a right Hilbert A-module, then $\mathcal{K}(A) = A$, and $\mathbb{B}(A) = \mathcal{M}(A)$.*

Proof Firstly, adjointable operators on the right Hilbert A-module A are precisely multipliers of A by definition of multiplier. So $\mathbb{B}(A) \cong \mathcal{M}(A)$. If $x, y \in A$, the rank-one operator $\theta_{x,y} : A \to A$ is the map $\theta_{x,y}(a) = xy^*a$, thus is left multiplication by $xy^* \in A$, so it is in the image of (the isometric *-homomorphism) $A \to \mathcal{M}(A) = \mathbb{B}(A)$. Hence the closed span of the $\theta_{x,y}$ is in the image, so $\mathcal{K}(A)$ is contained in the image. Conversely, let $x \in A$; then $xx^* \in \mathcal{K}(A)$ since it equals $\Theta_{x,x}$ as an operator. Hence x^*x is compact, whence so is x, by Exercise 5.5.12. $\square$

Exercise 5.5.14 Let A be a C*-algebra:

a) Generalize the Proposition 5.5.13 and prove that $\mathcal{K}(A^n) \cong M_n(A)$ for all $n = 1, 2, \ldots$.

b) Prove that $\mathbb{B}(A^n) \cong M_n(\mathcal{M}(A))$ (c.f. Exercise 5.5.3) for any $n = 1, 2, \ldots$. Hence if A is unital, then

$$\mathbb{B}(A^n) \cong \mathcal{K}(A^n) \cong M_n(A).$$

c) Prove that $\mathcal{K}(A^{\mathbb{N}}) \cong A \otimes \mathcal{K}$.

d) Prove that $\mathbb{B}(A^{\mathbb{N}}) \cong \mathcal{M}(A \otimes \mathcal{K})$.

The following result is very useful. Its proof is easy.

Lemma 5.5.15 *Let A and D be C*-algebras and let $\rho\colon A \to \mathcal{M}(D)$ be a nondegenerate *-homomorphism. Then ρ induces a canonical *-homomorphism*

$$(5.15) \qquad\qquad \mathbb{B}(A^{\mathbb{N}}) \to \mathbb{B}(D^{\mathbb{N}}),$$

which agrees with $\rho \otimes \mathrm{id}_{\mathcal{K}}$ on the closed ideal $A \otimes \mathcal{K} \cong \mathcal{K}(A^{\mathbb{N}})$. Moreover, (5.15) is injective if ρ is injective.

Example 5.5.16 Let $\mathcal{E}_{\mathbb{Z},\mathbb{R}}$ be the right Hilbert $C^*(\mathbb{Z})$-module of Example 5.4.9: the completion of $C_c(\mathbb{R})$ with respect to the $C^*(\mathbb{Z})$-valued Hermitian form

$$\langle \xi, \eta \rangle := \sum_{n \in \mathbb{Z}} \langle \xi \cdot n, \eta \rangle [n] \in \mathbb{C}\mathbb{Z} \subset C^*(\mathbb{Z})$$

(using group algebra notation). Here $\xi \cdot n$ denotes the function $(\xi \cdot n)(x) = \xi(x+n)$; this formula for $\xi \cdot n$ also determines the right $C^*(\mathbb{Z})$-module structure.

If $f \in C_c(\mathbb{R})$, let $\lambda(f)\colon \mathcal{E}_{\mathbb{Z},\mathbb{R}} \to \mathcal{E}_{\mathbb{Z},\mathbb{R}}$ be the map $\lambda(f)$ of convolution by f:

$$\lambda(f)\xi(x) := \int_{\mathbb{R}} f(y)\xi(x - y)dy.$$

Exercise 5.5.17 Prove the following:

a) Prove that if $f \in C_c(\mathbb{R}) \subset C^*(\mathbb{R})$, then $\lambda(f)$ is an adjointable right Hilbert $C^*(\mathbb{Z})$-module map, and that λ determines a C*-algebra homomorphism

$$C^*(\mathbb{R}) \to \mathbb{B}(\mathcal{E}_{\mathbb{Z},\mathbb{R}}).$$

b) Prove that $\lambda(f)$ acts by a *compact* Hilbert $C^*(\mathbb{Z})$-module operator on $\mathcal{E}_{\mathbb{Z},\mathbb{R}}$, for all $f \in C^*(\mathbb{R})$.

c) Let D be the densely defined, unbounded operator $D = i\frac{d}{dx}$, acting on compactly supported functions in $C_c^\infty(\mathbb{R})$. Prove that $1 + D^2$ extends to an *invertible* bounded operator on $\mathcal{E}_{\mathbb{Z},\mathbb{R}}$, and that $(1 + D^2)^{-1}$ is a compact Hilbert $C^*(\mathbb{Z})$-module map (it follows from part b).

A remark on f.g.p. Hilbert modules

Finitely generated projective modules (f.g.p.) have already been studied in the case of commutative C*-algebras $A = C(X)$; they correspond to complex vector bundles over X, by Swan's theorem. Recall that a finitely generated right A-module E is *finitely generated projective* (or f.g.p.) if there exists an idempotent $e \in M_n(A)$ such that E is isomorphic to eA^n. We have already proved that if A is unital, then] every idempotent is similar to a projection, and so the module is isomorphic as a module to pA^n, for some projection $p \in M_n(A)$, which has the structure of a right Hilbert A-module as a subset of A^n, and which is also orthogonally complemented in A^n with orthogonal complement $(1 - p)A^n$.

Lemma 5.5.18 *Let E be a right Hilbert A-module. Then if the identity operator* $\mathrm{id}_E \colon E \to E$ *is compact, then E is f.g.p.*

Proof If the identity is compact, there exist $x_1, \dots, x_n, y_1, \dots, y_n \in E$ such that $\|\mathrm{id}_E - \sum_i \theta_{x_i, y_i}\| < 1$. This makes the finite-rank operator $\sum_i \theta_{x_i, y_i}$ invertible. If $S \in \mathbb{B}(E)$ is its inverse, then $\mathrm{id}_E = S \sum_i \theta_{x_i, y_i} = \sum \theta_{Sx_i, y_i}$.

So we may assume, after replacing x_i with Sx_i for each i, that $\sum_i \theta_{x_i, y_i} = \mathrm{id}_E$ to begin with. Therefore, we have the identity

$$(5.16) \qquad\qquad x = \sum_i x_i \langle y_i, x \rangle \quad \forall x \in E.$$

Applying (5.16) to $x = x_j$ and taking the inner product with y_k give the identity

$$(5.17) \qquad\qquad \langle y_k, x_j \rangle = \sum_i \langle y_k, x_i \rangle \langle y_i, x_j \rangle.$$

Let $p \in M_n(A)$ the matrix $p_{ij} := \langle y_i, x_j \rangle$. The above identity can be written

$$(5.18) \qquad\qquad p_{kj} = \sum_i p_{ki} p_{ij}.$$

Hence p is an idempotent. We claim that the f.g.p. module pA^n is isomorphic to E. Let $T \colon E \to A^n$ be the map $T(x) = (\langle y_1, x \rangle, \dots, \langle y_n, x \rangle)$. Let $S \colon A^n \to E$ be the map

$$S(a_1, \dots, a_n) := \sum x_i a_i.$$

Then $TSa = pa$, as is easily checked, with p the projection matrix above, and any $a \in A$, and $STx = x$, for all $x \in E$, from (5.16). So T defines an isomorphism from E onto the direct summand pA^n of A^n. $\qquad\qquad\square$

Exercise 5.5.19 Give an example of an f.g.p. module over $C(S^2)$ which is not a free module.

Tensor products of Hilbert modules

Let E be a right Hilbert A-module, E' a right Hilbert B-module. Their tensor product in the category of complex vector spaces may be completed to a right Hilbert $A \otimes B$-module using the inner product

(5.19) $\langle x_1 \otimes y_1, x_2 \otimes y_2 \rangle := \langle x_1, x_2 \rangle \otimes \langle y_1, y_2 \rangle \in A \otimes B.$

The result is called the *external product* and denoted $E \otimes_{\mathbb{C}} E'$.

Another very important construction involves bimodules. We consider a right Hilbert B-module E together with a *-homomorphism $\pi : A \to \mathbb{B}(E)$. Algebraically, such an object is an A-B-bimodule (one has two multiplications, on the right by B and on the left by A, and these multiplications commute). Taking into account this extra structure, we might refer to this data as specifying a *right A-B Hilbert bimodule*, but the terminology is a bit cumbersome.

Algebraically, an A-B-bimodule defines a map from right A-modules to right B-modules by the tensor product construction described below. The procedure can be "Hilbert-ized" and is important, especially in KK-theory.

Suppose E is a right Hilbert A-module and that E' is an A-B Hilbert bimodule in the above sense, with $\pi : A \to \mathbb{B}(E')$. Form the quotient of the $\mathbb{C}$ vector space tensor product $E \otimes_{\mathbb{C}} E'$ (a right B-module), by the right B-submodule generated by the elements

(5.20) $xa \otimes y - x \otimes \pi(a)y.$

In the algebraic setting, the tensor product just defined is usually denoted $E \otimes_A E'$. We will retain the same notation for its completed version.

Definition 5.5.20 In the above notation, with E a right Hilbert A-module, E' a right Hilbert B-module, and $\pi : A \to \mathbb{B}(E')$ a representation, we will let $E \otimes_A E'$ denote the completion of the algebraic tensor product of modules described above, with respect to the semi-Hermitian B-valued form

(5.21) $\langle x_1 \otimes y_1, x_2 \otimes y_2 \rangle := \langle y_1, \pi(\langle x_1, x_2 \rangle) y_2 \rangle_B, \qquad x_1, x_2 \in E, \quad y_1, y_2 \in E'.$

Exercise 5.5.21 Check that (5.21) annihilates the relation (5.20) and descends to give a semi-Hermitian B-valued form on $E \otimes_{\mathbb{C}} E'$.

The completion $E \otimes_A E'$ is a right Hilbert B-module containing the algebraic version as a dense submodule.

We leave it to the reader to check the following:

Exercise 5.5.22 In the above notation, if $T \in \mathbb{B}(E)$, then there is a unique adjointable operator $T \otimes 1$ on $E \otimes_A E'$ such that

$$(T \otimes 1)(x \otimes y) = T(x) \otimes y, \quad \forall x \in E, y \in E'.$$

The adjoint of $T \otimes 1$ is $T^* \otimes 1$.

Remark 5.5.23 The map $T \mapsto T \otimes 1$ defines a *-homomorphism $A \to \mathbb{B}(E \otimes_A E')$. In particular, it is contractive, and hence automatic that $\|T \otimes 1\| \leq \|T\|$, where the norms are the respective operator norms.

Exercise 5.5.24 Let $\alpha \colon A \to B$ be a *-homomorphism. The right Hilbert B-module given by B itself has a representation of A by module maps on it using α:

$$a \cdot b := \alpha(a)b.$$

So a *-homomorphism $\alpha \colon A \to B$ determines a canonical A-B Hilbert bimodule E_α. Prove that if $\alpha \colon A \to B$ and $\beta \colon B \to C$ are *-homomorphisms, then $E_{\beta \circ \alpha} \cong E_\alpha \otimes_B E_\beta$ as right Hilbert C-modules, where the tensor product is over the homomorphism $\beta \colon B \to C \subset \mathbb{B}(E_\beta)$.

Exercise 5.5.25 Let $\mathcal{E}$ be a f.g.p. Hilbert A-module, i.e., an orthogonally complemented Hilbert submodule of A^n for some n. Let $\alpha \colon A \to \mathbb{B}(\mathcal{E})$ be a representative of A as bounded adjointable operators on a right Hilbert B-module $\mathcal{E}$. Show that the algebraic tensor product $\mathcal{E}_A \otimes_A \mathcal{E}$ is already complete with respect to the tensor product Hermitian form defined above, so that the tensor product in the category of Hilbert modules is the same as the algebraic tensor product.

Example 5.5.26 A particular case of the tensor product construction shows that if A and B are C*-algebras and $\varphi \colon A \to \mathcal{M}(B)$ a *-homomorphism, then, regarding φ as a representation of A in $\mathbb{B}(B)$ (by multipliers) we can form, for any right Hilbert A-module $\mathcal{E}$, the tensor product $\mathcal{E} \otimes_A B$. This results in a Hilbert B-module: the "pushforward" of $\mathcal{E}$ under φ.

Exercise 5.5.27 Suppose $\mathcal{E}_A$ is a right Hilbert A-module, and that $\pi \colon A \to \mathbb{B}(\mathcal{E}'_B)$ is a nondegenerate representation as bounded adjointable operators on $\mathcal{E}'_B$, a right Hilbert B-module. The nondegeneracy implies that the representation extends to the multiplier algebra $\mathcal{M}(A)$.

By Exercise 5.4.19, the right multiplication action of A on $\mathcal{E}_A$ also extends to an action of the multiplier algebra $\mathcal{M}(A)$.

Prove that in the tensor product of Hilbert modules

$$\mathcal{E}_A \otimes_A \mathcal{E}'_B,$$

that the vectors

$$\xi \cdot a \otimes_A \eta - \xi \otimes_A a \cdot \eta$$

are zero even if $a \in \mathcal{M}(A)$. (*Hint.* Compute, for $a_n \to a$ strictly, $a_n \in A$, $a \in \mathcal{M}(A)$,

$$\|\xi \cdot a_n \otimes_A \eta - \xi \cdot a \otimes_A \eta\|^2 \to 0$$

directly, for vectors $\xi, \eta \in \mathcal{E}_A, \mathcal{E}'_B$), and $n \to \infty$.)

Exercise 5.5.28 Consider the right $C^*(\mathbb{Z})$-module $\mathcal{E}_{\mathbb{Z},\mathbb{R}}$ of Example 5.4.9. Every point $\omega \in \mathbb{T}$ determines a *-homomorphism $C^*(\mathbb{Z}) \to \mathbb{C}$. Tensoring over this *-homomorphism gives the right Hilbert $\mathbb{C}$-module $H_\omega := \mathcal{E}_{\mathbb{Z},\mathbb{R}} \otimes_\omega \mathbb{C}$, that is, a Hilbert space.

Prove that this "bundle" of Hilbert spaces $\{H_\omega \mid \omega \in \mathbb{T}\}$ can be described as follows. For each ω, H_ω is the Hilbert space completion of the space of continuous functions φ on $\mathbb{R}$ such that $\varphi(x + n) = \omega^n \varphi(x)$ for all integers n, and inner product

$$\langle \varphi, \psi \rangle := \int_0^1 \overline{\varphi(t)} \psi(t) dt.$$

Although all the Hilbert spaces H_ω are isomorphic to each other as Hilbert spaces, one cannot find a *continuous* (in $\omega \in \mathbb{T}$) family of such isomorphisms.

Exercise 5.5.29 This exercise follows from Exercise 5.4.22 and gives a geometric interpretation of certain tensor products of Hilbert modules occurring in index theory.

Assume X, Y, Y' are smooth manifolds. Let $\pi : Y \to X$ be a submersion with orientable fibers. Let $\mathcal{E}_\pi$ the Hilbert $C_0(X)$-module constructed in the cited exercise. Let $\pi' : Y' \to Z$ be another such submersion, $E_{\pi'}$ the corresponding right Hilbert $C_0(Z)$-module. And suppose that $b : Y' \to X$ is a smooth map:

a) $Y \times_X Y' := \{(y, y') \mid \pi(y) = b(y')\}$ is a smooth manifold of dimension dim $Y +$ dim $Y' -$ dim X.
b) Let $\pi'' : Y \times_X Y' \to Z$ the second projection map, restricted to $Y \times_X Y'$, followed by $\pi' : Y' \to Z$. Show that π'' is a smooth submersion with oriented fibers.
c) The map b determines a *-homomorphism $C_0(X) \to C_b(Y')$. By construction, there is a natural representation of $C_0(Y')$ in $\mathcal{E}_{\pi'}$. Hence we can take the product of Hilbert modules $\mathcal{E}_\pi \otimes_{C_0(X)} \mathcal{E}_{\pi'}$. Show that

$$\mathcal{E}_\pi \otimes_{C_0(X)} \mathcal{E}_{\pi'} \cong \mathcal{E}_{\pi''}$$

as right Hilbert $C_0(Z)$-modules.

Exercise 5.5.30 The following construction amounts to a type of GNS construction for Hilbert modules.

Let A be a unital C*-algebra and $\varphi : A \to \mathbb{C}$ a state. Suppose that $\mathcal{E}$ is a right Hilbert A-module:

a) Prove that the sesquilinear form $(\xi, \eta) := \varphi(\langle \xi, \eta \rangle)$ makes $\mathcal{E}$ into a semi-Hermitian right Hilbert $\mathbb{C}$-module. So its completion is a Hilbert space $L^2(\mathcal{E}, \varphi)$.
b) Prove that any adjointable operator T on $\mathcal{E}$ determines a bounded operator on $L^2(\mathcal{E}, \varphi)$, so that we obtain a representation

$$\pi : \mathbb{B}(\mathcal{E}) \to \mathbb{B}(L^2(\mathcal{E}, \varphi))$$

of the C*-algebra $\mathbb{B}(\mathcal{E})$ on the Hilbert space $L^2(\mathcal{E}, \varphi)$.

c) Prove that if φ is a *trace*, then "scalar" multiplication by elements of A determines a representation $\rho: A \to \mathbb{B}\big(L^2(\mathcal{E}, \varphi),\big)$ of the C*-algebra A (on the right; that is, it is a representation of the opposite algebra A^{op} of A).

d) Check that with the assumption that φ is a trace, that the two actions (representations) of A and of $\mathbb{B}(\mathcal{E})$ commute.

Application to the theory of crossed products

In Section 1.8 we introduced crossed product C*-algebras. Let G be a discrete (countable, as usual) group and A a G-C*-algebra. To form the crossed product, one first forms the twisted group *-algebra $A[G]$. Previously, in order to make $A[G]$ into a C*-algebra, we chose an injective representation of A on a Hilbert space. This begs the question of the independence of the choice of representation. In fact there is a completely canonical representation of A as bounded (adjointable) operators on the right Hilbert A-*module* $l^2(G, A) := \{(a_g)_{g \in G} \mid \sum_{g \in G} a_g^* a_g \text{ converges in } A\}$, with A-valued inner product

$$\langle a, b \rangle = \sum_{g \in G} a_g^* b_g.$$

Define a *-homomorphism

$$\lambda_{A,G}: A[G] \to \mathbb{B}(l^2(G, A))$$

from the *-algebra $A[G]$ to the C*-algebra of adjointable operators on $l^2(G, A)$ by the covariant pair

(5.22) $\lambda_{A,G}(g)(a_h)_{h \in G} = (a_{gh})_{h \in G}, \quad \lambda_{A,G}(a)(a_h)_{h \in G} := \big(h^{-1}(a)a_h\big)_{h \in G}.$

It is left to the reader to check that

$$\lambda_{G,A}\big(h(a)\big) = \lambda_{A,G}(h)\lambda_{A,G}(a)\lambda_{A,G}(h)^*,$$

so that this is a covariant pair, and determines a *-homomorphism as required (see Exercise 1.8.4 e)).

We may now give a much better definition of the crossed product:

Definition 5.5.31 The *reduced crossed product* $A \rtimes G$ of a discrete group G acting by automorphisms of a C*-algebra A is the completion of the pre-C*-algebra $(A[G], \|\cdot\|_{\lambda_{A,G}})$.

Remark 5.5.32 As a general matter regarding cross products, the *-homomorphism

$$\lambda_{A,G}: A[G] \to \mathbb{B}(l^2(G, A))$$

maps the group algebra $A[G]$ into Hilbert A-module maps on $l^2(G, A)$. So we can represent the operator corresponding to a given element of $A[G]$ as a G-by-G matrix of elements of A, by trivializing the Hilbert A-module $l^2(G, A)$ in the standard way.

For example, if $a \in A$, its matrix is diagonal with $g^{-1}(a)$ in the (g, g)th coordinate. If $g \in G$ is a group element, and A is unital, so that we can view the elements of G as (unitary) elements of $A[G]$, then $\lambda_{A,G}(g)$ has constant g-diagonal (a collection of entries of the form (h, hg)) with 1's along it, so has the form

$$
S = \begin{bmatrix}
0 & 1 & 0 & \cdots\cdots\cdots \\
\cdots & 0 & 1 & 0 & \cdots \\
\cdots\cdots & 0 & 1 & 0 \\
\cdots\cdots\cdots & 0 & 1 \\
\cdots\cdots\cdots & 0 & 0 \\
\cdots\cdots\cdots\cdots & 0
\end{bmatrix}.
$$

Thus, the matrix representation of the general element of $A[G]$ (not of $A \rtimes G$, of course) has only finitely many nonzero diagonals.

Proposition 5.5.33 *If* $\alpha : A \rightarrow B$ *is a* G-*equivariant nondegenerate* *-*homomorphism, then* α *extends uniquely to a* *-*homomorphism* $j(\alpha) : A \rtimes G \rightarrow B \rtimes G$.*

Moreover, $j(\alpha)$ is injective if α is injective.

Proof The *-homomorphism

$$
\alpha \otimes \mathrm{id}_{\mathcal{K}} : A \otimes \mathcal{K} \rightarrow B \otimes \mathcal{K}
$$

is nondegenerate and extends to a *-homomorphism $\mathcal{M}(A \otimes \mathcal{K}) \rightarrow \mathcal{M}(B \otimes \mathcal{K})$. Since $\mathcal{M}(A \otimes \mathcal{K}) = \mathbb{B}(l^2(G, A))$ and similarly for B (by Exercise 5.5.14), we have a *-homomorphism, which we denote by $\tilde{\alpha}$ from $\mathbb{B}(l^2(G, A)) \rightarrow \mathbb{B}(l^2(G, B))$.

Let $j(\alpha) : A[G] \rightarrow B[G]$ denote the *-homomorphism determined by α. We leave it to the reader to check that

$$(5.23) \qquad \lambda_{G,B} \circ j(\alpha) = \tilde{\alpha} \circ \lambda_{G,A} : A[G] \rightarrow \mathbb{B}\left(l^2(G, B)\right).$$

Equation (5.23) implies that

(5.24)
$$\|j(\alpha)(x)\|_{B \rtimes G} := \|(\lambda_{G,B} \circ j(\alpha))(x)\| = \|(\tilde{\alpha}(\lambda_{G,A}(x))\| \le \|\lambda_{G,A}(x)\| = \|x\|_{A \rtimes G}$$

for all $x \in A[G]$, using the fact that any *-homomorphism is contractive. Hence $j(\alpha)$ extends continuously to a *-homomorphism $A \rtimes G \rightarrow B \rtimes G$ as claimed.

For the injectivity statement, injectivity of α implies that of $\alpha \otimes \mathrm{id}_{\mathcal{K}}$ and then that of its extension to a *-homomorphism $\mathbb{B}(l^2(G, A)) \rightarrow \mathbb{B}(l^2(G, B))$. $\square$

Exercise 5.5.34 Prove that if A is unital, then the inclusion $\mathbb{C}[G] \to A[G]$ extends to a *-homomorphism $C^*(G) \to A \rtimes G$.

We now relate the current definition of crossed product to an earlier one. Let A be a G-C^*-algebra and B a C^*-algebra (without any G-action). Let $\rho \colon A \to B$ be a nondegenerate *-homomorphism.

We produce a *-homomorphism

$$\operatorname{Ind}(\rho) \colon A[G] \to \mathbb{B}\big(l^2(G, B)\big)$$

by defining

$$\operatorname{Ind}(\rho)(a)(b \otimes e_h) := \rho\big(h^{-1}(a)\big)b \otimes e_h, \quad \operatorname{Ind}(g)(b \otimes e_h) := b \otimes e_{gh}.$$

This is a covariant pair. We obtain a *-algebra homomorphism

$$\operatorname{Ind}(\rho) \colon A[G] \to \mathbb{B}\big(l^2(G, B)\big).$$

As above, ρ determines a *-homomorphism $\tilde{\rho} \colon \mathbb{B}\big(A \otimes l^2(G)\big) \to \mathbb{B}\big(B \otimes l^2(G)\big)$.

Lemma 5.5.35 *In the above notation,* $\operatorname{Ind}(\rho) = \tilde{\rho} \circ \lambda_{G,A} \colon A[G] \to \mathbb{B}\big(l^2(G, B)\big)$.
Hence $\operatorname{Ind}(\rho)$ *extends to a C^*-algebra homomorphism* $A \rtimes G \to \mathbb{B}\big(l^2(G, B)\big)$.
It is injective if ρ is.

The proof follows the same pattern as that of Proposition 5.5.33.

Corollary 5.5.36 *If* $\rho \colon A \to \mathbb{B}(H)$ *is a representation of A on a Hilbert space, then the map* $\operatorname{Ind}(\rho) \colon A \rtimes G \to \mathbb{B}\big(l^2(G, H)\big)$ *is a representation of $A \rtimes G$ on* $l^2(G, H)$. *The latter is injective if ρ is injective.*

The proof consists merely of reinterpreting the target $\mathbb{B}\big(l^2(G, B)\big)$ of $\operatorname{Ind}(\rho)$ as the C^*-algebra $\mathbb{B}\big(l^2(G, H)\big)$, when $B := \mathbb{B}(H)$.

As a consequence:

Corollary 5.5.37 *Definitions 5.5.31 and 1.8.7 agree, that is, produce the same completion $A \rtimes G$ of $A[G]$, for any G, and any choice of injective representation involved in Definition 1.8.7.*

Example 5.5.38 Let $A = C_0(X)$, where G acts on X, locally compact. To each orbit $Gx_0 \subset X$ of the action, we associate a representation

$$\pi_{x_0} \colon C_0(X) \rtimes G \to \mathbb{B}(l^2 G),$$

of the crossed product $C_0(X) \rtimes G$ as follows. Let $\operatorname{ev}_{x_0} \colon C_0(X) \to \mathbb{C}$ be the *-homomorphism of evaluation of functions at x_0: a representation of $C_0(X)$ on a one-dimensional Hilbert space.

Applying the construction of $\text{Ind}(\text{ev}_{x_0})\colon C_0(X) \rtimes G \to \mathbb{B}(l^2 G)$ above we see that

$$\text{Ind}(\text{ev}_{x_0})(f)(e_h) := \text{ev}_{x_0}\big(h^{-1}(f)\big) := \text{ev}_{x_0}(f \circ h) = f\big(h(x_0)\big), \quad g(e_h) := e_{gh}.$$

Hence $\text{Ind}(\text{ev}_{x_0})$ is induced by the covariant pair

$$C_0(X) \to C_b(G) \subset \mathbb{B}(l^2 G), \quad \text{Gelfand dual to the orbit map} \ \ G \to X, g \mapsto gx_0,$$

$$\text{and} \ G \to \mathbf{U}(l^2 G), \text{the left regular representation.}$$

Exercise 5.5.39 Prove that the representation π_{x_0} associated to an orbit of G acting on X is injective if and only if the orbit Gx_0 of x_0 is dense in X.

Exercise 5.5.40 If $\rho\colon A \to B$ is a nondegenerate *-homomorphism and H is a Hilbert space, then the tensor product $(A \otimes H) \otimes_A B$ of Hilbert modules over ρ is canonically isomorphic to $B \otimes H$ as right Hilbert B-modules. Check that this identifies the map $\tilde{\rho}\colon \mathbb{B}(A \otimes l^2(G)) \to \mathbb{B}(B \otimes l^2(G))$ discussed in the proofs with the map $T \mapsto T \otimes 1$.

Chapter 6
Morita Equivalence

Modifying a C*-algebra A by replacing it with $M_2(A)$ or $M_n(A)$ or $A \otimes \mathcal{K}$ has little impact on the structure theory of the C*-algebra. It therefore seems reasonable to consider a pair of C*-algebras A and B to be equivalent if $A \otimes \mathcal{K}$ and $B \otimes \mathcal{K}$ are isomorphic. This equivalence relation is called Morita equivalence, due to having been defined by Kiiti Morita in 1958. Morita equivalent algebras have exactly the same spaces of irreducible representations, primitive ideal spaces, tracial state spaces, finitely generated projective modules, and other important invariants (see [29, 136–138] and [139], and the more recent [66] and the book [134]). Morita equivalent C*-algebras are for most purposes indistinguishable as objects of noncommutative geometry. Two commutative C*-algebras are Morita equivalent if and only if they are isomorphic, so Morita equivalence only becomes visible as a new concept for noncommutative algebras.

Morita equivalence reveals the fundamentally important fact that the crossed product construction for G-actions on spaces X, where G is a locally compact group, is actually a form of quotient space construction (producing, however, a potentially noncommutative space). Suppose that $G = \mathbb{Z}/2$ is the two-element group, acting on the 2-point space X by interchanging the two points. The quotient of the action $G \backslash X$ is the one-point space. As we discussed in Section 1.8 of Chapter 1, we may realize $C(X)$ as diagonal 2-by-2 matrices, and these and the off-diagonal unitary flip matrix generate the crossed product algebra $C(X) \rtimes G$, which is none other than $M_2(\mathbb{C})$. Thus the crossed product is Morita equivalent to $\mathbb{C} = C(\mathrm{pt}) = C(G \backslash X)$. More generally, if G acts properly and freely on X, then $C_0(X) \rtimes G$ is Morita equivalent to $C(G \backslash X)$. In [47] and [43] Connes speaks often of the C*-algebra $C^*(\mathcal{F})$ of a foliation $\mathcal{F}$ acting as a kind of replacement for the (topologically badly behaved) space of leaves $M/\mathcal{F}$ of the foliation.

One of the most general results along these lines is due to Muhly, Renault and Williams (see [125]), which applies to locally compact groupoids with Haar system (see [135]) which implies in particular that the restriction of the holonomy groupoid of a foliation to transversal results determines a Morita equivalence between the

© Springer Nature Switzerland AG 2024
H. Emerson, *An Introduction to C*-Algebras and Noncommutative Geometry*,
Birkhäuser Advanced Texts Basler Lehrbücher,
https://doi.org/10.1007/978-3-031-59850-0_6

holonomy groupoid and its (étale) restriction to a transversal. The C*-algebras of transversals are unital, if the transversal is compact, and are in many ways much easier to analyze.

In particular these results apply to Lie group actions like the Krönecker flow on the 2-torus, which we discuss here. As we do not discuss groupoids much in these notes, we restrict ourselves here to stating the special case of [125] for commuting group actions and only give an indication of why it is true by giving parts of the proof for special cases, see [125] for the full proof of Theorem 6.1.15 and its more general version for groupoids and their C*-algebras. The restriction of the Krönecker flow α along lines of slope $\hbar$ is Morita equivalent to the irrational rotation algebra $A_\hbar$, by restricting to an appropriate transversal. This reduces the dynamics of the Krönecker flow to the dynamics of irrational rotation on the circle.

A higher dimensional example of this general sort of discussion appears in classical two-dimensional hyperbolic geometry (see Section 6.7). If M is a hyperbolic surface, then weak stable equivalence of geodesic flow on the unit sphere bundle SM is Morita equivalent to the action of $G := \pi_1(M)$ on the boundary of hyperbolic space $\tilde{M} \cong \mathbb{H}^2$, again reflecting the common ergodic theory technique of studying geodesic and horocycle flow by analyzing the action of the fundamental group on the boundary of hyperbolic space (see e.g., [101, 151]), which coincides with the intrinsically defined Gromov boundary ∂G if M is compact (see Section 10.4). The Morita equivalence between the C*-algebra $C(\partial \mathbb{H}^2) \rtimes G$ of the boundary action of a surface group $G = \pi_1(M)$ on the boundary of the hyperbolic plane, and $C(SM) \rtimes \mathbb{R} \rtimes \mathbb{R}$ through the joint actions of geodesic and horocycle flow, also has consequences for the problem of computing the K-theory of $C(\partial \mathbb{H}^2) \rtimes G$, as we discuss at the end of this book. (See the Introduction of [71] for more discussion of the example.)

Finally, in this section we construct, using Morita equivalence, various examples of "noncommutative vector bundles": that is, f.g.p. (finitely generated projective) modules over various examples of noncommutative C*-algebras, from geometric ideas related to dynamics. Such modules are data for K-theory. For instance the famous Rieffel module over $A_\hbar$ is of this type (see Section 6.6). If G is a discrete group acting properly and cocompactly on X, then the space of sections of any G-equivariant vector bundle over X also determines a Morita equivalence bimodule (see Section 6.3), this time between a unital C*-algebra and an ideal in $C_0(X) \rtimes G$, and then an f.g.p. module over $C_0(X) \rtimes G$. This determines a bijective correspondence between isomorphism classes of G-equivariant vector bundles over X and isomorphism classes of f.g.p. modules over the crossed product $C_0(X) \rtimes G$ of some importance in K-theory (see [77] and references). We also discuss examples for finite group C*-algebras, which is the special case of X a point: If π is a finite-dimensional representation of a finite group, then the C*-algebra commutant of $\pi(G)$ is Morita equivalent to the ideal of $C^*(G)$ generated by the character and in particular determines an idempotent in $C^*(G)$. This correspondence between (equivalence classes of) representations and (equivalence classes of) projections in $C^*(G)$ is called the Green–Julg correspondence [103].

6.1   Morita Equivalence

The most basic example of Morita equivalent algebras is the pair A and $M_2(A)$, for any C*-algebra A. If $\pi : A \to \mathbb{B}(H)$ is an irreducible representation of A, then

$$\pi'\left(\begin{bmatrix} a & b \\ c & d \end{bmatrix}\right) := \begin{bmatrix} \pi(a) & \pi(b) \\ \pi(c) & \pi(d) \end{bmatrix}$$

is an irreducible representation of $M_2(A)$ on $H \oplus H$. To go in the reverse direction, consider the right Hilbert $M_2(A)$-module $\mathcal{E}$ given by $A \oplus A$, with the right $M_2(A)$-module structure by matrix multiplication, and inner product $\langle x, y \rangle := \theta_{x,y} \in \mathbb{B}(A \oplus A) \cong M_2(A)$. There is a natural (left) action of A by bounded Hilbert module operators on $\mathcal{E}$ by left scalar multiplication: Denote it $\rho : A \to \mathbb{B}(\mathcal{E})$. Now if π is an irreducible representation of $M_2(A)$, then the tensor product of Hilbert modules $\mathcal{E} \otimes_{M_2(A)} H$ is a Hilbert space, A is represented on it by $\pi'(a) := \rho(a) \otimes 1$, and this is an irreducible representation of A.

Exercise 6.1.1 Prove that the maps between representations of A and of $M_2(A)$ described above send irreducible representations to irreducible representations, equivalent representations to equivalent representations, and induce inverse maps on spectra.

Similar arguments should be that the representation theories of A and of $M_n(A) \cong A \otimes M_n(\mathbb{C})$ and even of $A \otimes \mathcal{K}$ are identical, where $\mathcal{K}$ is the C*-algebra of compact operators on any (separable) Hilbert space. We will not prove this general statement, but the reader should have it in the back of their mind. One way of defining Morita equivalent algebras is by saying A and B are Morita equivalent if $A \otimes \mathcal{K}$ is isomorphic to $B \otimes \mathcal{K}$.

Exercise 6.1.2 If G is a finite group, then $C^*(G)$ is Morita equivalent in the above sense to $C(X)$ where X is a finite set. What is X?

The previous exercise illustrates what one loses, and what one does not lose, in passing from a C*-algebra to its Morita equivalence class. In the case of $C^*(G)$ for a finite group G, the noncommutative space is the set of equivalence classes $\widehat{G}$ of irreducible representations of G. What is lost, are the dimensions of those representations.

One of the fascinating points of C*-algebra theory is that Morita equivalence, a concept from algebra, has a beautiful geometric interpretation in the context of group actions and crossed products.

We will not prove the following theorem but will discuss some aspects of the proof. Our main interest will be in examples where it can be applied.

Theorem 6.1.3 *Let G and H be locally compact groups acting properly and freely on X locally compact. Assume the actions commute. Then $C_0(G \backslash X) \rtimes H$ is strongly Morita equivalent to $C_0(H \backslash X) \rtimes G$.*

Notice that applying the theorem to a free and proper action of G on X and setting H to be the trivial group give that $C_0(X) \rtimes G$ is Morita equivalent to $C_0(G \backslash X)$.

Morita equivalence is best understood in terms of the existence of certain kinds of bimodules.

Definition 6.1.4 Let A and B be C*-algebras.

A *Morita equivalence A-B-bimodule* is a linear space $\mathcal{E}$ which is both a right Hilbert B-module, with form $\langle \cdot, \cdot \rangle_B \colon \mathcal{E} \times \mathcal{E} \to B$, and a Hermitian *left* A-module, with form $_A\langle \cdot, \cdot \rangle \colon \mathcal{E} \times \mathcal{E} \to A$, such that the two module structures commute

$$(ax)b = a(xb), \quad \forall a \in A, b \in B, x \in \mathcal{E},$$

the identities
(6.1)
$$_A\langle x, y \rangle z = x \langle y, z \rangle_B, \quad \text{and} \quad _A\langle x, yb \rangle = {_A\langle xb^*, y \rangle}, \quad \text{and} \quad \langle ax, y \rangle_B = \langle x, a^*y \rangle_B,$$

hold for all $x, y, z \in \mathcal{E}, a \in A, b \in B$, and such that the linear spans of $_A\langle \mathcal{E}, \mathcal{E} \rangle$ and of $\langle \mathcal{E}, \mathcal{E} \rangle_B$ are dense in A, B, respectively.

In this case we say A and B are *Morita equivalent*.

The term "Hermitian" used in reference to the inner products in Definition 6.1.4 means, in the case of $_A\langle \cdot, \cdot \rangle$, that the form must be A-linear in the first variable, A conjugate linear in the second. Similarly, $\langle \cdot, \cdot \rangle_B$ should be B-linear in the second variable and B conjugate linear in the first variable. See Definition 5.4.1 for the definition of "Hermitian."

The paper [29] contains a proof of the following theorem, relying on a result from [36]. The uniqueness part is Theorem 2.8 from [36].

Theorem 6.1.5 *Let A and B be separable C*-algebras. Then a Morita equivalence from A to B induces an isomorphism $A \otimes \mathcal{K} \to B \otimes \mathcal{K}$ which is canonical up to conjugation by a unitary in $\mathcal{M}(B \otimes \mathcal{K})$.*

Corollary 6.1.6 *A Morita equivalence between A and B induces a bijection between the sets of equivalence classes of irreducible representations of A and of B.*

Indeed, since the compact operators have a unique irreducible representation λ, the irreducible representations of $A \otimes \mathcal{K}$ are all of the form $\pi \otimes \lambda$, up to equivalence, for some irreducible representation of A.

Example 6.1.7 Let A be a C*-algebra. Then $M_n(A)$ is strongly Morita equivalent to $M_m(A)$ for any n, m.

For the proof, let $\mathcal{E} = M_{n \times m}(A)$. If $S, T \in \mathcal{E}$, then $ST^* \in M_n(A)$ and $S^*T \in M_m(A)$. This provides two inner products

$$_{M_n(A)}\langle \cdot, \cdot \rangle \colon \mathcal{E} \times \mathcal{E} \to M_n(A), \quad \text{and} \quad \langle \cdot, \cdot \rangle_{M_m(A)} \colon \mathcal{E} \times \mathcal{E} \to M_m(A).$$

We obtain a strong Morita equivalence bimodule.

Note that this contains the result that A is Morita equivalent to $M_n(A)$ for any A. More generally, A is Morita equivalent to $A \otimes \mathcal{K}(H)$, as we show below, for any A and any Hilbert space H.

Proposition 6.1.8 *Let $\mathcal{E}$ be any right Hilbert B-module and $J = \mathrm{supp}(\mathcal{E})$. Then $\mathcal{K}(\mathcal{E})$ is Morita equivalent to J.*

Proof For a strong Morita equivalence bimodule, we use $\mathcal{E}$ as a right Hermitian B-module. Its support is thus the ideal $J \subset B$. We set

$$\mathcal{K}(\mathcal{E}) \langle x, y \rangle := \theta_{x,y}.$$

with $\theta_{x,y}$ as in (5.14). If $T \in \mathcal{K}(\mathcal{E})$, or more generally if $T \in \mathbb{B}(\mathcal{E})$, then $T\theta_{x,y} = \theta_{Tx,y}$ so that $\mathcal{K}(\mathcal{E}) \langle \cdot, \cdot \rangle$ is $\mathcal{K}(\mathcal{E})$-linear in the first coordinate. Since $\theta_{x,y} = \theta_{y,x}^*$, we get $\mathcal{K}(\mathcal{E}) \langle x, y \rangle = \mathcal{K}(\mathcal{E}) \langle y, x \rangle^*$. Exercise 5.5.9 implies that $\mathcal{K}(\mathcal{E}) \langle x, x \rangle \geq 0$ for all $x \in \mathcal{E}$. Also, $\theta_{x,yb} = \theta_{xb^*,y}$ by an easy computation and hence $\mathcal{K}(\mathcal{E}) \langle x, yb \rangle = \mathcal{K}(\mathcal{E}) \langle xb^*, y \rangle$.

Finally, $\langle Tx, y \rangle_B = \langle x, T^*y \rangle_B$ since compact operators are adjointable, and the condition (6.1) follows from the definition of $\theta_{x,y}$. $\qquad\square$

Since $A \otimes \mathcal{K} \cong \mathcal{K}(A \otimes l^2)$, and since the Hilbert A-module $A \otimes l^2$ has full support in A, it follows that A is Morita equivalent to $A \otimes \mathcal{K}$, for any A, as we have mentioned previously.

Exercise 6.1.9 The following is a fibered version of Example 6.1.7. Let $\pi : V \to X$ be a complex vector bundle over X compact. Equip V with a Hermitian metric. Let $\mathrm{End}\, V$ denote the endomorphism bundle of V, and $A := \Gamma(\mathrm{End}\, V)$, the C*-algebra of sections of the endomorphism bundle. Let $B = C_0(X)$. Let $\mathcal{E} = \Gamma(V)$ with its standard structure of a right Hilbert $C(X)$-module

$$\langle s_1, s_2 \rangle_{C(X)}(x) = \langle s_1(x), s_2(x) \rangle_x.$$

Define a left $\Gamma(\mathrm{End}\, V)$-valued inner product by

$$\Gamma(\mathrm{End}\, V) \langle s_1, s_2 \rangle(x) = \theta_{s_1(x), s_2(x)}.$$

$\mathcal{E}$ into a Morita equivalence A-B-bimodule.

Returning to the general situation of a strong Morita equivalence A-B-bimodule, note that the proof of the Cauchy–Schwarz inequality for semi-Hermitian *right* A-modules (see Lemma 5.4.14) works equally well for semi-Hermitian *left* A-modules, so that in the setting of Definition 6.1.4 it holds that

$$\|_A\langle x, y \rangle\|^2 \leq \|_A\langle x, x \rangle\| \cdot \|_A\langle y, y \rangle\|,$$

as well as the corresponding statement

$$\|\langle x, y\rangle_B\|^2 \le \|\langle x, x\rangle_B\| \cdot \|\langle y, y\rangle_B\|$$

for the B-valued Hermitian form.

Secondly, note that (6.1) implies that left scalar multiplication by $a \in A$ acts on $\mathcal{E}$, as a right Hilbert B-module, as an adjointable operator. Its adjoint is scalar multiplication by a^*. That is, the left module structure is a C*-algebra homomorphism $A \to \mathbb{B}_B(\mathcal{E})$, with target the C*-algebra of adjointable operators on the right Hilbert B-module $\mathcal{E}$.

In the definitions above, there emerge two natural norms on $\mathcal{E}$: In order to show that they are the same, we temporarily denote

$$\|x\|_A := \|_A\langle x, x\rangle\|, \quad \|x\|_B := \|\langle x, x\rangle_B\|.$$

It follows from our remarks above that $\|ax\|_A \le \|a\| \cdot \|x\|_A$ and that $\|xb\|_B \le \|x\|_B \cdot \|b\|$, for all $x \in \mathcal{E}, a \in A, b \in B$.

Lemma 6.1.10 *If $\mathcal{E}$ is a strong Morita equivalence A-B-bimodule, then*

$$\|_A\langle x, x\rangle\| = \|\langle x, x\rangle_B\|$$

for all $x \in \mathcal{E}$.

Proof If $x \in \mathcal{E}$, then

$$\|x\|_A^4 = \|_A\langle x, x\rangle\|^2 = \|_A\langle x, x\rangle_A\langle x, x\rangle\| = \|_A\langle_A\langle x, x\rangle x, x\rangle\|$$

$$= \|_A\langle x\langle x, x\rangle_B, x\rangle\| = \|_A\langle x, x\langle x, x\rangle_B\rangle\| \le \|_A\langle x, x\rangle\| \cdot \|\langle x, x\rangle_B\| = \|x\|_A^2 \|x\|_B^2,$$

where we used the Cauchy–Schwarz inequality for $\langle \cdot, \cdot\rangle_A$, and the fact that $\|zb\|_A \le \|z\|\|b\|$ for all $z \in \mathcal{E}, b \in B$. Hence $\|x\|_A^2 \le \|x\|_B^2$, and the result follows by switching the roles of A and B. $\square$

Proposition 6.1.11 *Let $\mathcal{E}$ be a strong Morita equivalence A-B-bimodule. Then $A \cong \mathcal{K}(\mathcal{E})$ by the left multiplication action of A on $\mathcal{E}$, where $\mathcal{K}(E)$ is the C*-algebra of compact operators on $\mathcal{E}$ as a right Hilbert B-module.*

Proof Left multiplication by $a = \langle x, y\rangle_A \in A$ maps $z \in \mathcal{E}$ to $\langle x, y\rangle_A z = x\langle y, z\rangle_B = \theta_{x,y}(z)$. So left multiplication by $a \in A$ of this form is a compact operator. Since the span of the $_A\langle x, y\rangle$ is dense in A, the result follows. $\square$

Exercise 6.1.12 Suppose A and B are unital, commutative C*-algebras. Show that if they are Morita equivalent, they are isomorphic.

Exercise 6.1.13 Let A be a C*-algebra. A *full corner of* A is a C*-subalgebra of the form pAp, where p is a projection in A for which the ideal generated by p in A is all of A:

a) Prove that the ideal generated by p is the closed linear span of ApA.

b) Prove that if B is a full corner of A, $B = pAp$, then Ap with inner products

$$_A \langle ap, bp \rangle := apb^*, \quad \langle ap, bp \rangle_B := pa^*bp$$

and evident left A-module structure, and right B-module structure, is a strong Morita equivalence between A and B.
c) Prove that A can be realized as a full corner of $\mathcal{K}(A^{\mathbb{N}}) \cong A \otimes \mathcal{K}$ using any rank 1 projection $p \in \mathcal{K}$.

Finally, we address the adjective "equivalent" in the term "Morita equivalent" more carefully. Suppose that A and B are Morita equivalent by a Morita equivalence A-B-bimodule $\mathcal{E}$.

We define a "conjugate" bimodule as follows. Let $\mathcal{E}^*$ be $\mathcal{E}$ as an additive group, but with the conjugate $\mathbb{C}$-multiplication $\lambda x := \bar{\lambda} x$, making it a $\mathbb{C}$-vector space.

Denote elements of $\mathcal{E}^*$ by $\bar{x}$ (where $x \in \mathcal{E}$).

Proposition 6.1.14 *If $\mathcal{E}$ is a Morita equivalence A-B-bimodule, then $\mathcal{E}^*$ defined above together with the B-A-bimodule structure*

$$b \bar{x} a := \overline{a^* x b^*},$$

and inner products

$$_B \langle \bar{x}, \bar{y} \rangle := \langle x, y \rangle_B, \quad \langle \bar{x}, \bar{y}, \rangle_A :=_A \langle x, y \rangle,$$

is a strong Morita equivalence B-A-bimodule.
   Furthermore,

$$\mathcal{E}^* \otimes_A \mathcal{E} \cong B, \qquad \mathcal{E} \otimes_B \mathcal{E}^* \cong A,$$

as right Hilbert B-bimodules, A-modules, respectively.

Proof We leave it to the reader to check that $\mathcal{E}^*$ is a strong Morita equivalence B-A-bimodule. To see why that $\mathcal{E}^* \otimes_A \mathcal{E} \cong B$, recall that the tensor product $\mathcal{E}^* \otimes_A \mathcal{E}$ is defined as the completion of the algebraic tensor product over $\mathbb{C}$ with respect to the Hermitian B-valued form

$$(6.2) \quad \langle \bar{x}_1 \otimes y_1, \bar{x}_2 \otimes y_2 \rangle_B := \langle y_1, \langle \bar{x}_1, \bar{x}_2 \rangle_A \cdot y_2 \rangle_B. = \langle y_1, {}_A \langle x_1, x_2 \rangle y_2 \rangle_B.$$

Let

$$U : \mathcal{E}^* \otimes_A \mathcal{E} \to B, \quad U(x^* \otimes y) := \langle x, y \rangle_B.$$

Then U is a well-defined B-bimodule map: To see that it is a bimodule map, compute

$$U\left(b_1 \cdot (\bar{x} \otimes y)b_2\right) = U\left(\overline{xb_1^*} \otimes yb_2\right) = \langle xb_1^*, yb_2 \rangle_B = b_1^* \langle x, y \rangle_B b_2.$$

Finally,

$$(6.3) \quad U(\bar{x_1} \otimes y_1)^* U(\bar{x_2} \otimes y_2) = \langle y_1, x_1 \rangle_B \cdot \langle x_2, y_2 \rangle_B = \langle y_1, x_1 \langle x_2, y_2 \rangle_B \rangle_B$$
$$= \langle y_1, \ _A\langle x_1, x_2 \rangle \cdot y_2 \rangle_B,$$

which agrees with (6.2). Hence U is an isometry and is clearly surjective so is an isomorphism of right Hilbert B-modules.   □

We close by outlining the proof of Theorem 6.1.3 in the case of discrete group actions.

Let G and H be discrete groups acting properly and freely on X locally compact. Let $A = C_0(G\backslash X) \rtimes H$ and $B = C_0(H\backslash X) \rtimes G$.

If $\xi \in C_c(X)$ and $g \in G$ or H, let $g(\xi) := \xi \circ g^{-1}$. We consider functions on $G\backslash X$ to be G-periodic functions on X, and similarly, functions on $H\backslash X$ are H-periodic functions on X.

These comments supply $C_c(X)$ with a $C_0(G\backslash X) \rtimes H$-$C_0(H\backslash X) \rtimes G$-bimodule structure, at least for the subalgebras of finitely supported elements in the respective crossed products (elements of H act on the right by $(\xi \cdot h)(x) = h^{-1}(\xi)$ to give a right action). Let $\xi^*(x) = \overline{\xi(x)}$.

Theorem 6.1.15 *Let G and H be discrete groups acting properly and freely on X by commuting actions.*

On $C_c(X)$ define inner products

$$(6.4) \qquad _A\langle \xi, \eta \rangle := \sum_{g \in G, h \in H} g(\xi)(hg)(\eta)^*[h] \in A = C_0(G\backslash X) \rtimes H$$

and

$$(6.5) \qquad \langle \xi, \eta \rangle_B := \sum_{h \in H, g \in G} h(\xi)^*(gh)(\eta)[g] \in B = C_0(H\backslash X) \rtimes G.$$

Then the corresponding completion of $C_c(X)$ is a Morita equivalence $C_0(G\backslash X) \rtimes H$-$C_0(H\backslash X) \rtimes G$ bimodule.

Note that if $h \in H$, the sum $\sum_{g \in G} g(\xi)(gh)(\eta)^* = \sum_{g \in G} g(\xi \cdot h(\eta^*))$ is locally finite (its restriction to any compact $K \subset X$ is a finite sum of functions) because the G-action is proper. It defines a G-periodic function on X. So the coefficients of the group elements $[h]$ in (6.5) are indeed elements of $C_0(G\backslash X)$.

We will not prove the theorem here but will remark on a few aspects of the proof.

Exercise 6.1.16 In reference to the inner products above, prove the following, where $\xi, \eta \in C_c(X)$:

a) If $k \in H$, then $[k]_A \langle \xi, \eta \rangle =_A \langle k(\xi), \eta \rangle$.
   (*Hint.*

$$[k]_A \langle \xi, \eta \rangle = \sum_{g \in G, h \in H} (kg)(\xi)(khg)(\eta)^*[kh]$$

$$= \sum_{g \in G, h \in H} (kg)(\xi)(hg)(\eta)^*[h] =_A \langle k(\xi), \eta \rangle$$

by making the change of variables $h \mapsto k^{-1}h$ in the sum; remember the H- and G-actions commute.)
b) Prove that if $f \in C_0(G \backslash X)$, then $f_A \langle \xi, \eta \rangle =_A \langle f\xi, \eta \rangle$.
c) Prove that if $u \in G$, then $\langle \xi, \eta \rangle_B [u] = \langle \xi, \eta \cdot u \rangle$.
c) Prove that if $f \in C_0(G/H)$, then

$$\langle \xi, \eta \rangle_B f = \langle \xi, \eta f \rangle_B.$$

d) Prove that $_A \langle \xi, \eta \rangle \zeta = \xi \langle \eta, \zeta \rangle_B$.

We close with a discussion of the *positivity* of the B-valued inner product defined in Theorem 6.1.15 (the other inner product is similar) since it involves some ideas of importance.

To simplify things set H to be the trivial group, so we are looked at a $C_0(G \backslash X)$-$C_0(X) \rtimes G$-bimodule giving the important special case of the theorem asserting that $C_0(G \backslash X)$ and $C_0(X) \rtimes G$ are Morita equivalent, for a proper free action of G.

The $C_0(X) \rtimes G$-valued inner product is

(6.6)
$$\langle \varphi, \psi \rangle_{C_0(X) \rtimes G} = \sum_{g \in G} \overline{\varphi} \cdot g(\psi)[g].$$

The right $C_0(X) \rtimes G$-module structure is given by

(6.7)
$$(\xi f)(x) := \xi(x)f(x), \quad (\xi g)(x) := \xi(gx).$$

The left $C_0(G \backslash X)$-valued inner product is

(6.8)
$$_{C_0(G \backslash X)} \langle \varphi, \psi \rangle = \sum_{g \in G} \varphi \overline{g(\psi)}.$$

The left module structure lets $f \in C_0(G \backslash X)$ act by multiplication on $C_c(X)$ by considering f a G-periodic function on X.

The problem is to show that

$$\langle \varphi, \varphi \rangle_{C_0(X) \rtimes G} \geq 0, \quad \forall \varphi \in C_c(X),$$

the positivity of $C_0(G \backslash X) \langle \varphi, \varphi \rangle$ being clear.

Exercise 6.1.17 If $G \times X \rightarrow X$ is a proper action of a discrete group, a *cutoff function* for the action is a continuous function $\rho \in C_b(X)$ such that $0 \leq \rho \leq 1$ everywhere, and

$$\tag{6.9} \sum_{g \in G} g(\rho)^2 = 1.$$

a) Prove that if $K \subset X$ is any compact subset, then there exists a cutoff function ρ which equals 1 on K. If $G \backslash X$ is compact, such ρ may also be taken to have compact support.
b) Show that if ρ is any cutoff function, then

$$\tag{6.10} P_\rho := \sum_{g \in G} \rho g(\rho) [g]$$

is a projection in the multiplier algebra of $C_0(X) \rtimes G$. If ρ has compact support, then P_ρ is a projection in $C_0(X) \rtimes G$.
c) Prove that the collection of cutoff functions is a convex space and that a path between any two cutoff functions gives rise to a path of projections in $C_0(X) \rtimes G$ between the projections associated by (6.10) to them.

Remark 6.1.18 If G is finite and X is compact, then the constant function $\rho := \frac{1}{|G|}$ is a cutoff function. The associated projection (6.10) is the projection $\frac{1}{|G|} \sum_{g \in G} [g] \in \mathbb{C}[G] \subset C(X) \rtimes G$.

Lemma 6.1.19 *Let* $\varphi, \psi \in C_c(X)$. *Let* $K \subset X$ *be a compact subset containing* $\text{supp}(\varphi) \cup \text{supp}(\psi)$, *and* ρ *be a cutoff function which is 1 on* K.
  Then

$$\varphi^* P_\rho \psi = \langle \varphi, \psi \rangle_{C_0(X) \rtimes G}.$$

holds in $C_0(X) \rtimes G$, *where, on the left hand side, we are considering* φ, ψ *as elements of* $C_0(X) \rtimes G$, *by the inclusion* $C_c(X) \subset C_0(X) \rtimes G$.

Proof Since $\rho = 1$ on $\text{supp}(\varphi) \cup \text{supp}(\psi)$,

$$\overline{\varphi} P_\rho \psi = \sum_{g \in G} \overline{\varphi} \rho g(\rho) g(\psi) [g] = \sum_{g \in G} \overline{\varphi} g(\rho \psi) [g] = \sum_{g \in G} \overline{\varphi} g(\psi) [g]$$

$$\tag{6.11} = \langle \varphi, \psi \rangle_{C_0(X) \rtimes G}. \qquad \square$$

The Lemma immediately implies the positivity result we are looking for, since now for $\varphi \in C_c(X)$ and ρ a cutoff function which is 1 on the support of φ, we have

$$\langle \varphi, \varphi \rangle_{C_0(X) \rtimes G} = \overline{\varphi} P_\rho \varphi = (P_\rho \varphi)^* (P_\rho \varphi) \geq 0 \in C_0(X) \rtimes G.$$

The reader might have noticed that the verification of positivity of $\langle \cdot, \cdot \rangle_{C_0(X) \rtimes G}$ above does not use freeness of the G-action on X, which was one of the hypotheses of Theorem 6.1.15. In the case of isotropy, the range of this inner product is not dense but is an ideal in $C_0(X) \rtimes G$ (see Exercise 5.4.18).

Exercise 6.1.20 Deduce from Theorem 6.1.3 that if G is a locally compact group, then $C_0(G/H) \rtimes K$ is Morita equivalent to $C_0(K \backslash G) \rtimes H$, where K acts on G/H by left multiplication and H acts on $K \backslash G$ by right multiplication.

Exercise 6.1.21 Let $Z = \mathbb{R} \times \mathbb{T}$, with $\mathbb{T} = \mathbb{R}/\mathbb{Z}$, so Z is a cylinder. Let $\mathbb{R}$ act on Z by $\alpha_t(x, y) = (x + t, y + t\hbar)$, where $\hbar \in \mathbb{R}$ is a fixed constant, and the second coordinate is understood to be mod $\mathbb{Z}$. Let $\mathbb{Z}$ act on Z by $n \cdot (x, y) = (x + n, y)$. Clearly, the two actions commute:

a) Show that the given actions of $\mathbb{Z}$ and $\mathbb{R}$ are proper.
b) Show that the quotient $Z/\mathbb{Z}$ is the torus $\mathbb{T}^2$ and that the induced action of $\mathbb{R}$ on $\mathbb{T}^2$ is the Krönecker flow

$$\beta_t^h(x, y) = (x + t, y + \hbar t), \quad x, y \in \mathbb{R}/\mathbb{Z}.$$

c) Show that the quotient $Z/\mathbb{R}$ is the circle $\mathbb{T}$ and that the induced action of $\mathbb{Z}$ is the irrational rotation action $n \cdot x = x + n\hbar$.
d) Deduce that the crossed products $C(\mathbb{T}^2) \rtimes_{\beta^h} \mathbb{R}$ and $A_\hbar$ are Morita equivalent.

Example 6.1.22 Let G be a locally compact group, $H \subset G$ a closed subgroup, and Y an H-space. Let H act diagonally on $X := G \times Y$ by $h \cdot (g, y) := (gh^{-1}, hy)$. The quotient is denoted $G \times_H Y$. It carries a G-action by $g \cdot (g_1, y) := (gg_1, y)$. The G-space X is said to be *induced* from the H-space Y.

Since the given G- and H-actions commute, and since $G \backslash X \cong Y$ as H-spaces, an application of Theorem 6.1.3 gives that

$$C_0(G \times_H Y) \rtimes G \quad \text{is Morita equivalent to} \quad C_0(Y) \rtimes H.$$

This fact is important, as an arbitrary proper action of a Lie group G can be shown to be built locally from such induced spaces.

6.2 Finitely Generated Projective Modules and Morita Correspondences

An extremely important property of an A-B Morita equivalence bimodule is that it induces a map from (isomorphism classes of) finitely generated projective modules over A to f.g.p. modules over B. We discuss this below but begin with a special case.

Lemma 6.2.1 *Let A be unital, B a C*-algebra, $J \subset B$ a closed ideal, and $\mathcal{E}$ be a strong Morita equivalence A-J-bimodule. Then $\mathcal{E}$ is finitely generated projective (f.g.p.) as a right B-module: $\mathcal{E} = p \cdot B^n$ for some n, for some full projection $p \in M_n(B)$, and, moreover, $A \cong pM_n(B)p$.*

Proof By Exercise 5.4.18 the right scalar multiplication by $J \subset A$ on $\mathcal{E}$ extends to a right scalar multiplication by A, making $\mathcal{E}$ into a right Hilbert B-module with support J. Accordingly, we consider the given right J-valued inner product as being B-valued and denote it $\langle \cdot, \cdot \rangle_B$.

Now, I claim that there exist $x_1, \ldots x_n \in \mathcal{E}$ such that $\sum_{i=1}^n {}_A\langle x_i, x_i \rangle = 1_A$.

Indeed, since $\mathcal{E}$ is full as a left A-module, there exist a finite collection a_i, b_i in $\mathcal{E}$ such that $\sum {}_A\langle a_i, b_i \rangle = c$ where c is close to the unit $1 \in A$, and hence is invertible. Replacing each a_i by $c^{-1}a_i$, we may assume that we have found a_i, b_i such that

$$\sum {}_A\langle a_i, b_i \rangle = 1.$$

We have

$$\sum {}_A\langle a_i + b_i, a_i + b_i \rangle = \sum {}_A\langle a_i, a_i \rangle + \sum {}_A\langle b_i, b_i \rangle + 2 = h,$$

where $h \in A$ is strictly positive. The claim now follows from putting

$$x_i := h^{-\frac{1}{2}}(a_i + b_i).$$

Let

$$U \colon \mathcal{E} \to J^n \subset B^n, \quad Ux := (\langle x_1, x \rangle_B, \ldots \langle x_n, x \rangle_B).$$

The following calculation shows that U is an isometry of Hilbert B-modules:

$$(6.12) \quad \langle U(x), U(x) \rangle_B = \sum_i \langle x_i, x \rangle_B^* \langle x_i, x \rangle_B = \sum_i \langle x, x_i \rangle_B \langle x_i, x \rangle_B$$

$$= \sum_i \langle x, x_i \langle x_i, x \rangle_B \rangle_B = \sum_i \langle x, {}_A\langle x_i, x_i \rangle x \rangle_B = \langle x, x \rangle_B.$$

A similar calculation shows that U is adjointable with adjoint

$$U^* \colon B^n \to \mathcal{E}, \quad U(b_1, \ldots, b_n) := \sum x_i b_i.$$

The range projection of U is given by the matrix of elements of $J \subset B$

$$p := UU^* = (\langle x_i, x_j \rangle_B),$$

and we have shown that $\mathcal{E}$ is isomorphic as a right Hilbert B-module to $p \cdot B^n$. Finally, if

$$\Phi : A \to \mathbb{B}_B(\mathcal{E}), \quad \Phi(a) := (\langle a x_i, x_j \rangle_B),$$

then Φ maps A isomorphically to the corner $p M_n(B) p$ of B.

We verify that p is full if $n = 1$, and leave the general case to the reader. If $\eta, \eta' \in \mathcal{E}$, then since $_A\langle \xi, \xi \rangle \eta = \eta$, it follows that $\xi \langle \xi, \eta \rangle_B = \eta$. Similarly $\xi \langle \xi, \eta' \rangle_B = \eta'$. Hence

$$\langle \eta, \eta' \rangle_B = \langle \xi \langle \xi, \eta \rangle_B, \xi \langle \xi, \eta' \rangle_B = \langle \eta, \xi \rangle_B \, p \langle \xi, \eta' \rangle_B.$$

It follows that the ideal generated by p in B is exactly the support of $\mathcal{E}$, whence is B. If $n > 1$, there is no great further difficulty and we leave the extension to the reader.

$\square$

Exercise 6.2.2 Let J be an ideal in B and $p \in M_n(J)$ be a projection. Show that $p B^n = p J^n$ as right Hilbert B-modules.

We now deduce the following important result.

Theorem 6.2.3 *Let A and B be C^*-algebras and $\mathcal{E}$ an A-B Morita equivalence bimodule. Then if L is a finitely generated projective Hilbert A-module, then $L \otimes_A \mathcal{E}$ is a finitely generated projective Hilbert B-module.*

The assignment $L \mapsto L \otimes_A \mathcal{E}$ is compatible with isomorphism and a direct sum of f.g.p. modules and defines a semigroup isomorphism

$$\mathcal{E}_* : \mathcal{P}(A) \to \mathcal{P}(B),$$

where $\mathcal{P}(A)$, $\mathcal{P}(B)$ are the semigroups of isomorphism classes of f.g.p. modules over A, B.

Proof For simplicity let L be a f.g.p. module over A of the form pA, where p is a projection in A. The more general case of $p \in M_n(A)$ is left to the reader. We can identify $L \otimes_A \mathcal{E}$ with $p\mathcal{E} \subset \mathcal{E}$. Note that since p is a projection, $p\mathcal{E}$ is topologically closed in $\mathcal{E}$. Restricting the A- and B-valued inner products to $p\mathcal{E}$ gives inner products valued in pAp and B. The support of the B-valued inner product will be an ideal J in B. Thus, $p\mathcal{E}$ is a Morita equivalence bimodule between pAp and J. Now pAp is unital, with unit $p \in A$. By Lemma 6.2.1, $p\mathcal{E}$ is f.g.p. over $J \subset B$, and hence over B as well. This proves the first statement.

Clearly mapping L to $L \otimes_A \mathcal{E}$ respects isomorphisms and direct sums. So it defines a semigroup homomorphism $\mathcal{P}(A) \to \mathcal{P}(B)$.

Finally, if $\mathcal{E}^*$ is the conjugate bimodule, it is a Morita equivalence bimodule from B to A and induces a map $\mathcal{P}(B) \to \mathcal{P}(A)$. The composition of the two maps is tensoring with $\mathcal{E} \otimes_B \mathcal{E}^* \cong B$ and induces the identity map on $\mathcal{P}(A)$. $\square$

Remark 6.2.4 The semigroup homomorphism $\mathcal{E}_* : \mathcal{P}(A) \to \mathcal{P}(B)$ induced by an A-B Morita equivalence $\mathcal{E}$ thus has a simple description at the level of projections. In fact, if $p \in A$ is a projection, and if $\xi \in \mathcal{E}$ such that $_A\langle \xi, \xi \rangle = p$, then $\langle \xi, \xi \rangle_B =: q$ is a projection in B, and $\mathcal{E}_*([p]) = [q]$. More generally, if $p = \sum_A \langle \xi_i, \xi_i \rangle$, then the matrix q over B with entries $(\langle \xi_i, \xi_j \rangle_B$ is a projection in $M_n(B)$, and $\mathcal{E}_*([p]) = [q]$.

Definition 6.2.5 A *Morita correspondence* from A to B consists of the data in the following diagram:

$$A \xrightarrow{\alpha} C \xrightarrow{\mathcal{E}} B,$$

where C is a C*-algebra, α is a *-homomorphism, and $\mathcal{E}$ is a strong Morita equivalence between C and the ideal $\text{supp}(\mathcal{E}) \subset B$ of B.

Suppose now that

$$A \xrightarrow{\alpha} C \xrightarrow{\mathcal{E}} J \subset B$$

is a Morita correspondence. Set $\mathcal{E}_B := \mathcal{E} \otimes_J B$. Then $\mathcal{E}_B$ is a right Hilbert B-module. We leave it as an exercise to check that E_B is exactly the same as $\mathcal{E}$, but with the inner product regarded as valued in B.

If L is an f.g.p. module over A, then $\alpha_*(L) := L \otimes_A C$ is an f.g.p. right Hilbert C-module, since α is a *-homomorphism. By Lemma 6.2.1 we obtain the following:

Corollary 6.2.6 *Suppose that*

$$A \xrightarrow{\alpha} C \xrightarrow{\mathcal{E}} J \subset B,$$

is a Morita correspondence with C unital. Then in the notation above, if L is an f.g.p. right Hilbert A-module, then

$$\alpha_*(L) \otimes_C \mathcal{E}_B$$

is an f.g.p. module over B.

The Morita correspondence determines a semigroup homomorphism from the semigroup $\mathcal{P}(A)$ of isomorphism classes of f.g.p. modules over A, to the semigroup $\mathcal{P}(B)$ of isomorphism classes of f.g.p. modules over B.

Example 6.2.7 If G is a Lie group acting properly on X, the following method gives a geometric way of constructing f.g.p. modules over the crossed product $C_0(X) \rtimes G$, involving only *compact* group actions, over lower-dimensional subspaces.

Since the action is proper, if $x \in X$, $H = \text{Stab}_G(x)$, then H is compact. A "slice theorem" is available in this situation and implies (see [67]) that there exists an H-space Y such that the map $G \times_H Y \to X$, $[(g, y)] \mapsto gy$, is a G-equivariant homeomorphism $G \times_H Y \cong U$ with an open and G-invariant subset of X. Hence $C_0(U) \rtimes G$ is an ideal in $C_0(X) \rtimes G$. Since $C_0(Y) \rtimes H$ is Morita equivalent to

$C_0(G \times_H Y) \rtimes G$, by Example 6.1.22, and hence to $C_0(U) \rtimes G$, we obtain a Morita correspondence

$$C_0(Y) \rtimes H \xrightarrow{\text{id}} C_0(Y) \rtimes H \xrightarrow{\mathcal{E}} C_0(X) \rtimes G.$$

This shows that f.g.p. modules over the crossed product $C_0(Y) \rtimes H$ by the *compact* group H serve as a supply of f.g.p. modules, "noncommutative vector bundles," over $C_0(X) \rtimes G$.

6.3 Morita Correspondences from Equivariant Vector Bundles

Let G be a locally compact group acting properly on X. A *G-equivariant vector bundle* $\pi: V \to X$ over a G-space X is a vector bundle $\pi: V \to X$ which is also equipped with a group action for which:

a) $\pi: V \to X$ is a G-equivariant map.
b) The action of any $g \in G$ maps the fiber V_x linearly to the fiber V_{gx}, for any $x \in X$.

A vector bundle over a point acted on by a finite or compact group is simply a finite-dimensional representation of G.

Example 6.3.1 Let G be the group $\mathbf{SU}_2(\mathbb{C})$ of 2-by-2 unitary matrices with determinant 1. Clearly G acts (linearly) on $\mathbb{C}^2$, so there is an induced action of G on the space $\mathbb{CP}^1$ of lines in $\mathbb{C}^2$. Let H be the Hopf bundle on $\mathbb{CP}^1$ (see Example 5.1.7). If $L \subset \mathbb{C}^2$ is a line, spanned, say, by a vector $(z, w) \in \mathbb{C}^2$, then $g(L)$ is the line spanned by the vector $g \cdot (z, w)$. Moreover, g maps the line L to the line $g(L)$ linearly, in the obvious way, sending $\lambda \cdot (z, w) \in L$ to $\lambda \cdot g(z, w) \in g(L)$. So the Hopf bundle over $\mathbb{CP}^1$ carries a natural structure of a G-equivariant vector bundle over the G-space $\mathbb{CP}^1$.

Exercise 6.3.2 If G is a compact group acting on X and V is a G-equivariant vector bundle on X, prove that there is a Hermitian metric on V making the G-action fiberwise unitary. (*Hint.* Average an arbitrary Hermitian metric over G using the Haar measure.)

Extend this result to where G is possibly noncompact but acts properly, by using a cutoff function.

Definition 6.3.3 Let G be a discrete group acting properly on X and let $\pi: V \to X$ be a G-equivariant vector bundle on X. We define $\mathcal{E}_V$ to be the completion of the linear space of compactly supported sections of V with respect to the inner product valued in $C_0(X) \rtimes G$ and defined

$$\langle \xi, \eta \rangle_{C_0(X) \rtimes G}(x, g) := \langle \xi(x), g \cdot \eta(g^{-1}x) \rangle.$$

Note that in the formula, the g^{-1}-action maps the fiber $V_{g^{-1}x}$ to the fiber V_x because V is an equivariant vector bundle.

Give $\mathcal{E}_V$ the right $C_0(X) \rtimes G$-module structure by letting

$$(\xi \cdot g)(x) := g^{-1} \cdot \xi(gx), \quad (\xi \cdot f)(x) := \xi(x)f(x),$$

where the inner product on the right hand side is the Hermitian inner product on V.

These formulas satisfy the appropriate covariance condition and extend to a right module multiplication by $C_0(X) \rtimes G$.

Definition 6.3.4 If V is a G-equivariant vector bundle on X, we let A_V denote the C*-algebra of continuous, bounded, and G-equivariant sections of the endomorphism bundle $\mathrm{End}(V)$ over X.

Thus, a typical element of A_V is a bundle endomorphism a such that

$$a(gx) = g \cdot s(x) \cdot g^{-1} \quad \forall x \in X.$$

Note that A_V is unital.

If ξ is a compactly supported section of V, and $a \in A_V$, set

$$(a \cdot \xi)(x) := a(x)\xi(x).$$

Since $a(gx) = ga(x)g^{-1}$, it follows that if ξ is a compactly supported section of V, then

$$(6.13) \quad [(a \cdot \xi) \cdot h](x) = h^{-1} \cdot [(a \cdot \xi)(hx)] = (h^{-1} \cdot a(hx) \cdot h)h^{-1} \cdot \xi)(hx)$$

$$= a(x)h^{-1} \cdot \xi(hx) = [a \cdot (\xi \cdot h)](x),$$

and hence

$$(a \cdot \xi) \cdot h = a \cdot (\xi \cdot h).$$

Similarly

$$(a \cdot \xi) \cdot f = a \cdot (\xi \cdot f)$$

for any $f \in C_0(X) \rtimes G$. Hence the left multiplication action by $a \in A_V$ is a right $C_0(X) \rtimes G$-module map.

Next, we define a left A_V-valued inner product as follows. If ξ and η are two compactly supported sections of V, that is, elements of $\mathcal{E}_V$, they determine the rank-one module map $\theta_{\xi, \eta}$ acting on $\mathcal{E}_V$. Now if ξ is a section, let $(g\xi)(x) = g\left(\xi(g^{-1}x)\right)$. Then with this notation set

$$A_V \langle \xi, \eta \rangle (x) := \sum_{g \in G} \theta_{g\xi, g\eta}.$$

We leave it to the reader to check that we can find a finite collection $\xi_1, \ldots, \xi_n$ of compactly supported sections of V such that

$$\sum A_V \langle \xi_i, \xi_i \rangle = 1_{A_V},$$

from which it follows that id_V is a compact operator on $\mathcal{E}_V$ and that the projection p_V with matrix with the entries

$$\langle \xi_i, \xi_j \rangle_{C_0(X) \rtimes G}$$

satisfies

$$\mathcal{E}_V \cong p_V \cdot (C_0(X) \rtimes G)^n$$

as right Hilbert $C_0(X) \rtimes G$-modules.

Corollary 6.3.5 *If V is a G-equivariant vector bundle over X, a proper, cocompact G-space, then the right Hilbert $C_0(X) \rtimes G$-module $\mathcal{E}_V$ together with its left A_V action and inner product defined above is a Morita equivalence between the unital C^*-algebra A_V and the ideal generated by p_V. Thus,*

$$A_V \xrightarrow{\mathrm{id}} A_V \xrightarrow{\mathcal{E}_V} C_0(X) \rtimes G$$

is a Morita correspondence between A_V and $C_0(X) \rtimes G$.
   In particular

$$\mathbb{C} \to A_V \xrightarrow{\mathcal{E}_V} C_0(X) \rtimes G$$

is a Morita correspondence from $\mathbb{C}$ to $C_0(X) \rtimes G$, and $\mathcal{E}_V$ is an f.g.p. module over $C_0(X) \rtimes G$.

Example 6.3.6 A basic case where one can apply the theorem is to the trivial line bundle $\mathbf{1}$ over any (proper, cocompact) G-space X. Then $A_{\mathbf{1}} = C(G \backslash X)$. Our results above show that $C(G \backslash X)$ is Morita equivalent to an ideal of $C_0(X) \rtimes G$. Thus, one always has the correspondence

$$C(G \backslash X) \xrightarrow{\mathrm{id}} C(G \backslash X) \xrightarrow{\mathcal{E}_1} C_0(X) \rtimes G.$$

In particular, we see that any complex vector bundle V over $G \backslash X$ determines a canonical f.g.p. module $\mathcal{E}_V \otimes_{C(G \backslash X)} \mathcal{E}_1$ over $C_0(X) \rtimes G$. When the bundle is trivial, the projection corresponding to the f.g.p. module induced by this procedure is the same as the projection P_ρ of Exercise 6.1.17, manufactured from a cutoff function. It is full if and only if the action is free.

Actually, all isomorphism classes of f.g.p. modules over $C_0(X) \rtimes G$ arise from equivariant vector bundles from the above procedure. Theorem 6.5.10 of the next chapter provides the exact statement.

6.4   Morita Correspondences and Representations of Finite Groups

Let G be a finite group. A G-equivariant vector bundle over a point is a finite-dimensional representation $\pi: G \to \mathbb{B}(H_\pi)$. Applying the techniques of the previous section, we endow the Hilbert space H_π with the structure of a right Hilbert $C^*(G)$-module $\mathcal{E}_\pi$ with

$$(6.14) \qquad v \cdot g := \pi(g^{-1})v, \quad \langle v, w \rangle_{C^*(G)}(g) := \langle \pi(g^{-1})v, w \rangle.$$

Note that the inner product of two vectors is simply the corresponding matrix coefficient of the representation, as a function on G.

For $v, w \in \mathcal{E}_\pi = V_\pi$ let $\theta_{v,w}$ be the corresponding rank-one operator in $\mathbb{B}(H_\pi)$. The C*-algebra A_V from the vector bundle discussion above is then the commutant

$$\pi(G)' := \{T \in \mathbb{B}(H_\pi) \mid T\pi(g) = \pi(g)T, \ \forall g \in G\}$$

of the representation. The $C^*(G)'$-valued inner product is given by

$$_{C^*(G)'}\langle v, w \rangle := \sum_{g \in G} \theta_{gv, gw}.$$

The C*-algebra $\pi(G)'$ is by definition a subalgebra of $\mathbb{B}(H_\pi)$, but since elements of $\pi(G)'$ commute with $\pi(G)$, they define Hilbert $C^*(G)$-module maps on $\mathcal{E}_\pi$. Notice also that the span of the range of the $\pi(G)'$-valued inner product is all of $\pi(G)'$. Indeed, suppose $T \in \pi(G)'$. Let $e_1, \ldots, e_n$ be an orthonormal basis for H_π. We may write T as a matrix, that is

$$T = \sum_{i,j} T_{ij} \, \theta_{e_i, e_j}.$$

Since T commutes with $\pi(G)$, summing the above construction over G gives

$$(6.15) \qquad |G| \cdot T = \sum_{g \in G} \sum_{i,j} T_{ij} \cdot \theta_{ge_i, ge_j} = \sum_{i,j} T_{ij} \cdot_{A_\pi} \langle e_i, e_j \rangle.$$

This shows that E_π has the structure of a Morita equivalence between $A_\pi = \pi(G)'$ and an ideal of $C^*(G)$, namely the support of the inner product $\langle \cdot, \cdot \rangle_{C^*(G)}$. We discuss this ideal in more detail below. In any case we thus obtain a Morita correspondence

$$\pi(G)' \xrightarrow{\text{id}} \pi(G)' \xrightarrow{\mathcal{E}_\pi} C^*(G)$$

and so the following result.

Corollary 6.4.1 *Let* $\pi : G \to \mathbb{B}(V)$ *be a finite-dimensional representation of a finite group* G. *Let* $\pi(G)' \subset \mathbb{B}(H_\pi)$ *be the C*-algebra of operators on* H_π *which commute with* $\pi(G)$. *Then*

$$\pi(G)' \xrightarrow{\text{id}} \pi(G)' \xrightarrow{\mathcal{E}_\pi} C^*(G)$$

with $\mathcal{E}_\pi$ *defined above is a Morita correspondence from* $\pi(G)'$ *to* $C^*(G)$.

Example 6.4.2 Let $\rho : G \to \mathbf{U}(l^2 G)$ be the right regular representation. Applying the procedure above produces a Morita correspondence

$$\rho(G)' \xrightarrow{\text{id}} \rho(G)' \xrightarrow{\mathcal{E}_\rho} C^*(G).$$

But since $\rho(G)' = C^*(G)$, this is a Morita correspondence from $C^*(G)$ to itself. It is the "identity," as the following shows.

Exercise 6.4.3 In the above notation:

a) $\langle v, w \rangle_{C^*(G)} = v^* * w$, where $v^*(g) = \overline{v(g^{-1})}$ is the adjoint of v as an element of $C^*(G)$ and that $\mathcal{E}_\rho = C^*(G)$ as $C^*(G)$-bimodules.

b) If $v \in \mathcal{E}_\pi = l^2(G)$ is the point mass at the identity, then

$$_{C^*(G)}\langle v, v \rangle = 1_{C^*(G)}$$

and $\langle v, v \rangle_{C^*(G)} = 1_{C^*(G)}$.

Returning to a general finite group, let π be irreducible. Then $\pi(G)' \cong \mathbb{C}$ by Schur's Lemma. Let $v \in H_\pi$ be a vector. Then

$$\frac{1}{|G|} \sum_{g \in G} \theta_{\pi(g)v, \pi(g)v} = \frac{\langle v, v \rangle}{\dim H_\pi}$$

by (1.31). Therefore if v is a unit vector, then

$$_{\pi(G)'}\langle cv, cv \rangle = 1,$$

where $c = \sqrt{\frac{\dim H_\pi}{|G|}}$. We therefore obtain a projection

$$q_\pi \in C^*(G), \quad q_\pi := \langle cv, cv \rangle_{C^*(G)}.$$

We have

(6.16)
$$q_\pi(g) = \frac{\dim H_\pi}{|G|} \cdot \langle \pi(g^{-1})v, v \rangle,$$

a matrix coefficient of π. It is a projection in $C^*(G)$ which maps under π to the rank-one projection onto the span of v. Note that q_π really depends on v as well as π, so it might be better to use the notation $q_{\pi,v}$. However, Exercise 6.4.8 at the end of the section shows that q_π up to *Murray–von-Neumann equivalence* only depends on π and not v.

Exercise 6.4.4 In the above discussion, if v_1, v_2 are orthogonal vectors in H_π such that $_{\pi(G)'}\langle v_i, v_j \rangle = \delta_{ij}$, then $q_{ij} := \langle v_i, v_i \rangle_{C^*(G)}$, $i, j = 1, 2, i \neq j$, are orthogonal projections in $C^*(G)$: that is, $q_{\pi,v_i} \cdot q_{\pi,v_j} = 0$.

Proposition 6.4.5 *Let G be a finite group and $\pi: G \to \mathbb{B}(H_\pi)$ be an irreducible representation and $\mathcal{E}_\pi$ be the Morita correspondence from $\mathbb{C}$ to $C^*(G)$ defined as above. Then*

$$\mathcal{E}_\pi \cong q_\pi \cdot C^*(G)$$

as right f.g.p. $C^*(G)$-modules, with q_π the idempotent (6.16).

Exercise 6.4.6 In the above notation, if α and β are inequivalent irreducible representations of G, then the projections q_α and q_β are orthogonal projections in $C^*(G)$: $q_\alpha \cdot q_\beta = 0$.

Theorem 6.4.7 *Let G be a finite group. Then mapping a finite-dimensional representation $\pi: G \to \mathbf{U}(H)$ of G to the right Hilbert f.g.p. $C^*(G)$-module $\mathcal{E}_\pi$ determines an isomorphism between the semigroup $\mathrm{Rep}(G)$ of equivalence classes of finite-dimensional representations of G, and the semigroup $\mathcal{P}(C^*(G))$ of isomorphism classes of f.g.p. right $C^*(G)$-modules.*

This statement is generalized in the next section.

The use of Morita correspondences and the consequence of the Green–Julg isomorphism for finite groups may be interpreted in terms of KK-theory, which is discussed later in this book. The definition of Morita correspondence

$$A \xrightarrow{\alpha} C \xrightarrow{\mathcal{E}} B,$$

meshes perfectly with the definitions of Kasparov's KK and gives a cycle for $\mathrm{KK}_0(A, B)$. The cycle is based on the right Hilbert B-module $\mathcal{E}_B$, together with the representation $A \xrightarrow{\alpha} C \cong \mathcal{K}(\mathcal{E})$ of A on $\mathcal{E}$.

Exercise 6.4.8 Suppose that $\mathcal{E}$ is a Morita equivalence bimodule from a unital C*-algebra A to an ideal in B:

a) Show that if $\xi_1, \xi_2 \in \mathcal{E}$ such that $_A\langle \xi_i, \xi_i \rangle = 1, i = 1, 2$, then $p_i := \langle \xi_i, \xi_i \rangle_B$ are projections in $\mathrm{supp}(\mathcal{E}) \subset B$.

b) Show that p_1 and p_2 as in a) are *Murray–von-Neumann equivalent projections* by showing that if $u = \langle \xi_1, \xi_2 \rangle_B$, then u is a partial isometry in B such that $uu^* = p_1$ and $u^*u = p_2$. In particular, the f.g.p. modules determined by the projections p_i are isomorphic.

c) If ξ_i are as in a) and if $_A\langle \xi_1, \xi_2 \rangle = 0$, then p_1 and p_2 are orthogonal projections in B, that is, $p_1 \cdot p_2 = 0$.

d) Suppose that $\xi_1, \ldots, \xi_n \in \mathcal{E}$ constitute a *frame* for $\mathcal{E}$ in the sense that $_A\langle \xi_i, \xi_j \rangle = \delta_{ij}$ for all $i.j$ and that

$$\sum_{i=1}^{n} {_A\langle \xi_i, \xi \rangle} \cdot \xi_i = \xi$$

for all $\xi \in \mathcal{E}$. Show that if $p_i := \langle \xi_i, \xi_i \rangle_B$ then p_i are orthogonal projections in B and that $p := p_1 + \cdots + p_n$ is a unit for the ideal $\mathrm{supp}(\mathcal{E})$, that is, that $pb = bp = b$ for all $b \in \mathrm{supp}(\mathcal{E})$.

e) Suppose now that $\mathcal{E} = \mathcal{E}_\pi$ for an irreducible representation of G finite. The left $\pi(G)'$-valued inner product on $\mathcal{E}_\pi$ is then the (conjugate of the) Hilbert space inner product. If $\xi_1, \ldots, \xi_n$ is an orthonormal basis for H_π, then the projection $p \in C^*(G)$ obtained as in part d) is, up to a scalar, the character χ_π of π.

6.5 Remarks on Compact Noncommutative Spaces

Since it seems reasonable to consider unital C*-algebras as corresponding to compact (noncommutative spaces), and in noncommutative geometry, one should think up to Morita equivalence, we formalize the following definition.

Definition 6.5.1 A C*-algebra B *represents a compact noncommutative space* if B is Morita equivalent to a unital C*-algebra.

If A is unital, then $B := A \otimes \mathcal{K}$ is the simplest example, for B is Morita equivalent to A. As we will see below, certain crossed products $C_0(X) \rtimes G$ by proper actions of locally compact groups, and other interesting examples from geometry, have compact underlying noncommutative spaces. This has consequences for their f.g.p. module theory and K-theory.

Recall that a projection $p \in B$ in a C*-algebra is *full* if the ideal generated by p in B is B.

Exercise 6.5.2 If p is a rank-one projection in $\mathcal{K}$, then $1 \otimes p$ is a full projection in $B := A \otimes \mathcal{K}$ for any unital C*-algebra A.

Proposition 6.5.3 *A stable C*-algebra B = A represents a compact noncommutative space if and only if B contains a full projection.*

Proof Suppose that B contains a full projection. Since p is full, pBp is Morita equivalent to B. Since pBp is unital, with unit p, B is Morita equivalent to a unital C*-algebra. Hence B represents a compact noncommutive space. Conversely, if B is stable and is Morita equivalent to A unital, then $B \cong A \otimes \mathcal{K}$ and since $A \otimes \mathcal{K}$ contains a full projection (it can be taken to be $1 \otimes p$ where p is a rank-one projection in $\mathcal{K}$), it follows that B does as well.   □

Exercise 6.5.4 Let $A := \mathcal{T}$ be the Toeplitz algebra, and P_+ the Toeplitz–Szegö projection. Consider the C*-algebra crossed product $B. := C(\overline{\mathbb{Z}}) \rtimes \mathbb{Z}$, where $\overline{\mathbb{Z}} = \mathbb{Z} \cup \{\pm\infty\}$ is the 2-point compactification of the integers, on which $\mathbb{Z}$ acts by translation:

a) Prove that $P_+ B P_+ = A$, so that A is a corner in B.
b) Prove that the ideal in A generated by P_+ is the ideal $C_0((-\infty, +\infty]) \rtimes \mathbb{Z}$ of B.

Thus, $\mathcal{T}$ is Morita equivalent to $C_0((-\mathbb{N}) \rtimes \mathbb{Z}$. In particular, the latter (non-unital) C*-algebra represents a compact noncommutative space.

Exercise 6.5.5 Prove that $C_0(\mathbb{R}^2) \otimes \mathcal{K}$ contains no nonzero projection. Deduce that $C_0(\mathbb{R}^2)$ does not represent a compact noncommutative space.

More generally, if A and B are Morita equivalent, then their spectra: The spaces $\hat{A}$ and $\hat{B}$ of equivalence classes of irreducible representations are homeomorphic. Combining with the fact that if A is commutative, then $\widehat{A}$ is the usual Gelfand spectrum of A, one deduces:

Proposition 6.5.6 *If X is locally compact Hausdorff, then $C_0(X)$ represents a compact noncommutative space if and only if X is compact.*

We are going to show below the following.

Theorem 6.5.7 *If G discrete acts properly and cocompactly on X, then $C_0(X) \rtimes G$ represents a compact noncommutative space.*

Remark 6.5.8 Theorem 4.6.23 already suggests that the *spectrum* of such an example is compact. The theorem provided a parameterization of the spectrum: the set of equivalence classes of irreducible representations of $C_0(X) \rtimes G$, by the set $\sqcup_{x \in F} \widehat{\mathrm{Stab}_G(x)}$, where $F \subset X$ is a set of representatives of the orbits. In particular, the spectrum has the structure of a bundle over $G \backslash X$ with fiber over an orbit Gx the (finite) spectrum of the isotropy group $\mathrm{Stab}_G(x)$.

In order to prove that $C_0(X) \rtimes G$ represents a compact noncommutative space, we have to look more closely at the f.g.p. module theory of $C_0(X) \rtimes G$.

Let $\pi : V \to X$ be a G-equivariant vector bundle over X. Thus, G is a complex vector bundle, and G acts on V with an action for which the bundle projection $\pi : V \to X$ is G-equivariant, and so that the action of each $g \in G$ is fiberwise linear. We may assume without loss of generality that V possesses a G-invariant Hermitian metric.

Definition 6.3.3 describes a finitely generated projective right Hilbert $A :=$ $C_0(X) \rtimes G$-module $\mathcal{E}_V$, obtained by suitably completing the space of compactly supported sections of V. $\mathcal{E}_V$ is the right A-module underlying a Morita equivalence between an ideal J_V of $C_0(X) \rtimes G$ and a certain unital C*-algebra. In particular, if the support ideal J_V, for appropriate choice of V, can be arranged to be $C_0(X) \rtimes G$, then this would prove that $C_0(X) \rtimes G$ represents a compact noncommutative space.

Let $B = A \otimes \mathcal{K}$ with $A = C_0(X) \rtimes G$. The bundle therefore determines an f.g.p. module $\mathcal{E}_V$ and hence a projection $p_V \in M_n(A) \subset B$ for some n. Let us observe first that all such projections occur in this way. To see this, by Theorem 4.6.19 the crossed product $C_0(X) \rtimes G$ can be identified with the fixed-point algebra

$$C_0(X, \mathcal{K})^G := \{ f : X \to \mathcal{K}(l^2 G) \mid f \text{ is continuous, and } f(gx)$$
$$= \rho(g) f(x) \rho(g)^{-1} \, \forall g \in G, \, x \in X. \}$$

Now suppose that p is a projection in $M_n\big(C_0(X) \rtimes G\big)$. Thinking of p as a map $X \to \mathcal{K}((l^2 G)^n$ which is G-equivariant, the image $p(x) \subset l^2(G)$ for any $x \in X$ is a finite-dimensional, linear subspace $V_x \subset l^2(G)^n := l^2(G) \oplus \cdots l^2(G)..$

Exercise 6.5.9 In the above notation:

a) Show that the family of subspaces $\{V_x\}_{x \in X}$ are the fibers of a vector bundle V over X.

b) Show that if $g \in G$, then the right translation operator $\rho(g) \colon l^2(G) \to l^2(G)$ induces a linear isomorphism $V_x \to V_{gx}$ for any $x \in X$. Show that V has in this way the structure of a G-equivariant vector bundle over X, and whose associated f.g.p. module $\mathcal{E}_V$ over $C_0(X) \rtimes G$ is isomorphic to $p_V \cdot \big(C_0(X) \rtimes G\big)^n$.

The observations above lead to a bijection between isomorphism classes of f.g.p. modules over $C_0(X) \rtimes G$, and G-equivariant vector bundles over X and the following theorem (see [77]).

Theorem 6.5.10 *If G is a discrete group acting properly and cocompactly on X, then mapping a G-equivariant vector bundle V over X to the f.g.p. right $C_0(X) \rtimes G$-module $\mathcal{E}_V$ determines an isomorphism from the semigroup $\mathrm{Vect}_G(X)$ of isomorphism classes of G-equivariant vector bundles over X, to the semigroup $\mathcal{P}(C_0(X) \rtimes G)$ of isomorphism classes of f.g.p. modules over $C_0(X) \rtimes G$, equivalently, the semigroup of Murray–von-Neumann equivalences classes of projections in $C_0(X) \rtimes G \otimes \mathcal{K}$.*

Recall that if V is an equivariant vector bundle, then for each $x \in X$, the fiber V_x carries a representation of the isotropy $\mathrm{Stab}_G(x)$. Representations of compact groups decompose into finite direct sums of irreducibles, and so we can ask whether a given irreducible representation π of $H := \mathrm{Stab}_G(x)$ is contained (up to isomorphism) in the representation of H on V_x.

Definition 6.5.11 Let V be a G-equivariant vector bundle over X. We say V is *full* if for every $x \in X$, the representation of $\mathrm{Stab}_G(x)$ on the fiber V_x contains (up to isomorphism) every irreducible representation χ of $\mathrm{Stab}_G(x)$.

The following result is due to W. Lück and Bob Oliver [120].

Lemma 6.5.12 *If G is discrete and acts properly and cocompactly on X, then X has a full G-equivariant vector bundle.*

With these ideas we can now give the proof of Theorem 6.5.7, modulo the theorem of Lück and Oliver Lemma 6.5.12.

Proof Let V be a full, G-equivariant vector bundle on X, and $p_V \in C_0(X) \rtimes G \otimes \mathcal{K}$ a projection determining the corresponding f.g.p. module. As previously discussed, the C*-algebra $C_0(X) \rtimes G$ may be identified with the C*-algebra of continuous maps $f: X \to \mathcal{K}(l^2 G)$ which are G-equivariant, with G acting on $\mathcal{K}(l^2 G)$ by the action induced by the *right* regular representation of G on $L^2(G)$. In particular, if $p_V \in C_0(X) \rtimes G$ is a projection, then for each $x \in X$, with isotropy $H_x = \mathrm{Stab}_G(x)$, $p_V(x)$ is a finite-rank projection in $\mathcal{K}(l^2 G)$ which commutes with the right translation action $\rho(H_x)$: It projects to the fiber V_x of the corresponding equivariant vector bundle. Now, by fixing an H_x-equivariant bijection $G \cong G/H \times H$, one can H_x-equivariantly decompose $V_x = \oplus_i l^2(H_x)$ into a direct sum of (infinitely many) copies of $l^2(H_x)$, and $p_V(x)$ is a (finite) direct sum of projections $p_1, p_2, \ldots$ on $l^2(H_x)$, each $p_i \in \rho(H_x)' = C^*(H_x)$. Now suppose that $\chi \in \widehat{H_x}$. Then if χ does not appear in the representation V_x, it follows that $\chi(p_i) = 0$ for each i. Allowing now for projections p_V in matrix algebras over $C_0(X) \rtimes G$, we deduce that if $p_V \in \mathcal{P}(C_0(X) \rtimes G) \otimes \mathcal{K}$ is the projection associated with a G-equivariant vector bundle V, then V_x contains an irreducible representation χ of $\mathrm{Stab}_G(x)$ if and only if $\chi(p_V) \neq 0$.

In particular, if V is full, then $\pi(p_V) \neq 0$ for every irreducible representation π of $C_0(X) \rtimes G$, by Remark 6.5.8. Hence the ideal generated by p_V is all of $C_0(X) \rtimes G \otimes \mathcal{K}$, and the support ideal of $\mathcal{E}_V$ is hence $C_0(X) \rtimes G$. Thus $C_0(X) \rtimes G$ contains a full projection and is therefore Morita equivalent to a unital C*-algebra. $\square$

Note that the proof shows that $C_0(X) \rtimes G$ is Morita equivalent to the specific unital C*-algebra of G-equivariant endomorphisms of a full vector bundle over X.

We conclude by noting that there are many interesting examples of compact noncommutative spaces in the form of crossed products $C(X) \rtimes G$, with the G-action *not* assumed proper. Exercise 6.1.21 noted that the crossed product $C(\mathbb{T}^2) \rtimes_{\beta\hbar} \mathbb{R}$ by the Krönecker flow

$$\beta_t(x, y) = (x + t, y + t\hbar),$$

with $\hbar \in \mathbb{R}$, usually irrational, is Morita equivalent to the irrational rotation algebra $A_\hbar$.

Proposition 6.5.13 *For any $\hbar \in \mathbb{R}$, the C*-algebra $C(\mathbb{T}^2) \rtimes_{\beta\hbar} \mathbb{R}$ represents a compact noncommutative space.*

More generally:

Theorem 6.5.14 *If G is a locally compact group and $H \subset G$ is a closed subgroup with G/H compact, and if $\Gamma \subset G$ is a discrete subgroup, then $C_0(G/\Gamma) \rtimes H$ represents a compact noncommutative space.*

This follows from Exercise 6.1.20, which gives a Morita equivalence between $C_0(G/\Gamma) \rtimes H$ with $C(G/H) \rtimes \Gamma$; the latter is unital.

6.6   Morita Correspondences Between Irrational Tori

Morita equivalence techniques can be used very fruitfully in connection with the irrational rotation algebra, for example, for the construction of f.g.p. modules from geometric ideas.

Choose $\hbar \in \mathbb{R}$, and let $G = \mathbb{Z}\hbar$, $H = \mathbb{Z}$, discrete subgroups of $\mathbb{R}$, and $X = \mathbb{R}$, with each group acting by group translation on $\mathbb{R}$. We apply Theorem 6.1.15 to the two commuting group actions of G and H on X. The corresponding Morita equivalent crossed products $A = C_0(H\backslash X) \rtimes G$ and $B = C_0(G\backslash X) \rtimes H$ are respectively

$$A = C(\mathbb{R}/\hbar\mathbb{Z}) \rtimes \mathbb{Z}, \quad B = C(\mathbb{R}/\mathbb{Z}) \rtimes \hbar\mathbb{Z},$$

and the second crossed product we may identify with the crossed product $C(\mathbb{R}/\mathbb{Z}) \rtimes_\hbar \mathbb{Z}$ of the integers acting by translation by multiples of $\hbar$, so that $B = A_\hbar$, the irrational rotation algebra.

Exercise 6.6.1 The crossed product A is naturally isomorphic to $A_{1/\hbar}$.

(*Hint.* Show that addition of $+1$ on $[0, \hbar]$ (mod $\hbar$) is conjugate to addition of $1/\hbar$ on $[0, 1]$ (mod 1).)

In any case, A is unital. We obtain the following.

Corollary 6.6.2 *Define on $C_c(\mathbb{R})$ the inner products*

$$_{C(\mathbb{R}/\hbar\mathbb{Z})\rtimes\mathbb{Z}}\langle \xi, \eta\rangle(x, m) = \sum_{n\in\mathbb{Z}} \xi(x - n\hbar)\cdot \overline{\eta(x - n\hbar - m)}, \quad x \in \mathbb{R}/\hbar\mathbb{Z}, \; m \in \mathbb{Z},$$

and

$$\langle \xi, \eta\rangle_{C(\mathbb{R}/\mathbb{Z})\rtimes_\hbar\mathbb{Z}}(x, m) = \sum_{n\in\mathbb{Z}} \overline{\xi(x - n)}\eta(x - n - m\hbar), x \in \mathbb{R}/\mathbb{Z}, \; m \in \mathbb{Z}.$$

Give $C_c(\mathbb{R})$ the $C(\mathbb{R}/\hbar\mathbb{Z}) \rtimes \mathbb{Z}$- $C(\mathbb{R}/\mathbb{Z}) \rtimes_\hbar \mathbb{Z}$ bimodule structure with

$$(n\xi)(x) = \xi(x - n), \quad (f\xi)(x) = f(x)\xi(x), \qquad (\xi n)(x) = \xi(x + n\hbar),$$

$$(\xi f)(x) = f(x)\xi(x).$$

Then $C_c(\mathbb{R})$ completes to a Morita equivalence $C(\mathbb{R}/\hbar\mathbb{Z}) \rtimes \mathbb{Z}$- $C(\mathbb{R}/\mathbb{Z}) \rtimes_\hbar \mathbb{Z}$ bimodule $\mathcal{E}_\hbar$, that is, to a Morita equivalence $A_{1/\hbar}$-$A_\hbar$-bimodule.

From Proposition 6.2.1 we obtain:

Corollary 6.6.3 *The completion of $C_c(\mathbb{R})$ with respect to the right $A_\hbar$-valued inner product*

$$\langle \xi, \eta \rangle_{A_\hbar}(x, m) = \sum_{n \in \mathbb{Z}} \overline{\xi(x - n)} \eta(x - n - m\hbar)$$

and right module structure

$$(\xi n)(x) = \xi(x + n\hbar), \quad (\xi f)(x) = f(x)\xi(x),$$

is a finitely generated projective $A_\hbar$-module.

We will call $\mathcal{E}_\hbar$ the *Rieffel module*.

The proof of Proposition 6.2.1 gives a way of finding a projection $p_\hbar \in A_\hbar$ such that $\mathcal{E}_\hbar \cong p_\hbar A_\hbar$ as right $A_\hbar$-modules.

We are going to need a real-valued function ξ of compact support such that

$$(6.17) \qquad\qquad\qquad \sum_{n \in \mathbb{Z}} \xi(x + n\hbar)^2 = 1.$$

The resulting projections $\langle \xi, \xi \rangle_{A_\hbar} \in A_\hbar$ as ξ varies while still satisfying the given condition are all Murray–von-Neumann equivalent to each other (by Exercise 6.4.8).

Now we have

$$_{C(\mathbb{R}/\hbar\mathbb{Z}) \rtimes \mathbb{Z}}\langle \xi, \xi \rangle(x, m) = \sum_{n \in \mathbb{Z}} \xi(x - n\hbar)\xi(x - n\hbar - m), \quad x \in \mathbb{R}/\hbar\mathbb{Z}, \quad m \in \mathbb{Z}.$$

Since the support of ξ has diameter < 1, every term in the sum vanishes if $m \neq 0$. If $m = 0$, we have

$$_{C(\mathbb{R}/\hbar\mathbb{Z}) \rtimes \mathbb{Z}}\langle \xi, \xi \rangle(x, 0) = \sum_{n \in \mathbb{Z}} \xi(x - n\hbar)^2 = 1 \; \forall x \in \mathbb{R}/\hbar\mathbb{Z}.$$

Hence $_{C(\mathbb{R}/\hbar\mathbb{Z}) \rtimes \mathbb{Z}}\langle \xi, \xi \rangle = 1$.

The proof of Proposition 6.2.1 now shows that $p_\hbar := \langle \xi, \xi \rangle_{A_\hbar}$ is a projection, and

$$\Phi : \mathcal{E}_\hbar \to A_\hbar, \quad \Phi(\eta) = \langle \xi, \eta \rangle_{A_\hbar}$$

is a Hilbert module isometry with range $p_\hbar A_\hbar$.

Thus

$$p_\hbar(x, m) = \sum_{n \in \mathbb{Z}} \xi(x - n)\xi(x - n - m\hbar) = f + gU + (gU)^*,$$

in group algebra notation, where U corresponds to $1 \in \mathbb{Z}$, generates the action, and $f(x)$ is the function $\xi(x)^2$ (supported strictly inside $[0, 1]$ and thought of as a function on $\mathbb{T}$). The function g is $g(x) = \xi(x)\xi(x - \hbar)$.

One might think of ξ as taking the characteristic function χ of an interval in $\mathbb{T}$ of length $\hbar$ and rounding it off slightly at the edges to make it continuous but still satisfy (6.17).

Exercise 6.6.4 For $f \in C(\mathbb{T}) = C(\mathbb{R}/\mathbb{Z})$, let $R_\hbar(f)(x) = f(x - \hbar)$.

Suppose f and g are real-valued continuous functions on $\mathbb{T}$, supported strictly inside $[0, 1]$ (we consider $\mathbb{T} = \mathbb{R}/\mathbb{Z}$ as usual). Consider the self-adjoint

$$p := f + gu + (gu)^* \in A_\hbar.$$

Prove that:

a) Prove that p is a projection if and only if $0 \le f \le 1$, and

$$g = \sqrt{f - f^2}, \quad gR_\hbar(f) + fR_\hbar(g) = g, \quad gR_\hbar(g) = 0.$$

b) Now suppose that f is chosen, still compactly supported within $[0, 1]$ such that $R_\hbar^{-1}(f) + f + R_\hbar(f) = 1$ on supp(f). Prove that if we define $g := \sqrt{f - f^2}$, then the remaining two of the above conditions hold.

Thus, one gets a projection $p = f + gu + (gu)^*$ by defining $g := \sqrt{f - f^2}$ where f satisfies $f + R_\hbar(f) = 1$. Notice that with $f := \langle \xi, \xi \rangle_{A_\hbar} = \xi^2$ in the notation above previous to the exercise, then this condition corresponds exactly to (6.17). Such projections are often called *Rieffel projections*.

Exercise 6.6.5 Show that if τ is the trace on $A_\hbar$,

$$\tau \left(\sum_n f_n[n] \right) := \int_\mathbb{T} f d\mu$$

with μ normalized Lebesgue measure, then

$$\tau(p_\hbar) = \hbar.$$

Remark 6.6.6 Recall that if $\tau : A \to \mathbb{C}$ is a trace on a unital C*-algebra, then τ determines a map $\mathcal{P}(A) \to \mathbb{R}$: Indeed, if $p \in A$ is a projection, then $\tau(p) \in \mathbb{R}$ only depends on the Murray–von-Neumann equivalence class of p because of the tracial property of τ. The (positive) number $\tau(p)$ is sometimes referred to as the

Murray–von-Neumann dimension of the module pA, and denoted $\dim_\tau(pA)$. One
extends τ to projections in $M_n(A)$ by summing the diagonal entries of the matrix
and applying τ to the sum.

If τ is the unique unital trace on $\mathbb{C}$, then $\dim_\tau(L)$ is an integer for any f.g.p. $\mathbb{C}$-module, of course. Actually, it is a question of some interest to compute the range
of this construction, e.g., the set of possible values of $\tau(p)$, for p a projection in
A, or in $M_n(A)$, when $\tau : C^*(G) \to \mathbb{C}$ is the canonical trace on the C*-algebra
of a discrete group. If the group is torsion-free, this range is conjectured to be the
integers.

In any case, we see from the previous exercise that there are f.g.p. modules over
C*-algebras with arbitrary real numbers as their dimensions.

The above constructions can be generalized in the following way. Consider the
group $\mathbf{GL}_2(\mathbb{Z})$. It acts by linear automorphisms of $\mathbb{T}^2$. Suppose $\hbar \in \mathbb{R}$. Let $B_\hbar$ be
the crossed product

$$B_\hbar := C(\mathbb{T}^2) \rtimes_{\beta^h} \mathbb{R},$$

where β^h is the Krönecker flow on $\mathbb{T}^2$—see Exercise 6.1.21. By the same Exercise,
$A_\hbar$ is Morita equivalent to $B_\hbar$. Let $\mathcal{E}_\hbar$ denote the corresponding Morita $B_\hbar$-$A_\hbar$-bimodule. An element $g \in \mathbf{GL}_2(\mathbb{Z})$ defines an automorphism of $\mathbb{T}^2$, which
conjugates the Krönecker flow along lines of slope $\hbar$, to the Krönecker flow along
lines of slope $g(\hbar)$, where if $g = \begin{bmatrix} a & b \\ c & d \end{bmatrix}$, then $g(h) = \frac{a\hbar+b}{c\hbar+d}$. Hence g defines an
isomorphism $g : B_\hbar \to B_{g(\hbar)}$. We obtain a Morita correspondence

$$B_{g(\hbar)} \xrightarrow{g^{-1}} B_\hbar \xrightarrow{\mathcal{E}_\hbar} A_\hbar.$$

On the other hand, we have a correspondence

$$A_{g(\hbar)} \xrightarrow{\text{id}} A_{g(\hbar)} \xrightarrow{\mathcal{E}^*_{g(\hbar)}} B_{g(\hbar)}.$$

We can then, in effect, *compose* the two correspondences, as suggested by the
diagram

$$A_{g(\hbar)} \xrightarrow{\text{id}} A_{g(\hbar)} \xrightarrow{\mathcal{E}^*_{g(\hbar)}} B_{g(\hbar)} \xrightarrow{g^{-1}} B_\hbar \xrightarrow{\mathcal{E}_\hbar} A_\hbar.$$

To define this composition precisely, we pull back $\mathcal{E}_\hbar$ under the isomorphism g^{-1}.
This results in a Morita $B_{g(\hbar)}$-$A_\hbar$-bimodule $(g^{-1})^*\mathcal{E} = g_*(\mathcal{E})$. We then tensor the
bimodules, forming

(6.18) $$\mathcal{E}^*_{g(\hbar)} \otimes_{B_{g(\hbar)}} g_*(\mathcal{E}_\hbar),$$

an $A_{g(\hbar)}$-$A_\hbar$-bimdodule, and an equivalence.

This results in a Morita equivalence between $A_{g(\hbar)}$ and $A_\hbar$, yielding the following interesting fact:

Corollary 6.6.7 *If $\hbar$ and $\hbar'$ are in the same orbit of $\mathbf{GL}_2(\mathbb{Z})$ acting on $\mathbb{R}$ by Möbius transformations, then the rotation algebras $A_\hbar$ and $A_{\hbar'}$ are Morita equivalent, so represent the same noncommutative space.*

The corollary also implies something interesting about the f.g.p. module theory of $A_\hbar$.

Corollary 6.6.8 *If $g \in \mathbf{GL}_2(\mathbb{Z})$, then the right $A_\hbar$-module L_g given by forgetting the left $A_{g(\hbar)}$-structure on (6.18) is f.g.p. over $A_\hbar$.*

Exercise 6.6.9 Following Exercise 6.6.5, compute $\dim_\tau(L_g)$, for $g \in \mathbf{GL}_2(\mathbb{Z})$, L_g as in the Corollary above, and $\dim_\tau$ defined as in the discussion in Remark 6.6.6.

6.7 Morita Equivalence and Asymptotics in Hyperbolic Geometry

Endow $\mathbb{D}$ with the hyperbolic Riemannian metric $ds^2 = \frac{1}{1-x^2-y^2} \cdot (dx^2 + dy^2)$. Thus, infinitesimally, this metric is a positive scalar multiple of the usual Euclidean metric, but the scalar factor increases as one moves out toward the boundary of the disk.

Geodesics in this "Poincaré disk" model of the hyperbolic plane are straight lines through the origin in $\mathbb{D}$, together with arcs of circles perpendicular to $\partial \mathbb{D}$.

Of course the disk admits a compactification $\overline{\mathbb{D}}$ and boundary, $\partial \mathbb{D}$. What is particularly interesting about this boundary relative to the geometry is that if (x_n) and (y_n) are sequences in $\mathbb{D}$ which remain a bounded hyperbolic distance apart, at least one of them converging to a point of $\partial \mathbb{D}$, then the other does as well, and

$$\lim_{n \to \infty} x_n = \lim_{n \to \infty} y_n.$$

This is because the hyperbolic metric blows up near the boundary, relative to the usual Euclidean metric.

The isometry group of $\mathbb{D}$ with the Poincaré metric is the Lie group G of matrices of the form $g = \begin{bmatrix} a & b \\ \bar{b} & \bar{a}, \end{bmatrix}$ where $|a|^2 - |b|^2 = 1$. The action is by Möbius transformations

$$g(z) = \frac{az + b}{\bar{b}z + \bar{a}}.$$

Exercise 6.7.1 Such g maps $\mathbb{D}$ to itself and extends continuously to a self-map of $\overline{\mathbb{D}}$.

Inside G are many interesting lattice Γ, which are symmetry groups of various tessellations of $\mathbb{D}$ by hyperbolic polygons, like the following tessellation by ideal hyperbolic triangles.

(6.19)

The group Γ of the tessellation is isomorphic to $\mathbb{Z} * \mathbb{Z}/2$. One of the generators, say a, acts by a parabolic transformation which fixes the top vertex of the central dark triangle, and slides the central dark triangle to the right onto the adjacent white triangle. In the process of doing this, it maps the left edge of the central dark triangle onto the right edge. Likewise, a^{-1} slides the dark triangle left onto the left white triangle.

The generator b acts by elliptic element of order 2 which rotates the central dark triangle around the midpoint of its lower side, onto the white triangle below it.

In particular, if P is the central triangle, and Q is any other triangle which is adjacent to (intersects) P, then $Q = s(P)$ for one of these generators.

Take a geodesic in $\mathbb{D}$. Assume it starts somewhere in the central triangle P.

Now as t increases, the geodesic passes into another, adjacent triangle, which has the form $P_1 = s_1(P)$ for some generator $s_1 \in \{a, a^{-1}, b\}$ (unless it heads straight to one of the ideal vertices of the triangle). Next, it passes into a polygon P_2 adjacent to $s_1(P)$. It follows that $s_1^{-1}(P_2)$ meets P. Hence for some generator s_2, we have $s_1^{-1}(P_2) = s_2(P)$, giving

$$P_2 = s_1 s_2(P).$$

Continuing in this way, we "code" the geodesic by the sequence $s_1, s_2, \ldots$ of generators of Γ.

The sequence of group elements $g_1 = s_1, g_2 = s_1 s_2, g_3 = s_1 s_2 s_3, \ldots$ represents a path in the Cayley graph $X(\Gamma, S)$ of Γ with respect to the generating set S.

Note that if one started with a translate $g(r)$ of the given geodesic, they are the same as geodesics in M. The sequence $g_1, g_2, \ldots \in \Gamma$ obtained above, and converging to the boundary point $\xi \in \partial \Gamma$, would be replaced by the sequence $g g_1, g g_2, g g_3, \ldots$, which converges to $g(\xi)$. That is, different choices of initial lift of the geodesic on M correspond to boundary points in $\partial \Gamma$ in the same orbit of the Γ-action on $\partial \Gamma$.

Finally, observe that if one had another geodesic in M, lifting to a geodesic r' in $\mathbb{D}$ which is *asymptotic* to r in the sense that they remain as $t \to \infty$, a bounded hyperbolic distance apart, then the corresponding sequences of group elements (g_k) and (g'_k) also remain a bounded distance apart. Hence they converge to the same boundary point in $\partial \Gamma$.

The idea is then that a geodesic in $M = \Gamma \backslash \mathbb{D}$, up to the relation of being asymptotic, is equivalent to a Γ-orbit of the boundary point $\xi \in \partial \Gamma$, where the "boundary" $\partial \Gamma$ of the group Γ, with given generators, is the space of all sequences of generators in which no s_i is followed by s_i^{-1}. (This is a theorem: See [148] and related papers for exact statements.) Closed geodesics in M correspond to periodic sequences. The (chaotic) geodesic flow on M corresponds to the action of Γ on its boundary. This argumentation shows that the geodesic flow is effectively a subshift of finite type (these are discussed below).

We now show how Morita equivalence enters into this dictionary. To avoid discussing boundaries of groups at this stage, we work just with the geometry and natural boundary of the hyperbolic plane.

Fix then any lattice $\Gamma \subset G$. Let $M = G \backslash \mathbb{D}$, a hyperbolic manifold (possibly with some "marked points").

Let $S\mathbb{D}$ be the unit sphere bundle (in the hyperbolic metric) of the tangent bundle $T\mathbb{D}$ to the hyperbolic disk. A point of $S\mathbb{D}$ is a unit tangent vector v based at a point $z \in \mathbb{D}$. By the exponential map in Riemannian geometry, v determines a unique geodesic $r_v \colon (-\infty, +\infty) \to \mathbb{D}$ with $r_v(0) = z$ (with $r_v(t) = \exp_z(tv)$, exp the Riemannian exponential map).

This supplies a G-equivariant bijection between the set of geodesics in $\mathbb{D}$ and the three-dimensional manifold $S\mathbb{D}$, such that

$$g(r_v)(t) = r_{g(v)}(t),$$

for all g, v, t.

Likewise, the set of geodesics in $M := \Gamma \backslash \mathbb{D}$ is in 1-1 correspondence with points of the quotient manifold $SM := \Gamma \backslash S\mathbb{D}$.

Define an equivalence relation on geodesics in $M = \Gamma \backslash \mathbb{D}$ by

(6.20) $\qquad\qquad [r_1] \sim [r_2] \text{ iff } \lim_{t \to \infty} d(r_1(t), r_2(t)) = 0,$

for *some* lifts r_1, r_2 of the initial geodesics to geodesics in $\mathbb{D}$. We refer to equivalent geodesics as *strongly asymptotic*.

Let $(g_t)_{t \in \mathbb{R}}$ be the geodesic flow: $g_t(r)(u) = g(u - t)$. This defines an $\mathbb{R}$-action on $S\mathbb{D}$ and, since it commutes with the Γ-action on $S\mathbb{D}$, drops to an $\mathbb{R}$-action on SM.

Definition 6.7.2 Let d be the hyperbolic metric on M.
Let $[r_1]$ and $[r_2]$ be two geodesics in $M := \Gamma \backslash \mathbb{D}$.
We say then that $[r_1]$ and $[r_2]$ are *asymptotic* if there exists $s \in \mathbb{R}$ such that

$$g_s([r_1]) \sim [r_2].$$

If r_1 and r_2 are two geodesics in $\mathbb{D}$ which converge to the same boundary point of $\mathbb{D}$, then there exists s such that

$$\lim_{t \to \infty} d(r_1(t), r_2(t + s)) = 0,$$

so that the boundary points parameterize certain two-dimensional equivalence classes of geodesics, with one direction corresponding to the shift in parameterization (geodesic flow) and the other the horocycle direction.

Lemma 6.7.3 *There is a natural bijection between asymptotic equivalence classes of geodesics on M, and orbits of the Γ-action on $\partial\mathbb{D}$.*

The bijection associates to an asymptotic equivalence class $[r]$ of a geodesic on M obtained by projecting a geodesic r in $\mathbb{D}$, to the Γ-orbit of its endpoint $r(+\infty) = \lim_{t \to \infty} r(t) \in \partial\mathbb{D}$.

Proof If two geodesics on $M = \Gamma \backslash \mathbb{D}$ are asymptotic on M, then rescale one of them to make the pair strongly asymptotic. So they have lifts r_1, r_2 such that $d(r_1(t), r_2(t)) \to 0$ as $t \to \infty$. It follows that r_1 and r_2 converge to the same boundary point of $\partial\mathbb{D}$. Conversely, suppose that $[r_1]$ and $[r_2]$ are geodesics on M and that r_1 and r_2 are lifts converging to boundary points in the same Γ-orbit. Replacing r_2 by an appropriate Γ translate, we get a pair of lifts r_1, r_2 which converge to the same boundary point. $\qquad \square$

We are now going to construct a group action whose orbits are the asymptotic equivalence classes of geodesics on M.

To do so, we revert to our Lie group $G = \mathrm{Isom}(\mathbb{D})$ of Möbius transformations leaving the disk $\mathbb{D}$ invariant.

The G-action extends to $\overline{\mathbb{D}}$ and leaving $\partial\mathbb{D}$ invariant.

Choose a boundary point $\xi_0 \in \partial\mathbb{D}$, and let

$$P := \mathrm{Stab}_G(\xi_0) = \{g \in G \mid g(\xi_0) = \xi_0\}.$$

We will need two observations.

Exercise 6.7.4 Prove that for any $z \in \mathbb{D}$ there is a Möbius transformation $p_z : \mathbb{D} \to \mathbb{D}$ which fixes ξ_0 and maps 0 to z.

We now define an action of P on geodesics as follows.

Let $K = \text{Stab}_G(0)$, so K is the group of matrices in G of the form

$$k = k_\theta = \begin{bmatrix} e^{i\theta} & 0 \\ 0 & e^{-i\theta} \end{bmatrix},$$

acting by the G-action by Möbius transformations $k_\theta(z) = e^{2i\theta}z$. Clearly K acts simply transitively on $\partial\mathbb{D}$.

Definition 6.7.5 If r is a geodesic in $\mathbb{D}$, and $p \in P$, let $k \in K$ the unique element such that

$$r(+\infty) = k(\xi_0).$$

We define

$$p \cdot r := kpk^{-1}(r).$$

Note first that $p \cdot r$ has the same endpoint as r, for any r, since if $k(\xi_0) = r(+\infty)$, $p \cdot r = kpk^{-1}(r)$ and so

$$(p \cdot r)(+\infty) = kpk^{-1}(r(+\infty)) = kp(\xi_0) = k(\xi_0) = r(+\infty).$$

Suppose $p_1, p_2 \in P$. Let $k \in K$ such that $k(\xi_0) = r(+\infty)$, so $p_2 \cdot r = kp_2k^{-1}$. Now to compute $p_1 \cdot (p_2 \cdot r)$, we have that $p_2 \cdot r$ has the same endpoint as r, so $k(\xi_0) = (p_2 \cdot r)(+\infty)$ and hence

$$p_1 \cdot (p_2 \cdot r) = (kp_1k^{-1})(kp_1k^{-1}) = k(p_1p_2)(r) = (p_1p_2) \cdot r.$$

This shows that we have defined an action.

Lemma 6.7.6 *The P-action on the space $S\mathbb{D}$ of geodesics in $\mathbb{D}$ defined above commutes with the G-action on the space $S\mathbb{D}$ of geodesics in $\mathbb{D}$. Furthermore, two geodesics r_1, r_2 are asymptotic if and only if they are in the same P-orbit.*

Proof We have already noted that $p \cdot r \sim_a r$ since have the same endpoint. Conversely, suppose r ends at ξ, let $k(\xi_0) = \xi$, and suppose that $r' \sim_a r$. Since r' ends at ξ as well, and $k(\xi_0) = \xi$,

$$p \cdot r = kpk^{-1}(r), \quad p \cdot r' = (kpk^{-1}(r')$$

for any $p \in P$. Now since P acts simply transitively on $\mathbb{D}$, there exists unique $p \in P$ such that

$$p\left(pk^{-1}(r'(0))\right) = r(0).$$

Now the geodesics $kpk^{-1}(r')$ and r have the same starting point and initial point so

$$p \cdot r' = kpk^{-1}(r') = r$$

as claimed.   □

Since the P-action commutes with the G-action on $S\mathbb{D}$, it passes to an action on SM and we can form the crossed product $C(SM) \rtimes P$.

Corollary 6.7.7 *There is a natural bijection between orbits of the P-action on SM and orbits of the Γ-action on $\partial\mathbb{D}$.*

The C*-algebras $C(SM) \rtimes P$ and $C(\partial\mathbb{D}) \rtimes \Gamma$ are Morita equivalent, for any lattice $\Gamma \subset G$.

If $\Gamma \subset G$ is a cocompact lattice, then the inclusion $\Gamma \to \mathbb{D}$ is in this case a quasi-isometric equivalence in the sense of [91] and induces a homeomorphism $C(\partial\Gamma) \rtimes \Gamma \to C(\partial\mathbb{D}) \rtimes \Gamma$, where $\partial\Gamma$ is the Gromov boundary. We discuss hyperbolic groups later in this book. What is important for the present is that the Gromov boundary $\partial\Gamma$ with its Γ-action only depends on the abstract group Γ.

Corollary 6.7.8 *Let Γ be a cocompact torsion-free discrete group of isometries of the hyperbolic disk, $M = \Gamma\backslash\mathbb{D}$, a Riemann surface of genus ≥ 2, and SM the sphere bundle of M with the action of P defined above.*

Then $C(SM) \rtimes P$ and $C(\partial\Gamma) \rtimes \Gamma$ are Morita equivalent.

See Section 10.4 of Chapter 7 for more information on hyperbolic groups.

From a computational point of view, the Poincaré disk model of the hyperbolic plane can usefully be replaced by the upper half plane model $H = \{z \in \mathbb{C} \mid \mathrm{Im}(z) > 0\}$ with boundary $\partial H := \mathbb{R} \cup \{\infty\}$. The Cayley transform

$$\varphi: H \to \mathbb{D}, \quad \varphi(z) = \frac{z - i}{z + i}$$

conformally maps H to the disk. Pulling back the hyperbolic metric gives the Riemannian metric $ds^2 = \frac{1}{y^2}(dx^2 + dy^2)$ and conjugates the orientation-preserving isometry group G of the disk to the group $\mathbf{PSL}(\mathbb{R})$ of matrices $\begin{bmatrix} a & b \\ c & d \end{bmatrix}$, with $a, b, cd \in \mathbb{R}, ad - bc = 1$, acting by Möbius transformations leaving H invariant. The action continuously extends to $\overline{H}$ leaving ∂H invariant.

Exercise 6.7.9 In reference to the above discussion, prove the following:

a) $\mathbf{PSL}(2, \mathbb{R})$ acts transitively on H and $K := \mathrm{Stab}_{\mathbf{PSL}(2,\mathbb{R})}(i) \cong \mathbf{SO}(2, \mathbb{R})$. Hence $G/K \cong H$ by the orbit map at i.

b) $\mathbf{PSL}(2, \mathbb{R})$ acts transitively on ∂H and $\mathrm{Stab}_{\mathbf{PSL}(2,\mathbb{R})}(\infty)$ is the group B of upper triangular matrices in $\mathbf{PSL}(2, \mathbb{R})$. Hence $\partial H = G/B$ by the orbit map at ∞.

c) A unit tangent vector in H, or, equivalently, a geodesic in H, is determined by its base (the starting point of the geodesic), and the boundary point in ∂H converges to as $t \to \infty$. Moving the basepoint to i by an element of $\mathbf{PSL}(2, \mathbb{R})$ and then moving the boundary point of the resulting geodesic to ∞ by application of an element of $k \in K$ produce the unit tangent vector with base i and target ∞. Hence $\mathbf{PSL}(2, \mathbb{R})$ acts transitively on the space SH of geodesics in H. Moreover, the action is free. Hence $SH \cong \mathbf{PSL}(2, \mathbb{R})$.

d) Fix the boundary point $\xi_0 = \infty$, so B is its stabilizer. We refer to the definition of the P-action on geodesics in $\mathbb{D}$ in Definition 6.7.5, which now transplants to H with $P = B$ if ξ_0 was initially chosen to be $\varphi(\infty)$. Prove that if $p = \begin{bmatrix} a & b \\ 0 & a^{-1} \end{bmatrix} \in B$ then under the identification $SH \cong G$, we have

$$p \cdot g = gp.$$

That is, the B-action on G corresponding to the B-action on geodesics is simply right group multiplication.

Since right multiplication by B commutes with left multiplication by Γ, we obtain a somewhat more algebraic version of Corollary 6.7.8.

Corollary 6.7.10 *The C*-algebras $C(G/B) \rtimes \Gamma$ and $C(\Gamma \backslash G) \rtimes B$ are Morita equivalent.*

The group B of upper triangular matrices is isomorphic to the semi-direct product group $\mathbb{R} \rtimes \mathbb{R}$. To see this, note that

$$\begin{bmatrix} e^{\frac{t}{2}} & 0 \\ 0 & e^{-\frac{t}{2}} \end{bmatrix} \begin{bmatrix} 1 & s \\ 0 & 1 \end{bmatrix} \begin{bmatrix} e^{-\frac{t}{2}} & 0 \\ 0 & e^{\frac{t}{2}} \end{bmatrix} = \begin{bmatrix} 1 & e^t s \\ 0 & 1 \end{bmatrix}.$$

So the group N of nilpotent matrices in B is normal in B and the group A of diagonal matrices with positive entries acts by automorphisms of P, and, furthermore, $B = AN$ is clear, as is $B \cap A = \{e\}$. So there is a bijection between B and $A \times N$, and the group multiplication in $B = AN$ is exactly the same as in the semi-direct product group $N \rtimes A$.

Definition 6.7.11 Under the identification $G \cong SH$, *geodesic flow g_t on SH corresponds to right multiplication by* $\begin{bmatrix} e^{\frac{t}{2}} & 0 \\ 0 & e^{-\frac{t}{2}} \end{bmatrix}$. *Horocycle flow h_s is right multiplication by* $\begin{bmatrix} 1 & s \\ 0 & 1 \end{bmatrix}$.

Horocycle flow h_s applied to a unit tangent vector based at z and pointing to ∞ maps it to the tangent vector still pointing to ∞ but starting at $z + s$. In particular it leaves the lines $\text{Im}(z) = \text{const.}$ invariant. Möbius transforms fixing H send these lines to circles *tangent* to ∂H. They are called *horocycles*. Fix a geodesic r in H. Let $\xi = r(+\infty)$. There is a unique circle passing through $r(0)$ and tangent to ∂H, that is, a unique horosphere L, and notice that v is orthogonal to it. Actually, if w is any unit tangent vector based at points of L and orthogonal to L, then $\lim_{t\to\infty} r_w(t) = \xi$. Moreover, for any such w, $d(r_v(t), r_w(t)) \to 0$ as $t \to \infty$. The horocycles are the orbits under the horocycle flow. Therefore two unit tangent vectors in SH are strongly asymptotic if and only if they are in the same orbit of the horocycle flow.

They are asymptotic, if they are in the same B-orbit, which is if and only if after applying geodesic flow g_s to one of them, for some s, they are in the same orbit of the horocycle flow.

Exercise 6.7.12 Let $\Gamma \subset \mathbf{SL}_2(\mathbb{R})$ be a cocompact lattice, acting by isometries of the hyperbolic plane $H := \mathbb{H}^2$. Let $M = \Gamma \backslash H$, a compact Riemann surface with fundamental group Γ. Let

$$\partial^2 H := \{(a, b) \in \partial H \times \partial H \mid a \neq b\},$$

notation as in Gromov [91]. The Γ action extends to an action on ∂H:

a) Prove that if one quotients the sphere bundle SH by the geodesic flow action of $\mathbb{R}$, the result is exactly $\partial^2 \Gamma$.
b) By finding a commuting pair of group actions, prove that

$$C_0(\partial^2 H) \rtimes \Gamma \sim C(SM) \rtimes_g \mathbb{R},$$

where $\rtimes_g$ denotes the geodesic flow and $\sim$ Morita equivalence.

Horocycle and geodesic flow on unit tangent bundles of surfaces are prolifically studied in ergodic theory, see [112].

Chapter 7
Topological K-Theory and Clifford Algebras

Topological K-theory was invented by M. F. Atiyah and F.E.P. Hirzebruch. It forms a generalized cohomology theory on compact (or locally compact) spaces, and the K-theory $K^*(X)$ of a locally compact space is a ring, defined in terms of vector bundles. The book [8] is an excellent source. Topological K-theory is fairly easy to define but its key properties are based on the deep Bott Periodicity Theorem, which implies that $K^{-i}(X)$ is 2-periodic. The Bott Periodicity Theorem also gives that $K^0(S^2) = \mathbb{Z}[H]/([H] - 1)^2$ as a ring, where $[H] \in K^0(S^2)$ is the class of the Hopf bundle. The Bott Periodicity Theorem, due to R. Bott was originally phrased in terms of the homotopy groups of the unitary groups, see [30, 31] and Milnor's book [124], which contains a statement and proof of the Periodicity Theorem using Morse Theory. Famous applications of K-theory due to Adams include to the Hopf invariant problem classifying maps of Hopf invariant 1, [2], to upper bounds on the number of linearly independent vector fields on sphere [3] (a short proof of the Hopf Conjecture), and perhaps most importantly of all, to the index theorem of Atiyah and Singer. (see the survey [32].)

The Bott Periodicity theorem is proved in the context of C*-algebra K-theory in Chapter 8 and again by KK-methods in Chapter 11. Equivariant versions of the Bott Periodicity Theorem are essential for computing K-theory of crossed products. In this chapter we establish the basic structure of topological K-theory (functoriality, homotopy-invariance, long exact sequences, Bott Periodicity) without giving all the proofs, as these are covered in the chapter on C*-algebra K-theory. Clifford algebras (discussed extensively in the book [118]) naturally appear in connection with topological K-theory, as they do in analytic K-homology and KK-theory in connection with Dirac operators. Periodicities in the representation theories of Clifford algebras in fact underlie the respective periodicities (two-fold and eight-fold) which appear in complex and real K-theory (respectively). We give a brief discussion of this beautiful result of Atiyah, Bott and Shapiro [11] (the book [11] discusses this and is an excellent general source for Clifford algebras.) The Thom Isomorphism Theorem is stated and discussed, although we do not prove it.

© Springer Nature Switzerland AG 2024
H. Emerson, *An Introduction to C*-Algebras and Noncommutative Geometry*,
Birkhäuser Advanced Texts Basler Lehrbücher,
https://doi.org/10.1007/978-3-031-59850-0_7

7.1   The Definition of K-Theory, the K-Theory of the Circle

The collection of complex vector bundles over a fixed space X has a natural additive structure: given two vector bundles V, W, their direct sum $V \oplus W$ is another vector bundle of the same type.

It is rather easy to see that $V \oplus W$ only depends on the isomorphism classes of V and W. Therefore, the direct sum operation descends to an addition operation on the collection $\mathrm{Vect}(X)$ of isomorphism classes of complex vector bundles over X, and similarly, the collection of isomorphism classes $\mathrm{Vect}_{\mathbb{R}}(X)$ of real vector bundles over X has an addition operation. Thus, each of $\mathrm{Vect}(X)$ and $\mathrm{Vect}_{\mathbb{R}}(X)$ has a natural structure of *abelian semigroup with identity*. (the zero vector bundle is the identity.)

The *Grothendieck completion* of an abelian semigroup A (think of $A = \mathbb{N}$ the natural numbers (including zero) under addition, or $A = \mathbb{N}^*$ the nonzero natural numbers, under multiplication) is the group defined in the following manner.

Let $G(A) := A \times A / \sim$, modulo the equivalence relation $(a, b) \sim (c, d)$ if $a + d + \epsilon = b + c + \epsilon$ for some $\epsilon \in A$.

Denote the equivalence class of a pair (a, b) in $G(A)$ by $a - b$.

It is easy to check that the operation $(a - b) + (c - d) := (a + c) - (b + d)$ is well defined. There is a natural pair of semigroup homomorphisms $A \to G(A)$, in product notation, mapping $a \in A$ to the equivalence class of $(a, 0)$, and respectively, to the equivalence class of $(0, a)$. We write simply a for $a - 0$, and $-a$ for $0 - a$.

Then it is easy to verify that $-a$ is the additive inverse of a, and, more generally, the additive inverse of $a - b$ is $b - a$, in this notation.

Thus $G(A)$ is a group.

Remark 7.1.1 The condition, for $a, b \in A$, that $a = b$ as elements of $G(A)$, says that $a + \epsilon = b + \epsilon$ for some $\epsilon \in A$, which is weaker of course than to say $a = b$ as elements of A. If the semigroup has the property that this implies that $a = b$, we say it has the *cancelation property*. Many semigroups of interest for us do not have this property; for them, $G(A)$ does not contain A injectively, but rather only a homomorphic image of it. This is the case for $A = \mathrm{Vect}(X)$, for example, which fails cancelation in general. The real vector bundle TS^2 satisfies $TS^2 \oplus 1 \cong 1_3 = 1_2 \oplus 1$ but TS^2 is not isomorphic to 1_2 (by the Poincaré-Hopf Theorem, for example, since TS^2 has no non-vanishing section.)

Exercise 7.1.2 Prove that $G(A)$ has the following universal property. Let $f : A \to H$ be a semigroup homomorphism to an abelian group H mapping the zero element of A to the identity of H. Then f extends uniquely to a group homomorphism $\bar{f} : G(A) \to H$ such that $\bar{f} \circ i = f$, where $i : A \to G(A)$ is the canonical map discussed in part b).

Exercise 7.1.3 Prove that the Grothendieck completion of the natural numbers (including zero) $\mathbb{N}$ under addition, is the integers, and that the Grothendieck completion of the nonzero natural numbers $\mathbb{N}^*$ under multiplication, is the nonzero rational numbers $\mathbb{Q}^*$ under multiplication.

Definition 7.1.4 If X is a compact space, then $K^0(X)$ is the Grothendieck completion of the abelian semigroup $\text{Vect}(X)$ of isomorphism classes of complex vector bundles over X.

$KO^0(X)$ is the Grothendieck completion of $\text{Vect}_{\mathbb{R}}(X)$.

We generally denote by $[V]$ the class in $K^0(X)$ of a vector bundle V over X.

By Remark 5.2.21, the K-theory of X is a countable group for any compact second countable Hausdorff space X.

Exercise 7.1.5 Prove that if x is any point of a compact Hausdorff space X then the map $V \mapsto \dim(V_x)$ determines a group homomorphism $K^0(X) \to \mathbb{Z}$.

Deduce from this that if V is any nonzero real or complex vector bundle over a compact space X, then $[V] \neq 0 \in K^0(X)$. Similarly for real vector bundles.

In particular, $K^0(X)$ is not the zero group, for any compact X, because the subgroup generated by the one-dimensional trivial bundle $[1]$ over X generates an infinite cyclic subgroup. (and similarly $KO^0(X)$ is never zero.)

The simplest example of a space is the 1-point space pt. Clearly $\text{Vect}(\text{pt}) \cong \mathbb{N}$ as semigroups, by the map sending a vector bundle V over the point, which is exactly the same as a finite-dimensional vector space, to its rank.

Hence $K^0(\text{pt}) \cong \mathbb{Z}$. Similarly $KO^0(\text{pt}) \cong \mathbb{Z}$. The following easy exercise implies that for any finite space X, $K^0(X)$ is the free abelian group on the points of X.

Exercise 7.1.6 If a compact Hausdorff space X is the disjoint union of two clopen (both closed and open) subsets U and V, then $K^0(X) \cong K^0(U) \oplus K^0(V)$, the direct sum in the category of abelian groups.

We next compute $K^0([0, 1])$ and $KO^0([0, 1])$.

Theorem 7.1.7 *Any real or complex vector bundle over $[0, 1]$ is trivial. In particular,* $\text{Vect}([0, 1]) \cong \mathbb{N}$ *by the map* $V \mapsto \text{rank}(V)$, *and* $K^0([0, 1]) \cong \mathbb{Z}$, $KO^0([0, 1]) \cong \mathbb{Z}$.

Of course there is an analogous statement for real K-theory.

Proof By connectedness of $[0, 1]$ and Exercise 5.1.11, E has constant fiber dimension n, for some n.

The interval is covered by open subintervals on which E is trivializable, by definition of vector bundle. By compactness, there exists a finite subcover of $[0, 1]$ by such intervals. Thus, we can find open intervals $I_1, \ldots, I_m$, moving from left to right, overlapping, and such that E_{I_k} is trivializable for $k = 1, 2, \ldots m$.

Let $s_i^{(k)}$ be sections of E on I_k, everywhere linearly independent, $i = 1, 2, \ldots, n$. Moving from left to right along the interval we build n globally defined sections s_i which are everywhere linearly independent, as follows. Fix a point $t_0 \in I_1 \cap I_2$. We have two bases $s_i^{(1)}(t_0)$ and $s_i^{(2)}(t_0)$, $i = 1, \ldots, n$ for the fiber E_{t_0} of E at t_0. Let A be the matrix defined by

$$s_i^{(2)}(t_0) = \sum_j A_{ij} s_j^{(1)}(t_0).$$

Then A is invertible. And for each i the section $x \mapsto \sum_j (A^{-1})_{ij} s_j^{(2)}(x)$ on I_2 agrees with s_i at $t_0 \in I_1 \cap I_2$ and can thus be used to extend $s_i^{(1)}$ on $[0, t] \subset I_1$ to $I_1 \cup I_2$. We then choose a point $t_2 \in I_2 \cap I_3$, and continue this process until we have constructed n linearly independent global sections of E, showing that it is trivial.   □

Exercise 7.1.8 Let $v_1 \ldots, v_n$ and $w_1, \ldots, w_n$ be two bases for $\mathbb{C}^n$. Let p and q be two points of the interval $[0, 1]$. Prove that there are n everywhere linearly independent sections $s_1, \ldots, s_n$ of the trivial bundle $[p, q] \times \mathbb{C}^n$ such that $s_i(p) = v_i$, $s_i(q) = w_i$, $i = 1, \ldots, n$. (Hint. This is equivalent to showing that $GL_n(\mathbb{C})$ is path connected.)

Exercise 7.1.9 Is it possible to do Exercise 7.1.8 in the case of two bases for $\mathbb{R}^n$ instead of $\mathbb{C}^n$? What additional hypothesis on the bases is needed?

Extending the argument of Theorem 7.1.7 a little for *complex* vector bundles produces the following result—but it definitely does not work for real vector bundles, since the Möbius bundle is not trivial.

Proposition 7.1.10 *Any complex vector bundle over* S^1 *is trivial. Hence* $\mathrm{Vect}(S^1) \cong \mathbb{N}$ *and* $K^0(S^1) \cong \mathbb{Z}$.

Proof Cover the circle with a finite family $I_1, I_2, \ldots, I_m$ of open intervals (in the angular sense) such that $E_{|I_k}$ is trivial for $k = 1, \ldots, m$. By the argument of Theorem 7.1.7 we can take n linearly independent sections, call them $s_1, \ldots, s_n$ of $E_{|I_1}$ and extend them one interval at a time to $I_1 \cup I_2$, $I_1 \cup I_2 \cup I_3$ and so on, until they are defined on $I_1 \cup \cdots \cup I_{m-1}$. Proceeding to the next step produces two choices for our sections on $I_1 \cap I_m$, for by extending the constructed sections on I_{m-1} to sections on $I_{m-1} \cup I_m$ produces n sections $s_1', \ldots, s_n'$ which may not agree with the initially defined sections $s_1, \ldots, s_n$ defined on I_1, on the intersection $I_1 \cap I_m$.

To remedy this, choose two points $z, w \in S^1$ in $I_1 \cap I_m$, with w past z in the counter-clockwise direction. The bundle E is trivial over $I_1 \cap I_m$. Let $\varphi \colon E_{|I_1 \cap I_m} \to I_1 \cap I_m \times \mathbb{C}^n$ be a trivialization. Consider the basis $\varphi(s_1(z)), \varphi(s_2(z)), \ldots, \varphi(s_n(z))$, and the basis $\varphi(s_1'(w)), \varphi(s_2'(w)), \ldots, \varphi(s_n'(w))$, for $\mathbb{C}^n$. By Exercise 7.1.8, there is a family $t_1, \ldots, t_n$ of everywhere linearly independent sections of the trivial bundle $[z, w] \times \mathbb{C}^n$ such that $t_i(z) = \varphi(s_i'(z))$ and $t_i(w) = \varphi(s_i(w))$, $i = 1, 2, \ldots, n$. We can then glue the sections $\varphi^{-1}(t_i)$ to s_i' at z and to s_i at w. This produces the required family of n linearly independent sections of E on S^1.   □

Exercise 7.1.11 What goes wrong if one tries to run the same argument through for the Möbius bundle?

The determinant map $\det \colon GL_n(\mathbb{R}) \to \mathbb{R}^*$ is continuous for every n and so $GL_n(\mathbb{R})$ has two components, since $\mathbb{R}^*$ does.

Use this to show that if V is any orientable real vector bundle of rank n over the circle, then $V \cong 1_n$. Deduce that $[M] - [1] \in K^0_{\mathbb{R}}(S^1)$ is 2-torsion, where M is the Möbius bundle. Can you take a guess at the group $KO^0_{\mathbb{R}}(S^1)$ (we are still not in a position to prove it.)

The torsion class just discussed may be regarded as a special case of the following construction.

Suppose G is a finite group acting freely on $\tilde{X}$ compact, let $X := G\backslash X$. And suppose that

$$\alpha \colon G \to \mathbf{U}_n$$

is a finite-dimensional, unitary representation of G. Let $\tilde{X} \times_G \mathbb{C}^n := X \times \mathbb{C}^n \,/\, \sim$, where $\sim$ is the equivalence relation $(x, v) \sim (gx, \alpha(g)v)$. The first coordinate projection descends to a well-defined map

$$(7.1) \qquad\qquad \pi \colon E_\alpha := \tilde{X} \times_G \mathbb{C}^n \to G\backslash \tilde{X} = X.$$

Exercise 7.1.12 In the above notation, answer the following.

a) Prove that $\pi \colon E_\alpha \to X$ defines a complex n-dimensional vector bundle over X by showing that it is locally trivial.
b) Show that the transition functions $\varphi \colon W \to \mathbf{GL}(\mathbb{C}^n)$ are given by the action of elements of G on $\mathbb{C}^n$.)
c) Prove that the Möbius vector bundle of Example 5.1.4 is the bundle over $\mathbb{R}/\mathbb{Z} \cong S^1$ associated to the character $\chi(n) = (-1)^n$ of the integers $\mathbb{Z}$.
d) Since $\mathbb{RP}^n$ is the quotient of S^n by an action of the group $\mathbb{Z}/2$, exhibit a corresponding one-dimensional real vector bundle L_n over $\mathbb{RP}^n$ for all n. For a challenge, prove that $L_n \oplus L_n$ is a trivial bundle.

Proposition 7.1.13 *If $X = G\backslash X$ for a free action of a finite group on a compact space, and $\alpha \colon G \to \mathbf{U}_n$ is a representation of G on $\mathbb{C}^n$, E_α the bundle over X defined above, then*

$$(7.2) \qquad\qquad |G| \cdot ([E_\alpha] - [1_n]) = 0 \in K^0(X).$$

In particular, $[E_\alpha] - [1_n] \in K^0(X)$ is always a torsion class, of order a divisor of $|G|$.

The proof uses a simple device that is slightly more general, so we give this slightly more general statement.

Let $\pi \colon \tilde{X} \to X$ be a finite covering map. We define a *push-forward* operation on vector bundles as follows. If E is a vector bundle over $\tilde{X}$, define a vector bundle $\pi_\sharp(E)$ over X by setting the fiber at $x \in X$ to be

$$(7.3) \qquad\qquad \pi_\sharp(E)_x := \oplus_{y \in \pi^{-1}x} E_y.$$

Exercise 7.1.14 Prove that $\pi_\sharp(E)$ defined above is a vector bundle over X. If E and E' are isomorphic vector bundles over X, then $\pi_\sharp(E)$ and $\pi_\sharp(E')$ are isomorphic.

The push-forward construction therefore gives rise to a *group homomorphism*

$$(7.4) \qquad\qquad \pi_* : \mathrm{K}^0(\tilde{X}) \to \mathrm{K}^0(X).$$

Lemma 7.1.15 *In the above notation, let* $\pi^* : \mathrm{K}^0(X) \to \mathrm{K}^0(\tilde{X})$ *be the map induced by the finite covering map* $\pi : \tilde{X} \to X$. *Then*

$$(7.5) \qquad \pi^* \circ \pi_* = |G| \cdot \mathrm{id}_{\mathrm{K}^0(\tilde{X})}, \quad \pi_* \circ \pi^* = |G| \cdot \mathrm{id}_{\mathrm{K}^0(X)}$$

hold.

In particular, the push-forward rationally *inverts the pull-back map.*

Exercise 7.1.16 Verify that (7.5) holds.

We now prove Proposition 7.1.13.

Proof Let $\pi : \tilde{X} \to X$ the quotient map—a covering map. It is obvious that $\pi^*(E_\alpha) \cong \tilde{X} \times \mathbb{C}^n$, that is, the pull-back of E_α to $\tilde{X}$ is trivial. Hence

$$\pi^*([E_\alpha]) = [1_n].$$

Applying the push-forward map π_* gives

$$\pi_*\big(\pi^*([E_\alpha] - [1_n])\big) = 0 \in \mathrm{K}^0(X).$$

By (7.5)

$$|G| \cdot ([E_\alpha] - [1_n]) = 0 \in \mathrm{K}^0(X),$$

as required. $\square$

The above constructions show that torsion *may* exist in K^0-groups, but does not prove it. In order to prove it, one needs more methods of actually computing these groups.

Clutching constructions, a homotopy description of vector bundles over spheres

We finish this section with a discussion of "clutching." We restrict ourselves to complex vector bundles for simplicity; the analogous discussion goes through for real bundles.

Let $X = U \cup U'$ be the union of two open sets, let E be a complex vector bundle over U and E' a complex vector bundle over U', and let $\varphi : E|_{U \cap U'} \to E'|_{U \cap U'}$ be a bundle isomorphism.

Then the *clutching* of E and E' over φ is denoted $E \cup_\varphi E'$, is defined as follows. As a space, $E \cup_\varphi E'$ is the quotient of $E \sqcup E'$ by the equivalence relation which identifies $(x, v) \in E_x \subset E$ with $(x, \varphi(x)v)$ in E'. The projection maps $E \to U$ and $E' \to U'$ splice together to make a projection map $\pi : E \cup_\varphi E' \to U \cup U' = X$.

Each fiber of π has the structure of a vector space, since the glueing map $\varphi(x) : E_x \to E'_x$ is linear for all $x \in U \cap U'$, this is well defined on $E \cup_\varphi E'$, and the addition and scalar multiplication operators on $E \cup_\varphi E'$ are easily checked to be continuous, fiberwise, with respect to the quotient topology.

Exercise 7.1.17 In the above notation, prove that $\pi : E \cup_\varphi E' \to X$ is locally trivial and that the isomorphism class of $E \cup_\varphi E'$ only depends on the homotopy class of the vector bundle isomorphism $\varphi : E|_{U \cap U'} \to E'|_{U \cap U'}$. (*Hint.* A homotopy $(\varphi_t)_{t \in [0,1]}$ of bundle isomorphisms $E|_{U \cap U'} \to E'|_{U \cap U'}$ is equivalent to a single bundle isomorphism $\mathrm{pr}_X^*(E)|_{U \cap U' \times [0,1]} \to \mathrm{pr}_X^*(E')|_{U \cap U' \times [0,1]}$, where $\mathrm{pr}_X : X \times [0,1] \to X$ is the projection. Now prove that $E \cup_{\varphi_t} E' \cong f_t^*\left(\mathrm{pr}_X^* E \cup_\Phi \mathrm{pr}_X^* E'\right)$, with $f_t : X \to X \times [0,1]$ the map $f_t(x) = (x, t)$.)

Exercise 7.1.18 Prove that if U and U' are open in X, E and E' are vector bundles over U, U', and if V is a vector bundle over X whose restriction to U is isomorphic to E, and whose restriction to U' is isomorphic to E', then V is isomorphic to the clutching $E \cup_\varphi E'$, using the clutching function manufactured on $U \cap U'$ by using first the isomorphism $E' \cong V$ (on U) followed by the inverse of the isomorphism $V \cong E$ (on U').

Deduce from this that if a vector bundle is trivial over an open set $U \subset X$, then it is isomorphic to a vector bundle which is actually *equal* over U to a product bundle.

The following exercise generalizes the clutching idea over two open sets, to an arbitrary collection of them.

Exercise 7.1.19 (Clutching Using a Cocycle)
Suppose that $\{U_i\}_{i \in I}$ is a cover of X by open sets. And suppose we are given a family $\{\varphi_{ij} : U_i \cap U_j \to \mathbf{GL}(n, \mathbb{C}) \mid i, j \in I\}$ of maps satisfying the cocycle conditions

* $\varphi_{ii}(x) = \mathrm{id}$ for all i.
* $\varphi_{ij}(x)\varphi_{jk}(x) = \varphi_{ik}(x)$, $\forall i, j, k$.

Then the relation $\sim$ on $\bigsqcup_{i \in I} U_i \times \mathbb{C}^n$ defined $(x, v) \sim (x, \varphi_{ij}(x)v)$ for $x \in U_i \cap U_j$, is an equivalence relation, and the quotient space has a canonical structure of an n-dimensional vector bundle over X.

Exercise 7.1.20 Let $\pi : E \to X$ be an n-dimensional real or complex vector bundle. Suppose that $\{U_i, \varphi_i\}_{i \in I}$ is an atlas for E, i.e., $\varphi_i : E|_{U_i} \to U_i \times \mathbb{C}^n$ is a local trivialization for all $i \in I$. Let $\varphi_{ij} = \varphi_i \circ \varphi_j^{-1}$ the transition functions for the atlas, understood as maps $\varphi_{ij} : U_i \cap U_j \to \mathbf{GL}(n, \mathbb{C})$. Check that they satisfy the cocycle condition and that the vector bundle $\bigsqcup_{i \in I} U_i \times \mathbb{C}^n / \sim$ as in Exercise 7.1.19, is isomorphic to E.

Exercise 7.1.21 Let $\{\varphi_{ij} : U_i \cap U_j \to \mathbf{GL}(n, \mathbb{C})\}$ be a cocycle as in Exercise 7.1.19 which is a *coboundary* in the sense that there are maps $\psi_i : U_i \to \mathbf{GL}(n, \mathbb{C})$ for which $\varphi_{ij}(x) = \psi_j(x)\psi_i(x)^{-1}$ for $x \in U_i \cap U_j$. Prove that the "clutched" bundle $\bigsqcup_{i \in I} U_i \times \mathbb{C}^n \,/\, \sim$ described in the Exercise 7.1.19 is trivial.

Exercise 7.1.22 Let G be a finite group acting freely on $\tilde{X}$ compact. Show that the vector bundle E_α over $X := G\backslash X$ associated to a finite-dimensional representation $\alpha: G \to \mathbf{U}_n$, may be considered as being obtained by clutching in the following way. Cover X by the open images of sets $U_i \subset \tilde{X}$ for which $g(U_i) \cap U_i = \emptyset$ for $g \neq e$. The quotient map $\pi: \tilde{X} \to X$ restricts to a homeomorphism on each U_i, and the composition $(\pi|_{U_j})^{-1} \circ \pi|_{U_i}$ is a homeomorphism on $U_i \cap U_j$ onto an open subset of $\tilde{X}$. Show that this homeomorphism is the restriction of a group element $g_{ij} \in G$, and if we set $\varphi_{ij} := \alpha(g_{ij})$ then the system $\varphi_{ij}: U_i \cap U_j \to \mathbf{U}_n$ defines a cocyle, whose functions are locally constant.

Exercise 7.1.23 Prove that a pair of vector bundles can be "clutched" over two *closed* sets, as well as over two open sets. More precisely, let $A_1, A_2 \subset X$ be two closed subsets of X, let E_i be vector bundles over A_i, and let $\varphi: E_1|_{A_1 \cap A_2} \overset{\cong}{\to} E_2|_{A_1 \cap A_2}$ be a vector bundle isomorphism. Forming the quotient space of $E_1 \sqcup E_2$ by the equivalence relation which identifies $v \in E_1$ with $\varphi(v) \in E_2$ results in a vector bundle over $X = A_1 \cup A_2$, which is isomorphic to E.

We close this section with a homotopy-theoretic description of vector bundles over a sphere.

Let $\pi: V \to S^n$ be a vector bundle over the n-sphere. Let S_+^n be the (closed) upper hemisphere, S_-^n the lower hemisphere, so that $S_+^n \cap S_-^n \cong S^{n-1}$.

Let E be a k-dimensional complex vector bundle over S^n. Since $S_\pm^n$ are each contractible compact spaces, $E_{|S_\pm^n}$ is trivial. Fix trivializations

$$\alpha_\pm: E_{|S_\pm^n} \overset{\cong}{\longrightarrow} S_\pm^n \times \mathbb{C}^k.$$

The restriction of $\alpha_- \circ \alpha_+^{-1}$ to $S_+^n \cap S_-^n \cong S^{n-1}$ is a bundle map $S^{n-1} \times \mathbb{C}^n \to S^{n-1} \times \mathbb{C}^k$, which is equivalent to a map $\alpha: S^{n-1} \to \mathbf{GL}(k, \mathbb{C})$. Let $[\alpha] \in [S^{n-1}, \mathbf{GL}(k, \mathbb{C})]$ be the corresponding homotopy class of map.

Exercise 7.1.24 In the above notation, answer the following.

a) Prove that the homotopy class $[\alpha] \in [S^{n-1}, \mathbf{GL}(n, \mathbb{C})]$ does not depend on the choice of trivializations $\alpha_\pm$. (*Hint.* Due to contractibility of $S_\pm^n$, even the homotopy classes of the bundle maps $\alpha_\pm$ are uniquely defined.)

b) Using clutching to produce a map inverse to the construction above, prove that

$$\mathrm{Vect}_k(S^n) \cong [S^{n-1}, \mathbf{GL}(k, \mathbb{C})],$$

where $\mathrm{Vect}_k(S^n)$ is the set of isomorphism classes of k-dimensional complex vector bundles over the sphere.

Exercise 7.1.25 Let $S_\pm^2$ be the upper and lower closed hemispheres of the 2-sphere. Prove that the Hopf bundle is obtained by clutching two trivial bundles over $S_\pm^2$ using the $\mathbf{GL}(1, \mathbb{C}) \cong \mathbb{C}^*$-valued function

$$\varphi \colon S_+^2 \cap S_-^2 \cong S^1 \to \mathbb{C}^*, \quad \varphi(z) = \bar{z}.$$

Exercise 7.1.26 Let $\pi \colon V \to X$ be an n-dimensional vector bundle over a compact space. Let $\mathcal{F}(V)$ be the bundle of frames of V: a point of $\mathcal{F}$ is a pair $(x, \mathbf{v})$ where $\mathbf{v}$ is an n-tuple $(v_1, \dots, v_n)$ of linearly independent vectors in V_x. Topologize $\mathcal{F}(V)$ to be a compact space, and prove that the projection $p \colon \mathcal{F}(V) \to X$ pulls V back to a trivial bundle over $\mathcal{F}(V)$.

Orientations on vector bundles

Let $\pi \colon V \to X$ be any *real* vector bundle. Due to local triviality, V has local frames. Thus, for any point of X, there is a neighborhood U of the point, and a frame $\mathbf{e}$—i.e., a collection of sections $e_1, \dots, e_n$ of V defined on U, such that $e_1(x), \dots, e_n(x)$ is a basis for V_x for all $x \in U$.

If U' is another, intersecting open set, with another frame $\mathbf{e}'$ on it, then we say the frames are *compatibly oriented* on $U \cap U'$ if $\mathbf{e}(x)$ and $\mathbf{e}'(x)$ are compatibly oriented frames of V_x for all $x \in U$, equivalently, $e_1(x) \wedge \cdots e_n(x)$ is a *positive* multiple of $e_1'(x) \wedge \cdots \wedge e_n'(x)$ in $\Lambda^n(V_x)$ for all $x \in U \cap U'$.

A vector bundle is *orientable* if there is a cover of X by open sets U_i, and a collection of frames $\mathbf{e}_i$ on U_i such that if $U_i \cap U_j \neq \emptyset$ then $\mathbf{e}_i$ and $\mathbf{e}_i'$ are compatibly oriented frames. We call any such data an *orientation* on V.

Exercise 7.1.27 The following are equivalent for a real vector bundle $\pi \colon V \to X$.

a) V is orientable.
b) There exists an atlas $\{U_i, \varphi_i\}_{i \in I}$ for V, for which the transition functions $\varphi_{ij} = \varphi_i \circ \varphi_j^{-1} \colon U_i \cap U_j \to \mathbf{GL}(n, \mathbb{R})$ take values in the subgroup $\mathbf{GL}^+(n, \mathbb{R})$ of matrices of positive determinant.
c) There exists an atlas $\{U_i, \varphi_i\}_{i \in I}$ for V, for which the transition functions $\varphi_{ij} = \varphi_i \circ \varphi_j^{-1} \colon U_i \cap U_j \to \mathbf{GL}(n, \mathbb{R})$ take values in $\mathbf{SO}(n, \mathbb{R})$.

Exercise 7.1.28 Prove that if $E \to X$ is a complex vector bundle, then it is orientable, when regarded as a real vector bundle.

7.2 Vector Bundles on Smooth Manifolds

In this chapter we review some of the standard (real) vector bundles that come up in smooth manifold theory.

An *n-dimensional locally Euclidean space* M is a Hausdorff, second countable topological space with the property that every point $p \in M$ has a neighborhood U homeomorphic to an open subset of $\mathbb{R}^n$. A *smooth atlas* on an n-dimensional locally Euclidean space is a collection of pairs $\{U_i, \varphi_i\}$ with U_i an open subset of M and $\varphi_i : U_i \to \mathbb{R}^n$ a homeomorphism onto an open subset, such that $\cup U_i = M$ and $\varphi_i \circ \varphi_j^{-1} : \varphi_j(U_i \cap U_j) \to \varphi_i(U_i \cap U_j)$ is a *smooth* map, for all $i.j$ (where a map between open subsets of Euclidean space is *smooth* if it is infinitely differentiable.) Each such pair is called a (smooth) *local coordinate chart*. A *smooth map* between open subsets of Euclidean space is an infinitely differentiable map.

If $r_1, \ldots, r_n : \mathbb{R}^n \to \mathbb{R}$ are the usual coordinate projections, we can write $\varphi = (x_1, \ldots, x_n)$ where $x_i := r_i \circ \varphi$ and we can label points in U by their corresponding coordinate vectors $(x_1, \ldots, x_n)$.

A maximal smooth atlas is a *differentiable structure* on M. A continuous function $f : M \to \mathbb{R}$ is *smooth* if $f \circ \varphi^{-1} : \varphi(U) \to \mathbb{R}$ is smooth for every local coordinate chart on M. $C^\infty(M)$ denotes the real algebra of smooth functions on M.

If $p \in M$ is a point, a *point derivation* $X_p : C^\infty(M) \to \mathbb{R}$ is a linear map satisfying the Leibnitz rule

$$X_p(fg) = f(p)X_p(g) + g(p)X_p(f),$$

for f and g smooth functions on M.

The *tangent bundle* of M is the disjoint union $TM = \sqcup_{p \in M} T_p(M)$ where $T_p(M)$ is the real vector space of point derivations of $C^\infty(M)$ at p. There is an evident projection $\pi : TM \to M$; we want to show that TM can be given the structure of a vector bundle over M.

Exercise 7.2.1 If $X_p \in T_p(M)$ is a point derivation at p, and $f \in C^\infty(M)$ is a smooth function which vanishes in a neighborhood of p, then $X_p f = 0$. Deduce that $X_p f$, for $f \in C^\infty(M)$, really only depends on the *germ* of f at p. (germs are discussed below.)

Exercise 7.2.2 Suppose that $p \in M$ and $\gamma : (-\epsilon, \epsilon) \to M$ is a smooth curve such that $\gamma(0) = 0$. Show that γ determines a point derivation $\gamma'(0)$ at p by

$$\gamma'(0)f := (f \circ \gamma)'(0).$$

All point derivations at p arise in this way (Exercise 7.2.3.)

If p is in the domain of a coordinate chart $\varphi : U \to \mathbb{R}^n$ with coordinates $x_1, \ldots, x_n$, let $\frac{\partial}{\partial x_i}|_p$ denote the point derivation $f \mapsto \frac{\partial f}{\partial x_i}(p) := \frac{\partial (f \circ \varphi^{-1})}{\partial r_i}(p)$ at p.

Taylor's lemma asserts that if f is a smooth function in a neighborhood of a point $p \in \mathbb{R}^n$, then

$$f(r) = f(p) + \sum_{i=1}^{n} g_i(r)(r_i - p_i),$$

where $g_1, \ldots, g_n$ are smooth functions in a neighborhood of p satisfying $g_i(p) = \frac{\partial f}{\partial r_i}$.

Taylor's Lemma extends more or less verbatim to points $p \in M$ in a smooth manifold M, and to smooth functions on M. If (U, φ), $\varphi = (x^1, \ldots, x^n)$ is a local coordinate chart, then any $f \in C^\infty(M)$ may be written

$$f(x) = f(p) + \sum_{i=1}^{n} g_i(x)(x^i - p^i)$$

for a collection $g_1, \ldots, g_n$ of functions smooth in a neighborhood of p.

Now, if X_p is a point derivation at p, then by the Leibniz rule, and the fact that all the $x^i - p^i$ vanish at p,

$$X_p(f) = \sum_{i=1}^{n} g_i(p) X_p(x_i - p_i) = \sum_{i=1}^{n} a_i \frac{\partial f}{\partial x_i}(p)$$

with a_i the constants $a_i = X_p(x_i - r_i)$ obtained by applying X_p to the functions $x^i - p^i$, which vanish at p. That is, any X_p, for $p \in U$, can be expanded uniquely in the form

$$X_p = \sum_{i=1}^{n} a_i \frac{\partial}{\partial x^i}\big|_p, \text{ where } a_i = X_p(x^i - p^i).$$

It is also easy to check that the $\frac{\partial}{\partial x^i}\big|_p$ are linearly independent at each $p \in U$. Hence they form linearly independent and spanning sections of $TM|_U := \pi^{-1}(U) \subset TM$. This supplies local trivializations of TM, and a basis for a topology, and TM therefore becomes a real vector bundle over M of dimension $n = \dim(M)$.

Exercise 7.2.3 Show that all point derivations at a point $p \in X$ arise from the germ of a curve, as in Exercise 7.2.2.

The *dual* T^*M of the tangent bundle also has a nice geometric description. Fix a point $p \in M$. Let A_p be the algebra over $\mathbb{R}$ of germs (f, U) of smooth functions at p, which vanish at p. Thus, $f \in C_c^\infty(U)$, $p \in U$, $f(p) = 0$, and two such pairs (f, U) and (g, V) for which $f = g$ on $U \cap V$ are considered equivalent.

Now form $T_p^*(M) := A_p/A_p^2$. This is the algebra of germs which vanish to first order at p modulo the germs which vanish to second order at p.

Now if f is a smooth function defined in a neighborhood of p, we let $df(p)$ be the class in A_p/A_p^2 of the smooth $f - f(p)$, which vanishes at p.

By Taylor's Lemma, we can find smooth functions $g_1, \ldots, g_n$ in a neighborhood of p such that $g_i(p) = \frac{\partial f}{\partial x^i}(p)$. Applying Taylor's Lemma to each such g_i then yields smooth functions h_{ij} in a neighborhood of p such that

$$g_i(x) = \frac{\partial f}{\partial x^i}(p) + \sum_{j=1}^{n} h_{ij}(x)(x^j - p^j).$$

Substituting into the formula for $f - f(p)$ yields

$$f - f(p) = \sum_{i=1}^{n} \frac{\partial f}{\partial x^i}(p)(x^i - p^i) + \sum_{i.j} h_{ij}(x)(x^i - p^i)(x^j - p^j).$$

This immediately implies that

$$df(p) = \sum_{i=1}^{n} a_i dx^i(p), \quad \text{where } a_i = \frac{\partial f}{\partial x^i}(p)$$

since the germ $\sum_{i.j} h_{ij}(x)(x^i - p^i)(x^j - p^j)$ vanishes to order 2 at p.

It follows that $\sqcup_{p \in M} A_p / A_p^2$ can be given the structure of a real, n-dimensional vector bundle over M, with local sections given on the domains of coordinate charts by the cosets $dx^1, \ldots, dx^n$. In fact, this vector bundle can be naturally identified with the dual bundle T^*M of the tangent bundle. To prove this, define a map $A_p / A_p^2 \to T_p(M)^*$ by letting $df(p) \in A_p / A_p^2$ act on point derivations at p by

$$\langle df(p), X_p \rangle := X_p(f).$$

This formula is well defined, because any point derivation must vanish on functions which vanish to order 2 at p, so that $X_p(f)$ only depends on the coset of f modulo A_p^2.

The details are left as an exercise.

A section of the tangent bundle is called a *vector field* on M. A section of T^*M is called a *differential 1-form* on M. A section of the exterior algebra bundle $\Lambda^k T^*M$ is called a *differential k-form* on M. Any differential k-form ω on M can be locally expanded into a linear combination of the standard differential k-forms $dx^I := dx^{i_1} \wedge d^{i_2} \wedge \cdots \wedge dx^{i_k}$, where $I = (i_1, \ldots, i_k)$ is a multi-index: thus, on the domain of a local coordinate system, we can write

$$\omega = \sum_I a_I dx^I$$

for some collection of smooth functions a_I on the domain of the chart.

Exercise 7.2.4 A smooth manifold whose tangent bundle is trivial is called *parallelizable*. Prove that the n-torus $\mathbb{T}^n$ is parallelizable.

Exercise 7.2.5 Let M be any smooth manifold. Prove that the tangent bundle $T(TM)$, as a vector bundle over the space TM, is isomorphic to $\pi^*(TM) \oplus$

$\pi^*(TM)$, where $\pi : TM \to M$ is the projection map, and $\pi^*(TM)$ is the pull-back of the vector bundle TM over M, to a vector bundle over TM.

Deduce that the tangent bundle to TM has a complex structure.

Smooth structures on vector bundles

Definition 7.2.6 A vector bundle $\pi : E \to M$ over a manifold, is *smooth* if E is a smooth manifold, and if there exists an atlas for E consisting of smooth maps.

By an easy exercise, if E is smooth, then $\pi : E \to M$ is a smooth map, and M embeds by the zero section of E as a regular submanifold of E.

In this section, we show the important basic result that every vector bundle over a smooth manifold may as well be taken to be a smooth vector bundle. This idea is an important one in Noncommutative Geometry: it means that one may for purposes of K-theory computations, assume that all the K-theory data is smooth. The proof we give here is fairly C*-algebraic in nature.

Theorem 7.2.7 *Every real vector bundle* $\pi : E \to M$ *over a smooth compact manifold is isomorphic to a smooth vector bundle.*

Remark 7.2.8 The theorem can be phrased a bit more concretely as follows: any vector bundle E over smooth M be given a differentiable structure, and, moreover, one can find a system of local trivializations of E which are smooth.

Lemma 7.2.9 *Let a be a self-adjoint element of a C*-algebra of norm ≤ 1. Then if* $\|a - a^2\| < \frac{1}{4}$ *then* $\frac{1}{2} \notin \mathrm{Spec}(a)$.

Proof If a is self-adjoint then the functional calculus produces a *-isomorphism $C^*(a) \cong C(\mathrm{Spec}(a))$ mapping a to $f(t) = t$, so $\|a - a^2\| < \frac{1}{4}$ implies that $|t - t^2| < \frac{1}{4}$ for all $t \in \mathrm{Spec}(a)$, and hence that $\frac{1}{2} \notin \mathrm{Spec}(a)$ since $t - t^2$ assumes the value $-\frac{1}{4}$ there.   □

Lemma 7.2.10 *Let M be a smooth compact manifold and let $H \in C^\infty(M, M_n(\mathbb{C}))$ be a smooth element of the C*-algebra $C(M, M_n(\mathbb{C}))$. Let ψ be a continuous function on $\mathrm{Spec}(H)$. Then if ψ extends to a holomorphic function on a neighborhood in $\mathbb{C}$ of $\mathrm{Spec}(H)$, then $\psi(H)$ is also smooth.*

Proof Let $\tilde{\psi}$ be an extension of ψ to a holomorphic function in a neighborhood U of $\mathrm{Spec}(H)$, and let γ be a simple closed, positively oriented contour in U with $\mathrm{Spec}(H)$ contained in its interior. By the holomorphic functional calculus

$$\psi(H) = \frac{1}{2\pi i} \oint_\gamma \tilde{\psi}(w)(w - H)^{-1} dw.$$

As a function on X, thus,

$$\psi(H)(x) = \frac{1}{2\pi i} \oint_{\gamma} \tilde{\psi}(w)(w - H(x))^{-1} dw.$$

The usual technique of differentiating under the integral sign implies that this function of x is smooth, because H is assumed smooth.   □

Lemma 7.2.11 *If M is a smooth manifold and $p \colon M \to M_n(\mathbb{C})$ is a smooth projection-valued function, then the vector bundle $\mathrm{Im}(p)$ is a smooth vector bundle over M.*

Proof By definition, $\mathrm{Im}(p)$ is a *subset* of the smooth manifold $M \times \mathbb{R}^m$. So to show it has the structure of a smooth manifold, it suffices to show that it is a regular submanifold of $M \times \mathbb{R}^m$. Choose any point $a \in M$. We have already shown that if $v_1, \ldots, v_m$ is a basis for $\mathbb{R}^m$ with the first k vectors a basis for $\mathrm{Im}(p(a))$, then the sections $s_1(b) := p(b)v_1, \ldots, s_k(b) := p(b)v_k, s_{k+1}(b) := v_{k+1}, \ldots, s_m(q) := v_m$ form a basis for $\mathbb{R}^m$ for all q in a neighborhood U of a. After possibly shrinking U we may also assume it is the domain of a coordinate chart $(U, x^1, \ldots, x^n)$ for M.

Now if $v \in \mathbb{R}^m$ and $q \in U$ then we can find unique scalars $t^1(v, q), \ldots, t^m(v, q)$ such that $v = \sum_{i=1}^m t^i(v, q)v_i$. The functions t^i are smooth, and are linear in v for fixed q. We now make a local coordinate system for $M \times \mathbb{R}^m$ around $(a, 0)$ by

$$(q, v) \mapsto (x^1(q), \ldots, x^n(q), t^1(v, q), \ldots, t^m(v, q)).$$

This forms a coordinate system, and the last $m - k$ coordinates of (q, v) vanish if and only if v has the form $v = \sum_{i=1}^k t^i(v, q)P(q)v_i$ which lies in $\pi^{-1}(U) \subset \mathrm{Im}(p)$, so that locally $\mathrm{Im}(p)$ may be represented as

$$\{(x^1, \ldots, x^n, t^1, \ldots, t^m) \mid t_{k+1} = \cdots = t_m = 0\}$$

which is the condition for being a regular submanifold.

Note that with this differential structure, the bundle trivializations are smooth, indeed, in our local coordinates the bundle trivializations are, in the above notation,

$$\varphi(q, v) = (q, t^1(q, v), \ldots, t^k(q, v)) \in M \times \mathbb{R}^k, \quad \text{for } (q, v) \in \mathrm{Im}(p).$$

□

Proof *(Of Theorem 7.2.7)* Let $p \colon M \to M_n(\mathbb{C})$ be a continuous projection-valued function such that $\mathrm{Im}(p) \cong E$. Since smooth functions $M \to \mathbb{C}$ are dense in continuous functions, by the Stone-Weierstrass Theorem, it follows that smooth, matrix-valued functions $M \to M_n(\mathbb{C})$ are also dense in continuous matrix-valued functions. So there exists a sequence (H_n) of smooth functions $H_n \colon M \to M_n(\mathbb{C})$ with $H_n \to P$ in the C*-algebra $C(M, M_n(\mathbb{C}))$. Since $\frac{H_n + H_n^*}{2} \to P$ as well, we may as well assume that the H_n are also self-adjoint.

In particular, there exists a smooth self-adjoint element $H \in C^{\infty}(M, M_n(\mathbb{C}))$ such that $\|H - P\| < \frac{1}{4}$. By Lemma 7.2.9, $\frac{1}{2} \notin \mathrm{Spec}(H)$. The spectrum of

H is compact and does not contain $\frac{1}{2}$ and hence there exists a pair of open sets $U, V \subset \mathbb{C}$ such that $U \cap V = \emptyset$, V contains $\text{Spec}(H) \cap (\frac{1}{2}, +\infty)$, U contains $\text{Spec}(H) \cap (-\infty, \frac{1}{2})$. Let $\tilde{f}$ be the function assuming value 1 on V and 0 on U, then $\tilde{f}$ is clearly holomorphic on $U \cup V$, and if γ is a simple closed contour in V encircling $(\frac{1}{2}, +\infty) \cap \text{Spec}(H)$, then $f(H)$ is then smooth by Lemma 7.2.10, and is a projection, call it Q, since it is the image of a characteristic function on $\text{Spec}(H)$ under holomorphic (and hence continuous) functional calculus $C(\text{Spec}(H)) \rightarrow C^*(H)$.

Since $\|Q - P\| < 1$, $\text{Im}(Q) \cong \text{Im}(P) \cong E$ as vector bundles over M. Finally, an application of Lemma 7.2.11 gives that $\text{Im}(Q)$ is a smooth vector bundle, and we conclude that E is isomorphic to a smooth vector bundle as initially claimed.

$\square$

Exercise 7.2.12 Suppose that $A \subset X$ is a closed subspace of a locally compact space X and E a vector bundle over A. Prove that E can be extended to a vector bundle over a neighborhood of A. That is, prove that there exists an open neighborhood U containing A and a vector bundle $\tilde{E}$ over U whose restriction to A is E. (*Hint.* Find a projection-valued map $p \colon X \rightarrow M_n(\mathbb{C})$ such that $\text{Im}(p) \cong E$. Extend p to a map $\tilde{p} \colon X \rightarrow M_n(\mathbb{C})$, argue that for some neighborhood U of A, $\frac{1}{2} \notin \text{Spec}(\tilde{p}(x))$ for all $x \in U$, and use the functional calculus methods of the proof of Theorem 7.2.7 perturb $\tilde{p}$ so that it is projection-valued in a neighborhood of A.)

7.3 Functoriality and Homotopy-Invariance

Let $\varphi \colon X \rightarrow Y$ be a continuous map of compact spaces. The pull-back operation $V \mapsto \varphi^*(V)$ from vector bundles on Y to vector bundles on X, defined in Definition 5.1.13, can be easily checked to take isomorphic vector bundles to isomorphic vector bundles, and respects direct sums. Hence it induces a homomorphism $\varphi^* \colon \text{Vect}(Y) \rightarrow \text{Vect}(X)$ of abelian semigroups. This results in a pair of abelian group homomorphisms $\varphi^* \colon \text{K}^0(Y) \rightarrow \text{K}^0(X)$ and $\text{KO}^*(Y) \rightarrow \text{KO}^*(X)$.

It is routine to check that as maps on K-theory, or KO-theory, $(\varphi \circ \psi)^* = \psi^* \circ \varphi^*$, for $\varphi \colon Y \rightarrow Z$ and $\psi \colon X \rightarrow Y$, so that *the assignment $X \mapsto \text{K}^0(X)$, $\varphi \mapsto \varphi^*$ defines a contravariant functor from the category of compact Hausdorff spaces and continuous maps, to the category of abelian groups and group homomorphisms* (and similarly for KO^0-theory.)

The main result of this section is the homotopy-invariance of these K-theory functors: that is, that homotopic maps induce the same map on K-theory.

Lemma 7.3.1 *Let V be a vector bundle over a locally compact Hausdorff space X and $Y \subset X$ be a closed subspace. Then any section $s \colon Y \rightarrow V$ can be extended to a section $\bar{s} \colon X \rightarrow V$ of V on all of X.*

Proof For product bundles $X \times \mathbb{R}^n$, the result follows immediately from the Tietze Extension Theorem. For trivial bundles, it follows as well, since if $\varphi \colon V \xrightarrow{\cong} X \times \mathbb{R}^n$ is a bundle isomorphism, s a section of V on a closed subset, $Y \subset X$, then $\varphi \circ s$ is a section of a product bundle on Y, and if t is an extension of it to a section on X, then $\varphi^{-1} \circ t$ is a section of V which extends s on Y, as required.

Now let $\pi \colon V \to X$ be any vector bundle. Then X is covered by open sets on which V is trivial. Let $\{U_i\}_{i \in I}$ and $\{\rho_i\}_{i \in I}$ a partition of unity subordinate to this cover. Let $\varphi_i \colon V|_{U_i} \xrightarrow{\cong} U_i \times \mathbb{R}^n$ trivializations of V on each U_i.

Now using the partition of unity, it will be enough to extend $s|_{U_i \cap Y}$ to a section $s_i \colon U_i \to V$ on U_i. For then we may define $s(x) = \sum_{i \in I} \rho_i(x) s_i(x)$. The sum will be finite, for any fixed $x \in X$, because of local finiteness. And if $x \in Y$, it equals $s(x)$, because $s_i(x) = s(x)$ for every i, since we are assuming s_i extends s on $U_i \cap Y$, and since $\sum \rho_i(x) = 1$, for all $x \in X$.

So we are reduced to showing that any section of $V|_{U_i}$ can be extended from $U_i \cap Y$ to U_i. But by construction $V|_{U_i}$ is trivial, and hence, the extension property follows from our initial remarks. $\qquad \square$

Corollary 7.3.2 *If V_1 and V_2 are vector bundles over X, and if $Y \subset X$ a closed subset, then any isomorphism $V_1|_Y \cong V_2|_Y$ can be extended to an isomorphism $V_1|_U \cong V_2|_U$ on a neighborhood of Y.*

Proof Bundle maps $V_1 \to V_2$ are exactly sections of the vector bundle $\mathrm{HOM}(V_1, V_2)$ (see Exercise 5.1.19 of Chapter 5). So an isomorphism $V_1|_Y \to V_2|_Y$, since it is a section of $\mathrm{HOM}(V_1, V_2)$ on $Y \subset X$, extends, by Lemma 7.3.1 to a section on X, in other words, to a bundle map $T \colon V_1 \to V_2$ defined on all of X, and such that $T(y)$ is an isomorphism for all $y \in Y$. The result will then follow from the following

Claim. If $T \colon V_1 \to V_2$ is any vector bundle map, then the set

$$\{x \in X \mid T(x) \text{ an isomorphism}\}$$

is open in X.

To see this, let $x_0 \in X$ for which $T(x_0)$ is an isomorphism. We may find trivializations V_1 and V_2 on a neighborhood V of x_0, and hence we can find a frame for V_1 and a frame for V_2, defined on V, and write T in terms of these frames, as a matrix. Let $\tilde{T}$ be the corresponding map $U \to M_n(\mathbb{R})$—it is continuous, and takes an invertible value at x_0. Since the invertibles $\mathbf{GL}(n, \mathbb{R}) \subset M_n(\mathbb{R})$ are open in $M_n(\mathbb{R})$, it follows that $\tilde{T}$ takes invertible values in a neighborhood of x_0, so there exists U a neighborhood of x_0 on which $\tilde{T}$ takes invertible values. It follows immediately that T is an isomorphism on U. This completes the claim. $\qquad \square$

Lemma 7.3.3 *Let X be any compact Hausdorff space and let $i_0, i_1 \colon X \to X \times [0, 1]$ be the maps $i_0(x) := (x, 0)$, $i_1(x) := (x, 1)$. Then $i_0^* = i_1^* \colon \mathrm{KO}^0(X \times [0, 1]) \to \mathrm{KO}^0(X)$, and similarly $i_0^* = i_1^*$ as maps $\mathrm{K}^*(X \times [0, 1]) \to \mathrm{K}^*(X)$.*

Proof Let $\pi : V \to X \times [0, 1]$ be a vector bundle. For each t, let $i_t : X \to X \times [0, 1]$, $i_t(x) := (x, t)$ be the inclusion of X as the slice at t. Let $V_t := i_t^*(V)$, a vector bundle over X. We show the following

Claim. In the above notation, there exists $\epsilon > 0$ such that $V_s \cong V_t$ if $|s - t| < \epsilon$.

To prove the claim, choose any t and consider the bundle $\mathrm{pr}_1^*(V_t)$ on $X \times [0, 1]$, with $\mathrm{pr}_1 : X \times [0, 1] \to X$ the first projection map.

Obviously, by the definitions, $\mathrm{pr}_1^*(V_t)$ agrees on the nose with V on the slice $X \times \{t\}$, which is a closed subset of $X \times [0, 1]$. In particular, there is a bundle isomorphism $\mathrm{pr}_1^*(V_t) \to V$ defined on the slice. By Lemma this extends to a bundle isomorphism in a neighborhood of the slice. A routine compactness argument implies that any such neighborhood contains one of the form $X \times (t - \epsilon, t + \epsilon)$. In particular, if $|t - s| < \epsilon$, V_t is isomorphic to V_s, as claimed.

The result we are trying to prove—that V_0 is isomorphic to V_1—now follows from a routine compactness argument, producing a list of points $0 < t_1 < \cdots < \epsilon_n < 1$ of the interval close enough to each other that $V_{t_i} \cong V_{t_{i+1}}, i = 0, 1, \ldots n$. $\qquad\square$

Corollary 7.3.4 *Let φ_0 and φ_1 be homotopic maps $X \to Y$, where X and Y are compact. Then the induced group homomorphisms φ_1^* and $\varphi_2^* : \mathrm{KO}^*(Y) \to \mathrm{KO}^*(X)$ are equal. Similarly, $\varphi_1^* = \varphi_2^* : \mathrm{K}^*(Y) \to \mathrm{K}^*(X)$.*

Proof By definition of homotopy, there exists a map $F : X \times [0, 1] \to Y$ such that $F \circ i_0 = \varphi_0$ and $F \circ i_1 = \varphi_1$. By functoriality and Lemma 7.3.3, we get

$$\varphi_0^* = (F \circ i_0)^* = i_0^* \circ F^* = i_1^* \circ F^* = (F \circ i_1)^* = \varphi_1^*,$$

which completes the proof. $\qquad\square$

Ring and module structures on K^0

We close this section with a discussion of the very important *ring* structure on the K^0-group of a compact space.

If V_1 and V_2 are vector bundles over X, then their tensor product $V_1 \otimes V_2$ is a vector bundle over X. If $V_1 \cong V_1'$ and $V_2 \cong V_2'$ then $V_1 \otimes V_2 \cong V_1' \otimes V_2'$, so the tensor product operation descends to an operation on the semigroup of isomorphism classes $\mathrm{Vect}(X)$ (or on $\mathrm{Vect}_{\mathbb{R}}(X)$, if one is working with real bundles.) By the universal property of the Grothendieck completion, tensor products on real and respectively complex bundles descends to a pair of multiplication operations

$$\mathrm{K}^0(X) \times \mathrm{K}^0(X) \to \mathrm{K}^0(X), \quad \mathrm{KO}^0(X) \times \mathrm{KO}^0(X) \to \mathrm{KO}^0(X).$$

Proposition 7.3.5 *Under direct sum and tensor product, $\mathrm{K}^0(X)$ is a commutative ring with identity.*
Similarly for $\mathrm{KO}(X)$.

Exercise 7.3.6 If X is compact and $a = [E^1] - [E^2] \in \mathrm{K}^0(X)$, $b = [F^1] - [F^2] \in \mathrm{K}^0(X)$, then the ring product $a \cdot b \in \mathrm{K}^0(X)$ equals the difference $[(E^1 \otimes F^1) \oplus (E^2 \otimes F^2)] - [(E^2 \otimes F^1) \oplus (E^1 \otimes F^2)]$.

The multiplicative identity of $\mathrm{K}^0(X)$ is the class of the trivial line bundle over X. (and similarly in KO^0-theory.)

Exercise 7.3.7 If $\varphi \colon X \to Y$ is a map of compact spaces, the induced map abelian group homomorphism $\varphi^* \colon \mathrm{KO}^0(Y) \to \mathrm{KO}^0(X)$ is also a ring homomorphism. (Similarly for complex K-theory.)

Exercise 7.3.8 Let A be a closed, contractible subspace of a compact space X. Prove that $\mathrm{K}^0(X) \cong \mathrm{K}^0(X/A)$, where X/A is the quotient space obtained by crushing A to a point.

7.4   K-Theory for Noncompact Spaces, Higher K-groups

Everything we say in this chapter is equally valid for K-theory and KO-theory. We mainly focus on K-theory.

Let X be locally compact Hausdorff, X^+ its one-point compactification. Neighborhoods of the point ∞ at infinity are complements of compact subsets of X, with the point at ∞ added. See Exercise 1.1.17 of Chapter 1. The space X^+ is compact Hausdorff. Let $\epsilon_X \colon \mathrm{pt} \to X^+$ the inclusion of the one-point space as the point at infinity. It induces a map $\epsilon_X^* \colon \mathrm{K}^0(X^+) \to \mathrm{K}^0(\mathrm{pt}) \cong \mathbb{Z}$, and similarly induces a map $\mathrm{KO}^0(X^+) \to \mathrm{KO}^0(\mathrm{pt}) \cong \mathbb{Z}$.

Definition 7.4.1 If X is a locally compact Hausdorff space, we define $\mathrm{K}^0(X)$ to be the kernel of the map $\epsilon_X^* \colon \mathrm{K}^0(X^+) \to \mathbb{Z}$. Similarly, we define $\mathrm{KO}^0(X)$.

Exercise 7.4.2 Prove that $\mathrm{K}^0\big([0, 1)\big)$ is the zero group.

Remark 7.4.3 Elements in $\mathrm{K}^0(X^+)$ are differences $[V_1] - [V_2]$ of stable isomorphism classes $[V_1]$ and $[V_2]$ of vector bundles over X^+. At the level of vector bundles, the map ϵ_X^* just maps a vector bundle V over X^+ to its restriction V_∞ to the point at ∞; this results in a vector space, and the corresponding integer is its dimension. Thus, $\epsilon_X^*([V] - [W]) = \dim(V_\infty) - \dim(W_\infty)$.

In particular, $\mathrm{K}^0(X)$ is always an *ideal* in the ring $\mathrm{K}^0(X^+)$. In particular, it is of course a subring, and hence a (non-unital) ring in its own right.

Similarly $\mathrm{KO}(X)$-theory is a ring, with the ring structure inherited from $\mathrm{KO}^0(X^+)$.

Example 7.4.4 $\mathrm{K}^0(\mathbb{R}) = 0$. Indeed, by definition, $\mathrm{K}^0(\mathbb{R})$ is the kernel of the augmentation map $\epsilon_{\mathbb{R}}^* \colon \mathrm{K}^0(\mathbb{R}^+) \to \mathbb{Z}$, while $\mathbb{R}^+ \cong S^1$ is the circle, whose K^0 has already been computed (Proposition 7.1.10) to be infinite cyclic with generator the class $[1] \in \mathrm{K}^0(S^1)$ of the trivial line bundle on S^1. If $n[1] \in \mathrm{K}^0(S^1)$ is any element, then $\epsilon_{\mathbb{R}}^*(n[1]) = n$ so $\epsilon_{\mathbb{R}}^*$ is injective and $\mathrm{K}^0(\mathbb{R}) := \ker(\epsilon_{\mathbb{R}}^*)$ is the zero group.

Exercise 7.4.5 Let E be a vector bundle over X. Prove that if E is trivial outside a compact subset of X, then X is (isomorphic to) the restriction to X of a vector bundle over X^+ to X. (That is, if E is trivial outside a compact set, then E "extends" to a vector bundle over X^+.)

Exercise 7.4.6 From the previous exercise, check in detail that if X is noncompact, then $K^0(X)$ can be described as formal differences $[E^1] - [E^2]$ where E^i are each vector bundles over X, each trivial outside a compact subset of X, and, such that each have the same dimension outside some compact subset. Write down when two such formal differences correspond to the same element of $K^0(X)$.

Functoriality of K-theory for noncompact spaces involves a nuance. It is *not* functorial under arbitrary maps $f : X \to Y$, but only *proper* maps, for these are precisely the maps which extend continuously to maps $f_+ : X^+ \to Y^+$ mapping the point at infinity to the point at infinity. Due to this property, $f_+ \circ \epsilon_X = \epsilon_Y$, and hence by functoriality of KO^0 or K^0, $\epsilon_X^* \circ f_+^* = \epsilon_Y^*$ and hence f_+^* maps $\ker(\epsilon_Y^*)$ into $\ker(\epsilon_X^*)$.

Thus, a proper map $f : X \to Y$ induces maps $f^* : K^*(Y) \to K^*(X)$ and $f^* : KO^0(Y) \to KO^0(X)$.

Suppose that X is already compact. Then ∞ is isolated in X^+ (is an open set) and hence $K^0(X^+) \cong K^0(X) \oplus K^0(\{\infty\}) \cong K^0(X) \oplus K^0(\text{pt}) = K^0(X) \oplus \mathbb{Z}$, with ϵ_X^* corresponding to the second projection map (by Exercise 7.1.6). It is immediate that $\ker(\epsilon_X^*) = K^0(X)$, so we recover our old definition of K-theory for compact spaces. The same remarks go through verbatim for KO^0-theory.

Exercise 7.4.7 Prove that a proper map is a closed map.

Exercise 7.4.8 Prove that if $\varphi : X \to Y$ is a proper map and $f \in C_c(X)$ is a continuous, complex-valued function with compact support, then $f \circ \varphi$ has compact support.

Two proper maps $\varphi_0, \varphi_1 : X \to Y$ are *properly homotopic* if there is a proper map $F : X \times [0, 1] \to Y$ such that $F \circ i_0 = \varphi_0$ and $F \circ i_1 = \varphi_1$, with i_0, i_1 the inclusions of X at the endpoints, as usual.

Proposition 7.4.9 *If X and Y are locally compact Hausdorff and $\varphi_0, \varphi_1 : X \to Y$ are properly homotopic proper maps, then $\varphi_0^* = \varphi_1^*$ as maps $K^0(Y) \to K^0(X)$. Similarly, $\varphi_0^* = \varphi_1^* : KO^0(Y) \to KO^0(X)$*

Proof For the proof, we restrict ourselves to complex K-theory. The same proof works for the real version.

As before, it is enough to prove that the maps $i_0^*, i_1^* : K^0(X \times [0, 1]) \to K^0(X)$, are equal. (Note that they are each proper.) By the definitions, it is sufficient to show that i_0^+ and i_1^+ induce the same map $K^0((X \times [0, 1])^+) \to K^0(X^+)$.

Let $H : X^+ \times [0, 1] \to (X \times [0, 1])^+$ map any $(x, t) \in X \times [0, 1]$ to the image of (x, t) in $(X \times [0, 1])^+$, and let $H(\infty, t) = \infty$ for every $t \in [0, 1]$. The reader can easily verify that H is continuous. It gives a homotopy between i_0^+ and i_1^+, as maps between two compact spaces. Hence $(i_0^+)^* = (i_1^+)^*$ from Theorem 7.3.4. This proves the result. $\square$

We can now define the higher K-theory groups of a space.

Definition 7.4.10 For X locally compact Hausdorff, $K^{-n}(X)$ is defined to be $K^0(X \times \mathbb{R}^n)$, and likewise $KO^{-n}(X) := KO^0(X \times \mathbb{R}^n)$.

Example 7.4.11 Since $K^0(\mathbb{R}) = 0$, (Example 7.4.4) we have so far determined that $K^0(\text{pt}) \cong \mathbb{Z}$ and $K^{-1}(\text{pt}) = 0$. The computation of $K^{-2}(\text{pt})$ and the computation of the higher groups $K^{-3}(\text{pt})$, $K^{-4}(\text{pt})$, ..., which turn out to be 2-periodic, will require the Bott Periodicity Theorem.

Remark 7.4.12 $\mathbb{R}^n$ is obviously a contractible space, but it is not *properly* contractible. Hence there is no *a priori* reason to suppose the K-theory or KO-theory groups of $\mathbb{R}^n$, equivalently, the higher K-theory groups $K^{-n}(\text{pt})$, of a point, are uninteresting. (And similarly for KO-theory.)

Since $(\mathbb{R}^n)^+ \cong S^n$, $K^{-n}(\text{pt}) := K^0(\mathbb{R}^n)$ is the subgroup of $K^0(S^n)$ consisting of differences $[V_1] - [V_2]$ of vector bundles over the sphere, of the same dimension. The difference $[H] - [1]$ is an example of such a difference, where H is the Hopf bundle.

The Hopf bundle is nontrivial, so there is no immediate reason to conclude that this difference is zero in $K^0(S^n)$ (in fact it is not); the difference, in fact, measures exactly the nontriviality of the Hopf bundle.

K-*theory classes from triples;* K-*theory "germs"*

One of the key points in the construction of K-theory or KO-theory classes from geometric considerations is that they can be constructed on various interesting *non-compact* spaces, by considering pairs of bundles, isomorphic to each other off a compact set. Since many spaces (like manifolds) have interesting open subsets, one can often splice a K-theory class for the (noncompact) open subset, into a K-theory class for X, with interesting results.

We start with some basic observations about the K-theory of open subsets of a space.

Exercise 7.4.13 Let $U \subset X$ be an open subset. Show that mapping the complement of U in X^+ to the point at infinity of U^+ results in a continuous map $i^+ : X^+ \to U^+$ mapping the points at infinity to each other.

The following constructions work in either K-theory or KO-theory; for brevity we restrict ourselves to K-theory.

a) If $i_U : U \to X$ is the inclusion of an open set in X, and $i^+ : X^+ \to U^+$ the map described above, show that $(i^+)^* : K^0(U^+) \to K^0(X^+)$ maps $\ker(\epsilon_U^*)$ to $\ker(\epsilon_X^*)$. Let $i_U! : K^0(U) \to K^0(X)$ be the corresponding map.

b) Prove that if $U \xrightarrow{i_U} V$ and $V \xrightarrow{j_V} W$ are two open inclusions then $(j_V \circ i_U)! = j_V! \circ i_U! : K^0(U) \to K^0(W)$.

c) Prove that the groups $K^0(U)$, as U runs over the open subsets of X, directed by inclusion, and the group homomorphisms $i! : K^0(U) \to K^0(V)$, for $i : U \to V$ an inclusion, make up a directed system of groups, and prove that

$$K^0(X) \cong \varinjlim_U K^0(U).$$

d) Prove that (for X locally compact Hausdorff as usual), the result of c) holds if we restrict the directed system just to the collection of *pre-compact* open subsets of X.

As a consequence of the result in part d) of the Exercise, is that every K^0-class for X has the form $i_U!(a)$ for some K-theory class $a \in K^0(U)$, for an open and *pre-compact* subset $U \subset X$.

Definition 7.4.14 Let X be a locally compact space. A K-triple E for X (respectively a KO-triple) consists of a pair E^0 and E^1 of complex (respectively real) vector bundles over X, and a bundle map $\varphi\colon E^0 \to E^1$, which is an isomorphism on the complement of a compact subset of X.

Two triples $E = (E^0, E^1, \varphi)$ and $F = (F^0, F^1, \psi)$ are *isomorphic* if there are vector bundle isomorphisms $\alpha\colon E^0 \to F^0$ and $\beta\colon E^1 \to F^1$ such that the diagram

$$
\begin{array}{ccc}
E^0 & \xrightarrow{\;\varphi\;} & E^1 \\
\downarrow{\scriptstyle\alpha} & & \downarrow{\scriptstyle\beta} \\
F^0 & \xrightarrow{\;\varphi'\;} & F^1
\end{array}
$$

commutes.

A *homotopy* of triples is a triple (E^0, E^1, φ) over $X \times [0, 1]$; the inclusions $i_0\colon X \to X \times [0, 1]$ and $i_1\colon X \to X \times [0, 1]$ at the endpoints of the interval pull such a triple to a pair of triples for X, which we call *homotopic triples*.

The *support* of a triple (V, W, φ) is the set of points $x \in X$ for which $\varphi(x)$ is *not* an isomorphism. The support is compact, by the definitions.

A *degenerate triple* is one for which φ is an isomorphism everywhere.

On the collection of isomorphism classes of triples, we put the equivalence relation generated by homotopy and addition of degenerate triples. Let $L(X)$ denote the correspond semigroup, with addition operation direct sum of triples.

We are going to describe the map $L(X) \to K^0(X)$—it can be shown to be an isomorphism, but we will not need this fact. The *map* is of most importance.

Suppose $\tau = (E^0, E^1, \varphi)$ is a triple. Let $U \subset X$ be any neighborhood of its support, a compact subset of X (U could be X, for example). We define a K^0-class $[\tau_U] \in K^0(U)$ in the following way. Let W and V be open subsets of X with $\mathrm{supp}(E) \subset W \subset \overline{W} \subset V \subset \overline{V} \subset U$, and $\overline{V}$ compact. Since $\overline{V}$ is compact, there is a vector bundle F over $\overline{V}$ such that $E^1 \oplus F$ is trivial on $\overline{V}$. Adding the degenerate triple (F, F, id) to E results in a triple for $\overline{V}$ in which the second vector bundle is trivial.

Instead of introducing new notation for this, we just denote by $\tau = (E^0, E^1, \varphi)$ the triple we have constructed, for $\overline{V}$, in which now the bundle E^1 is a product bundle.

We now proceed as in the examples. Let $A = \overline{W}$ and $B \subset V^+$ be the complement of W in V, together with the point of infinity of V^+. Then A and B are closed in V^+.

Take the bundle E^0 on A, and clutch it to the trivial bundle $E^1 = B \times \mathbb{C}^n$ on B using the clutching function $E^0|_{A \cap B} \xrightarrow{\varphi} E^1_{A \cap B} = A \cap B \times \mathbb{C}^n = (B \times \mathbb{C}^n)|_{A \cap B}$. The clutching results in a vector bundle $\tilde{E}$ on V^+. The difference $[\tilde{E}] - [1_n]$ is in $K^0(V)$, where $n = \dim(E^0)$. We now set

Definition 7.4.15 $\tau_U := i_{U,V}!([\tilde{E}] - [1_n])) \in K^0(U)$ where $i_{U,V} \colon V \to U$ is the inclusion.

Exercise 7.4.16 In the above notation, if $j \colon U \to U'$ is an inclusion of open sets, then $[\tau_{U'}] = j!([\tau_U])$.

In particular, any triple τ over X determines a class $\tau_U \in K^0(U)$ for any neighborhood $U \subset X$ of its support, and in particular, determines a class $\tau_X \in K^0(X)$.

Example 7.4.17 (The Bott Element for $\mathbb{R}^2$) For $(x, y) \in \mathbb{R}^2$ let $c(x, y) \colon \mathbb{C} \to \mathbb{C}$ be multiplication by the complex number $x + iy$. We may interpret c as a vector bundle map from the trivial bundle $\mathbb{R}^2 \times \mathbb{C}$ over $\mathbb{R}^2$, to itself.

Now, c is a bundle isomorphism away from 0. Let U be any neighborhood of the origin (it could be all of $\mathbb{R}^2$), let $A \subset U$ be a small closed Euclidean ball contained in U and centered at the origin. Let B be the closure in U^+ of $U^+ \setminus A$. Thus, B consists of the closure of the complement of A in U, together with the point at infinity of U.

On A we put the product bundle $A \times \mathbb{C}$, on B we put the product bundle $B \times \mathbb{C}$, and we clutch them (see Exercise 7.1.23) using the function c on $A \cap B$. This results in a complex vector bundle H_U on U^+, and a class $\beta_U := [H_U] - [1] \in K^0(U)$, where $[1] \in K^0(U^+)$ is the class of the trivial line bundle on U^+, because $\epsilon^*_U(\beta_U) = 0$.

Exercise 7.4.18 In the above notation, prove that the complex vector bundle $H_{\mathbb{R}^2}$ over $(\mathbb{R}^2)^+ = S^2$ is isomorphic to the Hopf bundle.

Hence $\beta_{\mathbb{R}^2} \in K^0(\mathbb{R}^2) \subset K^0(S^2)$ is equal to the difference $[H] - [1]$, where 1 is the trivial complex line bundle over S^2 and H is the Hopf bundle over S^2.

Exercise 7.4.19 Verify that if $i \colon U \to V$ is an inclusion of neighborhoods of the origin in $\mathbb{R}^2$ then $i_U!(\beta_U) = \beta_V$.

It will be a consequence of Bott Periodicity that $K^0(\mathbb{R}^2)$ is an infinite cyclic group generated by $\beta_{\mathbb{R}^2}$.

Example 7.4.20 The system of "Bott elements" $\beta_U \in K^0(U)$, one attached to each neighborhood of the origin $\mathbb{R}^2$, suggests might be thought of as the specification of a kind of a "germ" of a K-theory class at the origin.

One can also get (one-dimensional, now) 'K-theory "germs" in this (informal) sense around smooth curves in the plane, as we now show. In order to make things topologically nontrivial, remove a finite set of points, let $X = \mathbb{R}^2 \setminus \{p_1, \dots, p_n\}$ from the plane.

We consider a smooth closed curve C in the plane looping around some of these points. By the Jordan Curve Theorem one can select a (smooth) field of unit vectors $\mathbf{n}(x)$ as $x \in C$, such that $\mathbf{n}(x)$ is perpendicular to the tangent of the curve at x. We can thus label points in a neighborhood of the curve by pairs (x, t) where $x \in C$ and $t \in \mathbb{R}$, but making this pair correspond to the point $x + t\mathbf{n}(x)$. Our labeling determines a natural diffeomorphism and system of coordinates on the neighborhood U of the curve consisting of points (x, t) where $|t| < \epsilon$. Let $V = U \times \mathbb{R}$. On U let $c(x, t, s) := t + is$ where (x, t) are the coordinates as explained above, of a point of U.

One finds suitable closed sets A and B to argue that the bundle map c determines, by clutching, a canonical class in $K^0(V^+)$ and then, by subtracting the class of a trivial line bundle over $A \cap B$, a class in $K^0(V) = K^0(U \times \mathbb{R}) = K^{-1}(U)$, which can then be pushed forward to a class $\beta_C \in K^{-1}(\mathbb{R}^2 \setminus \{p_1, \ldots, p_n\})$.

It can be shown that this "germ" is a nontrivial K-theory class for X if the curve loops around at least some of the points.

These examples of K-theory classes makes one think more of the fundamental group, or first homology group, of a space.

Graded ring structure on higher K-theory

A graded ring is *graded commutative* if $ab = (-1)^{\partial a \partial b} ba$ for any homogeneous elements a, b of degrees ∂a and ∂b. It turns out that $K^*(X) := \oplus_{n=0}^{\infty} K^{-n}(X)$ has a graded commutative ring structure extending, in an appropriate sense, the ring structure on $K^0(X)$ by tensor product of vector bundles.

Before proceeding, let X and Y be locally compact spaces and $Z = X \times Y$, $\pi_X : Z \to X$ and $\pi_Y : Z \to Y$ the projection maps. These will not be proper maps, if the spaces are not compact. So if $a \in K^0(X)$, it does not quite make sense to write $\pi_X^*(a) \in K^0(Z)$, as π being not proper, does not give a map on K-theory. However, $\pi_X^*(a) \cdot \pi_Y^*(b)$ does in fact make sense as an element of $K^0(X \times Y)$, its "support" is roughly speaking, the product of the support of a and the support of b, which will be compact.

Suppose $a = [E^1] - [E^2]$ for two vector bundles E^i on X, trivial and of the same dimension outside a compact subset $K \subset X$. Write $b = [F^1] - [F^2]$, F^i trivial and of the same dimension off $L \subset Y$.

Let $\tilde{E}^i := \pi_X^*(E^i)$, $\tilde{F}^i := \pi_Y^*(F^i)$. Then $\tilde{E}^i$ are trivial and isomorphic to each other outside $K \times Y$, and the $\tilde{F}^i$ are trivial and isomorphic to each other outside $X \times L$.

Consider the vector bundles

$$V^1 := (\tilde{E}^1 \otimes \tilde{F}^1) \oplus (\tilde{E}^2 \otimes \tilde{F}^2),$$

and

$$V^2 := (\tilde{E}^2 \otimes \tilde{F}^1) \oplus (\tilde{E}^1 \oplus \tilde{F}^2).$$

Now $\tilde{E}^1 \cong \tilde{E}^2$ outside $K \times Y$, so the first summand $\tilde{E}^1 \otimes \tilde{F}^1$ of V^1 is isomorphic to the first summand $\tilde{E}^2 \otimes \tilde{F}^1$ of V^2 outside $K \times Y$. By the same reasoning, the second summand of V^1 is isomorphic to the second summand of V^2 outside $K \times Y$. Therefore, V^1 is isomorphic to V^2 outside $K \times L$.

On the other hand, outside $X \times L$, the second summand of V^1 is isomorphic to the first summand of V^1, and, likewise, the first summand of V^1 is isomorphic to the second summand of V^2, so that in this case also, we see that V^1 is isomorphic to V^2.

We conclude therefore that V^1 is isomorphic to V^2 outside $K \times L$, a compact subset of $X \times Y$, and fixing the isomorphism, we obtain a triple (V^1, V^2, φ) representing an element of $\mathrm{K}^0(X \times Y)$, which we denote by $\pi^*(a) \cdot \pi^*(b)$.

Remark 7.4.21 The idea is that the product $\pi^*(a) \cdot \pi^*(b)$ should be represented by the product, formally speaking,

$$(7.6) \qquad\qquad [\tilde{E}^1] - [\tilde{E}^2]) \cdot ([\tilde{F}^1] - [\tilde{F}^2]),$$

the problem of course being that neither of the terms actually define K-theory classes for $X \times Y$.

However, the first term "vanishes" outside $K \times Y$, and the second term vanishes outside $X \times L$, so the idea is that the product should vanish outside $K \times L$, which of course is compact, making the product define a K-theory class.

In fact, if one multiplies, somewhat formally, the equality (7.6) out, one obtains the formal difference $[(\tilde{E}^1 \otimes F^1) \oplus (\tilde{E}^2 \otimes F^2)] - [(\tilde{E}^2 \otimes F^1) \oplus (\tilde{E}^1 \oplus F^2)]$, that is, one obtains $[V_1] - [V_2]$ with V_i defined as above.

Exercise 7.4.22 Construct an explicit formula for the isomorphism between V_1 and V_2 based on the assumed isomorphisms $E^1 \cong E^2$ and $F^1 \cong F^2$ (outside suitable compact sets.)

By similar arguments (see Exercise 7.4.24) one can argue that there is a multiplication operation between K-theory classes $a \in \mathrm{K}^0(X)$ and K-theory classes $c \in \mathrm{K}^0((X \times Y))$, with values in $\mathrm{K}^0(X \times Y)$, which we denote by $\pi^*(a) \cdot c$.

Thus, $\mathrm{K}^0(X \times Y)$ has the structure of a *module* over the ring $\mathrm{K}^0(X)$. Similarly, $\mathrm{K}^0(X \times Y)$ is a module over $\mathrm{K}^0(Y)$.

Proposition 7.4.23 *The pairings and module structures defined above, are all well defined, $\mathbb{Z}$-bilinear, and associative in the sense that*

$$\pi_X^*(a) \cdot \left(\pi_X^*(a') \cdot \pi_Y^*(b) \right) = \left(\pi_X^*(a \cdot a') \right) \cdot \pi_Y^*(b),$$

and

$$\left(\pi_X^*(a) \cdot \pi_Y^*(b) \right) \cdot \pi_Y^*(b') = \pi_X^*(a) \cdot \left(\pi_Y^*(b \cdot b') \right),$$

for $a, a' \in \mathrm{K}^0(X)$, $b, b' \in \mathrm{K}^0(Y)$.

Exercise 7.4.24 Let X and Y be locally compact spaces and let $f : Y \to X$ be any map (not necessarily proper). Then there is a well defined, $\mathbb{Z}$-bilinear multiplication operation $K^0(Y) \times K^0(X) \to K^0(Y)$ mapping a pair $a \in K^0(X)$ and $c \in K^0(Y)$ to an element $f^*(a) \cdot c \in K^0(Y)$, which makes $K^0(Y)$ into a module over the ring $K^0(X)$.

Exercise 7.4.25 Generalize the $\mathbb{Z}$-bilinear multiplication operation $K^0(X) \times K^0(Y) \to K^0(X \times Y)$ developed above to a multiplication operation $K^0(X) \times K^0(Y) \to K^0(Z)$, producing an element $\rho_1^*(a) \cdot \rho_2^*(b)$ from $a \in K^0(X)$, $b \in K^0(Y)$, whenever $\rho_1 : Z \to X$ and $\rho_2 : Z \to Y$ are two maps with the property that *for any pair of compact subsets $K \subset X$ and $L \subset Y$, $\rho_1^{-1}(K) \cap \rho_2^{-1}(L)$ is compact in Z.*

We may now define a *graded ring structure* on $K^*(X) = \oplus_{i=0}^{\infty} K^{-i}(X)$.

Choose $r, s \geq 0$ and let $\pi_1 : \mathbb{R}^{r+s} \to \mathbb{R}^r$, $\pi_2 : \mathbb{R}^{r+s} \to \mathbb{R}^s$ be the projection maps. Then for any locally compact space X, consider the maps $\rho_1 := \mathrm{id}_X \times \pi_1 : X \times \mathbb{R}^{r+s} \to X \times \mathbb{R}^r$, and $\rho_2 : X \times \mathbb{R}^{r+s} \to X \times \mathbb{R}^s$. It is easily checked that $\rho_1^{-1}(K) \cap \rho_2^{-1}(L)$ is compact in $X \times \mathbb{R}^{r+s}$, for any compact $K \subset X \times \mathbb{R}^r$ and any compact $L \subset X \times \mathbb{R}^s$. By Exercise 7.4.25, there is a well-defined product class $\rho_1^*(a) \cdot \rho_2^*(b) \in K^0(X \times \mathbb{R}^{k+s})$ for any $a \in K^0(X \times \mathbb{R}^r)$ and $b \in K^0(X \times \mathbb{R}^s)$.

Definition 7.4.26 If $a \in K^{-r}(X)$, $b \in K^{-s}(X)$, we let

$$a \wedge b \in K^{-(r+s)}(X)$$

denote the class $\rho_1^*(a) \cdot \rho_2^*(b)$ described above.

The wedge product notation is required to distinguish our graded multiplication from ordinary multiplication, when the situation is ambiguous. If, for example, $a, b \in K^{-2}(\mathrm{pt}) := K^0(\mathbb{R}^2)$, then since $K^0(X)$ is always a ring, for any X, and in particular for $X = \mathbb{R}^2$, we can form $a \cdot b \in K^0(\mathbb{R}^2)$, whereas, $a \wedge b \in K^{-4}(\mathrm{pt}) = K^0(\mathbb{R}^2 \times \mathbb{R}^2)$, lies, of course, in a different group. The two products are related by the following

Exercise 7.4.27 In the above notation, if $\delta : \mathbb{R}^2 \to \mathbb{R}^2 \times \mathbb{R}^2 = \mathbb{R}^4$ is the diagonal map, then $a \cdot b = \delta^*(a \wedge b)$.

The multiplication $a \wedge b$ is easily checked to be associative, and by the definitions, extends the usual ring structure on the summand $K^0(X)$. Recall that the latter ring structure is *commutative*. The more general multiplication turns out to be *graded commutative*.

Proposition 7.4.28 *If $a \in K^{-r}(X)$ and $b \in K^{-s}(X)$, then $a \wedge b = (-1)^{rs} b \wedge a$. That is, $K^*(X)$ is a graded commutative ring.*

We leave the proof as an exercise.

7.5   The Long Exact Sequence of a Pair

In ordinary cohomology, defined for spaces using chain complexes, the Snake Lemma implies that associated to a closed subspace $A \subset X$ is a long exact cohomology sequence.

The same is true of K-theory. We will just state the result here, as we will prove a more general version of it when we discuss K-theory for C*-algebras.

Theorem 7.5.1 *Let $A \subset X$ be a closed subspace of a locally compact space. Then there exist natural maps $\delta \colon K^{-i}(A) \to K^{-i+1}(X \setminus A)$ for which the sequence, infinite to the left,*

$$(7.7) \quad \cdots \to K^{-i-1}(A) \xrightarrow{\delta} K^{-i}(X \setminus A) \xrightarrow{i_!} K^{-i}(X) \xrightarrow{j^*} K^{-i}(A) \longrightarrow \cdots$$

$$\cdots \xrightarrow{\delta} K^0(X \setminus A) \to K^0(X) \to K^0(A)$$

is exact, where $i \colon X \setminus A \to X$ is the (open) inclusion, $j \colon A \to X$ the (closed) inclusion.

This long exact sequence is natural with respect to maps $(X, A) \to (X', A')$ of pairs of locally compact spaces.

We will know nothing about the range of the last map until Bott Periodicity. This makes the long exact sequence not hugely helpful for computations.

Exercise 7.5.2 Let X be a compact space and $Y \subset X$ a finite subset. Let X/Y be the *quotient* space obtained by identifying all the points of Y with each other. Let $\pi \colon X \to X/Y$ be the quotient map.

a) Prove that $\pi^* \colon K^0(X/Y) \to K^0(X)$ is always injective.
b) Prove that π^* is an isomorphism if all points of Y lie in the same connected component of X.

(*Hint.* X/Y can be identified with $(X \setminus Y)^+$. We get a long exact sequence from the pair $Y \subset X$. On the other hand, X/Y comes with a natural basepoint and this generates another long exact sequence. The second sequence maps naturally to the first; examine the corresponding commutative diagram, and use the fact that $K^{-1}(Y)$ is zero for any finite Y.)

Note that π^* does *not* induce an isomorphism on K^{-i} for $i > 0$; for example identifying the endpoints of $[0, 1]$ results in S^1, and $K^{-1}(S^1) \cong \mathbb{Z}$ will follow from Bott Periodicity. But $K^{-1}([0, 1]) \cong K^{-1}(\text{pt}) = 0$. Nor is π^* an isomorphism even on K^0, if the connectedness assumption is dropped. (consider the 2-point space X.)

Remark 7.5.3 The last exercise makes it a bit easier to visualize what space one is dealing with in computing $K^{-1}(X)$, for X compact (say). By the definitions, $K^{-1}(X) = K^0(X \times \mathbb{R}) := \ker \epsilon_X^* \colon K^0\big((X \times \mathbb{R})^+\big) \to \mathbb{Z}$. The space $(X \times \mathbb{R})^+$ is thus what is of interest here. We can consider it alternately as the quotient space

$X \times [0, 1] / \sim$ where the equivalence relation collapses $X \times \{0\} \cup X \times \{1\}$ to a single point. The exercise above shows that this results in the same K^0 group as for the quotient space obtained by each of $X \times \{0\}$ and $X \times \{1\}$ to (different) points.

Exercise 7.5.4 Deduce from Exercise 7.5.2 (and Remark 7.5.3) that

$$K^0(S^2) \cong K^0((S^1 \times \mathbb{R})^+)$$

by a natural isomorphism, and that as a consequence,

$$K^0(\mathbb{R}^2) \cong K^{-1}(S^1),$$

(*c.f.* Exercise 7.1.24). The same reasoning proves the more general result that $K^0(\mathbb{R}^n) \cong K^{-1}(S^{n-1})$ for all n.

Exercise 7.5.5 Give another proof that $K^{-1}(S^1) \cong K^0(\mathbb{R}^2)$ (even though at this stage, we are still not in a position to say what either of these groups are), using the following method. The closed subset $S^1 \subset \overline{\mathbb{D}}$ generates a long exact sequence

$$(7.8) \qquad \cdots \to K^{-1}(\overline{\mathbb{D}}) \to K^{-1}(S^1) \xrightarrow{\delta} K^0(\mathbb{R}^2) \to K^0(\overline{\mathbb{D}}) \to K^0(S^1).$$

Argue that the last map is an isomorphism and deduce that δ is an isomorphism.

Exercise 7.5.6 Let A be a contractible subspace (that is, contractible as a topological space in its own right) of a compact space X. Let X/A be the space obtained from X by identifying A to a point. Prove that the quotient map $\pi : X \to X/A$ induces an isomorphism $\pi^* : K^0(X/A) \to K^0(X)$. (*Hint.* Define an inverse map $K^0(X) \to K^0(X/A)$ as follows. If E is a vector bundle over X, find a trivialization $E \cong 1_n$ of E restricted to A; one exists, since A is contractible. If $u : E_{|A} \to 1_n = A \times \mathbb{C}^n$ is a trivialization, extend it to a bundle map $\bar{u}$ from E to 1_n in a neighborhood of A. Now clutch the bundle 1_n over a suitable (slightly smaller) neighborhood of A with E on the complement. This bundle now has a single fiber over A and can be considered a bundle over X/A.)

An alternative description of $K^{-1}(X)$ and the boundary map for the long exact sequence

We first discuss an interesting topological group, whose connected components are relevant to K-theory.

Let A be a unital C*-algebra. Let $\mathbf{U}_\infty(A)$ be the group of all $\mathbb{N}$-by-N-matrices with entries in A, which have a block-diagonal form $\begin{bmatrix} u & 0 \\ 0 & 1 \end{bmatrix}$ with u a (square) unitary matrix in $M_n(A)$, and 1 denoting the identity operator. There is an evident group structure on $\mathbf{U}_\infty(A)$ by multiplication, and we can regard, in the obvious way, all of the groups $\mathbf{U}(M_n(A))$ as subgroups of $\mathbf{U}_\infty(A)$.

We give $\mathbf{U}_\infty(A)$ the inductive limit topology): a subset $U \subset \mathbf{U}_\infty(A)$ is open if and only if $U \cap \mathbf{U}(M_n(A))$ is open for all n.

We are particularly interested in the path components $\pi_0(\mathbf{U}_\infty(A))$. If $u, v \in \mathbf{U}_\infty(A)$ let $u \sim v$ mean that u and v are in the same path component of $\mathbf{U}_\infty(A)$.

Assume that u and v are unitary matrices of a fixed size n, understood as elements of $\mathbf{U}_\infty(A)$. Form the matrix $\begin{bmatrix} u & 0 \\ 0 & v \end{bmatrix} \in \mathbf{U}_\infty(A)$. With respect to the same block decomposition put $R_t := \begin{bmatrix} \cos t & -\sin t \\ \sin t & \cos t \end{bmatrix}$. Then $R_0 = \begin{bmatrix} 1 & 0 \\ 0 & 1 \end{bmatrix}$ and $R_{\frac{\pi}{2}} = \begin{bmatrix} 0 & -1 \\ 1 & 0 \end{bmatrix}$.

We have:

(7.9)
$$\begin{bmatrix} 0 & 1 \\ -1 & 0 \end{bmatrix} \begin{bmatrix} u & 0 \\ 0 & v \end{bmatrix} \begin{bmatrix} 0 & -1 \\ 1 & 0 \end{bmatrix} = \begin{bmatrix} v & 0 \\ 0 & u \end{bmatrix}.$$

We obtain a path of unitaries $R_t^{-1} \begin{bmatrix} u & 0 \\ 0 & v \end{bmatrix} R_t$ between $\begin{bmatrix} u & 0 \\ 0 & v \end{bmatrix}$ and $\begin{bmatrix} v & 0 \\ 0 & u \end{bmatrix}$. That is,

(7.10)
$$\begin{bmatrix} u & 0 \\ 0 & v \end{bmatrix} \sim \begin{bmatrix} v & 0 \\ 0 & u \end{bmatrix}.$$

Now multiply both sides of this identity by $\begin{bmatrix} v^* & 0 \\ 0 & 1 \end{bmatrix}$. We obtain

(7.11)
$$\begin{bmatrix} v^*u & 0 \\ 0 & v \end{bmatrix} \sim \begin{bmatrix} 1 & 0 \\ 0 & u \end{bmatrix}.$$

Taking $u \doteq 1$ for example gives then that

(7.12)
$$\begin{bmatrix} v^* & 0 \\ 0 & v \end{bmatrix} \sim \begin{bmatrix} 1 & 0 \\ 0 & 1 \end{bmatrix}.$$

Multiplying (7.10) on both sides by the matrix $\begin{bmatrix} v^* & 0 \\ 0 & v \end{bmatrix}$, therefore, gives the identity

(7.13)
$$\begin{bmatrix} u & 0 \\ 0 & v \end{bmatrix} \sim \begin{bmatrix} 1 & 0 \\ 0 & uv \end{bmatrix}.$$

Of course, this is $\sim \begin{bmatrix} uv & 0 \\ 0 & 1 \end{bmatrix}$, the group product of u and v in $\mathbf{U}_\infty(A)$.

Proposition 7.5.7 *Let A be a unital C*-algebra. Then the group $\pi_0(\mathbf{U}_\infty(A))$ of path components of $\mathbf{U}_\infty(A)$ is abelian. Moreover, if $[u], [v] \in \pi_0(\mathbf{U}_\infty(A))$ are two elements of this group, with u, v unitary-valued matrices of the same size, then*

$$[u] \cdot [v] := [uv] = [\begin{bmatrix} u & 0 \\ 0 & v \end{bmatrix}] \in \pi_0\big(\mathbf{U}_\infty(A)\big).$$

Now suppose that X is compact Hausdorff and that $A = C(X)$. If $u: X \to \mathbf{U}_n = \mathbf{U}\big(M_n(\mathbb{C})\big)$ is a continuous map, then we may consider it alternatively as a unitary matrix in $M_n\big(C(X)\big)$, and then as an element of $\mathbf{U}_\infty\big(C(X)\big)$. We see that the group $\mathbf{U}_\infty\big(C(X)\big)$ is the same as the group $[X, \mathbf{U}_\infty]$ of continuous maps $X \to \mathbf{U}_\infty :=$ $\mathbf{U}_\infty(\mathbb{C})$, where such maps are multiplied pointwise in the obvious way.

Moreover, to say that two elements of $\mathbf{U}_\infty\big(C(X)\big)$ are in the same path component of the group, is equivalent to saying that the corresponding maps $X \to \mathbf{U}_\infty$ are homotopic.

Example 7.5.8 The $\mathbb{T} = \mathbf{U}_1$-valued map on the circle $S^1 \subset \mathbb{C}$ defined by the inclusion, determines a class $[z] \in \mathrm{K}^{-1}(S^1)$. It will be a consequence of Bott Periodicity that $\mathrm{K}^{-1}(S^1)$ is infinite cyclic, and $[z]$ generates it.

Theorem 7.5.9 *There is a canonical, natural isomorphism of abelian groups*

$$\mathrm{K}^{-1}(X) \cong [X, \mathbf{U}_\infty(\mathbb{C})].$$

Furthermore, under this identification, suppose that $A \subset X$ is a closed subspace of X. Then the boundary map $\delta: \mathrm{K}^{-1}(A) \to \mathrm{K}^0(X \setminus A)$ in the long exact sequence, sends the K-theory class corresponding to a homotopy class $u: X \to \mathbf{U}_n$, to the class of the K-theory triple $(1_n, 1_n, \bar{u})$, where $\bar{u}$ is any extension of u to a matrix-valued function $\bar{u}: X \to M_n(\mathbb{C})$.

Indeed, a vector bundle over $(X \times \mathbb{R})^+$ can be trivialized over the closure in $(X \times \mathbb{R})^+$ of $X \times (-\infty, 0]$, and similarly can be trivalized over the closure of $X \times [0, \infty)$. The difference of the two trivializations on the intersection $\cong X$ of these two closed subsets (neglecting the point at infinity) gives a unitary map $u: X \to \mathbf{U}_n$.

The above description of the boundary map is very helpful in doing computations.

Example 7.5.10 Consider the setting of Example 7.5.5, where we considered the pair $(\overline{\mathbb{D}}, S^1)$ and the associated long exact sequence. It was argued there (or rather left to the reader to argue) that $\delta: \mathrm{K}^{-1}(S^1) \to \mathrm{K}^0(\mathbb{D})$ is an isomorphism. In example 7.5.8 we pointed out the tautological class $[z] \in \mathrm{K}^{-1}(S^1)$. According to Theorem 7.5.9, the boundary map

$$\delta: \mathrm{K}^{-1}(S^1) \to \mathrm{K}^0(\mathbb{D})$$

maps $[z]$ to the class of the triple $(\mathbf{1}, \mathbf{1}, z)$, since z can be extended in the obvious way from a map S^1 to $\mathbb{C}^*$ to a map $\overline{\mathbb{D}}$ into $\mathbb{C}$. But this also describes the Bott class $\beta_{\mathbb{D}}$ described in Example 7.4.17 .

In other words, the boundary map $\delta: \mathrm{K}^{-1}(S^1) \to \mathrm{K}^0(\mathbb{D})$ maps the class $[z]$ to the Bott class $\beta_{\mathbb{R}^2}$ for the open disk.

7.6   Bott Periodicity, the 6-Term Exact Sequence

Let X be any locally compact space. In this section we describe Bott's celebrate Periodicity Theorem. The key character is the Bott class $\beta_{\mathbb{R}^2} \in K^{-2}(pt) = K^0(\mathbb{R}^2)$ described in Example 7.4.17.

The graded ring $K^*(X) := \oplus_{i=0}^{\infty} K^{-i}(X)$ is a (graded) module over the graded ring $K^*(pt)$. Therefore, multiplication by the Bott element defines a map

$$\beta_X \colon K^*(X) \to K^{*-2}(X),$$

shifting degrees by -2.

The Bott Periodicity theorem in complex K-theory is the following statement.

Theorem 7.6.1 *For every locally compact space X, the group homomorphism*

$$\beta_X \colon K^*(X) \to K^{*-2}(X)$$

of multiplication by the Bott element $\beta_{\mathbb{R}^2} \in K^{-2}(pt)$, is an isomorphism.

Periodicity says in particular that as an abstract group, $K^{-2}(pt) := K^0(\mathbb{R}^2)$ is isomorphic to the group $\mathbb{Z}$ of integers with generator the Bott element $\beta_{\mathbb{R}^2} \in K^0(\mathbb{R}^2)$. Furthermore, $\beta_{\mathbb{R}^2}^2$ generates $K^{-4}(pt)$, $\beta_{\mathbb{R}^2}^3$ generates $K^{-6}(pt)$ and so on.

On the other hand we have already established (it was relatively easy, based on a computation of $K^0(S^1)$) that $K^{-1}(pt) = 0$. Combining this observation with Bott Periodicity gives that all of the groups $K^{-3}(pt)$, $K^{-5}(pt)$, ... and so on, are zero.

Since we have already proved that $K^{-1}(S^1) \cong K^0(\mathbb{R}^2)$, we also get:

Corollary 7.6.2 $K^{-1}(S^1) \cong \mathbb{Z}$ *with generator the $\mathbf{GL}_1(\mathbb{C})$-valued function $u(z) = z$. Moreover, the boundary map $\delta \colon K^{-1}(S^1) \to K^0(\mathbb{D}) \cong K^0(\mathbb{R}^2)$ for the long exact sequence associated to $S^1 \subset \overline{\mathbb{D}}$ maps $[z] \in K^{-1}(S^1)$ to the Bott element $\beta_{\mathbb{R}^2} \in K^0(\mathbb{R}^2)$.*

Exercise 7.6.3 Verify that the system of maps β_X satisfies the following two naturality conditions: firstly, it is *natural* in X in the sense that if $f \colon X \to Y$ is a continuous, proper map, then the diagram

$$
\begin{array}{ccc}
K^0(Y) & \xrightarrow{\ f^*\ } & K^0(X) \\
{\scriptstyle \beta_Y}\big\downarrow & & \big\downarrow{\scriptstyle \beta_X} \\
K^{-2}(Y) & \xrightarrow{\ f^*\ } & K^{-2}(X)
\end{array}
$$

commutes. This says that β defines a *natural transformation* from the functor K^0 (from locally compact Hausdorff spaces, to abelian groups), to the functor K^{-2}.

Secondly, β_X, is compatible with the ring structure on K-theory in the sense that

(7.14) $$\beta_X(ab) = \beta_X(a)b, \quad a, b \in \mathrm{K}^*(X).$$

Exercise 7.6.4 Use the graded multiplicativity of K-theory and (7.14) to deduce that $\beta_X(ab) = a\beta_X(b)$ for all $a, b \in \mathrm{K}^*(X)$. (*Hint.* Prove it first for homogeneous elements $a \in \mathrm{K}^{-i}(X)$ and $b \in \mathrm{K}^{-j}(X)$. That is, β_X is a bimodule homomorphism of bimodules.)

Bott Periodicity also plays a role analogous to the essential Excision Theorem of cohomology in the sense that coupling it with the Long Exact sequence results in a periodic exact sequence of length 6, which makes it possible in principal to compute the K-groups of spaces which are inductively made up of simpler pieces (simplicial complexes.)

Let $A \subset X$ be a closed subset of X locally compact, and consider the associated long exact sequence

(7.15) $$\cdots \to \mathrm{K}^{-1}(A) \xrightarrow{\delta} \mathrm{K}^{-1}(X \setminus A) \xrightarrow{i!} \mathrm{K}^{-1}(X) \xrightarrow{j^*} \mathrm{K}^{-1}(A) \longrightarrow \cdots$$

$$\cdots \xrightarrow{\delta} \mathrm{K}^0(X \setminus A) \to \mathrm{K}^0(X) \to \mathrm{K}^0(A)$$

of Theorem 7.5.1. By Bott Periodicity, $\mathrm{K}^0(A) \cong \mathrm{K}^{-2}(A)$ by the Bott map β_A. Composing β_A with the boundary map $\delta: \mathrm{K}^{-2}(A) \to \mathrm{K}^{-1}(X \setminus A)$ thus produces a map

(7.16) $$\delta' := \delta \circ \beta_A: \mathrm{K}^0(A) \to \mathrm{K}^{-1}(X \setminus A).$$

To describe this map, let E be a vector bundle over A, assuming that A is compact, and $p: A \to M_n(\mathbb{C})$ a projection-valued map such that $\mathrm{Im}(p) \cong E$. Extend p to a continuous map $\bar{p}: X \to M_n(\mathbb{C})$ taking self-adjoint values. As a self-adjoint of the C*-algebra $C(X, M_n(\mathbb{C}))$, we have available functional calculus, and in particular the $\mathbb{T}$-valued function $e(x) = e^{2\pi i x}$. Applying this function, which is obviously continuous on the spectrum of $\bar{p}$, to $\bar{p}$ defines a unitary $e(\bar{p}) \in C(X, M_n(\mathbb{C}))$, that is, a unitary matrix-valued map on X.

Since the spectrum of $p(x)$ consists of 0 and 1 alone, for $x \in A$, the function $e(\bar{p})$ takes the constant value 1 (meaning the identity operator in $M_n(\mathbb{C})$) on A.

Hence it can be considered as a unitary matrix-valued function on $(X \setminus A)^+ \cong X/A$.

Corollary 7.6.5 *Let $A \subset X$ be a closed subspace of a locally compact space. Then there is a natural (with respect to maps of pairs) 6-term exact sequence*

$$
\begin{array}{ccccc}
\mathrm{K}^0(X \setminus A) & \longrightarrow & \mathrm{K}^0(X) & \longrightarrow & \mathrm{K}^0(A) \\
\delta \uparrow & & & & \downarrow \delta \\
\mathrm{K}^{-1}(A) & \longleftarrow & \mathrm{K}^{-1}(X) & \longleftarrow & \mathrm{K}^{-1}(X \setminus A)
\end{array}
$$

Moreover:

a) *The boundary map* $\delta\colon \mathrm{K}^{-1}(A) \to \mathrm{K}^0(X \setminus A)$ *maps the homotopy class of a map* $u\colon A \to \mathbf{GL}(n, \mathbb{C})$ *to the class of the* K-*theory triple* $(1_n, 1_n, \bar{u})$, *where* u *is any extension of* u *to a map* $X \to M_n$.
b) *The map* $\delta\colon \mathrm{K}^0(A) \to \mathrm{K}^{-1}(X \setminus A)$ *maps the class of a bundle* $E = Im(p)$, *for some projection-valued map* $p\colon A \to M_n(\mathbb{C})$, *to the (homotopy class of the)* $\mathbf{U}_n$-*valued map* $e(\bar{p})\colon X/A \to \mathbf{U}_n$, *described above, for any extension* $\bar{p}\colon X \to M_n(\mathbb{C})$ *of* p *to* X.

Exercise 7.6.6 Compute the K-theory groups of a (closed) annulus $a \le |z| \le b$, and of an open annulus $a < |z| < b$, respectively.

K-*theory of spheres*

We can view S^n as the one-point compactification $(\mathbb{R}^n)^+$ so that, the pair consisting of S^n together with the closed subspace $\{\infty\}$ consisting of the single point "at infinity," gives a 6-term exact sequence

$$\begin{array}{ccccc}
\mathrm{K}^0(\mathbb{R}^n) & \longrightarrow & \mathrm{K}^0(S^n) & \longrightarrow & \mathrm{K}^0(\mathrm{pt}) \\
\delta \uparrow & & & & \downarrow \delta \\
\mathrm{K}^{-1}(\mathrm{pt}) & \longleftarrow & \mathrm{K}^{-1}(S^n) & \longleftarrow & \mathrm{K}^{-1}(\mathbb{R}^n)
\end{array}$$

(7.17)

Let n be even. Plugging in what we know (the K-theory of $\mathbb{R}^n$, and of points), this boils down to the sequence

$$\begin{array}{ccccc}
\mathbb{Z} & \longrightarrow & \mathrm{K}^0(S^n) & \longrightarrow & \mathbb{Z} \\
\delta \uparrow & & & & \downarrow \delta \\
0 & \longleftarrow & \mathrm{K}^{-1}(S^n) & \longleftarrow & 0
\end{array}$$

(7.18)

from which it is immediate that $\mathrm{K}^{-1}(S^n) = 0$, and that there is an exact sequence

(7.19) $0 \to \mathbb{Z} \to \mathrm{K}^0(S^n) \to \mathbb{Z} \to 0,$

which can be described as follows. The first map corresponds to the using the open embedding of $\mathbb{R}^n$ as an open subset of S^n to map the Bott class $\beta_{\mathbb{R}^n} \in \mathrm{K}^0(\mathbb{R}^n)$ to $\mathrm{K}^0(S^n)$. If $i\colon \mathbb{R}^n \to S^n$ denotes this embedding, then, therefore, the first map $\mathbb{Z} \to \mathrm{K}^0(S^n)$ maps the generator $1 \in \mathbb{Z}$ to $i!(\beta_{\mathbb{R}^n})$, which we will call b.

On the other hand, the quotient map sends the class $[\mathbf{1}]$ of the trivial line bundle over S^n to the generator $1 \in \mathbb{Z}$. It follows that the map $\mathbb{Z} \oplus \mathbb{Z} \to \mathrm{K}^0(S^n)$, sending (n, m) to $nb + m[\mathbf{1}]$, is a group isomorphism.

Now suppose that n is odd. Then plugging what we know into (7.17) gives the sequence

$$0 \longrightarrow K^0(S^n) \longrightarrow \mathbb{Z} \, .$$

(7.20)

$$0 \longleftarrow K^{-1}(S^n) \longleftarrow \mathbb{Z}$$

with vertical maps δ on the left (upward) and δ on the right (downward).

The generator of the integers (the K^0 of a point) in the upper right corner, obviously is in the image of the map from $K^0(S^n)$, since it is the image of the class $[1]$ of the trivial line bundle on S^n. So the quotient map is surjective, and so both connecting maps δ vanish.

It follows that the map $K^{-1}(\mathbb{R}^n) \to K^{-1}(S^n)$, induced by the open embedding of $\mathbb{R}^n$ in S^n, is an isomorphism, and that $K^0(S^n) \cong \mathbb{Z}$ with generator $[1]$.

Remark 7.6.7 In view of the fact that, rather than suspending a space X to define its K^{-1}-group, we can instead use Theorem 7.5.9 and look for unitary-valued maps on the space, it would seem reasonable to look for such a description of $K^{-1}(S^n) \cong \mathbb{Z}$, when n is odd. We will do this once we have a bit of Clifford algebra theory in hand.

K-*theory of real projective spaces*

Real projective spaces $\mathbb{RP}^n$ are the quotient space obtained by identifying antipodal points x and $-x$ of the sphere S^n. The case $n = 1$ is slightly special; in this case the map

(7.21)
$$\mathbb{RP}^1 \to S^1, \quad [z] \mapsto z^2$$

is a homeomorphism of $\mathbb{RP}^1$ with S^1 (but the other projective spaces are not homeomorphic to spheres.)

Proposition 7.6.8 $K^1(\mathbb{RP}^n) = 0$ *and* $K^0(\mathbb{RP}^n) \cong \mathbb{Z}/2 \oplus \mathbb{Z}$ *with generators a certain "Bott element"* $b \in K^0(\mathbb{RP}^n)$, *of order 2, and described in the proof, and free generator* $[1]$, *the class of the trivial line bundle.*

Proof We can think of $\mathbb{RP}^2$ as obtained from the closed disk $\overline{\mathbb{D}}$ by the equivalence relation that identifies antipodal boundary points of the disk. Since no points of the interior $\mathbb{D} \cong \mathbb{R}^2$ of the disk are identified, $\mathbb{D}$ may be considered an open subset of $\mathbb{RP}^2$ with complement $\mathbb{RP}^1$.

This generates a six-term exact sequence

$$K^0(\mathbb{D}) \longrightarrow K^0(\mathbb{RP}^2) \longrightarrow K^0(\mathbb{RP}^1) \, .$$

(7.22)

$$K^1(\mathbb{RP}^1) \longleftarrow K^1(\mathbb{RP}^2) \longleftarrow K^1(\mathbb{D})$$

with vertical maps δ on the left (upward) and δ on the right (downward).

By Bott Periodicity, $K^0(\mathbb{D}) = \mathbb{Z}$ with generator the Bott element $\beta_{\mathbb{D}}$, and $K^1(\mathbb{D}) = 0$. By the remarks above, $K^0(\mathbb{RP}^1)$ and $K^1(\mathbb{RP}^2)$ are both infinite cyclic, with generators $[1] \in K^0(\mathbb{RP}^1)$ the class of the trivial line bundle, and the map (7.21), a unitary-valued map, the generator of $K^1(\mathbb{RP}^1)$. We denote its class $[z^2]$.

Inserting all this information into (7.22) we get the diagram

$$
\begin{array}{ccc}
\mathbb{Z} \longrightarrow K^0(\mathbb{RP}^2) \longrightarrow \mathbb{Z} \,. \\
\delta \big\uparrow \qquad\qquad\qquad\qquad \big\downarrow \\
\mathbb{Z} \longleftarrow K^1(\mathbb{RP}^2) \longleftarrow 0
\end{array}
$$

(7.23)

To compute the vertical map δ on the left, we take the known generator, the class of z^2, of $K^1(\mathbb{RP}^1)$, and we extend it from $\mathbb{RP}^1 \subset \mathbb{RP}^2$ to a map $\mathbb{RP}^2 \to \mathbb{C}$. One can clearly do this in a number of ways, for example, to simply take the extension to be defined by the function z^2, now defined on the whole closed disk.

Then we get a K-theory triple $(\mathbf{1}, \mathbf{1}, z^2)$ consisting of the trivial bundles $\mathbf{1}$ over $\mathbb{D}$ with z^2 the bundle map between them, and we have

$$\delta([z^2]) = [(\mathbf{1}, \mathbf{1}, z^2)].$$

As a triple, this is homotopic to the triple $(\mathbf{1}_2, \mathbf{1}_2, \begin{bmatrix} z & 0 \\ 0 & z \end{bmatrix})$. consisting of the direct sum of $(\mathbf{1}, \mathbf{1}, z)$ with itself. Since the latter represents the Bott element $\beta_{\mathbb{D}}$ for the open disk of radius 1 around 0, we get that

(7.24) $$\delta([z^2]) = 2\beta_{\mathbb{D}}.$$

Hence the vertical left map $\mathbb{Z} \to \mathbb{Z}$ induced by δ is multiplication by 2.

Since the kernel of this map is trivial, and since $K^1(\mathbb{RP}^2)$ injects to the kernel of this map, by the diagram, and since multiplication by 2 is injective, $K^1(\mathbb{RP}^2) = 0$.

The diagram now shows that the map $\mathbb{Z} \to K^0(\mathbb{RP}^2)$ induced by the open disk $\mathbb{D}$ sitting in $\mathbb{RP}^2$, and the corresponding map $K^0(\mathbb{D}) \to K^0(\mathbb{RP}^2)$, vanishes on the even integers. We obtain an injection $\mathbb{Z}/2\mathbb{Z} \to K^0(\mathbb{RP}^2)$. The corresponding element b of order 2 in $K^0(\mathbb{RP}^2)$ is simply obtained by mapping the open disk $\mathbb{D}$ into $\mathbb{RP}^2$, and in this way allowing us to view the Bott element $\beta_{\mathbb{D}} \in K^0(\mathbb{D})$ as a K^0-class, which we are calling b, for $\mathbb{RP}^2$.

Our calculations show thus that

$$2b = 0 \in K^0(\mathbb{RP}^2).$$

Finally, the class $[\mathbf{1}] \in K^0(\mathbb{RP}^2)$ of the trivial line bundle over $\mathbb{RP}^2$ is the other generator for $K^0(\mathbb{RP}^2)$; the two thus generate a copy of the group $\mathbb{Z}/2 \oplus \mathbb{Z}$. To check this, observe that we have produced a group extension

$$0 \to \mathbb{Z}/2 \to K^0(\mathbb{RP}^2) \to \mathbb{Z} \to 0,$$

and any such extension is split, since $\mathbb{Z}$ is free abelian. This makes $K^0(\mathbb{RP}^2) \cong \mathbb{Z}/2 \oplus \mathbb{Z}$ as claimed, and it is left to the reader to check that $[\mathbf{1}]$ can be taken to generate the copy of $\mathbb{Z}$. □

Exercise 7.6.9 Here is an alternative proof that the boundary map in (7.22) satisfies $\delta([z^2]) = 2\beta_{\mathbb{D}}$.

Consider $\mathbb{RP}^2$ as the quotient of the closed disk $\overline{\mathbb{D}}$ by identifying antipodal points. Let $\pi : \overline{\mathbb{D}} \to \mathbb{RP}^2$ be the quotient map. It restricts to a map $S^1 = \partial\overline{\mathbb{D}} \to \mathbb{RP}^1 \subset \mathbb{RP}^2$ and so can be considered as a map of pairs. By naturality of the connecting map, owe obtain a commutative diagram

(7.25)
$$
\begin{array}{ccc}
K^0(\mathbb{D}) & \xleftarrow{\ \delta\ } & K^1(\mathbb{RP}^1) \\
{\scriptstyle \pi^*}\downarrow & & \downarrow{\scriptstyle \pi^*} \\
K^0(\mathbb{D}) & \xleftarrow{\ \delta\ } & K^1(S^1)
\end{array}
$$

Now check that π^* maps the generator $[z^2]$ for $K^1(\mathbb{RP}^1)$ to $[z^2] \in K^1(S^1)$, *i.e.*, to $2[z]$. The lower map δ thus has $\delta \circ \pi^*([z^2]) = 2\beta_{\mathbb{R}^2}$, since we have already computed that $\delta([z]) = \beta_{\mathbb{R}^2}$. We conclude that $\pi^*(\delta([z^2])) = 2\beta_{\mathbb{R}^2}$. On the other hand, the restriction of π to $\mathbb{D}$ is the identity map. Hence $\delta([z^2]) = 2b$ follows immediately, with b the Bott element, considered as an element of $K^0(\mathbb{RP}^2)$.

Exercise 7.6.10 Recall the complex line bundle L over the 2-torus $\mathbb{T}^2$ defined in Exercise 5.1.10 (see also Exercise 5.2.13).

a) Prove that $K^0(\mathbb{T}^2)$ is generated by the classes $[\mathbf{1}]$ of the trivial line bundle, and the class $[L]$.
b) In Example 7.4.17 we discussed the K-theory "germ" around the origin $0 \in \mathbb{R}^2$, in the sense that in any neighborhood U of 0, there is an associated "Bott class" $\beta_U \in K^0(U))$. Since a small enough neighborhood of any point of $\mathbb{R}^2$ can be fit into $\mathbb{T}^2$, we obtain corresponding K-theory germs, which we still denote by β_U, at points of $\mathbb{T}^2$. Show that

$$\beta_U = [L] - [\mathbf{1}] \in K^0(\mathbb{T}^2),$$

that is, $[L] - [\mathbf{1}]$ is the K-theory germ of a point of the torus.

7.7 Spin Geometry and Clifford Algebras

For a comprehensive treatment of Clifford algebras, spin representations, and Dirac operators, containing far more detail than we are able to offer here, we refer the reader to the excellent book [118].

If V is a (finite-dimensional) real or complex vector space, the *tensor algebra* of V is the algebra $T(V) := \bigoplus_{n=0}^{\infty} V \otimes \cdots \otimes V$, where in the k-summand one is taking the tensor product of V with itself, k-times. (Tensor products are discussed in Section 4.5.) The direct sum here just indicates the algebraic direct sum (there is no completion.) The algebra structure is obtained by concatenating tensors

$$(v_1 \otimes \cdots \otimes v_k) \cdot (w_1 \otimes \cdots \otimes w_l) := v_1 \otimes \cdots v_k \otimes w_1 \otimes \cdots \otimes w_l.$$

This is well-defined and linear and gives an (associative) algebra structure. The tensor algebra is also "graded" by the natural numbers, and the multiplication respects the grading.

The tensor algebra construction converts linear maps into algebra homomorphisms in the sense that if $T \colon V \to V$ is a linear map then T extends uniquely to a grading-preserving algebra homomorphism $T(V) \to T(V)$ (proving this is an easy exercise, see also below.)

Definition 7.7.1 The *Clifford algebra* $\mathrm{Cliff}_{\mathbb{R}}(V)$ of a Euclidean vector space V is the unital algebra $T(V)/I$, where

- $T(V) := \bigoplus_{n=0}^{\infty} V^{\otimes n}$ is the tensor algebra of V.
- I is the ideal in $T(V)$ generated by the elements $v \otimes w + w \otimes v + 2\langle v, w \rangle$, for $v, w \in V$.

The *complex* Clifford algebra of V is $\mathrm{Cliff}_{\mathbb{R}}(V) \otimes_{\mathbb{R}} \mathbb{C}$ and will be simply denoted $\mathrm{Cliff}(V)$; it is the one we will be primarily working with.

By the definitions, $\mathrm{Cliff}(V)$ is generated as a unital complex algebra by the elements of V, with the relations

$$(7.26) \qquad\qquad v \cdot w + w \cdot v = -2\langle v, w \rangle, \quad v, w \in V.$$

The *grading* on a Clifford algebra is quite important. A $\mathbb{Z}/2$-*graded algebra* A is an algebra, either real or complex, which, as a vector space, decomposes into a direct sum of two subspaces $A = A^0 \oplus A^1$ in such a way that $A_i A_j \subset A_{i+j \bmod 2}$. The elements of A_0 are "even," those of A_1 "odd." One way of ensuring a $\mathbb{Z}/2$-grading on an algebra is to specify an automorphism $\epsilon \colon A \to A$ such that $\epsilon \circ \epsilon = \mathrm{id}$. Then A decomposes into a direct sum with even part $A^0 := \{a \in A \mid \epsilon(a) = a\}$ and odd part $A^1 := \{a \in A \mid \epsilon(a) = -a\}$. We call ϵ the *grading operator*.

Exercise 7.7.2 Let $T \colon V \to V$ be a linear map, where V is Euclidean (real or complex). Prove that T extends (canonically) to an algebra endomorphism of $\mathrm{Cliff}_{\mathbb{R}}(V)$ (or $\mathrm{Cliff}(V)$.)

Proposition 7.7.3 $\mathrm{Cliff}_{\mathbb{R}}(V)$ *and* $\mathrm{Cliff}(V)$ *are 2^n-dimensional, $\mathbb{Z}/2$-graded algebras over* $\mathbb{R}$ *and* $\mathbb{C}$ *respectively, with in each case the grading operator the automorphism of* $\mathrm{Cliff}_{\mathbb{R}}(V)$ *(or* $\mathrm{Cliff}(V)$*) induced by the map* $\epsilon(v) = -v$.

Proof We consider V as a subset of $\mathrm{Cliff}_{\mathbb{R}}(V)$ for the proof (and more generally), and write $v_1 \cdots v_k$ for the coset of $v_1 \otimes \cdots \otimes v_k$ in $\mathrm{Cliff}_{\mathbb{R}}(V) := T(V)/I$. The key relation is $v \cdot w + w \cdot v = -2\langle v, w \rangle$. Fix an orthonormal basis $e_1, \ldots, e_n$ for V. Then the algebra in $\mathrm{Cliff}_{\mathbb{R}}(V)$ generated by $e_1, \ldots, e_n$ already clearly contains V, and V, by the definitions, certainly generates $\mathrm{Cliff}_{\mathbb{R}}(V)$ as an algebra, so $\mathrm{Cliff}_{\mathbb{R}}(V)$ is generated by $e_1, \ldots, e_n$. Since they are orthonormal, $e_i \cdot e_j = -e_j \cdot e_i$ and $e_i^2 = -1$. It follows immediately that any product $e_{i_1} \cdots e_{i_k}$ in which $k > n$ contains at least two occurrences of some e_i, and moving them adjacent to each other in the monomial, and canceling them, results in a product of smaller length. So it follows that every element of $\mathrm{Cliff}_{\mathbb{R}}(V)$ can be written as a linear combination of monomials $e_{i_1} \cdots e_{i_k}$, with $k \leq n$, and $i_1, \ldots, i_k$ a set of k *distinct* indices. We leave it as an exercise to show that if one takes multi-indices $i_1 < i_2 < \cdots < i_k$ in increasing order, then the corresponding monomials $e_{i_1} \cdots e_{i_k}$ are linearly independent, and so form a basis for $\mathrm{Cliff}_{\mathbb{R}}(V)$. The basis is by definition in 1-1 correspondence with the collection of subsets of $\{1, 2, \ldots, n\}$ so the dimension of $\mathrm{Cliff}_{\mathbb{R}}(V)$ is 2^n.   □

Definition 7.7.4 Let V be a Euclidean vector space and $\mathbf{e} = \{e_1, \ldots, e_n\}$ an orthonormal basis for V The corresponding *volume element* is the element $\Gamma(\mathbf{e}) := e_1 \cdots e_n$.

Exercise 7.7.5 In the above notation, prove that $\Gamma(\mathbf{e})$ only depends on the *orientation class* of the orthonormal basis. That is, prove that $\Gamma(\mathbf{e}) = \pm\Gamma(\mathbf{e}')$ for any $\mathbf{e}, \mathbf{e}'$, and the sign is $+1$ if and only if $\mathbf{e}$ and $\mathbf{e}'$ determine the same *orientation* on V.

Exercise 7.7.6 The *transpose* map on $\mathrm{Cliff}_{\mathbb{R}}(V)$ is the unique $\mathbb{R}$-linear map $x \mapsto x^t$ such that $v^t := -v$ for all $v \in V$, and $(x \cdot y)^t = y^t x^t$ for $x, y \in \mathrm{Cliff}_{\mathbb{R}}(V)$. The *adjoint* on $\mathrm{Cliff}(V)$ is similarly defined: the unique $\mathbb{C}$-conjugate linear map $x \mapsto x^*$ such that $v^* = -v$ and $(vw)^* = w^*v^*$.

The transpose map extends to a *conjugate* linear map on the complex Clifford algebra $\mathrm{Cliff}(V)$ with the same properties, which defines an adjoint on $\mathrm{Cliff}(V)$ making it a *-algebra.

Lemma 7.7.7 *If V is any Euclidean vector space, then $\mathrm{Cliff}(V)$ has the structure of a ($\mathbb{Z}/2$-graded) C*-algebra with adjoint determined by conjugate linearity and the rules $(vw)^* = w^*v^*$ and $v^* = -v$, for $v \in V \subset \mathrm{Cliff}(V)$.*

Proof Let $e_1, \ldots, e_n$ be an orthonormal basis for V. The monomials $e_{i_1} \cdots e_{i_k}$ in $\mathrm{Cliff}(V)$, with $i_1 < i_< \cdots < i_k$ span $\mathrm{Cliff}(V)$ as a vector space, and they are linearly independent. We may therefore define an inner product $\langle \cdot, \cdot \rangle$ on $\mathrm{Cliff}(V)$ making it a ($\mathbb{Z}/2$-graded) Hilbert space for which the monomials $e_{i_1} \cdots e_{i_k}$ constitute an orthonormal basis, and on which $\mathrm{Cliff}(V)$ acts by left multiplication.

We leave it to the reader that this is an injective *-representation. We thus obtain a (finite-dimensional) C*-algebra structure on $\mathrm{Cliff}(V)$.   □

Lemma 7.7.8 *Let V be Euclidean and oriented, let $\Gamma \in \mathrm{Cliff}_{\mathbb{R}}(V)$ the volume element $\Gamma := e_1 \cdots e_n$, for a positively oriented orthonormal basis $e_1, \ldots e_n$.*

a) *If n is even then $\Gamma^2 = (-1)^{\frac{n}{2}}$, $\Gamma^t = (-1)^{\frac{n}{2}}\Gamma$, and Γ anti-commutes with the vectors e_i, $i = 1, \ldots n$.*

b) If n is odd then $\Gamma^2 = (-1)^{\frac{n+1}{2}}$, $\Gamma^t = (-1)^{\frac{n+1}{2}} \Gamma$ and Γ commutes with the e_i's, $i = 1, \ldots, n$.

We leave the proof as an exercise.

Exercise 7.7.9 Prove the following about Clifford algebras.

a) Prove that a bijective linear isometry $L \colon V \to V'$ of Euclidean vector spaces, extends uniquely to an algebra isomorphism $\mathrm{Cliff}_\mathbb{R}(V) \to \mathrm{Cliff}_\mathbb{R}(V')$.
b) Prove that if V is any Euclidean vector space, then the orthogonal group $\mathbf{O}(V)$ acts on $\mathrm{Cliff}_\mathbb{R}(V)$ by $\mathbb{Z}/2$-grading preserving C*-algebra automorphisms, extending the action of $\mathbf{O}(V)$ on $V \subset \mathrm{Cliff}_\mathbb{R}(V)$.

Example 7.7.10 Endowing $\mathbb{R}^n$ with its standard inner product. If $n = 1$, and $e \in \mathbb{R}^1$ is a unit vector, then $\mathrm{Cliff}_\mathbb{R}(\mathbb{R}^1)$ is spanned over $\mathbb{R}$ by 1 and e and the map $a + be \mapsto a + bi$ defines an isomorphism of $\mathbb{Z}/2$-graded real algebras $\mathrm{Cliff}_\mathbb{R}(\mathbb{R}^1) \cong \mathbb{C}$, where $\mathbb{C}$ (understood as a an algebra over $\mathbb{R}$) is graded by complex conjugation, $\epsilon(z) := \bar{z}$.

For $n = 2$ let $e_1 = (1, 0)$, $e_2 = (0, 1)$, say. Then $\mathrm{Cliff}_\mathbb{R}(\mathbb{R}^2)$ is linearly spanned over $\mathbb{R}$ by 1, e_1, e_2 and $e_1 e_2$, and is isomorphic as a real algebra to the algebra of quaternions $\mathbb{H}$, by the formula $\varphi(a \cdot 1 + b e_1 + c e_2 + d e_1 e_2) := a + bi + cj + dk$. We can realize $\mathrm{Cliff}(\mathbb{R}^2)$ as an algebra of matrices by mapping e_1, e_2 and $e_1 e_2$ to the matrices

$$c(e_1) := \begin{bmatrix} 0 & -1 \\ 1 & 0 \end{bmatrix}, \quad c(e_2) := \begin{bmatrix} i & 0 \\ 0 & -i \end{bmatrix}, \quad c(e_1 e_2) := \begin{bmatrix} 0 & i \\ i & 0 \end{bmatrix},$$

respectively.

Exercise 7.7.11 Check that the matrices $c(e_1)$ and $c(e_2)$ anti-commute, that $c(e_i)^* = -c(e_i))$ and $c(e_i)^2 = -1$, $i = 1, 2$. Produce an explicit formula for the image of $a \cdot 1 + b e_1 + c e_2 + d e_1 e_2 \in \mathrm{Cliff}(\mathbb{R}^2)$.

Exercise 7.7.12 Prove that the *even* part $\mathrm{Cliff}_\mathbb{R}(\mathbb{R}^n)^0$ of $\mathrm{Cliff}_\mathbb{R}(\mathbb{R}^n)$, is isomorphic to $\mathrm{Cliff}_\mathbb{R}(\mathbb{R}^{n-1})$. (*Hint.* $\mathrm{Cliff}(\mathbb{R}^{n-1})$ is generated by $e_1, \ldots, e_{n-1}$ with relations $e_i e_j + e_j e_i = -2\delta_{ij}$. Map generators to $\mathrm{Cliff}_\mathbb{R}(\mathbb{R}^n)$ by $e_i \mapsto e_n e_i$, and check that the relations are preserved.)

We define two subgroups of the group of invertibles in the algebra $\mathrm{Cliff}_\mathbb{R}(V)$ as follows.

$$\mathbf{Pin}(V) := \{v_1 \cdots v_k \mid v_1, \ldots, v_k \text{ unit vectors in } V\}$$

$$\mathbf{Spin}(V) := \{v_1 \cdots v_k \mid v_1, \ldots, v_k \text{ unit vectors in } V, \ k \text{ even.}\}$$

Clearly both are subgroups of the group of invertibles of $\mathrm{Cliff}_\mathbb{R}(V)$.

Exercise 7.7.13 Prove there is a group homomorphism $\mathbf{Pin}(V) \to \{\pm 1\} \subset \mathbb{T}$ mapping $v_1 \cdots v_k$ to $(-1)^k$. Deduce that $\mathbf{Spin}(V)$ is a subgroup of $\mathbf{Pin}(V)$ of index 2.

Exercise 7.7.14 Prove that $\mathbf{Spin}(V)$ is a compact, connected group. (*Hint.* To prove it is path connected, move a vector v along a path of unit vectors in V from v_n to $-v_{n-1}$, we obtain a path $v_1 \cdots v_{n-1} \cdot v$ in $\mathbf{Spin}(V)$ from $v_1 \cdots v_n$ to $v_1 \cdots v_{n-2}$. Now continue this process.)

Exercise 7.7.15 Prove that $\mathbf{Spin}_3$ is diffeomorphic to the 3-sphere S^3. (*Hint.* The elements of $\mathbf{Spin}_3$ may be uniquely written $a + be_2e_1 + ce_3e_2 + de_3e_1$ where $a^2 + b^2 + c^2 + d^2 = 1$.)

We now describe the *spin covering* $\rho \colon \mathbf{Spin}(V) \to \mathbf{SO}(V)$ of the special orthogonal group, V Euclidean.

A short calculation shows that for V Euclidean, and $v \in V$ a *unit* vector, and if $w \in V$, then

$$(7.27) \qquad v \cdot w \cdot v = v(-2\langle v, w\rangle - v \cdot w) = w - 2\langle v, w\rangle v = \mathrm{refl}_{v^\perp}(w),$$

where $\mathrm{refl}_{v^\perp}(w)$ is the orthogonal reflection of w through the hyperplane $v^\perp$ determined by v—thus $\mathrm{refl}_{v^\perp}$ is an orthogonal transformation of V. Inspired by this calculation, we define a group homomorphism

(7.28)

$$\rho \colon \mathbf{Pin}(V) \to \mathbf{O}(V), \quad \rho(\pm 1) = 1, \quad \rho(v_1 \cdots v_k) := \mathrm{refl}_{v_1^\perp} \cdots \mathrm{refl}_{v_n^\perp} \in \mathbf{O}(V).$$

which we will show below is well defined. Of special interest to us is the restriction of ρ to a homomorphism $\mathbf{Spin}(V) \to \mathbf{SO}(V)$.

Remark 7.7.16 If $\theta \in \mathbb{R}$ and v and w are orthogonal unit vectors in V, then $\cos\theta\, v + \sin\theta\, w$ is also a unit vector in V, and hence $(\cos\theta\, v + \sin\theta\, w) \cdot (-w) \in \mathbf{Spin}(V)$. A short calculation shows that this equals $\cos\theta + \sin\theta\, v \cdot w$, and that $\rho(\cos\theta + \sin\theta\, v \cdot w)$ is rotation through an angle of 2θ in the plane spanned by v and w (in the direction from v toward w.)

The main difficulty is that we need to prove that (7.28) is well defined, since although every element of $\mathbf{Pin}(V)$ is a product $v_1 \cdots v_k$ in some way, there may be different representations of it as such.

Lemma 7.7.17 *The expression (7.28) is well defined, and defines a surjective, 2-to-1 group homomorphism $\mathbf{Pin}(V) \to \mathbf{O}(V)$, and, its restriction to $\mathbf{Spin}(V)$ is a 2-to-1 group homomorphism onto $\mathbf{SO}(V)$.*

Since $\mathbf{Spin}(V)$ is connected, ρ is a nontrivial double cover of $\mathbf{SO}(V)$.

Proof If $x \in \mathrm{Cliff}_{\mathbb{R}}(V)$ is an invertible in the Clifford algebra, then $\mathrm{Ad}_\epsilon(x)y := xy\epsilon(x)^{-1}$ defines an algebra homomorphism $\mathrm{Cliff}_{\mathbb{R}}(V) \to \mathrm{Cliff}_{\mathbb{R}}(V)$, where ϵ is the grading automorphism. A further easy check shows that $\mathrm{Ad}_\epsilon(xy) =$

$\mathrm{Ad}_\epsilon(x)\mathrm{Ad}_\epsilon(y)$. Furthermore, if $x = v \in V \subset \mathrm{Cliff}_\mathbb{R}(V)$ then since $\epsilon(v) = -v$, and by the calculation (7.27) above, we get that $\mathrm{Ad}_\epsilon(v) = \mathrm{refl}_{v^\perp}$. The result follows: we have shown that (7.28) actually represents the value $\mathrm{Ad}_\epsilon(v_1 \cdots v_k)$, and hence only depends on $v_1 \cdots v_k$ as an element of the Clifford algebra, and not on its representation.

Surjectivity of $\rho \colon \mathbf{Pin}(V) \to \mathbf{O}(V)$ now follows from the classical result of Cartan and Dieudonné that the orthogonal group is generated by reflections. $\square$

Exercise 7.7.18 Prove that $\mathbf{Spin}_2 \cong \mathbb{T}$, and that the spin covering $\rho \colon \mathbf{Spin}_2 \to \mathbf{SO}(2, \mathbb{R}) \cong \mathbb{T}$ corresponds to the map $z \mapsto z^2$ on the circle.

Example 7.7.19 This example examines a bit further the geometry of, especially, $\mathbf{Spin}_3$, and the double covering $\mathbf{Spin}_3 \to \mathbf{SO}(3, \mathbb{R})$.

Firstly, $\mathbf{Spin}_3$ is generated as a group by the elements $v \cdot w$, where $v, w \in \mathbb{R}^3$ are unit vectors. Let e_1, e_2, e_3 be the standard orthonormal basis for $\mathbb{R}^3$.

If $v, v' \in \mathbb{R}^3$ are unit vectors, write $v = ae_1 + be_2 + ce_3$ where $a^2 + b^2 + c^2 = 1$, $v' = a'e_1 + b'e_2 + c'e_3$ where $(a')^2 + (b')^2 + (c')^2 = 1$.

Exercise 7.7.20 Multiply out $v \cdot v'$ and expand in the basis $1, e_1e_2, e_1e_3, e_2e_3$. Check that the first coordinate is the dot product $\langle v, v'\rangle$, or inner product of v and v', and that the next 3 coordinates are those of the crossed product $v \times v'$.

It follows from elementary properties of the dot and crossed products, that the map $\phi \colon \mathbf{Spin}_3 \to \mathbb{R}^4$, $\phi(v \cdot w) := (-\langle v, w\rangle, v \times w)$ takes values in unit vectors in $\mathbb{R}^4$, and, in fact, can be easily checked to define a diffeomorphism $\mathbf{Spin}_3 \cong S^3$.

Let us see what the double covering $\rho \colon \mathbf{Spin}_3 \to \mathbf{SO}(3, \mathbb{R})$ looks like in this picture.

Exercise 7.7.21 Three-dimensional real projective space $\mathbb{RP}^3$ is by definition the space $S^3/\sim$ obtained by identifying antipodal points of the 3-sphere. Since every equivalence class in $\mathbb{RP}^3$ is represented by a unit vector $v \in \mathbb{R}^4$ with z-coordinate ≥ 0, we may also consider $\mathbb{RP}^3$ as obtained by taking the closed upper hemisphere in S^3 and identifying antipodal points of its boundary. The map $(x, y, z) \mapsto (x, y, z, \sqrt{1 - x^2 - y^2 - z^2})$ identifies the upper hemisphere in S^3 with the closed 3-ball B^3. Thus, we can consider $\mathbb{RP}^3$ as the space obtained from the closed 3-ball B^3 by identified antipodal points on the boundary of the ball.

If $v \in B^3$ is a nonzero vector, with $t = \|v\|$, let $\alpha(v) \in \mathbf{SO}(3, \mathbb{R})$ be rotation in the axis determined by v, by an angle of πt, in the sense determined by the right hand rule. Since $\lim_{v \to 0} \alpha(v) = \mathrm{id}_{\mathbf{SO}(3,\mathbb{R})}$, as is easy to check, α extends to a continuous map $B^3 \to \mathbf{SO}(3, \mathbb{R})$. Prove this map is a homeomorphism. Deduce that $\mathbf{SO}(3, \mathbb{R}) \cong \mathbb{RP}^3$.

By the preceding exercise, $\mathbf{SO}(3, \mathbb{R})$ may be identified with $S^3/\sim$, where $\sim$ identifies antipodal points of the sphere. The diagram

$$\mathbf{Spin}_3 \xrightarrow{\ \rho\ } \mathbf{SO}(3,\mathbb{R})$$

$$\downarrow \cong \qquad\qquad \downarrow \cong$$

$$S^3 \xrightarrow{\ \pi\ } \mathbb{RP}^3$$

commutes where π is the quotient map.

Remark 7.7.22 Another interesting fact about the double cover $\rho \colon \mathbf{Spin}_3 \to$ $\mathbf{SO}(3,\mathbb{R})$, is that if it is followed by an orbit map $\mathbf{SO}(3,\mathbb{R}) \to S^2$, $A \mapsto Ae$, e a chosen fixed unit vector in $S^2 \subset \mathbb{R}^3$, then the resulting map $\rho_e \colon \mathbf{Spin}_3 \cong S^3 \to S^2$ gives a famous fibration,

$$S^1 \longrightarrow S^3$$
$$\downarrow \rho_e$$
$$S^2$$

called the *Hopf fibration*. It determines a nontrivial element of $\pi_3(S^2)$.

Remark 7.7.23 It can be checked that $\mathbf{Spin}(V)$ is a (compact, connected) *Lie* group in a natural way, and its Lie algebra $\mathfrak{spin}(V)$ can be identified with the Lie subalgebra of $\mathrm{Cliff}_{\mathbb{R}}(V)$ spanned by the $e_i \cdot e_j$, with $i \neq j$; the $e_i \cdot e_j$ with $i < j$ form a basis, giving a $\frac{n(n+1)}{2}$-dimensional space (and it is closed under commutators, as the reader can easily check.)

The exponential map $\exp \colon \mathfrak{spin}(V) \to \mathbf{Spin}(V)$ in this case has the form of the convergent series $\exp(X) = \sum_{n=0}^{\infty} \frac{X^n}{n!}$, where X is supposed an element of $\mathfrak{spin}(V) \subset \mathrm{Cliff}_{\mathbb{R}}(V)$ and X^n has its obvious meaning of X multiplied by itself n times. The series converges in the topology of $\mathrm{Cliff}_{\mathbb{R}}(V)$ (which as a linear space is just a finite-dimensional real vector space) to an element of $\mathrm{Cliff}_{\mathbb{R}}(V)$, and to an element of $\mathbf{Spin}(V)$ if $X \in \mathfrak{spin}(V)$.

It is an easy exercise (plug in $t e_i \cdot e_j$ into the power series for exp, use the basic properties of the Clifford multiplication) to see that $\exp(t e_i \cdot e_j) = \cos t + \sin t\, e_i \cdot e_j$. These elements form a closed subgroup of $\mathbf{Spin}(V)$ isomorphic to the circle $\mathbb{T}$. Thus, exp maps the line through $e_i \cdot e_j$ to this subgroup (in a $\mathbb{Z}$-to-1 fashion; it is a covering map.

To see the connection to the orthogonal group and the spin covering $\rho \colon \mathbf{Spin}_n \to$ $\mathbf{SO}(n,\mathbb{R})$, let E_{ij} be the n-by-n matrix with 1 in the i, jth coordinate, -1 in the j, ith coordinate, and zeros elsewhere. Then E_{ij} is an element of the Lie algebra $\mathfrak{so}(n,\mathbb{R})$ of $\mathbf{SO}(n,\mathbb{R})$ and for any $t \in \mathbb{R}$, $\exp(t E_{ij}) = \cos t + \sin t\, E_{ij}$ is rotation by t in the plane spanned by e_i and e_j, in the direction from e_i to e_j.

In particular, the identification of Lie algebras $\mathfrak{spin}(\mathbb{R}^n)$ and $\mathfrak{so}(n,\mathbb{R})$, maps the elements $e_i \cdot e_j \in \mathfrak{spin}(\mathbb{R}^n) \subset \mathrm{Cliff}_{\mathbb{R}}(\mathbb{R}^n)$ to the matrices E_{ij} of $\mathfrak{so}(n,\mathbb{R})$.

7.8   Representation Theory of Clifford Algebras

Let V be a Euclidean vector space, that is, a finite-dimensional real Hilbert space. We will usually refer to an C*-algebra representation $\pi : \mathrm{Cliff}(V) \to \mathrm{End}(W)$ on a (generally finite-dimensional)(Hilbert space W as a $\mathrm{Cliff}(V)$-*module*. Such representations we will always assume to be nondegenerate (*i.e.*, unital; *i.e.*, $1 \in \mathrm{Cliff}(V)$ acts by the identity map on W.)

We refer the reader to the book [118] for a proof of the following result.

Theorem 7.8.1 *For any even-dimensional Euclidean vector space V, there is, up to isomorphism, exactly one irreducible $\mathrm{Cliff}(V)$ module S of dimension $2^{\frac{\dim V}{2}}$. Furthermore, an orientation on V determines a $\mathbb{Z}/2$-grading on S, for which Clifford multiplication by vectors are odd operators.*

Furthermore, any $\mathrm{Cliff}(V)$-module of dimension $2^{\frac{\dim V}{2}}$ is irreducible.

If $\dim(V)$ is odd, there are, up to isomorphism, exactly two irreducible $\mathrm{Cliff}(V)$-modules of dimension $2^{\frac{\dim V - 1}{2}}$, and any $\mathrm{Cliff}(V)$-module of dimension $2^{\frac{\dim V - 1}{2}}$, is necessarily irreducible.

The isomorphism in the statement refers to the obvious notion of isomorphism of $\mathrm{Cliff}(V)$-modules.

In the case V is even-dimensional, if $c : \mathrm{Cliff}(V) \to \mathrm{End}(S)$ is an irreducible $\mathrm{Cliff}(V)$-module, and if $e_1, \ldots, e_n$ is an orthonormal basis of V, $\Gamma := e_1 \cdots e_n$ the volume (see Lemma 7.7.8) then the operator

$$\epsilon_S := i^{\frac{n}{2}} c(\Gamma)$$

satisfies $\epsilon_S^* = \epsilon_S$, $\epsilon^2 = 1$, and $\epsilon_S c(v) = -c(v)\epsilon_S$ for all $v \in V$, and so determines a $\mathbb{Z}/2$-grading on S which is a $\mathbb{Z}/2$-graded Clifford module.

Note that, conversely, if we assume from the beginning that S carries a $\mathbb{Z}/2$-graded Clifford module structure, with grading operator ϵ, then since $\epsilon_S \epsilon$ commutes with $\mathrm{Cliff}(V)$, it is a multiple of the identity, by irreducibility, and in fact must be ± 1.

Remark 7.8.2 One can always realize a given $\mathrm{Cliff}(V)$-module, up to isomorphism, by one in which the Clifford multiplication action by vectors $v \in V$ is skew-adjoint. We sometimes call these *Hermitian* $\mathrm{Cliff}(V)$-modules, if the context absolutely demands it.

Indeed, fix any Hermitian metric on V, and then average it over the compact group **Pin** $\subset \mathrm{Cliff}(V)$. This produces a metric with which **Pin**(V) acts by unitary maps. Since also $v^2 = -1$ as an endomorphism of W, we get $v^2 = -1 = -vv^*$ (since v is a unitary) and, canceling the v's, that $v^* = -v$.

Exercise 7.8.3 Prove that the averaging of the previous Remark could have been simply done over the *finite* subgroup of the invertibles in $\mathrm{Cliff}_{\mathbb{R}}(V)$ generated by a fixed orthonormal basis $e_1, \ldots, e_n$ for V. (Check that these vectors really do generate a finite subgroup.)

Example 7.8.4 (*The unique $\mathbb{Z}/2$-graded irreducible* $\text{Cliff}(\mathbb{R}^{2n})$-*module.*

If V is a Hermitian vector space, $\Lambda^*(V)$ its exterior algebra, a $\mathbb{Z}/2$-graded algebra, then $\Lambda^*(V)$ inherits an inner product from V by the formula

$$\langle v_1 \wedge \cdots \wedge v_k, v_1' \wedge \cdots \wedge v_k \rangle := \det(\langle v_i, v_j' \rangle).$$

Choose $v \in V$. The linea operator $\lambda_v: \Lambda^*(V) \to \Lambda^*(V)$ of exterior multiplication by v is odd with respect to the grading. The adjoint of this operator is discussed in the exercise below.

Exercise 7.8.5 The Hilbert space adjoint λ_v^* of the operator of exterior multiplication by v is given by the formula

$$(7.29) \quad \lambda_v^*(w_1 \wedge \cdots \wedge w_k) = \sum_{i=1}^{k} (-1)^{i+1} \langle w_i, v \rangle \, w_1 \wedge \cdots \wedge \widehat{w_i} \wedge \cdots \wedge w_k,$$

where the term with the hat is omitted.

Interior multiplication occurs in the famous identity of Cartan:

$$(7.30) \quad d\iota_X + \iota_X d = \mathcal{L}_X,$$

where X is a vector field on a manifold, ι_X and d the operators on differential forms on the manifold determined by (fiberwise) interior multiplication by X (a vector field is, fiberwise, an element of the dual of the cotangent space) and $\mathcal{L}_X$ is the Lie derivative.

Remark 7.8.6 The adjoint of exterior multiplication is closely related to *interior multiplication*. If $\varphi \in V^*$, then $\iota_\varphi: \Lambda^*(V) \to \Lambda^*(V)$ is the unique degree -1 operator such that

$$\iota_\varphi(w_1 \wedge \cdots \wedge w_k) = \sum_{i=1}^{k} (-1)^i \varphi(w_i) \, w_1 \wedge \cdots \wedge \hat{w}_i \wedge \cdots \wedge w_k.$$

Now for $v \in V$ let

$$c(v) := \lambda_v + \lambda_v^*.$$

Clearly $c(v) = c(v)^*$, and moreover, if $w_1 \ldots w_k \in V$, then

(7.31)

$$\lambda_v^* \lambda_v \, (w_1 \wedge \cdots \wedge w_k)$$

$$= \langle v, v \rangle \, w_1 \wedge \cdots \wedge w_k - \sum_{i=1}^{k} (-1)^{i+1} \langle w_i, v \rangle \, v \wedge w_1 \wedge \cdots \wedge \hat{w}_i \wedge \cdots \wedge w_k$$

$$= \|v\|^2 - \lambda_v \lambda_v^* \, (w_1 \wedge \cdots \wedge w_k)$$

giving that

$$(\lambda_v + \lambda_v^*)^2 = \|v\|^2.$$

Let

$$c_V : V \to \mathrm{End}(\Lambda^* V), \quad c_V(v) := i \, (\lambda_v + i_v),$$

then $c_V(v)^2 = -\|v\|^2$ and $c_V(v)^* = -c_V(v)$.

In particular, the above discussion applies to $V = \mathbb{C}^n$ with its standard Hermitian structure. Define now

$$c \colon \mathbb{R}^{2n} = \mathbb{R}^n \times \mathbb{R}^n \to \mathrm{End}(\Lambda^* V), \quad c(u, v) := c_{\mathbb{C}^n}(u + iv).$$

Then from the above discussion

$$c(u, v)^2 = -\|u + iv\|^2 = -\|u\|^2 - \|v\|^2 = -\|(u, v)\|^2.$$

Since $\dim \Lambda^*(\mathbb{C}^n) = 2^n$, this construction produces the unique irreducible representation of $\mathrm{Cliff}(\mathbb{R}^{2n})$.

Exercise 7.8.7 Show that if $n = 2$, then up to isomorphism, the $\mathrm{Cliff}(\mathbb{R}^2)$-module constructed above is the $\mathbb{Z}/2$-graded vector space $\mathbb{C}^2$, with grading operator $\begin{bmatrix} 1 & 0 \\ 0 & -1 \end{bmatrix}$, and Clifford representation in which a vector $(x, y) \in \mathbb{R}^2$ acts by

$$c(x, y) := \begin{bmatrix} 0 & ix - y \\ ix + y & 0 \end{bmatrix}.$$

Remark 7.8.8 How do we construct the irreducible representations of $\mathrm{Cliff}(V)$ when V is odd-dimensional? Then the orthogonal sum $V \oplus \mathbb{R}$ is even-dimensional. Let $e_1, \ldots e_n$ be an orthonormal basis for V, extend it by adding $e_{n+1} = (0, 1) \in V \oplus \mathbb{R}$. Let S be an irreducible $\mathrm{Cliff}(V \oplus \mathbb{R})$-module, with grading operator ϵ_S. Thus S is $2^{\frac{n+1}{2}}$-dimensional.

Now, since n is odd, the vectors $e_1, \ldots, e_n$ commute with $\Gamma := e_1 \cdots e_n \in \mathrm{Cliff}(V) \subset \mathrm{Cliff}(V \oplus \mathbb{R})$ Therefore, the element $\gamma := i^{\frac{n+1}{2}} c(\Gamma)$ satisfies $\gamma^* = \gamma$, $\gamma^2 = 1$, and γ commutes with $\mathrm{Cliff}(V)$. Hence $\mathrm{Cliff}(V)$ leaves each of its ± 1-eigenspaces invariant. These subspaces $S_\pm$ realize the two irreducible $\mathrm{Cliff}(V)$-modules.

For example, let $n = 1$, and the Cliff($\mathbb{R}^2$)-module $\mathbb{C}^2$ constructed in Exercise 7.8.7 with

$$c(x, y) = i \begin{bmatrix} 0 & x + iy \\ x - iy & 0 \end{bmatrix}.$$

The operator $\gamma' := i\, c(1, 0)$ is multiplication by the matrix $\begin{bmatrix} 0 & 1 \\ 1 & 0 \end{bmatrix}$ whose ± 1 eigenspaces are the subspaces $x = y$, and $x = -y$ in $\mathbb{C}^2$, and on these two one-dimensional subspaces, Clifford multiplication by $x \in \mathbb{R}$ is multiplication by the (real) scalar $x \in \mathbb{C}$, and respectively, the (real) scalar $-x$.

So we get the two irreducible, one-dimensional complex representations of Cliff($\mathbb{R}$).

The following argument shows that when dim(V) is even-dimensional, then an orientation on V supplies any irreducible Cliff(V)-module with a natural $\mathbb{Z}/2$-grading.

Proposition 7.8.9 *Let V be even-dimensional and oriented. Let $\Gamma \in$ Cliff(V) be the volume element associated to the orientation. Then $\Gamma' := i^{\frac{n}{2}} \cdot \Gamma$ is self-adjoint, $(\Gamma')^2 = 1$, and if S is an irreducible Cliff(V)-module, then the ± 1-eigenspaces of Γ' induce a $\mathbb{Z}/2$-grading on S with respect to which Clifford multiplication by a vector in V is an odd operator.*

If $n = $ dim(V) is odd, V oriented, Γ the volume element, then Γ acts by ± 1 in any irreducible Cliff(V)-module, and the sign depending on which of the two irreducible representations it is.

All of these statements follow from Lemma 7.7.8. Note that since v graded commutes with Γ, and hence Γ', for any vector $v \in V$, it follows that $c(v)$ interchanges the ± 1 eigenspaces of Γ'.

Exercise 7.8.10 Let V be odd-dimensional and suppose that $c:$ Cliff(V) $\rightarrow$ End(S) is an irreducible Cliff(V) module. Recall that there are two such modules, up to isomorphism. Show that letting $c'(v) := c(-v)$ we obtain the other one.

Exercise 7.8.11 Let A and B be $\mathbb{Z}/2$-graded *-algebras (over $\mathbb{C}$.) Their *graded tensor product* $A \hat{\otimes} B$ is, as a complex vector space, the same as their ordinary tensor product $A \otimes B$, but with the multiplication on homogeneous (*i.e.*, either even or odd) elements

$$(a \hat{\otimes} b) \cdot (c \hat{\otimes} d) := (-1)^{\partial b \partial c}\, ac \hat{\otimes} bd, \quad (a \hat{\otimes} b)^* := (-1)^{\partial a \partial b}\, a^* \hat{\otimes} b^*,$$

and grading having even part $(A \hat{\otimes} B)^0 := A^0 \hat{\otimes} B^0 \oplus A^1 \hat{\otimes} B^1$, and odd part $A^0 \hat{\otimes} B^1 \oplus A^1 \hat{\otimes} B^0$.

Prove that in this notation, there is a natural isomorphism

$$\text{Cliff}(V) \hat{\otimes} \text{Cliff}(W) \cong \text{Cliff}(V \oplus W)$$

if V and W are Euclidean vector space. (and $V \otimes W$ is the orthogonal direct sum.)

Exercise 7.8.12 Prove that any $\mathbb{Z}/2$-graded C*-algebra (*i.e.*, a C*-algebra, which is also $\mathbb{Z}/2$-graded as a *-algebra), is isomorphic to a closed C*-subalgebra of $\mathbb{B}(H)$ for some *graded* Hilbert space $H = H^0 \oplus H^1$, in such a way that A^0 acts by grading-preserving (*i.e.*, "even") operators on H, and A^1 acts by grading-reversing ('odd') operators. (*Hint.* Extend a faithful state for A^0 to A and apply the GNS representation to get a faithful $\mathbb{Z}/2$-graded representation.)

Exercise 7.8.13 Adapt the definition of the spatial tensor product of C*-algebras (Definition 4.5.16) to work for $\mathbb{Z}/2$-graded C*-algebras, using Exercise 7.8.12.

Exercise 7.8.14 This exercise introduces a bit more material on $\mathbb{Z}/2$-graded spaces and algebras.

A $\mathbb{Z}/2$-graded Hilbert space is of course a Hilbert space with a $\mathbb{Z}/2$-grading. Suppose that H_1 and H_2 are two $\mathbb{Z}/2$-graded Hilbert spaces. Their graded tensor product $H_1 \hat{\otimes} H_2$ is the same as $H_1 \otimes H_2$ as a Hilbert space, but $\mathbb{Z}/2$-graded with grading operator $\epsilon_1 \otimes \epsilon_2$.

Suppose that $T \in \mathbb{B}(H_1)$ and $S \in \mathbb{B}(H_2)$. Assume that T, S, v_1 and v_2 are all homogeneous elements for the gradings. Define

$$(T \hat{\otimes} S)(v_1 \hat{\otimes} v_2) := (-1)^{\partial S \, \partial v_1} \, T v_1 \hat{\otimes} S v_2.$$

Show that the above definition extends to a $\mathbb{Z}/2$-graded *-representation of the $\mathbb{Z}/2$-graded algebra $\mathbb{B}(H_1) \hat{\otimes} \mathbb{B}(H_2)$ on $H_1 \hat{\otimes} H_2$.

Lemma 7.8.15 *Assume that V and W are even-dimensional Euclidean spaces,*

$$c_V : \mathrm{Cliff}(V) \to \mathrm{End}(S_V), \quad c_W : \mathrm{Cliff}(W) \to \mathrm{End}(S_W)$$

are two $\mathbb{Z}/2$-graded irreducible Clifford modules. Set $S_{V \oplus W} := S_V \hat{\otimes} S_W$, the graded tensor product of graded vector spaces. Let $c_{V \oplus W} : V \oplus W \to \mathrm{End}(S_{V \oplus W})$,

$$c_{V \oplus W}(v, w) := c(v) \hat{\otimes} 1 + 1 \hat{\otimes} c(w).$$

Then $c_{V \oplus W} \, c_V \oplus W : \mathrm{Cliff}(V \oplus W) \to \mathrm{End}(S_{V \oplus W})$ is an irreducible $\mathbb{Z}/2$-graded representation of $\mathrm{Cliff}(V \oplus W)$.

Proof It is easily checked that $c_V(v) \hat{\otimes} 1 + 1 \hat{\otimes} c_W(w)$ is an odd operator on the $\mathbb{Z}/2$-graded $S_V \hat{\otimes} S_W$. On the other hand, by definition of the multiplication in $\mathbb{B}(S_V) \hat{\otimes} \mathbb{B}(S_W)$, we have

$$\left(c_V(v) \hat{\otimes} 11 + 1 \hat{\otimes} c_W(w) \right)^2$$

$$= c_V(v)^2 \hat{\otimes} 1 - c_V(v) \hat{\otimes} c_W(w) + c_V(v) \hat{\otimes} c_W(w) + 1 \hat{\otimes} c_W(w)^2$$

$$(7.32) \qquad = -\|v\|^2 - \|w\|^2 = -\|(v, w)\|^2$$

as required.

Since $\dim S_V \hat{\otimes} S_W = \dim S_V \dim S_W = 2^{\frac{\dim V + \dim W}{2}} = 2^{\frac{\dim(V \oplus W)}{2}}$, this is an irreducible, $\mathbb{Z}/2$-graded representation, as required. $\qquad\square$

When V is even-dimensional and W is odd-dimensional, we proceed as follows.

Lemma 7.8.16 *Suppose that V is even-dimensional, W is odd-dimensional,*

$$c_V : \mathrm{Cliff}(V) \to \mathrm{End}(S_V)$$

a $\mathbb{Z}/2$-graded, irreducible representation of $\mathrm{Cliff}(V)$, and

$$c_W : \mathrm{Cliff}(W) \to \mathrm{End}(S_W)$$

an irreducible representation.
With respect to the decomposition $S_V \otimes S_W = S_V^0 \otimes S_W \oplus S_V^1 \otimes S_W$, let

$$c_{V \oplus W}(v, w) := \begin{bmatrix} 0 & c_V(v) \otimes 1 + 1 \otimes i c_W(w) \\ c_V(v) \otimes 1 - 1 \otimes i c_W(w) & 0 \end{bmatrix}$$

Then $c_{V \oplus W} \, c_V \oplus W : \mathrm{Cliff}(V \oplus W) \to \mathrm{End}(S_{V \oplus W})$ is an irreducible representation of $\mathrm{Cliff}(V \oplus W)$.

Finally, we deal with the case when V and W are both odd-dimensional:

Lemma 7.8.17 *Suppose that V and W are odd-dimensional Euclidean vector spaces, and that $c_V : \mathrm{Cliff}(V) \to \mathrm{End}(S_V)$ is an irreducible representation of $\mathrm{Cliff}(V)$, and $c_W : \mathrm{Cliff}(W) \to \mathrm{End}(S_W)$ is an irreducible representation of $\mathrm{Cliff}(W)$.*
Set $S_{V \oplus W} := S_V \otimes S_W \oplus S_V \otimes S_W$, endow $S_{V \oplus W}$ with the $\mathbb{Z}/2$-grading with even part the first factor, odd part the second factor. Let

$$c_{V \oplus W}(v, w) := \begin{bmatrix} 0 & c_V(v) \otimes 1 + 1 \otimes i c_W(w) \\ c_V(v) \otimes 1 - 1 \otimes i c_W(w) & 0 \end{bmatrix}.$$

Then $c_{V \oplus W}$ extends to an irreducible, $\mathbb{Z}/2$-graded representation $c_{V \oplus W} : \mathrm{Cliff}(V \oplus W) \to \mathrm{End}(S_V \otimes S_W)$.

The two-out-of-three Lemma
Above we proved that a representation of $\mathrm{Cliff}(V)$ and one of $\mathrm{Cliff}(W)$ induced one of $\mathrm{Cliff}(V \oplus W)$. A converse holds in the form of an important "2-out-of-3 Lemma" for K-orientations:

Lemma 7.8.18 *Assume that V and W are even-dimensional Euclidean vector spaces, $V \oplus W$ their orthogonal direct sum.*
Suppose that $c_V : \mathrm{Cliff}(V) \to \mathrm{End}(S_V)$ is a $\mathbb{Z}/2$-graded irreducible representation of $\mathrm{Cliff}(V)$, and $c_{V \oplus W} : \mathrm{Cliff}(V \oplus W) \to \mathrm{End}(S_{V \oplus W})$ is a $\mathbb{Z}/2$-graded irreducible representation of $\mathrm{Cliff}(V \oplus W)$. Let

$$S_W := \mathrm{Hom}_{\mathrm{Cliff}(V)}(S_V, S_{V\oplus W}) := \{T : S_V \to S_{V\oplus W} \mid c_V(v)T(s)$$

$$= T\big(c_V(v)s\big) \; \forall s \in S_V\},$$

graded into even and odd operators. Define, for $w \in W$, *and* $T \in S_{V\oplus W}$ *homogeneous,*

(7.33) $$c_W(w)T := (-1)^{\partial T} \, c_{V\oplus W}(w) \circ T \circ \epsilon_V$$

Then this extends to a $\mathbb{Z}/2$-*graded, irreducible representation of* $\mathrm{Cliff}(W)$.

Proof We leave it to the reader to verify that (7.37) really is well defined (that the indicated action of $c_W(w)$ maps the space of invariant operators to itself), and extends to a $\mathbb{Z}/2$-graded Clifford representation.

Since there a unique irreducible representation of $\mathrm{Cliff}(V)$ up to equivalence, the evaluation map

(7.34) $$\mathrm{Hom}_{\mathrm{Cliff}(V)}(S_V, S_{V\oplus W}) \otimes_{\mathbb{C}} S_V \to S_{V\oplus W}$$

is an isomorphism. As a result, In particular,

$$\dim \mathrm{Hom}_{\mathrm{Cliff}(V)}(S_V, S_{V\oplus W}) \cdot 2^{\frac{\dim V}{2}} = 2^{\frac{\dim S_{V\oplus W}}{2}}$$

from which $\dim S_W = 2^{\frac{\dim W}{2}}$ follows, and hence irreducibility.   □

Lemma 7.8.19 *Assume that* V *and* W *are Euclidean vector spaces, one even-dimensional, one odd-dimensional.*

a) *If* W *is odd-dimensional,* V *even-dimensional,* $c_V : \mathrm{Cliff}(V) \to \mathrm{End}(S_V)$ *is a* $\mathbb{Z}/2$-*graded irreducible representation of* $\mathrm{Cliff}(V)$, *and* $c_{V\oplus W} : \mathrm{Cliff}(V\oplus W) \to \mathrm{End}(S_{V\oplus W})$ *is an irreducible representation of* $\mathrm{Cliff}(V \oplus W)$, *let*

$$S_W := \mathrm{Hom}_{\mathrm{Cliff}(V)}(S_V, S_{V\oplus W}) := \{T : S_V \to S_{V\oplus W} \mid c_V(v)T(s)$$

(7.35) $$= T\big(c_V(v)s\big) \; \forall s \in S_V\},$$

(with no grading). Define, for $w \in W$, *and* $T \in S_{V\oplus W}$ *homogeneous,*

(7.36) $$c_W(w)T := c_{V\oplus W}(w) \circ T \circ \epsilon_V,$$

where ϵ_V *is the grading operator on* V *Then* $c_W : \mathrm{Cliff}(W) \to \mathrm{End}(S_W)$ *is an irreducible representation of* $\mathrm{Cliff}(W)$.

b) *If* V *is odd-dimensional,* W *even-dimensional,* $c_V : \mathrm{Cliff}(V) \to \mathrm{End}(S_V)$ *is an irreducible representation of* $\mathrm{Cliff}(V)$, *and* $c_{V\oplus W} : \mathrm{Cliff}(V \oplus W) \to \mathrm{End}(S_{V\oplus W})$ *is an irreducible representation of* $\mathrm{Cliff}(V \oplus W)$, S_W *still as in (7.35), we let* ϵ_W *be the* $\mathbb{Z}/2$-*grading operator on* $S_{V\oplus W}$ *induced by the orientation on* W, *and*

$$c_W(w)T := \epsilon_W \circ c_{V \oplus W}(w) \circ T, \quad w \in W, T \in \mathrm{Hom}_{\mathrm{Cliff}(V)}(S_V, S_{V \oplus W}).$$

Then $c_W : \mathrm{Cliff}(W) \to \mathrm{End}(S_W)$ is an irreducible, $\mathbb{Z}/2$-graded representation of $\mathrm{Cliff}(W)$.

The verification of the above statements is a good exercise and is left to the reader. Finally, we finish with the last case:

Lemma 7.8.20 *Assume that V and W are odd-dimensional Euclidean vector spaces, $V \oplus W$ their orthogonal direct sum.*

Suppose that $c_V : \mathrm{Cliff}(V) \to \mathrm{End}(S_V)$ is an irreducible representation of $\mathrm{Cliff}(V)$, and $c_{V \oplus W} : \mathrm{Cliff}(V \oplus W) \to \mathrm{End}(S_{V \oplus W})$ is a $\mathbb{Z}/2$-graded irreducible representation of $\mathrm{Cliff}(V \oplus W)$, with grading operator ϵ. Let

$$S_W := \mathrm{Hom}_{\mathrm{Cliff}(V)}(S_V, S_{V \oplus W}) := \{T : S_V \to S_{V \oplus W} \mid c_{V \oplus W}(v)T(s)$$

$$= T\big(c_V(v)s\big) \; \forall s \in S_V\}.$$

Define, for $w \in W$, and $T \in S_{V \oplus W}$ homogeneous,

$$(7.37) \qquad\qquad c_W(w)T := i \, c_{V \oplus W}(w) \circ \epsilon \circ T.$$

Then this extends to a $\mathbb{Z}/2$-graded, irreducible representation of $\mathrm{Cliff}(W)$.

Exercise 7.8.21 In the notation of the Lemma, show that if one sets

$$S_W^- := \{T : S_V \to S_{V \oplus W} \mid c_V(v)T(s) = -T\big(c_V(v)s\big) \; \forall s \in S_V\},$$

then one also obtains another irreducible $\mathrm{Cliff}(W)$-module (recall that when W is odd-dimensional, there are, up to isomorphism, two irreducible $\mathrm{Cliff}(W)$-modules).

Representation theory of real Clifford algebras

We will be focusing on complex K-theory in the following, and correspondingly will not discuss much the representation theory of the *real* Clifford algebras $\mathrm{Cliff}_{\mathbb{R}}(V)$ of a Euclidean vector space. However, we give some examples of real Clifford modules below, and an application to the problem of vector fields on spheres. Real Clifford algebra theory is important in Riemannian geometry.

Example 7.8.22 $\mathrm{Cliff}_{\mathbb{R}}(\mathbb{R}^1) \cong \mathbb{C}$, so a $\mathrm{Cliff}_{\mathbb{R}}(\mathbb{R}^1)$-module W, disregarding the grading for the moment, corresponds to a real vector space, together with a linear operator $c(e)$, where $e = (1, 0)$, with $c(e)^2 = -\mathrm{id}_W$, which we can view as a complex structure on W. With this point of view, Clifford multiplication by e is multiplication by the complex scalar $i \in \mathbb{C}$.

In particular, W has real dimension a multiple of 2. Of course, 2 can be achieved (by $W := \mathbb{C}$, viewed as a module over $\mathrm{Cliff}_{\mathbb{R}}(\mathbb{R}^1) = \mathbb{C}$ by scalar multiplication.

There is a natural $\mathbb{Z}/2$-grading with even part the real axis $\mathbb{R}$ in $\mathbb{C}$, and odd part the imaginary axis $\mathbb{R}i$, so that $c(e)$ (multiplication by i) acts as an odd operator.

Example 7.8.23 We have already seen that $\mathrm{Cliff}_{\mathbb{R}}(\mathbb{R}^2) \cong \mathbb{H}$, the real algebra of quaternions. A $\mathrm{Cliff}_{\mathbb{R}}(\mathbb{R}^2)$-module is therefore the endowment of a real vector space with a "quaternionic" structure,' with i, j, k acting respectively as $c(e_1)$, $c(e_2)$ and $c(e_1 e_2)$.

Note that this forces W to have real dimension a multiple of 4. Using the standard embedding of the quaternions in $M_2(\mathbb{C})$ (sending the unit quaternions to $\mathbf{SU}_2$) results in a real representation of real dimension 4 of $\mathrm{Cliff}(\mathbb{R}^2)$.

Exercise 7.8.24 Suppose n is even. Let $c \colon \mathrm{Cliff}(\mathbb{R}^n) \to \mathrm{End}(S)$ be a $\mathbb{Z}/2$-graded irreducible $\mathrm{Cliff}(\mathbb{R}^n)$-module. Let A be an n-by-n orthogonal matrix. Show that the map $c'(v) := c(Av)$ extends to a $\mathbb{Z}/2$-graded irreducible $\mathrm{Cliff}(\mathbb{R}^n)$-module which is isomorphic, as a $\mathbb{Z}/2$-graded module, to c, if $\det(A) = 1$, and is isomorphic to c^{op} (the same Clifford module, but with grading reversed), if $\det(A) = -1$.

Some applications to vector fields over spheres

Some of the ideas above can be used to prove significant theorems about vector fields on spheres. We give a rather simple such result here.

Let $c \colon \mathbb{R}^n \to \mathrm{End}_{\mathbb{R}}(W)$ be a $\mathrm{Cliff}_{\mathbb{R}}(\mathbb{R}^n)$-module, with $\dim_{\mathbb{R}}(W) = m$. Endow W with an inner product with respect to which Clifford multiplication by vectors is orthogonal, whence skew-symmetric. Now, if $\xi \in \mathbb{R}^n$, let $V_\xi \colon S^{m-1} \to \mathbb{R}^m$ be $V_\xi(x) := c(\xi)x$. Since $c(\xi)$ is skew-adjoint, $\langle c(\xi)x, x \rangle = 0$. Hence $c(\xi)x \in T_x S^{m-1}$. The map $\xi \to V_\xi$ defines a linear injective map from $\mathbb{R}^n$ into the space of vector fields on S^{m-1}.

For example, the construction in Example 7.8.4 has a real counterpart, producing a real representation

$$\mathrm{Cliff}(\mathbb{R}^n) \to \mathrm{End}_{\mathbb{R}}\big(\Lambda_{\mathbb{R}}^*(\mathbb{R}^n)\big),$$

with $c(v) := \lambda_v + i_v$ (exterior and interior multiplication defined as in the complex case.) The space on which the representation occurs is thus of real dimension 2^n. So we get n linearly independent vector fields on S^{2^n-1}, *e.g.*, there are 2 on S^3 by this method, 3 on S^7 and so on. One can do much better. We refer the reader to Michaelson and Lawson's book, or to the paper of Adams [1] for more information.

7.9   Clifford Algebras and K-Theory: The Bott-Shapiro Theorem

Suppose that $c \colon \mathrm{Cliff}(\mathbb{R}^n) \to \mathrm{End}(S)$ is a $\mathbb{Z}/2$-graded Clifford module for $\mathbb{R}^n$. Let

$$\sigma_c \colon \mathbb{R}^n \to \mathrm{Hom}_{\mathbb{C}}(S_+, S_-), \quad \sigma_S(\xi) := c(\xi).$$

Then σ_c determines a vector bundle map $\mathbb{R}^n \times S_+ \to \mathbb{R}^n \times S_-$ between the trivial bundles $\mathbb{R}^n \times S_\pm$, which is invertible away from $\{0\}$. We get a K-theory triple $(\mathbb{R}^n \times S_+, \mathbb{R}^n \times S_-, \sigma_c)$ and a corresponding class $[\sigma_c] \in K^0(\mathbb{R}^n)$.

Lemma 7.9.1 *View* $\mathrm{Cliff}(\mathbb{R}^n) \subset \mathrm{Cliff}(\mathbb{R}^{n+1})$ *in the usual way.*

Then if a $\mathbb{Z}/2$-*graded Clifford module* $c: \mathbb{R}^n \to \mathrm{End}(S)$ *is the restriction to* $\mathrm{Cliff}(\mathbb{R}^n) \subset \mathrm{Cliff}(\mathbb{R}^{n+1})$ *of a representation* $c': \mathrm{Cliff}(\mathbb{R}^{n+1}) \to \mathrm{End}(S)$ *over* $\mathbb{R}^{n+1}$, *then the corresponding class* $[\sigma_c] \in K^0(\mathbb{R}^n)$ *is zero.*

Proof Let e_{n+1} be the $n+1$ th standard basis vector of $\mathbb{R}^{n+1}$. We show that there is a vector bundle map $\sigma': \mathbb{R}^n \times S_+ \to \mathbb{R}^n \times S_-$ which agrees with σ_c off a compact set, and is invertible everywhere. The result will follow. To accomplish this, if $v \in \mathbb{R}^n$, $\|v\| \leq 1$, set

$$\sigma'(v) := c'(v + \sqrt{1 - \|v\|^2}\, e_{n+1})$$

and if $\|v\| \geq 1$ let $\sigma'(v) = \sigma_c(v)$. $\square$

This gives a rather direct argument that at least *some* candidates for classes in $K^0(\mathbb{R}^n)$, when n is *odd*, are zero:

Proposition 7.9.2 *If n is odd then any* $\mathbb{Z}/2$-*graded Clifford module* $c: \mathrm{Cliff}(\mathbb{R}^n) \to \mathrm{End}(S)$ *is the restriction of a* $\mathbb{Z}/2$-*graded Clifford module for* $\mathbb{R}^{n+1}$, *and hence* $[\sigma_c] \in K^0(\mathbb{R}^n)$ *is the zero class.*

Proof Let $\epsilon: S \to S$ be the grading operator. Let $e_{n+1} \in \mathbb{R}^{n+1}$ act on S by $i\epsilon$. Since by hypothesis, ϵ graded commutes with Clifford multiplication by vectors in $\mathbb{R}^n$, we do indeed obtain a representation $\mathrm{Cliff}(\mathbb{R}^{n+1}) \to \mathrm{End}(S)$. $\square$

Now let $n = 2m$ be even. Let $c: \mathrm{Cliff}(\mathbb{R}^n) \to \mathrm{End}(\Lambda^*\mathbb{C}^m)$ be the $\mathbb{Z}/2$-graded Clifford module of Example 7.8.4. Let $\beta_n := \beta_{\mathbb{R}^2}^m \in K^0(\mathbb{R}^n)$ be the Bott generator of $K^0(\mathbb{R}^n) = K^{-n}(\mathrm{pt})$.

Proposition 7.9.3 *With c as above,*

$$[\sigma_c] = \beta_{\mathbb{R}^n} \in K^0(\mathbb{R}^n) = K^{-n}(pt),$$

where $\beta_{\mathbb{R}^n}$ *is the Bott generator for* $K^0(\mathbb{R}^n)$.

The following exercise gives a nice application to describing the K-theory of odd spheres.

Exercise 7.9.4 Let n be odd. Let $: \mathbb{R}^{n+1} \to \mathrm{End}(S)$ be a $\mathbb{Z}/2$-graded Clifford module over $\mathrm{Cliff}(\mathbb{R}^{n+1})$. Fix $v_0 \in S^n \subset \mathbb{R}^{n+1}$ and let

$$u_c: S^n \to \mathrm{End}(S_+)$$

be the map $u(v) := c(v)c(v_0)|_{S_+}$. Then u is a unitary-valued function on S^n and if c is irreducible, the

$$[u_c] \in [S^n, \mathbf{U}_\infty] \cong K^{-1}(S^n)$$

generates $K^{-1}(S^n)$.

For each n, the collection of isomorphism classes of $\mathbb{Z}/2$-graded $\mathrm{Cliff}(\mathbb{R}^n)$-modules forms a semigroup under addition of modules. By taking the Grothendieck completion of this semigroup, we obtain a $\mathbb{Z}/2$-graded abelian group, which we denote by $\widehat{\mathcal{M}}_n$ and let $\widehat{\mathcal{M}}_* := \oplus_{n=0}^\infty \widehat{\mathcal{M}}_n$.

Note that $\widehat{\mathcal{M}}_*$ has a natural ring structure, for if $c_i \colon \mathrm{Cliff}(\mathbb{R}^n) \to \mathrm{End}(S_i)$ is a pair of $\mathbb{Z}/2$-graded Clifford modules over $\mathbb{R}^n$ and $\mathbb{R}^m$ respectively, then their $\mathbb{Z}/2$-graded tensor product $S_1 \widehat{\otimes} S_2$ is a $\mathbb{Z}/2$-graded $\mathrm{Cliff}(\mathbb{R}^{n+m})$-module with $c(v_1, v_2) := c_1(v_1) \widehat{\otimes} 1 + 1 \widehat{\otimes} c_2(v_2)$. This multiplication descends to $\widehat{\mathcal{M}}_*$ to give a $\mathbb{Z}/2$-graded ring structure to $\widehat{\mathcal{M}}_*$. To check all this is left as an easy exercise for the reader.

Now, for each n let $i \colon \widehat{\mathcal{M}}_{n+1} \to \widehat{\mathcal{M}}_n$ be the group homomorphism induced by the inclusion $\mathrm{Cliff}(\mathbb{R}^n) \to \mathrm{Cliff}(\mathbb{R}^{n+1})$. We consider the quotient $\widehat{\mathcal{M}}_n / i^*(\widehat{\mathcal{M}}_{n+1})$. It maps, by Lemma 7.9.1 to $K^{-n}(\mathrm{pt})$.

Hence, by taking the direct sum of all these homomorphisms we obtain a homomorphism

(7.38)
$$\widehat{Q}_* := \widehat{\mathcal{M}}_* / i^*(\widehat{\mathcal{M}}_{*+1}) := \oplus_{n=0}^\infty \widehat{\mathcal{M}}_n / i^*(\widehat{\mathcal{M}}_{n+1}) \to \oplus_{n=0}^\infty K^{-n}(\mathrm{pt}) := K^*(\mathrm{pt}).$$

The Bott Periodicity Theorem now be stated in the following way.

Theorem 7.9.5 (Atiyah-Bott-Shapiro) *The map*

$$\widehat{Q}_* \to K^*(pt)$$

of (7.38) *is an isomorphism of* $\mathbb{Z}/2$-graded rings.

The irreducible $\mathrm{Cliff}(\mathbb{R}^2)$-module of Example 7.8.7, in which Clifford multiplication by $(x, y) \in \mathbb{R}^2$ acts by the matrix $\begin{bmatrix} 0 & x+iy \\ x-iy & \end{bmatrix}$, has thus an isomorphism class Λ in $\widehat{Q}_*$. Then in this ring, Λ^n is the class of Example 7.8.4 (by a routine exercise.) The ring $\widehat{Q}_*$ is isomorphic as a ring, to the polynomial ring $\mathbb{Z}[x]$ of polynomials in one variable, by the map sending a polynomial $f(x) = \sum_{k=0}^n a_k x^k$ to $= \sum_{k=0}^n a_k \Lambda^k$.

On the K-theory side, such a polynomial identifies with the $K^*(\mathrm{pt})$ element $\sum_{k=0}^n a_k \beta_{\mathbb{R}^2}^k$, with $\beta_{\mathbb{R}^2} \in K^{-2}(\mathrm{pt})$ the Bott generator.

7.10   K-Orientations and the Thom Isomorphism Theorem

The Thom isomorphism is a central result in K-theory (as is its analogue in ordinary cohomology) because it allows the construction in topological K-theory of so-called *wrong-way maps*.

Let $\pi : V \to X$ be a Euclidean vector bundle over a locally compact space X.

Since each fiber $V_x := \pi^{-1}(x)$ of V is a Euclidean vector space, we can form its Clifford algebra $\mathrm{Cliff}(V_x)$. The bundle of algebras $\{\mathrm{Cliff}(V_x)\}_{x \in X}$ is easily checked to be locally trivial as a bundle of algebras, and hence is, in particular, a complex vector bundle over X. We denote it $\mathrm{Cliff}(V)$.

There is an obvious bundle-version of the idea of a Clifford module. By a $\mathbb{Z}/2$-*graded vector bundle* we mean a vector bundle equipped with a fiberwise $\mathbb{Z}/2$-grading such that the even and odd parts $V^{\pm}$ of V are themselves (sub-) vector bundles (of V).

Definition 7.10.1 Let V be a Euclidean vector bundle over X.

By a $\mathrm{Cliff}(V)$-*module*, we shall mean a complex, Hermitian vector bundle S over X, together with a vector bundle map $c : \mathrm{Cliff}(V) \to \mathrm{End}(S)$, whose restriction $c_x : \mathrm{Cliff}(V_x) \to \mathrm{End}(S_x)$, to each fiber of $\mathrm{Cliff}(V)$, is a $\mathrm{Cliff}(V_x)$-module.

We sometimes require the bundle S to be $\mathbb{Z}/2$-graded as a vector bundle, and for the Clifford action to be by (fiberwise) odd operators. In this case we refer to a $\mathbb{Z}/2$-graded $\mathrm{Cliff}(V)$-module.

The case where the Clifford bundle is, fiberwise, an irreducible module, is particularly important, and we will generally call the S of the previous definition a *spinor bundle* in this case.

There is a fairly obvious notation of *isomorphism* of $\mathrm{Cliff}(V)$-modules: one requires a fiberwise unitary bundle map $S \to S'$ intertwining the two representations. In the case when the modules are graded, we generally require the isomorphism to be grading-preserving.

Definition 7.10.2 Suppose $\pi : V \to X$ is a a real vector bundle over X locally compact.

A K-*orientation* on V is a pair, consisting of

i) An inner product q on V.
ii) A fiberwise irreducible $\mathrm{Cliff}(V)$-module, with V equipped with the Euclidean structure from the metric in i). We require the module to be $\mathbb{Z}/2$-graded if V is even-dimensional.

Example 7.10.3 The zero vector bundle **0**, over any space, can be K-oriented in two different ways.

Indeed, the Clifford algebra of the zero vector space is $\mathbb{C}$, graded as an algebra with $\mathbb{C}^+ := \mathbb{C}$, $\mathbb{C}^- := \{0\}$. Of course the C*-algebra $\mathbb{C}$ has a unique irreducible representation, on the one-dimensional Hilbert space $\mathbb{C}$. Either choice of $\mathbb{Z}/2$-grading on this Hilbert space, setting $\mathbb{C}^+ := \mathbb{C}$, $\mathbb{C}^- := \{0\}$, or setting $\mathbb{C}^- := \{0\}$ and $\mathbb{C}^+ := \mathbb{C}$ yield $\mathbb{Z}/2$-graded Clifford modules.

The Clifford algebra of the zero bundle is the trivial rank-one bundle, so a K-orientation amounts to endowing the trivial line bundle over X with a $\mathbb{Z}/2$-grading. Note that such a grading might vary from component to component (of X).

Definition 7.10.4 A spin^c-*structure* on a smooth manifold X is a pair, consisting of a Riemannian metric on X, and a fiberwise irreducible Cliff(TX)-module, which is $\mathbb{Z}/2$-graded if $\dim X$ is even. That is, X is spin^c if the real vector bundle TX is K-oriented.

Remark 7.10.5 In the applications of Index Theory (of Dirac operators) to geometry, the slightly more rigid requirement of a *spin* structure is quite important, although we do not discuss it much in these notes. A spin structure on a Riemannian manifold X is the existence of an irreducible Cliff$_\mathbb{R}$(TX)-module, where Cliff$_\mathbb{R}$(TX) is the *real* Clifford algebra of the tangent bundle. If a spin structure exists, X is a *spin manifold*. The 2-sphere S^2 is a spin manifold.

It is completely obvious that a trivial Euclidean vector bundle $X \times \mathbb{R}^n$ bundle (with the standard Euclidean metric fixed on the fibers) is K-orientable.

Furthermore, a K-orientation on a vector bundle determines an orientation on it, as the following exercise shows.

Exercise 7.10.6 Suppose that V is an even-dimensional Euclidean vector bundle over X. and c: Cliff(V) $\to$ End(S) a $\mathbb{Z}/2$-graded irreducible Cliff(V)-module. If $e_1, \ldots, e_n$ is a local orthonormal frame for V, defined, say on a connected open set U, then we declare this local frame to be positively oriented if Clifford multiplication by $c(x, e_1(x) \cdots e_n(x)) \in$ End(S_x) equal to the grading operator ϵ on S_x, and call it negatively oriented if it equals $-\epsilon$ (one of these possibilities must occur, for all $x \in U$).

a) Check that the above prescription orients V.
b) Prove the analogous result (K-oriented implies oriented) for odd-dimensional vector bundles.

In particular, the Möbius bundle over the circle is not K-orientable.

Example 7.10.7 If $\pi: V \to X$ has a complex structure (*i.e.*, if it is isomorphic as a real vector bundle to a complex vector bundle E), and equipped with a Hermitian metric, and induced Euclidean metric, then V is K-oriented by letting

$$(7.39) \qquad S := \Lambda_{\mathbb{C}}^*(E), \quad c(v) := \lambda_v + i_v: \Lambda_x^* E \to \Lambda_x^* E, \quad v \in E_x,$$

where λ_v is external product with v and i_v interior product.

This all follows immediately from the definitions and the discussion in Example 7.8.4.

Exercise 7.10.8 Deduce from Example 7.10.7 that $V \oplus V$ is (canonically) K-oriented, for *any* real vector bundle V. (*Hint.* $V \oplus V \cong V \otimes_\mathbb{R} \mathbb{C}$ as real vector bundles, the latter is a complex bundle.)

Definition 7.10.9 Suppose that V is a vector bundle over X, and $c\colon \mathrm{Cliff}(V) \to$ $\mathrm{End}(S)$ is a $\mathbb{Z}/2$-graded $\mathrm{Cliff}(V)$-module, and that L is a $\mathbb{Z}/2$-graded Hermitian line bundle over X. Let $S' := S\hat{\otimes}L$ be the $\mathbb{Z}/2$-graded tensor product of these two vector bundles, endowed with the tensor product Hermitian structure, and the $\mathrm{Cliff}(V)$-module structure

$$(7.40) \quad c_L(v) := c(v) \otimes \mathrm{id} \in \mathrm{End}(S_x \otimes L_x) \quad x \in X, \; v \in V_x, s \in S_x, l \in L_x.$$

We call S' the $\mathrm{Cliff}(V)$-module obtained from S by *twisting by L*.

If S is an ungraded $\mathrm{Cliff}(V)$-module, we twist in exactly the same way, dropping all mentions of the gradings. The outcome is another non-graded $\mathrm{Cliff}(V)$-module.

Proposition 7.10.10 *Let V be a Euclidean vector bundle over X. Then if V admits one K-orientation S, then any other K-orientation S' on V is obtained by twisting S by some $\mathbb{Z}/2$-graded Hermitian line bundle L over X, as in (7.40)*

In the case V is odd-dimensional, the same statement holds, with L ungraded.

Proof Assume first that V is even-dimensional, and $c\colon \mathrm{Cliff}(V) \to \mathrm{End}(S)$ is a $\mathbb{Z}/2$-graded irreducible representation of $\mathrm{Cliff}(V)$, and $c'\colon \mathrm{Cliff}(V) \to \mathrm{End}(S')$ another. Let L be the Hermitian vector bundle with fibers

$$L_x := \mathrm{Hom}_{\mathrm{Cliff}(V_x)}(S_x, S'_x) := \{T \in \mathrm{Hom}_{\mathbb{C}}(S_x, S'_x) \mid c(v)T = Tc(v) \; \forall v \in V_x\},$$

graded into even and odd operators.

Then L is a complex line bundle over X and $S' \cong S \otimes L$ as $\mathbb{Z}/2$-graded Clifford modules.

In the odd-dimensional case, we simply drop the gradings and use the same argument. $\qquad\square$

Lemma 7.10.11 *If $f\colon X \to Y$ is a map and V is a vector bundle over Y, then a K-orientation on V pulls back to a K-orientation on $f^*(V)$.*

Proof Suppose $c\colon V \to \mathrm{End}(S)$ is a spinor bundle for V, equipped with some Euclidean metric. The pulled-back bundle f^*V has fiber at x the Euclidean vector space $V_{f(x)}$, so f^*V inherits a natural pulled-back Euclidean structure. And the associated bundle of Clifford algebras has fibers $\mathrm{Cliff}(V_{f(x)})$, which map to $\mathrm{End}(S_{f(x)})$. This provides the pull-back f^*S of the spinor bundle for V with an action of $\mathrm{Cliff}(f^*V)$ as required. $\qquad\square$

Lemma 7.10.12 (The 2-Out-of-3 Lemma) *Suppose that*

$$0 \to W \xrightarrow{i} V \xrightarrow{\pi} Q \to 0$$

is an exact sequence of real vector bundles over X. Then a K-orientation on any two of them, determines a canonical K-orientation on the third.

Proof Choosing a splitting $s: Q \to V$ determines an isomorphism of real vector bundles $V \cong Q \oplus W$. So need to prove that if V and Q is K-oriented, then so is Q, and if Q and W are K-oriented, then so is V. This all follows from the bundle versions of Lemmas 7.8.15, 7.8.16, 7.8.20 and 7.8.19.   □

Exercise 7.10.13 On the tangent bundle $T S^2$ to the 2-sphere, there are two natural K-orientations that spring to mind. The first is based on the complex structure on S^2, seen as $\mathbb{CP}^1$. The tangent bundle $T S^2$ thus has a complex structure, and one builds an irreducible Clifford module bundle accordingly (as in Example 7.10.7).

On the other hand, S^2 is the boundary of the closed ball $\bar{\mathbb{D}}^3$, with trivial normal bundle, so that one has an exact sequence of vector bundles over S^2 given by

$$0 \to T S^2 \to T \bar{\mathbb{D}}^3 \to S^2 \times \mathbb{C},$$

and $T \bar{\mathbb{D}}^3$ is also a trivial bundle and so canonically K-orientable.

So we get a canonical K-orientation on $T S^2$, by the 2-out-of-3 result, the *boundary K-orientation*.

Find the complex line bundle L which intertwines these two K-orientations.

Definition 7.10.14 Let $\pi: V \to X$ be a K-oriented, even-dimensional Euclidean vector bundle over X, with associated $\mathbb{Z}/2$-graded irreducible representation $c: \mathrm{Cliff}(V) \to \mathrm{End}(S)$.

On the total space V of V, define a K-theory triple by $(\pi^* S^+, \pi^* S^-, \sigma_V)$, where $\sigma_V(x, v) := c(x, v): S_x^+ \to S_x^-$.

The associated class $\xi_V \in \mathrm{K}^0(V)$ is the *Thom class of V*.

If V is odd-dimensional, identify $V \times \mathbb{R}$ with the total space of the vector bundle $V \oplus \mathbf{1}$, where $\mathbf{1}$ is the trivial line bundle over X.

Since V and $\mathbf{1}$ are K-oriented, so is their sum, which is even-dimensional. We let in this case the Thom class ξ_V be the class in $\mathrm{K}^{-1}(V) \cong \mathrm{K}^0(V \oplus \mathbf{1})$ associated to the even-dimensional K-oriented bundle $V \oplus \mathbf{1}$.

The Thom class ξ_V of a K-oriented bundle V thus lies in the group $\mathrm{K}^{-\dim V}(V)$, with the superscript $-\dim V$ to be interpreted mod 2.

For the purposes of the following central theorem, recall that if $\pi: V \to X$ is a vector bundle, then using pull-back by π and the ring structure on $\mathrm{K}^*(V)$ gives the $\mathbb{Z}/2$-graded abelian group $\mathrm{K}^*(V)$ the structure of a (graded) module over the $\mathbb{Z}/2$-graded abelian group $\mathrm{K}^*(X)$.

Theorem 7.10.15 (The Thom Isomorphism Theorem) *Let V be a K-oriented vector bundle over X locally compact. Then as a $\mathrm{K}^*(X)$-module, $\mathrm{K}^*(V)$ is free and rank-one, and is generated freely as a rank-one $\mathrm{K}^*(X)$-module by the Thom class $\xi_V \in \mathrm{K}^{-\dim V}(V)$. That is, the map*

(7.41) $\tau_V: \mathrm{K}^*(X) \to \mathrm{K}^{*-\dim V}(V), \quad \tau_V(a) := \pi^*(a)\xi_V,$

is a $\mathrm{K}^*(X)$-module isomorphism.

Furthermore, if $f: X \to Y$ is a map, then $\xi_{f^*V} = f^*(\xi_V)$, so that the Thom isomorphism is natural with respect to maps, and pull-backs of K-oriented vector bundles.

The last (naturality) statement means the following. Suppose that if $f: X \to Y$ is a map, V is a real vector bundle over Y, and f^*V is given the pull-back K-orientation, using f. There is an obvious extension of f to a map $\bar{f}: f^*V \to V$.

Then naturality of the Thom isomorphism is the statement that the diagram

(7.42)

$$
\begin{array}{ccc}
K^*(f^*V) & \xleftarrow{\;\bar{f}^*\;} & K^*(V) \\
\big\downarrow{\scriptstyle \tau_{f^*V)}} & & \big\downarrow{\scriptstyle \tau_V} \\
K^*(Y) & \xleftarrow{\;f^*\;} & K^*(X)
\end{array}
$$

commutes.

The Thom isomorphism is a "bundle" version of Bott Periodicity. If V is a K-oriented, n-dimensional vector bundle over X, with, initially, let us say, n even, then its restriction to a point $p \in X$ can be identified linearly with $\mathbb{R}^n$, and its K-orientation thus restricts to a K-orientation on the real vector space $\mathbb{R}^n$.

Since there are only two possible K-orientations of $\mathbb{R}^n$, for n even, up to equivalence, the standard one, and it's opposite, it follows that

$$
\xi_V \mid_{\text{pt}} = \pm \beta_{\mathbb{R}^n} \in K^0(\mathbb{R}^n) = K^{-n}(\text{pt}).
$$

The sign is positive if and only if the orientation determined on $\mathbb{R}^n$ by identifying it with V, and then taking the orientation determined by the K-orientation on V, agrees with the standard K-orientation on $\mathbb{R}^n$. One can tell the difference as follows. Take a positively oriented linear basis $e_1, \dots, e_n$ for $\mathbb{R}^n$, so that as vectors in V, they form an orthonormal set in V_p, and let $\text{sign}_p(V) = 1$ if $c(e_1) \cdots c(e_n) = \epsilon_p$, and $\text{sign}_p(V) := -1$ if $c(e_1) \cdots c(e_n) = -\epsilon_p \in \text{End}(S_p)$, where S_p is the spinor bundle at p, ϵ_p the grading operator at p.

Proposition 7.10.16 *If V is a K-oriented n-dimensional vector bundle over X, then the restriction of the Thom class for V to any point $p \in X$ is given by*

$$
(\xi_V)|_{\{p\}} = \text{sign}_p(V) \cdot \beta_{\mathbb{R}^n} \in K^{-n}(pt),
$$

where $\beta_{\mathbb{R}^n}$ is the Bott generator of $K^{-n}(\mathbb{R}^n)$ and sign_p is defined as above.

Thus, intuitively, the Thom isomorphism restricts to Bott periodicity on the fibers of any K-oriented vector bundle—but one must take care with the K-orientations.

Chapter 8
K-Theory for C*-Algebras

Due to Swan's Theorem 5.2.12, topological K-theory for compact Hausdorff spaces X may be equivalently defined in terms of isomorphism classes of finitely generated projective (f.g.p.) modules over $C(X)$, and this suggests a definition of K_0-theory of C*-algebras or indeed of any ring. At the latter level of generality the resulting theory is called *algebraic* K-theory, while when specialized to C*-algebras it is often called *operator* K-theory. Operator K-theory retains the Bott Periodicity phenomenon of topological K-theory, while algebraic K-theory does not, so the two differ in their treatment of higher K-groups. Operator K-theory is Morita invariant, and is the correct homology theory for studying the "noncommutative spaces" of Noncommutative Geometry.

K-theory has a number of applications in physics. In string theory, K-theory classification refers to a conjectured application of K-theory to superstrings, to classify the allowed Ramond–Ramond field strengths as well as the charges of stable D-branes (see [156], and [143].) In condensed matter physics K-theory has also found important applications, specially in the topological classification of topological insulators, superconductors and stable Fermi surfaces, see [113].

Some of the first interesting results in operator K-theory came from the study of C*-algebra inductive limits of finite-dimensional C*-algebras called AF-algebras. Such algebras were first introduced by Bratteli [35], who showed how to classify them by means of equivalence classes of certain graphs now called 'Bratteli diagrams.' However, this method of classification was unsatisfactory from a computational point of view. G. Elliott [69] showed that AF algebras are classified by their K_0- groups, together with the natural ordering on K_0 induced by the semigroup of finitely generated projective modules, and in the unital case, the order unit corresponding to the rank-one free module, all of this data comprising the *dimension group* of the algebra. Elliot's result generated several decades of activity on classification of simple C*-algebras by K-theory invariants (see [70]). An AF algebra is determined by successively refining partitions of a totally disconnected space, a procedure important in coding in dynamics. Following Elliot's work, I.

© Springer Nature Switzerland AG 2024
H. Emerson, *An Introduction to C*-Algebras and Noncommutative Geometry*,
Birkhäuser Advanced Texts Basler Lehrbücher,
https://doi.org/10.1007/978-3-031-59850-0_8

Putnam *et* al classified minimal homeomorphisms of Cantor sets using dimension groups [131] (and more recently $\mathbb{Z}^d$-actions, see [89].)

K-theory of group C*-algebras $C^*(G)$ for discrete groups G is a question of great interest due to ramifications in topology (e.g., the Novikov Conjecture, see [111]) and geometry (e.g., problem of existence of positive scalar curvature metrics [92]). The Baum–Connes Conjecture (see [20]) provides a (conjectural) method of computing the K-theory of group C*-algebras, or more generally groupoid C*-algebras, or C*-algebra crossed products. Finite group C*-algebras and crossed products by (some) proper actions of discrete groups are amenable to fairly direct methods and we treat them in this chapter. Computation of the K-theory groups of the irrational rotation algebra $A_\hbar$, which involves an *infinite* group and a non-proper action, requires more advanced methods (like KK-theory) , and we discuss it at the end of the book. In this chapter we summarize with proof all of the important properties of operator K-theory, following rather closely the efficient presentation of Higson and Roe in [98].

8.1   Basic Definitions of C*-Algebra K-Theory

If A is a unital ring, and L_1 and L_2 are f.g.p. modules over A, their sum $L_1 \oplus L_2$, as a right A-module, is also f.g.p. Obviously $L_1 \oplus L_2 \cong L_2 \oplus L_1$ as A-modules, and this addition operation is well defined on isomorphism classes of f.g.p. right A-modules. We write $[L]$ for the isomorphism class of L.

The collection of isomorphism classes of finitely generated projective modules over A defines a commutative semigroup $\mathcal{M}(A)$. If $A = C(X)$, this semigroup is in 1-1 correspondence with the semigroup $\mathrm{Vect}(X)$ of isomorphism classes of vector bundles over X, as we have already proved.

We may think of f.g.p. modules as kinds of "noncommutative vector bundles," and we correspondingly define

Definition 8.1.1   If A is a unital ring, and in particular, if A is a unital C*-algebra, $K_0(A)$ is the Grothendiek completion of the semigroup $\mathcal{M}(A)$ of finitely generated projective (f.g.p.) right A-modules.

An element of $K_0(A)$ is a formal difference $[L] - [M]$, of isomorphism classes of right f.g.p. A-modules L, M.

Two such formal differences $[L_1] - [L_2]$ and $[L'_1] - [L'_2]$ are equal in $K_0(A)$ if there exists an f.g.p. module L such that $L_1 \oplus L'_2 \oplus L \cong L'_1 \oplus L_2 \oplus L$.

Example 8.1.2 $\mathcal{M}(\mathbb{C}) \cong \mathbb{N}$ and $K_0(\mathbb{C}) \cong \mathbb{Z}$. Indeed, an f.g.p. module $\mathcal{E}$ over $\mathbb{C}$ is exactly the same as a finite-dimensional complex vector space, and isomorphism of f.g.p. modules over $\mathbb{C}$ corresponds to linear isomorphism of $\mathbb{C}$-vector spaces. Since a vector space over $\mathbb{C}$, or any field, is determined up to isomorphism by its dimension, $\mathcal{M}(\mathbb{C}) \cong \mathbb{N}$ by the map $\mathcal{E} \mapsto \dim_{\mathbb{C}}(\mathcal{E})$. Taking Grothendiek completions we get $K_0(\mathbb{C}) \cong \mathbb{Z}$ by the map $[\mathcal{E}_1] - [\mathcal{E}_2] \mapsto \dim_{\mathbb{C}}(\mathcal{E}_1) - \dim_{\mathbb{C}}(\mathcal{E}_2)$.

Example 8.1.3 Let $\pi : H \to \mathbb{CP}^1$ be the Hopf bundle. Then by the very definition, $H = \mathrm{Im}(P)$ where $p : \mathbb{CP}^1 \to M_2(\mathbb{C})$ is the projection-valued function $p : \mathbb{CP}^1 \to M_2(\mathbb{C})$ mapping a line $L \subset \mathbb{C}^2$ to orthogonal projection pr_L onto that line. In terms of homogeneous coordinates on $\mathbb{CP}^1$,

$$P([z, w]) = \frac{1}{|z|^2 + |w|^2} \begin{bmatrix} |z|^2 & \bar{w}z \\ \bar{z}w & |w|^2 \end{bmatrix}.$$

We can restrict p to $\mathbb{C} \subset \mathbb{CP}^1$, i.e., set

$$p : \mathbb{C} \to M_2(\mathbb{C}), \quad p(z) = p([z, 1]) = \frac{1}{|z|^2 + 1} \begin{bmatrix} |z|^2 & z \\ \bar{z} & 1 \end{bmatrix}$$

on $\mathbb{C}$. Note that

$$\lim_{z \to \infty} p(z) = \begin{bmatrix} 1 & 0 \\ 0 & 0 \end{bmatrix},$$

that is, p on $\mathbb{CP}^1$ takes the value $\begin{bmatrix} 1 & 0 \\ 0 & 0 \end{bmatrix}$ at "infinity"—the point with homogeneous coordinates $[1, 0]$.

The class

$$(8.1) \qquad\qquad \beta_{\mathbb{R}^2} := [p] - [1] \in \mathrm{K}_0(\mathbb{R}^2)$$

is the *Bott element* for $\mathbb{R}^2$, already discussed in the context of topological K-theory. Bott Periodicity implies that $\mathrm{K}^0(\mathbb{R}^2)$ is infinite cyclic with generator $\beta_{\mathbb{R}^2}$

If A is a C*-algebra, we let

$$M_\infty(A) := \bigcup_{n=1}^{\infty} M_n(A),$$

the *-algebra of infinite $\mathbb{N}$-by-$\mathbb{N}$ matrices with entries in A, which have only finitely many nonzero terms. Recall that two idempotents $p, q \in M_n(A)$ are *algebraically equivalent* if there exists $u, v \in M_n(A)$ elements such that $uv = p$, $vu = q$. We have already seen that if A is a unital C*-algebra, then any idempotent is algebraically equivalent to a projection, and two projections are algebraically equivalent if and only if they are *Murray–von-Neumann* equivalent: i.e., if and only if there is a partial isometry $u \in M_n(A)$ such that $uu^* = p$, $u^*u = q$.

If A is a unital ring, we set $\mathcal{P}(A)$ to be the collection of algebraic equivalence classes of idempotents in $M_\infty(A)$. If A is a unital C*-algebra, this is equivalent to looking at Murray–von-Neumann equivalence classes of projections, and we normally take this latter picture as defining $\mathcal{P}(A)$, when A is a C*-algebra.

Exercise 8.1.4 Let $p, q \in M_n(A)$ be projections, where A is a C*-algebra. Show that

a) The partial isometry

$$u := \begin{bmatrix} 0 & p \\ q & 0 \end{bmatrix}$$

gives a Murray–von-Neumann equivalence between the projections $\begin{bmatrix} p & 0 \\ 0 & q \end{bmatrix}$ and $\begin{bmatrix} q & 0 \\ 0 & p \end{bmatrix}$.

b) If $pq = qp = 0 \in M_n(A)$, so that $p + q$ is a projection, then the partial isometry

$$u := \begin{bmatrix} p & 0 \\ q & 0 \end{bmatrix}$$

gives a Murray–von-Neumann equivalence between $\begin{bmatrix} p+q & 0 \\ 0 & 0 \end{bmatrix}$ and $\begin{bmatrix} p & 0 \\ 0 & q \end{bmatrix}$.

c) Prove that if $p \sim p' \in M_n(A)$ and $q \sim q' \in M_n(A)$, then

$$\begin{bmatrix} p & 0 \\ 0 & q \end{bmatrix} \sim \begin{bmatrix} p' & 0 \\ 0 & q' \end{bmatrix} \in M_{2n}(A),$$

where $\sim$ means Murray–von-Neumann equivalence.

This shows that $\sim$-equivalence classes of projections in $\mathcal{P}(A)$ can be added.

From the exercise, the collection $\mathcal{P}(A)$ has the structure of a commutative semigroup under the addition operation

$$[p] + [q] := \begin{bmatrix} p & 0 \\ 0 & q \end{bmatrix},$$

corresponding to the direct sum operation on f.g.p. modules.

Proposition 8.1.5 *If A is a unital C*-algebra, then the map sending the Murray–von-Neumann equivalence class $[p]$ of a projection $p \in M_n(A)$ to the isomorphism class of the f.g.p. A-module pA^n in $K_0(A)$ defines an isomorphism*

$$\mathcal{P}(A) \cong \mathcal{M}(A).$$

Hence $K_0(A)$ *is naturally isomorphic to the Grothendiek completion of the abelian semigroup* $\mathcal{P}(A)$.

Exercise 8.1.6 Let p and q be projections in $M_\infty(A)$ such that $[p] = [q] \in K_0(A)$. Then for some k, $p \oplus 1_k$ is Murray–von-Neumann equivalent to $q \oplus 1_k$, as projections in $M_\infty(A)$, where 1_k is the k-by-k identity matrix.

(*Hint.* By the definitions, $p \oplus e \cong q \oplus e$ for some projection e. Now $e \oplus 1 - e$ is Murray–von-Neumann equivalent to 1_k for suitable k, by Exercise 8.1.4 b), and the result follows.)

Due to Swan's Theorem, K_0 for unital C*-algebras as described above generalizes topological K-theory for compact spaces:

Proposition 8.1.7 $K^0(X) \cong K_0(C(X))$ *for any second-countable compact Hausdorff space* X.

Proof The semigroup $\mathrm{Vect}(X)$ of isomorphism classes of vector bundles is isomorphic to the semigroup of isomorphism classes of f.g.p. modules by mapping the bundle $E \to X$ to the f.g.p. module $\Gamma(E)$ of sections of E. Lemma 5.2.10 proved that $E \cong E'$ if and only if $\Gamma(E) \cong \Gamma(E')$.   □

We next compute $K_0(\mathbb{B}(H))$, for any Hilbert space H.

Lemma 8.1.8 *If H is a Hilbert space, and p, q are projections in $\mathbb{B}(H)$, then p and q are Murray–von-Neumann equivalent in $\mathbb{B}(H)$ if and only if they have the same rank as operators on H.*

Proof if p and q have the same rank, their ranges are isomorphic as Hilbert spaces. Let $u : pH \to qH$ be such a unitary isomorphism, and extend it to $H = pH \oplus (pH)^\perp$ by setting it equal to the zero map on $(pH)^\perp$. Then $uu^* = q, u^*u = p$.   □

Corollary 8.1.9 *If H is finite-dimensional, then $K_0(\mathbb{B}(H)) \cong \mathbb{Z}$. If H is infinite-dimensional, $K_0(\mathbb{B}(H)) = \{0\}$.*

Proof First let H be infinite-dimensional. Let $p \in \mathbb{B}(H)$. If p has infinite rank, then apply the lemma to the projections $\begin{bmatrix} p & 0 \\ 0 & p \end{bmatrix}$ and $\begin{bmatrix} p & 0 \\ 0 & 0 \end{bmatrix}$ to get that they are Murray–von-Neumann equivalent in $\mathbb{B}(H \oplus H) \cong M_2(\mathbb{B}(H))$. Hence $[p]+[p] = [p] \in K_0$. Hence $[p] = 0$.

If H is finite-dimensional, $\mathbb{B}(H) \cong M_n(\mathbb{C})$ for some n, and a projection in $M_n(\mathbb{C})$ is determined up to equivalence by its rank.

□

We now continue with the general theory.

If $\alpha : A \to B$ is a *-homomorphism, then it induces, by applying α pointwise to the entries of a matrix, a C*-algebra homomorphism $M_n(A) \to M_n(B)$ for all n and a *-algebra homomorphism $M_\infty(A) \to M_\infty(B)$. Such a *-homomorphism maps projections to projections, and maps partial isometries to partial isometries, and hence induces a canonical semigroup homomorphism

$$\alpha_* : \mathcal{P}(A) \to \mathcal{P}(B)$$

which then determines a group homomorphism

$$\alpha_* \colon K_0(A) \to K_0(B).$$

It is obvious from the definitions that if $\beta \colon B \to C$ is another *-homomorphism, then

$$(\beta \circ \alpha)_* = \beta_* \circ \alpha_* \colon K_0(A) \to K_0(C).$$

Hence the assignment

$$A \to K_0(A), \quad \alpha \mapsto \alpha_*$$

defines a *functor* from the category of unital C*-algebras and *-homomorphisms, to the category of abelian groups, and abelian group homomorphisms.

Exercise 8.1.10 Let A be a unital C*-algebra and $p \in A$ be a projection. Then p determines a *-homomorphism $\alpha_p \colon \mathbb{C} \to A$. Such a *-homomorphism determines a group homomorphism $(\alpha_p)_* \colon K_0(\mathbb{C}) \to K_0(A)$. Check that $(\alpha_p)_*([1]) = [p]$. Formulate and prove a more general statement, where p is allowed to be in a matrix algebra $M_n(A)$ over A.

Exercise 8.1.11 If A is any unital C*-algebra, then the inclusion

$$A \to M_n(A), \quad i(a) := \begin{bmatrix} a & 0 \\ 0 & 0 \end{bmatrix}$$

of A as a corner of $M_n(A)$, determines an isomorphism

$$i_* \colon K_0(A) \cong K_0\big(M_n(A)\big).$$

Exercise 8.1.12 This exercise describes functoriality in terms of f.g.p. modules. Let A and B be C*-algebras and $\alpha \colon A \to B$ be a *-homomorphism.

a) If $\mathcal{E}_A$ is a f.g.p. module over A, then the algebraic tensor product

$$\alpha_*(\mathcal{E}_A) := \mathcal{E}_A \otimes_A B$$

of the right A-module $\mathcal{E}_A$ over the homomorphism $\alpha \colon A \to B$, with the left B-module B, gives a f.g.p. module over B.

b) If $\mathcal{E}_A$ and $\mathcal{E}'_A$ are isomorphic f.g.p. A-modules, then $\alpha_*(\mathcal{E}_A)$ and $\alpha_*(\mathcal{E}'_A)$ are isomorphic f.g.p. B-modules.

c) If $\mathcal{E}_A \cong pA^n$ for a projection $p \in M_n(A)$, then the pushed-forward module $\alpha_*(\mathcal{E}_A)$ is isomorphic to $\alpha(p)B^n$.

We are going to study the properties of the K_0-functor. Firstly, it is clearly *additive* in the following sense. Suppose A, B are unital C*-algebras. Firstly,

$$M_n(A \oplus B) \cong M_n(A) \oplus M_n(B), \quad \text{and} \quad M_\infty(A \oplus B) \cong M_\infty(A) \oplus M_\infty(B),$$

as *-algebras. Since *-algebra isomorphisms send projections to projections, and partial isometries to partial isometries, it follows that

$$\mathcal{P}(A \oplus B) \cong \mathcal{P}(A) \oplus \mathcal{P}(B)$$

as semigroups, and hence

$$K_0(A \oplus B) \cong K_0(A) \oplus K_0(B).$$

More exactly:

Lemma 8.1.13 *Let A, B be unital C*-algebras, $i_A \colon A \to A \oplus B$ and $i_B \colon B \to A \oplus B$, the inclusions and $\mathrm{pr}_A \colon A \oplus B \to A$ and $\mathrm{pr}_B \colon A \oplus B \to B$ the projections. Then*

$$(i_A)_* + (i_B)_* \colon K_0(A) \oplus K_0(B) \to K_0(A \oplus B)$$

is an isomorphism of groups, with inverse

$$(\mathrm{pr}_A)_* \oplus (\mathrm{pr}_B)_* \colon K_0(A \oplus B) \to K_0(A) \oplus K_0(B).$$

We now extend the K_0-functor to (possibly) non-unital C*-algebras. Let A be any C*-algebra, unital or not, and A^+ its unitization (see Definition 1.1.14). As a vector space $A^+ = A \oplus \mathbb{C}$, but with algebra multiplication and adjoint

$$(a, \lambda) \cdot (b, \mu) = (ab + \lambda b + \mu a, \lambda\mu), \quad (a, \lambda)^* = (a^*, \bar\lambda)$$

and the supremum norm makes A^+ a C*-algebra.

If A is already unital, then the map

(8.2) $$\gamma \colon A^+ \to A \oplus \mathbb{C}, \quad \gamma(a, \lambda) = (a + \lambda \cdot 1, \lambda)$$

is an isomorphism with the direct sum C*-algebra $A \oplus \mathbb{C}$.

Let

$$\epsilon \colon A^+ \to \mathbb{C}, \quad \epsilon(a, \lambda) = \lambda.$$

Then $\ker(\epsilon) = A$, embedded in the first copy of A^+. By functoriality of K_0 for unital C*-algebras, we obtain a group homomorphism

$$\epsilon_* \colon K_0(A^+) \to K_0(\mathbb{C}) \cong \mathbb{Z}.$$

Definition 8.1.14 If A is any C*-algebra, unital or not, we define $K_0(A)$ to be the kernel of the homomorphism $\epsilon_* \colon K_0(A^+) \to K_0(\mathbb{C}) \cong \mathbb{Z}$.

Lemma 8.1.15 *Definitions 8.1.1 and 8.1.14 agree for unital C*-algebras.*

Proof The isomorphism (8.2) identifies $\epsilon \colon A^+ \to \mathbb{C}$ with the projection $\mathrm{pr}_{\mathbb{C}} \colon A \oplus \mathbb{C} \to \mathbb{C}$. Since K_0 is additive (for unital C*-algebras), $K_0(A \oplus \mathbb{C}) \cong K_0(A) \oplus K_0(\mathbb{C})$ and under this isomorphism, $(\mathrm{pr}_{\mathbb{C}})_*$ becomes the second projection map of groups (see Lemma 8.1.13). The kernel of the second projection map $K_0(A) \oplus K_0(\mathbb{C}) \to K_0(\mathbb{C})$ is $K_0(A)$. $\qquad\qquad\square$

Remark 8.1.16 Suppose A is non-unital, let A^+ be its unitization, $\epsilon \colon A \to \mathbb{C}$ the augmentation.

Suppose that $\mathcal{E}$ is a f.g.p. module over A^+, with class $[\mathcal{E}] \in K_0(A^+)$. Let $[1] \in K_0(A^+)$ be the class of the free A^+-module given by A^+ itself. And suppose that

$$[\mathcal{E}] - m[1] \in K_0(A^+)$$

is in the kernel of

$$\epsilon_* \colon K_0(A^+) \to K_0(\mathbb{C}) = \mathbb{Z}.$$

Since

$$A^+ \otimes_{A^+} \mathbb{C} \cong \mathbb{C}$$

as right $\mathbb{C}$-modules, by the obvious map $b \otimes \lambda \mapsto \epsilon(b)\lambda$, application of ϵ_* to the given difference gives the formal difference in $K_0(\mathbb{C})$ of the isomorphism classes of $\mathcal{E} \otimes_{A^+} \mathbb{C}$ and $\mathbb{C}$. The isomorphism $K_0(\mathbb{C}) \to \mathbb{Z}$ is of course by taking the complex dimension of a complex vector space, and the difference is thus mapped to

$$\dim_{\mathbb{C}}(\mathcal{E} \otimes_{A^+} \mathbb{C}) - m.$$

For this to be zero means therefore that

$$m = \dim_{\mathbb{C}}(\mathcal{E} \otimes_{A^+} \mathbb{C}) - m.$$

So, rather generally, we can parameterize $K_0(A)$, for A non-unital, by isomorphism classes $[\mathcal{E}]$ of f.g.p. modules over A^+, the corresponding classes being

$$[\mathcal{E}] - \dim_{\mathbb{C}}(\mathcal{E} \otimes_{A^+} \mathbb{C}) \cdot [1] \in K_0(A) \subset K_0(A^+).$$

Before proceeding, we prove a useful lemma, which implies in particular that the K_0-group of a separable C*-algebra is always a *countable* abelian group.

Lemma 8.1.17 *If A is a unital C*-algebra and $p, q \in A$ are projections with $\| p - q \| < 1$, then p and q are unitarily equivalent.*

Proof Let $a = pq + (1 - p)(1 - q)$. Then

$$1 - a = (2p - 1)(p - q).$$

Since $\|pq\| < 1$ and $\|2p - 1\| \leq 1$, we see that $\|1 - a\| < 1$ and so a is invertible. Since $pa = pq = aq$, $a^{-1}pa = q$, and p, q are similar. From Proposition 5.2.16 they are unitarily equivalent. (using the unitary $u = a(a^*a)^{-\frac{1}{2}}$ in the polar decomposition of a.) $\qquad\square$

Corollary 8.1.18 *If A is a separable C*-algebra then $K_0(A)$ is countable.*

The proof is left as an exercise.

Two projections p, q in a C*-algebra A are *homotopic* if there is a continuous, projection-valued map $p \colon [0, 1] \to A$ with values p, q at the endpoints.

Corollary 8.1.19 *If A is a unital C*-algebra and $p, q \in A$ are homotopic projections, then p and q are unitarily equivalent, and in particular, are Murray–von-Neumann equivalent.*

Proof By compactness of the interval we can find $0 = t_0 < t_1 < \cdots < t_n = 1$ such that $\|p(t_k) - p(t_{k+1})\| < 1$ for all k. The result follows from Lemma 8.1.17. $\qquad\square$

The converse is almost true as well:

Proposition 8.1.20 *Let p, q be Murray–von-Neumann equivalent projections in a unital C*-algebra A. Then $\begin{bmatrix} p & 0 \\ 0 & 0 \end{bmatrix}$ and $\begin{bmatrix} q & 0 \\ 0 & 0 \end{bmatrix}$ are homotopic.*

Proof Assume v is a partial isometry in A such that $v^*v = p$, $vv^* = q$. By Exercise 5.2.19 of Chapter 4, the matrix

$$u := \begin{bmatrix} v & 1 - vv^* \\ v^*v - 1 & v^* \end{bmatrix}$$

is unitary, is connected by a path of unitaries to $\begin{bmatrix} 1 & 0 \\ 0 & 1 \end{bmatrix}$, and

$$upu^* = q.$$

The result follows. $\qquad\square$

Exercise 8.1.21 Suppose p is a projection in A (unital). Let $R_t = \begin{bmatrix} \cos t & -\sin t \\ \sin t & \cos t \end{bmatrix}$.

Show that $R_t^* \begin{bmatrix} p & 0 \\ 0 & 0 \end{bmatrix} R_t$, $0 \leq t \leq \frac{\pi}{2}$ is a path of projections between $\begin{bmatrix} p & 0 \\ 0 & 0 \end{bmatrix}$ and $\begin{bmatrix} 0 & 0 \\ 0 & p \end{bmatrix}$.

Exercise 8.1.22 Suppose p, q are projections in A and $pq = qp = 0$. Show that

$$\begin{bmatrix} q & 0 \\ 0 & 0 \end{bmatrix} + R_t^* \begin{bmatrix} p & 0 \\ 0 & 0 \end{bmatrix} R_t$$

gives a path of projections between $\begin{bmatrix} p+q & 0 \\ 0 & 0 \end{bmatrix}$ and $\begin{bmatrix} p & 0 \\ 0 & q \end{bmatrix}$, where R_t is the rotation

matrix $\begin{bmatrix} \cos t & -\sin t \\ \sin t & \cos t \end{bmatrix}$.

Exercise 8.1.23 Let $p \in M_2(C(\mathbb{T}))$ be the projection

$$p(z) := \frac{1}{2} \begin{bmatrix} 1 & z \\ \bar{z} & 1 \end{bmatrix}.$$

Find a loop $u(z)$ of unitaries in $M_2(\mathbb{C})$ such that

$$upu^* = \frac{1}{2} \begin{bmatrix} 1 & 1 \\ 1 & 1 \end{bmatrix} \in M_2(C(\mathbb{T})).$$

Note that p is the restriction to $\mathbb{T} \subset \mathbb{C}$ of the Bott projection of Example 8.1.3.

Definition 8.1.24 Two *-homomorphisms $\alpha \colon A \to B$ and $\beta \colon A \to B$ between unital C*-algebras are *homotopic* if there is a *-homomorphism

$$\gamma \colon A \to C([0, 1], B)$$

such that

$$\gamma(a)(0) = \alpha(a), \quad \gamma(a)(1) = \beta(a), \quad \forall a \in A.$$

Such a homotopy determines a 1-parameter family $(\alpha_t)_{t \in [0,1]}$ of *-homomorphisms $A \to B$ by $\alpha_t(a) := \gamma(a)(t)$, with $\alpha_0 = \alpha$, and $\alpha_1 = \beta$, which is continuous in the sense that for each $a \in A$, the map $[0, 1] \to A$, $t \mapsto \alpha_t(a)$, is continuous.

Corollary 8.1.25 *If $\alpha \colon A \to B$ and $\beta \colon A \to B$ are homotopic *-homomorphisms, then $\alpha_* = \beta_* \colon K_0(A) \to K_0(B)$.*

Proof Assume A and B are unital; the general case is dealt with by taking unitizations.

For any n, γ induces a *-homomorphism $M_n(A) \to C([0, 1], M_n(B))$, and a family of *-homomorphisms which we still denote by $\alpha_t \colon M_n(A) \to M_n(B)$. If $p \in M_n(A)$, then $q(t) := \alpha_t(p)$ is a path of projections in $M_n(B)$ between $\alpha(p)$ and $\beta(p)$ Hence these two projections determine the same class in $K_0(B)$.

$\square$

Exercise 8.1.26 A C*-algebra A is *contractible* if the identity homomorphism and zero homomorphisms $A \to A$ are homotopic.

a) Prove that if A is contractible (unital or not) then $K_0(A) = 0$.
b) Prove that if A is any C*-algebra, then the C*-algebra $C_0([0, 1), A)$ is contractible.

Exercise 8.1.27 Let $p \colon [0, 1] \to A$ be a path of projections in a unital C*-algebra A. Use the proof of Lemma 8.1.17 to show that there is a continuous path of unitaries $(u_t)_{t \in [0,1]}$ such that

$$p_t = u_t p_1 u_t^*.$$

(*Hint.* The proof of Lemma 8.1.17 shows that if $t_0 \in [0, 1]$ is any point, and

$$a_s := p_t p_s + (1 - p_t)(1 - p_s)$$

then a_s is invertible for $|t - s|$ sufficiently small, and

$$u_s := a_s (a_s^* a_s)^{-\frac{1}{2}}$$

is a unitary such that

$$u_s p_t u_s^* = p_s$$

holds.)

Exercise 8.1.28 Let $\pi \colon A \to B$ be a surjective C*-algebra homomorphism between unital C*-algebras. Prove that if $(p_t)_{t \in [0,1]}$ is a path of projections in B, $\tilde{p}_1 \in A$ is a projection such that $\pi(\tilde{p}_1) = p_1$, then there is a path $(\tilde{p}_t)_{t \in [0,1]}$ of projections in A, ending in $\tilde{p}_1$, and such that $\pi(\tilde{p}_t) = p_t$ for all t.

That is, show that the map $\mathcal{P}(A) \to \mathcal{P}(B)$ induced by a surjective *-homomorphism, has the path-lifting property. (*Hint.* Use Exercise 8.1.27 and the fact proved in Exercise 4.5.28 that the restriction of π to the unitary group $\mathbf{U}(A)$ has the path-lifting property.)

We next discuss a technique with broad applications in Noncommutative Geometry, in preparation for proving continuity of K_0 under inductive limits.

Definition 8.1.29 Let $\mathcal{A}$ be a unital *-subalgebra of a unital C*-algebra A. We say that $\mathcal{A}$ is *spectral* in A if $\mathcal{A}$ is dense in A, and whenever n is a positive integer and $a \in M_n(\mathcal{A})$ and f is a holomorphic function on $\mathrm{Spec}_A(a)$, then $f(a) \in M_n(\mathcal{A})$.

Notice that if $\mathcal{A}$ is spectral in A and $a \in \mathcal{A}$ is invertible in A, then $a^{-1} \in \mathcal{A}$, since $f(z) = \frac{1}{z}$ is holomorphic on $\mathrm{Spec}_A(a)$.

It follows that the spectrum of $a \in \mathcal{A}$, as an element of $\mathcal{A}$, is the same as its spectrum as an element of A.

Exercise 8.1.30 If M is a compact manifold, then $C^k(M)$ is spectral in $C(M)$ for every $k \in \mathbb{N} \cup \{\infty\}$.

Exercise 8.1.31 Suppose $A_1 \subset A_2 \subset \cdots \subset A$ is an increasing union of unital C*-subalgebras of a fixed unital C*-algebra A. Show that $\mathcal{A} := \bigcup_{n=1}^{\infty} A_n$ is a spectral *-subalgebra of A; in fact, show that it is even closed under *continuous* functional calculus.

We have already remarked that K_0, as defined in terms of idempotents in $M_\infty(\mathcal{A})$, is defined for any algebra (in fact any ring), and in particular for any *-algebra. Moreover, if $\mathcal{A} \subset A$ happens to be spectral, then since it is closed under holomorphic functional calculus in A, it follows easily that the idempotent picture of K_0 agrees with the projection picture, with algebraic equivalence corresponding to Murray von-Neumann equivalence, because any spectral $\mathcal{A}$ will be closed under the operation of taking square roots of strictly positive elements and hence polar decompositions of elements of $\mathcal{A}$ which are invertible in A, end up having their constituents in $\mathcal{A}$. (A spectral subalgebra is very nearly a C*-algebra).

Thus in the following, we will understand that K_0 as defined in terms of projections and Murray von-Neumann equivalence, extends directly to spectral *-subalgebras. With this convention:

Theorem 8.1.32 *Suppose A is a unital C*-algebra and $\mathcal{A} \subset A$ is a spectral *-subalgebra.*

Then the homomorphisms

$$i_* : \mathcal{P}(\mathcal{A}) \to \mathcal{P}(A), \quad i_* : K_0(\mathcal{A}) \to K_0(A)$$

induced from the inclusion $i : \mathcal{A} \to A$, are isomorphisms.

Lemma 8.1.33 *Let A be a unital C*-algebra and p a projection in A. If $a \in A$ is a normal element with $\|a - p\| < \delta$, then*

$$\mathrm{Spec}(a) \subset B_\delta(0) \cup B_\delta(1) \subset \mathbb{C},$$

with $B_\delta(\cdot)$ the disk of radius δ.

Proof Suppose $\lambda \notin B_\delta(0) \cup B_\delta(1)$, i.e., that $\min\{|\lambda|, |1 - \lambda|\} > \delta$. Then $\lambda - p$ is invertible in A. And

$$\|(\lambda - p)^{-1}\| = \max\{|\lambda|^{-1}, |1 - \lambda|^{-1}\} < \frac{1}{\delta}.$$

Consequently

$$(8.3) \quad \|(\lambda - p)^{-1}(\lambda - a) - 1\| = \|(\lambda - p)^{-1}(\lambda - a) - (\lambda - p)^{-1}(\lambda - p)\|$$

$$\leq \|(\lambda - p)^{-1}\| \cdot \|a - p\| < 1.$$

Hence $(\lambda - p)^{-1}(\lambda - a)$ is invertible, and hence so is $\lambda - a$. $\qquad\square$

Proof of Theorem 8.1.32 Let $p \in M_n(A)$ be a projection. By density of $M_n(\mathcal{A})$ in $M_n(A)$ there exists $h \in M_n(\mathcal{A})$ such that $\|h - p\| < \frac{1}{2}$. By replacing h with $\frac{h+h^*}{2}$, we may assume that h is self-adjoint. By Lemma 8.1.33, $\frac{1}{2} \notin \mathrm{Spec}(h)$, and hence the characteristic function $\chi := \chi_{[\frac{1}{2},1]}$ is holomorphic on $\mathrm{Spec}(h)$. Since $M_n(\mathcal{A})$ is spectral in $M_n(A)$, the projection $q := \chi(h)$ is in $M_n(\mathcal{A})$. Also, $\|\chi(h) - h\| < \frac{1}{2}$ since $\frac{1}{2} \notin \mathrm{Spec}(h)$, and hence

$$\|q - p\| \leq \|q - h\| + \|h - p\| < \frac{1}{2} + \frac{1}{2} = 1$$

so that p and q are unitarily equivalent and hence Murray–von-Neumann equivalent in A by Lemma 8.1.17, and hence $i_*([q]) = [p] \in \mathrm{K}_0(A)$.

We have shown that $i_* : \mathcal{P}(\mathcal{A}) \to \mathcal{P}(A)$ is surjective.

Injectivity is similar. Suppose p, q are projections in $M_n(\mathcal{A})$ which become Murray–von-Neumann equivalent in $M_n(A)$, by a partial isometry u such that

$$u^*u = p, \quad uu^* = q.$$

It follows from density of $M_n(\mathcal{A})$ in $M_n(A)$, that we can find a contraction $x \in M_n(\mathcal{A})$ such that $\|x - u\| < \frac{1}{4}$. As

$$\|x^*x - p\| = \|x^*x - x^*u + x^*u - u^*u\| \leq (\|x\| + \|u\|)\|x - u\| \leq 2\|x - u\|,$$

we have that $\|x^*x - p\| < \frac{1}{2}$. Similarly $\|xx^* - q\| < \frac{1}{2}$.

By Lemma 8.1.33, $\mathrm{Spec}(x^*x) \cup \mathrm{Spec}(xx^*) \subset B_{\frac{1}{2}}(0) \cup B_{\frac{1}{2}}(1)$. So $\chi = \chi_{[0,\frac{1}{2}]}$ as in the proof of the same Lemma, is continuous on the spectra both of x^*x and xx^*, and $\|\chi(x^*x) - x^*x\| < \frac{1}{2}$, whence $\|\chi(x^*x) - p\| < 1$ so p is unitarily equivalent to $p' := \chi(x^*x)$ in A by Lemma 8.1.17. Similarly, q is unitarily equivalent to $q' := \chi(xx^*)$. We show that p' and q' are Murray von-Neumann equivalent in $\mathcal{A}$. Let $f(t) = \sqrt{\frac{\chi(t)}{t}}$, a continuous function on $\mathrm{Spec}(x^*x) \cup \mathrm{Spec}(xx^*)$ satisfying

$$tf(t)^2 = \chi(t) \quad \forall t \in \mathrm{Spec}(x^*x) \cup \mathrm{Spec}(xx^*).$$

Set

$$v := xf(x^*x).$$

Since $\mathcal{A}$ is spectral and $x \in \mathcal{A}$, $v \in \mathcal{A}$ as well.

Since x^*x commutes with $f(x^*x)$,

$$v^*v = f(x^*x)x^*xf(x^*x) = f(x^*x)^2x^*x = \chi(x^*x) = p'.$$

Now notice that

$$x(x^*x)^k x = (xx^*)^k xx^*$$

for any positive integer k. It follows that

$$xh(x^*x)x^* = h(xx^*)xx^*$$

for any polynomial h and hence for any continuous function. In particular, taking $h = f^2$, we see

$$vv^* = xf(x^*x)^2 x^* = xg(x^*x)^2 x^* = g(xx^*)^2 xx^* = \chi(xx^*) = q',$$

so that $p' \sim q'$ in $\mathcal{A}$ as claimed.   □

In the course of the proof, we proved the following, which will be used again:

Lemma 8.1.34 *If A is a unital C*-algebra, $x \in A$, p, q are projections in A, then if there exists $x \in A$ such that $\|x^*x - p\| < \frac{1}{2}$ and $\|xx^* - q\| < \frac{1}{2}$, then p and q are Murray–von-Neumann equivalent in A.*

We next verify the continuity of K_0 under inductive limits. Suppose that $\{A_i, \ \varphi_{ij}\}_{i \in I}$ is an inductive system of C*-algebras, $A := \varinjlim A_i$ the inductive limit. Let $\varphi_i : A_i \to \varinjlim A_i$ be the associated C*-algebra homomorphisms. The homomorphisms φ_i and the homomorphisms φ_{ij} all determine maps $(\varphi_i)_*$, etc between the appropriate K-theory groups. By functoriality of K_0,

$$(\varphi_{ij})_* \circ (\varphi_{jk})_* = (\varphi_{ik})_*$$

for all relevant indices $i.j, k$ so we obtain an inductive system $\{K_0(A_i), \ (\varphi_{ij})_*\}$ of abelian groups.

Similarly, $\varphi_i \circ \varphi_{ij} = \varphi_j$ implies $(\varphi_i)_* \circ (\varphi_{ij})_* = (\varphi_j)_*$. Hence, by the universal property of inductive limits of groups, we obtain a unique group homomorphism

$$\Phi : \varinjlim K_0(A) \cong K_0\big(\varinjlim A_i\big).$$

Theorem 8.1.35 *If $\{A_i, \ \varphi_{ij}\}_{i \in I}$ is an inductive system of C*-algebras, and $\varphi_i : A_i \to \varinjlim A_i$ the associated C*-algebra homomorphisms, then $\{K_0(A_i), \ (\varphi_{ij})_*\}$ is an inductive system of abelian groups, and*

$$\varinjlim K_0(A) \cong K_0\big(\varinjlim A_i\big).$$

The isomorphism is induced from the coherent family of group homomorphisms

$$(\varphi_i)_* : K_0(A_i) \to K_0(\varinjlim A_i)$$

induced by the φ_i.

Proof For the proof we will assume first that all of the C*-algebras A_i are unital, and that the structure maps of the system are unital *-homomorphisms. This implies that the inductive limit A is also unital. The general case is dealt with by unitizations.

Observe first that $\mathcal{A} := \cup_{i \in I} \varphi_i(A_i)$ is spectral in A, and the map Φ factors through the isomorphism $K_0(\mathcal{A}) \to K_0(A)$. So we are reduced to showing that

$$\Phi : \varinjlim K_0(A_i) \to K_0(\mathcal{A})$$

is an isomorphism.

If p is a projection in $M_n(\mathcal{A})$, then $p \in M_n\big(\varphi_i(A_i)\big)$ for some i. Let $h \in M_n(A_i)$ self-adjoint with $\varphi_i(h) = p$. Then $h^2 - h \in \ker(\varphi_i) = \{a \in A_i \mid \lim_{j \to \infty} \|\varphi_{ji}(a)\| = 0\}$, so that for j large enough $\|\varphi_{ji}(h)^2 - \varphi_{ji}(h)\| < \frac{1}{4}$. Since $\varphi_j \circ \varphi_{ji}(h) = \varphi_i(h) = p$, due to the definitions, by replacing $h \in A_j$ by $\varphi_{ji}(h) \in A_j$, we may as well have assumed from the beginning that

$$h \in A_j, \; h = h^*, \; \varphi_j(h) = p, \; \|h^2 - h\| < \frac{1}{4}.$$

Then $\frac{1}{2} \notin \mathrm{Spec}(h)$, and $q := \chi(h)$ is a projection in $M_n(A_j)$ such that $\|q - h\| < \frac{1}{2}$. Hence $\|\varphi_j(q) - p\| = \|\varphi_j q) - \varphi_j(h)\| \le \|q - h\| < \frac{1}{2}$ so that $\varphi_j(q)$ is unitarily equivalent to p in $M_n(A)$ and

$$\Phi([q]) = (\varphi_j)_*([q]) = [\varphi_j(q)] = [p].$$

This shows that Φ is surjective.

Injectivity: suppose that p and q are in $M_n(A_i)$ whose φ_i-images are Murray–von-Neumann equivalent by a partial isometry u in $M_n\big(\varphi_i(A_i)\big) \subset \mathcal{A}$, with, say, $u^*u = p$ and $uu^* = q$. We can lift u to an $x \in A_i$. As previously, we have

$$\lim_{j \to \infty} \|\varphi_{ji}(x)^*\varphi_{ji}(x) - \varphi_{ji}(p)\| = 0$$

and similarly

$$\lim_{j \to \infty} \|\varphi_{ji}(x)\varphi_{ji}(x)^* - \varphi_{ji}(q)\| = 0.$$

Moreover, the image in $\varinjlim K_0(A_i)$ of the class of $\varphi_{ji}(p)$ in $K_0(A_i)$ is the same as the image of $[p]$ in this inductive limit of groups, and similarly for q. So we may as well have assumed from the beginning that

$$x \in M_n(A_j), \; \|x^*x - p\| < \frac{1}{2}, \; \|xx^* - q\| < \frac{1}{2}.$$

By Lemma 8.1.34, p and q are Murray–von-Neumann equivalent in A_i. Hence $[p] = [q] \in \varinjlim K_0(A_i)$, as required. $\qquad\qquad\square$

Exercise 8.1.36 Complete the proof of Theorem 8.1.35 by showing that if $\varinjlim A_i = A$ then $\varinjlim A_i^+ = A^+$ with unital structure maps, and using the unital case proved above to complete the argument.

Remark 8.1.37 The proof Theorem 8.1.35 is really just an amplification of that of Theorem 8.1.32. In fact, if an inductive system has injective structure maps, then $K_0(\varinjlim A_i) \cong \varinjlim K_0(A_i)$ is a direct consequence of Theorem 8.1.32, because then the union $\mathcal{A} := \cup_{i \in I} \varphi_i(A_i)$ is spectral in the C*-algebra limit A, and the following basic exercise in the definitions:

Corollary 8.1.38 *If $\mathcal{K}$ is the C*-algebra of compact operators on a Hilbert space H, then $K_0(\mathcal{K}) \cong \mathbb{Z}$, under an isomorphism sending $1 \in \mathbb{Z}$ to the class $[p] \in K_0(\mathcal{K}(H))$ of any rank-one projection.*

Proof Indeed, $\mathcal{K}(H) \cong \varinjlim \mathbb{B}(V)$, the inductive limit of the C*-algebras $\mathbb{B}(V)$, as V ranges over the directed set of finite-dimensional subspaces of H. The structure maps $\mathbb{B}(V) \to \mathbb{B}(V')$, for $V \subset V'$, are by setting $T \in \mathbb{B}(V)$ to be zero on the orthogonal complement $V^\perp$ of V in V'.

For each finite-dimensional Hilbert space V, $K_0(\mathbb{B}(V)) \cong \mathbb{Z}$. A generator for the K_0-group can be taken to be the class of any rank-one projection p_V. With V fixed, any two such projections are unitarily equivalent in $\mathbb{B}(V)$ and so determine the same K_0-class for $\mathbb{B}(V)$.

Now if if $V \subset V'$, then the image of p_V in $\mathbb{B}(V')$ under the structure map is the projection $\begin{bmatrix} p_V & 0 \\ 0 & 0 \end{bmatrix}$, (with respect to the decomposition $V' = V \oplus V^\perp$, and this image is clearly also a minimal projection in $\mathbb{B}(V')$. Hence the induced maps $K_0(\mathbb{B}(V)) \to K_0(\mathbb{B}(V'))$ are the identity maps $\mathbb{Z} \to \mathbb{Z}$, for every inclusion of subspaces. The result follows. $\square$

More generally:

Corollary 8.1.39 *Let $p \in \mathcal{K}(H)$ be any rank-one projection. Then the C*-algebra homomorphism $\alpha_p : A \to A \otimes \mathcal{K}$, $\alpha(a) := a \otimes p$, descends to an isomorphism $K_0(A) \cong K_0(A \otimes \mathcal{K})$.*

Proof Write $A \otimes \mathcal{K} \cong \varinjlim M_n(A)$ as an inductive limit. Since $M_n(A)$ and A have the isomorphic K-theory for all n, $K_0(A \otimes \mathcal{K}) \cong \varinjlim K_0(M_n(A)) \cong \varinjlim K_0(A) \cong K_0(A)$. The statement regarding the projection is left to the reader. $\square$

Corollary 8.1.40 (Morita Invariance of K-Theory) *If A and B are Morita equivalent then $K_0(A) \cong K_0(B)$.*

This follows immediately from the fact that if A and B Morita equivalent then they are stably isomorphic: that is, $A \otimes \mathcal{K} \cong B \otimes \mathcal{K}$ as C*-algebras, and the fact observed above that $K_0(A) \cong K_0(A \otimes \mathcal{K})$ for any A.

On the other hand, this argument gives no hint as to *what* map a Morita equivalence might induce on K-theory, since (the proof of) Corollary 8.1.40 appeals to the theorem (not proved in these notes) that Morita equivalent C*-algebras are stably isomorphic. Actually the proof of this theorem is not very constructive and knowing it would not help much anyway.

We give a more thorough discussion of the Morita invariance of K-theory in the next section.

Corollary 8.1.41 *For the UHF algebra $U(d^\infty)$,*

$$\mathrm{K}_0\big(U(d^\infty)\big) \cong \mathbb{Z}\left[\frac{1}{d}\right],$$

the additive subgroup underlying the subring of $\mathbb{Q}$ generated by $\mathbb{Z}$ and $\frac{1}{d}$.
If $\mathcal{N}$ is the universal UHF algebra of Example 1.7.10, then

$$\mathrm{K}_0(\mathcal{N}) \cong \mathbb{Q},$$

with $\mathbb{Q}$ as a group under addition.

Proof We just do the case $d = 2$, since it is typographically simpler. The UHF is the inductive limit of the system

$$\mathbb{C} \subset M_2(\mathbb{C}) \to M_{2^2}(\mathbb{C}) \to M_{2^3}(\mathbb{C}) \to \cdots$$

where the structure maps between adjacent C*-algebras are of the form

$$A \mapsto \begin{bmatrix} A & 0 \\ 0 & A \end{bmatrix}.$$

A rank-one projection in $M_{2^n}(\mathbb{C})$ is thus identified in the inductive limit with a rank-two projection in $M_{2^{n+1}}(\mathbb{C})$. The K_0-group of each $M_{2^k}(\mathbb{C})$ is $\mathbb{Z}$ with $1 \in \mathbb{Z}$ corresponding to the class of a rank-one projection. It follows that the maps $\mathbb{Z} \cong \mathrm{K}_0\big(M_{2^n}(\mathbb{C})\big) \to \mathrm{K}_0\big(M_{2^{n+1}}(\mathbb{C})\big) \cong \mathbb{Z}$ induced by the corresponding structure map, is multiplication by 2.

Thus

$$\mathrm{K}_0\big(U(2^\infty)\big) \cong \varinjlim \mathbb{Z},$$

where in the inductive limit on the right hand side, the structure maps are given between adjacent groups by multiplication by 2. The result follows from Example 1.7.3.

The second assertion follows from similar arguments and Exercise 1.7.4. $\qquad\square$

Exercise 8.1.42 Model the Cantor set X as $X := \prod_{n=1}^{\infty}\{0, 1\}$. For an element $\mu := (i_1, \ldots, i_k) \in \{0, 1\} \times \cdots \times \{0, 1\} = \{0, 1\}^n$, let $U_\mu \subset X$ be the clopen set of all sequences (x_n) starting with μ. Let χ_μ be the characteristic function of U_μ. Then X is the inverse limit of the spaces $\{0, 1\}^n \leftarrow \{0, 1\}^{n+1}$, with $S^n = S \times \cdots \times S$ where the maps drop the last coordinate, and $C(X) \cong \varinjlim C(\{0, 1\}^n)$, in such a way that a delta function at a point $\mu \in \{0, 1\}^n$, defining an element of $C(\{0, 1\}^n)$, corresponds under the map into the inductive limit, to the characteristic function χ_μ.

Prove, using compactness of X, that if $f \in C(X, \mathbb{Z})$ is any continuous, integer-valued function on X, then f is a finite, $\mathbb{Z}$-linear combination of the χ_μ's. (*Hint.* Start with f a characteristic function of a clopen set.)

Deduce that the natural map

$$C(X, \mathbb{Z}) \to K^0(X)$$

is an isomorphism, with $C(X, \mathbb{Z})$ the group, under addition, of integer-valued, continuous functions on X.

Exercise 8.1.43 This exercise explores the K_0-group of the C*-algebra $C^*(G)$ of a finite group.

For any such group, the *representation ring* $\mathrm{Rep}(G)$ of G, as a group, is the Grothendieck completion of the semigroup of unitary isomorphism classes of finite-dimensional unitary representations of G.

Elements of the representation ring can be designated $[\pi_1] - [\pi_2]$ where π_i are finite-dimensional unitary representations of G.

Prove that the Green-Julg correspondence determines an isomorphism

$$K_0\big(C^*(G)\big) \cong \mathrm{Rep}(G).$$

And if π is irreducible, then this isomorphism maps the class $[e_\pi] \in K_0(C^*(G))$ of the projection $e_\pi := \frac{\dim H_\pi}{|G|} \cdot \chi_\pi^* \in C^*(G)$, with χ_π^* the (conjugate) character of π, to the class $[\pi] \in \mathrm{Rep}(G)$ of the representation π.

In particular, $K_0\big(C^*(G)\big)$ is a finitely generated free abelian group with free generators the classes of the projections e_π, with π an irreducible representation of G.

See Theorem 6.5.10.

Exercise 8.1.44 Let G be the group $\mathbb{Z}/2$ acting on the interval $I := [-1, 1]$ by $\sigma(x) = -x$. Compute the K-theory of the crossed product $C(I) \rtimes \mathbb{Z}/2$. (*Hint.* The interval is $\mathbb{Z}/2$-equivariantly contractible. Use the homotopy-invariance of K-theory.)

Exercise 8.1.45 Let A be unital and $B = A \otimes \mathcal{K}$. Let B^+ be the unitization of B, $\epsilon \colon B^+ \to \mathbb{C}$ the augmentation, and $\underline{1}$ the unit of B^+. Prove that if p is a projection in B^+ such that $\epsilon(p) = 1$ then $[p] - [\underline{1}] = [p'] - [1_n]$ for some n, where $p' \in M_n(A)$ and $1_n \in M_n(A) \subset B^+$ is the unit of $M_n(A)$.

(This follows from Theorem 8.1.35 but try to prove it as directly as possible yourself.)

Prove the generalization of this statement where p is allowed to be in a matrix algebra $M_m(B^+)$.

8.2 Morita Invariance and Applications

The Morita invariance of K-theory (Corollary 8.1.40) means that $K^0(A)$ only depends on the noncommutative space (the Morita equivalence class) determined by A. Thus, K^0 is effectively a theory on noncommutative spaces. Actually, as we show below, K_0 has a simpler definition for the class of *compact* noncommutative spaces.

Recall that for a C*-algebra A, not necessarily unital, $\mathcal{P}(A)$ denotes the semigroup of isomorphism classes of f.g.p. modules over A, equivalently, of Murray–von-Neumann equivalence classes of projections in the *-algebra $M_\infty(A)$.

Definition 8.2.1 If A is a C*-algebra we define $K_{00}(A)$ to be the Grothendiek completion of $\mathcal{P}(A)$.

Thus, K_{00} is defined similarly to K_0, but without the unitization. Observe that due to $\mathcal{K} \otimes \mathcal{K} \cong \mathcal{K}$, K_{00} is stable: $K_{00}(A) \cong K_{00}(A \otimes \mathcal{K})$. There is clearly a canonical map $K_{00}(A) \to K_0(A)$. Note that if $A = C_0(\mathbb{R}^2)$ then $K_{00}(A) = \{0\}$ by Exercise 6.5.5. So K_{00} definitely differs from K_0. The following shows that the discrepancy is due to $C_0(\mathbb{R})$ representing a noncompact noncommutative space.

Theorem 8.2.2 *If a C*-algebra B represents a compact noncommutative space, then* $K_{00}(B) \cong K_0(B)$.

Proof It suffices to prove the statement for $B = A \otimes \mathcal{K}$ where A is unital, but then $B = \lim M_n(A)$, and so $K_0(B) \cong \lim K_0(M_n(A))$, the isomorphism being that induced by the system of *-homomorphisms $\varphi_n \colon M_n(A) \to B$, placing matrices in the top left corner as usual. The imagine of a K-theory class $[p]-[q] \in K_0(M_n(A))$ in $K_0(B)$ is then a difference $[p] - [q]$ of classes of projections in B (and not in B^+). Similar remarks hold for projections in matrix algebras over $M_n(A)$, of course this essentially changes nothing.

□

Example 8.2.3 We have already encountered the following examples of compact noncommutative spaces.

a) If G discrete acts properly and co-compactly on X then $C_0(X) \rtimes G$ represents a noncommutative space by Theorem 6.5.7.
b) If G is a locally compact group and $\Gamma \subset G$ a discrete subgroup and $H \subset G$ a co-compact subgroup, then $C_0(G/\Gamma) \rtimes H$ represents a compact noncommutative space by Theorem 6.5.14. This contains the example $C(SM) \rtimes P$ of Corollary 6.7.8, of the geodesic/horocyclic flow on the unit tangent sphere of a compact Riemann surface. It also contains the crossed products $B_\hbar = C(\mathbb{T}^2) \rtimes_{\beta^\hbar} \mathbb{R}$ of the Krönecker flow on $\mathbb{T}^2$ (Proposition 6.5.13).

We have already noted that if L is an f.g.p. module over A and $\mathcal{E}$ is a Morita equivalence between A and B then $L \otimes_A \mathcal{E}$ is an f.g.p. module over B. Tensoring with $\mathcal{E}$ in this way gives a semigroup isomorphism $\mathcal{P}(A) \to \mathcal{P}(B)$ (Theorem 6.2.3). Combining this with Theorem 8.2.2 we get:

Corollary 8.2.4 *Let $\mathcal{E}$ be a Morita A-B equivalence bimodule. If A (equivalently, B) represents a compact noncommutative space, then the map*

$$\mathcal{E}_* : K_0(A) \to K_0(B), \quad \mathcal{E}_*([L] - [L']) := [L \otimes_A \mathcal{E}] - [L' \otimes_A \mathcal{E}]$$

is an isomorphism.

By combining Theorems 6.5.10 and 6.5.7 we obtain the following alternative description of the K_0-theory of the orbifold C*-algebras $C_0(X) \rtimes G$, G discrete, acting properly and co-compactly.

Corollary 8.2.5 *If G locally compact acts properly and co-compactly on X, let $K_G^0(X)$ be the Grothendiek completion of the semigroup $\mathrm{Vect}_G(X)$ of isomorphism classes of G-equivariant vector bundles over X.*

Then if G is discrete, then $K_G^0(X) \cong K_0(C_0(X) \rtimes G)$, induced by the map $\mathrm{Vect}_G(X) \cong \mathcal{P}(C_0(X) \rtimes G)$ of Theorem 6.5.10.

Exercise 8.2.6 Let G be a finite group acting *freely* on X compact. Define a map $\tau_* : K^0(X) \to K^0(G \backslash X)$ by the composition of the map

$$K^0(X) = K_0(C(X)) \to K_0(C(X) \rtimes G)$$

induced by the inclusion $C(X) \to C(X) \rtimes G$, and the isomorphism

$$K_0(C(X) \rtimes G) \cong K^0(G \backslash X)$$

induced by the Morita equivalence $C(X) \rtimes G \sim C(G \backslash X)$.

a) Show that if V is a vector bundle over X then $\tau_*([V])$ is the class of the vector bundle $\tilde{V}$ on $G \backslash X$ with fibers

$$\tilde{V}_{Gx} := \oplus_{g \in G} V_{gx}.$$

b) Let $\pi : X \to G \backslash X$ the quotient map. Show that $(\tau_* \circ \pi^*)([V]) = |G| \cdot [V]$ for any vector bundle V over $G \backslash X$.

c) Suppose α is a finite-dimensional representation of G on H_α and let $V_\alpha := X \times_G H_\alpha = X \times H_\alpha / (x, v) \sim (gx, \alpha(g)v)$ be the induced vector bundle on $G \backslash X$. Show that $\pi^*(V_\alpha)$ is trivial. Deduce that

$$|G| \cdot ([V_\alpha] - \dim(V_\alpha)[1]) = 0,$$

where 1 is the trivial line bundle over $G \backslash X/$ Hence $[V_\alpha] - \dim(V_\alpha) \cdot [1] \in K^0(G \backslash X)$ is a torsion class.

8.3 Higher K-Theory, Loops and Unitaries

Higher K-theory groups are described as in topological K-theory by suspension.

Definition 8.3.1 If A is a C*-algebra, $i = 0, 1, 2, \ldots$, then we define $K_i(A) := K_0(S^n A)$, where $S^n(A) = C_0(\mathbb{R}^n) \otimes A$.

It is clear that K_i is functorial with respect to *-homomorphisms, and stable (that is, Morita invariant), for all i. Bott Periodicity will tell us that the groups $K_i(A)$ are actually automatically 2-periodic, and hence there are in effect only two of them.

In topological K-theory, we noted that a vector bundle over $(X \times \mathbb{R})^+$ can be trivialized over the closure in $(X \times \mathbb{R})^+$ of $X \times (-\infty, 0]$, and similarly can be trivialized over the closure of $X \times [0, \infty)$. The difference of the two trivializations on the intersection $\cong X$ of these two closed subsets (neglecting the point at infinity) gives a unitary map $u \colon X \to \mathbf{U}_n$. The argument shows that homotopy classes of such u's give a group isomorphic to $K^{-1}(X)$. (See Proposition 7.5.7). We now extend the construction to noncommutative C*-algebras.

Let A be a unital C*-algebra. Recall (see Proposition 7.5.7 and the that that $\mathbf{U}_\infty(A)$ denotes the group of all $\mathbb{N}$-by-$\mathbb{N}$-matrices with entries in A, which have a block-diagonal form $\begin{bmatrix} u & 0 \\ 0 & 1 \end{bmatrix}$ with u a (square) unitary matrix in $M_n(A)$, and 1 denoting the identity operator. There is an evident group structure on $\mathbf{U}_\infty(A)$ by multiplication, and we can regard, in the obvious way, all of the groups $\mathbf{U}(M_n(A))$ as subgroups of $\mathbf{U}_\infty(A)$.

Clearly $\mathbf{U}_\infty(A)$ is the inductive limit of the groups

$$\mathbf{U}_n(A) := \{u \in M_n(A) \mid u \text{ is unitary}\}.$$

We give it the corresponding inductive limit topology): a subset $U \subset \mathbf{U}_\infty(A)$ is open if and only if $U \cap \mathbf{U}(M_n(A))$ is open for all n.

We write $u, v \in \mathbf{U}_\infty(A)$ let $u \sim v$ if u and v are in the same path component of $\mathbf{U}_\infty(A)$. The quotient $\pi_0(\mathbf{U}_\infty(A))$ is an abelian group, by Proposition 7.5.7.

Proposition 8.3.2 If A is any unital C*-algebra, then $K_1(A)$ is naturally isomorphic to the abelian group $\pi_0(\mathbf{U}_\infty(A))$.

Proof By definition, $K_1(A) := K_0(S(A))$ is the kernel of the augmentation homomorphism $\epsilon_* \colon K_0((S(A)^+) \to \mathbb{Z}$, and in particular is a subgroup of $K_0((S(A)^+)$. The latter consists of equivalence classes of projections in $S(A)^+$, and a projection in $S(A)^+$ is a loop

$$p \colon [0, 1] \to M_n(A)$$

of projections in A, such that $p(0) = p(1) \in M_n(\mathbb{C}) \subset M_n(A)$. Any projection in $M_n(\mathbb{C})$ is unitarily equivalent to $\begin{bmatrix} 1_k & 0 \\ 0 & 0_{n-k} \end{bmatrix}$ for some k, and conjugating the loop by this unitary gives an equivalent loop with

$$p(0) = p(1) = \begin{bmatrix} 1_k & 0 \\ 0 & 0 \end{bmatrix}$$

so we replace the original loop with this one without change in notation.

By Exercise 8.1.27, there is a path of unitaries $u : [0, 1] \to M_n(A)$ with $u(1) = 1_n$ and

$$p(t) = u(t)p(1)u(t)^*, \quad t \in [0, 1].$$

We have

$$p(0) = u(0)p(1)u(0)^* = u(0)p(0)u(0)^*$$

and hence $u(0)$ commutes with $p(0) = p(1) = \begin{bmatrix} 1_k & 0 \\ 0 & 0 \end{bmatrix}$. Therefore it has a block-diagonal form

$$u(0) = \begin{bmatrix} v & 0 \\ 0 & w \end{bmatrix}$$

for some pair of unitaries $v \in M_k(A)$ and $w \in M_{n-k}(A)$.

We map

$$K_1(A) \to \pi_0\big(\mathbf{U}_\infty(A)\big)$$

by sending $[p]$ to $[v]$. We leave it to the reader to check that this assignment is well defined on K-theory classes $[p]$, and that it is a group homomorphism. To construct an inverse, let $v \in \mathbf{U}_m(A) \subset M_m(A)$ be a unitary. Then

$$\begin{bmatrix} v & 0 \\ 0 & v^* \end{bmatrix}$$

is a unitary which is in the same path component of $\mathbf{U}_\infty(A)$ as the identity $1_{2m} \in M_{2m}(\mathbb{C}) \subset M_n(A)$, by (7.13). Let, therefore,

$$u : [0, 1] \to \mathbf{U}_{2m}(A)$$

be a unitary-valued function such that

$$u(0) = \begin{bmatrix} v & 0 \\ 0 & v^* \end{bmatrix}, \quad u(1) = \begin{bmatrix} 1_m & 0 \\ 0 & 1_m \end{bmatrix}.$$

Now let

$$p\colon [0, 1] \to M_{2m}(A), \quad p(t) := u(t) \begin{bmatrix} 1_m & 0 \\ 0 & 0_m \end{bmatrix} u(t)^*.$$

Note that

$$p(0) = \begin{bmatrix} 1_m & 0 \\ 0 & 0_m \end{bmatrix} = p(1),$$

so that p is an element of $S(A)^+$. If now $q\colon [0, 1] \to M_{2m}(A)$ is the constant loop $q(t) = \begin{bmatrix} 1_m & 0 \\ 0 & 0_m \end{bmatrix}$, then

$$[p] - [q] \in K_0\big(S(A)\big)$$

and maps to $[u]$ under our construction above. □

To define K_1 for non-unital C*-algebras in terms of unitaries, let A be possibly non-unital. Then the augmentation homomorphism

$$\epsilon\colon A^+ \to \mathbb{C}$$

induces a group homomorphism $\mathbf{U}_\infty(A^+) \to \mathbf{U}_\infty(\mathbb{C})$ and then an induced group homomorphism

$$\pi_0\big(\mathbf{U}_\infty(A^+)\big) \to \pi_0\big(\mathbf{U}_\infty(\mathbb{C})\big).$$

Proposition 8.3.3 *For any C*algebra A,*

$$K_1(A) \cong \ker(\epsilon_*\colon \pi_0\big(\mathbf{U}_\infty(A^+)\big) \to \pi_0\big(\mathbf{U}_\infty(\mathbb{C})\big).$$

We leave the proof to the reader, using the ideas and constructions from the unital case.

In fact, since $K_1(\mathbb{C}) = K^{-1}(\mathrm{pt}) = 0$, the augmentation homomorphism ϵ_* is actually the zero map, so actually

$$K_1(A) \cong K_1(A^+) \cong \pi_0\big(\mathbf{U}_\infty(A^+)\big)$$

holds for any A.

One should take a bit of care with this statement, however. It is not true in KO-theory. (since it is no longer true that $KO^{-1}(\mathrm{pt}) = 0$.)

Remark 8.3.4 The discussion above shows that we now have two different ways of describing $K_2(\mathbb{C}) = K^{-2}(\mathrm{pt})$. The first is the definition

$$K_2(\mathbb{C}) := K_0\big(C_0(\mathbb{R}^2)\big),$$

defined as a certain subgroup of $K^0((\mathbb{R}^2)^+) = K^0(S^2)$ (the kernel of the augmentation homomorphism.)

On the other hand, $K^0(\mathbb{R}^2) = K^{-1}(\mathbb{R}) = K_1\big(C_0(\mathbb{R})\big)$ and according to the discussion above,

$$K_1\big(C_0(\mathbb{R})\big) \cong \pi_0\big(\mathbf{U}_\infty(C(\mathbb{T}))\big)$$

since $C_0(\mathbb{R})^+ = C(\mathbb{T})$.

Therefore, the "unitary description" in this case produces a group isomorphism

$$(8.4) \qquad \ker(\epsilon_*) \subset K^0(S^2) \to \pi_0\big(\mathbf{U}_\infty(C(\mathbb{T}))\big) \cong [\mathbb{T}, \mathbf{U}_\infty],$$

the last group being homotopy classes of maps $\mathbb{T} \to \mathbf{U}_\infty$, a group under pointwise multiplication of homotopy classes (another and briefer way of describing $\pi_0\big(\mathbf{U}_\infty(C(\mathbb{T}))\big)$.)

It is not difficult to check that this isomorphism is the clutching construction of Theorem 7.5.9. If $E \to S^2$ is any complex vector bundle, it can be trivialized over the top and bottom $S^2_\pm$ of the sphere. On the equator $\mathbb{T} = S^2_- \cap S^2_+$, one obtains, by following the inverse of the one trivialization, followed by the other, a map

$$\mathbb{T} \times \mathbb{C}^n \to \mathbf{GL}_n(\mathbb{C}),$$

and such a map determines a unique homotopy class $u(E) \in [\mathbb{T}, \mathbf{GL}_n(\mathbb{C})] \cong [\mathbb{T}, \mathbf{U}_\infty]$.

Notice that $u([1])$ is the zero element of the group $[\mathbb{T}, \mathbf{U}]$. However, $[1] \in K^0(S^2)$ is not zero. Thus, clutching directly describes a group homomorphism

$$K^0(S^2) \to [\mathbb{T}, \mathbf{U}_\infty]$$

which annihilates the class $[1]$ of the trivial bundle, and maps the class $[H] \in K^0(S^2)$ of the Hopf bundle $H \to S^2$ to the class of the unitary $z \colon \mathbb{T} \to \mathbb{C}$. In particular, it maps

$$\beta := [H^*] - [1]$$

to the class $[\bar{z}]$ of the unitary $\bar{z} \in C(\mathbb{T})$.

Exercise 8.3.5 Let $u \in \mathbf{U}(A)$ be a unitary in a unital C*-algebra A. It determines, by functional calculus, a *-homomorphism

$$C(\mathbb{T}) \to C(\mathrm{Spec}(u)) \to A,$$

where the first map is restrictions of functions on the circle $\mathbb{T}$ to the spectrum of u, and second is functional calculus for u.

The Cayley transform $T: \mathbb{R} \to \mathbb{T}$ is $T(x) = \frac{x-i}{x+i}$. It maps the point at infinity of $\mathbb{R}$ to $1 \in \mathbb{T}$, and gives a natural identification of $\dot{\mathbb{R}}$ with the open subset $\mathbb{T} - \{1\}$ of the circle.

Since there is a *-algebra inclusion of the ideal $C_0(\mathbb{T} - \{1\}) \subset C(\mathbb{T})$ we obtain a *-homomorphism

$$\alpha_u: C_0(\mathbb{R}) \subset C(\mathbb{T}) \to C\big(\mathrm{Spec}(u)\big) \to A,$$

with the last map functional calculus.

By functoriality of K-theory α_u determines a group homomorphism

$$(\alpha_u)_*: \mathrm{K}_1\big(C_0(\mathbb{R})\big) \to \mathrm{K}_1(A).$$

Prove that

$$(\alpha_u)_*(\beta) = [u],$$

where $\beta \in \mathrm{K}_1\big(C_0(\mathbb{R})\big) = \mathrm{K}^0(\mathbb{R}^2)$ is the Bott element.

Deduce that if $\mathrm{Spec}(u) \subset \mathbb{T}$ is a *proper* subset of the circle, then $[u] = 0 \in \mathrm{K}_1(A)$. (*Hint*. The unitary $z|_{\mathrm{Spec}(u)} \in C\big(\mathrm{Spec}(u)\big)$ is connected by a path of unitaries in $C\big(\mathrm{Spec}(u)\big)$ to 1.)

8.4 The Long Exact Sequence

We now develop the *long exact sequence* in K-theory associated to an ideal $J \subset A$ of a C*-algebra. For this it will be convenient to describe a "relative" version of K_0, similar to the 'K-theory triples' discussed previously.

Definition 8.4.1 Let A be a unital C*-algebra and $J \subset A$ an ideal. Let $\pi: A \to A/J$ be the quotient map

A *relative triple* is a triple (p, q, x) where p, q are projections in $M_n(A)$, for some n, $x \in M_n(A)$, and

$$\pi(x)^*\pi(x) = p, \quad \pi(x)\pi(x)^* = q.$$

A triple is *degenerate* if $x^*x = p$ and $xx^* = q$ in A.

A homotopy of triples is a triple of continuous paths p_t, q_t and x_t, in $M_n(A)$, $t \in [0, 1]$, such that (p_t, q_t, x_t) is a triple for all t. We say the endpoints (p_0, q_0, x_0) and (p_1, q_1, x_1) are *homotopic* triples.

Exercise 8.4.2 Show that (p, q, x) is a relative triple, and $x' \in A$ with $x - x' \in J$, then (p, q, x') is a relative triple which is homotopic to (p, q, x). (*Hint*. Straight line homotopy.)

The reader might recognize the similarity to the K-theory triples (E^0, E^1, φ) we introduced in connection with K-theory of noncompact spaces X. Let $U \subset X$ be an open subset and $\varphi \colon E^0 \to E^1$ be a vector bundle map which is an isomorphism on $X \setminus U$.

Then the triple defines a relative triple for the ideal $C_0(U)$ in $C_0(X)$, since $\varphi|_{X \setminus U}$ is an isomorphism between the two bundles, and hence determines a Murray–von-Neumann equivalence between projections p_0, p_1 such that $E^i \cong \operatorname{Im}(p_i)$.

For example, let $X = \overline{\mathbb{D}}$ the closed unit disk in $\mathbb{C}$, and $z \in C(\overline{\mathbb{D}})$ the usual complex variable, here considered as a bundle map from the trivial line bundle $\mathbf{1}$ on $\overline{\mathbb{D}}$, to itself.

Then its restriction to $\partial \mathbb{D}$ is unitary, and hence $(\mathbf{1}, \mathbf{1}, z)$ defines a relative triple for the ideal $C_0(\mathbb{D})$ of $C(\overline{\mathbb{D}})$.

Definition 8.4.3 The relative group $\mathrm{K}_0(A, A/J)$ is defined to be the free abelian group with one generator for each *homotopy class* of relative triple (p, q, x), subject to the relations

a) $(p_0, q_0, x_0) + (p_1, q_1, x_1) = (p_0 \oplus p_1, q_0 \oplus q_1, x_0 \oplus x_1)$, for any pair of triples (p_0, q_0, x_0) and (p_1, q_1, x_1);
b) Degenerate triples are zero.

Remark 8.4.4 Let A be any C*-algebra, A^+ its unitization. Then A is an ideal in A^+, with quotient map $\epsilon \colon A^+ \to A^+/A \cong \mathbb{C}$. By definition $\mathrm{K}_0(A) = \ker(\epsilon_*)$ with $\epsilon_* \colon \mathrm{K}_0(A^+) \to \mathrm{K}_0(\mathbb{C}) \cong \mathbb{Z}$ the induced group homomorphism.

Define a map

(8.5) $$\mathrm{K}_0(A^+, A^+/A) \to \mathrm{K}_0(A)$$

by sending a triple (p, q, x) of elements of $M_n(A^+)$ to $[p] - [q] \in \mathrm{K}_0(A^+)$. The element $\epsilon(x) =: v \in M_n(\mathbb{C})$ is a partial isometry with $v^*v = \epsilon(p)$, $vv^* = \epsilon(q)$, by the definitions, whence $[p] - [q] \in \ker(\epsilon_*)$, so our map has range in $\mathrm{K}_0(A)$.

Conversely, if $[p] - [q] \in \ker(\epsilon_*) = \mathrm{K}_0(A)$ with $p, q \in M_n(A^+)$, then the projections $\epsilon(p)$ and $\epsilon(q)$ in $M_n(\mathbb{C})$ determine the same K-theory class for $\mathbb{C}$ and hence have the same rank. There is then a partial isometry $z \in M_n(\mathbb{C})$ such that $z^*z = \epsilon(p)$ and $zz^* = \epsilon(q)$, so if $x \in M_n(A^+)$ is any lift of x under ϵ then (p, q, x) is a relative triple for the ideal A in A^+, which maps to $[p] - [q]$ under (8.5). Hence the map is surjective. Injectivity is left to the reader as an exercise; the conclusion is that

(8.6) $$\mathrm{K}_0(A) \cong \mathrm{K}_0(A^+, A^+/A).$$

So ordinary K-theory is a special case of the relative theory.

Exercise 8.4.5 Let A, A' be unital C*-algebras, $J \subset A$ and $J' \subset A'$ ideals. Then a *-homomorphism

$$\alpha \colon A \to A'$$

which maps J into J', induces a group homomorphism

$$K_0(A, A/J) \to K_0(A', A'/J').$$

Let $J \subset A$, A unital. Then J^+ can be identified with a C*-subalgebra of A. The inclusion maps the ideal J into A. So we have an induced map

(8.7) $$K_0(J) \cong K_0(J^+, J^+/J) \to K_0(A, A/J)$$

which we call the *excision map*. (The first isomorphism is a case of (8.6)).

We are going to show that the excision map is an isomorphism. We prove it in several steps.

As in our discussion of K-theory triples for noncompact spaces, we start by showing that the first projection in a relative triple can be taken to be "trivial," without changing the class of the relative triple.

Lemma 8.4.6 *Any relative triple for the ideal J of A is equivalent to a relative triple in which*

$$p = \begin{bmatrix} 1_k & 0 \\ 0 & 0 \end{bmatrix} \in M_n(\mathbb{C}) \subset M_n(A),$$

for some k and n.

Proof Suppose that (p, q, x) be a relative triple with $p, q, x \in A$ for simplicity. Note that $(1 - p, 1 - p, 1 - p)$ is a degenerate triple. Adding it to the original triple gives the equivalent triple $(p \oplus 1 - p, q', x')$ for some q', x'.

Using rotation matrices $R_t = \begin{bmatrix} \cos t & -\sin t \\ \sin t & \cos t \end{bmatrix}$ we construct the homotopy of triples

$$\left(\begin{bmatrix} p & 0 \\ 0 & 0 \end{bmatrix} + R_t \begin{bmatrix} 0 & 0 \\ 0 & 1 - p \end{bmatrix} R_t^*, R_t q' R_t^*, R_t x R_t^* \right).$$

As $R_{\frac{\pi}{2}} = \begin{bmatrix} 0 & -1 \\ 1 & 0 \end{bmatrix}$ conjugates $\begin{bmatrix} 0 & 0 \\ 0 & 1 - p \end{bmatrix}$ to $\begin{bmatrix} 1 - p & 0 \\ 0 & 0 \end{bmatrix}$, at the $t = \frac{\pi}{2}$ end of the path, we get a triple of the required kind, and at the $t = 0$ end of the path we get the cycle $(p \oplus 1 - p, q', x')$. The result follows. $\quad\square$

Lemma 8.4.7 *Any relative triple is equivalent to one of the form (p, q, x), where $x = up$ for some unitary $u \in M_n(A)$ connected to the identity in $M_n(A)$ by a path of unitaries, and satisfying $upu^* = q$ mod J.*

Proof As in Exercise 5.2.19, let (p, q, x) be a triple. Let

$$w = \begin{bmatrix} \pi(x), & 1 - \pi(x)\pi(x)^* \\ \pi(x)^*\pi(x) - 1 & \pi(x)^* \end{bmatrix}.$$

Then w is unitary in $M_2(A/J)$ and is connected to the identity $\begin{bmatrix} 1 & 0 \\ 0 & 1 \end{bmatrix}$ by a path of unitaries in $M_2(A/J)$. Since a path of unitaries starting at the identity can be lifted under $\pi : A \to A/J$ to a path of unitaries starting at the identity in $M_2(A)$, we get a unitary $u \in A$, connected to the identity, and such that $\pi(u) = w$. Then, working mod J, we compute

$$u \begin{bmatrix} p & 0 \\ 0 & 0 \end{bmatrix} = \begin{bmatrix} x & 1 - xx^* \\ x^*x - 1 & x^* \end{bmatrix} \begin{bmatrix} p & 0 \\ 0 & 0 \end{bmatrix} = \begin{bmatrix} x & 0 \\ 0 & 0 \end{bmatrix}$$

since $xp = x$ mod J and $(x^*x - 1)p = 0$ mod J, due to $x^*x = p$ mod J. The triple

$$\left(\begin{bmatrix} p & 0 \\ 0 & 0 \end{bmatrix}, \begin{bmatrix} q & 0 \\ 0 & 0 \end{bmatrix}, \begin{bmatrix} x & 0 \\ 0 & 0 \end{bmatrix} \right)$$

is a degenerate perturbation of the cycle we started with, and $\begin{bmatrix} x & 0 \\ 0 & 0 \end{bmatrix} - u \begin{bmatrix} p & 0 \\ 0 & 0 \end{bmatrix}$ is in J by the computation just done. So the triple we started with is equivalent to

$$\left(\begin{bmatrix} p & 0 \\ 0 & 0 \end{bmatrix}, \begin{bmatrix} q & 0 \\ 0 & 0 \end{bmatrix}, u \begin{bmatrix} p & 0 \\ 0 & 0 \end{bmatrix} \right).$$

The fact that $upu^* = q$ mod J follows from the construction of u, which equals w mod J.

This proves the Lemma.   □

Lemma 8.4.8 *Any triple (p, q, pu) where u is a unitary in $M_n(A)$ connected to the identity by a path of unitaries, and such that $upu^* = q$, is equivalent to one of the form (p, q, p).*

Note that if one has a triple of the form (p, q, p) then by the definitions, it must be that $p - q \in J$.

Proof Let $(u_t)_{t \in [0,1]}$ be a path of unitaries with $u_1 = u$ and $u_0 = 1$. Then

$$(p, u_t^* u q u^* u_t, pu_t)$$

is a path of triples. When $t = 0$ we get (p, uqu^*, p). When $t = 1$ we get (p, q, pu), the original triple. The result is proved.   □

Theorem 8.4.9 *For any ideal J in a unital C*-algebra A, the excision map (8.7) is an isomorphism.*

Proof We prove surjectivity and leave injectivity as an exercise.

Suppose that (p, q, x) is a relative triple. By Lemma 8.4.6, it is equivalent to a triple in which $p = \begin{bmatrix} 1_k & 0 \\ 0 & 0 \end{bmatrix}$, and in particular, p is in the range of the inclusion $J^+ \to A$. Next, by Lemma 8.4.7, the triple can be replaced by one in which $x = up$, where u is a unitary in A, connected to the identity by a path of unitaries, and by Lemma 8.4.8 the unitary u can be removed by a homotopy, to get a triple now of the form (p, q, p). The projection p remains of the form $\begin{bmatrix} 1_m & 0 \\ 0 & 0 \end{bmatrix}$ for some m, since p has not been changed through any of the previous steps, except having zeros added to it. In particular, since $p - q \in J$. Now since $p = \begin{bmatrix} 1_m & 0 \\ 0 & 0 \end{bmatrix}$, both p and q are in the range of the inclusion $J^+ \to A$. The excision map sends $[p] - [q]$ to the class of the triple (p, q, x) as claimed, and so excision is surjective.   □

8.5 The Long Exact Sequence

As a consequence of the excision isomorphism, we deduce the existence of the long exact sequence in C*-algebra K-theory, as follows.

Let A be unital, $J \subset A$ an ideal. Let $\alpha \colon K_0(A/A/J) \to K_0(A)$ be the group homomorphism

$$\alpha([(p, q, x)]) := [p] - [q] \in K_0(A).$$

Lemma 8.5.1 *The sequence of groups*

$$K_0(A, A/J) \xrightarrow{\alpha} K_0(A) \xrightarrow{\pi_*} K_0(A/J)$$

is exact in the middle.

Proof Clearly $\pi_* \circ \alpha$ is the zero homomorphism, so $\mathrm{ran}(\alpha) \subset \ker(\pi^*)$. To show the other equality, let $[p] - [q] \in K_0(A)$, where p, q are projections in $M_n(A)$ for some n, such that $\pi_*([p] - [q]) = 0$. Then by Exercise 8.1.6, $\pi(p) \oplus 1_k$ is Murray–von-Neumann equivalent to $\pi(q) \oplus 1_k$, for some k, so there exists $z \in M_{n+k}(A/J)$ such that $z^* z = \pi(p) \oplus 1_k$, $zz^* = \pi(q) \oplus 1_k$. Let $x \in M_{n+k}(A)$ be any lift of z. Then (p, q, x) is a triple such that $\alpha([(p, q, x)]) = [p] - [q]$.   □

Corollary 8.5.2 *Let J be an ideal in a unital C*-algebra A, Then the sequence of groups*

$$K_0(J) \to K_0(A) \to K_0(A/J)$$

is exact in the middle.

The proof is easy, by the Excision theorem.

Let F be any functor from the category of C*-algebras and C*-algebra homomorphisms, to the category of abelian groups. Since F is a functor, the inclusion $i : J \to A$ of an ideal in a C*-algebra determine a sequence of group homomorphisms

$$F(J) \xrightarrow{F(i)} F(A) \xrightarrow{F(\pi)} F(A/J).$$

The functor is called *half-exact* if this sequence of group homomorphisms is exact in the middle. The functor is *homotopy invariant* if $F(\alpha) = F(\beta)$ for any pair of homotopic *-homomorphisms $\alpha, \beta : A \to B$.

We have proved so far that the functor K_0 is both homotopy invariant and half-exact.

We are going to show that *any* half-exact, homotopy invariant functor determines a long exact sequence of the form

$$F(J) \xrightarrow{F(i)} F(A) \xrightarrow{F(\pi)} F(A/J)$$

$$\xrightarrow{\delta} F(SJ) \xrightarrow{F(Si)} F(SA) \xrightarrow{F(S\pi)} F\big(S(A/J)\big)$$

(8.8) $$\xrightarrow{\delta} F(S^2 J) \xrightarrow{F(S^2 i)} F(S^2 A) \xrightarrow{F(S^2 \pi)} F\big(S^2(A/J)\big) \xrightarrow{\delta} \cdots$$

This will in particular hold for the functor K_0.

Let $\pi : A \to A/J$ be the quotient map. We define two auxiliary C*-algebras. Let

$$C_\pi := \{(a, f) \in A \oplus C([0, 1], A/J) \mid f(0) = 0, \ f(1) = \pi(a)\},$$

(8.9) $$Q := \{f : [0, 1] \to A \mid f(0) \in J\}$$

called the *mapping cone* of π.

There is an obvious inclusion $i : J \to Q$, as constant functions. There is also a map $\rho : Q \to J$, defined $\rho(f) := f(0)$. Clearly $\rho : i = \mathrm{id}_J$. On the other hand,

$$\iota_t(f)(s) := f(ts)$$

gives a homotopy between $i \circ \rho$ and the identity homomorphism $Q \to Q$.

That is, i and ρ are homotopy-inverses of each other.

Let $k : J \to C_\pi$ be the inclusion $j(a) := (a, 0)$.

Lemma 8.5.3 k induces an isomorphism $k_* : K_*(J) \to K_*\big(C_\pi\big)$.

Proof Consider the map

(8.10) $$\alpha : Q \to C_\pi, \quad \alpha(f) = (f(1), \pi \circ f).$$

Note that $(\pi \circ f)(0) = 0$ since $f(0) \in J$. The kernel of α is the ideal $\{f \in Q \mid f(1) = 0, \ \pi \circ f = 0\}$ and such f map into J. It follows that

$$\ker(\alpha) \cong C_0\big((0, 1], J\big),$$

which is a contractible C*-algebra (Exercise 8.1.26). The sequence of groups

$$(8.11) \qquad\qquad \mathrm{K}_0\big(\ker(\alpha)\big) \to \mathrm{K}_0(Q) \to \mathrm{K}_0(C_\pi)$$

is exact in the middle, and $\mathrm{K}_0\big(\ker(\alpha)\big)$ is the zero group by the above discussion. Hence

$$\alpha_* : \mathrm{K}_0(Q) \to \mathrm{K}_0(C_\pi)$$

is injective.

Now the composition

$$J \xrightarrow{i} Q \xrightarrow{\alpha} C_\pi$$

equals k. Hence $k_* = \alpha_* \circ i_*$ and i_* is an isomorphism and α_* is injective. Hence k_* is injective.

For surjectivity of k_* observe that k actually embeds J as an *ideal* in C_π. It is the kernel of the map

$$\beta : C_\pi \to C\big((0, 1], A/J\big), \quad \beta(a, f) := f.$$

We obtain a sequence of groups

$$(8.12) \qquad\qquad \mathrm{K}_0(J) \xrightarrow{k_*} \mathrm{K}_0(C_\pi) \to \mathrm{K}_0\big[C\big((0, 1], A/J\big)\big]$$

and $\mathrm{K}_0\big[C_0\big((0, 1], A/J\big)\big]$ is the zero group, since $C_0\big((0, 1], A/J\big)$ is contractible. This shows that k_* is surjective. $\square$

Note that $S(A/J) := \{f : [0, 1] \to A/J \mid f(0) = f(1) = 0\}$ is an ideal in C_π. We let $s : S(A/J) \to C_\pi$ be the inclusion.

Definition 8.5.4 The *connecting homomorphism*, or boundary homomorphism,

$$\partial : \mathrm{K}_1(A/J) \to \mathrm{K}_0(J)$$

is defined to be the composition

$$(8.13) \qquad \mathrm{K}_1(A/J) := \mathrm{K}_0\big(S(A/J)\big) \xrightarrow{s_*} \mathrm{K}_0(C_\pi) \xrightarrow{k_*^{-1}} \mathrm{K}_0(J),$$

where the first map is induced from the inclusion $s\colon S(A/J) \to C_\pi$ and the second map is the inverse of the group isomorphism $k_*\colon K_0(J) \to K_0(C_\pi)$ induced by the inclusion $k\colon J \to C_\pi$.

We now take our extension $0 \to J \to A \to A/J \to 0$. It generates the sequence of groups and group homomorphisms

$$\text{(8.14)} \qquad K_0(J) \xrightarrow{j_*} K_0(A) \xrightarrow{\pi_*} K_0(A/J),$$

which is exact in the middle.

The sequence $0 \to C_0(\mathbb{R}) \otimes J \to C_0(\mathbb{R}) \otimes A \to C_0(\mathbb{R}) \otimes A/J \to 0$ is still exact, and writing $C_0(\mathbb{R}) \otimes J = SJ$ and so on, we get a sequence of groups

$$\text{(8.15)} \qquad K_0(SJ) \to K_0(SA) \to K_0\big(S(A(A/J))\big)$$

which by the definitions can be written

$$\text{(8.16)} \qquad K_1(J) \xrightarrow{j_*} K_1(A) \xrightarrow{\pi_*} K_1(A/J).$$

It is exact in the middle.

We then connect the end of (8.16) with (8.14) using the boundary map (8.13). to get the spliced-together sequence

$$\text{(8.17)} \quad K_1(J) \xrightarrow{j_*} K_1(A) \xrightarrow{\pi_*} K_1(A/J) \xrightarrow{\partial} K_0(J) \xrightarrow{j_*} K_0(A) \xrightarrow{\pi_*} K_0(A/J).$$

Lemma 8.5.5 *In reference to* (8.17), *we have* $\ker(\partial) = \operatorname{ran}(\pi_*)$ *and* $\operatorname{ran}(\partial) = \ker(j_*)$.

Proof Consider the diagram

$$K_1(A/J) \xrightarrow{s_*} K_0(C_\pi) \xrightarrow{p_*} K_0(A)$$

$$\partial \searrow \quad \nearrow_{k_*} \quad \nearrow_{j_*}$$

$$K_0(J)$$

Here s is the map induced by the inclusion $s\colon S(A/J) \to C_\pi$ of $S(A/J)$ as an ideal in C_π, p_* is induced by the map $C_\pi \to A$, whose kernel is the image of s, and $k\colon J \to C_\pi$ is the inclusion as constant functions.

The diagram commutes by the definitions. The top row is exact in the middle because it comes from a short exact sequence of C*-algebras. It is now apparent that $\operatorname{ran}(\partial) = \ker(j_*)$, as claimed.

Now consider the exact sequence

$$\text{(8.18)} \qquad 0 \to S(A/J) \xrightarrow{s} C_\pi \xrightarrow{q} A \to 0,$$

where $q(a, f) = a$. It generates a connecting map ∂' fitting into a sequence

(8.19)

$$K_1\big(S(A/J)\big) \to K_1(C_\pi) \to K_1(A) \xrightarrow{\partial'} K_0\big(S(A/J)\big) \xrightarrow{s_*} K_0(C_\pi) \to K_0(A).$$

and $\mathrm{ran}(\partial') = \ker(s_*)$ by what has already been proved, while $\ker(s_*) = \ker(\partial)$ by the definitions, so we get that

$$\ker(\partial) = \mathrm{ran}(\partial').$$

But the map ∂' is simply equal to π_* as a map $K_1(A) \to K_1(A/J)$. The result follows.   □

Theorem 8.5.6 *Let J be an ideal in a C*-algebra A. Then there exist connecting homomorphisms $\partial \colon K_{i+1}(A/J) \to K_i(A)$, for each $i = 0, 1, \ldots$, making the sequence of groups and group homomorphisms*

$$\cdots \to K_2(J) \xrightarrow{i_*} K_2(A) \xrightarrow{\pi_*} K_0(A/J)$$

$$\xrightarrow{\partial} K_1(J) \xrightarrow{i_*} K_1(A) \xrightarrow{\pi_*} K_1(A/J)$$

(8.20) $$\xrightarrow{\partial} K_0(J) \xrightarrow{i_*} K_0(A) \xrightarrow{\pi_*} K_0(A/J)$$

exact (with nothing known about the right endpoint), and with the following naturality property.

If $\varphi \colon A_1 \to A_2$ is a *-homomorphism, mapping an ideal $J_1 \subset A_1$ to an ideal J_2 in A_2, then the diagram

(8.21)
$$
\begin{array}{ccccccccc}
\cdots & K_{i+1}(J_1) & \longrightarrow & K_{i+1}(A_1) & \longrightarrow & K_{i+1}(A_1/J_1) & \longrightarrow & K_{i-1}(J_1) & \longrightarrow & K_{i-1}(A_1) & \longrightarrow & \cdots \\
& \downarrow & & \downarrow & & & & \downarrow \\
\cdots & K_{i+1}(J_2) & \longrightarrow & K_{i+1}(A_2) & \longrightarrow & K_{i+1}(A_2/J_2) & \longrightarrow & K_{i-1}(J_2) & \longrightarrow & K_{i-1}(A_2) & \longrightarrow & \cdots
\end{array}
$$

commutes, with the top and bottom being the long exact sequences associated to the ideals $J_1 \subset A_1$ and $J_2 \subset A_2$.

An explicit description of the connecting homomorphism

We finish this section with a fairly specific description of the connecting homomorphism

$$\delta \colon K_1(A/J) := K_0\big(S(A/J)\big) \to K_0(J),$$

of Definition 8.5.4, associated to an ideal $J \subset A$.

Assume A is unital, so $\mathbb{C} \subset A$ naturally, by multiplying against the unit of A. The quotient mapping is unital. So we also have a copy of $\mathbb{C}$ in A/J.

We consider $S(A/J)$ to be the C*-algebra of continuous $f : [0, 1] \to A/J$ with $f(0) = f(1) = 0$. Then its unitization $S(A/J)^+$ is the C*-algebra of continuous $f : [0, 1] \to A/J$ such that $f(0) = f(1) \in \mathbb{C} \subset A/J$.

On the other hand, the mapping cone C_π is the C*-algebra of pairs (a, f) in the direct sum $A \oplus C([0, 1], A/J)$ such that $f(0) = 0$ and $f(1) = \pi(a)$. Its unitization C_π^+ is pairs (a, f) where $f : [0, 1] \to A/J$ with $f(0) \in \mathbb{C} \subset A/J$ and $f(1) = \pi(a)$. As above, we have a natural injective *-homomorphism $s : S(A/J) \to C_\pi$ and it extends canonically to a *-homomorphism

$$(8.22) \qquad s : S(A/J)^+ \to C_\pi^+, \quad s(f) = \big(f(0), f\big).$$

Note that $f(0) = f(1) \in \mathbb{C}$ in this formula, so $s(f)$ lies in C_π^+. The other ingredient in the boundary map is the inclusion $k : J \to C_\pi$. It extends to a *-homomorphism

$$(8.23) \qquad k : J^+ \to C_\pi^+, \quad k(a) := (a, \pi(a)),$$

for $a \in J^+$ understood as a C*-subalgebra of A. Note that the restriction of the quotient map π to J^+, has kernel exactly equal to J.

Now take a projection $p \in S(A/J)^+$. It is a loop of projections: a continuous, projection-valued map $p : [0, 1] \to A/J$ with $p(0) = p(1) = \lambda \in \mathbb{C}$. By the path-lifting property of projections, p lifts under the quotient map $\pi : A \to A/J$ to a path

$$\tilde{p} : [0, 1] \to A$$

of projections in A such that $\tilde{p}(1) = \lambda$. We know that $\pi\big(\tilde{p}(0)\big) = p(0) = \lambda$, so that $\tilde{p}(0) - \lambda$ maps to zero under the quotient map $\pi : J^+ \to A$ and hence $\tilde{p}(0) - \lambda \in J$, whence $\tilde{p}(0) \in J^+$.

We set

$$(8.24) \qquad \mathrm{Twist}(p) := [\tilde{p}(0)] \in \mathrm{K}_0(J^+).$$

Now consider the image of $[p]$ under $s_* : \mathrm{K}_0\big(S(A/J)^+\big) \to \mathrm{K}_0(C_\pi^+)$. By the definitions, see (8.5.4), the class $s_*([p])$ is the class of the projection in C_π^+ given by the element

$$\big(p(0), p\big) \in C_\pi^+.$$

On the other hand, consider the map $k : J^+ \to C_\pi^+$, given by (8.23). By its definition, $k_*([\tilde{p}(0)]) \in \mathrm{K}_0(C_\pi^+)$ is the class of the projection $\big(\tilde{p}(0), p(0)\big) \in C_\pi^+$.

Lemma 8.5.7 *In the above notation, the projections $\big(\tilde{p}(0), p(0)\big)$ and $\big(p(0), p\big)$ in C_π^+ are homotopic.*

In particular,

$$k_*([\tilde{p}(0)]) = s_*([p]) \in K_0(C_\pi^+).$$

Proof Let $q_s \in C_\pi^+$ be the projection $q_s := (\tilde{p}(s), p_s)$, where $p_s : [0, 1] \to A/J$ is the projection-valued map $p_s(t) := p(ts)$. For each $s \in [0, 1]$, $\pi(\tilde{p}(s)) = p(s) = p_s(1)$, and, moreover, $p_s(0) = p(0) \in \mathbb{C}$, so q_s really is in C_π^+. It is clearly a projection. When $s = 0$, since p_0 is the constant function $p(0)$ we obtain the projection

$$(\tilde{p}(0), p(0)))$$

and when $s = 1$ we get, since $p_1 = p$,

$$(\tilde{p}(1), p).$$

Moreover, $\tilde{p}(1) = p(1) = p(0) \in \mathbb{C}$, by construction, and so the two endpoints of our path are the two given projections, as claimed.   □

Now let $p, q \in S(A/J)^+$ be projections such that $[p] - [q] \in K_0(S(A/J)) = \ker(\epsilon_*)$, with $\epsilon : S(A/J)^+ \to \mathbb{C}$ the usual augmentation.

I claim that

$$\delta([p] - [q]) = \text{Twist}(p) - \text{Twist}(q) \in K_0(J).$$

Indeed, from Lemma 8.5.7,

$$s_*([p] - [q]) = s_*([p]) - s_*([q]) = k_*([\tilde{p}(0)]) - k_*([\tilde{q}(0)])$$
$$= k_*(\text{Twist}(p) - \text{Twist}(q))$$

and hence

$$\delta([p] - [q]) := (k_*^{-1} \circ s_*)([p] - [q]) = \text{Twist}(p) - \text{Twist}(q)$$

follows.

The extension of this argument to where the projections are matrix-valued, is routine and is left to the reader. In fact it essentially follows from simply replacing A by $M_n(A)$, J by $M_n(J)$, etc., in the given argument.

Theorem 8.5.8 *Let $J \subset A$ be an ideal, A unital. Let $p : [0, 1] \to M_n(A/J)$ be a continuous, projection-valued map with $p(0) = p(1) \in M_n(\mathbb{C}) \subset M_n(A/J)$. Let $\tilde{p}(1)$ be a lift of $p(1) \in M_n(\mathbb{C})$ to $M_n(A)$, and let $\tilde{p} : [0, 1] \to M_n(A)$ be a lifting of the path p with $\tilde{p}(1)$ prescribed as in the previous sentence. Then $\tilde{p}(0) \in J^+ \subset A$, and if*

$$\text{Twist}(p) := [\tilde{p}(0)] \in K_0(J^+),$$

then the connecting map

$$\delta \colon K_1(A/J) \to K_0(J)$$

satisfies

$$\delta([p] - [q]) = \text{Twist}(p) - \text{Twist}(q),$$

for any group element $[p] - [q] \in K_1(A/J)$.

Exercise 8.5.9 The Mischenko element β defines a canonical loop of vector bundles over the circle $\mathbb{T}$, which is the boundary of the closed disk $\overline{\mathbb{D}}$. Since $\mathbb{D}^+ \cong S^2$, the 2-sphere,

$$\text{Twist}(\beta) \in K^0(S^2).$$

Show that $\text{Twist}(\beta) = [H]$ is the class of the Hopf bundle.

Finally, we complete this section by describing the connecting map

$$\delta \colon K_1(A/J) \to K_0(J)$$

in terms of the description of $K_1(A/J)$ as equivalence classes of unitaries in (matrix algebras over) A/J.

Lemma 8.5.10 *If A is a unital C*-algebra and $a \in A$ with $\|a\| \leq 1$, then*

$$(8.25) \qquad w := \begin{bmatrix} a & -(1 - aa^*)^{\frac{1}{2}} \\ (1 - a^*a)^{\frac{1}{2}} & a^* \end{bmatrix}$$

is a unitary in $M_2(A)$.

Proof This follows from a direct calculation using the fact that

$$a(1 - a^*a)^{\frac{1}{2}} = (1 - aa^*)^{\frac{1}{2}}a$$

(see Exercise 3.5.11.)   □

Now let $w \in M_n(A/J)$ be a unitary. We need to describe the corresponding cycle for $K_0(SA)$ and describe its twist, which will be a cycle for $K_0(J)$, so there will be two steps in the calculation.

Firstly, looking back at the proof of Proposition 8.3.2, we need to find a path of unitaries $w \colon [0, 1] \to M_{2n}(A/J)$ such that

$$(8.26) \qquad w(0) = \begin{bmatrix} w & 0 \\ 0 & w^* \end{bmatrix}, \quad w(1) = \begin{bmatrix} 1_n & 0 \\ 0 & 1_n \end{bmatrix}.$$

From this is obtained a loop of projections in $M_{2n}(A)$ by setting

$$p_t := w(t) \begin{bmatrix} 1_n & 0 \\ 0 & 0_n \end{bmatrix} w(t)^*.$$

In the second step, the twist $\mathrm{Twist}(p)$ is by definition obtained by lifting this path of projections in A/J to a path of projections in A, starting at $\begin{bmatrix} 1_n & 0 \\ 0 & 0_n \end{bmatrix} \in M_n(J^+) \subset M_n(A)$.

We can carry out both steps at once quite efficiently, however. Let $a \in M_n(A)$ be any lift of $u \in M_n(A/J)$ to an element of A with $\|a\| \leq 1$. Set

$$a_t := t \cdot 1_n + (1 - t) \cdot a \in M_n(A).$$

Then $\|a_t\| \leq 1$, and hence

(8.27)
$$u(t) := \begin{bmatrix} a_t & -(1 - a_t a_t^*)^{\frac{1}{2}} \\ (1 - a_t^* a_t)^{\frac{1}{2}} & a_t^* \end{bmatrix}$$

is unitary in $M_{2n}(A)$. Note that $\pi(a_0) = \pi(a) = u$ and that $u_1 = \begin{bmatrix} 1_n & 0 \\ 0 & 1_n \end{bmatrix} = 1_{2n}$.

Therefore the path $w := \pi \circ u \colon [0, 1] \to M_{2n}(A/J)$ is as required by (8.26). We obtain the loop of projections in $M_{2n}(A/J)$,

$$p_t := w(t) \begin{bmatrix} 1_n & 0 \\ 0 & 0_n \end{bmatrix} w(t)^*.$$

This loop, however, has a ready-make lift to a path of projections in $\tilde{p}_t \in M_{2n}(A)$ starting at $\begin{bmatrix} 1_n & 0 \\ 0 & 0_n \end{bmatrix}$ since we may set

$$\tilde{p}(t) := u(t) \begin{bmatrix} 1_n & 0 \\ 0 & 0_n \end{bmatrix} u(t)*$$

with $u(t)$ the path of unitaries in (8.27) .

This lifted path has endpoints

$$\tilde{p}(1) = \begin{bmatrix} 1_n & 0 \\ 0 & 0_n \end{bmatrix}$$

and

$$\tilde{p}(0) = \begin{bmatrix} a & -(1-aa^*)^{\frac{1}{2}} \\ (1-a^*a)^{\frac{1}{2}} & a^* \end{bmatrix} \begin{bmatrix} 1_n & 0 \\ 0 & 0_n \end{bmatrix} \begin{bmatrix} a^* & (1-a^*a)^{\frac{1}{2}} \\ -(1-aa^*)^{\frac{1}{2}} & a \end{bmatrix}$$

$$(8.28) \qquad = \begin{bmatrix} aa^* & a(1-a^*a)^{\frac{1}{2}} \\ (1-a^*a)^{\frac{1}{2}} & 1-aa^* \end{bmatrix},$$

which is a projection in $M_{2n}(A)$. By the definitions,

$$\mathrm{Twist}(p) = [\tilde{p}(0)] - [1_n] \in \mathrm{K}_0(J^+).$$

We have proved the following.

Theorem 8.5.11 *Let $u \in M_n(A/J)$ be a unitary representing a class $[u] \in \mathrm{K}_1(A/J)$. Let $a \in M_n(A)$ such that $\|a\| \le 1$ and $\pi(a) = u$.*
   Set

$$(8.29) \qquad q := \begin{bmatrix} aa^* & a(1-a^*a)^{\frac{1}{2}} \\ (1-a^*a)^{\frac{1}{2}}a^* & 1-a^*a \end{bmatrix} \in M_{2n}(A).$$

Then q is a projection in $M_{2n}(J^+)$ such that $\pi(q) = \begin{bmatrix} 1_n & 0 \\ 0 & 0_n \end{bmatrix} \in M_{2n}(A/J)$, and

$$\delta([u]) = [q] - [1_n] \in \mathrm{K}_0(J),$$

holds, where $\delta \colon \mathrm{K}_1(A/J) \to \mathrm{K}_0(J)$ is the connecting map in the long exact sequence.

8.6   Examples of the Connecting Homomorphism

We start by considering an extremely important instance of an exact sequence of C*-algebras: namely the sequence

$$0 \to \mathcal{K}(H) \to \mathbb{B}(H) \to \mathcal{Q}(H) \to 0,$$

where $\mathcal{Q}(H) = \mathbb{B}(H)/\mathcal{K}(H)$ is the Calkin algebra. This exact sequence generates a long exact sequence of K-theory groups, and noting that $\mathrm{K}_i(\mathcal{K}) \cong \mathrm{K}_i(\mathbb{C})$, it has the form

$$\cdots \to \mathrm{K}_2(\mathbb{C}) \to \mathrm{K}_2(\mathbb{B}) \to \mathrm{K}_2(\mathcal{Q}) \xrightarrow{\delta}$$

$$\mathrm{K}_1(\mathbb{C}) \to \mathrm{K}_1(\mathbb{B}) \to \mathrm{K}_1(\mathcal{Q}) \xrightarrow{\delta}$$

(8.30) $$K_0(\mathbb{C}) \to K_0(\mathbb{B}) \to K_0(Q)$$

At this stage, we are interesting in computing the connecting maps δ. Using the isomorphism $K_0(\mathbb{C}) \cong \mathbb{Z}$, the first of these connecting maps boils down to a group homomorphism

$$\delta \colon K_1(Q) \to \mathbb{Z}.$$

In order to compute it explicitly, we start with a general lemma.

Lemma 8.6.1 *Suppose that A is unital and $J \subset A$ is an ideal.*
Suppose that $u \in M_n(A/J)$ and that u lifts to a partial isometry $v \in M_n(A)$. Then

$$\delta([u]) = [1 - v^*v] - [1 - vv^*] \in K_0(J),$$

where $\delta \colon K_1(A/J) \to K_0(J)$ is the connecting map of the previous section.

Proof From Theorem 8.5.11,

$$\delta([u]) = [q] - [1_n]$$

where $q = \begin{bmatrix} vv^* & v(1 - v^*v)^{\frac{1}{2}} \\ (1 - v^*v)^{\frac{1}{2}}v^* & 1 - v^*v \end{bmatrix} \in M_{2n}(A)$. Since v is a partial isometry,

$(1 - v^*v)^{\frac{1}{2}} = 1 - v^*v$ is projection to the kernel of v and so the off-diagonal entries in the matrix q are zero. We get

$$[q] - [1_n] = [vv^*] + [1 - v^*v] - [1_n] = [1 - v^*v] - [1 - vv^*]$$

as claimed.   □

Lemma 8.6.2 *If $u \in Q$ is a unitary, then u has a lift to $\mathbb{B}(H)$ which is a partial isometry.*

Proof Let $a \in \mathbb{B}$ be any lift of u. Then it has a polar decomposition $a = v|a|$ where v is a partial isometry from $\ker(a)^{\perp}$ to $\overline{\mathrm{ran}(a)}$. If $\pi \colon \mathbb{B} \to Q$ is the quotient map, then

$$u = \pi(a) = \pi(v)\pi(|a|).$$

Now it follows from the definitions that the projections vv^* and v^*v satisfy

$$a \cdot v^*v = a, \quad vv^* \cdot a = a.$$

Hence, projecting these equations to the Calkin algebra we get

$$u \cdot \pi(v)^* \pi(v) = u, \quad \pi(v)\pi(v)^* \cdot u = u \in Q.$$

Since u is unitary, we get

$$\pi(v)^* \pi(v) = 1, \quad \pi(v)\pi(v)^* = 1.$$

Hence $\pi(v)$ is unitary in the Calkin algebra, and hence $\pi(|a|)$ is also unitary, and is also positive. Any positive unitary in a C*-algebra must be equal to 1. Hence

$$u = \pi(v).$$

So v provides the required lift of u to a partial isometry in $\mathbb{B}$.   □

Corollary 8.6.3 *Let H be a separable Hilbert space, $Q = \mathbb{B}(H)/\mathcal{K}(H)$ the Calkin algebra of H. Let u be a unitary in $M_n(Q)$ for some n, representing a class $[u] \in K_1(Q)$. Lift u to a bounded operator*

$$T : H \oplus \cdots H \to H \oplus \cdots \oplus H.$$

Then T is Fredholm, and

$$\delta([u]) = \mathrm{Index}(T) \in K_0(\mathcal{K}) \cong \mathbb{Z},$$

where

$$\delta : K_1(Q) \to K_0(\mathcal{K}) \cong \mathbb{Z}$$

is the connecting homomorphism of the exact sequence

$$0 \to \mathcal{K} \to \mathbb{B} \to Q \to 0.$$

Proof Since $M_n(Q(H)) \cong Q(H \oplus \cdots H)$ we assume for simplicity that $u \in Q$ from the start. If T is a lift of u to $\mathbb{B}(H)$ then T is essentially unitary and hence Fredholm. Its Fredholm index $\mathrm{Index}(T) \in \mathbb{Z}$ is therefore well defined , and independent of the lift.

By Lemma 8.6.2, we can find a lift T which is a partial isometry. By Lemma 8.6.1,

$$\delta([u]) = [1 - T^*T] - [1 - TT^*] = [\mathrm{pr}_{\ker(T)}] - [\mathrm{pr}_{\ker(T^*)}] \in K_0(\mathcal{K}),$$

the isomorphism $K_0(\mathcal{K}) \cong \mathbb{Z}$ maps $[\mathrm{pr}_{\ker(T)}] - [\mathrm{pr}_{\ker(T^*)}]$ to the difference of integers

$$\dim \ker(T) - \dim \ker(T^*) = \mathrm{Index}(T)$$

as required.   □

Corollary 8.6.4 *Let*

$$\delta \colon K^{-1}(\mathbb{T}) \to K_0(\mathcal{K}) \cong \mathbb{Z}$$

be the connecting homomorphism of the Toeplitz extension

$$0 \to \mathcal{K} \to \mathcal{T} \to C(\mathbb{T}) \to 0.$$

Then if $u \in M_n(C(\mathbb{T}))$ is a unitary, lift u to a matrix of Toeplitz operators on $l^2(\mathbb{N})$ and let

$$T_u \colon l^2(\mathbb{N}) \oplus \cdots \oplus l^2(\mathbb{N})$$

be the associated generalized Toeplitz operator. Then T is Fredholm and

$$\delta([u]) = \text{Index}(T_u),$$

where $\text{Index}(T_u)$ *is the Fredholm index of T_u.*

Proof We have a commutative diagram of C*-algebras and homomorphisms

$$
\begin{array}{ccccccccc}
0 & \longrightarrow & \mathcal{K} & \longrightarrow & \mathcal{T} & \longrightarrow & C(\mathbb{T}) & \longrightarrow & 0 \\
& & \downarrow{\scriptstyle \text{id}} & & \downarrow{\scriptstyle i} & & \downarrow{\scriptstyle \tau} & & \\
0 & \longrightarrow & \mathcal{K} & \longrightarrow & \mathbb{B} & \longrightarrow & Q & \longrightarrow & 0
\end{array}
$$

(8.31)

The map $i \colon \mathcal{T} \to \mathbb{B}$ is the natural inclusion; the map $\tau \colon C(\mathbb{T}) \to Q$ maps $f \in C(\mathbb{T})$ to the image of

$$T_f \in \mathbb{B}\big(l^2(\mathbb{N})\big)/\mathcal{K}\big(l^2(\mathbb{N})\big) \cong Q\big(l^2(\mathbb{N})\big).$$

By naturality of connecting maps with respect to *-homomorphisms, the diagram

$$
\begin{array}{ccc}
K^{-1}(\mathbb{T}) & \overset{\delta}{\longrightarrow} & K_0(\mathcal{K}) = \mathbb{Z} \\
\downarrow{\scriptstyle \tau_*} & & \downarrow{\scriptstyle \text{id}} \\
K_1(Q) & \overset{\delta}{\longrightarrow} & K_0(\mathcal{K}) = \mathbb{Z}
\end{array}
$$

(8.32)

commutes, where the δ's on the top and bottom are associated to the two exact sequences. The result follows from Corollary 8.6.4.   □

The connecting homomorphism for the boundary extension of the disk.

We next consider a more purely topological instance of the connecting homomorphism, i.e., one which involves only topological K-theory of spaces.

Consider the exact sequence of C*-algebras

(8.33) $0 \to C_0(\mathbb{D}) \to C(\overline{\mathbb{D}}) \to C(\partial\mathbb{D}) = C(\mathbb{T}) \to 0$

of $C(\mathbb{T})$ by $C_0(\mathbb{D})$, with $\mathbb{D}$ the open disk. Since $\overline{\mathbb{D}}$ is compact and contractible, $K^{-1}(\overline{\mathbb{D}}) = K^{-1}(\text{pt}) = 0$ and $K^0(\overline{\mathbb{D}}) = K^0(\text{pt}) = \mathbb{Z}$. Since $K^0(\mathbb{T}) = \mathbb{Z}$ with generator the class of the trivial line bundle over $\mathbb{T}$, the restriction map $C(\overline{\mathbb{D}}) \to C(\mathbb{T})$ induces a surjection on K^0. Hence we get an exact sequence of groups

(8.34) $0 \to K^{-1}(\mathbb{T}) \xrightarrow{\delta_t} K^0(\mathbb{D}) \to \mathbb{Z} \to 0,$

where δ_t denotes the connecting homomorphism; we subscript it by t (standing for 'topological') to distinguish it from the Toeplitz connecting map.

The map $K^0(\mathbb{D}) \to \mathbb{Z}$ is induced by the C*-algebra homomorphism $C_0(\mathbb{D}) \to C(\overline{\mathbb{D}})$ and the isomorphism $K^0(\overline{\mathbb{D}}) \cong \mathbb{Z}$ due to contractibility of the closed disk.

The following exercise is a good one, and does not require Bott Periodicity to solve it.

Exercise 8.6.5 Prove that if $U \subset X$ is an open subset of a compact, contractible space X (so that $K_0\big(C(X)\big) = K^0(X) \cong \mathbb{Z}$), then the C*-algebra inclusion $C_0(U) \to C(X)$ induces the zero homomorphism

$$K^0(U) \to K^0(X).$$

(*Hint.* Let $i : \text{pt} \to X$ be an inclusion of the one-point space in X mapping the point to $x_0 \in X$; contractibility of X implies that $i^* : K^0(X) \to \mathbb{Z}$ is an isomorphism, for any choice of x_0. But the point x_0 can be moved to be disjoint from the support of any K-theory class for U, since any such class has a compact support inside U.)

From the exercise,

$$\delta_t : K^{-1}(\mathbb{T}) \to K^0(\mathbb{D})$$

is an isomorphism of groups.

Now let $z : \mathbb{T} \to \mathbb{C}$ the inclusion, so $z \in C(\mathbb{T})$ is unitary and defines a class $[z] \in K^{-1}(\mathbb{T})$. It lifts to the inclusion $z : \overline{\mathbb{D}} \to \mathbb{C}$, an element of $C(\overline{\mathbb{D}})$ of norm ≤ 1.

By Theorem 8.5.11,

(8.35) $q := \begin{bmatrix} |z|^2 & z\sqrt{1-|z|^2} \\ \bar{z}\sqrt{1-|z|^2} & 1-|z|^2 \end{bmatrix} \in M_2\big(C(\overline{\mathbb{D}})\big)$

is a projection in $M_2(C_0(\mathbb{D})^+)$ such that and

(8.36) $\delta_t([z]) = [q] - [1] \in K^0(\mathbb{D}).$

Exercise 8.6.6 Let $\varphi \colon \mathbb{R}^2 = \mathbb{C} \to \mathbb{D}$ be the diffeomorphism

$$\varphi(z) := \frac{z}{\sqrt{1 + |z|^2}}.$$

Show that

$$q\big(\varphi(z)\big) = p(z) := \frac{1}{1 + |z|^2} \begin{bmatrix} |z|^2 & z \\ \bar{z} & 1 \end{bmatrix},$$

the projection-valued map $p \colon \mathbb{C} \to M_2(\mathbb{C})$ defined in Example 8.1.3; the class $[p] - [1]$ is a representation of the *Bott element* $\beta \in \mathrm{K}^0(\mathbb{R}^2)$. This shows that the isomorphism

$$\varphi^* \colon \mathrm{K}^0(\mathbb{D}) \to \mathrm{K}^0(\mathbb{R}^2)$$

satisfies

$$\varphi^*\big(\delta_t([z])\big) = \beta_{\mathbb{R}^2} \in \mathrm{K}^0(\mathbb{R}^2).$$

Thus, up to the canonical isomorphism $\mathrm{K}^0(\mathbb{D}) \cong \mathrm{K}^0(\mathbb{R}^2)$, the class $\delta([z])$ is the Bott element.

Since it is going to play a significant role in what follows, we review the construction of the "Bott element." Its initial source was the Hopf bundle H over $\mathbb{CP}^1$. The Hopf bundle has a rather convenient representation in terms of a projection: the map $p \colon \mathbb{CP}^1 \to M_2(\mathbb{C})$ is the projection-valued function $p \colon \mathbb{CP}^1 \to M_2(\mathbb{C})$ mapping a line $L \subset \mathbb{C}^2$ to orthogonal projection pr_L onto that line. In terms of homogeneous coordinates on $\mathbb{CP}^1$,

$$P([z, w]) = \frac{1}{|z|^2 + |w|^2} \begin{bmatrix} |z|^2 & \bar{w}z \\ \bar{z}w & |w|^2 \end{bmatrix}.$$

Under the standard embedding of $\mathbb{C}$ as an open subset of $\mathbb{CP}^1$, $z \mapsto [z, 1]$ we obtain the restriction p of P to $\mathbb{C}$, given by the formula

$$p(z) = \frac{1}{|z|^2 + 1} \begin{bmatrix} |z|^2 & z \\ \bar{z} & 1 \end{bmatrix}.$$

Thinking of $\mathbb{CP}^1$ as $(\mathbb{R}^2)^+$, note that p takes the value

$$p(\infty) = \begin{bmatrix} 1 & 0 \\ 0 & 0 \end{bmatrix},$$

and at the origin $0 \in \mathbb{C}$, takes the value

$$p(0) = \begin{bmatrix} 0 & 0 \\ 0 & 1 \end{bmatrix}.$$

The Bott element is by definition

$$\beta := [p] - [1] \in K^0(\mathbb{R}^2).$$

As noted above, the particular choice of homeomorphism $\mathbb{R}^2 \cong \mathbb{D}$ maps p (and β) into a corresponding projection, and K-theory class, for $C_0(\mathbb{D})$, with $\mathbb{D}$ the unit disk; the formula of this new, projection-valued function on the disk, is conveniently given by

$$(8.37) \qquad q(z) = \begin{bmatrix} |z|^2 & z\sqrt{1-|z|^2} \\ \bar{z}\sqrt{1-|z|^2} & 1-|z|^2 \end{bmatrix} \in M_2(C(\overline{\mathbb{D}}))$$

– conveniently because this is precisely the formula for the projection involved in the formula produced by us for

$$\partial([z]) \in K_0(C_0(\mathbb{R}^2)),$$

with δ the connecting homomorphism for the disk and the closed disk.

Note that on the boundary of the disk $q(z) = \begin{bmatrix} 1 & 0 \\ 0 & 0 \end{bmatrix}$. In particular, since it is constant on the boundary, we can extend it (by the same constant matrix value to a function $\tilde{q} \colon \mathbb{C} \to M_2(\mathbb{C})$, and even further to $\mathbb{C}^+ = (\mathbb{R}^2)^+ M_2(\mathbb{C})$, with value $\begin{bmatrix} 1 & 0 \\ 0 & 0 \end{bmatrix}$ at ∞.

Exercise 8.6.7 The projections $\tilde{q}$ and p are homotopic as projections in $C_0(\mathbb{R}^2)^+$. (*Hint.* Argue that the composition of the map $\varphi \colon \mathbb{R}^2 \to \mathbb{D}$ of Exercise 8.6.6 and the open embedding $\mathbb{D} \to \mathbb{R}^2$ is homotopic to the identity map $\mathbb{R}^2 \to \mathbb{R}^2$, through a homotopy of open embeddings fixing the point at infinity.

From the Exercise above, $[p] - [1]$ and $[q] - [1]$ define the same element of $K^0(\mathbb{R}^2)$.

Note that q (or $\tilde{q}$) can be rescaled into an arbitrarily small disk around $0 \in \mathbb{C}$, or, of course, moved into a disk centered at another point. One thus obtains varieties of formulas for projection-valued functions on $(\mathbb{R}^2)^+$, all taking the constant value $\begin{bmatrix} 1 & 0 \\ 0 & 0 \end{bmatrix}$ outside of a small open disk in the plane. The classes $[q] - [1]$ are equal to the Bott element β in the group $K^0(\mathbb{R}^2)$. We have already discussed classes defined in this manner; we call them K-theory "germs," and the one under discussion was referred to as the "K-theory germ of a point in $\mathbb{R}^2$." The general outcome of that discussion was that one can produce a K-theory class for $\mathbb{R}^2$ in the following way.

Take a point $p \in \mathbb{R}^2 = \mathbb{C}$ and let $\varphi(z) = z - p$, which is non-vanishing away from p. Now let B be any closed ball around p and $B' = \mathbb{C} \setminus \text{int}(B) \cup \{\infty\}$. Each are closed, contractible subsets of $(\mathbb{R}^2)^+$ and we can clutch the trivial bundles $B \times \mathbb{C}$ and $B' \times \mathbb{C}$ over $B \cap B' = \partial B$ using the clutching function $z - p$. This results in a complex line E bundle over S^2, and the difference $[E] - [1] \in \mathrm{K}^0(\mathbb{R}^2)$ equals the Bott element β.

One can obviously use more general clutching functions than $z - p$. Any complex-valued continuous function with p as isolated zero determines enough data to use it to clutch two trivial bundles, one defined over a neighborhood of p, one defined over its complement in S^2, to produce a vector bundle over S^2.

Let us fix $p = 0$ and clutch using the closed disk $D = \overline{\mathbb{D}}$ and D', its complement in $(\mathbb{R}^2)^+$. The intersection $D \cap T' = \mathbb{T}$ is the circle. If $u \colon \mathbb{T} \to \mathbf{U}_n$ is any unitary-valued function, let E_φ be the vector bundle over S^2 defined by clutching $D \times \mathbb{C}^n$ and $D' \times \mathbb{C}^n$ using φ. Let

$$b(\varphi) := [E_\varphi] - [1_n] \in \mathrm{K}^0(\mathbb{R}^2)$$

be the corresponding "Bott-type" element. It is an element of $\mathrm{K}^0(S^2)$ which is in the kernel of the augmentation homomorphism

$$\epsilon^* \colon \mathrm{K}^0(S^2) = \mathrm{K}^0((\mathbb{R}^2)^+) \to \mathrm{K}^0(\text{pt}) = \mathbb{Z}$$

and hence defines a class in $\mathrm{K}^0(\mathbb{R}^2)$.

It is an easy exercise to check that if u and u' are homotopic $\mathbf{U}_n$-valued maps, amongst such maps, then the vector bundles E_u and $E_{u'}$ are homotopic vector bundles on S^2, and hence are isomorphic. In particular, since the clutching function

$$\begin{bmatrix} z & 0 \\ 0 & \bar{z} \end{bmatrix}$$

is homotopic to the constant clutching function $\begin{bmatrix} 1 & 0 \\ 0 & 1 \end{bmatrix}$ we obtain that

$$E_{z \oplus \bar{z}} \sim S^2 \times \mathbb{C}^2,$$

isomorphic as vector bundles, and thus in $\mathrm{K}^0(S^2)$ it follows that

$$[E_z] + [E_{\bar{z}}] = [1_2] \in \mathrm{K}^0(S^2)$$

holds, so that

$$[E_z] - [1] + [E_{\bar{z}}] - [1] = 0 \in \mathrm{K}^0(S^2).$$

Hence

$$b(z) = -b(\bar{z}) \in \mathrm{K}^0(S^2).$$

Exercise 8.6.8 Let E_u be the bundle over S^2 obtained from $u \colon \mathbb{T} \to \mathrm{U}_n$, as in the above discussion. Let $\alpha \colon S^2 \to S^2$ be the extension of a linear, isometric map (an element of $\mathrm{O}(2, \mathbb{R})$.) Such a map restricts to a map $\alpha \colon \mathbb{T} \to \mathbb{T}$, and $u \circ \alpha$ is another clutching function determining a bundle $E_{u \circ \alpha}$ over S^2.

Show that

$$E_{u \circ \alpha} \cong \alpha^*(E_u)$$

as vector bundles over S^2.

Combining the discussion above with the exercise we obtain the following simple result.

Proposition 8.6.9 *If $\alpha \colon \mathbb{R}^2 \to \mathbb{R}^2$ is an orthogonal map,*

$$\alpha^* \colon \mathrm{K}^0(\mathbb{R}^2) \to \mathrm{K}^0(\mathbb{R}^2)$$

the induced map, then

$$\alpha^*(\beta) = \det(\alpha) \cdot \beta,$$

with $\det(\alpha)$ the determinant of the matrix α, and β is the Bott element.

Proof The Bott element in the notation of the discussion above is given by

$$\beta = b(z),$$

where $z \colon \mathbb{T} \to \mathbb{C}$ is the usual coordinate. Since α is homotopic to either the identity map $\mathbb{R}^2 \to \mathbb{R}^2$ or to the complex conjugation map, through elements of $\mathrm{O}(2, \mathbb{R})$, the result follows from the above discussion. $\square$

A K-theoretic perspective on the Toeplitz index theorem.

A pseudo-Toeplitz operator $T = T_u + S$, for $u \colon \mathbb{T} \to \mathbb{C}^*$ smooth, say, and S a smoothing operator on the circle, is Fredholm, and has a Fredholm index

$$\mathrm{Index}(T) := \dim \ker(T) - \dim \ker(T^*).$$

We have shown that this "analytic index" admits a K-theoretic interpretation involving the Toeplitz algebra and the Toeplitz extension:

$$\mathrm{Index}(T) = \delta([u]),$$

where

$$\delta \colon \mathrm{K}^{-1}(\mathbb{T}) \to \mathrm{K}_0(\mathcal{K}) = \mathbb{Z}$$

is the connecting map for the Toeplitz extension.

The essential idea leading to the Atiyah–Singer index theorem, is that this index also has a purely *topological* K-theoretic interpretation. That is, it can be described purely in terms of a certain *topological* K-theoretic invariant of the symbol, which involves no noncommutative C*-algebras.

We describe this more general statement now, assuming Bott Periodicity, which implies that $K^0(\mathbb{D}) = \mathbb{Z} \cdot \beta_{\mathbb{D}}$, where $\beta_{\mathbb{D}}$ is the Bott element of the open disk. Consider the connecting homomorphism

$$\delta_t : K^{-1}(\mathbb{T}) \to K^0(\mathbb{D})$$

associated to the exact sequence

$$0 \to C_0(\mathbb{D}) \to C(\overline{\mathbb{D}}) \to C(\mathbb{T}) \to 0.$$

Now for u the symbol of T as above, set

(8.38) $\text{Index}_t(T) := n \iff \delta_t([u]) = n \cdot \beta_{\mathbb{D}} \in K^0(\mathbb{D}).$

Then

Theorem 8.6.10 *If T is a pseudo-Toeplitz operator on $\mathbb{T}$ with symbol u, then*

$$\text{Index}(T) = \text{Index}_t([u]),$$

where $[u] \in K^{-1}(\mathbb{T})$ is the K-theory class determined by u.

Exercise 8.6.11 Let $i: \mathbb{D} \to \mathbb{R}^2$ be the inclusion of $\mathbb{D}$ as an open subset of $\mathbb{R}^2$. It determines a map

$$i! : K^0(\mathbb{D}) \to K^0(\mathbb{R}^2).$$

Let $\beta_{\mathbb{D}} \in K^0(\mathbb{D})$ be the Bott element of the disk (8.36). Let $\varphi: \mathbb{R}^2 \to \mathbb{D}$ the diffeomorphism of Exercise 8.6.6, $\beta_{\mathbb{R}^2}$ the Bott element for $\mathbb{R}^2$. Prove that

$$i!(\beta_{\mathbb{D}}) = \varphi^*(\beta_{\mathbb{D}}).$$

The proof only involves some simple homotopies.

Exercise 8.6.12 If u is a non-vanishing function on $\mathbb{T}$, define $\text{Index}_t([u])$ by (8.38). Let E_u be the vector bundle over S^2 obtained by clutching the trivial vector bundles $S_{\pm}^2 \times \mathbb{C}$ over the top and bottom hemispheres, using $u: S_+^2 \cap S_-^2 \to \mathbb{C}^*$. Prove that

$$[E_u] - [1] = \text{Index}_t([u]) \cdot [H^*] \in K^0(S^2),$$

where H^* is the dual of the Hop bundle.

This gives another way of looking at the topological index.

8.7   The External Product Operation on K-Theory

While K-theory group of a commutative C*-algebras has a natural ring structure (induced by the tensor product of vector bundles, fiberwise), the K-theory of a noncommutative C*-algebra has in general no natural ring structure.

Formally, the fiberwise tensor product of two vector bundles $E \to X$ and $E' \to X$ may be interpreted in the following way. First, one forms the *external tensor product* of the two bundles, forming the bundle over $X \times X$ whose fiber at (x, y) is $E_x \otimes E'_y$, which can easily be checked to be a vector bundle over $X \times X$.

Then one *restricts* this vector bundle over $X \times X$ to the diagonal, a copy of X inside $X \times X$.

This results, obviously, in precisely the (fiberwise) tensor product bundle $E \otimes E'$.

The first step makes sense for noncommutative C*-algebras. The second step involves the diagonal map $\delta \colon X \to X \times X$, whose Gelfand dual is the multiplication map $C(X) \otimes C(X) \to C(X)$. The multiplication map makes sense for general C*-algebras but is not a *-homomorphism, unless they are commutative, and so it does not induce a product at the level of K-theory.

In this section we describe the extension of the first step to general C*-algebras. We will show that tensor product (of f.g.p. modules) gives rise to a natural bilinear map

$$\mathrm{K}_i(A) \times \mathrm{K}_j(B) \to \mathrm{K}_{i+j}(A \otimes B), \quad (x, y) \mapsto x \hat{\otimes} y$$

for any A, B, which we will call the *external product*.

We start with A and B unital.

We have already defined (see Equation (5.19)) the external product of two Hilbert modules over A, B, respectively, which is then a Hilbert module over $A \otimes B$.

Observe that the external product of a finitely generated *free* right A-module E_A, and a finitely generated free right B-module E_B, in this sense, is a free right $A \otimes B$-module. Indeed, choosing an isomorphism $E_A \cong A^n$ and an isomorphism $E_B \cong B^m$, we obtain an isomorphism on the algebraic tensor product $E_A \otimes_{\mathbb{C}} E_B$ with $A^n \otimes B^m \cong (A \otimes B)^{nm}$, where the tensor product is algebraic, that is, to the direct sum of nm copies of the algebraic tensor product $A \otimes_{\mathbb{C}} B$ of A and B.

Now completing this direct sum with respect to the Hermitian form above, results in the direct sum nm copies of the C*-algebraic tensor product $A \otimes B$ of A and B.

If E_A and E'_A are isomorphic, then they have Hermitian forms and an isometric isomorphism between the two corresponding Hilbert modules. It follows that $E_A \otimes_{\mathbb{C}} E_B \cong E'_A \otimes_{\mathbb{C}} E_B$, if E_B is a right B-module. By similar such simple arguments, one verifies that one obtains a well-defined "external product" operation on K-theory, as summarized by the following

Proposition 8.7.1 *Let A and B be unital C^*-algebras. If E_A and E_B be finitely generated projective A, B-modules, respectively, $E_A \otimes E_B$ their external tensor product, then $E_A \otimes_{\mathbb{C}} E_B$ is finitely generated projective over $A \otimes B$.*

1. Defining

$$[E_A] \hat{\otimes} [E_B] := [E \otimes_{\mathbb{C}} E_B]$$

gives a well defined , $\mathbb{Z}$-bilinear map

(8.39) $\hat{\otimes}: \mathcal{P}(A) \times \mathcal{P}(B) \to \mathcal{P}(A \otimes B),$

on the semigroups of isomorphism classes of f.g.p. modules, and consequent $\mathbb{Z}$-bilinear map

$$K_0(A) \times K_0(B) \to K_0(A \otimes B).$$

2. The pairing (8.46) is natural in the sense that if $\alpha: A \to A'$ and $\beta: B \to B'$ are $*$-homomorphisms between unital C^*-algebras, $\alpha \otimes \beta: A \otimes B \to A' \otimes B'$ their tensor product, then

(8.40) $\alpha_*([E_A]) \hat{\otimes} \beta_*([E_B]) = (\alpha \otimes \beta)_*([E_A] \hat{\otimes} [E_B]) \in K_0(A' \otimes B')$

holds, for all f.g.p. modules E_A over A, and E_B over B.

3. If $x \in K_0(A)$ and $[1] \in K_0(\mathbb{C})$ denotes the positive generator of $K_0(\mathbb{C})$, then under the identifications $A \otimes \mathbb{C} \cong A$, $\mathbb{C} \otimes A \cong A$,

$$x \hat{\otimes} [1] = x = [1] \hat{\otimes} x$$

holds for all $x \in K_0(A)$.

Remark 8.7.2 If one is thinking of projections, rather than f.g.p. modules, let $p \in M_n(A)$ and $q \in M_m(B)$ be two projections. Then $p \otimes q \in M_n(A) \otimes M_m(B)$. Choosing any bijection between $\{1, \ldots, n\} \times \{1, 2, \ldots, m\} \cong \{1, 2, \ldots, nm\}$ gives an isomorphism $M_n(A) \otimes M_m(B) \cong M_{nm}(A \otimes B)$. The projection $p \otimes q$ so defined, orthogonally projections $(A \otimes B)^{nm}$ to an isomorphic copy of the external product of modules $pA^n \otimes_{\mathbb{C}} qB^m$.

Exercise 8.7.3 Suppose $E \to X$ and $E' \to X'$ are vector bundles over X, X' (compact). Prove that the section module of the external product of the two bundles, a bundle over $X \times X'$, defined at the beginning of this section, is isomorphic to the external product $\Gamma(E) \otimes_{\mathbb{C}} \Gamma(E')$.

Exercise 8.7.4 If $1_n = \mathbb{C}^n$ is the trivial rank n free $\mathbb{C}$-module, then

$$[1_n] \hat{\otimes} x = nx$$

for any A unital, any $x \in K_0(A)$.

Exercise 8.7.5 Recall that $\mathcal{P}(\mathbb{C}) \cong \mathbb{N}$ and $K_0(\mathbb{C}) \cong \mathbb{Z}$. Show directly that the external product

$$\mathcal{P}(\mathbb{C}) \times \mathcal{P}(\mathbb{C}) \to \mathcal{P}(\mathbb{C} \otimes \mathbb{C}) \cong \mathcal{P}(\mathbb{C})$$

defined above corresponds to multiplication of natural numbers.

Analogously, show that

$$K_0(\mathbb{C}) \times K_0(\mathbb{C}) \to K_0(\mathbb{C} \otimes \mathbb{C}) \cong K_0(\mathbb{C})$$

corresponds to multiplication of integers.

You can use Exercise 8.7.4 to show, more generally, that the external product map

$$K_0(\mathbb{C}) \times K_0(A) \to K_0(\mathbb{C} \otimes A) = K_0(A)$$

identifies with the obvious $\mathbb{Z}$-multiplication map

$$\mathbb{Z} \times K_0(A) \to K_0(A).$$

The external product on the K_0-groups of a pair of possibly non-unital algebras is slightly more complicated to define. Suppose A_1 and A_2 are two, possibly non-unital algebras. Let $\epsilon_1 : A_1^+ \to \mathbb{C}$ and $\epsilon_2 : A_2^+ \to \mathbb{C}$ the usual augmentation *-homomorphisms. They induce *-homomorphisms

$$\epsilon_1 \otimes 1_{A_2^+} : A_1^+ \otimes A_2^+ \to \mathbb{C} \otimes A_2^+ \cong A_2^+, \quad \text{and} \quad 1_{A_1^+} \otimes \epsilon_2 : A_1^+ \otimes A_2^+ \to A_1^+.$$

Let

$$(8.41) \quad \pi : A_1^+ \otimes A_2^+ \to A_1^+ \oplus A_2^+,$$
$$\pi(a_1 \otimes a_2) := \big((1_{A_1^+} \otimes \epsilon_2)(a_1 \otimes a_2), (\epsilon_1 \otimes 1_{A_2^+})(a_1 \otimes a_2)\big).$$

be the direct sum of the *-homomorphisms $1_{A_1^+} \otimes \epsilon_2$ and $\epsilon_1 \otimes 1_{A_2^+}$.

Note that $(A_1 \otimes A_2)^+$ embeds in $A_1^+ \otimes A_2^+$ by extending the obvious embedding $A_1 \otimes A_2 \to A_1^+ \otimes A_2^+$ and then extending it to the unitization by mapping the unit to the unit $1 \otimes 1$ of $A_1^+ \otimes A_2^+$.

Lemma 8.7.6 *For π as in (8.41), the map* $K_0((A_1 \otimes A_2)^+) \to K_0(A_1^+ \otimes A_2^+)$ *induced by the inclusion* $(A_1 \otimes A_2)^+ \to A_1^+ \otimes A_2^+$, *maps* $K_0(A_1 \otimes A_2)$ *to the subgroup* $\ker(\pi_*) \subset K_0(A_1^+ \otimes A_2^+)$ *of* $K_0(A_1^+ \otimes A_2^+)$.

Note that (omitting subscripts)

$$\ker(\pi) = \ker\,(\epsilon_1 \otimes 1)_* \cap \ker\,(1 \otimes \epsilon_2)_*.$$

Proof The restriction $\epsilon_1 \otimes 1_{A_2}$ of $\epsilon_1 \otimes 1_{A_2^+}$ to $A_1^+ \otimes A_2$ has kernel $A_1 \otimes A_2$. Hence the sequence

$$0 \to A_1 \otimes A_2 \to A_1^+ \otimes A_2 \to A_2 \to 0$$

is exact. It is actually split exact, using the splitting $A_2 \to A_1^+ \otimes A_2$, $a_2 \mapsto 1 \otimes a_2$. Hence we obtain an exact sequence of K_0-groups

$$0 \to K_0(A_1 \otimes A_2) \to K_0(A_1^+ \otimes A_2) \to K_0(A_2) \to 0.$$

The quotient map is the map induced on K_0 from $\epsilon_1 \otimes 1_{A_2}$. Therefore, $K_0(A_1 \otimes A_2)$ embeds in $K_0(A_1^+ \otimes A_2)$ as the kernel of $(\epsilon_1 \otimes 1_{A_2})_*$.

On the other hand, $K_0(A_1^+ \otimes A_2)$ injects in $K_0(A_1^+ \otimes A_2^+)$ as the kernel of $(1_{A_1^+} \otimes \epsilon_2)_*$. by arguing similarly with the exact sequence

$$0 \to A_1^+ \otimes A_2 \to A_1^+ \otimes A_2^+ \to A_1^+ \to 0,$$

which is also easily checked to be split exact.

Putting these two observations together, we conclude that $K_0(A_1 \otimes A_2)$ injects naturally into $K_0(A_1^+ \otimes A_2^+)$ with kernel

(8.42) $$\ker\left((\epsilon_1 \otimes 1_{A_2})_*\right) \cap \ker\left((1_{A_1^+} \otimes \epsilon_2)_*\right).$$

Note that the (injective) map on K_0-theory induced by the inclusion $A_2 \to A_2^+$ identifies the kernels of $(\epsilon_1 \otimes 1_{A_2})_*$ and $(\epsilon_1 \otimes 1_{A_2^+})_*$. Hence (8.42) is the same as the subgroup $\ker(\pi_*)$, by the additivity property of K_0. $\qquad\square$

From the Lemma, we obtain the following recipe for taking external products in the non-unital case.

Let A_1, A_2 be two, possibly non-unital algebras, A_i^+ their unitizations, the map π defined as above. Suppose $x \in K_0(A_1)$, $y \in K_0(A_2)$. So $x \in K_0(A_1^+)$ is in the kernel of $(\epsilon_1)_* \colon K_0(A_1^+) \to \mathbb{Z}$, and $y \in K_0(A_2^+)$ is in the kernel of $(\epsilon_2)_* \colon K_0(A_2^+) \to \mathbb{Z}$, and

Since $(\epsilon_2)_*(y) = 0$,

$$(1_{A_1^+} \otimes \epsilon_2)_*(x \hat{\otimes} y) = x \hat{\otimes} (\epsilon_2)_*(y) = 0,$$

where we have used Proposition 8.7.1 2). Similarly,

$$(\epsilon_1 \otimes 1_{A_2^+})_*(x \hat{\otimes} y) = 0.$$

Hence

$$\pi_*(x \hat{\otimes} y) = 0$$

and therefore $x \hat{\otimes} y \in K_0(A_1^+ \otimes A_2^+)$ is in the kernel of $\pi_* \colon K_0(A_1^+ \otimes A_2^+) \to K_0(A_1^+) \oplus K_0(A_2^+)$. Applying the identification of this subgroup with $K_0(A_1 \otimes A_2)$, we obtain therefore a map

(8.43) $$K_0(A_1) \times K_0(A_2) \to K_0(A_1 \otimes A_2)$$

for arbitrary C*-algebras A, B. Furthermore, for any n if we replace in (8.43) the C*-algebra A_1 by $S^n(A_1) = C_0(\mathbb{R}^n) \otimes A_1$, then we obtain a bilinear pairing

$$K_n(A_1) \times K_0(A_2) := K_0\big(S^n(A_1)\big) \times K_0(A_2) \to K_0(S^n(A_1) \otimes A_2)$$

(8.44) $$\cong K_0\big(C_0(\mathbb{R}^n) \otimes A_1 \otimes A_2\big) \cong K_n(A_1 \otimes A_2),$$

which plays a role in part 3) of the Theorem below.

Theorem 8.7.7 *Let A and B be C*-algebras.*

1. *There is a $\mathbb{Z}$-bilinear pairing*

   (8.45) $$K_0(A) \times K_0(B) \to K_0(A \otimes B)$$

   mapping $(x, y) \in K_0(A) \times K_0(B)$ to their external product $x \hat{\otimes} y$.
2. *The external product (8.45) is natural in the sense that if $\alpha \colon A \to A'$ and $\beta \colon B \to B'$ are *-homomorphisms $\alpha \otimes \beta \colon A \otimes B \to A' \otimes B'$ their tensor product *-homomorphism, then*

   (8.46) $$\alpha_*(x) \hat{\otimes} \beta_*(y) = (\alpha \otimes \beta)_*(x \hat{\otimes} y) \in K_0(A' \otimes B')$$

   holds, for all $x \in K_0(A)$ and $y \in K_0(B)$.
3. *If*

   (8.47) $$0 \to J \to A \to A/J \to 0$$

   is a c.p. split exact sequence of C^*-algebras and B is any C^*-algebra, so that

   (8.48) $$0 \to J \otimes B \to A \otimes B \to A/J \otimes B \to 0$$

   is also short exact, then

   (8.49) $$\partial(x \hat{\otimes} y) = \partial(x) \hat{\otimes} y$$

for all $x \in K_1(A/J) = K_0\big(S(A/J)\big)$ and $y \in K_0(B)$. The boundary map on the left hand side is the K-theory connecting map for the exact sequence (8.55), and the boundary map on the right hand side is the K-theory map associated to the exact sequence (8.47).

Proof The statement 2 follows routinely from Proposition 8.7.1. For 3, we recall that definition of the boundary map

$$(8.50) \qquad\qquad \partial \colon K_1(A/J) \to K_0(J).$$

By definition, $K_1(A/J) = K_0\big(S(A/J)\big)$, and $S(A/J)$ is naturally isomorphic to the ideal $\{f \in C_\pi \mid f(1) = 0\}$, where C_π is the mapping cone of the quotient map. The inclusion $s \colon S(A/J) \to C_\pi$ induces a map

$$(8.51) \qquad\qquad s_* \colon K_0\big(S(A/J)\big) \to K_0(C_\pi).$$

On the other hand the inclusion

$$(8.52) \qquad\qquad k \colon J \to C_\pi, \quad k(a) := (a, 0),$$

induces an isomorphism

$$(8.53) \qquad\qquad k_* \colon K_0(J) \to K_0(C_\pi).$$

The boundary map ∂ is defined

$$(8.54) \qquad\qquad \partial := k_*^{-1} \circ s_* \colon K_1(A/J) \to K_0(J).$$

On the other hand, the sequence

$$(8.55) \qquad\qquad 0 \to J \otimes B \to A \otimes B \to A/J \otimes B \to 0.$$

This implies by an easy exercise that

$$A/J \otimes B \cong A \otimes B \,/\, J \otimes B.$$

Associated therefore to the ideal $J \otimes B$ in $A \otimes B$, we have the quotient map, which we denote

$$\pi_B \colon A \otimes B \to A \otimes B/ J \otimes B,$$

and its mapping cone, which we denote by C_{π_B}, the inclusions $k_B \colon J \otimes B \to C_{\pi_B}$ and $s_B \colon S(A/J) \otimes B) \to C_{\pi_B}$.

On the other hand, in a natural way

$$C_{\pi_B} \cong C_\pi \otimes B,$$

and under this identification, the inclusion

$$s_B : S(A \otimes B / J \otimes B) \to C_{\pi_B}$$

identifies with $s \otimes 1_B : S(A/J) \otimes B \to C_\pi \otimes B$. In particular, by part 2) of the Theorem, if $x \in K_0\big(S(A/J)\big)$ and $y \in K_0(B)$, the

$$(8.56) \qquad\qquad (s_B)_*(x \hat\otimes y) = (s \otimes 1_B)_*(x \hat\otimes y) = s_*(x) \hat\otimes y.$$

Similarly, under the identification $C_{\pi_B} \cong C_\pi \otimes B$, the inclusion $k_B : J \otimes B \to C_{\pi_B}$ for the ideal $J \otimes B \subset A \otimes B$ identifies with $k \otimes 1_B$. Hence

$$(k_B)_* \cong (k \otimes 1)_* : K_0(J \otimes B) \to K_0(C_\pi \otimes B)$$

and thus

$$(8.57) \qquad\qquad (k_B)_*^{-1} \cong (k \otimes 1)_*^{-1} : K_0(C_\pi \otimes B) \to K_0(J \otimes B)$$

Now functoriality of $\hat\otimes$ with respect to *-homomorphisms gives

$$(8.58) \qquad\qquad (k \otimes 1)_* \big(k_* s_*(x) \hat\otimes y\big) = k_* k_*^{-1} s_*(x) \hat\otimes y = s_*(x) \hat\otimes y.$$

Since $(k \otimes 1)_*$ is an isomorphism, we get

$$(8.59) \qquad\qquad (k \otimes 1)_*^{-1} \big(s_*(x) \hat\otimes y\big) = k_* s_*(x) \hat\otimes y$$

and under our standard identifications this says that

$$(k_B)_*^{-1} \big((s_B)_*(x \hat\otimes y)\big) = k_* s_*(x) \hat\otimes y$$

giving

$$\partial(x \hat\otimes y) = \partial(x) \hat\otimes y$$

as required.   □

We conclude this section with an extension of the external product

$$(8.60) \qquad K_0(A) \times K_0(B) \to K_0(A \otimes B), \quad (x, y) \in K_0(A) \times K_0(B) \mapsto x \hat\otimes y$$

to a bilinear pairing

$$K_i(A) \times K_j(B) \to K_{i+j}(A \otimes B).$$

This is very easily done. If

$$x \in K_i(A) := K_0(C_0(\mathbb{R}^i) \otimes A), \quad y \in K_j(B) := K_0(C_0(\mathbb{R}^j) \otimes B),$$

then the product already defined gives an element

$$x \hat{\otimes} y \in K_0(C_0(\mathbb{R}^i) \otimes A \otimes C_0(\mathbb{R}^j) \otimes B).$$

Rearranging factors gives a canonical isomorphism

$$C_0(\mathbb{R}^i) \otimes A \otimes C_0(\mathbb{R}^j) \otimes B \cong C_0(\mathbb{R}^i \times \mathbb{R}^j) \otimes A \otimes B.$$

Furthermore, we can identify $C_0(\mathbb{R}^i \times \mathbb{R}^j)$ with $C_0(\mathbb{R}^{i+j})$ by identifying a pair $x \in \mathbb{R}^i$ and $y \in \mathbb{R}^j$ with the element $(x, y) = (x_1, \ldots, x_i, y_1, \ldots, y_j)$ of $\mathbb{R}^{i+j}$. This gives a further isomorphism with $C_0(\mathbb{R}^{i+j}) \otimes A \otimes B$, so that we may interpret the product $x \hat{\otimes} y$ already defined for K_0-classes as lying in $K_{i+j}(A \otimes B)$.

If one identifies a pair $(x, y) \in \mathbb{R}^i \times \mathbb{R}^j$ with the element $(y, x) = (y_1, \ldots, y_j, x_1, \ldots, x_i)$ instead, the two differ by a permutation of the coordinates of sign $(-1)^{ij}$. This accounts for an important *graded* commutativity of the external product, acting on higher K-theory:

Theorem 8.7.8 *The external product of Theorem 8.7.7 extends to a more general bilinear, natural pairing*

$$K_i(A) \times K_j(B) \to K_{i+j}(A \otimes B)$$

mapping x, y to $x \hat{\otimes} y$. The generalized external product is graded commutative *in the sense that if $x \in K_i(A)$ and $y \in K_j(B)$, then*

$$x \hat{\otimes} y = (-1)^{ij} \, \sigma_*(y \hat{\otimes} x),$$

where $\sigma : B \otimes A \to A \otimes B$ is the flip isomorphism.

The sign is of course material only when both classes x and y involved in the product, are odd-dimensional classes.

Exercise 8.7.9 Let A, B be unital C*-algebras and $u \in A$ be a unitary. Let $p \in B$ be a projection. Show that the external product

$$[u] \hat{\otimes} [p] \in K_1(A \otimes B)$$

is represented by the class of the unitary

$$u \otimes p + 1 \otimes (1 - p) \in A \otimes B.$$

8.8   The Bott Periodicity Theorem

Let $\beta \in K^0(\mathbb{R}^2) = K_0\big(C_0(\mathbb{R}^2)\big)$ be the Bott element.

External product with β, discussed in the previous section, defines a map
(8.61)
$$\beta_A \colon K_0(A) \to K_2(A) := K_0(C_0(\mathbb{R}^2) \otimes A), \quad \beta_A(x) := \beta\hat{\otimes}x, \quad x \in K_0(A),$$

and for any C*-algebra A. We aim to show that it is an isomorphism.

We start with some general remarks about β.

The first point is that the map

$$\beta_A \colon K_0(A) \to K_2(A)$$

is natural in A in the sense that if $\alpha \colon A \to B$ is a C*-algebra homomorphism, then the diagram of groups and group homomorphisms

$$
\begin{array}{ccc}
K_0(A) & \xrightarrow{\ \beta_A\ } & K_2(A) \\
\alpha_* \downarrow & & \downarrow \alpha_* \\
K_0(B) & \xrightarrow{\ \beta_B\ } & K_2(B)
\end{array}
$$

(8.62)

commutes.

In the language of functors, β is a *natural transformation* between the functors K_0 and K_2 (each is a functor from the category of C*-algebras and C*-algebra homomorphisms, to the category of abelian groups, and group homomorphisms.)

The second important point is that β *commutes with external products* in the sense that

$$\beta_{A\otimes B}(x\hat{\otimes}y) = \beta_A(x)\hat{\otimes}y.$$

This statement is merely the associativity of the external product, since the left hand side is $\beta\hat{\otimes}(x\hat{\otimes}y)$ and the right hand side is $(\beta\hat{\otimes}x)\hat{\otimes}y$.

In order to invert Bott Periodicity, we will define a similar, natural transformation of functors: a group homomorphism, for each A,

$$\alpha_A \colon K_2(A) \to K_0(A),$$

using the Toeplitz extension.

Let A be a unital C*-algebra. The first step in defining α_A is the rather trivial one of identifying

(8.63) $$K_2(A) := K_0(C_0(\mathbb{R}^2) \otimes A) = K_1(C_0(\mathbb{R}) \otimes A).$$

Now let $\mathcal{T}$ be the Toeplitz algebra. Form the exact sequence of C*-algebras

$$0 \to \mathcal{K} \otimes A \to \mathcal{T} \otimes A \to C(\mathbb{T}) \otimes A \to 0.$$

There is an associated connecting homomorphism

$$\delta_A \colon \mathrm{K}_1(C(\mathbb{T}) \otimes A) \to \mathrm{K}_0(\mathcal{K} \otimes A) \cong \mathrm{K}_0(A).$$

Now, identify $\mathbb{R}$ with the open subset $\mathbb{T} \setminus \{1\}$ of the circle, using (say) the Cayley transform. This gives an embedding

$$i \colon C_0(\mathbb{R}) \otimes A \subset C(\mathbb{T}) \otimes A.$$

Putting things together we obtain the map

(8.64) $\qquad \alpha_A \colon \mathrm{K}_2(A) = \mathrm{K}_1\big(C_0(\mathbb{R}) \otimes A)\big) \xrightarrow{i_*} \mathrm{K}_1(C(\mathbb{T}) \otimes A) \xrightarrow{\delta_A} \mathrm{K}_0(A).$

Lemma 8.8.1 α *is a natural transformation* $\mathrm{K}_2 \to \mathrm{K}_0$, *which commutes with external products in the sense that*

(8.65) $\qquad\qquad\qquad \alpha_{A \otimes B}(x \hat{\otimes} y) = \alpha_A(x) \hat{\otimes} y$

for any $x \in \mathrm{K}_2(A)$ *and* $y \in \mathrm{K}_0(A)$.
  Furthermore,

$$\alpha_{\mathbb{C}}(\beta) = [1] \in \mathrm{K}_0(\mathbb{C}),$$

where $\beta \in \mathrm{K}_2(\mathbb{C}) = \mathrm{K}_0\big(C_0(\mathbb{R}^2)\big)$ *is the Bott element, and* $[1]$ *is the class of the unit* $1 \in \mathbb{C}$, *the generator of* $\mathrm{K}_0(\mathbb{C}) \cong \mathbb{Z}$.

Proof The first two statements are obvious. For the last one, recall that the unitary in $C_0(\mathbb{R})^+ \cong C(\mathbb{T})$ corresponding to the projection $p \in C_0(\mathbb{R}^2)^+$ defining the Bott element is the inclusion $\bar{z} \colon \mathbb{T} \to \mathbb{C}$. We have already proved that

$$\delta_{\mathbb{C}}([\bar{z}]) = \mathrm{Index}(T_{\bar{z}}) = -\mathrm{Index}(T_z) = 1,$$

where $T_{\bar{z}}$ is the Toeplitz operator with symbol $\bar{z}$, and Index is the Fredholm index. This proves the Lemma.

$\qquad\qquad\qquad\qquad\qquad\qquad\qquad\qquad\qquad\qquad\qquad\qquad\qquad\qquad\qquad\qquad$ $\square$

Corollary 8.8.2 *The transformations* α *and* β *satisfy*

$$\alpha_A \circ \beta_A = \mathrm{id}_{\mathrm{K}_0(A)},$$

for any C*-algebra A.

Proof Both transformations are natural with respect to external products, and we may write any $x \in \mathrm{K}_0(A)$ as the external product

$$x = [1] \hat{\otimes} x,$$

where $[1] \in K_0(\mathbb{C})$ is the generator. We get

$$(\alpha_A \circ \beta_A)(x) = \alpha_A \big(\beta_{\mathbb{C}}([1]) \hat{\otimes} x\big)$$

since β commutes with external products. By the same reasoning with α

$$= \alpha_{\mathbb{C}} \big(\beta_{\mathbb{C}}([1])\big) \hat{\otimes} x = \alpha_{\mathbb{C}}(\beta) \hat{\otimes} x$$

and by the Lemma

$$= [1] \hat{\otimes} x = x.$$

$\square$

Theorem 8.8.3 (Bott Periodicity) *The Toeplitz transformation α and the Bott transformation β are inverse to each other. That is,*

$$\alpha_A \circ \beta_A = \mathrm{id}_{K_0(A)}, \qquad \beta_A \circ \alpha_A = \mathrm{id}_{K_2(A)},$$

for any C-algebra A.*

Therefore, K-theory is Bott periodic: $K_i(A) \cong K_{i+2}(A)$ for any C-algebra A, any nonnegative integer i.*

The proof boils down to a "rotation trick" initially devised by Atiyah, to reduce left-invertibility of the transformation β, to right invertibility, which has already been proved in Corollary 8.8.2.

Proof Let $x \in K_2(A) := K_0(C_0(\mathbb{R}^2) \otimes A)$. We want to show that

$$(\beta_A \circ \alpha_A)(x) = x \in K_2(A) = K_0(C_0(\mathbb{R}^2) \otimes A).$$

By definition, $\alpha_A(x) \in K_0(A)$ and

$$(\beta_A \circ \alpha_A)(x) = \beta \hat{\otimes} \alpha_A(x).$$

By commutativity of the external product this equals

$$\sigma_* \big(\alpha_A(x) \hat{\otimes} \beta\big),$$

where

$$\sigma : A \otimes C_0(\mathbb{R}^2) \to C_0(\mathbb{R}^2) \otimes A$$

is the flip homomorphism. Moreover, since α commutes with external products, our product can be rewritten as

$$\text{(8.66)} \qquad \sigma_*\big(\alpha_{A\otimes C_0(\mathbb{R}^2)}(x\hat{\otimes}\beta)\big).$$

Now, the naturality property of α with respect to *-homomorphisms is that if $\nu\colon B \to B'$ is a *-homomorphism, then

$$\alpha_{B'}\big((\mathrm{id}_{C_0(\mathbb{R}^2)} \otimes \nu)_*(x)\big) = \nu_*\big(\alpha_B(x)\big).$$

Applying naturality to (8.66) gives that it equals

$$\text{(8.67)} \qquad \alpha_{A\otimes C_0(\mathbb{R}^2)}\big((\mathrm{id}_{C_0(\mathbb{R}^2)} \otimes \sigma)_*(\beta\hat{\otimes}x)\big).$$

By commutativity of $\hat{\otimes}$, this can be rewritten

$$\text{(8.68)} \qquad \alpha_{A\otimes C_0(\mathbb{R}^2)}\big((\mathrm{id}_{C_0(\mathbb{R}^2)} \otimes \sigma)_* \circ \sigma'_*(x\hat{\otimes}\beta)$$

where

$$\sigma'\colon C_0(\mathbb{R}^2) \otimes A \otimes C_0(\mathbb{R}^2) \to C_0(\mathbb{R}^2) \otimes C_0(\mathbb{R}^2) \otimes A$$

is the flip homomorphism given permuting the factors cyclically:

$$\sigma'(f \otimes a \otimes f') := f' \otimes f \otimes a.$$

By functoriality of K-theory we can write (8.68) as

$$\text{(8.69)} \quad \alpha_{A\otimes C_0(\mathbb{R}^2)}\big[\big((\mathrm{id}_{C_0(\mathbb{R}^2)} \otimes \sigma) \circ \sigma'\big)_*(\beta\hat{\otimes}x)\big] = \alpha_{A\otimes C_0(\mathbb{R}^2)}\big((\sigma'')_*(\beta\hat{\otimes}x)\big)$$

with σ'' the composition

$$C_0(\mathbb{R}^2)\otimes A\otimes C_0(\mathbb{R}^2) \xrightarrow{\sigma'} C_0(\mathbb{R}^2)\otimes C_0(\mathbb{R}^2)\otimes A \xrightarrow{\mathrm{id}_{C_0(\mathbb{R}^2)}\otimes\sigma} C_0(\mathbb{R}^2)\otimes A\otimes C_0(\mathbb{R}^2)$$

– which just flips the first and third factors (of $C_0(\mathbb{R}^2)$). It is therefore homotopic to the identity *-homomorphism because the flip homomorphism

$$C_0(\mathbb{R}^2) \otimes C_0(\mathbb{R}^2) \to C_0(\mathbb{R}^2) \otimes C_0(\mathbb{R}^2)$$

is already homotopic to the identity homomorphism, since $C_0(\mathbb{R}^2) \otimes C_0(\mathbb{R}^2) \cong C_0(\mathbb{R}^4)$ and our homomorphism is induced by the corresponding map $\mathbb{R}^4 \to \mathbb{R}^4$, which is matrix multiplication by $\begin{bmatrix} 0 & 1_2 \\ 1_2 & 0 \end{bmatrix}$, an orthogonal matrix with determinant $+1$.

Since σ'' is homotopic to the identity and since, by definition, $\beta \hat{\otimes} x = \beta_{C_0(\mathbb{R}^2) \otimes A}(x)$ we can write (8.69) as

$$(8.70) \qquad \alpha_{A \otimes C_0(\mathbb{R}^2)}(\beta \hat{\otimes} x) = (\alpha_{A \otimes C_0(\mathbb{R}^2)} \circ \beta_{C_0(\mathbb{R}^2) \otimes A})(x) = x,$$

the last step by Corollary 8.8.2.   □

This concludes the proof of Bott Periodicity.

The 6-term exact sequence

Suppose now that $J \subset A$ is an ideal. The associated long exact sequence

$$\cdots \to K_2(J) \xrightarrow{i_*} K_2(A) \xrightarrow{\pi_*} K_2(A/J)$$

$$\xrightarrow{\partial} K_1(J) \xrightarrow{i_*} K_1(A) \xrightarrow{\pi_*} K_1(A/J)$$

$$(8.71) \qquad \xrightarrow{\partial} K_0(J) \xrightarrow{i_*} K_0(A) \xrightarrow{\pi_*} K_0(A/J))$$

give no information about the final map $\pi_*: K_0(A) \to K_0(A/J)$. However, by naturality of the Bott Periodicity isomorphism β, the diagram

$$(8.72) \qquad \begin{array}{ccc} K_0(A) & \xrightarrow{\pi_*} & K_0(A/J) \\ \cong \downarrow \beta_A & & \cong \downarrow \beta_{A/J} \\ K_2(A) & \xrightarrow{\pi_*} & K_2(A/J) \end{array}$$

and the range of $\pi_*: K_0(A) \to K_0(A/J)$ identifies under Bott Periodicity with the range of

$$\pi_*: K_2(A) \to K_2(A/J).$$

which equals the kernel of the connecting homomorphism

$$\delta: K_2(J) \to K_1(J),$$

by exactness of the long exact sequence. Let

$$\delta': K_0(A/J) \xrightarrow{\beta_{A/J}} K_2(A/J) \xrightarrow{\delta} K_1(J)$$

be the indicated composition.

Then the *periodic* sequence

$$K_0(J) \xrightarrow{\;i_*\;} K_0(A) \xrightarrow{\;\pi_*\;} K_0(A/J)$$

$$\delta \uparrow \qquad\qquad\qquad\qquad\qquad \downarrow \delta'$$

$$K_1(A/J) \xleftarrow[\;\pi_*\;]{} K_1(A) \xleftarrow[\;i_*\;]{} K_1(J)$$

(8.73)

is exact.

Theorem 8.8.4 *For any ideal J in a C*-algebra A, the sequence (8.73) is exact. Furthermore, the sequence is natural with respect to *-homomorphisms, and the boundary maps commute with external products.*

The map $\delta': K_0(A/J) \to K_1(J)$ has a particularly simple description. Assume that A is unital. Let $p \in M_n(A/J)$ be a projection. Lift p to a self-adjoint $H \in M_n(A)$. Note that $H^2 - H \in J$. Using functional calculus for self-adjoints, we form the unitary

$$e^{2\pi i H} \in M_n(A).$$

Identifying J^+ with the C*-subalgebra of A generated by J and the unit $1 \in A$, I claim that $e^{2\pi i H} \in J^+$. Indeed, for $\pi: M_n(A) \to M_n(A/J)$ the quotient map, since $p = \pi(H)$ is a projection, its spectrum consists of 0 and 1, and hence $e^{2\pi i p} = 1$. Hence

$$e^{2\pi i H} = 1 \bmod J,$$

proving the claim.

We call the map associating to a projection p the corresponding exponentiated unitary $e^{2\pi i H}$, H a lift of p to a self-adjoint in A, the *exponential map*

Proposition 8.8.5 *In terms of the description of $K_1(J)$ as $\pi_0(U_\infty(J^+))$, the homomorphism*

$$\delta': K_0(A/J) \to K_1(J)$$

is the exponential map.

Example 8.8.6 Let $A = C_0([0, 1))$, a contractible C*-algebra containing $C_0((0, 1))$ as an ideal with quotient $\mathbb{C}$, under the map evaluating a function at 0. The 6-term exact sequence looks like

$$K^0((0,1)) \xrightarrow{\;i_*\;} K^0([0,1)) \longrightarrow K^0(\mathrm{pt})$$

$$\delta \uparrow \qquad\qquad\qquad\qquad\qquad \downarrow \delta'$$

$$K^{-1}(\mathrm{pt}) \longleftarrow K^{-1}([0,1)) \longleftarrow K^{-1}((0,1))$$

(8.74)

Substituting into the sequence the identities

$$K^0\big((0, 1)\big) = K^0\big([0, 1)\big) = K^1\big([0, 1)\big) = K^{-1}(\text{pt}) = 0, \quad K^0(\text{pt}) = \mathbb{Z}$$

gives the exact sequence

(8.75) $$0 \to K^0(\text{pt}) = \mathbb{Z} \xrightarrow{\delta'} K^{-1}\big((0, 1)\big) \to 0,$$

so that δ' is an isomorphism.

The generator of $K^0(\text{pt})$ is the projection $1 \in \mathbb{C}$, and lifting it to $C_0\big([0, 1)\big)$ in this case amounts to extending the function 1 at 0 to a continuous, real-valued function $f(t)$ on $[0, 1)$. There is an obvious explicit such extension:

$$f(t) := 1 - t.$$

Applying the exponential map to f we get the unitary function

$$u(t) = e^{2\pi i(1-t)} = e^{-2\pi it}$$

on the interval $[0, 1]$; notice that u assumes the same value 1 at each endpoint, so it is a unitary in $C_0\big((0, 1)\big)^+$. Of course, under the standard identification

$$C_0\big((0, 1)\big)^+ \cong C(\mathbb{T}),$$

this unitary is nothing but the usual complex (conjugate) coordinate $\bar{z} \colon \mathbb{T} \to \mathbb{C}$.

Exercise 8.8.7 Consider the extension of C*-algebras

(8.76) $$0 \to C_0(\mathbb{T} - \{1\}) \to C(\mathbb{T}) \to \mathbb{C} \to 0$$

obtained by removing a point (say, the point $1 \in \mathbb{T}$).

By direct computation using the "exponential map" description, as in the discussion above, show that the connecting map

$$\delta' \colon K_0(\mathbb{C}) \to K_1\big(C_0(\mathbb{T} - \{1\})\big) = K^{-1}(\mathbb{R})$$

is the zero map.

This is also clear from the 6-term exact sequence. Why?

A *homology theory* on the category $\mathbf{C}^*$ of C*-algebras and homomorphisms, is a sequence of functors

$$F_n \colon \mathbf{C}^* \to \mathbf{Ab}$$

to the category of abelian groups satisfying the following two axioms.

a) Homotopy-invariance: if $\alpha, \beta \colon A \to B$ are homotopic *-homomorphisms then $F_n(\alpha) = F_n(\beta)$ for all n.

b) Long exact sequences: if

$$0 \to J \xrightarrow{i} A \xrightarrow{\pi} B \to 0$$

is an exact sequence of C*-algebras then there exist group homomorphisms

$$\delta_n \colon F_n(B) \to F_{n-1}(J)$$

for all n so that the sequence

$$\cdots F_n(J) \xrightarrow{F_n(i)} F_n(A) \xrightarrow{F_n(\pi)} F_n(B) \xrightarrow{\delta_n} F_{n-1}(J) \to \cdots$$

The homomorphisms δ_n are required to be natural with respect to maps between short exact sequences—we leave it to the reader to formulate the condition exactly.

A homology theory is *stable* if the following holds. Let $p \in \mathcal{K}$ be a rank-one projection, it induces a *-homomorphism $i \colon \mathbb{C} \to \mathcal{K}$, and more generally a *-homomorphism $i_A \colon A \to A \otimes \mathcal{K}$ for any A. Stability is the condition that $F_n(i_A)$ is an isomorphism for all A, and all such projections p.

Joachim Cuntz has given a proof (see [56]) that any stable homology theory on the category of C*-algebras automatically satisfies Bott Periodicity. This is a typical example of a *rigidity theorem* about K-theory. Another, due to Nigel Higson (see [95]) asserts that any functor $F \colon \mathbf{C}^* \to \mathbf{Ab}$ exhibiting homotopy-invariance, stability and split exactness (maps split exact sequences of C*-algebras to short exact sequences of abelian groups), factors through KK: that is, KK is the *universal* such functor.

Such very strong rigidity results are in stark contrast with the situation with homology theories on *spaces*.

8.9   Some Orbifold K-Theory Computations

In this section we will draw from the material developed in Section 4.6 about the basic structure of crossed products $C_0(X) \rtimes G$ of discrete groups, acting *properly* on spaces X, and the strong Morita equivalence results of Section, together with the Morita invariance of K-theory, to compute the K-theory of some of these examples.

If G acts properly on X then the C*-algebra $C_0(X) \rtimes G)$ is isomorphic to the fixed-point algebra $C(X \times_G \mathcal{K})$, where $\mathcal{K} := \mathcal{K}(l^2(G))$, and the 'fixed-point algebra' $C(X \times_G \mathcal{K})$ is by definition the C*-algebra of all bounded $f \colon X \to \mathcal{K}$ such that $f(gx) = \rho(g)f(x)\rho(g)^{-1}$ for all $x \in X$, where ρ is the right-regular representation of G.

A good way to think of these functions is as sections of a bundle of C*-algebras over $G \backslash X$. The fiber of this bundle at an orbit Gx is $\mathcal{K}(l^2 G)^H$, where $H = \mathrm{Stab}_G(x)$, that is, compact operators on $l^2(G)$ which commute with the right translation action of H on $l^2(G)$.

As we have shown, actually

$$\mathcal{K}(l^2 G)^H \cong C^*(H) \otimes \mathcal{K}(l^2(G/H)).$$

Given our results on finite groups, $C^*(H)$ decomposes into a direct sum

$$C^*(H) \cong \oplus_{[\sigma] \in \widehat{H}} \mathcal{K}(V_\sigma)$$

of matrix algebras, with the summands parameterized by the points $\widehat{H}$, the irreducible representations of H. This induces a direct sum decomposition of $L^2(G)$ respected by the action of $\mathcal{K}(l^2 G)^H$. Putting everything together we obtain

$$(8.77) \qquad \mathcal{K}(l^2 G)^H \cong \oplus_{[\sigma] \in \widehat{H}} \mathcal{K}(l^2(G/H) \otimes V_\sigma).$$

We can summarize all of this in terms of the idea of a bundle of C*-algebras. if $A = C_0(X) \rtimes G$ and A_{Gx} is the "fiber" at Gx, then

$$A_{Gx} \cong \mathcal{K}(l^2 G)^H \cong \oplus_{[\sigma] \in \widehat{H}} \mathcal{K}(l^2(G/H) \otimes V_\sigma), \quad \text{where } H = \mathrm{Stab}_G(x).$$

In particular, in each fiber, one can single out the ideal corresponding to the ϵ-coordinate in the direct sum. This gives an ideal J_x in A_{Gx}, and the quotient is given by

$$(8.78) \qquad A_x/J_x \cong \mathcal{K}(l^2 G)^H \cong \oplus_{\widehat{H} \ni [\sigma] \neq \epsilon} \mathcal{K}(l^2(G/H) \otimes V_\sigma).$$

Of course $A_x/J_x = 0$ if x has no nontrivial isotropy.

The bundle of ideals $\{J_x\}$ corresponds to the ideal J_X discussed in the section on Morita equivalence, by the definitions: it is the ideal corresponding to the range of a certain inner product involved in a Morita equivalence between J_X and $C(G \backslash X)$.

Proposition 8.9.1 *Suppose that G acts properly on X, with only a finite set of orbits with nontrivial isotropy. Let J_X the ideal of $C_0(X) \rtimes G$ discussed above. Then $C_0(X) \rtimes G \,/\, J_X$ is isomorphic to a direct sum of compact operators. More exactly, if $F \subset G \backslash X$ denotes the set of points with nontrivial isotropy, then*

$$(8.79) \qquad C_0(X) \rtimes G \,/\, J_X \cong \oplus_{Gx \in F} \oplus_{\widehat{\mathrm{Stab}_G(x)} \ni [\sigma] \neq \epsilon} \mathcal{K}(l^2(Gx) \otimes V_\sigma),$$

where $\widehat{\mathrm{Stab}_G(x)}$ denotes, as usual, the collection of irreducible representations of the finite group $\mathrm{Stab}_G(x)$.

In particular, the K-theory groups of the quotient $C_0(X) \rtimes G \,/\, J_X$ are very easy: the K_0-group is

$$\bigoplus_{Gx \in F} \bigoplus_{\widehat{\mathrm{Stab}_G(x)} \ni [\sigma] \neq \epsilon} \mathbb{Z}$$

and K_1-group of the quotient is the zero group.

The 6-term exact sequence associated with the exact sequence

$$0 \to J_X \to C_0(X) \rtimes G \to C_0(X) \rtimes G \,/\, J_X \to 0$$

has therefore the form

$$0 \to K^0(G/X) \to K_0(C_0(X) \rtimes G) \to \bigoplus_{Gx \in F} \bigoplus_{\widehat{\mathrm{Stab}_G(x)} \ni [\sigma] \neq \epsilon} \mathbb{Z}$$

(8.80) $$\xrightarrow{\delta} K^{-1}(G \backslash X) \to K_1(C_0(X) \rtimes G) \to 0.$$

Now we recall Corollary 8.2.5, which asserts that in this situation, $K_0(C(X) \rtimes G)$ can be described completely in terms of G-equivariant vector bundles on X.

Suppose that $E \to X$ is a G-equivariant vector bundle over X, and $x \in X$, then the fiber E_x carries, by the assumptions, a representation of the compact group $\mathrm{Stab}_G(x)$. This results in a canonical group homomorphism

$$K_G^0(X) \to \mathrm{Rep}\big[(C^*(\mathrm{Stab}_G(x)))\big].$$

On the other hand, in our discussion above of the structure of $C_0(X) \rtimes G$, there is, for any $x \in X$, a natural *-homomorphism

$$K_0\big(C_0(X) \rtimes G\big) \to K_0\big(C^*(\mathrm{Stab}_G(x)))\big)$$

by restriction to the orbit of x. These two maps fit into a diagram

$$
\begin{array}{ccc}
K_G^0(X) & \longrightarrow & \mathrm{Rep}\big(\mathrm{Stab}_G(x)\big) \\
\downarrow & & \downarrow \\
K_0\big(C_0(X) \rtimes G\big) & \longrightarrow & K_0\big(C^*(\mathrm{Stab}_G(x))\big)
\end{array}
$$

(8.81)

where the vertical map on the left was discussed above, the vertical map on the right is the Green-Julg isomorphism.

Interpreting $K_0\big(C_0(X) \rtimes G\big)$ in this way as $K_G^0(X)$ allows us to describe the exact sequence (8.80) as follows.

Theorem 8.9.2 *Let G be a locally compact group acting properly on X with only finitely many points in $G \backslash X$ having nontrivial isotropy. Then if $F \subset X$ is a set of representatives of these points, then there is an exact sequence*

$$0 \to \mathrm{K}^0(G/X) \to \mathrm{K}_0\big(C_0(X) \rtimes G\big) \cong \mathrm{K}^0_G(X) \xrightarrow{r_*} \oplus_{x \in F} \mathrm{Rep}^*\big(\mathrm{Stab}_G(x)\big)$$

(8.82) $\xrightarrow{\delta} \mathrm{K}^{-1}(G \backslash X) \to \mathrm{K}_1(C_0(X) \rtimes G) \to 0,$

where, for any finite group H, $\mathrm{Rep}^*(H)$ denotes the free abelian group with one generator for each nontrivial *irreducible representation of H.*

What can we say about the map δ in the above sequence? In fact, it is rather subtle. It turns out that the question has to do with *torsion* in the K-theory of $G \backslash X$, and reflects a somewhat more general result, to the effect that it is much easier to compute *rationalized* K-theory of crossed products of the kind we are discussing, than it is to compute ordinary, integral K-theory.

Theorem 8.9.3 *The connecting homomorphism δ vanishes rationally. In particular, if the group $\mathrm{K}^{-1}(G \backslash X)$ has no torsion, then δ is the zero map.*

We will actually show, more precisely, that

$$m \cdot \delta(x) = 0, \quad \forall x \in \mathrm{K}_0(C_0(X) \rtimes G),$$

where m is the least common multiple of the cardinalities of the subgroups $\mathrm{Stab}_G(x)$.

Proof This is equivalent to showing that mx *lifts* to an element of $\mathrm{K}_0(C_0(X) \rtimes G)$ under the map r_* of (8.82),

(8.83) $m \cdot x = r_*(y), \quad y \in \mathrm{K}_0(C_0(X) \rtimes G) = \mathrm{K}^G_0(X).$

In order to do this, fix a point x with nontrivial isotropy. Denote

$$H := \mathrm{Stab}_G(x).$$

Let U be an H-slice at x: thus for some neighborhood V of x, the natural map

$$G \times_H V \to U$$

is a homeomorphism. We have already discussed that there is a natural "induction" map

$$\mathrm{Vect}_H(V) \to \mathrm{Vect}_G(G \times_H V) = \mathrm{Vect}(U),$$

applying in this situation. To induce an H-equivariant vector bundle on V to a G-equivariant vector bundle on U, we form

$$\tilde{E}_V := G \times_H E,$$

defined similarly as with $G \times_H V$. Let 1_H denote in this argument the trivial H-equivariant vector bundle, over whatever space, say, W, is under discussion. Thus, $1_H = W \times \mathbb{C}$ with the trivial action of H on the factor $\mathbb{C}$. Similarly for 1_G, the trivial G-equivariant vector bundle. Induction clearly maps $1_H \in \text{Vect}_H(V)$ to $1_G \in \text{Vect}_G(U)$.

Let $\rho: H \to U(V_\rho)$ be a unitary representation of H. We can view V_ρ as a H-equivariant vector bundle over the 1-point H-space $\{x\}$. Inducing it results in a G-equivariant vector bundle $\tilde{V}_\rho$ over the orbit Gx. Now consider the restriction of $\tilde{V}_\rho$ to $U - G \cdot x$. I claim that for some positive integer m, $m \cdot \tilde{V}_\rho = \tilde{V}_\rho \oplus \cdots \oplus \tilde{V}_\rho$, is isomorphic, as a G-equivariant vector bundle over $U - G \cdot x$, to a multiple $k \cdot 1_G$ of the trivial G-equivariant vector bundle over $U - Gx$.

Note that if we can prove this, the extension problem has been solved for $x := [V_\rho] \in \text{Rep}(H)$. Indeed, take the G-equivariant vector bundle $\tilde{V}_\rho$ over the G-invariant open set $W_1 := U = G \times_H V$ obtained by inducing the H-equivariant vector bundle $W_1 \times V_\rho$ (with diagonal H-action.) On the G-invariant open set $W_2 := X \backslash Gx$ take the trivial G-vector bundle $k \cdot 1_G = W_2 \times \mathbb{C}^k$ with $k = \dim(V_\rho)m$. Now glue these two G-equivariant vector bundles together to form a G-equivariant vector bundle over $W_1 \cup W_2 = X$. The its class $y \in K^0_G(X)$ is the required lift of x.

In order to prove the claim, we only need to observe that H acts freely on $V^* := V \setminus \{y\}$, and due to this,

$$K^0_H(V) \cong K^0(H \backslash V^*)$$

by a map sending the class of the H-equivariant vector bundle $V \times V_\rho$ to the class of the induced bundle $[V \times_H V_\rho] \in K^0(G \backslash V^*)$. We have already proved (Exercise 8.2.6) that in this situation, exists m so that $m \cdot [V \times_H V_\rho] = k \cdot [1_H]$, with $[1] \in K^0(H \backslash V)$ the class of the trivial line bundle. (And m is a divisor of $|H|$.) It follows that the bundles $V^* \times V_\rho$ and $V^* \times \mathbb{C}^k$, with H acting trivially on $\mathbb{C}^k$, are H-equivariantly isomorphic, for some k, over $V^* = V \setminus \{y\}$. Now inducing this result to $K^0_G(G \times_H V \setminus G \cdot x) = K^0_G(U \setminus G \cdot x)$ gives the required statement. $\square$

Exercise 8.9.4 Compute $K_*(C(I) \rtimes \mathbb{Z}/2)$, where the generator u of $\mathbb{Z}/2$ acts on $I := [-1, 1]$ by $u(x) = -x$.

Chapter 9
The Index Theorem of Atiyah and Singer

The elliptic operator $D = d/dx$ acting on the circle $\mathbb{T} = \mathbb{R}/2\pi i\mathbb{Z}$ has a discrete spectrum of the integers. In particular, the dimensions of the kernels of the continuously varying family of operators $D + \lambda$, for $\lambda \in \mathbb{R}$, jump discontinuously as λ crosses an integer point. But the *Fredholm index* given by the *difference* $\dim\ker(D + \lambda) - \dim\ker(D^* + \lambda)$ is constant in λ. An observation going back to Gelfand is that, more generally, elliptic operators on compact manifolds are Fredholm and that the Fredholm index (see Chapter 2) $\dim\ker(D) - \dim\ker(D^*)$ of an elliptic operator on a compact manifold is invariant under small perturbation of D: It is a homotopy invariant. Gelfand inquired whether there was a formula for the Fredholm index of any elliptic operator involving topological data of the manifold.

The Index Theorem was announced in [12] in 1963 with a proof sketch using cobordism; a full proof using K-theory appeared in 1968 in the papers [13–15, 17]. Theorems of this type had already appeared: namely the Riemann–Roch Theorem (see [140, 141]) and the Hirzebruch Signature Theorem. One of the questions answered by the Index Theorem has to do with the $\hat{A}$-genus of a manifold. This is a characteristic class which can expanded into an infinite series involving the Pontryagin classes $p_0, p_1, p_2, \ldots$ (see [142]), which are certain de Rham cohomology classes of even degrees (see [34]). The first few terms in the $\hat{A}$-genus are $\hat{A}_0 = 1$, $\hat{A}_1 = -\frac{1}{24}p_1$, $\hat{A}_2 = \frac{1}{5760}(-4p_2 + 7p_1^2)$. A fact proved earlier by Borel and Hirzebruch was the curious fact that the integral of the $\hat{A}$-genus of X over X, which is always a rational number, is an *integer* if X is a spin manifold (see Remark 7.10.5 for the definition of spin manifold). The Index Theorem explains this integrality: $\int_X \hat{A}(X)$ is equal to the Fredholm index of the spin Dirac operator on X.

Explicit characteristic class formulas for the index are discussed further in Section 10.8, but in this chapter we establish the basic Fredholm and spectral theory of Dirac-type operators and state the K-theory version of the Index Theorem. A lot of geometry and analysis goes into defining the "analytic index" (Fredholm index) of an elliptic operator. But the Dirac operator on a spin^c-manifold, twisted by a

© Springer Nature Switzerland AG 2024

H. Emerson, *An Introduction to C*-Algebras and Noncommutative Geometry*,
Birkhäuser Advanced Texts Basler Lehrbücher,
https://doi.org/10.1007/978-3-031-59850-0_9

vector bundle, has as its ultimate source a spin^c-structure, and a vector bundle, and each of these are bits of topological data, which can be combined in a completely different manner using primarily the Thom Isomorphism in topological K-theory, to produce an integer called the *topological index*. The equality of the topological index and the analytic index of the Dirac operator twisted by the bundle is the K-theory statement of the Index Theorem.

The most important lesson to be drawn from the Atiyah–Singer Theorem, for the purposes of this book, is that elliptic operators on manifolds define, by a process of twisting and the extraction of an analytic index from the twisted operator, group homomorphisms from the K-theory of the manifold to the integers. As pointed out by Atiyah, it thus seems that suitably defined equivalence classes of elliptic operators are in a relation of duality with suitably defined equivalence classes of vector bundles, i.e., K-theory classes. This is the idea which led to the formulation of K-*homology* by Kasparov [111], as is discussed in the next chapter.

9.1   Differential Operators on Euclidean Space

A *differential operator* (of order $\leq m$) on $\mathbb{R}^n$ is a linear operator on the complex vector space $C_c^\infty(\mathbb{R}^n)$, of the form

$$(9.1) \qquad\qquad (Df)(x) = \sum_{|\alpha| \leq m} a_\alpha(x) D^\alpha f(x).$$

Our notation is that

$$D^\alpha := (-i)^{|\alpha|} \frac{\partial^{\alpha_1}}{\partial x_1^{\alpha_1}} \frac{\partial^{\alpha_2}}{\partial x_2^{\alpha_2}} \cdots \frac{\partial^{\alpha_n}}{\partial x_1^{\alpha_n}}, \qquad |\alpha| := \sum_{i=1}^{n} \alpha_i.$$

The Fourier transform allows us to rewrite this as follows. If $f \in C_c(\mathbb{R}^n)$, or, more generally, if $f \in L^1(\mathbb{R}^n)$, then its Fourier transform $\hat{f}$ is the function on the dual group $\widehat{\mathbb{R}^n}$ defined by

$$\hat{f}(\xi) := \frac{1}{(2\pi)^n} \int_{\mathbb{R}^n} f(x) e^{-ix \cdot \xi} dx.$$

Let $S(\mathbb{R}^n)$ denote the Schwartz space of infinitely differentiable functions on $\mathbb{R}^n$ such that for all α, β there exists a constant $C_{\alpha\beta}$ such that $|D^\alpha f(x)| \leq C_{\alpha\beta}(1+|x|)^\beta$ for all $x \in \mathbb{R}^n$. It is routine to check that $f \in S(\mathbb{R}^n)$ implies $\hat{f}$ is in $S(\widehat{\mathbb{R}^n})$. So Fourier transform defines a linear map

$$F \colon S(\mathbb{R}^n) \to S(\widehat{\mathbb{R}^n}).$$

The Fourier inversion formula says that $f \in S(\mathbb{R}^n)$ and then

$$(9.2) \qquad\qquad f(x) = \int_{\mathbb{R}^n} \hat{f}(\xi) e^{ix \cdot \xi} d\xi.$$

Differentiating (9.2) under the integral sign and using that $D_x^\alpha (e^{ix \cdot \xi}) = \xi^\alpha e^{ix \cdot \xi}$, we see that

$$Df(x) = \int_{\mathbb{R}^n} \sigma(x, \xi) e^{ix \cdot \xi} \hat{f}(\xi) d\xi,$$

where $\sigma(x, \xi) = \sum_{|\alpha| \leq m} a_\alpha(x) \xi^\alpha$, an infinitely differentiable function of (x, ξ) which is polynomial in ξ of order $\leq m$. The top-order term of the symbol is called the *principal symbol* given by

$$\sigma_p(x, \xi) = \sum_{|\alpha| = m} a_\alpha(x) \xi^\alpha.$$

Directly in terms of σ, using a limit:

$$\sigma_p(x, \xi) = \lim_{t \to \infty} \frac{\sigma(x, t\xi)}{t^m}.$$

The symbol is *elliptic* if its principal symbol satisfies $\sigma_p(x, \xi) \neq 0$ for $\xi \neq 0$. The operator D is an *elliptic differential operator of order m* if its principal symbol is an elliptic symbol of order m.

Example 9.1.1 $\Delta = \sum_{i=1}^n -\frac{\partial^2}{\partial x^2}$ is an elliptic operator of order 2 on $\mathbb{R}^n$, the Laplacian. It has constant coefficients, and $\sigma(x, \xi) = -|\xi|^2$.

Exercise 9.1.2 Let $\mathrm{Diff}(\mathbb{R}^n)$ denote the collection of operators of the form (9.1):

a) Show that if S and T are differential operators of orders m, m', then ST is a differential operator of order $m + m'$.
b) Check that the commutator of differential operators $[\frac{\partial}{\partial x_k}, a]$ has order zero, for any coordinate x_k, and any smooth function a.
c) Extend the result of b) to show that if S and T are as in a), then the commutator $[S, T] = ST - TS$ is a differential operator of order $m + m' - 1$.
d) Deduce from c) that if S and T are as above, then the principal symbol of the differential operator ST of order $m + m'$ is the pointwise product of the principal symbols of S and T:

$$\sigma_p^{ST}(x, \xi) = \sigma_p^T(x, \xi) \cdot \sigma_p^S(x, \xi).$$

An important special case of part c) of the Exercise above is that if $f \in C^\infty(\mathbb{R}^n)$ is a smooth function, acting by multiplication on $C_c^\infty(\mathbb{R}^n)$, and if D is a differential operator of order m on $\mathbb{R}^n$, then the commutator

$$[f, D]$$

is a differential operator of order $m - 1$. In the important special case $m = 1$, this means $[D, f]$ is order zero and thus is a *bounded* operator, as long as f' is bounded.

In the applications of elliptic operator theory, one usually works not with Hilbert spaces of scalar-valued functions a_α, but Hilbert spaces of *matrix*-valued functions. Fix a positive integer m. For $f : \mathbb{R}^n \to \mathbb{C}^m$ a smooth, vector-valued function, with entries $f(x) = (f_1(x), \ldots, f_m(x))$, and α a multi-index, set $D^\alpha f := (D^\alpha f_1, \ldots D^\alpha f_m)$. Now, as above, if we are given a family of smooth functions

$$a_\alpha : \mathbb{R}^n \to M_m(\mathbb{C}),$$

for various multi-indices α, with $|\alpha| \leq m$ as before, we can define an operator

$$D : C_c^\infty(\mathbb{R}^n, \mathbb{C}^m) \to C_c^\infty(\mathbb{R}^n, \mathbb{C}^m)$$

by setting,

$$(Df)(x) = \sum_{|\alpha| \leq m} a_\alpha(x)(D^\alpha f)(x), \quad f \in C_c^\infty(\mathbb{R}^n, \mathbb{C}^m).$$

The *symbol* of such an operator and the principal symbol are defined just as in the scalar case and now are matrix-valued functions

$$\sigma : \mathbb{R}^n \times \widehat{\mathbb{R}^n} \to M_n(\mathbb{C})$$

(and similarly for σ_p). The symbol is *elliptic* if $\sigma_p(x, \xi)$ is *invertible* in $M_n(\mathbb{C})$ for all nonzero ξ, and all $x \in \mathbb{R}^n$.

Slightly more abstractly, if V is any finite-dimensional Hilbert space, then the partial differentiation operators $\frac{\partial}{\partial x_j}$ act on $C_c^\infty(\mathbb{R}^n, V)$; to see this one can define the action directly by a limit, in the usual way, or one can fix a basis for V and identify $C_c^\infty(\mathbb{R}^n, V)$ with functions valued in $\mathbb{C}^m$ for some m, by expanding vectors into their coefficients. Functions valued in $\mathbb{C}^m$ can then be differentiated component, and it is easily checked that the result is independent of the choice of basis for V. A differential operator D on $C_c^\infty(\mathbb{R}^n, V)$ is then one of the form

(9.3) $$(Df)(x) = \sum_{|\alpha| \leq m} a_\alpha(x)(D^\alpha f)(x), \quad f \in C_c^\infty(\mathbb{R}^n, V).$$

The coefficients a_α are smooth functions $\mathbb{R}^n \to \mathrm{End}(V)$, the symbol $\sigma : \mathbb{R}^n \times \widehat{\mathbb{R}^n} \to \mathrm{End}(V)$ is defined as in the scalar case by

$$\sigma(x, \xi) = \sum_{|\alpha| \leq m} a_\alpha(x)\xi^\alpha,$$

and the principal symbol is the top-order part as before. Ellipticity means that $\sigma_p(x, \xi)$ is invertible if $\xi \neq 0$.

Example 9.1.3 Let $m = 2$, $n = 2$ and define a matrix-valued function

$$(9.4) \qquad \sigma: \mathbb{R}^2 \times \widehat{\mathbb{R}^2} \to M_2(\mathbb{C}), \quad \sigma(x, \xi) = i \begin{bmatrix} 0 & \xi_1 - i\xi_2 \\ \xi_1 + i\xi_2 & 0 \end{bmatrix}.$$

Then σ is elliptic, for $\det \sigma(x, \xi) = \|\xi\|^2 \neq 0$ if $\xi \neq 0$.
The associated operator on $C_c^\infty(\mathbb{R}^2, \mathbb{C}^2)$ is

$$\bar{\partial} = \begin{bmatrix} 0 & \frac{\partial}{\partial x} - i\frac{\partial}{\partial y} \\ \frac{\partial}{\partial x} + i\frac{\partial}{\partial y} & 0 \end{bmatrix},$$

called the *Dolbeault* operator. Note that if $f_1, f_2 \in C^\infty(\mathbb{R}^2)$, then $\bar{\partial} \begin{bmatrix} f_1 \\ f_2 \end{bmatrix} = 0$ if and only if f_2 is anti-holomorphic, and f_1 is holomorphic, by the Cauchy–Riemann equations.

The previous example might remind the reader of Clifford algebras. In fact the map $\sigma: \mathbb{R}^2 \to \mathbb{B}(\mathbb{C}^2)$ in (9.4) satisfies $\sigma(\xi) = -\sigma(\xi)^*$, and $\sigma(\xi)^2 = -\|\xi\|^2$, for ξ a vector in $\mathbb{R}^2$. This is no accident. Let

$$c: \mathrm{Cliff}(\mathbb{R}^n) \to \mathrm{End}(S)$$

be a $\mathrm{Cliff}(\mathbb{R}^n)$-module; we may as well assume that it is one of the irreducible representations of Theorem 7.8.1—the unique one, if n is even. We will build an associated *Dirac operator* D. The operator will act on the space $C_c^\infty(\mathbb{R}^n, S)$ of smooth, compactly supported functions $s: \mathbb{R}^n \to S$.

The Clifford module structure gives, for each unit vector $\xi \in \mathbb{R}^n$, a linear operator $c(\xi)$ on the Hilbert space S, which is $\mathbb{Z}/2$-graded, if n is even, and with the properties that

$$c(\xi)^2 = -1, \quad c(\xi)^* = -c(\xi),$$

and $c(\xi)$ is odd with respect to the grading, in the case n is even.
Let $\xi_1, \ldots, \xi_n$ be the standard orthonormal basis for $\mathbb{R}^n$. For each i we form the composition of the partial differentiation operator $\frac{\partial}{\partial x_i}$ and the Clifford multiplication operator $c(\xi_i)$. Adding them up gives a differential operator with constant coefficients

$$(9.5) \qquad D := \sum_{i=1}^n c(\xi_i) \frac{\partial}{\partial x_i} = \sum_{i=1}^n -ic(\xi_i) \cdot D_i$$

on $C_c^\infty(\mathbb{R}^n, S)$.

Note that D is elliptic. Indeed, its symbol is given by the self-adjoint operator

$$\sigma(x, \xi) = -i \sum_{i=1}^{n} c(\xi_i) \cdot \xi_i \in \text{End}(S).$$

Hence

$$\sigma(x, \xi)^2 = - \sum_{i,j} \xi_i \xi_j c(\xi_i) c(\xi_j).$$

Since

$$c(\xi_i) c(\xi_j) = -c(\xi_j) c(\xi_i$$

for all $i \neq j$, and $c(\xi_i)^2 = -1$, this equals

$$\sum_{i=1}^{n} \xi_i^2 = \|\xi\|^2,$$

a nonzero scalar multiple of the identity operator on S, provided that $\xi \neq 0$.

Although D is an unbounded operator, it is formally self-adjoint. Indeed, $c(\xi_i)$ obviously commutes with $\frac{\partial}{\partial x_j}$ as linear operators on $C_c^\infty(\mathbb{R}^n, S)$. The spin representation space S is, by assumption, a Hilbert space, with an inner product. We endow $C_c^\infty(\mathbb{R}^n, S)$ with the inner product

$$\langle s, s' \rangle := \int_{\mathbb{R}^n} \langle s(x), s'(x) \rangle, \quad s, s' \in C_c^\infty(\mathbb{R}^n, S).$$

In this notation, integration by parts gives that

$$\left\langle \frac{\partial s}{\partial x_j}, s' \right\rangle = -\left\langle s, \frac{\partial s'}{\partial x_j} \right\rangle,$$

so that partial differentiation is a skew-adjoint operator. Since the operator $c(\xi_i)$ on $C_c^\infty(\mathbb{R}^n, S)$ is also skew-adjoint, for any vector $\xi \in \mathbb{R}^n$, we get that D, a real linear combination of compositions of two skew-adjoint operators, is formally self-adjoint, i.e.,

$$\langle Ds, s' \rangle = \langle s, Ds' \rangle, \quad s, s' \in C^\infty(\mathbb{R}^n, S).$$

Finally, we note that if we give the linear space $C_c^\infty(\mathbb{R}^n, S)$ the $\mathbb{Z}/2$-grading induced by the $\mathbb{Z}/2$-grading on S, then the Dirac operator D is an *odd* operator $C_c^\infty(\mathbb{R}^n, S) \to C_c^\infty(\mathbb{R}^n, S)$, i.e., interchanges the even and odd parts of $C_c^\infty(\mathbb{R}^n, S)$.

Example 9.1.4 The simplest example of all of a Dirac operator is $D = -i\frac{d}{dx}$ acting on $C_c^\infty(\mathbb{R})$. This is associated to the Clifford module

$$c\colon \text{Cliff}(\mathbb{R}) \to \mathbb{C}$$

of its positive irreducible representation on the one-dimensional spinor space $S = \mathbb{C}$, which maps the unit vector $1 \in \mathbb{R} \subset \text{Cliff}(\mathbb{R})$ to the scalar i.

Example 9.1.5 We conclude this section with some general remarks regarding changes of coordinates.

Differential operators on $\mathbb{R}^n$ are *local*, in the sense that if D is such an operator, thus, of the form (9.3), and if $\rho \in C_c^\infty(\mathbb{R}^n)$ is supported in an open set $W \subset \mathbb{R}^n$, then $D\rho$ is also supported in W.

It follows that any such operator D restricts to an operator

$$D|_U\colon C_c^\infty(U) \to C_c^\infty(U)$$

for any open set $U \subset \mathbb{R}^n$. We let $\text{Diff}(U)$ be the algebra of differential operators on U: It is generated by the partial differentiation operators $\frac{\partial}{\partial x_j}$ and the multiplication operators by smooth functions $f \in C^\infty(U)$.

Now, suppose that $\phi\colon U \to V$ is a diffeomorphism between two open subsets of $\mathbb{R}^n$. Let $T_\phi\colon C_c^\infty(V) \to C_c^\infty(U)$ be the linear map of composition with ϕ,

$$(T_\phi f)(x) = f\big(\phi(x)\big).$$

Let $\phi_1, \ldots, \phi_n$ be the coordinate functions of ϕ. Then by the Chain Rule

$$\frac{\partial}{\partial x_j}(f \circ \phi)(x) = \sum_i \frac{\partial f}{\partial x_i}(\phi(x)) \cdot \frac{\partial \phi_i}{\partial x_j}(x),$$

from which it follows that

$$T_\phi^{-1} \frac{\partial}{\partial x_j} T_\phi = \sum_i \left(\frac{\partial \phi_i}{\partial x_j} \circ \phi^{-1} \right) \cdot \frac{\partial}{\partial x_i}$$

as operators on $C_c^\infty(V)$.

Since

$$T_\phi^{-1} \circ f \circ T_\phi = f \circ \phi$$

as (multiplication) operators on $C_c^\infty(V)$, for any $f \in C^\infty(U)$, and since such functions, and the partial differentiation operators, generate $\text{Diff}(U)$ as an algebra, it follows that conjugation $L \mapsto T_\phi^{-1} L T_\phi$ by T_ϕ maps $\text{Diff}(U)$ to $\text{Diff}(V)$, in a canonical manner.

Exercise 9.1.6 If $L \in \text{Diff}(U)$, $\phi \colon U \to V$ a diffeomorphism, T_ϕ the operator of composition with ϕ as above, and if $\sigma_L^p \in C^\infty(U)$ is the principal symbol of L, then the principal symbol of $T_\phi^{-1} L T_\phi \in \text{Diff}(V)$ is given by

$$(9.6) \qquad \sigma_{T_\phi^{-1} L T_\phi}^p (x, \xi) = \sigma_T^p(\phi^{-1}x, {}^t D_{\phi^{-1}x}\phi \cdot \xi), \quad x \in V, \quad \xi \in \widehat{\mathbb{R}^n},$$

where ${}^t D_{\phi^{-1}x}\phi \cdot \xi$ is shorthand for

$$\sum_i \frac{\partial \phi_i}{\partial x_j}(\phi^{-1}x) \cdot \xi_i.$$

9.2   Differential Operators on Manifolds

See Chapter 6 Section 7.2 for background on vector bundles on smooth manifolds, like the tangent and cotangent bundles.

Definition 9.2.1 Let M be a smooth manifold, and $L \colon C_c^\infty(M) \to C_c^\infty(M)$ be a linear, *local* operator: That is, L leaves the subspaces $C_c^\infty(U)$ invariant, for every $U \subset M$ open.

We say that L is a *differential operator of order m* on M if for every $p \in M$, there exists a coordinate chart

$$\phi \colon U \to \mathbb{R}^n,$$

such that the operator

$$T_\phi^{-1} L T_\phi \colon C_c^\infty(\mathbb{R}^n) \to C_c^\infty(\mathbb{R}^n)$$

is a differential operator of order m on $\mathbb{R}^n$, where

$$T_\phi \colon C_c^\infty(\mathbb{R}^n) \to C_c^\infty(U)$$

is the operator of composition with the chart ϕ.

The discussion preceding the definition shows that L is differential of order m on M if and only if *for every* manifold chart $\phi \colon U \to \mathbb{R}^n$, the operator $T_\phi^{-1} L T_\phi$ is differential of order m.

Example 9.2.2 $D = -i\frac{d}{d\theta}$ is a differential operator of order 1 on the circle $\mathbb{T} = \mathbb{R}/\mathbb{Z}$, with θ the usual angular coordinate.

Example 9.2.3 One of the most important differential operators in Riemannian geometry is the *Laplacian* operator on a Riemannian manifold. Let M be a manifold, equipped with a Riemannian metric. Let g be the coefficient matrix in a chart with coordinates $x_1, \ldots, x_n$, then locally

$$(9.7) \qquad \Delta f = \frac{-1}{\sqrt{\det(g)}} \sum_{i,j} \frac{\partial}{\partial x_i} \left(\sqrt{\det g} \cdot (g^{-1})_{ij} \frac{\partial f}{\partial x_j} \right).$$

It has order 2.

Suppose L is a differential operator of order m on M and that in the domain $U \subset M$ of a coordinate chart, with coordinates $x_1, \ldots, x_n$, L can be represented in the form

$$(9.8) \qquad L|_U = \sum_\mu a_\mu D^\mu,$$

with a_μ smooth functions on U, $D_i := -i\frac{\partial}{\partial x_i}$, differentiation in the coordinate direction x_i, and D_μ the corresponding product of such operators, according to the multi-index μ. We are assuming that the top-order part of this operator is in degree m.

Recall that a smooth function f in a neighborhood of a point $p \in \mathbb{R}^n$ *vanishes to order* 1 at p if $f(p) = 0$. It vanishes to order 2 at p if it is a product of two functions each vanishing to order 1 at p, and so on.

Let J_p^k be the algebra of germs of smooth functions at p which vanish to order k at p. By definition, $J_p^1 \supset J_p^2 \supset \cdots$.

Then the cotangent bundle T^*M has fiber J_p^1/J_p^2, by definition. The differential 1-form df of a germ of a smooth function at p is by definition the class modulo J_p^2 of $f - f(p)$. The duality with the tangent bundle pairs a tangent vector v at p and an element df of T_p^*M, to produce the derivative by v of f at p. This process annihilates constant functions, and functions which vanish to order 2 at p.

In fact, vanishing to order 2 at p might be rephrased in terms of differential operators by observing that $f \in J_p^2$ is *equivalent* to saying that $(Df)(p) = 0$ for all differential operators D of order 1, defined in a neighborhood of p.

Lemma 9.2.4 *Let D be a differential operator of order m on $\mathbb{R}^n$, and $f \in C_c^\infty(\mathbb{R}^n)$ be a smooth function which vanishes to order $m + 1$ at p. Then $(Df)(p) = 0$.*

Proof Suppose first that $m = 1$. If f vanishes to order 2 at p, then $f = f_1 f_2$ with f_i vanishing to order 1 at p. Hence for any i,

$$\frac{\partial f}{\partial x_i}(p) = \frac{\partial f_1}{\partial x_i}(p) \cdot f_2(p) + f_1(p) \cdot \frac{\partial f_2}{\partial x_i}(p) = 0,$$

since each of f_1, f_2 vanishes at p.

This implies the result for $m = 1$, and the general result follows from induction.

□

Lemma 9.2.5 *If f_1 and f_2 each vanish to order 1 at p and $f_1 - f_2$ vanishes to order 2 at p, then $f_1^m - f_2^m$ vanishes to order $m + 1$ at p.*

Proof We use the identity

$$(9.9) \qquad f_1^m - f_2^m = (f_1 - f_2) \cdot f_1^{m-1} + f_2 \cdot (f_1^{m-1} - f_2^{m-1}).$$

If $f_i \in J_p^1$ and $f_1 - f_2 \in J_p^2$, then it follows that the first term is in J_p^{m+1}. By induction, $f_1^{m-1} - f_2^{m-1} \in J_p^m$. Hence the second term is also in J_p^{m+1}. $\qquad \square$

As a corollary:

Lemma 9.2.6 *If D is a differential operator of order m in a neighborhood of p, and if $f \in J_p^1$ is a germ of smooth function at p, then the value of $D(f^m)$ at p only depends on the coset of f in J_p^1/J_p^2—that is, depends only on $df \in J_p/J_p^2 = T_p^*M$.*

As a consequence, the symbol can be defined in the following natural way.

Definition 9.2.7 If D is a differential operator of order m, its principal symbol σ is the function on T^*M whose value at a point $df \in T_p^*M$ is the value

$$\frac{i^m}{m!} \cdot D(f^m)(p).$$

Example 9.2.8 Suppose D is Lie derivative with respect to a vector field V on M, or an open subset. Then for $p \in M, f$ a smooth germ of a function at p, then $(Df)(p) := V(f)(p)$. This is a linear function of the coset $df \in T_p^*M$. If in a coordinate system near p, centered at 0 for convenience, the vector field is $V = \sum_i a_i \cdot \frac{\partial}{\partial x_i}$, and if $f = x_j$, so $df = dx_j$ and then

$$(Vf)(p) = \sum_i a_i \cdot \frac{\partial x_j}{\partial x_i}(p) = a_j$$

so that in cotangent coordinates $\xi_1, \ldots \xi_n$, the symbol is given by

$$\sigma_D(x, \xi) = i \cdot \sum a_i \xi_i,$$

which is (except for the multiplication by i) precisely the pairing between T_p^*M and T_pM, applied to the value of the vector field $V(p)$ and the cotangent vector df.

Exercise 9.2.9 Show that the symbol of the Laplacian is given in local coordinates by

$$\sigma_\Delta(x, \xi) = -\sum_{i,j} (g^{-1})_{ij}(x) \cdot \xi_i \xi_j, \quad \xi \in T_x^*M.$$

That is, the symbol of the Laplacian on M is the Riemannian metric

$$\sigma_\Delta(x, \xi) = -\|\xi\|^2$$

on the cotangent bundle.

Our discussion of differential operators on $\mathbb{R}^n$ contained variants involving an auxiliary (finite-dimensional) Hilbert space. In the context of manifolds, the interesting examples of elliptic differential operators related to geometry are operators not on $C_c^\infty(M)$ but on the spaces of smooth sections of a smooth vector bundle $\pi: S \to M$ over M: We use the notation $C_c^\infty(M, S)$ for this linear space of smooth, compactly supported sections. More generally, we consider pairs of vector bundles, and maps between their spaces of smooth sections.

Definition 9.2.10 A *differential operator of order m*

$$D: C_c^\infty(M, S^+) \to C_c^\infty(M, S^-)$$

between sections of a pair of bundles S^+, S^- over M is a linear operator which is, firstly, local, and secondly, such that every point of M has a neighborhood U such that the restriction $D: C_c^\infty(U, S^+) \to C_c^\infty(U, S^-)$ can be written in local coordinates on M in the form

$$(9.10) \qquad (Ds)(x) = \sum_{|\mu| \le m} a_\mu(x) \cdot (A^{-1}D^\mu)(As)(x), \quad s \in C_c^\infty(U, S^+|_U)$$

for some (any) smooth trivialization $A: S^+|_U \to U \times \mathbb{C}^n$ of the bundle S^+ over U, and a family $a_\mu \in C_c^\infty(U, \mathrm{Hom}(S^+, S^-))$ of bundle maps $S^+ \to S^-$.

Of course any section $T \in C^\infty(M, \mathrm{End} S)$ of the endomorphism bundle of a single bundle S, that is, any bundle map $S \to S$, defines a differential operator of order zero. Such an operator commutes with multiplication by smooth functions on M.

Example 9.2.11 Suppose that $\pi: S \to M$ is a vector bundle over M and that ∇ is a connection on S. Then ∇ restricts to a connection on $S|_U$ for any open subset. Picking U with $S|_U$ is trivial, with $A: S|_U \to U \times \mathbb{C}^n$ a trivialization. On the trivial bundle $U \times \mathbb{C}^n$ we always have the trivial connection

$$\nabla_X^{\mathrm{triv}}(s_1, \ldots, s_n) := (X(s_1), \ldots X(s_n)),$$

where $(s_1, \ldots, s_n)$ is a section of $U \times \mathbb{C}^n$.

Hence $A^{-1} \cdot \nabla^{\mathrm{triv}} \cdot A$ is another connection on $S|_U$.

Now any two connections ∇^1 and ∇^2 on $S|_U$ differ by an $\mathrm{End}(S)$-valued 1-form: that is, a bundle map $T^M \to \mathrm{End}(S)$, given by the pairing

$$\langle X, s \rangle_p := \nabla_X^1 s - \nabla_X^2 s,$$

where s is a section of $S|_U$ and X is a tangent vector. This expression is $C^\infty(M)$-linear in the variable s since

$$\nabla_X^1(fs) - \nabla_X^2(fs) = X(f)s + f\nabla_X^1 s - X(f)s + f\nabla_X^2 s,$$

by the connection property. Hence $\langle X, s \rangle(p)$, for any $p \in U$, only depends on the value of X at p and the value of s at p.

In particular, any covariant derivative ∇_V, for V a vector field on M, is locally the sum of a section of $\mathrm{End}(S)$, and a conjugate, as above, of Lie differentiation by V, acting on sections of a trivial bundle. In particular, it locally has the form specified in (9.10). Therefore, ∇_V is a differential operator of order 1 on $C_c^\infty(M, S)$.

Exercise 9.2.12 Prove that if D is an order 1 differential operator $D: C_c^\infty(M, S^+)$ $\to C_c^\infty(M, S^-)$ between sections of a pair of bundles over M, and if $f \in C^\infty(M)$ acts on sections of each of these bundles by multiplication, then the commutator $[f, D]$ is an operator of order zero, and in particular, is a bounded operator.

Definition 9.2.13 Let $D: C_c^\infty(M, S^+) \to C_c^\infty(M, S^-)$ be a differential operator of order m between section spaces of a pair of bundles $S^\pm$ over a smooth manifold M.

The *symbol* of D is the smooth bundle map

$$\sigma_D: \pi^*(S^+) \to \pi^*(S^-)$$

mapping a covector $\xi := df \in T_p^*M$, where f is smooth and vanishes to order 1 at x, and an element $w \in S_x^+$, to

$$(9.11) \qquad\qquad \sigma_D(x, df) \cdot w := D(f^m s)(x) \in S_x^-,$$

where s is any smooth extension of s to a smooth section defined near p.

D is *elliptic* if $\sigma_D(x, \xi): S_x^+ \to S_x^-$ is invertible for every $\xi \neq 0$ in T_x^*M.

The formula (9.11) is well defined, since if s and s' are sections that agree at x, then $s - s'$ vanishes at x, and hence the section $f^m(s - s')$ vanishes to order $m + 1$ at x, and by a slight generalization of Lemma 9.2.4, we deduce that $D\big(f^m(s - s')\big)$ vanishes at x.

The de Rham and Laplace operators

The de Rham operator is the simplest example of an elliptic differential operator encoding topological information about the manifold on which it is defined. Let X be an n-dimensional compact, oriented manifold, $\Lambda^*(TX) \otimes_\mathbb{R} \mathbb{C}$ the complexified exterior bundle of the cotangent bundle, $\Omega^*(X)$ the space of smooth sections of this bundle, i.e., the space of smooth differential forms on X.

The de Rham differential

$$d: \Omega^*(X) \to \Omega^*(X)$$

is determined uniquely by the following two conditions, the first local, and described in a local coordinate system, and the second algebraic, and applying to forms α, β of some degrees, perhaps different:

$$df = \sum \frac{\partial f}{\partial x_i} dx_i, \quad f \in C^\infty(X), \quad d(\alpha \wedge \beta) = d\alpha \wedge \beta + (-1)^{\partial \alpha} \alpha \wedge d\beta, \quad \alpha, \beta \in \Omega^*(X).$$

The Riemannian metric on X is a Euclidean metric on TX and has an associated matrix in each local coordinate system, with entries $g_{ij} := \langle \frac{\partial}{\partial x_i}, \frac{\partial}{\partial x_j} \rangle$. Let g^{ij} be the i, jth entry of the inverse matrix and g the determinant of the matrix. If $x_1, \ldots, x_n$ is a positively oriented coordinate system, then the n-form $\sqrt{g} dx_1 \cdots dx_n$ defined on the coordinate patch does not depend on the (oriented) coordinate system used, so using an oriented atlas one pieces these forms together to give the *volume form* dvol on X.

A metric on the bundle of k-forms is defined in a local coordinate patch by

$$(9.12) \qquad \langle \alpha, \beta \rangle := \frac{1}{k!} \sum_{\mu, \nu} g^{\mu_1 \nu_1} \cdots g^{\mu_k \nu_k} A_{\mu_1 \cdots \mu_k} B_{\nu_1 \cdots \nu_k},$$

where $\alpha = \sum A_\mu dx_\mu$, $\beta = \sum B_\nu dx_\nu$, with μ, ν multi-indices of length k.

The Hodge $*$ maps k-forms to n-forms and is uniquely defined by the requirement

$$\langle \alpha, \beta \rangle \text{dvol} = \beta \wedge *\alpha.$$

Note that if α, β are k-forms, and we take their fiberwise inner product by (9.12) and integrate, we obtain exactly

$$\langle \alpha, \beta \rangle := \int_X \alpha \wedge *\beta.$$

This is extremely convenient, since by Stoke's theorem and calculation, one sees that

$$(9.13) \quad 0 = \int_X d(\beta \wedge *\alpha) = \int_X d\beta \wedge *\alpha + (-1)^k \int_X \beta \wedge d(*\alpha)$$

$$= \langle \alpha, d\beta \rangle - \langle \delta\alpha, \beta \rangle$$

if δ is defined by

$$\delta(\alpha) := (-1)^{nk+n+1} * d * \alpha,$$

a map from k-forms to $k - 1$-forms. Hence the operator

$$D_{\text{dR}} := d + \delta,$$

acting on $\Omega^*(X)$, is formally self-adjoint. Giving $\Omega^*(X)$, and its ambient L^2-completion to a Hilbert space, using the inner product above, the $\mathbb{Z}/2$-grading into even and odd degree forms, we see that D_{dR} is grading–reversing.

The *Laplacian* on X is $\Delta := (d + \delta)^2 = d\delta + \delta d$.

Exercise 9.2.14 In the notation above:

a) Show that if $\alpha = \sum A_i dx_i$ is a 1-form, then $\delta(\alpha) = -\frac{1}{\sqrt{g}} \sum_i \frac{\partial}{\partial x_i} (A_i g^{ij} \sqrt{g})$.

b) Show that $\Delta f = -\frac{1}{\sqrt{g}} \sum_{i,j} \frac{\partial}{\partial x_i} \left(\sqrt{g} g^{ij} \frac{\partial f}{\partial x_j} \right)$.

c) If $f \in C^\infty(X)$, then the commutator $[f, D_{\mathrm{dR}}]$ is the operator of left wedge product by the 1-form df, which is, in particular, a bounded operator.

d) Show that the value of the symbol of d at a cotangent vector $\xi \in T_x^* X$ is the endomorphism of $\Lambda^*(T_x^* X)$ of degree $+1$ of exterior multiplication (wedge product) λ_ω by ω.

e) The symbol of δ at a point $\xi \in T_x^* X$ is given by λ_ω^*, an endomorphism of $\Lambda^*(T_x^* X)$ of degree -1 (otherwise known as "interior multiplication"—this depends on the metric).

9.3   Analytic Aspects of Elliptic Operators

An unbounded operator $D \colon \mathrm{dom}(D) \subset H \to K$ between Hilbert spaces consists of a domain $\mathrm{dom}(D) \subset H$, typically a dense subspace of H, and a linear operator $D \colon \mathrm{dom}(D) \to K$. We say D is *closed* if its graph

$$\{(v, w) \in H \oplus K \mid w = Dv\}$$

is closed in $H \oplus K$.

The *adjoint* of D has domain consisting of all vectors $\xi \in H$ such that $\eta \mapsto \langle \xi, D\eta \rangle$ extends from $\mathrm{dom}(D)$ to a bounded linear functional on H. If $\xi \in \mathrm{dom}(D^*)$, then by the characterization of bounded linear functionals on a Hilbert space, there exists unique $\zeta \in H$ such that $\langle \xi, D\eta \rangle = \langle \zeta, \eta \rangle$ for all $\eta \in \mathrm{dom}(D)$. We define $D^*\xi := \zeta$.

Example 9.3.1 In the language of distributions, if $f \in L^2(\mathbb{R})$, then the distributional derivative f' of f is the unique continuous linear functional $S \to \mathbb{C}$ mapping g to $-\int g' f$. To say that the distributional derivative is in L^2 means that this linear functional extends to a bounded linear functional on H and hence is given by pairing with an L^2-function (denote f') on $\mathbb{R}$. Hence $-\int g' f = \int g f'$, for all $g \in S$.

By the definitions, it follows that $f \in \mathrm{dom}(D^*)$ if and only if the distributional derivative f' of f is in L^2, and in this case $D^* f = -f'$.

Example 9.3.2 If $\{\alpha_n\}$ is a sequence of complex numbers, then D defined on the subspace

$$\text{dom}(D) := \{\xi \in l^2(\mathbb{N}) \mid \sum |\xi_n \cdot \alpha_n|^2 < \infty$$

by multiplication by the sequence defines an unbounded operator on $l^2(\mathbb{N})$. The domain of D^* is the same as the domain of D (exercise), and D^* multiplies by the conjugate sequence $\{\overline{\alpha_n}\}$.

Some unbounded operators are not closed; indeed some admit no closed extension. If the closure of the graph of D is a graph, then D is said to be *closable*. It then extends to a closed operator on a potentially larger domain.

A useful class of closable operators is the class of symmetric operators. D is *symmetric* if $\langle D\xi, \eta \rangle = \langle \xi, D\eta \rangle$ for all $\xi, \eta \in \text{dom}(D)$.

Exercise 9.3.3 If D is symmetric, then D is closable.

The following exercise gives a more general class of closable operators.

Exercise 9.3.4 A "formal adjoint" of a densely defined operator $D\colon \text{dom}(D) \subset H \to H$ is a linear operator $D^*\colon \text{dom}(D) \to H$ such that $\langle Dv, w \rangle = \langle v, D^\sharp w \rangle$ for all $v, w \in \text{dom}(D)$:

a) If D has a formal adjoint, then D is closable.
b) Let M be a manifold with a Borel measure μ which is locally smoothly equivalent to Lebesgue measure. If X is a vector field on M, then X defines an unbounded operator $C^\infty(M) \to L^2(M)$. As such X has a formal adjoint X^* and $X^* = -X + \varphi$ where $\varphi \in C^\infty(M)$.
c) Any differential operator $D\colon C^\infty(M) \to L^2(M)$, M as in b), admits a formal adjoint and hence is closable.

The operator $i\frac{d}{dx}$ with domain $C_c^\infty(\mathbb{R})$ is symmetric. Hence it is closable. Its closure has domain the first Sobolev space

$$H^1(\mathbb{R}) = \{\xi \in L^2(\mathbb{R}) \mid \int_{\mathbb{R}} |\hat{u}(\xi)|^2 (1 + |\xi|^2)\, d\xi < \infty\},$$

the space of L^2-functions whose distributional derivative is in L^2.

Definition 9.3.5 A densely defined unbounded operator D is *self-adjoint* if $D = D^*$.

An operator $D\colon \text{dom}(D) \subset H \to K$ is *invertible* if there is a *bounded* operator $Q\colon K \to H$ such that DQ and QD are the identities on K, $\text{dom}(D)$, respectively.

The *spectrum* of an unbounded operator D is the collection of $\lambda \in \mathbb{C}$ such that $\lambda - D$ is not invertible.

The following exercise gives a criterion for a closed, symmetric operator to be self-adjoint.

Exercise 9.3.6 Let D be symmetric and $\lambda = \alpha + i\beta \in \mathbb{C}$:

a) If $v \in \text{dom}(D)$, then

$$\|(D - \lambda)v\|^2 = \|(D - \alpha)v\|^2 + \beta^2\|v\|^2.$$

b) If $\beta \neq 0$, then $\ker(D - \lambda) = \{0\}$.

c) If D is closed and $\beta \neq 0$, then $D - \lambda$ has closed range.

d) If D is also closed, then $\ker(D^* \pm i) = \{0\}$ and then D is self-adjoint.

d) If D is closed and symmetric and if $\mathrm{Spec}(D) \subset \mathbb{R}$, then D is self-adjoint.

Proposition 9.3.7 *A densely defined symmetric and closed operator is self-adjoint if its spectrum does not contain $\mathbb{R}$.*

We omit the proof (see [54]).

Self-adjoint unbounded operators have functional calculus, as with bounded operators, as we now demonstrate.

Lemma 9.3.8 *Let D be a self-adjoint operator on H. Then $D \pm i$ are invertible and $U := (D - i)(D + i)^{-1}$ extends continuously to a unitary operator on H.*

Exercise 9.3.9 Let D be multiplication by x on $L^2(\mathbb{R})$. Show that $U := (D - i)(D + i)^{-1}$ is multiplication by the Cayley transform $C : \mathbb{R} \to \mathbb{T}$, defined $C(t) = \frac{t-i}{t+i}$ and that $1 \in \mathrm{Spec}(U)$ but 1 is not an eigenvalue of U.

Suppose now that f is a bounded, continuous function on $\mathbb{R}$. If $C : \mathbb{R} \to \mathbb{T}$ is the Cayley transform, then $f \circ C^{-1}$ is a bounded, continuous function on $\mathbb{T} \setminus \{1\}$ and extends to a bounded Borel function on $\mathbb{T}$ with a single point of discontinuity at $1 \in \mathbb{T}$, which implies that $\{1\}$ has spectral measure zero, and making $f(D) := (f \circ C^{-1})(U)$ well defined, and giving a method of doing functional calculus $f \mapsto f(D)$ for bounded continuous functions on $\mathbb{R}$. If f vanishes at ∞, then $f \circ C^{-1}$ vanishes at 1 and is continuous on $\mathbb{T}$ so one can avoid the Borel theory (not discussed in these notes) for $f \in C_0(\mathbb{R})$. To deal with general $f \in C_b(\mathbb{R})$, let

$$\alpha : C_0(\mathbb{R}) \to \mathbb{B}(H), \quad \alpha(f) := f(D)$$

be the C*-algebra homomorphism just described. It is based on the Cayley transform and functional calculus for the unitary U, and the latter defines a unital *-homomorphism $C(\mathbb{T}) \to \mathbb{B}(H)$, and it follows that α is nondegenerate and so extends by strict continuity to a *-homomorphism $C_b(\mathbb{R}) = \mathcal{M}(C_0(\mathbb{R})) \to \mathbb{B}(H)$. This gives meaning to $f(D)$ for $f \in C_b(\mathbb{R})$ as required.

Theorem 9.3.10 *If D is a densely defined self-adjoint operator on H, there is a unique unital C*-algebra homomorphism $f \mapsto f(D)$, from $C_b(\mathbb{R})$ to $\mathbb{B}(H)$, which maps $\frac{1}{x \pm i}$ to $(D \pm i)^{-1}$.*

We will for the most part be using the theorem in connection with formally self-adjoint elliptic operators on compact manifolds. The closures of such operators have *discrete* spectra with finite multiplicities and are in particular self-adjoint, and even orthogonally diagonalizable, so that $f(D)$, for f bounded on $\mathbb{R}$, has an obvious meaning in this case, and one does not have to appeal to the Cayley transform to define it.

Returning to elliptic operators (Definition 9.2.10), such an operator acts by definition between the spaces $\Gamma^\infty(S^\pm)$ of smooth sections of a pair of Hermitian vector bundle. These linear spaces can be completed to Hilbert spaces $L^2(S^\pm)$ by fixing a measure μ on X and defining

$$\langle \xi, \eta \rangle := \int_X \langle \xi(x), \eta(x) \rangle \, d\mu(x).$$

In the following, we assume that $S^- = S^+$ so that one has a densely defined operator on a single Hilbert space $L^2(S)$.

Theorem 9.3.11 *Let X be a compact manifold, let μ be a probability measure on X, and $S \to X$ a Hermitian vector bundle. Let D be an order 1 elliptic operator defined initially on the space of smooth sections $\Gamma^\infty(X, S)$ of S. Assume that D is formally self-adjoint (or symmetric):*

$$\langle Ds_1, s_2 \rangle = \langle s_1, Ds_2 \rangle, \quad \forall s_1, s_2 \in C^\infty(X, S).$$

Then D has a canonical extension to an unbounded, self-adjoint operator on the Hilbert space $H := L^2(X, S)$ of L^2-sections of S. Furthermore, the spectrum of D is a discrete subset of $\mathbb{R}$, consisting of eigenvalues λ such that:

a) *Each λ-eigenspace H_λ is finite-dimensional and consists of smooth sections of S.*
b) *A section $s \in L^2(X, S)$ is smooth if and only its Fourier coefficients with respect to an orthogonal basis of H of eigenvectors of D are a Schwartz function on Spec(D).*

The simplest example is of course $D = -i \frac{d}{d\theta}$ on $C^\infty(\mathbb{T})$. It has spectrum $\mathbb{Z}$ and eigenfunctions z^n corresponding to n.

Exercise 9.3.12 Let

$$\bar{\partial} = \frac{1}{2\pi} \cdot \begin{bmatrix} 0 & \frac{\partial}{\partial x} - i \frac{\partial}{\partial y} \\ \frac{\partial}{\partial x} + i \frac{\partial}{\partial y} & 0 \end{bmatrix},$$

be the Dolbeault operator of Example 9.1.3:

a) Prove that $\bar{\partial}$ commutes with translations of $\mathbb{R}^2$ and determines a symmetric elliptic differential operator on sections of the trivial two-dimensional complex bundle over the 2-torus $\mathbb{T}^2 = \mathbb{R}^2/\mathbb{Z}^2$.
b) Show that the spectrum of $\bar{\partial}$ consists of the numbers $\pm\sqrt{n^2 + m^2}$, for $n, m \in \mathbb{Z}$.
c) The (scalar) Laplacian on $\mathbb{T}^2$ with the flat metric is $-\frac{\partial^2}{\partial x^2} - \frac{\partial^2}{\partial y^2}$. What are the eigenvalues of Δ? If one lists them in increasing order $\mu_0 \leq \mu_1 \leq \mu_2 \leq \cdots$, show that $\mu_n \sim n^{\frac{1}{2}}$.

Exercise 9.3.13 Let X be a smooth manifold and S_1, S_2 Hermitian vector bundles over X. Fix a Borel probability measure of full support to define L^2-spaces of sections.

Show that the Hilbert spaces $L^2(X, S_1 \oplus S_2)$ and $L^2(X, S_1) \oplus L^2(X, S_2)$ are identical, and deduce that the Hilbert space direct sum of an elliptic operator on sections of S_1 and an elliptic operator on sections of S_2 is an elliptic operator on sections of $S_1 \oplus S_2$.

9.4 Dirac Operators

In this section, we specialize to what is for us the most important class of elliptic operators, the *Dirac operators*. These are built from Clifford algebras on manifolds. See Sections 7.7, 7.8, and 7.9 of Chapter 5 for background on Clifford algebras and their representations (Clifford modules).

Let X be a Riemannian manifold. The Fundamental Theorem of Riemannian geometry asserts that there is a unique connection ∇^{LC} on the tangent bundle TX of X, called the *Levi-Cevita connection*, which is both torsion-free, and compatible with the metric. Torsion-free means that $\nabla_X^{\mathrm{LC}} Y - \nabla_Y^{\mathrm{LC}} X = [X, Y]$, and compatibility with the metric asserts that

$$\langle \nabla_X^{\mathrm{LC}} Y, Z \rangle + \langle Y, \nabla_X^{\mathrm{LC}} Z \rangle = X(\langle Y, Z \rangle),$$

for vector fields X, Y, Z.

Let $e_1, \ldots, e_n$ be a (pointwise) orthogonal frame for the tangent bundle TX, defined on an open subset $U \subset X$. For each vector field V on U there is a family of smooth real-valued functions $\omega_{ij}(V)$ such that

$$(9.14) \qquad\qquad \nabla_V^{\mathrm{LC}} e_i = \sum_{i,j} \omega_{ij}(V) e_j.$$

This expression is $C^\infty(U)$-linear in the variable V. Note that $\omega_{ij}(V) = \langle \nabla_V^{\mathrm{LC}} e_i, e_j \rangle$. Compatibility of ∇^{LC} with the metric gives

$$\langle \nabla_V^{\mathrm{LC}} e_i, e_j \rangle + \langle e_i, \nabla_V^{\mathrm{LC}} e_j \rangle = V(\langle e_i, e_j \rangle) = 0$$

for all i, j. Hence

$$\omega_{ij}(V) + \omega_{ji}(V) = 0,$$

so that the matrix $\omega(V)$ defined by its coordinates $\omega_{ij}(V)$ is skew-symmetric and lies in the Lie algebra $\mathfrak{so}(n, \mathbb{R})$ of $\mathbf{SO}(n, \mathbb{R})$. The *connection 1-form* of ∇^{LC} is the map

(9.15) $$\omega \colon TU \to \mathfrak{so}(n, \mathbb{R}),$$

defined on U, and depending on the initial choice of frame, and determined by the ω_{ij}. More exactly, if E_{ij} denotes the matrix in $\mathfrak{so}(n, \mathbb{R})$ with $+1$ in entry (i, j), -1 in entry (j, i), then

$$\omega(V) = \sum_{i<j} \omega_{ij}(V) \cdot E_{ij} \in \mathfrak{so}(n, \mathbb{R}).$$

Note that one can recover the connection ∇^{LC} from the 1-form ω in (9.15). The frame $e_1, \ldots, e_n$ for TU gives an identification of smooth vector fields $\Gamma^\infty(TU)$ on U, with $C^\infty(U, \mathbb{R}^n)$, and in terms of this identification, we have

(9.16) $$\nabla_V(\sigma) := \big(V(\sigma_1), \ldots, V(\sigma_n)\big) + \omega(V) \cdot \begin{bmatrix} \sigma_1 \\ \cdots \\ \cdots \\ \sigma_n \end{bmatrix},$$

for $\sigma = (\sigma_1, \ldots, \sigma_n)$ a smooth $\mathbb{R}^n$-valued function on U.

Exercise 9.4.1 Suppose the frame $e_1, \ldots, e_n$ is transformed into a frame $e'_1, \ldots, e'_n$ by the action of an orthogonal matrix $g \in \mathbf{SO}(n, \mathbb{R})$, where

(9.17) $$e'_i = \sum_j g_{ij} e_j.$$

Show that the 1-form in terms of the new basis is given by

$$\omega'(v) = \mathrm{Ad}_g\big(\omega(v)\big),$$

where

$$\mathrm{Ad} : \mathbf{SO}(n, \mathbb{R}) \to \mathrm{End}\big(\mathfrak{so}(n, \mathbb{R})\big)$$

is the adjoint representation.

Now suppose that

$$\rho \colon \mathrm{Cliff}(\mathbb{R}^n) \to \mathrm{End}(\Delta)$$

is a representation for $\mathrm{Cliff}(\mathbb{R}^n)$. It restricts to a representation of $\mathbf{Spin}_n$ and differentiating gives a representation

$$\rho_* \colon \mathfrak{spin}_n \to \mathrm{End}(\Delta)$$

of the Lie algebra $\mathfrak{spin}_n$.

The Lie algebras $\mathfrak{spin}_n$ and $\mathfrak{so}(n, \mathbb{R})$ of the Lie groups $\mathbf{Spin}_n$ and of $\mathbf{SO}(n, \mathbb{R})$ are naturally isomorphic, by differentiating the standard double covering $\mathbf{Spin}_n \to \mathbf{SO}(n, \mathbb{R})$. On the other hand, there is a natural embedding of $\mathfrak{spin}_n$ in $\mathrm{Cliff}(\mathbb{R}^n)$. Putting things together, one checks that under the resulting embedding of $\mathfrak{so}(n, \mathbb{R})$ in $\mathrm{Cliff}(\mathbb{R}^n)$, the matrix E_{ij} with $+1$ in entry (i, j), -1 in entry (j, i), and zeros elsewhere, corresponds to the Clifford algebra element

$$e_i e_j \in \mathrm{Cliff}(\mathbb{R}^n).$$

With these preliminary remarks aside, we define a connection 1-form

$$\omega_\Delta : \Gamma^\infty(TU) \to \mathrm{End}(\Delta)$$

depending on our initial choice of frame, by setting

$$(9.18) \qquad \omega_\Delta(V) := \frac{1}{2} \sum_{i<j} \omega_{ij}(V) \cdot c(e_i e_j) \in \mathrm{End}(\Delta),$$

for a vector field V on U.

Definition 9.4.2 Let $U \subset X$ be an open subset of a Riemannian manifold on which an orthonormal framing $e_1, \ldots, e_n$ of TX is defined. If $\rho : \mathrm{Cliff}(\mathbb{R}^n) \to \mathrm{End}(\Delta)$ is the spin representation, then the composition

$$\mathrm{Cliff}(TU) \cong U \times \mathrm{Cliff}(\mathbb{R}^n) \to \mathrm{End}(U \times \Delta),$$

where the first map is induced by the frame, defines a $\mathrm{Cliff}(TU)$-module. The *spin connection* on the product spinor bundle $S := U \times \Delta$, whose sections we understand as smooth maps $\sigma : U \to \Delta$, is given by

$$(9.19) \qquad \nabla_V^S(\sigma) = V(\sigma) + \omega_\Delta(V) \cdot \sigma,$$

where ω_Δ is the 1-form valued in $\mathrm{End}(\Delta)$ given by (9.18), and $V(\sigma)$ is the usual Lie derivative of a vector-valued function.

The crucial property of the spin connection constructed locally above is the following.

Lemma 9.4.3 *If ∇^S is the spin connection on sections of $U \times \Delta$ as above, $w : U \to \mathbb{R}^n \subset \mathrm{Cliff}(\mathbb{R}^n)$ is a smooth map, and $\sigma : U \to \Delta$ is a smooth section of $U \times \Delta$, then*

$$(9.20) \qquad \nabla_X^S\big(c(w) \cdot \sigma\big) = c(w) \cdot \nabla_X^S(\sigma) + c\big(\nabla_X^{LC} w\big) \cdot \sigma,$$

where ∇^{LC} is the Levi-Civita connection.

The proof is left as an exercise.

Definition 9.4.4 Let S be a Cliff(TX)-module, a Hermitian vector bundle, with Clifford multiplication $c\colon \text{Cliff}(TX) \to \text{End}(S)$. We say that a connection ∇^S on S is *compatible with the Clifford module structure* if

$$(9.21) \qquad \nabla^S_X\big(c(w) \cdot \sigma\big) = c(w) \cdot \nabla^S_X(\sigma) + c\big(\nabla^{\text{LC}}_X w\big) \cdot \sigma$$

holds for all smooth vector fields w on X, and smooth sections s of S. The connection ∇^S will be called a *Dirac connection* if it is compatible with the Clifford multiplication, and compatible with the Hermitian metric on S.

We aim to prove the following.

Proposition 9.4.5 *Every fiberwise irreducible* Cliff(TX)-*module has a Dirac connection.*

Lemma 9.4.6 *Suppose* $c\colon \text{Cliff}(TX) \to \text{End}(S)$ *and* $c'\colon \text{Cliff}(TX) \to \text{End}(S')$ *are two Clifford modules over* Cliff(TX) *and that* $U\colon S \to S'$ *is a unitary bundle isomorphism intertwining the two Clifford multiplications.*

Then if ∇ *is a Dirac connection on* S, *then the conjugate* $\nabla' := U\nabla U^*$ *connection is a Dirac connection on* S'.

Proof The conjugate connection is defined

$$(9.22) \qquad \nabla'_X(s) := U\nabla_X(U^*s)$$

for a vector field X and smooth section s of S'. It is easily checked that ∇' is a connection. If s_1, s_2 are smooth sections of S', X a vector field, then

$$(9.23) \quad \langle \nabla'_X(s_1), s_2 \rangle + \langle s_1, \nabla'_X(s_2) \rangle = \langle U\nabla'_X(U^*s_1), s_2 \rangle + \langle s_1, U\nabla'_X(U^*s_2) \rangle$$

$$= \langle \nabla'_X(U^*s_1), U^*s_2 \rangle + \langle U^*s_1, \nabla'_X(U^*s_2) \rangle = X(\langle U^*s_1, U^*s_2 \rangle) = X(\langle s_1, s_2 \rangle)$$

shows that it is compatible with the metric. Finally, the assumption $Uc(w)^* = c'(w)$ for a tangent vector field w, and the assumed Clifford compatibility of ∇, gives

$$U\nabla_X U^*\big(c(w)s\big) = c\big(\nabla^{\text{LC}}_X(w)\big)s + c(w)U\nabla_X(U^*s)$$

so the conjugate connection is Clifford multiplication compatible as well. $\qquad\square$

Lemma 9.4.7 *If* S *is a fiberwise irreducible* Cliff(TX)-*module and* L *is a complex Hermitian line bundle over* X, *then* $S \otimes L$ *is a fiberwise irreducible* Cliff(TX)-*module with module structure*

$$c(w)(s \otimes l) := c(w)s \otimes l,$$

for $w \in T_x X$, $s \in S_x$ *and* $l \in L_x$, $x \in X$.

Finally, we recall the following result observed earlier.

Lemma 9.4.8 *If* S *and* S' *are two fiberwise irreducible* $\mathrm{Cliff}(TX)$*-modules, then there is a Hermitian line bundle* L *and a unitary isomorphism of Clifford modules* $S \cong S' \otimes L$.

Proof Set $L := \mathrm{Hom}_{\mathrm{Cliff}(TX)}(S, S')$, the bundle of bundle maps $S \to S'$ which commute with the Clifford module structures. Then L is a complex line bundle, with a natural Hermitian structure, and the obvious map $S \otimes L \to S'$ sending $s \otimes T$ to $T(s)$ is a bundle isomorphism intertwining the Clifford multiplications. $\square$

Proof (Of Proposition 9.4.5) Let S be a fiberwise irreducible $\mathrm{Cliff}(TX)$-module. If $U \subset X$, with the restricted Riemannian metric, then $S|_U$ is also a $\mathrm{Cliff}(TU)$-module, which is fiberwise irreducible. Suppose that U has a globally defined orthonormal frame. The frame gives a unitary bundle isomorphism $TU \cong U \times \mathbb{R}^n$ and an induced isomorphism $\mathrm{Cliff}(TU) \cong U \times \mathrm{Cliff}(\mathbb{R}^n)$. Composing this isomorphism with the product $\mathrm{Cliff}(U \times \mathbb{R}^n)$-module $U \times \Delta$ gives a new, fiberwise irreducible $\mathrm{Cliff}(TU)$-module. Therefore, there is a complex Hermitian line bundle L over U such that $S|_U \otimes L \cong U \times \Delta$. If U is also contractible, every line bundle is trivial, and hence we get a unitary isomorphism

$$S|_U \cong U \times \Delta$$

of $\mathrm{Cliff}(TU)$-modules.

On the other hand, we have already shown in Lemma 9.4.3 that $U \times \Delta$ has a $\mathrm{Cliff}(U \times \mathbb{R}^n)$-compatible connection, and hence a $\mathrm{Cliff}(TU)$-compatible connection. Hence $S|_U$ has a compatible connection as well.

This shows that compatible connections exist locally. We may then piece them together using a partition of unity $\{\rho_i\}$, setting

$$\nabla := \sum_i \rho_i \nabla_i,$$

where $\{U_i\}_{i \in I}$ is a locally finite open cover of X by contractible open sets U_i each of which has a globally defined orthonormal frame, and ∇_i is a Clifford-compatible connection on $S|_{U_i}$. $\square$

We are finally in a position to define the Dirac operator associated to a spin^c-manifold.

Definition 9.4.9 Let X be a spin^c Riemannian manifold. Let $c \colon \mathrm{Cliff}(TX) \to \mathrm{End}(S)$ be an irreducible Clifford module, and ∇^S be a Dirac connection on S. The *Dirac operator* is the differential operator on sections $C_c^\infty(M, S)$ of the spin bundle, given locally in terms of a local orthonormal frame $e_1, \ldots, e_n$ of TX by the formula

(9.24) $$D = \sum_i c(e_i) \cdot \nabla_{e_i} : C_c^\infty(X, S) \to C_c^\infty(X, S),$$

where ∇_{e_i} is covariant differentiation by e_i.

Exercise 9.4.10 The expression (9.26) is independent of the frame.

Exercise 9.4.11 D is elliptic (see Definition 9.2.13).

Exercise 9.4.12 If $f \in C^\infty(X)$ is a smooth function, acting on smooth sections of S, then the commutator $[f, D]$ is the endomorphism of S given by Clifford multiplication by the tangent vector dual under the metric to df (that is, to the gradient ∇f of f). (*Hint*. Compute using a local orthonormal frame.)

Proposition 9.4.13 *The Dirac operator is an order 1 elliptic, differential operator on sections $C_c^\infty(X, S)$ of the spinor bundle. The symbol of D is the composition of the bundle isomorphism $T^*X \cong TX$ given by the Riemannian metric, and the Clifford multiplication $c: TX \to \mathrm{End}(S)$.*

In the case X is even-dimensional, D is $\mathbb{Z}/2$-grading–reversing and so maps $C_c^\infty(X, S^+)$ to $C_c^\infty(X, S^-)$.

In the case when X is even-dimensional, the fact that D is grading–reversing means that, with respect to the decomposition $C_c^\infty(X, S) = C_c^\infty(X, S^+) \oplus C_c^\infty(X, S^-)$, D has a 2-by-2 matrix decomposition

$$D = \begin{bmatrix} 0 & D_+ \\ D_- & 0 \end{bmatrix},$$

and the (formal) self-adjointedness of D implies that $D_+^* = D_-$.

In particular, we may apply Theorem 9.3.11 to the Dirac operator D, as it is formally self-adjoint (and elliptic).

Definition 9.4.14 Let X be a compact spin^c-manifold of even dimension. Let D be the Dirac operator, acting on smooth sections in $L^2(X, S)$, for some probability measure μ on X, and $\mathbb{Z}/2$-graded spinor bundle $S \to X$.

We define the *analytic index* of D by

(9.25) $$\mathrm{Index}_\epsilon(D) := \dim(\ker D_+) - \dim(\ker D_-),$$

where D_+ is the restriction of D to $H_+ = C^\infty(X, S^+)$, D_- the restriction to H_-.

The symbol ϵ refers to the grading.

We relate this index to the ordinary Fredholm index of bounded operators in the next section.

The index defined above may be described in the following equivalent way. The operator D is self-adjoint with kernel $\ker D = \ker D^2$, and squaring D gives the direct sum of $D_+ D_+^*$ and $D_+^* D_+$. Hence $\ker D$ is the direct sum of $\ker D_+$ and $\ker D_+^*$. The Fredholm index (9.25) is the *graded* dimension of $\ker D$. Indeed, with respect to the $\mathbb{Z}/2$-grading, $\ker D_+$ is contained in the even part H_+ of the Hilbert space of sections, $\ker D_-$ in the odd part. If in a graded Hilbert space we define the *graded dimension* of a subspace W to be

$$\dim_s := \dim W \cap H_+ - \dim W \cap H_-,$$

then by these remarks

$$\dim_s \ker D = \dim \ker D_+ - \dim \ker D_+^* = \mathrm{Index}(D_+) = \mathrm{Index}_\epsilon(D),$$

the same as (9.25).

Example 9.4.15 Let $X = \mathbb{T}^2$, the 2-torus, which we regard as $\mathbb{R}^2/\mathbb{Z}^2$.

The Dirac–Dolbeault operator

$$\bar{\partial} = \frac{1}{2\pi} \cdot \begin{bmatrix} 0 & \frac{\partial}{\partial z} \\ \frac{\partial}{\partial \bar{z}} & 0, \end{bmatrix}$$

where $\frac{\partial}{\partial \bar{z}} = \frac{\partial}{\partial x} + i \frac{\partial}{\partial y}$, acts on sections of the spinor bundle coming from the complex structure. The bundle is trivial, isomorphic to $\mathbb{T}^2 \times \mathbb{C}^2$, so that the spinor grading corresponds to grading the first factor of $\mathbb{C}^2$ even and the second odd. An exercise in Clifford algebras shows that the corresponding Dirac operator is given by the above matrix.

Under Fourier transform, $L^2(\mathbb{T}^2) \cong l^2(\mathbb{Z}^2)$ and with respect to the associated canonical orthonormal basis $\{e_{n,m}\}_{n,m\in\mathbb{Z}}$, with $e_{n,m}(x, y) = \exp(2\pi i(nx + my))$ we compute that

$$\frac{1}{2\pi} \cdot \left(\frac{\partial}{\partial x} + i \frac{\partial}{\partial y} \right) (e_{n,m}) = (-m + in)e_{n,m}.$$

Therefore, up to unitary equivalence, the Dirac operator on $\mathbb{T}^2$ is the operator

$$\begin{bmatrix} 0 & M \\ M & 0 \end{bmatrix}$$

acting on $l^2(\mathbb{Z}^2) \oplus l^2(\mathbb{Z}^2)$, where M is the diagonal operator in the standard basis with entries the $m + in$.

The kernel of $\frac{\partial}{\partial \bar{z}}$ is the holomorphic functions on $\mathbb{T}^2$ and the cokernel the antiholomorphic functions, and each space is one-dimensional, consisting of constants. Hence

$$\mathrm{Index}_\epsilon(\bar{\partial}) = \dim \ker \frac{\partial}{\partial \bar{z}} - \dim \ker \frac{\partial}{\partial z} = 0.$$

We end this section with a definition of a more general class of operators, but which are defined in the same way as *the* Dirac operator. Only the irreducibility requirement on the Clifford module structure has been dropped.

Definition 9.4.16 Let X be a Riemannian manifold. Let $c \colon \mathrm{Cliff}(TX) \to \mathrm{End}(S)$ be a Clifford module, and ∇^S be a Hermitian connection on S satisfying the compatibility condition of Definition 9.4.4 (a Dirac connection).

The associated *Dirac operator* is the differential operator on sections $C_c^\infty(M, S)$ of the spin bundle, given locally in terms of a local orthonormal frame $e_1, \ldots, e_n$ of TX by the formula

$$(9.26) \qquad D = \sum_i c(e_i) \cdot \nabla_{e_i} \colon C_c^\infty(X, S) \to C_c^\infty(X, S),$$

where ∇_{e_i} is covariant differentiation by e_i.

We call such operators *Dirac-type operators*, or simply, Dirac operators.

An advantage of the more general definition is that if D_1 and D_2 are Dirac-type operators on sections of spinor bundles S_1 and S_2, respectively, determined by two Clifford module structures and connections, then the direct sum of the bundles $S_1 \oplus S_2$ has a direct sum Clifford module structure, direct sum connection, and hence Dirac-type operator.

Similarly, if D is a Dirac operator associated to a spinor bundle S and connection ∇^S, and if E is a vector bundle over X, then the tensor product $S \otimes E$ of vector bundles over X has an obvious Clifford module structure and connection (see Definition 9.6.1) making a new operator typically denoted D_E and called D "twisted" by E. This procedure is essential in understanding the connection between K-theory and the index theory of the Dirac operator.

Exercise 9.4.17 Let D_i be Dirac-type operators on sections of Clifford modules S_i, $i = 1, 2$ over a compact Riemannian manifold X, determined by choice of compatible connections on S_i.

Show that under the identification $L^2(X, S_1 \oplus S_2) \cong L^2(X, S_1) \oplus L^2(X, S_2)]$ (see Exercise 9.3.13), the orthogonal direct sum $D_1 \oplus D_2$ of the two operators identifies with the Dirac operator associated to the direct sum Clifford module $S_1 \oplus S_2$, with the direct sum connection.

9.5   Bounded Transforms of Dirac Operators

It is convenient to translate the index of a Dirac operator on an even-dimensional compact manifold, defined in the last section as $\mathrm{Index}(D) := \dim \ker D_+ - \dim \ker D_-$, or, equivalently, as $\dim \ker D_+ - \dim \ker D_+^*$, of the operator acting on smooth sections (of the spinor bundle) into a Fredholm index of an ordinary bounded operator on a Hilbert space. This allows us to make use of the tools of functional analysis in index theory. The general procedure is the motivation for the definitions of KK-theory.

By Theorem 9.3.11 (or Theorem 9.3.10) we can apply functional calculus to a Dirac operator D. If D is elliptic and self-adjoint, acting on sections $\Gamma^\infty(X, S) \subset L^2(X, S)$ of a spinor bundle, and if χ is a bounded function on $\mathbb{R}$, then the spectrum of D is discrete and

$$\chi(D) = \sum_{\lambda \in \text{Spec}(D)} \chi(\lambda) \cdot \text{pr}_\lambda,$$

where pr_λ is projection to the λ-eigenspace. The sum converges in the strong operator topology, and all the eigenspaces are finite-dimensional.

An important example for our purposes will be $\chi(0) = 0$ and $\chi(n) = \text{sign}(n) = \frac{n}{|n|}$, $n \neq 0$, so

$$\chi(D) =: F = \oplus_{\lambda \in \text{Spec}(D), \lambda \geq 0} \text{pr}_\lambda - \oplus_{\lambda \in \text{Spec}(D), \lambda < 0} \text{pr}_\lambda.$$

Note that $\ker(F_D) = \ker(D)$ and that $F^2 - 1$ has finite rank equal to the dimension of the kernel of D.

Now suppose that S is a $\mathbb{Z}/2$-graded bundle, such as for example happens if S is the spinor bundle for an even-dimensional spin^c-manifold, and D is the Dirac operator. Let $\epsilon \colon L^2(X, S) \to L^2(X, S)$ be the grading operator, and assume that D anti-commutes with $\epsilon \colon D\epsilon = -\epsilon D$. If $s \in H_\lambda$, then

$$D(\epsilon s) = -\epsilon D s = -\lambda \cdot \epsilon s.$$

Hence $\epsilon s \in H_{-\lambda}$ if $s \in H_\lambda$. In particular, ϵ maps $\ker(D)$ to itself, and F thus anti-commutes with ϵ. That is, F is an odd operator with respect to the grading.

We may describe things as follows. The Hilbert space H decomposes as $H = H^+ \oplus .H^-$. The restriction of the Dirac operator D to H^+ gives a densely defined unbounded operator $D_+ \colon H^+ \to H^-$, its restriction to H^- an operator $D_- \colon H^- \to H^+$, and so we may write D as a 2-by-2 matrix

$$D = \begin{bmatrix} 0 & D_- \\ D_+ & 0 \end{bmatrix}.$$

Since D is self-adjoint, $D_- = D_+^*$.

For example, squaring D gives

$$D^2 = \begin{bmatrix} D_+^* D_+ & 0 \\ 0 & D_+ D_+^* \end{bmatrix}.$$

If D happens to be actually invertible, that is, if $\ker(D) = \{0\}$, or in other words, if $0 \notin \text{Spec}(D)$, then it follows that

$$F = D|D|^{-1} = \begin{bmatrix} 0 & D_-(D_-^* D_-)^{-\frac{1}{2}} \\ D_+(D_+^* D_+)^{-\frac{1}{2}} & 0 \end{bmatrix}$$

(recall that $D_- = D_+^*$).

In the general case, $D + \mathrm{pr}_{\ker(D)}$ is invertible, and the above formula applies to describe the operator $\chi(D + \mathrm{pr}_{\ker D})$. On the other hand

$$\chi(D) = \chi(D + \mathrm{pr}_{\ker D}) - \mathrm{pr}_{\ker(D)},$$

and one gets a corresponding explicit formula for F.

We obtain the following.

Proposition 9.5.1 *Let $\chi \in C_b(\mathbb{R})$ be the normalizing function defined $\chi(n) = \pm 1$ according to the sign of the nonzero integer n, and $\chi(0) = 0$.*

Let D be the Dirac operator on a compact, even-dimensional spin^c-manifold M, with a corresponding self-adjoint extension on $H := L^2(M, S)$. Let $F := \chi(D)$, using functional calculus for the self-adjoint unbounded operator D.

Then F is self-adjoint, odd (with respect to the grading on H), the restriction $F_+: H_+ \to H_-$ of F to H_+ is Fredholm, and $\mathrm{Index}(F_+) = \mathrm{Index}(D_+) := \dim \ker(D_+) - \dim \ker(D_-)$.

Describing the index of D in this way is more flexible, because we know for example that the index of a Fredholm operator does not change under compact perturbation. Such compact perturbations can appear if one changes the normalizing function. Suppose that $\chi : \mathbb{R} \to [-1, 1]$ is any continuous, odd function, such that

$$\lim_{t \to \pm\infty} \chi(t) = \pm 1.$$

We call any such χ a *normalizing function*. An example is the sign function used above to define F from D.

If D is a self-adjoint operator such that $\psi(D)$ is compact for all $\psi \in C_0(\mathbb{R})$, and if χ and χ' are any two normalizing functions, observe that

$$\chi(D) - \chi'(D) = (\chi - \chi')(D)$$

and $\chi - \chi' \in C_0(\mathbb{R})$. The following easy exercise shows that $\chi(D) - \chi'(D) \in \mathcal{K}(H)$.

Exercise 9.5.2 Let M be a compact manifold, S a complex vector bundle over M, and D be the self-adjoint extension of an elliptic operator on $\Gamma^\infty(M, S)$, on the Hilbert space $L^2(M, S)$.

Then if $\psi \in C_0(\mathbb{R})$, then the (bounded) operator $\psi(D)$ obtained by functional calculus for D is a compact operator on $L^2(M, S)$.

Corollary 9.5.3 *Let D, acting on the graded Hilbert space $H = L^2(X, S)$, be as in Theorem 9.3.11. Then if χ is any normalizing function, then $F := \chi(D)$ is a bounded, self-adjoint Fredholm operator on H such that $F^2 - 1 \in \mathcal{K}(H)$ and F is* *odd with respect to the grading on H. Moreover, if χ, χ' are any two such functions, then $\chi(D) - \chi'(D)$ is a compact operator.*

In particular, as the Fredholm index does not change under compact perturbation, the index of the bounded Fredholm operator $\chi(D)$ does not depend on the choice of normalizing function.

Exercise 9.5.4 For a self-adjoint unbounded operator D on a Hilbert space H, the condition

$$(1 + D^2)^{-1} \in \mathcal{K}(H)$$

is equivalent to the condition

$$\psi(D) \in \mathcal{K}(H)$$

for all $\psi \in C_0(\mathbb{R})$. (*Hint.*

$$(1 + D^2)^{-1} = (i + D)^{-1}(-i + D)^{-1} = TT^*,$$

where $T = (i + D)^{-1}$, and compactness of TT^* implies that of T, for general bounded operators T. This shows that $\psi_\pm(D)$ is compact for the particular functions $\psi_\pm(t) := (\pm i + t)^{-1}$, and these generate $C_0(\mathbb{R})$ as a C*-algebra.)

Exercise 9.5.5 Suppose D is a self-adjoint unbounded operator on a Hilbert space. Show that

$$(1 + \lambda_1 + D^2)^{-1} - (1 + \lambda_2 + D^2)^{-1} = (1 + \lambda_1 + D^2)^{-1}(\lambda_2 - \lambda_1)(1 + \lambda_2 + D^2)^{-1}$$

for all $\lambda \geq 0$. Deduce that $(1 + \lambda_1 + D^2)^{-1}$ is compact if and only if $(1 + \lambda_2 + D^2)^{-1}$ is compact.

We can use the above observations to prove that making different choices of some of the data (like the connection on S) when constructing D from a given spin^c-structure results in F's which are compact perturbations of each other. This in particular shows the fact that changing such data does not change the analytic index. To prove this we will use the particular normalizing function $\chi(t) = t(1 + t^2)^{-\frac{1}{2}}$, which has the integral formula

$$(9.27) \qquad (1 + t^2)^{-\frac{1}{2}} = \frac{1}{\pi} \int_0^\infty \lambda^{-\frac{1}{2}}(1 + \lambda + t^2)^{-1} d\lambda$$

—the integral converges absolutely for each t because

$$\lambda^{-\frac{1}{2}}(1 + \lambda + t^2)^{-1} \leq \lambda^{-\frac{3}{2}}.$$

Lemma 9.5.6 *Let $\chi(t) = t(1 + t^2)^{-\frac{1}{2}}$ and $F = \chi(D)$, where D is a self-adjoint unbounded operator on a Hilbert space H such that $(1 + D^2)^{-1}$ is compact.*

Then for $v \in dom(D)$, the integral $\frac{1}{\pi} \int_0^\infty \lambda^{-\frac{1}{2}} D(1 + \lambda + D^2)^{-1} v \, d\lambda$ converges in the topology of H to Fv.

We write accordingly, sometimes:

$$(9.28) \qquad F := \chi(D) = \frac{1}{\pi} \int_0^\infty \lambda^{-\frac{1}{2}} D(\lambda + 1 + D^2)^{-1} \, d\lambda$$

with it understood that the integral converges in the strong topology.

Proof Since $|(1 + \lambda + x^2)^{-1}| \le \frac{1}{1+\lambda}$ for all $x \in \mathbb{R}$, it follows from the properties of functional calculus that $\|(1 + \lambda + D^2)^{-1}\| \le \frac{1}{1+\lambda}$. Hence the integral

$$(9.29) \qquad (1 + D^2)^{-\frac{1}{2}} = \frac{1}{\pi} \int_0^\infty \lambda^{-\frac{1}{2}} (1 + \lambda + D^2)^{-1} d\lambda$$

converges norm absolutely in $\mathcal{K}(H)$. In particular, if $v \in \mathrm{dom}\, D$, then

$$\|D(1 + \lambda + D^2)^{-1} v\| = \|(1 + \lambda + D^2)^{-1} Dv\| \le \|Dv\| \cdot (1 + \lambda)^{-1}$$

and so the integral

$$(9.30) \qquad Fv := \frac{1}{\pi} \int_0^\infty \lambda^{-\frac{1}{2}} D(1 + \lambda + D^2)^{-1} v \, d\lambda$$

converges norm absolutely in the Hilbert space to Fv. $\qquad\qquad \square$

Corollary 9.5.7 *Suppose D_1, D_2 are unbounded self-adjoint operators on a Hilbert space H such that $D_1 - D_2$ is bounded. Then $(1 + D_1^2)^{-1}$ is compact if and only if $(1 + D_2^2)^{-1}$ is compact. Furthermore, if this is the case, and $F_i := D_i(1 + D_i^2)^{-\frac{1}{2}}$, then $F_1 - F_2$ is compact.*

Proof We use the integral formula of Lemma 9.5.6. One first computes that if $\lambda \ge 0$, then

$$(9.31) \quad (\lambda + 1 + D_1^2)^{-1} - (\lambda + 1 + D_2^2)^{-1} = (\lambda + 1 + D_1^2)^{-1}(D_2^2 - D_1^2)(\lambda + 1 + D_2^2)^{-1}.$$

Applying the Lemma gives

$$(9.32) \quad (1 + D_1^2)^{-\frac{1}{2}} - (1 + D_2^2)^{-\frac{1}{2}}$$
$$= \int_0^\infty \lambda^{-\frac{1}{2}} (1 + \lambda + D_1^2)^{-1}(D_2^2 - D_1^2)(1 + \lambda + D_2^2)^{-1} d\lambda$$

$$= \int_0^\infty \lambda^{-\frac{1}{2}}(1+\lambda+D_1^2)^{-1}(D_1(D_1-D_2)+(D_1-D_2)D_2)(1+\lambda+D_2^2)^{-1}d\lambda.$$

$$= \int_0^\infty \lambda^{-\frac{1}{2}}(1+\lambda+D_1^2)^{-1}D_1(D_1-D_2)(1+\lambda+D_2^2)^{-1}d\lambda$$

$$+ \int_0^\infty \lambda^{-\frac{1}{2}}(1+\lambda+D_1^2)^{-1}((D_1-D_2)D_2)(1+\lambda+D_2^2)^{-1}d\lambda.$$

The operator $(1+\lambda+D_1^2)^{-1}D_1$ is bounded with norm $\leq \frac{1}{2}(1+\lambda)^{-\frac{1}{2}}$, since $\frac{x}{1+\lambda+x^2} \leq \frac{1}{2}(1+\lambda)^{-\frac{1}{2}}$ for all $x \in \mathbb{R}$. The operator $(1+\lambda+D_2^2)^{-1}$ is bounded with norm $\leq (1+\lambda)^{-1}$. And $D_1 - D_2$ is bounded. Hence the first term is a norm absolutely convergent integral of bounded operators. Similarly for the second term.

Now suppose that $(1+D_1^2)^{-1}$ is compact. It follows that $(1+\lambda+D_i^2)^{-1}$ is compact for all λ (Exercise 9.5.5) and $i = 1, 2$. Moreover, $((D_1-D_2)D_2)(1+\lambda+D_2^2)^{-1}$ is bounded, for all λ. Hence the integrand in the second term is compact operator valued, so the second term is compact.

For the first term, factor out the compact operator $(1+\lambda+D_1^2)^{-\frac{1}{2}}$ from the integrand. The operator $(1+\lambda+D_1^2)^{-\frac{1}{2}}D_1(D_1-D_2)(1+\lambda+D_2^2)^{-1}$ is bounded since $(1+\lambda+D_1^2)^{-\frac{1}{2}}D_1$ is bounded (a contraction), and $D_1 - D_2$ is bounded. Hence the integrand in the second term is the product of a bounded operator and $(1+\lambda+D_1^2)^{-\frac{1}{2}}$, so is compact.

We have proved that if $(1+D_1^2)^{-1}$ is compact and $D_1 - D_2$ is bounded, then $(1+D_2^2)^{-1}$ is compact, as claimed.

Next, we multiply (9.33) by D_1 to get

$$(9.33) \quad D_1(1+D_1^2)^{-\frac{1}{2}} - D_1(1+D_2^2)^{-\frac{1}{2}}$$

$$= \int_0^\infty \lambda^{-\frac{1}{2}}(1+\lambda+D_1^2)^{-1}D_1^2(D_1-D_2)(1+\lambda+D_2^2)^{-1}d\lambda$$

$$+ \int_0^\infty \lambda^{-\frac{1}{2}}D_1(1+\lambda+D_1^2)^{-1}((D_1-D_2)D_2)(1+\lambda+D_2^2)^{-1}d\lambda,$$

where the integrals converge strongly. The operator $(1+\lambda+D_1^2)^{-1}D_1^2(D_1-D_2)$ is bounded with norm $\leq \|D_1 - D_2\|$. The operator $(1+\lambda+D_2^2)^{-1}$ is compact with norm $\leq (1+\lambda)^{-1}$. It follows that the first term is an absolutely convergent integral of compact operators.

Consider the second term. The operator $D_1(1+\lambda+D_1^2)^{-1}$ is compact with norm $\leq \frac{1}{2}(1+\lambda)^{-\frac{1}{2}}$ as observed above. The operator $(D_1-D_2)D_2)(1+\lambda+D_2^2)^{-1}$ is also compact, with norm $\leq \frac{\|D_1-D_2\|}{2}(1+\lambda)^{-\frac{1}{2}}$. It follows that the second term above is a compact operator, as claimed.

We have therefore showed that $D_1(1+D_1^2)^{-\frac{1}{2}} - D_1(1+D_2^2)^{-\frac{1}{2}}$ is compact. Since $D_1(1+D_2^2)^{-\frac{1}{2}} - D_2(1+D_2^2)^{-\frac{1}{2}} = (D_1 - D_2)(1+D_2^2)^{-\frac{1}{2}}$ is a compact operator, we deduce that $D_1(1+D_1^2)^{-\frac{1}{2}} - D_2(1+D_2^2)^{-\frac{1}{2}}$ is a compact operator, as claimed.   □

We record that we have established the following fact.

Corollary 9.5.8 *Let D be a self-adjoint, unbounded operator on a Hilbert space H, with $(1 + D^2)^{-1}$ compact.*

Then $\ker(D_{\pm})$ *are finite-dimensional vector spaces. Furthermore, if D and D′ are two such operators with a dense common domain and* $D - D'$ *bounded, then* Index(D) = Index(D').

Remark 9.5.9 As an easy consequence, the index of a Dirac operator associated to a spin^c-structure on M compact does not depend on the choice of connection on the spinor bundle, since changing the connection only changes the Dirac operator by a bounded perturbation.

The next result we establish here is particularly important for index theory in odd dimensions. The key point about the interaction between the operators of multiplication by smooth $f \in C^\infty(M)$ on sections of a spinor bundle and the action of the Dirac operator is that the commutator $[D, f]$ is *bounded*. The next result shows that this implies the commutator $[F, f]$ is *compact*.

Lemma 9.5.10 *Let D be a densely defined self-adjoint operator on a Hilbert space H such that* $(1 + D^2)^{-1} \in \mathcal{K}(H)$.

Let a be a bounded operator on H, leaving the domain of D invariant, and such that the commutator $[a, D]$ is bounded.

Then, for any normalizing function χ, the commutator

$$[a, \chi(D)]$$

is compact.

Proof By Corollary 9.5.3 it suffices to prove the second assertion for the particular normalizing function

$$\chi(t) := t(1 + t^2)^{-\frac{1}{2}}.$$

Using the integral formula (9.28) and some algebra we get

$$(9.34) \quad [a, F] = \frac{1}{\pi} \int_0^\infty \lambda^{-\frac{1}{2}} [a, D](\lambda + 1 + D^2)^{-1} d\lambda$$

$$+ \frac{1}{\pi} \int_0^\infty \lambda^{-\frac{1}{2}} D(\lambda + 1 + D^2)^{-1}[a, D^2](\lambda + 1 + D^2)^{-1} d\lambda$$

all a priori in the strong topology. However, all of these integrals are absolutely operator norm convergent integrals of compact operators. Indeed, the first integral converges since $(1 + \lambda + D^2)^{-1}$ is compact, $\|(1 + \lambda + D^2)^{-1}\| \leq \frac{1}{1+\lambda}$, and $[a, D]$ is bounded.

Consider the second term. Using $[a, D^2] = [a, D]D + D[a, D]$, we see it breaks into the sum

$$(9.35) \quad \frac{1}{\pi} \int_0^\infty \lambda^{-\frac{1}{2}} D(\lambda + 1 + D^2)^{-1}[a, D]D(\lambda + 1 + D^2)^{-1} d\lambda$$

$$+ \frac{1}{\pi} \int_0^\infty \lambda^{-\frac{1}{2}} D(\lambda + 1 + D^2)^{-1} D[a, D](1 + \lambda + D^2)^{-1} d\lambda.$$

Now $\|D(1 + \lambda + D^2)^{-1}\| \leq (1 + \lambda)^{-\frac{1}{2}}$ whence $\|D(1 + \lambda + D^2)^{-1}[a, D]D(1 + \lambda + D^2)^{-1}\| \leq \|[a, D]\| \cdot (1 + \lambda)^{-1}$ for some constant and so the first term is a norm convergent integral of compact operators and hence is a compact operator, and the same remarks apply to the second term. $\square$

Some of the technical results of this section are used in the next to define maps on K-theory groups, whose computation in two different ways constitutes the Index Theorem(s) of Atiyah and Singer.

9.6   The K-Theory Index Theorem(s)

An essential observation of Atiyah was that an elliptic operator can be "twisted" by a vector bundle to produce another elliptic operator D_E. Moreover, since the twisted operator is again elliptic, it is Fredholm, and has an index. Following this reasoning, one concludes that a Dirac operator D on an even-dimensional manifold determines a group homomorphism $K^0(M) \to \mathbb{Z}$. A slightly different procedure defines a group homomorphism $K^1(M) \to \mathbb{Z}$ if M is odd-dimensional, again by a construction with the Dirac operator on M, but a slightly different one.

These facts suggested to Atiyah that it might be possible to organize elliptic operators (e.g., Dirac operators) into a *homology* theory dual to the cohomolology theory K-theory, at least for compact manifolds. The idea led to Kasparov's K-homology, discussed in the next chapter.

In this section we describe the Index Theorem in these terms. We start by defining twisting by vector bundles.

Index maps in even dimensions: twisting

Let D be a Dirac (type) operator (Definition 9.4.16)) on sections of a spinor bundle $S \to X$, where X is a Riemannian manifold. Thus, for a Dirac connection ∇^S on S, and a Clifford module structure c on S, D acts on smooth sections of S by

$$(Ds)(x) = \sum_i c(e_i)(\nabla_{e_i} s)(x), \quad s \in \Gamma^\infty(S),$$

with (e_i) a local orthonormal frame for TX.

We show that D can be "twisted" by any complex vector bundle $E \to X$ as follows. Choose a Hermitian metric and compatible connection ∇^E on E.

The bundle $S \otimes E$ admits the Clifford module structure $c_E(\xi) := c(\xi) \otimes 1_{E_x} : S_x \otimes E_x \to S_x \otimes E_x$, for $x \in X, \xi \in T_x X$.

The bundle $S \otimes E$ admits a tensor product connection $\nabla^{S \otimes E} := \nabla^S \otimes 1 + 1 \otimes \nabla^E$ as well, defined as follows. On the algebraic tensor product $\Gamma^\infty(X, S) \otimes_{C^\infty(X)} \Gamma^\infty(X, E)$ of the two $C^\infty(X)$-modules of sections, and for X a vector field on X, the formula

$$\nabla_X^{S \otimes E} \left(\sum s_i \otimes t_i \right) := \sum \nabla_X^S s_i \otimes t_i + s_i \otimes \nabla_X^E t_i$$

is well defined (that is, satisfies $\nabla^{S \otimes E}(sf \otimes t) = \nabla(s \otimes ft)$ for $f \in C^\infty(X)$). Now sections of this form are dense in all sections, and the connection condition is easily checked. Moreover, it is compatible with the tensor product Hermitian metric.

Therefore $S \otimes E$ has a structure of Clifford module and has a compatible connection. From this data one builds as discussed above, the associated Dirac (-type) operator.

Definition 9.6.1 Let X be a compact Riemannian manifold, D a Dirac operator on X, and $E \to X$ a vector bundle over X. The Dirac operator D *twisted* by the complex Hermitian vector bundle $E \to X$ is the Dirac-type operator defined with respect to a local orthonormal frame by

$$D_E = \sum_i c_E(e_i) \nabla_{e_i}^{S \otimes E}, \quad s \in \Gamma^\infty(X, S \otimes E),$$

with $\nabla^{S \otimes E}$ and c_E as above.

Locally, if $s \otimes v$ is a simple tensor section of $S \otimes E$, then

$$D_E(s \otimes v) = \sum_i c(e_i) \, \nabla_{e_i}^S s \otimes v + c(e_i) s \otimes \nabla_{e_i}^E v.$$

The Dirac operator twisted by E is again of Dirac type. Hence D_E is elliptic and extends to a self-adjoint, unbounded, grading–reversing operator on the $\mathbb{Z}/2$-graded Hilbert space $L^2(X, S \otimes E)$, and has a Fredholm index $\mathrm{Index}_\epsilon(D_E) := \dim \ker (D_E)_+ - \dim \ker (D_E)_-$.

Exercise 9.6.2 In the above notation:

a) A different choice of connection on $E \to X$ differs from the original one by an $\mathrm{End}(E)$-valued one-form, and the corresponding twisted Dirac operators differ by a bounded operator. Deduce (see Lemma 9.5.7) that the analytic index $\text{Index}_\epsilon(E)$ does not depend on the choice of connection on E.

b) If E, E' are two bundles, then $\text{Index}_\epsilon(D_{E \oplus E'}) = \text{Index}_\epsilon(D_E) + \text{Index}_\epsilon(D_{E'})$. (*Hint.* Exercise 9.4.17.)

c) If E and E' are stably isomorphic vector bundles (that is if $E \oplus F \cong E' \oplus F$ for some vector bundle F), then $\text{Index}_\epsilon(D_E) = \text{Index}_\epsilon(D_{E'})$. (*Hint.* Start by assuming $E \cong E'$ and show that D_E and $D_{E'}$ are unitarily conjugate modulo order zero operators.)

The results of the previous exercise allow us to make the following:

Definition 9.6.3 Let X be a compact Riemannian, even-dimensional spin^c-manifold. The *analytic index map* $K^0(X) \to \mathbb{Z}$ determined by the spin^c-structure is the group homomorphism $[E] \to \text{Index}_\epsilon(D_E)$, for $E \to X$ a complex vector bundle over X.

Example 9.6.4 The Dirac–Dolbeault operator $\bar{\partial}$ on $L^2(\mathbb{T}^2, \mathbb{C}^2)$ has index zero. But twisting $\bar{\partial}$ by the Poincaré line bundle over $\mathbb{T}^2$ gives an elliptic operator on $\mathbb{T}^2$ with index 1.

We conclude with some further remarks about twisting.

Let $p \in C^\infty(X, M_n(\mathbb{C}))$ be a smooth, projection-valued function, and $E := \text{Im}(p)$, the (Hermitian) vector bundle with

$$E_x := \text{range}\,(p(x)) \subset \mathbb{C}^n, \quad x \in X.$$

Associated with p is a canonical connection on E, the *Grassmann* connection, defined by

$$\nabla_X s\,(x) := p(x)\,d_X s\,(x),$$

if $s \in C^\infty(X, \mathbb{C}^n)$ is a smooth section of the trivial bundle in the image of p. The symbol d_X denotes the trivial, or de Rham connection

$$d_X(s_1, \dots, s_n) := (X(s_1), \dots, X(s_n)).$$

Exercise 9.6.5 Show that the Grassmann connection is a connection compatible with the metric on E.

Suppose now that D is a Dirac-type operator on a Clifford module S equipped with a compatible connection. Following the discussion described above, using the Grassmann connection on E, we obtain an operator D_E on $L^2(X, S \otimes E)$.

We can describe the same operator as a compression.

Exercise 9.6.6 In the above notation, let P be the projection on the Hilbert space $L^2(X, S)^n \cong L^2(X, S^n) = L^2(X, S \otimes \mathbf{1}_n)$ obtained by fiberwise projecting $S_x \otimes \mathbb{C}^n$ to $S_x \otimes E_x$ using $1 \otimes p(x)$. Then the range of P is the subspace $L^2(X, S \otimes E)$, and

$$PDP = D_E$$

up to lower order terms, as densely defined operators on $L^2(X, S \otimes E)$.

In particular, $\text{Index}_\epsilon(D_E) = \text{Index}_\epsilon(PDP)$.

Index maps in odd dimensions: Toeplitz operators

In the case of an *odd*-dimensional spin^c-manifold with Dirac operator D acting on sections $L^2(X, S)$ of a spinor bundle, there is no $\mathbb{Z}/2$-grading available, with respect to which D is odd, so as D is moreover self-adjoint, there is no reasonable sense in which D has a Fredholm index, unless it be zero.

The right kind of index theory to do in odd dimensions is instead that involved in the Toeplitz Index Theorem.

Let X be odd-dimensional, Riemannian, spin^c, and D the Dirac operator, acting on sections of S. We use the same notation for its extension to a self-adjoint unbounded operator on $H := L^2(X, S)$. Since we are in odd dimensions, there is no grading. Being self-adjoint, the spectrum of D is real and splits into its positive and negative spectrum (and zero), and accordingly H splits into an orthogonal direct sum $H_{\lambda \geq 0} \oplus H_{\lambda < 0}$ of two closed subspaces.

Definition 9.6.7 The *Dirac–Szegö projection* p_D is the projection to the closed spectral subspace $H_{\lambda \geq 0}$ spanned by the nonnegative eigenvectors of D.

Alternatively,

$$p_D = \chi_{[0,\infty)}(D),$$

in the sense of functional calculus.

Note that $p_D = \frac{F+1}{2}$ where $F = \chi(D)$, χ the normalizing function $2\chi_{[0,\infty)} - 1$.

Lemma 9.6.8 *Let* $f \in C^\infty(X)$, *regarded as a multiplication operator on* $H = L^2(X, S)$.

Then the commutator $[f, p_D]$ *is compact.*

Proof By Lemma 9.5.10, since the commutator $[f, D]$ is bounded, the commutator $[f, \chi(D)]$ is compact, for any normalizing function χ. Since $2p_D - 1 = \chi(D)$ up to a compact operator for any normalizing function χ, $[f, p_D]$ is compact. □

Definition 9.6.9 Let X be odd-dimensional Riemannian spin^c, let D be the (self-adjoint) Hilbert space Dirac operator on $H = L^2(X, S)$, $p_D := \chi_{[0,\infty)}(D)$ the Szegö projection.

The Szegö projection extends to $H^n = L^2(X, S \otimes \mathbb{C}^n)$ by applying it coordinate-wise. We use the same notation. Now let u be a smooth, unitary-valued function in $C(X, M_n(\mathbb{C}))$, and it acts by a unitary multiplication operator on H^n by applying it fiberwise.

Then the associated *Dirac–Toeplitz operator* is the operator $T_u := p_D u p_D + (1 - p_D)$, acting on $H := L^2(X, S \otimes \mathbb{C}^n)$.

Lemma 9.6.10 *If u and v are smooth functions in $C^\infty (X, M_n(\mathbb{C}))$, then the operators*

$$T_{uv} - T_u T_v, \quad T_u^* - T_{u^*}$$

are compact.

In particular, T_u is an essentially unitary operator on H^n, for any u taking unitary values.

This follows from Lemma 9.6.8; we leave the details to the reader.

Definition 9.6.11 Let X be a compact Riemannian, odd-dimensional spin^c-manifold. The *analytic index map* $\mathrm{K}^1(X) \to \mathbb{Z}$ determined by the spin^c-structure is the group homomorphism $[u] \to \mathrm{Index}(T_u)$, for u a smooth, unitary-valued function in $C(X, M_n(\mathbb{C}))$, and T_u the associated Dirac–Toeplitz operator of Definition 9.6.9.

Example 9.6.12 The simplest example is of course the Dirac operator on the circle. The corresponding theory of Toeplitz operators was discussed in Chapter 1.

The results above imply that we may define an index map

$$\mathrm{K}^1(X) \to \mathbb{Z}, \quad [u] \mapsto \mathrm{Index}(T_u),$$

where T_u is the Dirac–Toeplitz operator associated to $u \in C^\infty (X, M_n(\mathbb{C}))$, and, of course, Index is the Fredholm index.

We now give the K-theory statement of the Index Theorem.

If $E \to X$ is a complex vector bundle over a compact even-dimensional spin^c-manifold X, then the integer $\mathrm{Index}(D_E)$ defines an invariant which depends on solving some differential equations. The Atiyah–Singer Index Theorem describes this integer in purely topological terms.

As observed in the previous section, the Fredholm index $\mathrm{Index}(D_E)$ only depends on the isomorphism class of the bundle E. Moreover, it is additive,

$$\mathrm{Index}(D_{E \oplus E}) = \mathrm{Index}(D_E) + \mathrm{Index}(D_{E'}).$$

It follows that our analytic invariant $\mathrm{Index}(D_E)$ only depends on the K-theory class $[E] \in \mathrm{K}^0(X)$.

Exactly parallel remarks hold for the odd-dimensional case. If X is odd-dimensional, T_u the Dirac–Toeplitz operator associated to a unitary in $C^\infty(X, M_n(\mathbb{C}))$, then $\mathrm{Index}(T_u)$ only depends on the path component of u, in $C^\infty(X, U_n)$, since the Fredholm index itself is homotopy invariant in this sense.

Recall that Bott Periodicity states that for every $n = 0, 1, 2, \ldots$, there is an isomorphism $\mathrm{K}^{-n}(\mathbb{R}^n) \cong \mathbb{Z}$, with $1 \in \mathbb{Z}$ corresponding to a certain Bott generator β_n. Now suppose that

$$X \subset \mathbb{R}^n$$

is a closed submanifold, with spin^c-structure. Let $\pi : \nu \to X$ be the normal bundle to TX in $\mathbb{R}^n$, so that $TX \oplus \nu = T\mathbb{R}^n$. By the 2-out-of-3 Lemma for K-orientations (Lemma 7.10.12) the given K-orientation on TX and the standard K-orientation on $\mathbb{R}^n$ induce a unique K-orientation on ν. Therefore, the Thom Isomorphism of Section 7.10 applies and gives an isomorphism

$$K^*(X) \cong K^{*-\dim \nu}(\nu).$$

Finally, let

$$\hat{\varphi} : \nu \to \mathbb{R}^n$$

be the open embedding onto a tubular neighborhood of X in $\mathbb{R}^n$, associated with the normal bundle. Then $\hat{\varphi}$ induces a group homomorphism

$$\hat{\varphi}! : K^*(\nu) \to K^*(\mathbb{R}^n).$$

For the following, recall that every smooth compact manifold can be embedded in $\mathbb{R}^n$.

Definition 9.6.13 Let X be a smooth compact spin^c-manifold, and choose an embedding $X \subset \mathbb{R}^n$. Then in the above notation, if $a \in K^*(X)$, then the *spin number* spin$^c(a)$ *of* a is defined by the equation

$$\hat{\varphi}! \left(\pi^*(a)\xi_\nu \right) = \text{spin}^c(a) \cdot \beta_n,$$

with $\beta_n \in K^{-n}(\mathbb{R}^n)$ the Bott generator, $\pi : \nu \to X$ the projection for the normal bundle, K-oriented by the K-orientation on X, and $\xi_\nu \in K^{-\dim \nu}(\nu)$ the Thom class.

If X is even-dimensional, only classes in $K^0(X)$ have nonzero spin^c numbers, and if X is odd-dimensional, only classes in $K^1(X)$ have nonzero spin^c numbers. This follows immediately from the definitions. Proving that the spin number is well defined requires of course more work than we have given. We refer the reader to the papers of Atiyah and Singer, or the more recent [78].

Exercise 9.6.14 Prove that if $u \in C^\infty(\mathbb{T})$ is a unitary, then spin$^c([u]) = -\text{wind}(u)$.

Theorem 9.6.15 (The Atiyah–Singer Index Theorem) *Let X be a compact, spin^c-manifold. Then if X is even-dimensional, and $E \to X$ is a complex vector bundle, D_E the Dirac operator on X twisted by E, then*

$$(9.36) \qquad\qquad \text{Index}(D_E) = \text{spin}^c([E]).$$

If X is odd-dimensional and T_u is a Dirac–Toeplitz operator associated to $u \in C^\infty(X, U_n)$, then

$$(9.37) \qquad\qquad \text{Index}(T_u) = \text{spin}^c([u]).$$

The Atiyah–Singer K-theoretic formula for the index leads to "local formulas," like the formula

$$\text{Index}(T_u) = -\frac{1}{2\pi i} \int_{\mathbb{T}} \frac{du}{u}$$

for the index of a Toeplitz operator.

If $\bar{\partial} = \begin{bmatrix} 0 & \frac{\partial}{\partial x} - i\frac{\partial}{\partial y} \\ \frac{\partial}{\partial x} + i\frac{\partial}{\partial y} & \end{bmatrix}$ acting on its domain in $L^2(\mathbb{T}^2, \mathbb{C}^2)$, and $E \to \mathbb{T}^2$
is a complex vector bundle over $\mathbb{T}^2$, then

$$\text{Index}(D_E) = \int_{\mathbb{T}^2} c_1(E),$$

where $c_1(E)$ is the first Chern class of E. (For more general Riemann surfaces there is a further, "curvature" term, which vanishes on the flat $\mathbb{T}^2$.)

We will discuss local formulas more in connection with our discussion of the Heat Equation proof of the Index Theorem in Section 10.7. Although the original statement of the Index Theorem [12] involved such formulas, that is, was formulated in terms of cohomology, the paper [13] states and proves it in terms of K-theory (as above); the cohomology version, using the Chern character isomorphism, is deduced from it in [15].

Chapter 10
K-Homology and Noncommutative Geometry

This chapter deals with some of the key aspects of the geometric part of the subject of Noncommutative Geometry. Connes' program for a Noncommutative (Riemannian) Geometry is based on using ideas from classical Index Theory to endow potentially noncommutative C*-algebras with further geometric structure. For example, the C*-algebra $C(M)$ for a compact smooth manifold M contains the dense and holomorphically closed subalgebra $C^\infty(M)$, and the geometry of M leads to many interesting functionals on $C^\infty(M)$, like $\tau(f, g) = \int_\gamma f dg$, where γ is a closed curve in M. Such γ defines a *closed* 1-*current*: a continuous linear functional $\Omega^1(M) \to \mathbb{C}$. From the point of view of the algebra $C^\infty(M)$, it defines a *cyclic* 1-*cocycle* (see Example 10.2.3). Such cocycles pair with K-theory classes, in this case with classes in $K^{-1}(M)$. In [40, 41, 43], and [47], A. Connes developed the theory of *cyclic cohomology*, a "noncommutative" analog of de Rham homology of currents, and this was one of the steps initiating Noncommutative Geometry as a subject. The irrational rotation algebra $A_\hbar$, with its certain natural subalgebra $A_\hbar^\infty$ of "smooth" elements, is the source of interesting examples of such cocycles. The differential structure on $A_\hbar$ is specified by an $\mathbb{R}^2$-action determining two transverse flows and derivations δ_1, δ_2, and if τ is the standard trace, then

$$\psi_2(b^0, b^1, b^2) = \tau \left(a^0 \delta(a^1)\delta_2(a^2) - a^0 \delta_2(a^1)\delta_1(a^2) \right)$$

is a cyclic 2-cocycle analogous to the curvature tensor of Riemannian geometry (see [47]), and it enters into several index formulas for elliptic operators over the noncommutative torus (see Section 10.12).

The idea of studying Riemannian geometry by studying spectra of appropriate geometrically defined operators goes back to Hermann Weyl (see [24] or the book [145]) who showed that the volume of a bounded domain in $\mathbb{R}^n$ can be determined from the spectral asymptotics of the Dirichlet boundary value problem of the Laplacian (see Kac's famous paper [106]). A refinement of Weyl's asymptotic

© Springer Nature Switzerland AG 2024
H. Emerson, *An Introduction to C*-Algebras and Noncommutative Geometry*,
Birkhäuser Advanced Texts Basler Lehrbücher,
https://doi.org/10.1007/978-3-031-59850-0_10

formula due to Pleijel and Minakshisundaram provides a series of local spectral invariants for compact Riemannian manifolds involving derivatives of the curvature tensor. Their method was based on certain asymptotic expansions. Asymptotic expansions of the heat kernel $e^{-t\Delta}$ where $\Delta = D^2$ for a Dirac operator D on a compact spin^c-manifold M give rise to an almost instant "local" proof of the Index Theorem insofar as it identifies the index with the integral of an appropriate smooth function on M, and the only remaining problem being to identify the function explicitly. This is the *Heat Equation* proof of the index theorem. This proof is quite different from the K-theory proof. The heat equation proof is discussed in Section 10.7. See the books [142] or [25] for detailed expositions of the heat equation proof. What is most important for our purposes is that the heat equation argument also applies in "noncommutative" settings and leads eventually to the Local Index Theorem of Connes and Moscovici [52] discussed in Section 10.8. This very general result applies to spectral triples, or cycles, and zeta functions of the type $\mathrm{Tr}(a\Delta^{-s})$, with a an element of a *-algebra playing a big role. The index formula assigns to a spectral triple a certain *cyclic cocycle* for periodic cyclic cohomology, the noncommutative analog invented by Connes of de Rham cohomology (see [45, 47]). Cyclic cocycles pair with K-theory classes, and the work of Connes and Moscovici amounts to an explicit description of a Chern character for K-homology. We discuss only the minimum of cyclic cohomology in this book, essentially enough to formulate the Connes–Moscovici theorem, and study simple examples. The book [119] is a good treatment of cyclic cohomology.

Topological invariants of elliptic operators and spectral triples are organized by K-*homology*, which is a generalized cohomology theory on C*-algebras dual to K-theory defined in terms of cycles and relations: The cycles in K-homology are "abstract elliptic operators" or *Fredholm modules*. K-homology defined this way is a special case of KK-theory and was developed by Kasparov [111], but it was preceded historically by another model involving equivalence classes of C*-algebra extensions by Brown, Douglas, and Filmore (see [64]), which has a beautiful application to the classification of essentially normal operators. The book [98] gives a complete and detailed account of both of these theories, and the important connections between K-homology of noncompact spaces, and coarse geometry. The paper [21] gives an excellent modern treatment of Dirac operators and their role in K-homology.

Connes and coauthors have shown that spectral triples give a new perspective on the geometric interpretation of the detailed structure of the Standard Model in particle physics and of the Brout–Englert–Higgs mechanism. See [37–39]. The book [84] discusses extensively the use of spectral triples in physics, and [49] is an essay on the interplay between Noncommutative Geometry and physics. The book [84] is concerned almost entirely with spectral triples and their use in physics. See [22] for connections between Noncommutative Geometry and Solid State Physics.

10.1   Fredholm Modules and Their Pairing with K-Theory

Suppose $\chi : \mathbb{R} \to [-1, 1]$ is a normalizing function: χ is odd, and $\lim_{t \to \pm\infty} \chi(t) = \pm 1$.

Lemma 9.5.10 (and Exercise 9.5.4) of Chapter 8 implies the following.

Lemma 10.1.1 *Suppose H is a Hilbert space and D is a densely defined self-adjoint operator on H such that $(1 + D^2)^{-1} \in \mathcal{K}(H)$. Suppose A is a C*-algebra represented on H by bounded operators containing a dense subalgebra $A^\infty \subset A$ leaving the domain of D invariant and such that $[a, D]$ is bounded for $a \in A^\infty$. Then if χ is a normalizing function, then $F := \chi(D)$ is self adjoint, and*

$$[a, F] \in \mathcal{K}(H), \quad \forall a \in A, \quad F^2 - 1 \in \mathcal{K}(H).$$

If H carries a $\mathbb{Z}/2$-grading with respect to which D is an odd operator, then F is also odd.

The lemma applies to elliptic differential operators D on sections of vector bundles over smooth manifolds, due to the results of the previous chapter.

Exercise 10.1.2 Suppose D is a densely defined self-adjoint operator with discrete spectrum, and assume D odd with respect to a $\mathbb{Z}/2$-grading with (self-adjoint) grading operator ϵ. Suppose that χ is an odd normalizing function. Show that $\chi(D)$ anti-commutes with ϵ.

Atiyah's idea, motivated by the Index Theorem and specifically Lemma 10.1.1, was that one should be able to define a theory *dual* to K-theory (called K-homology) with cycles specified by the following data.

Definition 10.1.3 Let A be a C*-algebra. An *even Fredholm module over A* is a triple consisting of a $\mathbb{Z}/2$-graded Hilbert space H, a representation

$$\pi : A \to \mathbb{B}(H),$$

as even operators (preserving grading), and a self-adjoint, odd (reversing grading) operator F on H such that

$$\pi(a) \cdot (F^2 - 1) \in \mathcal{K}(H)$$

for all $a \in A$.

An *odd* Fredholm module is defined in the same way, except we drop the assumption of a $\mathbb{Z}/2$-grading.

Example 10.1.4 For any normalizing function χ, let $F = \chi|_\mathbb{Z}$, acting as a multiplication operator on $l^2(\mathbb{Z})$. Then the triple $(l^2(\mathbb{Z}), \lambda, F = \chi_\mathbb{Z})$ is an odd Fredholm module over $A := C * (\mathbb{Z})$, where $\lambda : C^*(\mathbb{Z}) \to \mathbb{B}(H)$ is the regular representation.

Example 10.1.5

1. If X is an even-dimensional spin^c-manifold and D is the Dirac operator on sections of the Hermitian$\mathbb{Z}/2$-graded bundle $S \to M$, then the triple $(H := L^2(X, S), \pi, F_D := \chi(D))$ with the representation $\pi: C(X) \to \mathbb{B}(H)$ by multiplication operators, and χ a normalizing function, defines an even Fredholm module and corresponding class, the *Dirac class* $[D] \in \mathrm{KK}_0(C(X), \mathbb{C})$. The Hilbert space H is graded by the grading on S.
2. If X is an odd-dimensional spin^c-manifold and D is the Dirac operator on sections of the Hermitian bundle $S \to M$, then the triple $(H := L^2(X, S), \pi, F_D := \chi(D))$ with the representation $\pi: C(X) \to \mathbb{B}(H)$ by multiplication operators defines an odd Fredholm module and corresponding class, the Dirac class $[D] \in \mathrm{KK}_1(C(X), \mathbb{C})$.

There is an obvious notation of unitary equivalence of a pair of Fredholm modules, and unitary equivalence classes of Fredholm modules admit a direct sum operation.

Definition 10.1.6 Equivalence of Fredholm modules is the equivalence relation generated by the following two relations:

1. A Fredholm module (H, π, F) is *degenerate* if $\pi(a)(F^2 - 1)$, $[\pi(a), F]$ are both zero for all $a \in A$.
2. Two Fredholm modules (H_i, π_i, F_i) are *operator homotopic* if $H_1 = H_2 = H$, $\pi_1 = \pi_2 = \pi$, and F_1, F_2 are connected by a path in $\mathbb{B}(H)$ through operators F_t for which (H, π, F_t) are Fredholm modules.

Exercise 10.1.7 Prove that if (H, π, F) is a Fredholm module and (H, π, F') is another, such that $F - F' \in \mathcal{K}(H)$, then the two Fredholm modules are operator homotopic.

They are called *compact perturbations* of each other.

Remark 10.1.8 Any Fredholm module over unital A can be replaced by an equivalent one, in which the representation is nondegenerate. If $\pi(1) = p \in \mathbb{B}(H)$, then we compress the Fredholm module by the projection p

$$(pH, \pi, pFp).$$

This results in an equivalent Fredholm module, because the original Fredholm module differs from this one by a compact perturbation of a degenerate Fredholm module, namely

$$((1 - p)H, 0, (1 - p)F(1 - p)),$$

where 0 denotes the zero representation.

Definition 10.1.9 The *analytic K-homology*

$$KK_*(A, \mathbb{C}) := KK_0(A, \mathbb{C}) \oplus KK_1(A, \mathbb{C})$$

of a C*-algebra A is the $\mathbb{Z}/2$-graded abelian group generated by the equivalence classes of even/odd Fredholm modules over A, modulo the relation $x + y = [x \oplus y]$, where $\oplus$ denotes direct sum of Fredholm modules.

The following result is useful for theoretical purposes, e.g., for cyclic cohomology.

Proposition 10.1.10 *Every Fredholm module is equivalent to one of the form* (H, π, F) *where* $F^2 = 1$.

Proof Let (H, π, F) be an even Fredholm module over A. Then $F = \begin{bmatrix} 0 & u^* \\ u & 0 \end{bmatrix}$ for $u \colon H_+ \to H_-$ with $uu^* - 1$ and $u^*u - 1$ each compact. Applying the trick of Lemma 8.5.10, consider the operator $w \colon H_+ \oplus H_- \to H_- \oplus H_+$ given by matrix multiplication by

$$w := \begin{bmatrix} u & -(1 - uu^*)^{1/2} \\ (1 - u^*u)^{1/2} & u^* \end{bmatrix}.$$

Let $V_+ := H_+ \oplus H_-$ and $V_- := H_- \oplus H_+$ and A be represented on V_+ by $\pi_+ \oplus 0$ on V_+ and $\pi_-(a) \oplus 0$ on V_-. Let $\tilde{\pi}$ denote the direct sum of these two representations. Let $\tilde{H} = V_+ \oplus H_-$, with grading as notation indicates, $\tilde{F} = \begin{bmatrix} 0 & w^* \\ w & 0 \end{bmatrix}$, then $(\tilde{H}, \tilde{\pi}, \tilde{F})$ is a Fredholm module over A with $\tilde{F}^2 = 1$, and it is a compact perturbation of the direct sum of (H, π, F) and a degenerate.

The odd case is dealt with similarly. $\qquad\square$

It is a special case of Kasparov's KK-theory, and we will discuss its general functoriality properties in that context in the next chapter.

The reader will find it easy however to do the following exercise.

Exercise 10.1.11 Prove that a *-homomorphism $\alpha \colon A \to B$ between C*-algebras induces a pair of group homomorphisms $KK_i(B, \mathbb{C}) \to KK_i(A, \mathbb{C})$ by replacing the representation π in a Fredholm module by $\pi \circ \alpha$.

Pairing of Even Fredholm Modules with Even K-theory Classes

Suppose that (H, π, F) is an even Fredholm module over A.

Consider the operator F. We know that $F = F^*$ and that $F^2 - 1$ is compact. Furthermore, F is odd with respect to the grading. So with respect to the orthogonal decomposition $H = H_+ \oplus H_-$, the operator F and representation have the forms

$$F = \begin{bmatrix} 0 & U^* \\ U & 0 \end{bmatrix}, \quad \pi(a) = \begin{bmatrix} \pi^+(a) & 0 \\ 0 & \pi^-(a) \end{bmatrix},$$

where $U\colon H^+ \to H^-$ is an essentially unitary operator, with the property that

$$\pi^-(a)U - U\pi^+(a) \in \mathcal{K}(H^+, H^-)$$

is a compact operator for all $a \in A$.

Now suppose that p is a projection in A. Since $\pi(p)$ commutes mod compact operators with F, the operator

$$(10.1) \qquad \pi^-(p)U\pi^+(p)\colon \pi^+(p)H^+ \to \pi^-(p)H^-$$

is a Fredholm operator. Then we make the definition

$$(10.2) \qquad \langle(H, \pi, F), p\rangle := \mathrm{Index}\big(\pi^-(p)U\pi^+(p)\big).$$

Exercise 10.1.12 Show that this "index pairing" is invariant under changing the Fredholm module to either a homotopic one or perturbing it by a degenerate. Show that if A is unital and π is nondegenerate, then pairing (H, π, F) with $1 \in A$ gives the ordinary Fredholm index of U.

To define the pairing of (H, π, F) with a projection not necessarily in A, but in $M_n(A)$, for some n, observe that $(H \oplus \cdots H, \pi \oplus \cdots \oplus \pi, F \oplus \cdots F)$ is also a Fredholm module over A, where the $\mathbb{Z}/2$-grading is induced by $\epsilon \oplus \cdots \oplus \epsilon$. Hence, if $p \in M_n(A)$ is a projection, then we can pair p with the original Fredholm module by first forming a direct sum of n copies of (H, π, F), compressing the operator with the projections $\pi^\pm(p)$ as in (10.1), and taking its Fredholm index, that is, the index of

$$\pi^-(p)(U \oplus \cdots \oplus U)\pi^+(p)\colon \pi^+(p)H^+ \oplus \cdots \oplus \pi^+(p)H^+$$
$$\to \pi^-(p)H^- \oplus \cdots \oplus \pi^-(p)H^-.$$

We obtain a bilinear pairing

$$\mathrm{K}_0(A) \times \mathrm{KK}_0(A, \mathbb{C}) \to \mathbb{Z},$$

generalizing the pairing

$$\langle[D], [E]\rangle := \mathrm{Index}(D_E)$$

between Dirac operators and (K-theory classes of) vector bundles over X (i.e., generalizing the *analytic index map* of Definition 9.6.3).

Pairing of Odd Fredholm Modules with Odd K-theory Classes

The pairing

$$K_1(A) \times KK_1(A, \mathbb{C}) \to \mathbb{Z}$$

between $K_1(A)$-classes and *odd* Fredholm modules generalizes the pairing

$$\langle [D], [u] \rangle := \mathrm{Index}(T_u)$$

between Dirac operators on *odd*-dimensional compact manifolds and (K-theory classes of) unitaries in $C(X, M_n(\mathbb{C}))$.

Assume that (H, π, F) is a nondegenerate odd Fredholm module. Then $P := \frac{F+1}{2} \in \mathbb{B}(H)$ satisfies

$$P^* = P, \quad P^2 - P, \quad [P, \pi(a)] \in \mathcal{K}(H).$$

It follows that if $u \in A$ is a unitary, then

$$PuP + (1 - P) \in \mathbb{B}(H)$$

is an essentially unitary operator, and we can set

(10.3) $\langle (H, \pi, F), u \rangle := \mathrm{Index}(P\pi(u)P + 1 - P).$

Exercise 10.1.13 In the above notation, show that

a) $P\pi(u)P + (1 - P)$ is an essentially unitary operator.
b) Show that the pairing (10.3) does not change if the Fredholm module is replaced by an operator homotopic one or under adding a degenerate (odd) Fredholm module to it.
c) Show that if u is replaced by another unitary, connected by a path of unitaries to u, then the pairing (10.3) does not change.

The pairing above extends to unitaries in $M_n(A)$ by the same procedure as in the even case.

We close this section with a beautiful example of a Fredholm module due to Julg and Valette [102].

Example 10.1.14 The following example is based on the free group $\mathbb{F}_2$ on two generators a, b. The Cayley graph of this finitely generated group is (by definition) the graph with vertices labelled by elements of $\mathbb{F}_2$, and for each generator $s \in S = \{a, a^{-1}, b, b^{-1}\}$, there is an (oriented) edge from g to gs. For the group $\mathbb{F}_2$, this results in a tree:

$\mathbb{F}$

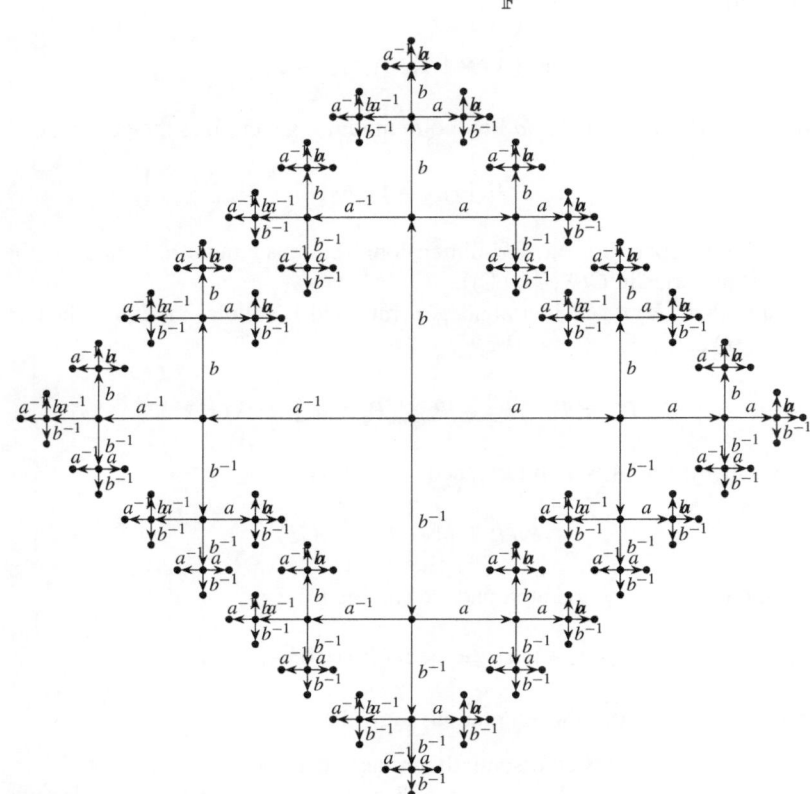

Let X^0 be the set of vertices of the tree and X^1 the set of geometric (unoriented) edges. Let $\{e_v\}_{v \in X^0}$ be the corresponding orthonormal basis of $l^2(X^0)$ by point masses at vertices and similarly $\{e_s\}_{s \in X^1}$ an orthonormal basis for $l^2(X^1)$.

Choose any basepoint $w \in X^0$. Define

$$b_w : l^2(X^0) \to l^2(X^1),$$

by setting $b(e_v) = 0$ if $v = w$ is the basepoint. Otherwise, $v \neq w$, then there is a unique path of geometric edges between v and w, and we let $s(v)$ be the first edge in this edge path and set $b(e_v) = e_{s(v)}$.

Thus, $b_w(e_v) = e_s$ means that one of the vertices of s is v, and the other is a vertex one unit closer to w than v.

Exercise 10.1.15 $b_w b_w^* = 1$ and $b_w^* b_w = 1 - p_w$, where p_w is projection to the basis vector e_w.

In particular, $F := \begin{bmatrix} 0 & b_w^* \\ b_w & 0 \end{bmatrix}$ satisfies $F^2 - 1$ has rank 1 and so is compact.

The group $\mathbb{F}_2$ permutes the vertices and edges of the tree, and we obtain two unitary representations

$$\pi_0 \colon \mathbb{F}^2 \to \mathbb{B}\left(l^2(X^0)\right), \quad \pi_1 \colon \mathbb{F}^2 \to \mathbb{B}\left(l^2(X^1)\right).$$

By Exercise 1.3.17 of Chapter 2, these representations extend to representations of $C^*(\mathbb{F}_2)$.

Exercise 10.1.16 Show that

$$\pi_1(g) b_w \pi_0(g)^{-1} = b_{g(w)}$$

for any vertex w and any $g \in \mathbb{F}_2$ and that if w, w' are any two vertices, then $(b_w - b_{w'})(e_v) = 0$ unless v lies on the path between w and w'.

Now we let w be the basepoint of the tree and just denote by b the corresponding operator b_w.

Use the observations above to show that for $\pi = \pi_0 \oplus \pi_1$ and $F = \begin{bmatrix} 0 & b* \\ b & 0 \end{bmatrix}$

$$[\pi(a), F] \in \mathcal{K}\left(l^2(X^0) \oplus l^2(X^1)\right), \quad \forall a \in C^*(\mathbb{F}_2).$$

Therefore, $\left(l^2(X^0) \oplus l^2(X^1), \pi, F\right)$ is an even Fredholm module over $C^*(\mathbb{F}_2)$. Note that b_w has Fredholm index 1. Hence pairing (classes of) projections in $C^*(\mathbb{F}_2)$ with the Fredholm module $\left(l^2(X^0) \oplus l^2(X^1), \pi, F\right)$ determines a nonzero group homomorphism $K_0(C^*(\mathbb{F}_2)) \to \mathbb{Z}$.

Exercise 10.1.17 Let F be the Fredholm operator above and γ the grading operator on $l^2(X^0) \oplus l^2(X^1)$ ($\gamma = 1$ on the first factor, and $\gamma = -1$ on the second factor).

Verify that the commutator $[\pi(a), F]$ has finite rank for $a \in \mathbb{C}[\mathbb{F}_2]$ and that

(10.4) $$\frac{1}{2}\mathrm{Trace}(\gamma F[F, \pi(a)]) = \tau(a),$$

where τ is the trace on $C^*(\mathbb{F}_2)$, $\tau(\sum_g a_g[g]) = a_e$.

In combination with Connes' character formula (Theorem 10.3.3), this implies that the group homomorphism $\tau_* \colon K_0(C^*(\mathbb{F}_2)) \to \mathbb{R}$ induced by the trace takes integer values. Since the trace is also faithful, this implies the remarkable result that $C^*(\mathbb{F}_2)$ has no nontrivial projections (the Kadison Conjecture).

10.2 Cyclic Cohomology

Definition 10.2.1 Let A be an algebra.

a) An $n + 1$-linear function $\phi \colon A \times \cdots A \to \mathbb{C}$ is a *cyclic n-cochain* if

$$\phi(a^0, \ldots, a^n) = (-1)^n \phi(a^n, a^0, a^2, \ldots, a^{n-1}).$$

b) The *Hochschild coboundary* $b\phi$ of an $n + 1$-linear functional ϕ (not necessarily cyclic) on A is the $n + 2$-linear functional

$$b\phi\,(a^0, \ldots, a^{n+1}) := \sum_{j=0}^{n}(-1)^j \phi(a^0, \ldots, a^j a^{j+1}, \ldots, a^{n+1})$$

$$+ (-1)^{n+1} \phi(a^{n+1} a^0, a^1, \ldots, a^n).$$

A Hochschild n-cocycle is an $n + 1$-linear functional ϕ on A such that $b\phi = 0$. A *cyclic n-cocycle* is a cyclic cochain which is also a Hochschild n-cocycle, i.e., $b\phi = 0$.

Example 10.2.2 A Hochschild 0-cocycle, equivalently, a cyclic 0-cocycle, since the cyclic condition is trivial, is a linear map $\tau \colon A \to \mathbb{C}$ such that $0 = b\tau(a^0, a^1) = \tau(a^0 a^1) - \tau(a^1 a^0)$. This shows that cyclic 0-cocycles are traces on the algebra A.

Example 10.2.3 A p-dimensional de Rham current on M is a continuous linear functional on the Fréchet space $\Omega^p(M)$. If C is a p-dimensional current, then the $p + 1$-linear functional

$$(10.5) \qquad \varphi_C(f^0, f^1, \ldots, f^p) := \int_C f^0 df^1 \wedge \cdots \wedge df^p$$

is a Hochschild p-cocycle which is a cyclic p-cocycle if C is *closed*: That is, if $C \circ d \colon \Omega^*(M) \to \mathbb{C}$ is zero, d is the de Rham differential.

In particular, if ω is a closed $n - p$-dimensional form on an oriented compact manifold M, then

$$(10.6) \qquad \varphi(f^0, \ldots, f^p) := \int_M f^0 df^1 \cdots df^p \wedge \omega$$

defines a cyclic p-cocycle on $C^\infty(M)$.

Definition 10.2.4 The *Hochschild cohomology* of A is the quotient of the linear space of Hochshild coycles by the subspace of coboundaries. The *cyclic cohomology* of A is the space of cyclic cocycles modulo the subspace of cyclic coboundaries.

The cyclic cohomology of A is denoted $\mathrm{HC}^*(A)$.

Remark 10.2.5 A converse to Example 10.2.3 holds. For example, observe that any Hochschild 1-cocycle ϕ on $C^\infty(M)$ determines a one-dimensional current

$$\langle C_\phi, f\,dg \rangle := \phi(f, g).$$

By an easy exercise the current is closed if ϕ is a cyclic cocycle. The Connes–Hochschild–Kostant–Rosenberg Theorem [41] extends this result to an embedding of the Hochschild cohomology of $C^\infty(M)$ into the space of currents (continuous linear functionals on the space of differential forms) on M, mapping cyclic cocycles to closed currents.

Example 10.2.6 Let A be an algebra and $\delta: A \to A$ be a derivation: $\delta(ab) = a\delta(b) + \delta(a)b$. Let $\tau: A \to \mathbb{C}$ be a δ-invariant trace in the sense that $\tau \circ \delta = 0$. Then

(10.7)
$$\phi(a^0, a^1) := \tau\left(a^0\delta(a^1)\right)$$

is a cyclic 1-cocycle on A. Indeed, the cyclic condition $\tau(a^0, a^1) = -\tau(a^1, a^0)$ follows from

$$\tau(a^0\delta(a^1) + \tau(a^1\delta(a^0)) = \tau(a^0\delta(a^1) + \tau(\delta(a^0)a^1) = (\tau \circ \delta)\,(a^0a^1) = 0.$$

The Hochschild cocycle condition $b\phi = 0$ is checked similarly and is left to the reader.

A concrete example is proved by a *flow* $\{\alpha_t\}_{t\in\mathbb{R}}$ on a compact manifold M and an α-invariant measure. Let A be the algebra $C^\infty(M)$. The flow generates a vector field X and corresponding derivation

$$\delta: C^\infty(M) \to C^\infty(M), \quad \delta(f) = X(f).$$

Suppose that μ is an α-invariant probability measure on M. Let $\tau(f) := \int_M f\,d\mu$, a trace on $C^\infty(M)$. Differentiating the equality

$$\int_M f \circ \alpha_t \, d\mu = \int_M f \, d\mu$$

true for arbitrary $f \in C^\infty(M)$ gives that

(10.8)
$$\int_M X(f)\,d\mu = 0.$$

Hence $\tau \circ \delta = 0$.

The *Ruelle–Sullivan current* is the continuous linear functional on 1-forms

$$\langle C_\alpha, \omega \rangle := \int_M \omega(X)\, d\mu.$$

It is a closed current since

$$\langle C_\alpha, df \rangle = \int_M X(df)\, d\mu = \int_M X(f)\, d\mu = 0$$

by (10.8).

Every 1-form is a linear combination of 1-forms of the kind $f\, dg$, for f, g smooth. Evaluating C_α on $f\, dg$ gives $\int_M f X(g)\, d\mu$, or in other notation,

$$\langle C_\alpha, f\, dg \rangle = \tau\left(f \delta(g) \right),$$

the associated cyclic cocycle (10.7).

For purposes of K-theory, there is an important variant of cyclic cohomology called *periodic* cyclic cohomology. This is defined as follows.

Definition 10.2.7 Let A be an algebra and $C^n(A)$ all $n + 1$-linear maps $\varphi \colon A \otimes \cdots \otimes A \to \mathbb{C}$. The B-operator $B \colon C^n(A) \to C^{n-1}(A)$ is defined by $B = AB_0$ where

$$(A_0\varphi)(a^0, \dots, a^{n-1}) := \sum (-1)^{(n-1)j} \varphi(a^j, a^{j+1}, \dots, a^{j-1})$$

and $(B_0\varphi)(a^0, \dots, a^{n-1}) := \varphi(1, a^0, \dots, a^{n-1})$.

Combining B with the Hochschild operator b of Definition 10.2.1, we obtain a complex of length 2: Set

$$C^{\mathrm{ev}}(A) := \oplus_{k=0}^{\infty} C^{2k}(A), \quad C^{\mathrm{od}}(A) := \oplus_{k=0}^{\infty} C^{2k+1}(A).$$

Then $b + B$ maps $C^{\mathrm{ev}}(A)$ to $C^{\mathrm{od}}(A)$ and $C^{\mathrm{od}}(A)$ to $C^{\mathrm{ev}}(A)$.

Definition 10.2.8 The *periodic cyclic cohomology of A* is the cohomology of the complex

$$C^{\mathrm{ev}}(A) \leftrightarrow C^{\mathrm{od}}(A),$$

under the operator $b + B$.

Periodic cyclic cohomology is the direct sum of the even periodic cyclic cohomology $\mathrm{HCP}^0(A)$ and odd periodic cyclic cohomology $\mathrm{HCP}^1(A)$. A class in $\mathrm{HCP}^0(A)$ is represented by a finitely supported tuple $(\phi_{2k})_{k=0}^{\infty}$, where $\phi_{2k} \in$

$C^{2k}(A)$, and $(b + B)(\phi_0 + \phi_2 + \phi_4 + \cdots = 0)$, meaning that $b\phi_{2k} + B\phi_{2k+2} = 0$ for $k = 0, 1, 2, \ldots$. A class in $\mathrm{HCP}^1(A)$ is described in a similar way.

Remark 10.2.9 If φ is a cyclic n-cycle, then $(0, \ldots, 0, \varphi, 0, \cdots)$ defines a cocycle in the $b + B$ complex, supported in the single dimension n because by definition $b\varphi = B\varphi = 0$ for cyclic cocycles.

Exercise 10.2.10 In the notation of Example 10.2.3, if C is a p-dimensional current and φ_C the associated $p + 1$-linear map of Example 10.2.3, then $B\varphi_C = p\,\varphi_{d^t C}$ (giving another proof that φ_C is a cyclic cocycle if C is closed).

The following theorem shows that the periodic cyclic cohomology of $C^\infty(M)$ is exactly the even/odd de Rham cohomology:

Theorem 10.2.11 *The map sending the class of a closed, k-dimensional current $C \in \Omega^k(M)'$ to the class in $\mathrm{HCP}^*(C^\infty(M))$ of the cyclic cocycle $\frac{1}{k!}\varphi_C$, where ϕ_C is as in (10.5), induces a grading-preserving vector isomorphism of $\mathbb{Z}/2$-graded vector spaces*

$$\mathrm{H}^*_{\mathrm{dR}}(M) \cong \mathrm{HCP}^*\left(C^\infty(M)\right),$$

where $H^*_{\mathrm{dR}}(M) := \oplus_{p=0}^{\dim M} \mathrm{H}^p_{\mathrm{dR}}(M)$.

We describe the pairing between periodic cyclic cocycles and idempotents and unitaries in matrix algebras over A.

If φ is an $n + 1$-linear functional $A \otimes \cdots \otimes A \to \mathbb{C}$, it induces an $n + 1$-linear functional $\varphi \,\sharp\, \mathrm{Trace} : M_k(A) \otimes \cdots M_k(A) \to \mathbb{C}$ by

$$(\varphi \,\sharp\, \mathrm{Trace})\,(a^0 \otimes m^0, \ldots, a^n \otimes m^n) := \mathrm{Trace}(m^0 \cdots m^n)\varphi(a^0, \ldots, a^n).$$

Theorem 10.2.12 *Let A be a unital algebra and $[\varphi] \in \mathrm{HCP}^{\mathrm{ev}}(A)$ be an even class, represented by a mixed-degree (finitely supported) cochain $(\varphi_{2k})_{k=0}^\infty$. Let $e \in M_k(A)$ be an idempotent. Then*

$$(10.9) \qquad \langle \varphi, e \rangle := \sum_{k=0}^\infty (-1)^k \frac{(2k)!}{k!}(\varphi_{2k} \,\sharp\, \mathrm{Trace})\,(e, e, \cdots, e)$$

only depends on the Murray–von Neumann equivalence class of e and the periodic cyclic cohomology class of φ.

Let $[\varphi] \in \mathrm{HCP}^{\mathrm{od}}(A)$ be an odd class, represented by a mixed-degree (finitely supported) cochain $(\varphi_{2k+1})_{k=0}^\infty$. Let $u \in M_k(A)$ be an invertible. Then

$$(10.10) \qquad \langle \varphi, u \rangle := \frac{1}{\sqrt{2\pi i}} \sum_{k=0}^\infty (\varphi_{2k+1} \,\sharp\, \mathrm{Trace})\,(u^{-1}, u, \cdots, u^{-1}, u)$$

only depends on the class of u in the abelianization of $GL_1(A)$ *and the periodic cyclic cohomology class of* φ.

See [87] and [88] for the proofs.

Remark 10.2.13 The constants in the pairings of Theorem 10.2.12 are chosen so that the following holds: Recall that the Chern character $Ch(E)$ or $Ch(e)$ of the class of a vector bundle, or equivalently, of an idempotent $e \in C^\infty(M) \otimes M_n(\mathbb{C})$, is represented by the closed differential form $Trace(\exp(-R_E))$, where R_E is the curvature of a Hermitian connection on E. The Chern character of a unitary $U \in C^\infty(M) \otimes M_n(\mathbb{C})$ is represented by the closed differential form

$$\sum_{k=0}^{\infty}(-1)^k \frac{2k!}{(2k+1)!} Trace(u^{-1}du)^{2k+1}.$$

Then,

$$\langle[\varphi_C],[e]\rangle = \langle[C], Ch(e)\rangle \quad \text{and} \quad \langle[\varphi_C],[u]\rangle = \frac{1}{\sqrt{2\pi i}}\langle[C], Ch(u)\rangle$$

hold for any closed current C on M.

10.3   Finitely Summable Fredholm Modules and Connes Character Formula

Connes' character formula supplies a Chern character map from K-homology to periodic cyclic cohomology. This results in a pairing between K-homology classes which are finite dimensional in a suitable sense and K-theory classes, which is one of the seminal results of Connes' early development of Noncommutative Geometry.

Definition 10.3.1 Let $p \in [1, \infty)$. A Fredholm module (H, π, F) over A is *p-summable* if

$$[\pi(a), F] \in \mathcal{L}^p(H)$$

for dense $a \in A$, where $\mathcal{L}^p(H)$ is the p-Schatten ideal (Section 2.1 of Chapter 1).

One sometimes refers, in the definition, more explicitly to the *-subalgebra $A^\infty \subset A$ of elements $a \in A$ such that $[\pi(a), F] \in \mathcal{L}^p(H)$, which may of course be strictly smaller than A (but is required to be dense by the definitions). Thus, one might speak of a Fredholm module over A, which is p-summable over A^∞.

Finitely summable, nondegenerate Fredholm modules are generally delicate to construct and depend on geometric constructions. The *classical* examples of them are elliptic pseudodifferential operators T of order 0 on sections of vector bundles over a compact manifold. We have not discussed general pseudodifferential operators in this book. However, the Toeplitz projection $T = P_+$, projecting $L^2(\mathbb{T})$ to the Hardy subspace $\mathbf{H}^2$, is an example of one, as are, more generally, the operators $T = \chi(D)$ for suitable normalizing functions χ and D a Dirac operator. The main relevant properties of pseudodifferential operators are the following:

a) The *-algebra of pseudodifferential operators contains the class of differential operators and is closed under holomorphic functional calculus. Since an elliptic self-adjoint operator D on a compact manifold M has discrete spectrum, we can always find a normalizing function χ which is actually holomorphic on $\mathrm{Spec}(D)$, and hence the corresponding sign operator $F := \chi(D)$ is pseudodifferential of order 0.

b) If T is a pseudodifferential operator of order α on a compact, d-dimensional manifold, then the singular values of T are $O(n^{\frac{\alpha}{d}})$. In particular, on a d-dimensional compact manifold, a pseudodifferential operator of order $\alpha < -d$ is trace-class (*Weyl* asymptotics).

c) The pseudodifferential operators are filtered by order, and

$$\mathrm{order}([S, T]) = \mathrm{order}(S) + \mathrm{order}(T) - 1.$$

This implies in particular that the commutators $[f, T]$ are also pseudodifferential operators of order -1, for $f \in C^\infty(M)$ and T of order 0, because multiplication by a smooth function is also a pseudodifferential operator of order 0.

Combining these facts gives that the principal values of $[f, T]$ are $O(n^{-\frac{1}{d}})$ if T has order 0. It follows immediately that $[f, T] \in L^p(H)$ for $p > d$.

Theorem 10.3.2 *Let M be a compact manifold and $H = L^2(M, E)$ be a (possibly graded) Hilbert space of sections of a bundle over M. Let $\pi \colon C(M) \to \mathbb{B}(H)$ be the representation of $C(M)$ by multiplication operators, and let F be a pseudodifferential operator on H of order 0 such that $F^2 - 1$ is compact. Then for f smooth on M, the commutator $[f, F]$ is in the Schatten class $L^p(H)$ for all $p > n$.*

In particular, (H, π, F) is a p-summable Fredholm module for $p > n$.

Such Fredholm modules are generic in a sense, for smooth manifolds: every K-homology class for $C(M)$, where M is a compact smooth manifold, is represented by a pseudodifferential operator of order 0 in the sense of the theorem above and hence is represented by a p-summable Fredholm module for $p > \dim(M)$. Therefore, the *dimension* of the manifold M is reflected to some degree in the order of the "infinitesimals" $[f, F]$, as F ranges over all pseudodifferential operator representatives of all K-homology classes.

The situation for general C*-algebras is much more delicate. M. Puschnigg [130] has proved that the group C*-algebras $C^*(\Gamma)$ of lattices $\Gamma \subset G$ in higher rank Lie groups have *no* nondegenerate finitely summable Fredholm modules over them.

However, we discuss some constructions connected to hyperbolic groups (like lattices in rank 1 Lie groups) in the next section, which yield (interesting examples of) finitely summable, nondegenerate Fredholm modules in connection with boundary actions of hyperbolic groups.

Suppose that (H, π, F) is a p-summable Fredholm module, p-summable over a $*$-subalgebra $A^\infty \subset A$, with $F^2 = 1$. If $a^0, \dots, a^n \in A^\infty$, and if $n \geq p$, consider the expression:

$$(10.11) \qquad \tau(a^0, \cdots, a^n) := \frac{1}{2} \cdot \mathrm{Trace}'(F[F, a^0][F, a^1][F, a^2] \cdots [F, a^n]),$$

where $\mathrm{Trace}'(T) := \mathrm{Trace}(\epsilon T)$ if H is $\mathbb{Z}/2$-graded, with grading operator ϵ, and $\mathrm{Trace}(T) = \mathrm{Trace}'(T)$ otherwise. The commutators $[F, a^i]$ for $i = 1, \dots, n$ are all in $\mathcal{L}^p(H)$, and there are at least p of them, so their product is trace-class, and so the expression is well defined provided that $n \geq p$.

Theorem 10.3.3 (Connes' Character Formula) *Let (H, π, F) be a p-summable Fredholm module over a C*-algebra A, p-summable over a subalgebra $A^\infty \subset A$. Then if $n = 2k + 1$ is an odd integer greater than or equal to p, then*

$$(10.12) \qquad \phi(a^0, \dots, a^n) := \frac{1}{2}\mathrm{Trace}(F[F, a^0] \cdots [F, a^n])$$

defines a cyclic cocycle on A^∞, and

$$(10.13) \qquad \phi(u, u^{-1}, \dots, u, u^{-1}) = (-1)^{k+1} 2^{k+1} \cdot \langle [u], [(H, \pi, F)] \rangle$$

for any unitary $u \in A$, with the right hand side the index pairing (10.3) in KK-theory between the class $[u] \in K_1(A)$ (where $u \in A^\infty \subset A$) and the class of the odd Fredholm module (H, π, F) in $\mathrm{KK}_1(A, \mathbb{C})$.

If $n = 2k$ is even and ϵ the grading of the Fredholm module, then

$$(10.14) \qquad \phi(a^0, \dots, a^n) := \frac{1}{2}\mathrm{Trace}(\epsilon F[F, a^0] \cdots [F, a^n])$$

defines a cyclic n-cocycle, and if $e \in A^\infty$ is an idempotent, then

$$\phi(e, \dots, e) = (-1)^k \cdot \langle [e], [(H, \pi, F)] \rangle,$$

where the right hand side is the analytic index pairing in KK-theory of (10.1) between the class in $\mathrm{K}_0(A)$ of the idempotent and the class of the Fredholm module (H, π, F) in $\mathrm{KK}_0(A, \mathbb{C})$.

Definition 10.3.4 The *Chern character* of (H, π, F) is the class in periodic cyclic cohomology of the cyclic cocycles (10.12) (odd case) and (10.14) (even case).

Example 10.3.5 A good example of the character formula applies to the Fredholm module over $C^*(\mathbb{F}_2)$ of Julg and Valette, described in Example 10.1.14. As in Exercise 10.1.17, one computes that

$$\frac{1}{2}\mathrm{Trace}'((F[F, a]) = \tau(a),$$

where $\tau \colon C^*(\mathbb{F}_2) \to \mathbb{C}$ is the standard trace $\tau(\sum_g a_g[g]) = a_e$, and $a \in \mathbb{C}[\mathbb{F}_2] \subset C^*\mathbb{F}_2)$.

Thus, Connnes' character formula shows that

$$\tau(e) = \frac{1}{2}\mathrm{Trace}'((F[F, e]) = \langle[e], [F]\rangle,$$

where $[F] \in \mathrm{KK}_0(C^*(\mathbb{F}_2), \mathbb{C})$ is the class of the Fredholm module of Julg and Valette, and $e \in \mathbb{C}[\mathbb{F}_2]$ is a projection in the group algebra. By Exercise 10.3.6 below, this statement can be improved to allow $a \in \mathcal{A}$, where $\mathcal{A}$ is an isospectral dense subalgebra and thus one having the same K-theory as $C^*(\mathbb{F}_2)$.

As a consequence, the trace τ *takes integer values on projections*. Since τ is faithful, this implies that the famous conjecture of R. Kadison holds for $C^*(\mathbb{F}_2)$: the C*-algebra $C^*(\mathbb{F}_2)$ contains no idempotents other than 0 and 1. (We can think of this result as asserting the connectedness of the "noncommutative space" $\widehat{\mathbb{F}_2}$).

Exercise 10.3.6 Use the results of Exercise 3.5.22 to prove that there is a dense subalgebra $\mathcal{A} \subset C^*(\mathbb{F}_2)$ which is isospectral in $C^*(\mathbb{F}_2)$, has a norm with respect to which it is complete, and in addition satisfies $[\pi(a), F] \in \mathcal{L}^1(H)$ for all $a \in \mathcal{A}$.

Deduce that the group ring $\mathbb{C}[\mathbb{F}_2]$ may be replaced in Example 10.3.5 by $\mathcal{A}$, which has the same K-theory as A.

Example 10.3.7 The simplest one-dimensional example of the Chern character involves Toeplitz operators on the circle, discussed in Section 2.3 of Chapter 1 where $F = 2P_+ - 1$, with P_+ the Szegö projection, an operator on $L^2(\mathbb{T})$. The following exercise shows that the cocycle involved in the Chern character can be simplified as follows:

Exercise 10.3.8 Show that the cyclic 1-cocycle ϕ in the formula (10.12) when $n = 1$ and $F = 2P_+ - 1$ with P_+ the Szegö projection satisfies

$$\varphi(f^0, f^1) = \mathrm{Trace}(f^0[P_+, f^1]),$$

for $f^0, f^1 \in C^\infty(\mathbb{T})$. Deduce (we did this in Chapter 2, see (2.17)) that

$$\varphi(f, g) = \int_{\mathbb{T}} f(\theta)g'(\theta)d\theta,$$

for f, g smooth on $\mathbb{T}$.

The reader will now see that our proof in Chapter 2 of the Toeplitz index theorem was in effect an application of Connes' Chern character formula for the Toeplitz 1-summable Fredholm module over $C(\mathbb{T})$.

We end with an exercise about p-summable odd Fredholm modules which will be used in the following section.

Exercise 10.3.9 Let A be a unital C*-algebra, let $\pi : A \to \mathcal{B}(H)$ be a representation, and let P be a projection in $\mathcal{B}(H)$. Denote by $s(a) := P\pi(a)P$ the corresponding compression: a completely positive linear map.

a) $(H, \pi, F := 2P-1)$ is a Fredholm module for A if and only if $s(|a|^2) - |s(a)|^2 \in \mathcal{K}(H)$ for all $a \in A$.
b) $(H, \pi, F := 2P - 1)$ is p-summable (Definition 10.3.1) over a dense subalgebra $\mathcal{A} \subseteq A$ if $\left(s(|a|^2) - |s(a)|^2\right)^{1/2} \in L^p(H)$ for all $a \in \mathcal{A}$.

(*Hint.* Let $\Pi(a) = (1 - P)\pi(a)P$; this is the lower left corner of the 2-by-2 matrix defined by the decomposition of π with respect to P

$$\pi(a) = \begin{bmatrix} P\pi(a)P & P\pi(a)(1 - P) \\ (1 - P)\pi(a)P & (1 - P)\pi(a)P) \end{bmatrix}.$$

Using the relations

$$[\pi(a), P] = \Pi(a) - \Pi(a^*)^*, \qquad \Pi(a) = (1 - P)[\pi(a), P],$$

we see that (π, P) is a Fredholm module if and only if $\Pi(a) \in \mathcal{K}(H)$ for all $a \in A$ and that (π, P) is p-summable over $\mathcal{A}$ if and only if $\Pi(a) \in L^p(H)$ for all $a \in \mathcal{A}$. Now

$$\Pi(a)^*\Pi(a) = P\pi(a^*)(1 - P)\pi(a)P$$

$$= P\pi(a^*a)P - \left(P\pi(a)^*P\right)\left(P\pi(a)P\right) = s(a^*a) - s(a)^*s(a)$$

shows that $|\Pi(a)| = \sqrt{s(|a|^2) - |s(a)|^2}$.

10.4   Fredholm Modules from Boundary Actions of Hyperbolic Groups

The examples of (finitely summable) Fredholm modules we have seen so far come from bounded transforms of Dirac operators and other pseudodifferential elliptic operators on compact manifolds. The module of Julg and Valette (Example 10.1.14) is a 1-summable Fredholm module over the C*-algebra $C^*(\mathbb{F}_2)$. The free group is a special case of a *Gromov hyperbolic group*, which we define below. A hyperbolic group is a finitely generated group G with a kind of large-scale negative curvature property. To such a group, or more generally, to a Gromov hyperbolic space, one can associate an asymptotic boundary, which can be used to compactify the space or group, and in the case of a group the boundary ∂G, which is a compact metrizable space, carries an action of G. The example of the crossed product $C(\partial \mathbb{F}_2) \rtimes \mathbb{F}_2$ of the free group $\mathbb{F}_2$ acting on its boundary $\partial \mathbb{F}_2$ has already been discussed in Example 1.8.16. Any discrete, co-compact group of isometries of the hyperbolic plane is Gromov hyperbolic, and the boundary of such a group is the boundary of the disk, the circle. The crossed products $C(\partial G) \rtimes G$ for such G are Morita equivalent to foliation C*-algebras (discussed in Section 6.7) encoding asymptotics of geodesic flow.

In general, the crossed products $C(\partial G) \rtimes G$ of hyperbolic groups acting on their boundaries are of significant interest in dynamics. If G is torsion-free, then every $g \in G$ acts by a homeomorphism with exactly two fixed points, and under the g-action, points of ∂G are attracted to the one fixed point and repelled by the other. If G is not cyclic, the action is always minimal. A result of [1] implies that the action is also amenable and that the crossed products are nuclear C*-algebras.

For current purposes what is interesting about these examples is that they have a very rich Fredholm module theory. For compact manifolds, every K-homology class (equivalence class of Fredholm module) is represented by a zero order elliptic pseudodifferential operator (indeed, in the spin^c-case, by a Dirac-type operator), see Theorem 10.3.2 and remarks following it, and in particular by a p-summable Fredholm module for $p > \dim M$. As we discuss below, the noncommutative spaces $C(\partial G) \rtimes G$ have an analogous property. This depends on the existence of one particular Fredholm module over $C(\partial G) \rtimes G$ which is constructed from a certain metric-measure structure on the boundary.

We now explain some of this in detail.

Let G be a finitely generated group with generating set S, which we assume is closed under inverses. The *word length* $|g|$ of $g \in G$ is the minimal n such that $g = s_1 \cdots s_n$ for $s_i \in S$. The *distance* between two group elements is $d(g_1, g_2) := |g_1^{-1} g_2|$. This gives G the structure of a metric space.

The *Gromov product* is defined as

$$\langle g_1, g_2 \rangle := \frac{1}{2} \left(|g_1| + |g_2| - d(g_1, g_2) \right).$$

Definition 10.4.1 G is δ-*hyperbolic* for $\delta \geq 0$ if

$$(10.15) \qquad \langle g_1, g_2 \rangle \geq \inf\{\langle g_1, h \rangle, \langle g_2, h \rangle\} - \delta.$$

See [93] for an excellent exposition of the theory of hyperbolic groups.

The basic example is the free group $\mathbb{F}_2$ on two generators, say a, b, whose Cayley graph we drew in the previous section. The Cayley graph gives a way of visualizing the group. Let X be its realization. It has a natural graph metric. It is also connected in the strong sense that if $x, y \in X$, then there is a geodesic between them: an

isometric map $r: [0, l] \to X$ such that $r(0) = x$, $r(l) = y$. Here $l = d(x, y)$ in
the graph metric. If $g \in \mathbb{F}_2$, considered as a vertex of X, the word length $|g|$ agrees
with the distance from g to the basepoint $e \in X$.

Notice also that the group $\mathbb{F}_2$ acts naturally by translation on the vertices of the
tree and isometrically on X (with quotient a wedge of two circles).

Suppose that $g_1, g_2 \in \mathbb{F}_2$, thought of as points of the graph X. The rays in X from
e to g_1 and g_2, respectively, agree on some initial segment and then diverge: Observe
then that $\langle g_1, g_2 \rangle$ is the length of the initial segment, as in the picture below. Group
theoretically, the initial segment corresponds to reduced expressions $g_1 = pu_1$ and
$g_2 = pu_2$, for some u_1, u_2, and $\langle g_1, g_2 \rangle = |p|$.

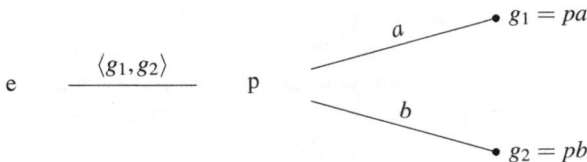

From this discussion, the reader should be able to easily verify that $\mathbb{F}_2$ is 0-
hyperbolic.

Exercise 10.4.2 Prove that $\mathbb{F}_2$ with the given generating set is 0-hyperbolic.

Remark 10.4.3 The same kind of "tripod" interpretation of the Gromov product
exists for general hyperbolic groups, up to an error controlled by δ.

Every hyperbolic group has a boundary, a compact, metrizable space giving a
kind of (compact) "geometry at infinity" invariant under the group. The dynamics of
this interesting action, and the structure of the crossed product C*-algebra $C(\partial G) \rtimes$
G, encode large-scale geometric information about the group.

In the case of the free group, the *boundary* of the realization of the Cayley graph
X may be defined to be the collection of "rays": isometric maps $r: [0, \infty) \to X$
such that $r(0) = e$, where $e \in X$ is the basepoint, corresponding to the identity of
the group. Reducing such a ray to the vertices gives a bijection with the collection
of infinite reduced words $s_1 s_2 s_3 \cdots$ in the generators $S = \{a, a^{-1}, b, b^{-1}\}$ and has
a natural Cantor set topology. We denote the boundary by $\partial \mathbb{F}_2$.

The boundary can be glued to X, or to $\mathbb{F}_2$ itself, to compactify either of them. Let
$\overline{X} = X \cup \partial \mathbb{F}_2$. Suppose that $\xi = s_1 s_2 s_3 \cdots \in \partial \mathbb{F}_2$, corresponding to a ray r_ξ with
$r(n) = s_1 \cdots s_n$. If $x \in X$, let r_x be the geodesic in X from e to x. Now if N is a
large positive integer, in X let $U_N(\xi)$ be the collection of points $x \in X$, or boundary
points $\eta \in \partial \mathbb{F}_2$, such that $r_x(t) = r_\xi(t)$ for $t \geq N$, respectively, $r_\eta(t) = r_\xi(t)$ for
$t \geq N$. Then $\{U_N \mid N = 1, 2, 3, \ldots\}$ gives a system of neighborhoods of ξ in $\overline{X}$. We
may similarly compactify the discrete space $\mathbb{F}_2$ itself by this method: Set $U_N(\xi)$ to
be all boundary points or group elements which are represented as finite or infinite
words $t_1 t_2 \cdots$, such that $t_i = s_i$ for $i = 1, 2, \ldots, n$.

Then we obtain a system of neighborhoods of $\xi \in \overline{\mathbb{F}}_2 := \mathbb{F}_2 \cup \partial \mathbb{F}_2$.

Exercise 10.4.4 Let (x_n) and (y_n) be sequences in $\mathbb{F}_2 \subset X$ which remain a bounded distance apart. Show that if $x_n \to \xi$ for some boundary point ξ, then $y_n \to \xi$ as well.

We may alternatively define the boundary using *all* geodesic rays, modulo a suitable equivalence relation, instead of just rays emanating from the basepoint e. Indeed, if $r : [0, \infty) \to X$ is a geodesic ray starting at any point (vertex, say) $p \in X$, then we can construct a ray starting at the basepoint $e \in X$ by concatenating the unique path $[e, p]$ from p to x with the ray and eliminating cancellation (which results if the original ray passing through e). The new ray remains at a bounded distance from the original one.

Since if $g \in \mathbb{F}_2$, acting isometrically by translation on X and r is any ray, then $g(r)$ is a ray, and we obtain the following.

Lemma 10.4.5 *The $\mathbb{F}_2$ action by translation on X extends to an action by homeomorphisms of $\overline{\mathbb{F}}_2$, leaving the boundary $\partial \mathbb{F}_2$ invariant.*

Note that the action of $\mathbb{F}_2$ is simple to describe in terms of "infinite words." If $\xi = s_1 s_2 \cdots$, $g \in \mathbb{F}_2$, write g as a reduced word, form the concatenation $g s_1 s_2 \cdots$, and make any necessary cancellations to get an infinite reduced word.

Exercise 10.4.6 Let $g \in \mathbb{F}_2$. Show that the sequence $(g^n)_{n \in \mathbb{Z}}$ converges as $\to \pm \infty$ to a pair ξ_g^+ and ξ_g^- of distinct points in $\partial \mathbb{F}_2$ which are fixed by the g-action on $\partial \mathbb{F}_2$.

Next, we observe that the boundary $\partial \mathbb{F}_2$ comes equipped with a natural family of metrics, and this is true for general hyperbolic groups as well. To define these metrics geometrically, we extend the Gromov product $\langle \cdot, \cdot \rangle$ to the boundary, as in the picture below, where ξ_1, ξ_2 are rays starting at $e \in X$ and diverge at $p \in \mathbb{F}_2$.

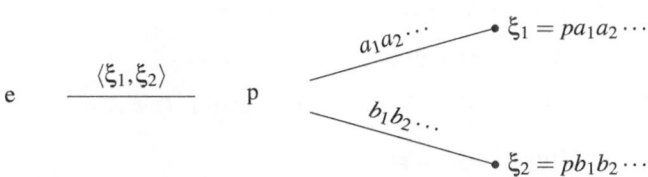

In the notation of the picture, $\langle \xi_1, \xi_2 \rangle = |p|$.

Exercise 10.4.7 For any $\epsilon > 0$, the formula

$$d(\xi_1, \xi_2) := e^{-\epsilon \langle \xi_1, \xi_2 \rangle}$$

gives a metric on $\partial \mathbb{F}_2$ generating the topology on $\partial \mathbb{F}_2$. The $\mathbb{F}_2$ action on $\partial \mathbb{F}_2$ is by bi-Lipschitz homeomorphisms with respect to this metric.

The theory of boundaries of general hyperbolic groups proceeds along similar lines.

Definition 10.4.8 Let G be a Gromov hyperbolic group, with respect to a generating set S.

a) A sequence $(x_i) \subset G$ *converges to infinity* if $\langle x_i, x_j \rangle \to \infty$ as $i, j \to \infty$.
b) Two sequences (x_i), (y_i) converging to infinity are *asymptotic* if $\langle x_i, y_i \rangle \to \infty$ as $i \to \infty$. The asymptotic relation is an equivalence on sequences converging to infinity.
c) The *boundary* of X, denoted ∂G, is the set of asymptotic classes of sequences converging to infinity. A sequence $(x_i) \subseteq G$ *converges to* $\xi \in \partial G$ if (x_i) converges to infinity, and the asymptotic class of (x_i) is ξ.
d) The Gromov product on $\partial G \times \partial G$ is defined as follows:

$$\langle \xi, \xi' \rangle := \inf \left\{ \liminf \langle x_i, x_i' \rangle : \ x_i \to \xi, \ x_i' \to \xi' \right\}.$$

Lemma 10.4.9 *In the above notation, if $\xi = \xi'$, then $\langle \xi, \xi' \rangle = \infty$. If $\xi \neq \xi'$, then the sequence $\langle x_i, x_i' \rangle$ is bounded whenever $x_i \to \xi$ and $x_i' \to \xi'$, hence $\langle \xi, \xi' \rangle < \infty$. Moreover,*

(10.16)
$$\langle \xi, \xi' \rangle \leq \liminf \langle x_i, x_i' \rangle \leq \limsup \langle x_i, x_i' \rangle \leq \langle \xi, \xi' \rangle + 2\delta \qquad (x_i \to \xi, x_i' \to \xi').$$

Definition 10.4.10 A *visual metric* on ∂G is a metric d_ϵ satisfying $d_\epsilon \asymp \exp(-\epsilon \langle \cdot, \cdot \rangle)$ for some $\epsilon > 0$, called the *visual parameter* of d_ϵ.

We have noted that $\mathbb{F}_2$ has visual metrics of any visual parameter $\epsilon > 0$. In general, the parameter is bounded above by $\frac{1}{5\delta}$.

Lemma 10.4.11 *Let $\epsilon > 0$ be such that $\epsilon\delta < 1/5$. Then there exists a visual metric d_ϵ on ∂X, having visual parameter ϵ.*

See [157, Prop.5.16] for the proof.

Lemma 10.4.12 *Let d_ϵ and $d_{\epsilon'}$ be two visual metrics. Then,*

a) *d_ϵ and $d_{\epsilon'}$ are Hölder equivalent: $d_\epsilon^{1/\epsilon} \asymp d_{\epsilon'}^{1/\epsilon'}$.*
b) *The metric space $(\partial G, d_\epsilon)$ has finite Hausdorff dimension. For ϵ, ϵ' in the allowed visual range, the relation $\epsilon \cdot \mathrm{hdim}(\partial G, d_\epsilon) = \epsilon' \, \mathrm{hdim}(\partial G, d_{\epsilon'})$ holds between the corresponding Hausdorff dimensions.*
c) *The corresponding Hausdorff measures are comparable: $\mu_\epsilon \asymp \mu_{\epsilon'}$.*

Putting the above results together, we conclude that any visual metric on the boundary of a hyperbolic group G determines a compact metrizable topology on ∂G and these topologies are all equivalent because any two visual metrics are Hölder equivalent. The group G acts by Lipschitz maps with respect to any visual metric and hence by homeomorphisms of ∂G. Therefore the crossed product $C(\partial G) \rtimes G$ is defined.

The source of the Fredholm module we are going to construct is a remarkable convergence property of Hausdorff measure with respect to a visual metric. We first describe this in the case of $G = \mathbb{F}_2$.

Let μ be Hausdorff measure with respect to the metric on $\mathbb{F}_2$ with visual parameter $\epsilon = \log 3$. Then μ is equidistributed around the boundary, as one looks out at the boundary from the basepoint e. If $w \in \mathbb{F}_2$, let U_w be all boundary points, considered as infinite reduced words $s_1 s_2 \cdots$, which start with the reduced form of w. Equivalently, in terms of the Gromov product, if $|w| = k$, then

$$U_w = \{\xi \in \partial \mathbb{F}_2 \mid \langle \xi, w \rangle \geq k\},$$

and geometrically, U_w consists of endpoints of geodesic rays r starting at e which pass through the vertex w.

Exercise 10.4.13 In the notation above, if $\eta \in U_w$, show that U_w is the metric ball $B_{3^{-|w|}}(\eta)$, with respect to the metric d on $\partial \mathbb{F}_2$ defined above.

As w ranges over $S_k := \{g \in \mathbb{F}_2 \mid |g| = k\}$, we obtain a partition

$$\partial \mathbb{F}_2 = \sqcup_{|w|=k} U_w.$$

As $|S_k| = 4 \cdot 3^{k-1}$ and the measure of all the U_w's will be the same,

$$\mu(U_w) = \frac{3}{4} \cdot 3^{-k}.$$

We may think of μ in terms of probability: If one starts at $e \in \mathbb{F}_2$ and walks (at unit speed) in a random way along edges X without every going back along the same edge just traversed, then the walk defines a unique geodesic ray. In particular, any such infinite walk will terminate at a boundary point $\xi \in \mathbb{F}_2$.

A path starting at e then involves a choice of four possible initial edges to step along, then three possible edges, then three again, and so on. The probability a walk will pass through a given vertex w is $\frac{3}{4} \cdot 3^{-k}$, and this is the same as the probability the walk will eventually terminate in U_w.

The measure μ depends on choice of the origin as basepoint. If $x \in \mathbb{F}_2$. let μ_x be the measure defined in the same way, but with basepoint $x \in \mathbb{F}_2$, and $\mu_x(U)$ for $U \subset \partial \mathbb{F}_2$ is the probability that a random geodesic ray starting at x ends in a point of U. Note that if x is a group element (i.e. a vertex of the graph), then

$$\mu_x = x_*(\mu),$$

where

$$\int_{\partial \mathbb{F}_2} f(\xi) d\mu_x := \int_{\partial \mathbb{F}_2} f(x\xi) \, d\mu(\xi)$$

is the pushed-forward measure. The measures μ_x are all equivalent, and in particular μ is quasi-invariant under $\mathbb{F}_2$, so

$$\int_{\partial \mathbb{F}_2} f(x\xi)\, d\mu(\xi) = \int_{\partial \mathbb{F}_2} f(\xi)\, \frac{d(x_*\mu)}{d\mu}(\xi)\, d\mu(\xi).$$

We compute below the probability distribution

$$\sigma(x,\xi) = \frac{d(x_*\mu)}{d\mu}(\xi)$$

and show that as $x \to \xi$ approaches a boundary point $\xi \in \mathbb{F}_2$, the measures μ_x accumulate at a delta distribution at ξ.

Let $x \in \mathbb{F}_2$. Let $|x| = n$. Partition the boundary into the sets

$$C_k(x) = \{\xi \in \partial \mathbb{F}_2 \mid \langle \xi, x \rangle = k\}, \qquad k = 0, 2, \ldots, n-1,$$

and $C_n(x) := \{\xi \in \partial \mathbb{F}_2 \mid \langle x, \xi \rangle \geq n\}$.

Lemma 10.4.14 *If $x \in \mathbb{F}_2 \subset X$, then*

$$\sigma(x,\xi) = 3^{-|x|+2\langle \xi, x \rangle}.$$

In particular, $\sigma(x,\xi) = 3^{-n+2k}$ for $\xi \in C_k(x)$).

Proof We work in terms of reduced words. So if $x = x_1 \cdots x_n$ is reduced, then $C_n(x)$ is all $\xi = \xi_1 \xi_2 \cdots$ beginning in x, which has measure $\frac{3}{4} \cdot 3^{-n}$. On the other hand,

$$x^{-1}\left(C_n(x)\right) = \partial \mathbb{F}_2 \setminus C_1(x_n),$$

as one easily checks and hence has measure $1 - \frac{1}{4} = \frac{3}{4}$. From this we see that

$$\frac{\mu\left(x^{-1}(C_n(x))\right)}{\mu\left(C_n(x)\right)} = 3^n,$$

and the same argument holds for smaller subsets. Hence

$$\sigma(x,\xi) = 3^n = 3^{-|x|+2\langle x, \xi \rangle}, \qquad \forall \xi \in C_k(x).$$

On the other hand,

$$x^{-1}\left(C_0(x)\right) = C_N(X^{-1},$$

as one easily checks. So an inversion of the previous argument shows that x^{-1} is uniformly expanding by a factor of 3^n. That is,

$$\sigma(x, \xi) = 3^n, \quad \forall \xi \in C_0(x).$$

The intermediate cases are left to the reader to check.   □

We have shown that if $|x| = n$, then $\sigma(x)$ is a discrete Gaussian probability distribution on the boundary taking the values 3^{-n+2k} for $k = 0, 1, \ldots, n$.

We state the general result.

Corollary 10.4.15 *Let G be a Gromov hyperbolic group and μ Hausdorff measure with respect to a visual metric. Then the orbit $G\mu$ of μ in $\mathrm{Prob}(\partial G)$ accumulates only at point masses on ∂G. Moreover, as $g \to \xi$ in $\overline{G}$, $g_*\mu \to \delta_\xi$ in the space of probability measures.*

Proof We verify this for the free group only, where the proof is quite easy.

Choose a test function χ_{U_w} for some w, with $|w| = n$. It suffices to show that

$$\lim_{x \to \xi} \mu_x(U_w) = 0 \quad \text{if } \xi \notin U_w.$$

So choose $\xi \notin U_w = C_n(w)$. Then $\langle \xi, w \rangle = k$ for some $0 \le k \le n - 1$. Since $x \to \xi$, for large enough x, $\langle x, w \rangle = k$ as well, and then $\langle x, \eta \rangle = k$ for all $\eta \in U_w$: Hence $\sigma(x, \eta) = 3^{-|x|+2\langle x, \eta \rangle} = 3^{-|x|+2k}$. This shows that

$$\mu_x(U_w) = \int_{U_w} \sigma(x, \eta) d\mu(\eta) \to 0$$

as $x \to \infty$, as claimed.

□

Fix a hyperbolic group G with Hausdorff measure μ with respect to a visual metric on the boundary.

We now show that the convergence property of the corollary gives rise to an odd Fredholm module over the crossed product $C(\partial G_2) \rtimes G$. For the representation, we use the regular representation of the crossed product $C(\partial G) \rtimes G$ on $l^2(G, L^2(\partial G, \mu))$, defined by the covariant pair

(10.17)
$$\lambda_\mu(\phi)\Big(\sum \psi_h \delta_h\Big) = \sum (h^{-1}.\phi)\psi_h \delta_h, \qquad \lambda_\mu(g)\Big(\sum \psi_h \delta_h\Big) = \sum \psi_h \delta_{gh},$$

where $\phi \in C(\partial G)$, $g \in G$, and $\sum \psi_h \delta_h \in \ell^2(G, L^2(\partial G, \mu))$.

We introduce some statistical concepts related to the construction.

Definition 10.4.16 The *G-expectation* and the *G-deviation* of $\phi \in C(\partial G)$ with respect to μ are the functions $E\phi : G \to \mathbb{C}$ and $\sigma\phi : G \to [0, \infty)$ given as follows:

$$E\phi(g) = \int_{\partial G} \phi \circ g \, d\mu = \int_{\partial G} \phi \, dg_*\mu, \qquad \sigma\phi = \sqrt{E(|\phi|^2) - |E\phi|^2}.$$

Due to Corollary 10.4.15, the G-deviation $\sigma\phi$ of any $\phi \in C(\partial G)$ is a C_0-function on G. In fact one can make a stronger statement. We compute the deviation in the free group example for characteristic functions and give the general statement without proof, see [81] for the general result and proof.

Lemma 10.4.17 Let $w \in \mathbb{F}_2$ and U_w be all boundary points $\xi = \xi_1\xi_2\cdots$ which start with w. Then

$$\sigma\chi_{U_w} \in l^p(\mathbb{F}_2)$$

for all $p > 2$, where σ is the $\mathbb{F}_2$-deviation.
   More generally, if G is a Gromov hyperbolic group, μ is a visual measure from a visual metric d_ϵ, and $\phi \in C(\partial G)$ is Lipschitz then $\sigma\phi \in l^p(G)$ for $p > \max\{2, \mathrm{hdim}(\partial G, d_\epsilon)\}$.

Proof Choose any $x \in \mathbb{F}_2 \subset X$. We compute $\mu_x(U_w)$. Suppose first that $\langle x, w \rangle < |w|$, that is, suppose x does not belong to the sector of the graph determined by w. A random path starting at x will then enter U_w if and only if it passes through w. There are $\frac{3}{4}3^{-d(x,w)}$ paths of length $d(x, w)$ emanating from x, and exactly one of them passes into U_w. It follows that

$$\mu_x(U_w) = \frac{4}{3}3^{-d(x,p)}.$$

Therefore

$$\sigma\chi_{U_w}(x) = \sqrt{\mu_x(U_w) - \mu_x(U_w)^2}$$

$$= \sqrt{\frac{4}{3}3^{-d(x,w)} - (\frac{4}{3})^2 3^{-2d(x,w)}} \leq \mathrm{const} \cdot 3^{-\frac{1}{2}d(x,p)}.$$

On the other hand suppose $\langle x, w \rangle \geq |w|$, so that x lies in the sector determined by w. In this case, a random path starting at x will *leave* the sector if and only if it passes through w. So by a similar calculation, we find that $\sigma\chi_{U_w}(x) \leq \mathrm{const}.3^{-\frac{1}{2}d(x,w)}$ in this case as well. Hence

$$\sum_{x \in \mathbb{F}_2} \sigma \chi_{U_w}(x)^p \le \text{const.} \sum_{x \in \mathbb{F}_2} 3^{-\frac{p}{2}d(x,p)} . \text{const.} \sum_{n=0}^{\infty} \sum_{|x|=n} 3^{-\frac{p}{2}d(x,p)} .$$

$$\le \text{const.} \sum_{n=0}^{\infty} \sum_{|x|=n} 3^{-\frac{np}{2}} \le \text{const.} \sum_{n=0}^{\infty} 3^{-\frac{np}{2}+n},$$

and the series converges if $p > 2$.

$\square$

Proceeding with construction of a Fredholm module, we identify $\ell^2 G$ with the constant-coefficient subspace of $\ell^2(G, L^2(\partial G, \mu))$. Let $P_{\ell^2 G}$ be orthogonal projection to this subspace. Then $P_{\ell^2 G}$ is given by coefficient-wise integration:

$$P_{\ell^2 G}\Big(\sum \psi_h \delta_h\Big) = \sum \Big(\int \psi_h \, d\mu\Big)\delta_h.$$

We form the triple

(10.18) $$\Big(l^2\big(G, L^2(\partial G, \mu)\big), \lambda_\mu, F := 2P_{l^2 G} - 1\Big),$$

where λ_μ is the representation (10.17). The condition we need to get a Fredholm module is that the commutators $[F, \lambda_\mu(a)] = [P_{l^2 G}, \pi_\mu(a)]$ are compact for $a \in C(\partial G) \rtimes G$. Since $P_{l^2 G}$ clearly commutes with $\lambda_\mu(G)$, it suffices to analyze the commutators with functions $\phi \in C(\partial G)$, and even Lipschitz functions, since they are dense. But the projection $P_{\ell^2 \mathbb{F}_2}$ compresses the space restriction $\lambda_\mu|_{C(\partial G)}$ to multiplication by the G-expectation on $\ell^2 G$:

$$P_{\ell^2 G}\lambda_\mu(\phi)P_{\ell^2 G} = \mathrm{M}(E\phi)$$

for all $\phi \in C(\partial G)$. Referring to Exercise 10.3.9, let $s_\mu(\phi) := P_{\ell^2 G}\lambda_\mu(\phi)P_{\ell^2 G}$. Then

$$\sqrt{s_\mu(|\phi|^2) - |s_\mu(\phi)|^2} = \mathrm{M}(\sigma\phi).$$

The result now follows from Exercise 10.3.9 and Lemma 10.4.17.

Theorem 10.4.18 *Let G be a Gromov hyperbolic group and μ a visual measure from a visual metric d_ϵ. Then the triple of (10.18) defines a p-summable odd Fredholm module over the crossed product $C(\partial G) \rtimes G$ for any $p > \max\{2, \mathrm{hdim}(\partial G, d_\epsilon)\}$.*

The significance for K-theory of these Fredholm modules will be discussed in the next section.

We conclude with some exercises illustrating further structure in these "Type III" examples of noncommutative spaces.

Exercise 10.4.19 Let G be a Gromov hyperbolic group and μ a visual probability measure on ∂G such that $\frac{dg_*\mu}{d\mu}$ is continuous for all $g \in G$ (such measures exist), where $g_*\mu$ is the pushed-forward measure $(g_*\mu)(A) := \mu(g^{-1}A)$ or, equivalently, satisfying the change of variables

$$\int_{\partial G} f \circ g \, d\mu = \int_{\partial G} f \, dg_*\mu$$

for $f \in C(\partial G)$.

a) The formula $\sigma_t(\sum f_g[g]) := \sum f_g \lambda_g^{it} \, [g]$ defines a 1-parameter group of automorphisms of $C(\partial G) \rtimes G$, a time evolution.

b) Let $\widehat{H} := \oplus_{g \in G} L^2(\partial G, g_*\mu)$. For $\phi \in C(\partial G)$, let $\pi(\phi)(v \otimes e_h) := \phi v \otimes e_h$, and for $g \in G$ let $\pi(g)(v \otimes e_h) := v \circ g^{-1} \otimes e_{gh}$. Show that $\pi(g)$ is unitary and that these formulas specify a covariant pair and induced representation $\pi : C(\partial G) \rtimes G \to \mathbb{B}(\widehat{H})$.

c) Let $U : l^2(G) \otimes L^2(\partial G, \mu) \to \widehat{H}$ be the unitary $U(v \otimes e_h) := v \circ h^{-1} \otimes e_h$ (prove it is unitary). Show that U conjugates the representation λ_μ of the Fredholm module (10.18), to the representation π, and conjugates P_{l^2G} to the projection P on $\widehat{H}$ which is the direct sum of the projection operators to the constant functions in $L^2(\partial G, g_*\mu)$ for every $g \in G$.

d) (See also Exercise 1.8.29). Show that the time evolution $(\sigma_t)_{t \in \mathbb{R}}$ analytically extends to $\mathbb{C}$ in the sense of Exercise 1.8.29 and that the KMS condition $\tau(\sigma_{-i}(a)b) = \tau(ba)$ for $a, b \in C(\partial G)[G]$ holds for $\tau : C(\partial G) \rtimes G \to \mathbb{C}$ the state $\tau(\sum f_g[g]) = \int f_e d\mu$.

e) Let δ be the densely defined derivation of $C(\partial G) \rtimes G$ defined by differentiating the time evolution $(\sigma_t)_{t \in \mathbb{R}}$ (see Section 3.5). Show that

$$\delta \left(\sum f_g[g] \right) = i \sum f_g \log \lambda_g \, [g].$$

d) Let B be the densely defined operator $B := \oplus_{g \in G} \log \lambda_g$ on $\widehat{H}$. Show that $(e^{itB})_{t \in \mathbb{R}}$ is a 1-parameter group of unitaries inducing the time evolution $(\sigma_t)_{t \in \mathbb{R}}$ of a) in the sense that $\pi(\sigma_t(a)) = e^{itB}\pi(a)e^{-itB}$ for all $a \in C(\partial G) \rtimes G$. The operator B is a direct sum of *Busemann functions*, which are important in hyperbolic geometry.

f) Show that $(\sigma_t)_{t \in \mathbb{R}}$ analytically extends to a family $(\sigma_z)_{z \in \mathbb{C}}$ of algebra homomorphisms $C(\partial G)[G]$ and that $\tau(ab) = \tau(b\sigma_{-i}(a)$, where τ is the state $\tau(\sum f_g[g]) := \int_{\partial G} f_e d\mu$ (τ satisfies a KMS$_\beta$ condition at inverse temperature $\beta = 1$).

10.5 Fredholm Modules from Extensions

There is a general method associating an odd Fredholm module over A with a (completely positively split) C*-algebra extension of A by $\mathcal{K}$. The method is called the *Stinespring construction*.

Definition 10.5.1 A linear, unital map $s: A \to B$ between two unital C*-algebras is *completely positive* if $s \otimes \mathrm{id}: A \otimes M_n(\mathbb{C}) \to B \otimes M_n(\mathbb{C})$ is a positive linear map for all n.

Exercise 10.5.2 Let $s: A \to B$ be a positive unital, linear map, where A is commutative. Show that s is completely positive.

The following result is discussed in detail in the book of [98]

Theorem 10.5.3 *Assume A is a separable, nuclear C*-algebra. Then (1) if $0 \to \mathcal{K} \to B \xrightarrow{\pi} A \to 0$ is an extension of A by the compact operators, then there exists a completely positive map $s: A \to B$ such that $\pi \circ s = \mathrm{id}_A$.*

(2) If $\tau: A \to Q(H)$ is a *-homomorphism, then there exists a c.p. map $s: A \to \mathbb{B}(H)$ such that $\pi \circ s = \mathrm{id}_A$, where $\pi: \mathbb{B}(H) \to Q(H)$ is the quotient map.

The two statements are equivalent, as the reader can check without much difficulty.

The reason that c.p. split extensions are important is that combining a c.p. split extension of A by the compact operators results in a Fredholm module. The procedure is called the Stinespring construction.

Lemma 10.5.4 *Let H be a Hilbert space and $s: A \to \mathbb{B}(H)$ a completely positive map. Define a sesquilinear form $\langle \cdot, \cdot \rangle_A$ on the algebraic tensor product $A \otimes H$ by*

$$(10.19) \qquad \langle \sum_i a_i \otimes \xi_i, \sum_j b_j \otimes \eta_j \rangle_A := \sum_{i,j} \langle \xi_i, s(a_i^* b_j) \eta_j \rangle.$$

Then,

a) *The completion of $A \otimes H$ with respect to (10.19) is a Hilbert space $\widehat{H}$.*

b) *The left multiplication action of A on $A \otimes H$ determines a representation $\pi: A \to \mathbb{B}(\widehat{H})$.*

c) *The map $P(\sum_i a_i \otimes \xi_i) := \sum_i 1 \otimes s(a_i)\xi_i$ extends continuously to a self-adjoint projection on $\widehat{H}$ with range unitarily isomorphic to H.*

d) *$s(a) = P\pi(a)P$, up to the unitary identification of part c).*

e) *The commutators $[\pi(a), P]$ are compact (respectively, in the Schatten class $\mathcal{L}^p$) for dense $a \in A$ iff the elements $\left(s(|a|^2) - |s(a)|^2\right)^{1/2}$ are in $\mathcal{K}(H)$ (respectively, in $\mathcal{L}^p(H)$) for dense $a \in A$.*

Proof a)–d) are routine. For e) proceed as in Exercise 10.3.9, write $[\pi(a), P] = \Pi(a) - \Pi(a^*)^*$, where $\Pi(a) = (1-)\pi(a)P$, then $\Pi(a)^*\Pi(a) = s(a^*a) - s(a)^*s(a)$, and the result follows.

$\square$

Definition 10.5.5 A *B.D.F.-cycle* for a unital C*-algebra A is a nondegenerate *-homomorphism $\tau : A \to Q(H)$, where $Q(H) = \mathbb{B}(H)/\mathcal{K}(H)$ is the Calkin algebra of a Hilbert space H, together with a completely positive map $s : A \to \mathbb{B}(H)$ such that

$$\tau(a) = \pi \circ s,$$

where $\pi : \mathbb{B}(H) \to Q(H)$ is the quotient map.

The Stinespring construction shows the following:

Proposition 10.5.6 *If $\tau : A \to Q(H)$, with splitting s, is a B.D.F. cycle for A, then the triple $(\widehat{H}, \pi, F := 2P - 1)$ is an odd Fredholm module over A. It is p-summable if and only if there is a dense *-subalgebra $\mathcal{A} \subset A$ such that $\sqrt{s(|a|^2) - |s(a)|^2} \in L^p(H)$ for all $a \in \mathcal{A}$.*

The last part follows as in Exercise 10.3.9.

Remark 10.5.7 Actually the class in $KK_1(A, \mathbb{C})$ of the Fredholm module of a B.D.F. cycle does not depend on the splitting s, but only on its existence. This is because of the homotopy-invariance properties of KK and the convexity of the space of c.p. splittings. Furthermore, if A is *nuclear*, then every $\tau : A \to Q$ is c.p. split. See [98] for an extension discussion of nuclearity and completely positive maps.

Example 10.5.8 B.D.F. cycles arise from completely positively split extensions by the compact operators. Suppose

$$0 \to \mathcal{K}(H) \overset{i}{\to} B \overset{p}{\to} A \to 0$$

is an extension of A by the compact operators on H. Since $\mathcal{M}(\mathcal{K}(H)) \cong \mathbb{B}(H)$, the inclusion i extends to a *-homomorphism $\tilde{i} : B \to \mathbb{B}(H)$.

Let $a \in A$, lift a to $b \in B$ under p, and set

$$\tau(a) := \pi\left(\tilde{i}(b)\right) \in Q(H).$$

Exercise 10.5.9 τ is a *-homomorphism.

The map τ is called the *Busby invariant* of the extension.

Now suppose that $s' : A \to B$ is a completely positive map such that $p \circ s' = \mathrm{id}_A$. Let $s = \tilde{i} \circ s' : A \to \mathbb{B}(H)$. Then the map τ and the splitting s comprise a B.D.F cycle for A.

The following proposition shows that the boundary map associated with a c.p. split extension of A by the compacts is an instance of a Kasparov product.

Proposition 10.5.10 *Let* $0 \to \mathcal{K} \to B \to A \to 0$ *be a c.p. split extension of A by the compact operators. Then the map* $K_1(A) \to \mathbb{Z}$ *induced by pairing with the odd Fredholm module in Proposition 10.5.6 agrees with the boundary map*

$$\delta \colon K_1(A) \to K_0(\mathcal{K}) \cong \mathbb{Z}$$

associated with the extension.
  That is,

$$\langle [\tau], a \rangle = \delta(a),$$

where $[\tau] \in KK_1(A, \mathbb{C})$ *is the class of the extension (of the associated B.D.F. cycle), and* $\delta \colon K_1(A) \to K_0(\mathcal{K}) \cong \mathbb{Z}$ *is the connecting homomorphism of the extension.*

Exercise 10.5.11 Prove Theorem 10.5.14 by comparing the description of the boundary operator δ in Theorem 8.5.11 of Chapter 6, with the description of the pairing between K-theory and Fredholm modules given in pairing (10.3).

We now illustrate these ideas by revisiting boundary actions of hyperbolic groups. Let G be a Gromov hyperbolic group, as in the last section, with boundary ∂G. The boundary ∂G can be glued to G to make a compactification $\overline{G} = G \cup \partial G$ of G: a compact Hausdorff space containing the (discrete) subspace G as an open subset, with complement ∂G.
  We obtain a C*-algebra extension

(10.20) $$0 \to C_0(G) \to C(\overline{G}) \to C(\partial G) \to 0.$$

Since all of the maps in this sequence are G-equivariant, it generates an extension of C*-algebras

$$0 \to C_0(G) \rtimes G \to C(\overline{G}) \rtimes G \to C(\partial G) \rtimes G \to 0.$$

Using the left regular representation of G on $l^2(G)$ and the usual representation of $C_0(G)$ on $l^2(G)$ by multiplication operators, we obtain a canonical isomorphism $C_0(G) \rtimes G \cong \mathcal{K}(l^2 G)$. So with this identification of the ideal in the sequence above, we get an extension

(10.21) $$0 \to \mathcal{K} \to C(\overline{G}) \rtimes G \to C(\partial G) \rtimes G \to 0.$$

We call it the *boundary extension*.

Lemma 10.5.12 *Let G be hyperbolic and μ normalized Hausdorff measure with respect to a visual metric d_ϵ on ∂G.*

1. The G-expectation

$$E \colon C(\partial G) \to C_b(\overline{G}), \quad E(\phi)(g) = \int_{\partial G} \phi \circ g \, d\mu$$

is a (completely) positive, G-equivariant map which splits the restriction *-homomorphism $C(\overline{G}) \to C(\partial G)$.

2. The G-equivariant expectation E induces a completely positive map

$$C(\partial G) \rtimes G \to C(\overline{G}) \rtimes G \subset \mathbb{B}(l^2 G)$$

by

$$(10.22) \qquad s \colon C(\partial G) \rtimes G \to \mathbb{B}(l^2 G), \quad s\Big(\sum_g \phi_g[g]\Big) := \sum_g M_{E\phi_g} \lambda(g),$$

with λ the left regular representation, M_ψ multiplication by ψ. The c.p. map s splits the restriction *-homomorphism

$$r \colon C(\overline{G}) \rtimes G \to C(\partial G) \rtimes G,$$

i.e., $r \circ s = \mathrm{id}_{C(\partial G) \rtimes G}$.

The lemma follows from Corollary 10.4.15, which asserts that

$$\lim_{g \to \xi \in \partial G} \int_{\partial G} \phi \circ g \, d\mu = \phi(\xi)$$

for all $\phi \in C(\partial G)$.

Proposition 10.5.13 *The boundary extension (10.21) of a hyperbolic group is c.p. split by the expectation map s of (10.22). The Fredholm module obtained by applying the Stinespring construction (Proposition 10.5.6) is unitarily isomorphic to the Fredholm module (10.18).*

Proof The c.p. map $s \colon C(\partial G) \rtimes G \to \mathbb{B}(l^2 G)$ G-equivariant in the sense that $s = E$ on functions in $C(\partial G) \rtimes G$ and for them $E(g(\phi)) = \lambda(g) E(\phi) \lambda(g)^*$, where λ is the left regular representation and E is the G-expectation. It follows that the vectors

$$\phi[g] \otimes e_k - \phi \otimes e_{gk} \in C(\partial G) \rtimes G \otimes l^2(G)$$

are null vectors in the inner product (10.19). Hence they are zero in the Stinespring Hilbert space $\widehat{H}$. On the other hand, if $\phi \otimes e_g, \psi \otimes e_h \in \widehat{H}$, then their inner product is given by

$$\langle \phi \otimes e_g, \psi \otimes e_h \rangle := \langle e_g, E(\bar{\phi}\psi) e_h \rangle = \delta_{g,h} \, E(\bar{\phi}\psi)(h).$$

So $\widehat{H}$ breaks into an orthogonal direct sum of the subspaces obtained, for each $g \in G$, by completing $C(\partial G)$ with respect to the inner product

$$\langle \phi, \psi \rangle := \int_{\partial G} (\bar{\phi}\psi) \circ g \, d\mu = \int_{\partial G} \bar{\phi}\psi \, d(g_* \mu).$$

The Stinespring projection is the direct sum of the projections to the constant functions in each $L^2(\partial G, g_* \mu)$. The representation of $C(\partial G) \rtimes G$ is the representation π of Exercise 10.4.19, so the Stinespring construction has produced precisely the Fredholm module of Exercise 10.4.19 c), which is unitarily equivalent to (10.18).

$\square$

Corollary 10.5.14 *Let G be a Gromov hyperbolic group, then the connecting homomorphism δ of the boundary extension (10.21) is given by pairing with the Fredholm module $(l^2(G, L^2(\partial G, \mu)), \pi_\mu, F := 2P_{l^2 G} - 1)$ in the sense of (10.3). That is,*

$$\langle [u], (l^2(G, L^2(\partial G, \mu)), \pi_\mu, F := 2P_{l^2 G} - 1) \rangle = \delta([u])$$

for any $[u] \in K_1(C(\partial G)] \rtimes G)$.

One can show (see [81]) that this pairing is nontrivial (nonzero) for groups G of zero Euler characteristic, such as discrete, co-compact groups of isometries of hyperbolic 3-space $\mathbb{H}^3$.

It would be a very nice thing if one could find a (finite-dimensional) *spectral* representative (in the sense of the next chapter) of the boundary extension class, rather than just a finitely summable Fredholm module. However, this seems to be unlikely or impossible with the usual definition of spectral cycle because the C*-algebras $C(\partial G) \rtimes G$ have no nonzero tracial states, because the boundary action leaves no probability measure invariant.

Exercise 10.5.15 Let χ be the characteristic function of all boundary points in $\partial \mathbb{F}_2$ which are endpoints of infinite words $s_1 s_2 \cdots$ with $s_1 = a$ (where $\mathbb{F}_2$ is generated by $\{a, b\}$).

Show that χ is Murray–von-Neumann equivalent in the group algebra $C(\partial \mathbb{F}_2)[\mathbb{F}_2] \subset C(\partial \mathbb{F}_2) \rtimes \mathbb{F}_2$ to a proper subprojection of itself, i.e., $\chi \sim \chi'$, where $\chi' + \chi'' = \chi$, some χ'' a projection orthogonal to χ'.

Deduce that if τ is a tracial state on $C(\partial \mathbb{F}_2) \rtimes \mathbb{F}_2$, then $\tau(\chi'') = 0$.

Extend the argument to prove that $C(\partial \mathbb{F}_2) \rtimes \mathbb{F}_2$ has no tracial states.

(*Hint.* Experiment with partial isometries of the form $S = \chi \cdot [a] \in C(\partial \mathbb{F}_2)[\mathbb{F}_2]$, for x a generator.)

A Fundamental Class for Boundary Actions of Hyperbolic Groups

If M is a compact, n-dimensional oriented manifold, so that one can integrate n-forms on M, the cohomology and homology of M exhibit *Poincaré duality*. In de Rham theory this duality is the nondegeneracy of the pairing

$$\langle [\alpha], [\beta] \rangle := \int_M \alpha \wedge \beta$$

between classes of closed forms in dimensions k and $n - k$. One may phrase this alternatively as the fact that the map $[\alpha] \mapsto \text{PD}([\alpha])$, where $\text{PD}([\alpha])$ is the class of the closed current $\beta \mapsto \int_M \alpha \wedge \beta$, is an isomorphism $H^k(M) \to H_{n-k}(M)$.

If M is a compact spin^c-manifold, then the class $[D] \in \text{KK}_0(C(M), \mathbb{C})$ of the Dirac operator on M turns out to generate a similar kind of duality. If $[E] \in \text{K}^0(M)$ is the class of a complex vector bundle, then twisting D by E and applying the methods of the previous chapter give a class $[D_E] \in \text{KK}_0(C(X), \mathbb{C})$. With some work (explained in the chapter on KK-theory), one can upgrade this to a map

$$\text{K}^*(X) \to \text{KK}_*(C(X), \mathbb{C}),$$

which is a K-theoretic version of Poincaré duality.

Note that the *orientation* assumption on M has been slightly upgraded to *K-orientation*.

One of the interesting features of Noncommutative Geometry is that this special feature of (oriented) manifolds among locally compact spaces appears in a quite wide variety of situations in operator algebras as it applies to dynamics. It appears in the situation of hyperbolic groups G acting on their boundaries, for example.

The main ingredient to this duality for the noncommutative spaces $C(\partial G) \rtimes G$ is a certain Fredholm module obtained by enriching the Fredholm module for the boundary extension class in $\text{KK}_1(C(\partial G) \rtimes G, \mathbb{C})$ which we have discussed above, using some further structure.

Let G be a hyperbolic group.

From the definition of the Gromov boundary if (x_i) and (y_i) are two sequences in the metric space G (with the word metric from a finite generating set) such that for some $C \geq 0$, we have $d(x_i, y_i) \leq C$; then by the definition of the Gromov product

$$(10.23) \qquad \langle x_i, y_i \rangle \geq \frac{1}{2} \langle x_i, x_i \rangle + \frac{1}{2} \langle y_i, y_i \rangle,$$

and hence if (x_i) converges to a boundary point, then (y_i) is asymptotic to (x_i) and hence converges to the same boundary point.

Lemma 10.5.16 *Let G be Gromov hyperbolic. Then if v_g is the unitary induced by right translation by $g \in G$, then the commutator $[M_\phi, v_g]$ is compact for any $\phi \in C(\overline{G})$.*

Proof If $h \in G$, $e_h \in l^2(G)$ the corresponding standard basis element, then

$$[M_\phi, v_g](e_h) = (\phi(hg) - \phi(h)) e_{hg} = (M_\psi v_g)(e_h),$$

where $\psi(h) = \phi(hg) - \phi(h)$, so it suffices to show that ψ is a C_0-function on G. Since $\overline{G}$ is compact and ϕ is continuous on $\overline{G}$, it suffices to show that if (h_i) is a sequence in G converging to a boundary point ξ, then $h_i g \to \xi$, which is immediate from (10.23) and surrounding discussion, since $d(g_i, g_i h) = |h|$ is bounded as $i \to \infty$.

$\square$

The following important result is obtained by combining the results of [1] and [63], which establish a notion of amenability for group actions and prove that the boundary action of a hyperbolic group is amenable and hence that the crossed product $C(\partial G) \rtimes G$ is nuclear.

Lemma 10.5.17 *For any Gromov hyperbolic group, the C*-algebra $C(\partial G) \rtimes G$ is nuclear.*

Corollary 10.5.18 *A *-homomorphism $\lambda \colon C(\partial G) \rtimes G \to Q(l^2 G)$ is obtained by mapping $\phi \in C(\partial G)$ to the class mod $\mathcal{K}(l^2 G)$ of the multiplication operator $M_{\tilde{\phi}}$, where $\tilde{\phi}$ is any extension of ϕ to a continuous function on $\overline{G}$, and mapping $g \in G$ to the class mod $\mathcal{K}$ of the left translation operator u_g.*

Let $J \colon l^2(G) \to l^2(G)$ be the unitary induced by inversion on G, $J(e_h) = e_{h^{-1}}$. Then a *-homomorphism $\rho \colon C(\partial G) \rtimes G \to Q(l^2)$ is defined by $\rho(a) = J\lambda(a)J$.

The homomorphisms λ and ρ commute and determine a *-homomorphism

$$\tau \colon C(\partial G) \rtimes G \otimes C(\partial G) \rtimes G \to Q(l^2 G), \qquad \tau(a \otimes b) := \lambda(a)\rho(b).$$

The *-homomorphism is c.p. split and therefore determines a B.D.F. cycle for $C(\partial G) \rtimes G \otimes C(\partial G) \rtimes G$ and class

$$\Delta \in \mathrm{KK}_1(C(\partial G) \rtimes G \otimes C(\partial G) \rtimes G, \mathbb{C}).$$

Remark 10.5.19 The amenability of the action on G on ∂G implies (see [63]) that any covariant pair determines a *-homomorphism defined on the crossed product $C(\partial G) \rtimes G$, whence the given covariant pairs determine *-homomorphisms λ and ρ from $C(\partial G) \rtimes G \to Q$. The fact that λ and ρ commute follows from Lemma 10.5.16. The fact that two commuting *-homomorphisms with domain $C(\partial G) \rtimes G$ combined to a *-homomorphism τ on the tensor product also follows from amenability of the action.

The class Δ acts as a "fundamental class" in K-homology of the crossed products, as we discuss in Section 11.2 of Chapter 10 in connection with duality in KK-theory.

10.6 Spectral Cycles and Fredholm Modules

The disadvantage of the formulas of Theorem 10.3.3 describing the index pairing of a finitely summable Fredholm module with projections or unitaries is that the formulas are not *local*. In fact, if one applies them to the classical situation of a zero order pseudodifferential operator F on a compact manifold, then one does not obtain an analog of the Atiyah–Singer formulas: The expressions $\text{Trace}(F[F, f^0] \cdots [f, f^n])$ involve taking the trace of a product of a sufficient number of operators (of order -1, so that the result is trace-class), and this produces a quite different formula from one involving only the integral over the manifold of a single explicit function. (In very simple situations, like that of Toeplitz operators, one may take $n = 1$, there is no product of operators involved, and the character formula is already "local," as we have already established in Section 2.3).

Remark 10.6.1 Section 10.4 discussed an example of a finitely summable Fredholm module over $C(\partial G) \rtimes G$ coming from measure-theoretic aspects of the action of a hyperbolic group G on its boundary. The Chern character of this Fredholm module can be computed by Connes' character formula Theorem 10.3.3 to be a cyclic cocycle over a dense subalgebra of $C(\partial G) \rtimes G$, which is a *sum over the group G*, of the statistical covariance functions

$$c(g, \phi, \psi) = \int_{\partial G} (\phi \cdot \psi) \circ g - \left(\int_{\partial G} \phi \circ g \right) \cdot \left(\int_{\partial G} \psi \circ g \right), \quad g \in G, \phi, \psi \in C(\partial G).$$

The function c decays as $g \to \infty$ to zero due to the convergence property of Hausdorff measure (sufficiently rapidly that it is in $l^p(G)$ for $p > 2$). The resulting formula is not computable (not "local") because it involves a sum over the group. Thus, the situation for this example is roughly similar to that of pseudodifferential operators. What would be desired, in this case, is a formula which does not involve any sums over the group.

The Atiyah–Singer formula, in contrast to these examples, translated into the language of cyclic cohomology, involves cyclic cocycles which have the form (10.6), which are of an entirely local, geometric nature.

The difference between the Atiyah–Singer formula and the formula provided by Connes' Chern character is due to the fact that there is a great deal of loss of spectral and geometric information in passage from a self-adjoint elliptic operator D, like a Dirac operator to its sign $F := \chi(D)$. Although some important information is retained, e.g., the boundedness of the interaction, through commutators, $[F, a]$, of the operator F with the algebra of observables, the concept of *length* has been in a sense lost when one replaces D by $\chi(D)$ (although the notion of *angle* has not been lost, which is why Fredholm operators can sometimes be thought of as corresponding to a conformal structure, rather than a Riemannian structure).

The Local Index Formula is based on enriching the notion of Fredholm module to involve a suitable notion of length, by requiring an unbounded self-adjoint D whose sign is F. Suitable hypotheses on D and A then give rise to a good noncommutative analog of integration on a manifold, using the *residue trace* (or the *Dixmier trace*, which we discuss less here, see [47]).

We fix the standard branch of λ^{-s} defined for $\text{Re}(s) > 0$. Suppose that D is a densely defined self-adjoint operator with discrete spectrum $\sim n^{\frac{1}{d}}$, with finite spectral multiplicities. If D is invertible, then $|D| > 0$, and we may apply λ^{-s} to $|D|$ using functional calculus.

Convention 10.6.2 *If D contains zero in the spectrum, there are some difficulties about defining 0^s, so we define $\mathrm{Tr}(|D|^{-s})$ to be $\mathrm{Tr}((|D| + \epsilon)^{-s})$ for $\epsilon > 0$ small. If $\epsilon' > 0$ is another choice, the resulting functions will differ by a function which extends analytically to $\mathbb{C}$. The same remarks apply to $\mathrm{Tr}(a|D|^{-s})$ for a a bounded operator. Therefore, making such minor alternations to $|D|$ makes no difference to the pole structure of the zeta functions. (An alternative fix is to simply add the projection $p_{\ker D}$ to $|D|$. The new operator $|D| + p_{\ker D}$ has no kernel, and it makes sense to define complex powers of it).*

To avoid cluttering up notation, we will simply write $\mathrm{Tr}(a|D|^{-s})$, with the above remarks in mind.

Definition 10.6.3 Let A be a unital C*-algebra and $A^\infty \subset A$ a dense *-subalgebra. Let $d \geq 1$.

A *d-dimensional even spectral cycle for* $A^\infty \subset A$ is a triple consisting of a $\mathbb{Z}/2$-graded Hilbert space H, a representation

$$\pi : A \to \mathbb{B}(H)$$

by even operators, and a densely defined self-adjoint operator D on H which is odd with respect to the grading, whose domain is invariant under A^∞, and such that $[\pi(a), D]$ is bounded for all $a \in A^\infty$, and furthermore that

a) $(1 + D^2)^{-1/2} \in \mathcal{L}^p(H)$ for all $p > d$.
b) The analytic function

$$\mathrm{Tr}(\pi(a)|D|^{-s}), \quad \mathrm{Re}(s) > d$$

extends to a meromorphic function on $\mathbb{C}$, for every $a \in A^\infty$.

An *odd* spectral cycle is defined the same way, except we drop the assumption of a $\mathbb{Z}/2$-grading.

Remark 10.6.4 (1) Definition 10.6.3 is similar, but not identical, to Connes' definition of *spectral triple*, which consists of a triple (A^∞, H, D), where A^∞ is a *-algebra, H is a Hilbert space carrying a representation of A^∞, and D is an operator, satisfying the conditions of Definition 10.6.3. (2) Often in examples of d-dimensional spectral cycles, the principal values μ_n of D are $O(n^{\frac{1}{d}})$, $n \to \infty$, a condition slightly stronger than (a). Condition (a) is often called *finite summability*.

Exercise 10.6.5 Let D be self-adjoint and invertible, and $(1 + D^2)^{-1}$ is compact. Show that if $\mathrm{Re}(s) \geq 0$, then $(1 + D^2)^{-s/2} \in \mathcal{L}^1(H)$ if and only if $|D|^{-s} \in \mathcal{L}^1(H)$, and if this holds for $\mathrm{Re}(s) > n$, then $\mathrm{Trace}((1 + D^2)^{-s/2}) - \mathrm{Trace}(|D|^{-s})$ extends to an analytic function on $\mathrm{Re}(s) > n - 2$.

The methods of Section 9.5 show the following:

Proposition 10.6.6 *If $\mathcal{D} = (H, \pi, D)$ is a spectral cycle for $A^\infty \subset A$ and χ is a normalizing function, then $(H, \pi, \chi(D))$ is a Fredholm module over A.*

Hence a spectral cycle (H, π, D) determines a class in $\mathrm{KK}_j(A, \mathbb{C})$ and an induced group homomorphism $\mathrm{K}_j(A) \to \mathbb{Z}$, where $j = 0$ if the cycle is even and $j = 1$ if the cycle is odd.

The group homomorphisms are induced by the corresponding Fredholm modules, by the method discussed in Section 10.1.

We generally denote by $[(H, \pi, D)]$, or sometime just $[D]$, the class of a spectral Fredholm triple in $\mathrm{KK}_*(A, \mathbb{C})$, but of course we really mean the class of $(H, \pi, \chi(D))$.

Example 10.6.7 Let $D = -i\frac{d}{d\theta}$, acting on $C^\infty(\mathbb{T})$ initially. It extends to a self-adjoint operator with domain the first Sobolev space $H^1(\mathbb{T}) = \{f \in L^2(\mathbb{T}) \mid \sum_{n \in \mathbb{Z}} |\hat{f}(n)|^2 (1 + n^2) < \infty\}$. Its spectrum is $2\pi\mathbb{Z}$. Hence $(1 + D^2)^{-1}$ is compact. The principal values of $|D|^{-1}$ are $\mu_n \sim n$, and hence the triple is one-dimensional.

The C*-algebra $C(\mathbb{T})$ and the smooth subalgebra $C^\infty(\mathbb{T}) \subset C(\mathbb{T})$ act by multiplication operators on $L^2(\mathbb{T})$, $C^\infty(\mathbb{T})$ leaves $\mathrm{dom}(D)$ invariant, and $[D, f] = f'$ is bounded for $f \in C^\infty(\mathbb{T})$.

We prove in Corollary 10.9.4 that if $f \in C^\infty(\mathbb{T})$, then the zeta function $\mathrm{Tr}(f|D|^{-s})$ extends mermorphically to $\mathbb{C}$, with a simple pole at $s = 1$, and residue there equal to $\int_\mathbb{T} f(x) d\mu$ where μ is a normalized Haar measure.

The corresponding Fredholm module is $(L^2(\mathbb{T}), \pi, F := 2P_+ - 1)$, where π is the representation by multiplication operators and P_+ the Szegö projection.

In particular, the pairing of this spectral cycle with K-theory is the pairing with the odd Fredholm module $(L^2(\mathbb{T}), \pi, P_+)$, and we have already observed that this pairing is given by the Toeplitz index map

$$u \mapsto \mathrm{Index}(T_u),$$

where T_u is the Toeplitz operator with symbol u.

Exercise 10.6.8 Suppose that (H, π, D) is an odd spectral cycle for $A^\infty \subset A$ and that T is a bounded operator on H leaving $\mathrm{dom}(D)$ invariant and such that T commutes with $\pi(A)$. Prove that $(H, \pi, D + T)$ is a spectral cycle for $A^\infty \subset A$ and that the Fredholm modules associated with D and $D + T$ are compact perturbations of each other. (See Lemma 9.5.7.)

We now show how a spectral cycle over $A^\infty \subset A$ gives rise to a trace on A, called the *residue trace*.

The Local Index Formula of Connes and Moscovici describes the Chern character of a spectral cycle (and hence its pairing with K-theory classes) in terms of residue traces of a more general kind; these more general residue functionals behave more like distributions, as they do not necessarily extend from A^∞ to A.

Before discussing the Residue Trace, we prove several technical results.

To shorten notation slightly, let $\Delta := |D|$. If Δ is not invertible, perturb it by a small positive constant to make it so, or add the projection to its kernel.

Lemma 10.6.9 *Let Δ be a strictly positive unbounded operator on a Hilbert space with discrete spectrum.*

Let C be the contour in the complex plane given by the straight line, $\mathrm{Re}(\lambda) = \epsilon$, for $\epsilon > 0$ small enough to miss the spectrum of Δ. C is oriented straight downward. Then for all s with $\mathrm{Re}(s) > 0$,

$$(10.24) \qquad \Delta^{-s} = \frac{1}{2\pi i} \oint_C \lambda^{-s}(\lambda - \Delta)^{-1} d\lambda.$$

The integral converges absolutely in $\mathcal{K}(H)$.

More generally, if $\binom{n}{p}$ is the usual binomial coefficient, $p = 0, 1, 2, \ldots$, then

$$(10.25) \qquad \binom{s}{p} \Delta^{-s-p} = \frac{1}{2\pi i} \oint_C \lambda^{-s}(\lambda - \Delta)^{-p-1} d\lambda.$$

Proof The function $\lambda^{-s}(\lambda - \Delta)^{-1} d\lambda$ is valued in $\mathcal{K}(L^2\mathbb{R})$. Since

$$\|(\lambda - \Delta)^{-1}\| = O(|\lambda|^{-1}), \quad |\lambda| \to \infty,$$

$\|\lambda^{-s}(\lambda - \Delta)^{-1}\| = O(|\lambda|^{-s-1})$ as $|\lambda| \to \infty$, and hence the integral converges absolutely if $\mathrm{Re}(s) > 0$. Now the remaining statements follow from the following application of Cauchy's formula: I claim that if $\mathrm{Re}(z) > 0$ and $\mathrm{Re}(s) > 0$, then

$$z^{-s} = \frac{1}{2\pi i} \oint_C \frac{\lambda^{-s}}{\lambda - z} \, dz.$$

For the proof let C_R be the right half-circle of radius R centered at 0, oriented clockwise, and C'_R the union of C_R with the segment of the imaginary axis from Ri to $-Ri$, then by Cauchy's formula if $0 < |z| < R$, then

$$z^{-s} = \frac{1}{2\pi i} \oint_{C'_R} \frac{\lambda^{-s}}{\lambda - z} \, d\lambda.$$

On the other hand $|\frac{\lambda^{-s}}{\lambda - z}| = O(R^{-s-1})$ for $\lambda \in C_R$ and $R \to \infty$, and since the length of C_R is πR, $\oint_{C_R} \frac{\lambda^{-s}}{\lambda - z} \, d\lambda = O(R^{-s})$ as $R \to \infty$, and in particular $\oint_{C_R} \frac{\lambda^{-s}}{\lambda - z} \, d\lambda \to 0$ as $R \to \infty$ and $\mathrm{Re}(s) > 0$. $\square$

Lemma 10.6.10 *In the notation of Lemma 10.6.9, suppose $a \in A^\infty$. Then $[\pi(a), \Delta^{-s}] \cdot \Delta^{s+1}$ is bounded, $\mathrm{Re}(s) > 0$.*

Proof To lighten notation, we just write a for $\pi(a)$.

For $a \in A^\infty$, let $\delta(a) = [a, \Delta]$.
By (10.24),

(10.26) $\qquad [a, \Delta^{-s}] = \dfrac{1}{2\pi i} \displaystyle\int_C \lambda^{-s} \cdot [a, (\lambda - \Delta)^{-1}] d\lambda.$

We have

$$[a, (\lambda - \Delta)^{-1}] = (\lambda - \Delta)^{-1}[a, \Delta](\lambda - \Delta)^{-1} = (\lambda - \Delta)^{-1}\delta(a)(\lambda - \Delta)^{-1}.$$

Putting this into (10.26) gives

(10.27) $\qquad [a, \Delta^{-s}] = \dfrac{1}{2\pi i} \displaystyle\int_C \lambda^{-s}(\lambda - \Delta)^{-1}\delta(a)(\lambda - \Delta)^{-1} d\lambda.$

Now,

$$(\lambda - \Delta)^{-1}\delta(a) = \delta(a)(\lambda - \Delta)^{-1} - [\delta(a), (\lambda - \Delta)^{-1}]$$

(10.28) $\qquad\qquad\qquad = \delta(a)(\lambda - \Delta)^{-1} - (\lambda - \Delta)^{-1}\delta^2(a)(\lambda - \Delta)^{-1}).$

Plugging into (10.32) and using Cauchy's formula (10.25) give

(10.29) $2\pi i[a, \Delta^{-s}] = \delta(a) \displaystyle\int_C \lambda^{-s}(\lambda - \Delta)^{-2}d\lambda + R_s = \delta(a)\dbinom{-s}{1}\Delta^{-s-1} + R_s,$

where the remainder term R_s is

$$R_s = \oint_C \lambda^{-s}(\lambda - \Delta)^{-1}\delta^2(a)(\lambda - \Delta)^{-2}d\lambda.$$

Notice that $R_s \Delta^2$ is bounded.
Now iterate the argument with R_s to get

(10.30)
$$R_s = \dfrac{\delta^2(a)}{2\pi i} \int_C \lambda^{-s}(\lambda - \Delta)^{-3}d\lambda + \dfrac{1}{2\pi i}\int_C \lambda^{-s}(\lambda - \Delta)^{-1}\delta^3(a)(\lambda - \Delta)^{-3}d\lambda$$

$$= \dfrac{1}{2\pi i}\dbinom{-s}{2}\delta^2(a)\Delta^{s-2} + R_s',$$

where $R_s'\Delta^3$ is bounded.
Proceeding inductively, we get an expansion

$$(10.31)\quad 2\pi i[a, \Delta^{-s}] = \binom{-s}{1}\delta(a)\Delta^{-s-1} + \binom{-s}{2}\delta^2(a)\Delta^{-s-2} + \cdots$$

$$+ \binom{-s}{k}\delta^k(a)\Delta^{-s-k} + R^{(k)},$$

where $R_s^{(k)}\Delta^{k+1}$ is bounded, valid for $\mathrm{Re}(s) > 0$. Multiplying the expansion by Δ^{s+1} gives

$$(10.32)\quad 2\pi i[a, \Delta^{-s}]\Delta^{s+1} = \binom{-s}{1}\delta(a) + \binom{-s}{2}\delta^2(a)\Delta^{-1} + \cdots$$

$$+ \binom{-s}{k}\delta^k(a)\Delta^{-k} + R_s^{(k)}\Delta^{s+1},$$

and all the terms up to the remainder term are bounded, while if $k > \mathrm{Re}(s)$, then the last term is bounded as well.

$\square$

We now show that a d-dimensional spectral cycle on A^∞ gives rise to a (positive) trace on A.

Theorem 10.6.11 *Suppose that (H, π, D) is a d-dimensional spectral cycle for $A^\infty \subset A$. Then*

$$\mathrm{Res}\,\mathrm{Tr}(a) := \mathrm{Res}_{s=d}\mathrm{Tr}(\pi(a)|D|^{-s})$$

is a positive trace on A^∞.

In particular, $\mathrm{Res}\,\mathrm{Tr}$ extends to a trace on A.

Proof We have

$$ab|D|^{-s} - a|D|^{-s/2}b|D|^{-s/2} = a[b, |D|^{-s/2}]\,|D|^{-s/2}.$$

By the lemma, $a[b, |D|^{-s/2}] \cdot |D|^{1+s/2} =: T$ is bounded. We have thus shown that $ab|D|^{-s} - a|D|^{-s/2}b|D|^{-s/2} = T|D|^{-s-1}$ with T bounded. Now $|D|^{-s-1}$ is trace-class for $\mathrm{Re}(s) > n - 1$. Therefore,

$$(10.33)\qquad \mathrm{Tr}(ab|D|^{-s}) - \mathrm{Tr}(a|D|^{-s/2}b|D|^{-s/2})$$

extends to a function ψ analytic on $\mathrm{Re}(s) > n - 1$.

Hence $\mathrm{Tr}(ab|D|^{-s}) = \mathrm{Tr}(a|D|^{-s/2}b|D|^{-s/2}) + \psi$, and reversing the roles of a and b gives

$$\mathrm{Tr}(ab|D|^{-s}) = \mathrm{Tr}(ba|D|^{-s}) + \psi'$$

for ψ' analytic for $\text{Re}(s) > n - 1$. Taking residues at $s = n$ gives the tracial property $\text{Res Tr}(ab) = \text{Res Tr}(ba)$.

This proves the tracial property. An analogous argument shows that

$$\text{Res}_{s=n} \text{Tr}(aa^*|D|^{-s}) = \text{Res}_{s=n} \text{Tr}(a^*|D|^{-s}a^*),$$

and since $\text{Tr}(a^*|D|^{-s}a) \geq 0$ for all s with $\text{Re}(s) > n$,

$$\text{Res}_{s=n} \text{Tr}(a^*|D|^{-s}a) = \lim_{s \to 1^+} (s - n)\text{Tr}\left((a^*|D|^{-s}a\right) \geq 0,$$

we deduce positivity.

□

The idea of constructing a trace using residues is essential in Noncommutative Geometry. An alternative approach uses the *Dixmier trace*. Both methods produce functionals which encode geometric information. In the classical situation, the following theorem (for the statement see [100], and for proof see [142]) summarizes why this idea is important. It is based on a theorem of Weyl.

Example 10.6.12 (Weyl) Let M be an n-dimensional compact Riemannian spin^c-manifold and D be a twisted Dirac operator on M, i.e., a twist of the Dirac operator on M by a vector bundle.

Then if $f \in C^\infty(M)$, then the analytic function

$$\text{Trace}(f|D|^{-s}), \quad \text{Re}(s) > n$$

extends meromorphically to $\mathbb{C}$ and has a simple pole at $s = n$, and

$$\text{Res}_{s=n} \text{Trace}(f|D|^{-s}) = c_n \cdot \int_M f \, d\mu,$$

with $d\mu$ volume measure on M and c_n the constant

$$c_n = \frac{\dim(S)}{(2\sqrt{\pi})^n \cdot \Gamma(1 + \frac{n}{2})},$$

where S is the bundle on which D acts (so being an irreducible Clifford module has dimension depending only on n).

The reason for the meromorphic extendibility property lies in certain asymptotic expansions of the heat kernel, discussed in the next section.

Note that taking $f = 1$ in the theorem and D the Dirac operator, we see that the spectrum of $|D|$ determines the Riemannian volume.

Weyl's result suggests the philosophical idea that a spectral triple over $A^\infty \subset A$ over a dense subalgebra of a C*-algebra may endow, in a sense, the corresponding "noncommutative space" with an analog of Riemannian geometric structure.

10.7   The Heat Equation Proof of the Atiyah–Singer Index Theorem

The meromorphic extension property for zeta functions $\mathrm{Tr}(a|D|^{-s})$ affiliated with spectral cycles may seem quite mysterious. In this section, we describe the classical situation and explain why this extension property follows from the heat equation method and gives a local formula for the index of a Dirac operator.

Suppose that D is a self-adjoint operator with singular values $\lambda_n \sim O(n^{\frac{1}{d}})$. For example, D could be the Dirac operator on a compact Riemannian spin^c-manifold of dimension d.

Let $\Delta = D^2$. The eigenvalues μ_n of Δ grow like $n^{\frac{2}{d}}$ as $n \to \infty$. Hence $\mathrm{Tr}(T\Delta^{-s})$ is finite for $\mathrm{Re}(s) > \frac{d}{2}$ and arbitrary bounded operators T.

By definition of the Γ-function and making a change of variables $t \to \lambda t$, for any $\lambda > 0$, we have

$$\Gamma(s) = \int_0^\infty t^{s-1} e^{-t} dt = \lambda^s \int_0^\infty t^{s-1} e^{-\lambda t} dt.$$

from which

$$\lambda^{-s} \Gamma(s) = \lambda^s \int_0^\infty t^{s-1} e^{-\lambda t} dt.$$

Summing over $\lambda \in \mathrm{Spec}(\Delta)$ gives

$$(10.34) \qquad \Gamma(s) \cdot \mathrm{Tr}(\Delta^{-s}) = \int_0^\infty t^{s-1} \mathrm{Tr}(e^{-t\Delta}) dt.$$

The semigroup of compact operators $e^{-t\Delta}$ is called the "heat semigroup." It satisfies the *heat equation*: the partial differential equation

$$(10.35) \qquad \left(\frac{\partial}{\partial t} + \Delta_x\right) k_t(x, y) = 0,$$

with initial condition $k_t(x, y) \to \delta_{x-y}$ as $t \to 0$.

In physical terms, if $u \in C^\infty(X)$ is a smooth function, describing the temperature at points of a Riemannian manifold, then

$$(e^{-t\Delta} u)(x) = \int_X k_t(x, y) u(y) dy,$$

where Δ is the scalar Laplacian, describes the distribution of temperature at time $t > 0$. As $t \to \infty$, as heat tends to flow into cooler areas, the temperature becomes evenly distributed across the manifold.

Exercise 10.7.1 Let Δ be any densely defined self-adjoint positive operator on a Hilbert space H with $(1 + \Delta^2)^{-1}$ compact. Assume the eigenvalues of Δ grow at most $O(n^\mu)$ for some $\mu \in \mathbb{R}_+$. Let A be any bounded operator. Prove that $A\Delta^{-s}$ is trace class for $\mathrm{Re}(s) > \frac{1}{\mu}$ and that

$$(10.36) \qquad \Gamma(s) \cdot \mathrm{Tr}(A\Delta^{-s}) = \int_0^\infty t^{s-1} \mathrm{Tr}(Ae^{-t\Delta}) dt.$$

In the situation of Dirac operators, one has a spinor bundle S and D is the Dirac operator acting on a dense subspace of $L^2(X, S)$, with respect to a measure μ on X. The operator $\Delta := D^2$ is an *even* operator with respect to the gradings, and the operator $e^{-t\Delta}$ acts as an integral operator with kernel k_t:

$$(e^{-t\Delta}u)(x) = \int_X k_t(x, y)u(y)d\mu(y),$$

where here k_t takes values in endomorphisms of the bundle S, so that $k_t(x, y) \in \mathbb{B}(S_x, S_y)$ for all x, y.

Since the trace of a smoothing operator like $fe^{-t\Delta}$ is the integral of its kernel along the diagonal, we obtain the following lemma:

Lemma 10.7.2 *If D is the Dirac operator on X, $f \in C^\infty(X)$, acting on $L^2(X, S)$ by multiplication, then*

$$\mathrm{Tr}(fe^{-t\Delta}) = \int_X f(x) \cdot \mathrm{Tr}(k_t(x, x)) d\mu(x),$$

where k_t is the heat kernel for Δ.

Now assume X is even-dimensional and that S is $\mathbb{Z}/2$-graded. Then

$$D = \begin{bmatrix} 0 & D_+^* \\ D_+ & 0 \end{bmatrix}$$

and

$$\Delta = \begin{bmatrix} D_+^* D_+ & 0 \\ & D_+ D_+ * \end{bmatrix}$$

so that

$$\mathrm{Ind}(D) = \dim \ker(D_+) - \dim \ker(D_-),$$

while on the other hand

$$e^{-t\Delta} = \begin{bmatrix} e^{-tD_+^* D_+} & 0 \\ & e^{-tD_+ D_{+*}} \end{bmatrix}.$$

If $T = \begin{bmatrix} A & B \\ C & D \end{bmatrix}$ is a compact operator on the $\mathbb{Z}/2$-graded Hilbert space $L^2(X, S)$, let $\mathrm{Tr}_s(T) = \mathrm{Tr}(A) - \mathrm{Tr}(D)$, the *graded trace*. Then with this notation,

$$\mathrm{Tr}_s(e^{-t\Delta}) = \mathrm{Tr}(e^{-tD_+^* D_+}) - \mathrm{Tr}(e^{-tD_+ D_+^* \cdot}).$$

Since $D_+ D_+^*$ and $D_+^* D_+$ have the same *nonzero* eigenvalues and multiplicities, expressing the above trace as a sum of eigenvalues gives through the resulting cancellation that

$$\mathrm{Tr}_s(e^{-t\Delta}) = \mathrm{Tr}(\mathrm{proj}_{\ker D}) - \mathrm{Tr}(\mathrm{proj}_{\ker(D^*)}) = \mathrm{Index}(D).$$

We obtain the *McKean–Singer formula*:

Theorem 10.7.3 *If D is the Dirac operator on a spinor bundle S over X compact Riemannian, then*

$$\mathrm{Index}(D) = \mathrm{Tr}_s(e^{-t\Delta}) = \int_X \mathrm{Tr}_s\left(k_t(x, x)\right) d\mu(x)$$

for any $t > 0$.

This formula is useless for actual computation, since

$$\mathrm{Tr}(e^{-tD_\pm^* D_\pm^*}) = \int_X \mathrm{Tr}\left(k_t^\pm(x, x)\right) d\mu(x),$$

each *individually* diverges to ∞ as $t \to 0$, while only their difference converges.

What is perhaps remarkable is that one may precisely quantify these divergences by means of an asymptotic expansion, in which the coefficients are geometrically defined distributions on the manifold X.

The result goes back to Minakshisundaram and Plejel; see the book [145] for an exposition.

Theorem 10.7.4 *Let X be a compact Riemannian manifold with volume measure $\mu = (\det g)^{\frac{1}{2}} dx$ (locally) the volume, and let Δ be the scalar Laplacian, a densely defined operator on $L^2(X, \mu)$.*

Then there exist A_k smooth functions on X such that for any $f \in C^\infty(X)$, there is an asymptotic expansion

$$\mathrm{Tr}(f e^{-t\Delta}) \sim (4\pi t)^{-\frac{d}{2}} \cdot \sum_{n=0}^{\infty} \left(\int_X f \cdot A_k \, d\mu\right) t^k,$$

as t → 0.

For the exact meaning of "asymptotic expansion," we refer the reader to more specialized texts, e.g., [145] or [142]. We will look at some specific examples in the sections to follow, where the meaning of ~ will be made precise.

We now give an indication of why asymptotic expansions imply meromorphic extendibility of certain zeta functions. Recall the formula (10.36) for the zeta function in terms of the Mellin transform of the heat trace. Let $f \in C^\infty(X)$. Then

(10.37) $$\Gamma(s) \cdot \text{Tr}(f \cdot \Delta^{-s}) = \int_0^\infty t^{s-1}\text{Tr}(f \cdot e^{-t\Delta})dt.$$

Observe first that

$$\int_1^\infty t^{s-1}\text{Tr}(f \cdot e^{-t\Delta})dt$$

is an *entire* function of s, due to uniform exponential decay of $\text{Tr}(f \cdot e^{-t\Delta})$ for $t \geq 1$. So, we have, up to an entire function, since the asymptotic expansion for $f e^{-t\Delta}$ is based on the kernel $k_t^f(x, y) = f(x)k_t(x, y)$,

(10.38)

$$\Gamma(s) \cdot \text{Tr}(f \cdot \Delta^{-s}) = \int_0^1 t^{s-1}\text{Tr}(f \cdot e^{-t\Delta})dt = \int_X \int_0^1 t^{s-1} f(x)k_t(x, x)dx$$

$$\sim \int_X \int_0^1 t^{s-\frac{d}{2}-1} A_0(x) f(x)dx + \int_X \int_0^1 t^{s-\frac{d}{2}} A_1(x) f(x0dx$$

$$+ \int_X \int_0^1 t^{s-\frac{d}{2}+1} A_2(x) f(x)dx + \cdots$$

$$= \left(s - \frac{d}{2}\right)^{-1} \cdot \int_X A_0(x) f(x)dx$$

$$+ \left(s - \frac{d}{2} + 1\right)^{-1} \cdot \int_X A_1(x) f(x)dx + \cdots,$$

where we have used Lemma 2.1.15 of Chapter 1 to compute the trace of an integral operator with smooth kernel in terms of the integral of the kernel over the diagonal.

A similar reasoning applies to the Dirac operator D on sections of a spinor bundle S.

Theorem 10.7.5 *Let D be a Dirac-type operator on sections of a bundle $S \to X$, X compact K-oriented, and d-dimensional. Let $\Delta = D^2$.*
Let k_t be the heat kernel, solving the heat equation $\frac{\partial}{\partial t} + \Delta$.
Then there is an asymptotic expansion

$$f(x)k_t(x, y) \sim f(x)A_0(x, y) \cdot t^{-\frac{d}{2}} + f(x)A_1(x, y) \cdot t^{-\frac{d}{2}+1}$$
$$+ f(x)A_2(x, y)t^{-\frac{d}{2}+2} + \cdots,$$

where $A_i(x, y)\colon S_x \to S_y$ are smooth linear operator-valued maps, defined in a neighborhood of the diagonal in $X \times X$.

Hence, the zeta function $\mathrm{Tr}(f\Delta^{-s})$ extends to a meromorphic function on $\mathbb{C}$ with poles at $\frac{d}{2}, \frac{d}{2} - 1, \cdots$ and residue at $\frac{d}{2} - k$ given by

$$\int_X f(x) \cdot \mathrm{Tr}(A_k(x, x))dx.$$

In particular,

$$\mathrm{Res}_{s=\frac{d}{2}} \mathrm{Trace}(f\Delta^{-s}) = \int_X f(x) \cdot \mathrm{Tr}(A_{d/2}(x, x)) \, dx.$$

Furthermore, if X is even-dimensional, S graded, then the endomorphisms A_k are all even with respect to the grading. Let $A_k = A_k^+ \oplus A_k^-$. Then

$$\mathrm{Index}(D) = \int_X \mathrm{Tr}_s(A_{d/2}(x, x)) \, dx = \mathrm{Res}_{s=\frac{d}{2}} \mathrm{Tr}_s(\Delta^{-s}),$$

where Tr_s is the fiberwise graded trace.

The previous result is a sort of "in-principal" index theorem, as it produces a formula for the index which is a difference of integrals of a pair of functions $A_{d/2}^{\pm}$. But it requires more work to compute $A_{d/2}^{\pm}$ explicitly. (See [10].) We will state the result below for the spin Dirac operator.

Exercise 10.7.6 This exercise is about the method of translating "asymptotic expansions" into meromorphic functions with pole structure determined by the expansion. The basic example of an asymptotic expansion is a Taylor series. Let f be a smooth function in a neighborhood of $0 \in \mathbb{R}$.

a) Let $f \sim \sum_{n=0}^{\infty} a_n t^n$ be the Taylor series of f at $t = 0$. By definition, $\sim$ means that $f - \sum_{k=0}^{n} a_k t^k = O(|t|^{n+1})$ as $t \to 0$, for all n. Show that the function

$$(10.39) \qquad\qquad \gamma(s) := \int_0^1 t^{s-1} f(t)dt,$$

defined initially and analytically for $\mathrm{Re}(s) > 0$, can be meromorphically extended to $\mathbb{C}$, that the poles are simple, parameterized by the natural numbers, and that the residues are the coefficients a_n of the power series.

b) Suppose that $f(t) \rightarrow 0$ exponentially quickly as $t \rightarrow 0$ in the sense that $t^{-k} f(t) \rightarrow 0$ for every positive integer k. Prove that (10.39) extends *analytically* to $\mathbb{C}$.

We now describe the Atiyah–Singer formula for the Dirac operator on a *spin* manifold, twisted by a vector bundle. The answer should involve topological information about the vector bundle. The case of a spin^c-manifold is similar, but slightly more complicated, and we omit it.

The Bare Bones of Chern–Weil Theory

A topic we have not discussed in this book, for reasons of space, is Chern–Weil theory, which is used to construct "local" models for vector bundles. The Chern–Weil theory can be used to define an explicit isomorphism from the K-theory of a compact manifold M, tensored with $\mathbb{R}$, to the sum of the de Rham cohomology vector spaces of M. The Chern character appears explicitly in the Atiyah–Singer Index Theorem in order to record the effect on the index of twisting by a vector bundle. Despite the importance of Chern–Weil theory, we will describe it only very briefly.

Suppose $E \rightarrow M$ is a complex vector bundle. Endow it with a Hermitian structure. Let R_E be the curvature of a Hermitian connection on E, an End(E)-valued 2-form, and by locally trivializing E we may represent R locally as a matrix of 2-forms. There are of course many natural constructions with matrices with entries in a ring, like the ring of 2-forms. One may form Trace(R_E), for example, locally, giving a 2-form. This is actually independent of the choice of trivialization of E used, since the trace is invariant under conjugation. So Trace(R_E) is a globally defined 2-form. Similarly Trace(R^2) is a globally defined even-degree form, and so on.

Suppose now that $f(z) = \sum_{n=0}^{\infty} a_n z^n$ is a power series defining a function analytic in a neighborhood of zero in $\mathbb{C}$. Note that the ring of matrices with entries differential forms on any manifold M is nilpotent, because there are no differential k-forms on M if $k > \dim M$. In particular, if R_E is the curvature of a complex vector bundle E, then $R_E^k = 0$ for $k > \frac{n}{2}$. We may thus apply f to R_E, defining it as $f(R_E) := \sum_{k=0}^{\infty} a_k R_E^k$. The sum is actually finite and is a differential form.

Let $g(z) = \frac{z/2}{\sinh(z/2)}$. Then g is analytic at $z = 0$ and expands into a power series. We may apply the discussion above to the function

$$f(A) := \exp\left(\frac{1}{2} \text{Trace} \, \log g(A)\right).$$

If R_M is the curvature of M, we obtain a differential form

$$\hat{A}(M) := f(R_M).$$

The Chern character ch(E) of E is the class of the differential form

$$\text{Trace}(e^{-R_E}) := \sum_{n=0}^{\infty}(-1)^n \cdot \frac{R_E^n}{n!}.$$

This is a finite sum of even-dimensional closed differential forms.

Theorem 10.7.7 (The Atiyah–Singer Index Theorem) *If D is the Dirac operator on a spin manifold M and $E \to M$ is a complex vector bundle, then*

$$\text{Index}(D_E) = \int_M \hat{A}(M) \wedge \text{ch}(E).$$

The index has been expressed in terms of the integral over M of certain canonical differential forms associated with the bundle and the spin structure. To compute the index explicitly therefore is reduced to computation of these differential forms, which is a local problem.

10.8 The Atiyah–Singer and Connes–Moscovici Local Index Theorems

The proof of the following basic lemma is quite easy using KK-theory, and we defer it.

Proposition 10.8.1 *If D is a Dirac operator on an even-dimensional compact* spin^c-manifold M, then $\text{Index}(D_E) = \langle [E], [D] \rangle$, the pairing between the K-theory class $[E]$ for C(M) and the K-homology class $[D] \in KK_0(C(M), \mathbb{C})$ in the sense of Section 10.1.

The Local Index formula of Connes and Moscovici gives a formula for the K-theory/K-homology pairing between the class in $KK_*(A, \mathbb{C})$ of a spectral cycle for a C*-algebra and the class of a projection in $K_*(A)$. The formula is "local" in the sense that it involves the distributions which appear at poles of zeta functions (like $\text{Trace}(a|D|^{-s})$) and is compatible with the K-theory–Fredholm module pairing we have already defined.

In this section we state the Local Index Theorem, starting with a translation of the Atiyah–Singer formula into the Connes–Moscovici framework of cyclic cohomology. The elegant short paper [128] by R. Ponge was one of our main sources, and we refer the reader to the article for further information and proofs.

Let M be an even-dimensional spin manifold and $\hat{A}(M)$ the corresponding genus, (represented by) a mixed-degree differential form on M which is nonzero only in even degrees (discussed in the previous section). The spin structure therefore defines a current on M of mixed degrees

$$\alpha \mapsto \int_M \hat{A}(M) \wedge \alpha.$$

It is closed, because $\hat{A}(M)$ is a closed differential form.

Theorem 10.8.2 *Let*

$$\varphi_{2k}(f^0, \ldots, f^{2k}) := \frac{1}{(2k)!} \int_M f^0 df^1 \cdots df^n \wedge \hat{A}^{(n-2k)}(M).$$

Then $\varphi := (\varphi_{2k})_{k=0}^{\infty}$ *defines a class in* $\mathrm{HCP}^0(C^{\infty}(M))$, *and if* $e \in C^{\infty}(M) \otimes M_n(\mathbb{C})$ *is a smooth idempotent, defining a smooth complex vector bundle* $E \to M$, *then*

$$\mathrm{Index}(D_E) = \langle [E], [D] \rangle = \langle [\varphi], [e] \rangle = \sum_{k=0}^{\infty} \varphi_{2k}(e, e, \ldots, e).$$

The general Local Index Theorem of Connes and Moscovici runs along similar lines.

Definition 10.8.3 Let (H, π, D) be a spectral cycle for $A^{\infty} \subset A$.

a) By definition dom(δ) is all T mapping dom(δ) to itself such that $\delta(T)$ extends to a bounded operator on H. If $T \in \mathbb{B}(H)$ maps dom(D) to itself, let $\delta(T) = [T, |D|]$. We say that the spectral Fredholm module is *regular* over A^{∞} if A^{∞} and $[D, A^{\infty}]$ both are in $\cap_{n=1}^{\infty} \mathrm{dom}(\delta^n)$.
b) Let $\Psi_D^0(A^{\infty})$ denote the algebra generated by the $\delta^k(a)$'s, $a \in A^{\infty}$. Say (H, π, D) has a *simple dimension spectrum* if there exists a discrete subset $\Gamma \subset \mathbb{C}$ such that $\mathrm{Trace}(T|D|^{-s})$ meromorphically extends to $\mathbb{C}$ with simple poles in Γ for every $T \in \Psi_D^0(A^{\infty})$.

Theorem 10.8.4 *Let* (H, π, D) *be an even, finite-dimensional spectral cycle for* $A^{\infty} \subset A$ *with the meromorphic continuation property and simple dimension spectrum. Let* $\Delta = D^2$.
    Set

$$\varphi_0(a^0) = \mathrm{Res}_{s=0} \left(\Gamma(s) \cdot \mathrm{Tr}(\epsilon a^0 \Delta^{-s}) \right).$$

For $k > 0$, *set*

$$\varphi_{2k}(a^0, \ldots, a^{2k})$$

(10.40)

$$= \sum_{\alpha \geq 0} c_{k,\alpha} \cdot \mathrm{Res}_{s=0} \mathrm{Tr} \left(\epsilon a^0 [D, a^1]^{(\alpha_1)} \cdots [D, a^{2k}]^{(\alpha_{2k})} |D|^{-2(|\alpha|+k)-s} \right),$$

where

$$\Gamma(|\alpha| + k) c_{k,\alpha}^{-1} = 2(-1)^{|\alpha|} \alpha!(\alpha_1 + 1)(\alpha_1 + \alpha_2 + 2) \cdots (\alpha_1 + \cdots + \alpha_{2k} + 2k).$$

Then the sequence $(\varphi_{2k})_{k=0}^{\infty}$ defines an even cochain in the (b, B) bicomplex and a class in $\mathrm{HCP}^0(A^{\infty})$. It is cohomologous to the Chern character (Definition 10.3.4) of (H, π, F), where $F = \chi(D)$ and χ a normalizing function.

Let (H, π, D) be an odd, regular, n-dimensional spectral cycle for $A^{\infty} \subset A$ with the meromorphic continuation property and simple dimension spectrum. Let $\Delta = D^2$.

For $k \geq 0$, set

$$\varphi_{2k+1}(a^0, \ldots, a^{2k+1})$$

$$(10.41)$$

$$= \sqrt{2\pi i} \sum_{\alpha \geq 0} c_{k,\alpha} \cdot \mathrm{Res}_{s=0} \mathrm{Tr}\left(a^0[D, a^1]^{(\alpha_1)} \cdots [D, a^{2k+1}]^{(\alpha_{2k+1})}|D|^{-2(|\alpha|+k)-1-s}\right),$$

where

$$\Gamma(|\alpha| + k + 1/2)c_{k,\alpha}^{-1} = (-1)^{|\alpha|}\alpha!(\alpha_1 + 1)(\alpha_1 + \alpha_2 + 2) \cdots$$

$$(\alpha_1 + \cdots + \alpha_{2k} + 2k + 1).$$

Then the sequence $(\varphi_{2k+1})_{k=0}^{\infty}$ defines an odd cochain in the (b, B) bicomplex and a class in $\mathrm{HCP}^1(A^{\infty})$. It is cohomologous to the Chern character (Definition 10.3.4) of (H, π, F), where $F = \chi(D)$ and χ a normalizing function.

10.9 Zeta Functions and the Local Index Theorem for the Circle

The Local Index Theorem provides a formula for the index pairing of a K-homology class for a C*-algebra, represented by a suitably regular spectral cycle. The formula involves certain functionals arising as residues from zeta functions, making them have a "local" character. These functions agree in the commutative case of a smooth manifold with the distributions which appear in the Atiyah–Singer formula in the heat equation approach.

In this section we work out this framework in detail from scratch, for some very basic examples: the crossed products $C(\mathbb{T}) \rtimes \Gamma$ for a finite group of rotations of $\mathbb{T}$. We set aside the group for the moment and focus on $C(\mathbb{T})$.

We identify $\mathbb{T} \cong \mathbb{R}/\mathbb{Z}$ (it has volume 1). Set $D = -i\frac{d}{dx}$. The Laplacian on the circle is $\Delta = -\frac{\partial^2}{\partial x^2}$, densely defined on $L^2(\mathbb{T})$ and diagonalizable with eigenvalues $4\pi^2n^2, n \in \mathbb{Z}$.

The first problem is to prove the meromorphic extendibility of the analytic functions $\mathrm{Tr}(f\Delta^{-s})$ for $f \in C^{\infty}(\mathbb{T})$. If $f = 1$, this is equivalent to the problem of

meromorphically extending the Riemann zeta function since $\mathrm{Tr}(\Delta^{-s}) = \sum_n n^{-2s}$.
We show how the heat equation method produces such a continuation.

The solution to the heat equation (10.35) on $\mathbb{R}$ is

$$k_t(x, y) = (4\pi t)^{-\frac{1}{2}} \exp\left(-\frac{(x-y)^2}{4t}\right),$$

as the reader may verify. Thus the operator $e^{-t\Delta}$ is an integral operator with kernel
k_t. Notice that this is a group convolution operator on the group $\mathbb{R}$:

$$e^{-t\Delta}u = f_t * u, \quad \text{where} \quad f_t(x) = (4\pi t)^{-\frac{1}{2}} \exp\left(-\frac{x^2}{4t}\right).$$

That is, $e^{-t\Delta} \in C^*(\mathbb{R})$ for all $t > 0$: The heat semigroup is contained in the C*-
algebra of $\mathbb{R}$.

For the Laplacian on the circle, we have

(10.42)
$$e^{-t\Delta} = \sum_{n\in\mathbb{Z}} e^{-4\pi^2 t n^2} p_n,$$

where p_n is the projection to the span of z^n, the integral operator with kernel

$$p_n(x, y) = e^{2\pi i n(x-y)}.$$

Substituting this into the series gives that

$$k_t(x, y) = \sum_{n\in\mathbb{Z}} e^{-4\pi^2 t n^2 + 2\pi i n(x-y)},$$

Therefore, $e^{-t\Delta} \in C^*(\mathbb{T})$ is again given by a convolution, this time on the group
$\mathbb{T} = \mathbb{R}/\mathbb{Z}$:

$$e^{-t\Delta}u = k_t * u, \quad k_t(x) = \sum_{n\in\mathbb{Z}} e^{-4\pi t n^2 + 2\pi i n x}.$$

This description of the heat kernel $k_t(x, y)$ on $\mathbb{T}$, by the formula

$$k_t(x, y) = \sum_{n\in\mathbb{Z}} e^{-4\pi^2 t n^2 + 2\pi i n(x-y)},$$

is "spectral" rather than geometric, producing a function of t in the class of functions
known as θ-functions and without any closed form.

However, we can argue in this case that the heat kernel on the circle, being locally Euclidean, agrees with the heat kernel on Euclidean space up to an exponentially small error. This will produce a rather trivial asymptotic expansion of the form

$$\mathrm{Tr}(fe^{-t\Delta}) = (\pi t)^{-\frac{1}{2}} \cdot \int_{\mathbb{T}} f(x)dx + r(t),$$

where $r(t) \to 0$ exponentially fast as $t \to 0$: that is, $t^{-k}r(t) \to 0$ as $t \to 0$ for any positive k. We may then apply Exercise 10.7.6.

The formula $(4\pi t)^{-\frac{1}{2}} \exp(-\frac{(x-y)^2}{4t})$ for the heat kernel on $\mathbb{R}$ can be periodized because it decays rapidly at infinity.

Lemma 10.9.1 *If k_t is the heat kernel on $\mathbb{T}$, then*

$$(10.43) \qquad k_t(x, y) = \sum_{n \in \mathbb{Z}} (4\pi t)^{-\frac{1}{2}} \exp\left(-\frac{(x - y + n)^2}{4t}\right).$$

Proof By an easy calculation, the formula satisfies the heat equation. As $t \to 0$, it converges to the distribution δ_{x-y}, so it satisfies the initial condition as well. □

Remark 10.9.2 In the above constructions there appeared the identity

$$(10.44) \qquad k_t(x) = \frac{1}{\sqrt{4\pi t}} \sum_{n \in \mathbb{Z}} e^{-\frac{(x+n)^2}{4t}} = \sum_{n} e^{-4\pi^2 t n^2 + 2\pi i n x},$$

where $k_t(x, y) = k_t(x - y)$, with k_t the heat kernel on the circle. The left hand side of (10.44) is geometric and is obtained by forming the periodization $\sum_{n \in \mathbb{Z}} k_t^{\mathbb{R}}(x+n)$ of the heat kernel on $\mathbb{R}$. The right hand side is spectral in nature: from the point of view of the right hand side, k_t is the element of $L^2(\mathbb{T})$ with Fourier series

$$\sum e^{-4\pi^2 t n^2} \cdot e^{2\pi i n x} = \sum \hat{f}_t(n) e^{2\pi i n x},$$

where $\hat{f}(\xi) = e^{-4\pi^2 t \xi^2}$. From our discussion of Fourier transforms, $f_t(x) = \frac{1}{\sqrt{4\pi t}} e^{-\frac{x^2}{4t}}$. Hence the identity (10.44) is nothing but the Poisson summation formula (Theorem 1.9.22)

$$\sum_{n \in \mathbb{Z}} f_t(x + n) = \sum_{n \in \mathbb{Z}} \hat{f}_t(n) e^{2\pi i n x}.$$

Corollary 10.9.3 *If k_t is the heat kernel on $\mathbb{T}$, then*

$$k_t(x, x) = (4\pi t)^{-\frac{1}{2}} + r(t),$$

where $r(t) \to 0$ exponentially fast as $t \to 0$.

Proof From (10.43) when $x = y$, we have

$$(10.45) \qquad k_t(x, x) = \sum_{n \in \mathbb{Z}} (4\pi t)^{-\frac{1}{2}} e^{-\frac{4\pi^2 n^2}{4t}} = (\pi t)^{-\frac{1}{2}} + r(t),$$

where

$$r(t) = 2 \sum_{n=1}^{\infty} (4\pi t)^{-\frac{1}{2}} e^{-\frac{4\pi^2 n^2}{4t}}.$$

An application of the integral test shows that $r(t)$ vanishes to infinite order at $t = 0$.

$\square$

Following the methods discussed in the previous sections (see Exercise 10.7.1), we obtain a continuation of the zeta function $\mathrm{Tr}(f\Delta^{-s})$ for $f \in C^{\infty}(\mathbb{T})$:

$$\Gamma(s) \cdot \mathrm{Tr}(f\Delta^{-s}) = \int_0^1 t^{s-1} \int_{\mathbb{T}} f(x)k_t(x, x)\,dx\,dt$$

$$= \frac{1}{2\sqrt{\pi}} \int_0^1 \int_{\mathbb{T}} f(x)t^{s-1} \cdot \left(t^{-\frac{1}{2}} + r(t)\right) dx\,dt$$

$$= \frac{1}{2\sqrt{\pi}} \int_0^1 \int_{\mathbb{T}} f(x)t^{s-\frac{3}{2}}\,dx\,dt + \frac{1}{2\sqrt{\pi}} \int_0^1 \int_{\mathbb{T}} f(x)t^{s-1}r(t)\,dx\,dt$$

$$= \left(\frac{1}{2\sqrt{\pi}} \int_{\mathbb{T}} f(x)\,dx\right) \cdot \left(\frac{1}{s - 1/2}\right) + e(s),$$

where $e(s)$ is entire.

Since $\Gamma(\frac{1}{2}) = \sqrt{\pi}$, we get the following corollary:

Corollary 10.9.4 *Let Δ be the Laplacian on $\mathbb{T} = \mathbb{R}/\mathbb{Z}$. Then if $f \in C(\mathbb{T})$, then the function $\mathrm{Tr}(f\Delta^{-s})$, $\mathrm{Re}(s) > 1$, extends meromorphically to $\mathbb{C}$ with a single simple pole at $s = \frac{1}{2}$, and*

$$(10.46) \qquad \mathrm{Res}_{s=\frac{1}{2}} \mathrm{Tr}(f\Delta^{-s}) = \frac{1}{2\pi} \int_{\mathbb{T}} f(x)\,dx,$$

where dx is Lebesgue measure on $\mathbb{T}$.

Corollary 10.9.5 *The triple consisting of the Hilbert space $L^2(\mathbb{T})$, the operator $D = -i\frac{d}{dx}(x)$, and the representation of $C(\mathbb{T})$ by multiplication operators defines an odd, one-dimensional regular spectral cycle over $C^{\infty}(\mathbb{T}) \subset C(\mathbb{T})$ with the meromorphic continuation property and simple dimension spectrum consisting of the single point $\{1/2\} \subset \mathbb{C}$.*

Now to get a mildly noncommutative example, let $\Gamma \cong \mathbb{Z}/n$ be a finite group of rotations, generated by U. We consider the crossed product $C(\mathbb{T}) \rtimes \Gamma$. The group Γ is represented on $L^2(\mathbb{T})$ by the unitary action induced by the given action on $L^2(\mathbb{T})$.

Since differentiation on $\mathbb{T}$ commutes with rotations, the group Γ of unitaries on $L^2(\mathbb{T})$ commutes with $D = -i\frac{d}{dx}$, and we obtain therefore a spectral cycle over the *crossed product* $C(\mathbb{T}) \rtimes \Gamma$, with only one issue remaining: the meromorphic continuation property for elements of $C^\infty(\mathbb{T}) \rtimes \Gamma$.

Let $\hbar \neq 0$ be any nonzero real number, and suppose that $U_\hbar$ is a rotation unitary on $L^2(\mathbb{T})$: i.e., $(U_\hbar \xi)(x) = \xi(x - \hbar)$. The Fourier transform of $U_\hbar$ acting on $l^2(\mathbb{Z})$ is the diagonal operator with entries $\omega^n = e^{-2\pi i n \hbar}$, and hence looking in the Fourier transform picture we see that $U_\hbar \Delta^{-s}$ is, up to a constant, the diagonal operator with entries $\omega^n |n|^{-2s}$. Hence,

$$\operatorname{Tr}(U_\hbar \Delta^{-s}) = 2 \sum_{n=1}^\infty \omega^n n^{-2s}.$$

But we will not try to understand this series, but rather use the heat equation approach.

Lemma 10.9.6 *If $\hbar \neq 0$ and $f \in C(\mathbb{T})$, then the analytic function $\operatorname{Tr}(fU_\hbar \Delta^{-s})$ initially defined for $\operatorname{Re}(s) > \frac{1}{2}$ extends analytically to $\mathbb{C}$.*

Proof As discussed above, $\Gamma(s) \cdot \operatorname{Tr}(fU_\hbar \Delta^{-s}) \sim \int_0^1 t^{s-1} \operatorname{Tr}(fe^{-t\Delta}) dt$. The operator $fU_\hbar e^{-t\Delta}$ is an integral operator on $L^2(\mathbb{T})$ with kernel $g_t(x, y) := f(x)k_t(x + \hbar, y)$, where k_t is the heat kernel. By Lemma 10.9.1,

$$g_t(x, y) = (4\pi t)^{-1/2} \cdot \sum_{n \in \mathbb{Z}} f(x) \cdot \exp\left(-\frac{(x + \hbar - y + n)^2}{4t}\right)$$

and hence

$$g_t(x, x) = (4\pi t)^{-1/2} \cdot \sum_{n \in \mathbb{Z}} f(x) \cdot \exp\left(-\frac{(\hbar + n)^2}{4t}\right).$$

We have, thus,

$$\Gamma(s) \cdot \operatorname{Tr}(fU_\hbar \Delta^{-s}) \sim \int_0^1 t^{s-1} \int_{\mathbb{T}} g_t(x, x) dx \, dt.$$

The function $\int_{\mathbb{T}} g_t(x, x) dx \to 0$ exponentially fast as $t \to 0$. The result then follows from Exercise 10.7.6.

$\square$

Corollary 10.9.7 *Let $\Gamma \subset \mathbb{T}$ be a finite group acting on $\mathbb{T}$ by the group multiplication, and let $\pi: C(\mathbb{T}) \rtimes \Gamma \to \mathbb{B}(L^2(\mathbb{T}))$ the representation in which*

$f \in C(\mathbb{T})$ *acts by multiplication, and group elements act by translations. Then the triple*

$$\left(L^2(\mathbb{T}), \pi, D = -i\frac{d}{dx} \right)$$

is a one-dimensional regular spectral cycle for $C^\infty(\mathbb{T}) \rtimes \Gamma \subset C(\mathbb{T}) \rtimes \Gamma$ *with the meromorphic continuation property and a simple dimension spectrum consisting of the point* $1/2$.

If $a = \sum_\gamma a_\gamma[\gamma] \in C^\infty(\mathbb{T}) \rtimes \Gamma$, *then*

$$\frac{1}{2} \cdot \mathrm{Res}_{s=1/2}\mathrm{Tr}(a\Delta^{-s}) = \int_{\mathbb{T}} a_e(x)dx = \int_{\mathbb{T}} a_e d\mu = \tau(a),$$

where $\tau \colon C(\mathbb{T}) \rtimes \Gamma \to \mathbb{C}$ *is the trace induced by* normalized *Lebesgue measure on* $\mathbb{T}$.

What does the Local Index Theorem 10.8.4 say about this situation?

In general, if (H, π, D) is a one-dimensional regular spectral cycle for $A^\infty \subset A$ with simple dimension spectrum, then inspection of the Local Index Formula shows that there is a nonzero residue only when $\alpha = 0 = k$, where we see the class of the cyclic 1-cocycle

$$\varphi(a^0, a^1) = \mathrm{const.} \cdot \mathrm{Res}_{s=1/2}\mathrm{Tr}(a^0[D, a^1]\Delta^{-s}),$$

where $\Delta = D^2$, for an appropriate constant.

As differentiation D on the circle commutes with the group action, the commutator $\delta(a) := [D, a]$ for any $a \in C^\infty(\mathbb{T})[\Gamma]$ is given explicitly by

$$(10.47) \qquad \delta\left(\sum_{g \in G} a_g\,[g] \right) = \sum_{g \in G}[D, a_g]\,[g] = \sum_{g \in G} -ia'_g\,[g],$$

with g' the angular derivative of g. In particular, δ maps $C^\infty(\mathbb{T})[\Gamma]$ to itself. Let τ be the trace

$$(10.48) \qquad \tau \colon C(\mathbb{T}) \rtimes \Gamma \to \mathbb{C}, \qquad \tau\left(\sum_{g \in G} a_g[g] \right) := \int_{\mathbb{T}} a_e(x)dx.$$

Sorting out the constants and applying Corollary 10.9.7, we get the following theorem:

Theorem 10.9.8 *Let* Γ *be a finite subgroup of* $\mathbb{T} = \mathbb{R}/\mathbb{Z}$ *acting on* $C(\mathbb{T})$ *by translation. Let* $\delta \colon C^\infty(\mathbb{T}) \rtimes \Gamma \to C^\infty(\mathbb{T}) \rtimes \Gamma$ *be the derivation* (10.47) *and* τ *be the normalized trace* (10.48) *on* $C(\mathbb{T}) \rtimes \Gamma \to \mathbb{C}$. *Then*

(10.49) $$\varphi(a^0, a^1) := \tau\left(a^0\delta(a^1)\right)$$

defines a cyclic 1-cocycle on $C^\infty(\mathbb{T}) \rtimes \Gamma$ and

$$\langle[u], [D]\rangle = \frac{-1}{2\pi i} \cdot \varphi(u^{-1}, u) = \frac{-1}{2\pi i} \cdot \tau(u^{-1}\delta(u))$$

for any unitary $u \in C^\infty(\mathbb{T}) \rtimes \Gamma$, where $\langle\cdot\rangle$ is the index pairing

$$K_1(C(\mathbb{T}) \rtimes \Gamma) \times KK_1(C(\mathbb{T}) \rtimes \Gamma, \mathbb{C}) \to \mathbb{Z}.$$

Remark 10.9.9   Setting the group Γ to be trivial we recover the Toeplitz Index Theorem 2.6.3 since $\frac{-1}{2\pi i} \cdot \varphi(u^{-1}, u) = \frac{-1}{2\pi i} \cdot \int_\mathbb{T} \frac{du}{u} = -\text{wind}_u(0)$ for $u \in C^\infty(\mathbb{T})$ a smooth unitary, and the index pairing $\langle[u], [D]\rangle$ is by the definitions equal to the Fredholm index of the Toeplitz operator T_u.

10.10   Heisenberg Spectral Cycles and Irrational Rotation

In the last section we examined an example of a one-dimensional spectral cycle over the crossed product $C(\mathbb{T}) \rtimes \Gamma$ of $C(\mathbb{T})$ by a finite order group. Since such actions are free, the crossed product is Morita equivalent to the quotient $\Gamma\backslash\mathbb{T}$, which is $\mathbb{T}$ again. On the other hand, the irrational rotation algebra $A_\hbar := C(\mathbb{T}) \rtimes_\hbar \mathbb{Z}$, the crossed product by a dense subgroup of $\mathbb{T}$, is a *simple* C*-algebra and not commutative even up to Morita equivalence because it is simple. It has a two-dimensional, rather than one-dimensional, nature, containing the two one-dimensional C*-algebras $C(\mathbb{T})$ and $C^*(\mathbb{Z}) \cong C(\widehat{\mathbb{Z}}) \cong C(\mathbb{T})$, which give it a sort of twisted (noncommutative) two-dimensional product structure. $A_\hbar$ is often referred to in Noncommutative Geometry as a *noncommutative torus*.

In order to exhibit an example of a two-dimensional spectral cycle over $A_\hbar$, we perturb the differentiation operator $\frac{d}{dx}$ on $L^2(\mathbb{R})$ by adding x to it, which produces a two-dimensional, noncommutative "Heisenberg" geometry based on the Dirac–Schrödinger operators $x \pm d/dx$. This Heisenberg geometry is actually another way of thinking of the irrational tori $A_\hbar$.

The Heisenberg group is the matrix group

$$H = \left\{ \begin{bmatrix} 1 & x & z \\ 0 & 1 & y \\ 0 & 0 & 1 \end{bmatrix} \mid x, y, z \in \mathbb{R} \right\}.$$

The Lie algebra $\mathfrak{h}$ of H is the tangent space at the identity of the group and is the Lie algebra of 3-by-3 strictly upper triangular matrices. Let X, Y be the elements

$$X = \begin{bmatrix} 0 & 1 & 0 \\ 0 & 0 & 0 \\ 0 & 0 & 0 \end{bmatrix}, \quad Y = \begin{bmatrix} 0 & 0 & 0 \\ 0 & 0 & 1 \\ 0 & 0 & 0 \end{bmatrix},$$

of $\mathfrak{h}$. Then

$$[X, Y] = Z := \begin{bmatrix} 0 & 0 & 1 \\ 0 & 0 & 0 \\ 0 & 0 & 0 \end{bmatrix},$$

while Z is central in $\mathfrak{h}$.

Now let π be any *irreducible* representation of H. Since it is irreducible and $\pi(Z)$ commutes with $\pi(\mathfrak{h})$, $\pi(Z)$ is a multiple of the identity operator:

$$\pi(Z) = \hbar,$$

for some $\hbar \in \mathbb{R}$, a "Planck constant."

The name "Heisenberg group" is motivated by these relations, which have the same form as the canonical commutation relations in quantum mechanics, where x and $\frac{d}{dx}$ model position and momentum operators.

From the above remarks, we obtain a classification of irreducible representations of H. Either $\hbar = 0$, in which case $\pi(Z) = 0$ and hence $\pi(X)$ and $\pi(Y)$ commute, which implies the representation is one-dimensional and is completely determined by the pair of real numbers $(\pi(X), \pi(Y))$, or $\hbar \neq 0$, in which case one can show that the representation is isomorphic to the following interesting representation $\pi_{\hbar}$ of $\mathfrak{h}$ by unbounded operators on $L^2(\mathbb{R})$. Let

$$\pi_{\hbar}(X) = x, \quad \text{and} \quad \pi_{\hbar}(Y) = \hbar \frac{d}{dx}.$$

Then $[x, \hbar \frac{d}{dx}] = \hbar$, so the required identity is satisfied to give a representation.

To see what the canonical anti-commutation relations have to do with rotation algebras, observe that the application of functional calculus to the operators x and $\frac{d}{dx}$ produces the operators

$$u = e^{2\pi i x}, \quad v_{\hbar} := e^{2\pi \hbar \frac{d}{dx}},$$

where u is multiplication by the periodic function $e^{2\pi i x}$ and

$$(v_{\hbar})\xi(x) = \xi(x - \hbar).$$

We have

$$u v_{\hbar} = e^{-2\pi i \hbar} v_{\hbar} u.$$

Now let

$$A_\hbar := C(\mathbb{T}) \rtimes_\hbar \mathbb{Z},$$

where $\mathbb{Z}$ acts on the circle $\mathbb{T} := \mathbb{R}/\mathbb{Z}$ with generator the automorphism induced by translation by $\hbar$ mod $\mathbb{Z}$. If $U \in A_\hbar = C(\mathbb{T}) \rtimes_h \mathbb{Z}$ is the generator $U(t) = e^{2\pi i t}$ of $C(\mathbb{T})$ and V_h the generator of the $\mathbb{Z}$ action in the crossed product, then a quick computation shows that

$$UV = e^{-2\pi i \hbar} VU \ \in A_\hbar,$$

and it follows that we obtain, for each $\hbar$, a representation

$$\pi_\hbar \colon A_\hbar \to \mathbb{B}(L^2(\mathbb{R}))$$

of $A_\hbar$ on $L^2(\mathbb{R})$.

Before proceeding, we make note that $A_\hbar$ has a natural "smooth structure" and corresponding subalgebra $A_\hbar^\infty \subset A_\hbar$ analogous to the algebra of smooth functions on $\mathbb{T}^2$.

Definition 10.10.1 The *-subalgebra $A_\hbar^\infty \subset A_\hbar$ is given by the collection of all $a = \sum_n f_n[n] \in A_\hbar$ such that f_n is smooth for all n and $\sup_{n \in \mathbb{Z}} \|f_n^{(l)}\| n^k < \infty$ for all $k, l = 1, 2, \ldots$.

Remark 10.10.2 The notation $a = \sum_n f_n[n]$ has a specific meaning for crossed products such as $C(\mathbb{T}) \rtimes \mathbb{Z}$ and does not imply convergence in the sense of series. But if $a \in l^1(G, A)$, then the series converges norm absolutely in $A \rtimes \mathbb{Z}$. This in particular holds if $a \in A_\hbar^\infty$, so that the expansion $a = \sum f_n[n]$ is a norm absolutely convergent series in $A \rtimes G$ for smooth a.

Exercise 10.10.3 Show that δ_1 and δ_2 defined on $A_\hbar^\infty$

$$\delta_1 \left(\sum_n f_n[n] \right) = \sum_n f_n'[n], \qquad \delta_2 \left(\sum_n f_n[n] \right) = \sum_n i n f_n[n]$$

are well defined, and define derivations of the algebra $A_\hbar^\infty$.

We are going to fit the Heisenberg representations discussed above into a family of spectral cycles using the properties of the *harmonic oscillator*

$$H := -\frac{d^2}{dx^2} + x^2,$$

a second-order elliptic operator on $\mathbb{R}$, whose domain we will take initially to be the Schwartz space $\mathcal{S}(\mathbb{R})$.

Let $A = x + \frac{d}{dx}$, with initial domain the Schwartz space $S(\mathbb{R})$—the "annihilation" operator. The operator $A^* = x - \frac{d}{dx}$ is a "creation operator." The reason for these terms will emerge shortly. Firstly, some routine computations show that

(10.50)
$$AA^* = H + 1, \quad A^*A = H - 1, \quad [A, A^*] = 2, \quad [H, A] = -2A, \quad [H, A^*] = 2A^*.$$

Exercise 10.10.4 Let $\varphi_0(x) = e^{-\frac{x^2}{2}}$ and $\varphi_k(x) = A^*\varphi_{k-1}, k = 1, 2, \ldots$. Using the relation $HA^* = A^*H + 2A^*$, from (10.50), prove that $H\varphi_k = (2k + 1)\varphi_k$, for $k = 0, 1, \ldots$.

The operator A^* (sometimes also called a "ladder operator" in physics) shifts along the eigenspaces of the harmonic oscillator, corresponding to the eigenvalues, the odd integers $1, 2, 3, \ldots$.

As in the exercise above, let $\varphi_k = A^*(\psi_{k-1})$ define a family of vectors starting with $\varphi_0(x) = e^{-\frac{x^2}{2}}$, with φ_k an eigenvector of H with eigenvalue $2k + 1$.

Observe that $\langle \varphi_k, \varphi_l \rangle = 0$ if $k \neq l$ since φ_k, φ_l belong to different eigenspaces of the self-adjoint operator H. Furthermore, we can rescale the φ_n to get unit vectors $\psi_0(x) = \frac{1}{\sqrt{\pi}}e^{-\frac{x^2}{2}}$ and for $k = 1, 2, \ldots, \psi_k = \frac{1}{\sqrt{2k}}A^*(\psi_{k-1})$. Induction implies ψ_k is a unit vector:

$$\|\psi_k\|^2 = \frac{1}{2k}\|A^*\psi_{k-1}\|^2 = \frac{1}{2k}\langle AA^*\psi_{k-1}, \psi_{k-1}\rangle = \frac{1}{2k}\langle (H + 1)\psi_{k-1}, \psi_{k-1}\rangle$$

(10.51) $$= \langle \psi_{k-1}, \psi_{k-1}\rangle = 1.$$

We show that the functions φ_n are given explicitly by $\varphi_n(x) = H_n(x)e^{-\frac{x^2}{2}}$, where H_n are the classical *Hermite polynomials*. They can be defined by

$$H_n(x) = (-1)^n e^{x^2}\frac{d^n}{dx^n}(e^{-x^2})$$

(with $H_0(x) = 1$).

Exercise 10.10.5 Prove that:

a) $H_3(x) = 8x^3 - 12x$.
b) $H_n(x)$ is a polynomial for all n.
c) The identity

$$H_{n+1}(x) = 2x H_n(x) - H'_n(x)$$

holds.

For $n = 0, 1, 2, \ldots$, let $\phi_n(x) = H_n(x)e^{-\frac{x^2}{2}}$. Applying the operator A^* to φ_n gives

$$A^*(\phi_n) = x H_n(x)e^{-\frac{x^2}{2}} - H_n'(x)e^{-\frac{x^2}{2}} + x H_n(x)e^{-\frac{x^2}{2}} = 2x H_n(x)e^{-\frac{x^2}{2}} - H_n'(x)e^{-\frac{x^2}{2}},$$

and the identity of part c) of the exercise above gives

$$A^*(\phi_n) = \phi_{n+1}.$$

This shows that $\phi_n = \varphi_n$, where $\varphi_n := (A^*)^n \varphi_0$ is as in the prior discussion.

Exercise 10.10.6 Deduce from $[H, A] = -2A$ that

$$A\psi_k = \sqrt{2k} \cdot \psi_{k-1}, \qquad A^*\psi_k = \sqrt{2k+2} \cdot \psi_{k+1}.$$

The vectors $\{\psi_k\}$ form an orthonormal basis for $L^2(\mathbb{R})$, as the following exercise demonstrates.

Exercise 10.10.7 Prove...

a) If $f \in L^2(\mathbb{R})$ and f is orthogonal to every L^2-function of the form $p(x)e^{-\frac{x^2}{2}}$, where $p(x)$ is a polynomial, then $f = 0$.
b) The linear span of the φ_k defined above (or the ψ_k) is equal to the linear span of the functions of the form $p(x)e^{-\frac{x^2}{2}}$, $p(x)$ a polynomial.
c) If $f \in L^2(\mathbb{R})$ is orthogonal to every $p(x)e^{-\frac{x^2}{2}}$, with $p(x) \in \mathbb{R}[x]$, then $f = 0$ in L^2. (*Hint.* Exercise 1.9.17.)
d) The ψ_k form an orthonormal basis for $L^2(\mathbb{R})$.

Each ψ_k is in the Schwartz class $\mathcal{S}(\mathbb{R})$. The discussion above shows that $\{\psi_k\}_{k=0}^{\infty}$ is an orthonormal basis for $L^2(\mathbb{R})$ and that with respect to this basis, H is diagonal with eigenvalues the odd integers $1, 2, 3, \ldots$:

$$H = \begin{bmatrix} 1 & 0 & \cdots & \cdots \\ 0 & 3 & 0 & \cdots \\ 0 & 0 & 5 & \cdots \\ 0 & \cdots & \cdots & \cdots \end{bmatrix}.$$

In particular, H has a canonical extension to a self-adjoint operator on $L^2(\mathbb{R})$, and $f(H)$ is a compact operator for all $f \in C_0(\mathbb{R})$ and a bounded operator for all $f \in C_b(\mathbb{R})$.

The *-algebra $\mathcal{D}$ generated by the unbounded operator A is called the *Weyl algebra*. It contains A^* and hence H. It contains every differential operator

$$f_0 + f_1 \cdot \frac{d}{dx} + \cdots + f_m \cdot \frac{d^m}{x^m}$$

on $\mathbb{R}$ with polynomial coefficients.

If $f \in L^2(\mathbb{R})$, let $(\hat{f}(n))$ denote the sequence of its Fourier coefficients with respect to the eigenbasis $\{\psi_k\}_{k \in \mathbb{Z}}$ for $L^2(\mathbb{R})$ for H discussed above.

Lemma 10.10.8 *If $f \in L^2(\mathbb{R})$, then $f \in S$ if and only if $(\hat{f}(n))$ is a rapidly decreasing sequence of integers:*

$$|\hat{f}(n)| = O(n^{-k})$$

for any k.

Proof If $f \in S$, then Hf is in S, as is clear from the definition of H. Similarly, $H^k f \in S$ for all k. Since

$$\widehat{H^k f}(n) = (2n + 1)^k \cdot \hat{f}(n)$$

and since this is an L^2-sequence (since $H^k f \in S$ as already observed), and hence bounded, we get for each k a constant C such that

$$(2n + 1)^k \cdot |\hat{f}(n)| \leq C_k$$

and so

$$|\hat{f}(n)| = O(n^{-k})$$

follows.

Conversely, suppose that $f \in L^2(\mathbb{R})$ and that $(\hat{f}(n))$ is a rapidly decreasing sequence. The eigenvectors ψ_k are Schwartz functions. It follows easily that $f = \sum \hat{f}(n) \psi_n \in S$ as well.   □

Let D be the unbounded operator

$$D = \begin{bmatrix} 0 & A^* \\ A & 0 \end{bmatrix}$$

on $L^2(\mathbb{R}) \oplus L^2(\mathbb{R})$, defined initially on Schwartz functions; it admits a canonical extension to a densely defined self-adjoint operator. We have

$$D^2 = \begin{bmatrix} H - 1 & 0 \\ 0 & H + 1 \end{bmatrix},$$

and hence $1 + D^2 = \begin{bmatrix} H & 0 \\ 0 & H + 2 \end{bmatrix}$ which is diagonal with respect to the basis $\{\psi_n\}_{n=0}^{\infty}$ and invertible.

Next, writing just a rather than $\pi_{\hbar}(a)$ in the notation temporarily, since $\hbar$ is fixed, we compute

$$\left[\begin{bmatrix} a & 0 \\ 0 & a \end{bmatrix}, \ D\right] = \begin{bmatrix} 0 & [a, A^*] \\ [a, A] & 0 \end{bmatrix}.$$

If $a = \sum f_n[n]$, then

$$[a, A] = \sum_n [f_n, A][n] + f_n[A, [n]] = \sum_n f_n'[n] + n f_n[n] = \delta_1(a) - i\delta_2(a),$$

where $\delta_1, \delta_2 \colon A_\hbar^\infty \to A_\hbar^\infty$ are the derivations of Exercise 10.10.3. Similarly, $[a, A^*] = \delta_1(a) + i\delta_2(a)$. These expressions are all well defined on the smooth subalgebra $A_\hbar^\infty$. Hence

$$\left[\begin{bmatrix} a & 0 \\ 0 & a \end{bmatrix}, \ D\right] = \begin{bmatrix} 0 & \delta_1(a) + i\delta_2(a) \\ \delta_1(a) - i\delta_2(a) & 0 \end{bmatrix}.$$

Definition 10.10.9 Let $A_\hbar^\infty \subset A_\hbar := C(\mathbb{T}) \rtimes_\hbar \mathbb{Z}$ as in Definition 10.10.1.

The *Heisenberg cycle* is the even, two-dimensional spectral cycle over $A_\hbar^\infty \subset A_\hbar$, given by

$$\left(L^2(\mathbb{R}) \oplus L^2(\mathbb{R}), \ \pi_\hbar \oplus \pi_\hbar, \ D = \begin{bmatrix} 0 & A^* \\ A & 0 \end{bmatrix} \right),$$

where $A = x + \frac{d}{dx}$.

We denote by $[D_\hbar]$ the class in $KK_0(A_\hbar, \mathbb{C})$ of the corresponding Fredholm module, given by

$$\left(L^2(\mathbb{R}) \oplus L^2(\mathbb{R}), \ \pi_\hbar \oplus \pi_\hbar, \ F := \chi(D) = \begin{bmatrix} 0 & A(H+2)^{-\frac{1}{2}} \\ AH^{-\frac{1}{2}} & 0 \end{bmatrix} \right),$$

using the normalizing function $\chi(x) = x(1 + x^2)^{1/2}$.

Note that the parameter $\hbar$ only appears in the representation.

The spectrum of $\Delta := D^2 = \begin{bmatrix} H - 1 & 0 \\ 0 & H + 1 \end{bmatrix}$ grows linearly, so the principal values of D, i.e., the eigenvalues of $|D| = H^{1/2}$, are $O(n^{1/2})$. So the cycle is two-dimensional. Establishing the meromorphic continuation property and simple dimension spectrum will be the task of the next section.

We close with noting that the $x \pm d/dx$ construction determines a spectral cycle over a much larger algebra than $A_\hbar$. Continuous periodic functions, e.g., elements of $C(\mathbb{T})$, are special cases of bounded, uniformly continuous functions on $\mathbb{R}$. Other examples are Lipschitz functions or functions with continuous, bounded first derivative.

Definition 10.10.10 Let $C_u(\mathbb{R})$ be the C*-algebra of all bounded uniformly continuous functions on $\mathbb{R}$.

The following exercise shows that the Heisenberg spectral cycle is defined over $C_u(\mathbb{R})$, modulo the issue of the meromorphic continuation property.

Exercise 10.10.11 Let $C_b^\infty(\mathbb{R})$ be all $f \in C_b^\infty(\mathbb{R})$ with bounded derivatives of all orders. Show that $C_b^\infty(\mathbb{R})$ is dense in $C_u(\mathbb{R})$.

If D is as in Definition 10.10.9, then $[D, f]$ is bounded for $f \in C_b^\infty(\mathbb{R})$. We have already noted that $[D, U_\alpha]$ is bounded for the unitary U_α induced by translation by $\alpha \in \mathbb{R}$.

Let $\Gamma \subset \mathbb{R}$ be any countable subgroup (possibly dense) of the real numbers. Inside the crossed product $C_u(\mathbb{R}) \rtimes \Gamma$, let $(C_u(\mathbb{R}) \rtimes \Gamma)^\infty$ denote all elements $\sum_{\gamma \in \Gamma} f_\gamma[\gamma]$ in the crossed product such that $\sum_{\gamma \in \Gamma} \|f_\gamma^{(k)}\| \cdot |\gamma|^l < \infty$ for all k, l positive integers. Then the derivations

$$\delta_1\left(\sum_\gamma f_\gamma[\gamma]\right) := \sum_\gamma f_\gamma'[\gamma], \quad \delta_2\left(\sum_\gamma f_\gamma[\gamma]\right) = \sum_\gamma |\gamma| \cdot f_\gamma[\gamma]$$

extend to $(C_u(\mathbb{R}) \rtimes \Gamma)^\infty$.

Proposition 10.10.12 *For any countable subgroup $\Gamma \subset \mathbb{R}$, let π be the standard representation of the crossed product $C_u(\mathbb{R}) \rtimes \Gamma \subset \mathbb{B}(L^2\mathbb{R})$; then the triple*

$$\left(L^2(\mathbb{R}) \oplus L^2(\mathbb{R}), \pi \oplus \pi, \quad D = \begin{bmatrix} 0 & A^* \\ A & 0 \end{bmatrix}\right)$$

defines a two-dimensional spectral cycle for $(C_u(\mathbb{R}) \rtimes \Gamma)^\infty \subset C_u(\mathbb{R}) \rtimes \Gamma$.

Exercise 10.10.13 Let α be a smooth flow on a compact manifold M and $p \in M$. If $f \in C(M)$, let $f_p(t) = f(\alpha_t(p))$.

a) Show that f_p is uniformly continuous.
b) Show that $f_p \in C_b^\infty(\mathbb{R})$ if $f \in C^\infty(M)$.
c) Show that the Heisenberg triple of Proposition 10.10.12 restricts to a two-dimensional spectral cycle for $(C(M) \rtimes [\Gamma])^\infty \subset C(M) \rtimes \Gamma$ for any subgroup $\Gamma \subset \mathbb{R}$, acting on M by the flow (an action of $\mathbb{R}$).

The next exercise provides an interesting example of the method of the former exercise, where the compact manifold M is $\mathbb{T}^2$.

Exercise 10.10.14 Let $\hbar, \mu \in \mathbb{R}$ be rationally independent and $(\alpha_t)_{t \in \mathbb{R}}$, $(\beta_t)_{t \in \mathbb{R}}$ the Krönecker flows $\alpha_t(x, y) = (x + t, y + \hbar t)$, $\beta_t(x, y) = (x + t, y + \mu t)$.

a) α and β determine connected dense subgroups K_α and K_β of the compact Lie group $\mathbb{T}^2$.

b) Suppose $\alpha_s(0) = \beta_t(0)$ for some $s, t \in \mathbb{R}$. That is, suppose $(s, s\hbar) = (t, t\mu)$. Show there exist integers n, m such that $s = t + n$ and $t = \frac{m}{\hbar-\mu} + \frac{n\hbar}{\hbar-\mu}$. Deduce that $K_\alpha \cap K_\beta$ is countable.

c) Show that the rank 2two free abelian subgroup Λ of $\mathbb{R}$ generated by $\frac{1}{\hbar-\mu}$ and $\frac{\hbar}{\hbar-\mu}$ is dense in $\mathbb{R}$ and is isomorphic to $K_\alpha \cap K_\beta$.

d) Prove that the crossed product $C(\mathbb{T}^2) \rtimes \Lambda$ is (canonically) isomorphic to $A_\hbar \otimes A_\mu$.

The spectral triples over $A_\hbar \otimes A_\mu$ obtained by using part d) and the method of Exercise 10.10.13 seem to play the role of KK-duality classes (our construction produces in particular a cycle for $KK_0(A_\hbar \otimes A_\mu, \mathbb{C})$).

10.11 The Harmonic Oscillator Residue Trace

In this section we establish the meromorphic continuation property of Definition 10.6.3 of spectral cycle, for the Heisenberg cycles of Definition 10.10.9. If $\Delta = D^2$ where D is as in the definition, then $\Delta = \begin{bmatrix} H-1 & 0 \\ 0 & H+1 \end{bmatrix}$, where H is the harmonic oscillator $H = -\frac{d^2}{dx^2} + x^2$.

Since $\mathrm{Tr}(a\Delta^{-s}) = 2\mathrm{Tr}(aH^{-s})$ up to an entire function, the meromorphic property of the Heisenberg cycles depends on proving that $\mathrm{Tr}(aH^{-s})$ extends to a meromorphic function on $\mathbb{C}$, where a is in a suitable class of operators on $L^2(\mathbb{R})$, and H is the harmonic oscillator. We will do this in this section. The only pole is at $s = 1$ and the residue functional $\mathrm{Res}_{s=1}\mathrm{Tr}(fH^{-s})$ defined there as some interesting properties.

Let $f \in C_u(\mathbb{R})$ be a bounded, uniformly continuous function.

We consider the zeta function $\mathrm{Tr}(fH^{-s})$, where H is the harmonic oscillator, which is analytic for $\mathrm{Re}(s) > 1$.

Theorem 10.11.1 *If $f \in C_u(\mathbb{R})$ and $\mathrm{Re}(s) > 1$, then*
(10.52)
$$\Gamma(s) \cdot \mathrm{Tr}(fH^{-s}) = \frac{1}{2\sqrt{\pi}} \cdot \int_0^1 \int_{\mathbb{R}} t^{s-1} \operatorname{csch} t \cdot f(x\sqrt{\coth t}) \cdot e^{-x^2} dx dt + \psi(s),$$

where $\psi(s)$ extends to an entire function.

Remark 10.11.2 We make some remarks about the statement.

a) Recall that $\operatorname{csch} t = 1/\sinh t$.

b) If $f = 1$ is constant, (10.52) gives that

$$\Gamma(s) \cdot \mathrm{Trace}(H^{-s}) \sim \frac{1}{2\sqrt{\pi}} \int_0^1 \int_{\mathbb{R}} t^{s-1} \operatorname{csch} t \, e^{-x^2} \, dx dt = \frac{1}{2} \int_0^1 t^{s-1} \operatorname{csch} t \, dt.$$

Now csch t has a Laurent series expansion at $t = 0$, with a simple pole at $t = 0$ and residue 1. So csch $t = \frac{1}{t} + h$, where h is analytic at $t = 0$. Hence

$$\int_0^1 t^{s-1} \operatorname{csch} t \, dt = \int_0^1 t^{s-2} \, dt + \int_0^1 t^{s-1} h(t) \, dt.$$

The first term equals $(1/2) \cdot \frac{1}{s-1}$ for $\operatorname{Re}(s) > 1$ and evidently extends to a meromorphic function. The second term extends analytically to $\mathbb{C}$. Hence we get

$$\operatorname{Res}_{s=1} \operatorname{Tr}(H^{-s}) = \frac{1}{2}.$$

c) The slightly awkward factor of $1/2$ disappears if we use Δ instead of H; thus, $\operatorname{Res}_{s=1} \operatorname{Tr}(\Delta^{-s}) = 1$.

Before proving the theorem, we show how we can use the integral formula of the theorem to produce meromorphic continuations of some zeta functions.

Lemma 10.11.3 *If $f \in C_u(\mathbb{R})$ admits n successive bounded anti-derivatives, $^1 f, ^2 f, \ldots, ^n f$, then $\operatorname{Trace}(f H^{-s})$ extends analytically to $\operatorname{Re}(s) > 1 - \frac{n}{2}$.*

Proof By Theorem 10.11.1,

$$(10.53) \quad \Gamma(s) \cdot \operatorname{Tr}(f H^{-s}) \sim \frac{1}{2\sqrt{\pi}} \cdot \int_0^1 t^{s-1} \operatorname{csch} t \cdot \int_\mathbb{R} f(x\sqrt{\coth t}) \cdot e^{-x^2} dx dt.$$

Let $F =^1 f$, then integration by parts gives

$$(10.54) \quad = \frac{1}{\sqrt{\pi}} \cdot \int_0^1 t^{s-1} \operatorname{csch} t \sqrt{\tanh t} \int_\mathbb{R} F(x\sqrt{\coth t}) \, x e^{-x^2} dx dt$$

$$= \frac{1}{\sqrt{\pi}} \int_0^1 t^{s-1} \operatorname{csch} t \sqrt{\tanh t} \cdot \phi(t) dt,$$

with $\phi(t) = \int_\mathbb{R} F(x\sqrt{\coth t}) \, x e^{-x^2} dx$. The function $t^{s-1} \operatorname{csch} t \sqrt{\tanh t} \cdot \phi(t)$ is $\sim t^{s-3/2} \cdot \phi(t)$ as $t \to 0$ and is integrable over $[0, 1]$ for $\operatorname{Re}(s) > \frac{1}{2}$ if ϕ is continuous and bounded as $t \to 0$. This is indeed the case for us since by assumption, F is bounded.

So we have verified analyticity for $\operatorname{Re}(s) > \frac{1}{2}$. Repeating the argument, if $^2 f =^1 F$ is the second anti-derivative, then the previous expression can be written

$$(10.55) \quad \frac{1}{\sqrt{\pi}} \cdot \int_0^1 t^{s-1} \operatorname{csch} t \tanh t \cdot \int_\mathbb{R} {}^2 f(x\sqrt{\coth t}) \, (1 - 2x) e^{-x^2} dx dt,$$

which is analytic now for $\text{Re}(s) > 0$ if $^2 f$ is also bounded. One repeats this argument n times and the statement follows.

□

The *cohomological equation* in dynamics refers to the differential equation

$$Xu = f,$$

where X is a generating vector field for a smooth flow α on a compact manifold M. Let $p \in M$, $f \in C(M)$, and

(10.56) $$f_p(t) = f(\alpha_t(p)).$$

Then $f_p \in C_u(\mathbb{R})$. If $f \in C^\infty(M)$, then $f_p \in C_u^\infty(\mathbb{R})$.

An obstruction to solving the cohomological equation for given f is the mean of f with respect to any α-invariant probability measure μ. This follows from differentiating the equation $\int_M u \circ \alpha_t \, d\mu = \int_M u d\mu$, which gives that $\int_M Xu \, d\mu = 0$, that is, $\int_M f d\mu = 0$ if $Xu = f$ has a solution.

Lemma 10.11.3 implies a connection to the harmonic oscillator zeta functions $\text{Tr}(f_p H^{-s})$, for any $p \in M$. Suppose the flow has *trivial cohomology* in this sense: thus, suppose that if $f \in C^\infty(M)$ and $\int_M f d\mu = 0$ implies $f = Xu$ for some smooth u.

Now suppose that $f \in C^\infty(M)$ is arbitrary. Then $g := f - \int_M f \, d\mu$ has μ-integral zero. Hence the obstruction vanishes and $g = Xu$ for some u. It follows immediately that $g_p = u'_p$ and u_p is bounded on $\mathbb{R}$ since u is smooth on M, and so $f_p = \int_M f d\mu + u'_p$, where u is bounded, and applying Lemma 10.11.3 and repeating the argument give the following proposition:

Proposition 10.11.4 *Let α be a smooth flow on M with generator X and μ any α-invariant measure with trivial cohomology. Then for any $f \in C^\infty(M)$, $\text{Trace}(f_p H^{-s})$ extends meromorphically to $\mathbb{C}$ and*

$$\text{Res}_{s=1}\text{Trace}(f_p H^{-s}) = \frac{1}{2}\int_M f d\mu$$

for any $p \in M$.

Only fairly simple flows have trivial cohomology in this sense. The standard periodic flow on the circle has it. In fact, let f be continuous and ρ-periodic on $\mathbb{R}$ with zero mean: $\int_0^\rho f(t)dt = 0$. Then $F(T) := \int_0^T f(t) \, dt$ is also continuous, ρ-periodic, with zero mean. This shows that the cohomological equation is solvable even for *continuous* functions, if they have zero mean, and so we obtain the following corollary.

Corollary 10.11.5 *If f is continuous and ρ-periodic, then $\text{Trace}(f H^{-s})$ meromorphically extends to $\mathbb{C}$ with a simple pole at $s = 1$ and*

$$\operatorname{Res}_{s=1} \operatorname{Trace}(f H^{-s}) = \frac{1}{2\rho} \int_0^\rho f(t)dt.$$

Proof $\bar{f} := f - \mu(f)$ has zero mean. Applying the previous lemma gives that f has bounded anti-derivatives of all orders; the result follows from Lemma 10.11.3 and Remark 10.11.2   □

Examples of $f \in C_u(\mathbb{R})$ with the meromorphic extension property but which are not periodic are given in the following exercise.

Exercise 10.11.6 Suppose $\hbar$ is an irrational number satisfying a *Diophantine condition*: There exists $0 < \gamma < 1$ such that $|n\hbar + m| \geq (n^2 + m^2)^{-\frac{\gamma}{2}}$, for all $n, m \in \mathbb{Z}$. Let α be the Krönecker flow on $\mathbb{T}^2$:

$$\alpha_t(x, y) = (x + t, y + t\hbar), \quad (x, y) \in \mathbb{R}^2/\mathbb{Z}^2.$$

If X is the generating vector field, show that the cohomological equation $Xu = f$ has a smooth solution for every smooth $f \in C^\infty(\mathbb{T}^2)$. If f_0 is the restriction of f to the orbit of $(0, 0) \in \mathbb{T}^2$, $f_0(t) := f(t, \hbar t)$, deduce that $\operatorname{Tr}(f_0 H^{-s})$ meromorphically extends to $\mathbb{C}$ and that

$$\operatorname{Res}_{s=1}(f H^{-s}) = \frac{1}{2} \int_{\mathbb{T}^2} f d\mu,$$

with μ normalized Lebesgue measure. Such f_0 are of course not periodic.

We now proceed to the proof of Theorem 10.11.1. A computation of the heat kernel of e^{-tH} follows from solving a differential equation: The heat kernel k_t satisfies $(\frac{\partial}{\partial t} + H) \cdot \phi_t = 0$, where $\phi_t(x) = \int_{\mathbb{R}} k_t(x, y)\phi(y)dy$, for $\phi \in S(\mathbb{R})$, and $t \geq 0$, together with the initial condition $\lim_{t \to 0} \phi_t = \phi$. Consider the *ansatz*

$$k_t(x, y) = \exp\left(\frac{a_t}{2}x^2 + b_t xy + \frac{a_t}{2}y^2 + c_t\right).$$

Setting this equal to 0 and solving for coefficients give the ordinary differential equations

$$\frac{\dot{a}_t}{2} = a_t^2 - 1 = b_t^2, \quad \dot{c}_t^2 = a_t.$$

Solving these gives

$$a_t = -\coth(2t + C), \quad b_t = \operatorname{csch}(2t + C), \quad c_t = -\frac{1}{2}\log\sinh(2t + C) + D.$$

Using the initial conditions, we get $C = 0$ and $D = \log(2\pi)^{-\frac{1}{2}}$. See [25].
    We obtain the following, called *Mehler's formula* [123].

Lemma 10.11.7

$$(10.57)\quad k_t(x, y) = \frac{1}{\sqrt{2\pi \sinh 2t}} \exp\left(-\tanh t \cdot \frac{(x + y)^2}{4} - \coth t \cdot \frac{(x - y)^2}{4}\right).$$

Proof of Theorem 10.11.1 The operator H^{-s} is trace-class for $\mathrm{Re}(s) > 1$, and the operator-valued integral $\int_0^\infty t^{s-1} e^{-tH} dt$ converges in norm to $\Gamma(s) \cdot H^{-s}$. Hence if $a \in \mathbb{B}(L^2\mathbb{R})$,

$$(10.58)\qquad\qquad \Gamma(s) \cdot aH^{-s} = \int_0^\infty t^{s-1} ae^{-tH} dt,$$

and taking traces gives

$$(10.59)\qquad \Gamma(s) \cdot \mathrm{Trace}(aH^{-s}) = \int_0^\infty t^{s-1}\mathrm{Trace}(ae^{-tH}) \, dt.$$

Furthermore, if a is any bounded operator, then

$$\int_1^\infty t^{s-1}\mathrm{Trace}(ae^{-tH}) dt$$

extends to an analytic function on $\mathbb{C}$. Hence

$$(10.60)\qquad \Gamma(s) \cdot \mathrm{Trace}(aH^{-s}) - \int_0^1 t^{s-1}\mathrm{Trace}(ae^{-tH}) \, dt$$

extends analytically to $\mathbb{C}$.

Now let $f \in C_u(\mathbb{R})$, and set $a = f$. Then fe^{-tH} is an integral operator with kernel $f(x)k_t(x, y)$, and hence $\mathrm{Trace}(fe^{-tH}) = \int_\mathbb{R} f(x)k_t(x, x) \, dx$. Applying Mehler's formula, Lemma 10.11.7 gives

$$(10.61)\quad \Gamma(s) \cdot \mathrm{Trace}(aH^{-s}) = \int_0^\infty t^{s-1} \frac{1}{\sqrt{2\pi \sinh 2t}} \int_\mathbb{R} f(x)e^{-x^2 \tanh t} \, dxdt.$$

Making the change of variables $x \mapsto \frac{x}{\sqrt{\tanh(t)}}$ gives

$$(10.62)\ \ \Gamma(s) \cdot \mathrm{Trace}(fH^{-s}) = \int_0^\infty t^{s-1} \frac{\sqrt{\coth t}}{\sqrt{2\pi \sinh 2t}} \int_\mathbb{R} f(x\sqrt{\coth t})e^{-x^2} \, dxdt.$$

The result follows from the identity $\frac{\coth t}{\sinh 2t} = \mathrm{csch}^2 t$.

□

If $\text{Trace}(f H^{-s})$ meromorphically extends past $\text{Re}(s) = 1$, then the residue of the pole at $s = 1$ defines, up to the factor of $1/2$, a kind of "mean" of $f \in C_u(\mathbb{R})$. In certain examples of flows where $f = g_p$ for $p \in M$, $f(t) := g(\alpha_t(p))$, we have noted (Proposition 10.11.4) that this spectrally defined mean agrees with the geometric mean $\int_M f d\mu$ over the manifold.

The spectrally defined mean does not require meromorphic continuation, but only existence of the limit $\lim_{s \to 1^+}(s - 1) \text{Trace}(f H^{-s})$, which is a weaker condition which we now discuss.

Definition 10.11.8 Let $\mathcal{D} \subset C_u(\mathbb{R})$ be the closed linear subspace of all f such that

$$(10.63) \qquad \text{Res Tr}(f) := 2 \lim_{s \to 1^+} (s - 1) \text{Trace}(f H^{-s})$$

exists.

Theorem 10.11.9 *If $f \in C_u(\mathbb{R})$, then $f \in \mathcal{D}$ if and only if*

$$(10.64) \qquad \mu_u(f) := \lim_{\lambda \to \infty} \frac{1}{\sqrt{\pi}} \int_0^1 \int_{\mathbb{R}} f(x \, t^{-\lambda}) e^{-x^2} dx dt$$

exists, and if this holds, then $\text{Res Tr}(f) = \mu_u(f)$.

Proof Choose $\epsilon > 0$. Since $\Gamma(1) = 1$ by Theorem 10.11.1, we have for $\text{Re}(s) > 1$,
$$(10.65)$$

$$\lim_{s \to 1^+} \text{Trace}(f H^{-s}) = \lim_{s \to 1^+} \frac{s - 1}{2\sqrt{\pi}} \cdot \int_0^1 \int_{\mathbb{R}} t^{s-1} \operatorname{csch} t \, f(x\sqrt{\coth t}) \, e^{-x^2} dx dt.$$

In the proof, we noted that the part of the integral corresponding to $t \geq \delta$ extends analytically to $\mathbb{C}$. Hence it contributes zero to the limit, and we may choose $\delta > 0$ small enough that $|t \operatorname{csch} t - 1| < \epsilon$ for $0 < t < \delta$, so that $|\operatorname{csch} t - \frac{1}{t}| < \frac{\epsilon}{t}$ for $t < \delta$. Let $\phi_f(t) = \int_{\mathbb{R}} f(x\sqrt{\coth t}) \cdot e^{-x^2} dx$, then

$$(10.66) \qquad \left| \int_0^\delta t^{s-1} \left(\operatorname{csch} t - \frac{1}{t} \right) \phi(t) dt \right| < \frac{\epsilon}{s - 1} \cdot \|f\|$$

by a brief computation. Letting $\epsilon \to 0$, we see that the limit on the right hand side of (10.65), if it exists, equals the limit

$$(10.67) \qquad \lim_{s \to 1^+} \frac{s - 1}{2\sqrt{\pi}} \cdot \int_0^1 \int_{\mathbb{R}} t^{s-2} \, f(x\sqrt{\coth t}) \cdot e^{-x^2} dx dt.$$

Let $\lambda = \frac{1}{s-1}$ and substitute $t \to t^\lambda$ in the above expression, and, noting $\delta^{\frac{1}{\lambda}} \to 1$ as $\lambda \to \infty$, we deduce that

(10.68)

$$\lim_{s \to 1+} (s-1)\text{Trace}(fH^{-s}) = \lim_{\lambda \to \infty} \frac{1}{2\sqrt{\pi}} \cdot \int_0^1 \int_{\mathbb{R}} f(x\sqrt{\coth t^{\lambda}}) \cdot e^{-x^2} dx dt.$$

Since Lipschitz functions are dense in $C_u^{\infty}(\mathbb{R})$ and $\mathcal{D}$ is closed, we may assume f is Lipschitz, and it follows that

(10.69) $$\left| \int_0^1 \int_{\mathbb{R}} \left(f(x\sqrt{\coth t^{\alpha}}) - f(xt^{-\frac{\alpha}{2}}) \right) e^{-x^2} dx t \right|$$

$$\leq \text{const.} \lim_{\alpha \to \infty} \int_0^1 |\sqrt{\coth t^{\alpha}} - t^{-\frac{\alpha}{2}}| \, dt,$$

which $\to 0$ as $\lambda \to \infty$. This proves the result. $\square$

The theorem can be expressed this way:

Theorem 10.11.10 *Let* $\mu_0 = \frac{1}{\sqrt{\pi}} e^{-x^2} dx$, *the Gaussian probability measure on* $\mathbb{R}$. *For* $t \in \mathbb{R}_+^*$, *let* $\rho_t \colon \mathbb{R} \to \mathbb{R}$, $\rho_t(x) = tx$, *and* $\mu_t := (\rho_t)_* \mu_0$. *Then*

$$\lim_{\lambda \to \infty} \int_0^1 \mu_{t^{\lambda}} \, dt = \text{Res Tr} \in \mathcal{D}$$

in the weak topology on $\mathcal{D}$.

We deduce the following.

Corollary 10.11.11 *If* $f \in C_u(\mathbb{R})$ *and* $\lim_{T \to \pm\infty} \frac{1}{T} \int_0^T f(t) \, dt$ *exists, then* $f \in \mathcal{D}$ *and*

(10.70) $$\text{Res Tr}(f) = \lim_{T \to \pm\infty} \frac{1}{T} \int_0^T f(t) \, dt.$$

In particular, if α *is an ergodic flow on a compact smooth manifold* M, μ *an* α-*invariant probability measure,* $f \in C(M)$, $f_p(t) := f(\alpha_t p)$, *then for a.e.* $p \in M$, $f_p \in \mathcal{D}$, *and*

$$\text{Res Tr}(f_p) = \int_M f \, d\mu$$

for a.e. $p \in M$.

Proof Integration by parts, the change of variables $u \to ut^{\lambda}$, and a slight rearrangement, gives

$$\int_0^1 \int_{\mathbb{R}} f(xt^{-\lambda}))e^{-x^2}dxdt = 2\int_0^1 \int_{\mathbb{R}} \int_0^x f(ut^{-\lambda})xe^{-x^2}dudxdt$$

$$= 2\int_0^1 \int_{\mathbb{R}} t^{\lambda} \int_0^{xt^{-\lambda}} f(u)xe^{-x^2}dudxdt$$

(10.71)
$$= 2\int_0^1 \int_{\mathbb{R}} \frac{1}{xt^{-\lambda}} \int_0^{xt^{-\lambda}} f(u)x^2 e^{-x^2}dudxdt.$$

Now letting $\lambda \to \infty$ and using the hypothesis that $L := \lim_{T \to \pm\infty} \frac{1}{T}\int_0^T f(t)\,dt$ exists give

(10.72)
$$\lim_{\lambda \to \infty} \int_0^1 \int_{\mathbb{R}} f(xt^{-\lambda}))e^{-x^2}dxdt = L\sqrt{\pi}.$$

By Theorem 10.11.9,

(10.73)
$$\mathrm{Res\,Tr}(f) = \lim_{\lambda \to \infty} \frac{1}{2\sqrt{\pi}} \int_0^1 \int_{\mathbb{R}} f(x\,t^{-\lambda})e^{-x^2}dxdt$$

giving (10.70).

The second statement follows from combining the first with the Birkhoff Ergodic Theorem. □

We next produce some estimates related to group translation operators on $L^2(\mathbb{R})$.

Lemma 10.11.12 *Let U_α be the unitary induced by translation on the real line by $\alpha \neq 0$. Then if $f \in C_u(\mathbb{R})$ and $a = fU_\alpha$, then*
(10.74)
$$\Gamma(s) \cdot \mathrm{Tr}(fU_\alpha H^{-s}) = \frac{1}{2\sqrt{2\pi}} \int_0^\infty t^{s-1} \mathrm{csch}\, t \exp\left(-\frac{\alpha^2}{4}\coth t\right) \mu_{\alpha,t}(f)dt,$$

where $\mu_{\alpha,t}(f) = \int_{\mathbb{R}} f(x\sqrt{\coth t} + \alpha)e^{-x^2}dx$.

Proof The argument proceeds as in the proof of Lemma 10.11.7. The operator $fU_\alpha e^{-tH}$ is a compact integral operator with kernel

$$k'_t(x, y) = f(x)k_t(x - \alpha, y).$$

Hence for $\mathrm{Re}(s) > 1$,

$$\Gamma(s) \cdot \mathrm{Tr}(fU_\alpha H^{-s}) = \int_0^\infty \int_{\mathbb{R}} t^{s-1}k_t(x - \alpha, x)dxdt$$

(10.75)
$$= \int_0^\infty \int_{\mathbb{R}} t^{s-1} (2\pi \sinh 2t)^{-\frac{1}{2}} f(x) \exp\left(-\frac{(2x-\alpha)^2}{4} \tanh t - \frac{\alpha^2}{4} \coth t\right) dx\, dt.$$

Basic manipulations yield (10.74).

□

Lemma 10.11.13 *If $f \in C_u(\mathbb{R})$ and $\alpha \in \mathbb{R}$ nonzero, then the function $\phi_{f,\alpha}(s) := \Gamma(s) \cdot \mathrm{Trace}(f U_\alpha H^{-s})$, $\mathrm{Re}(s) > 1$, extends to an analytic function on $\mathbb{C}$. There are constants C'_s and C''_s depending holomorphically on s such that*

(10.76)
$$|\phi_{f,\alpha}(s)| \leq \left(C'_s \alpha^{-2\mathrm{Re}(s)} + C''_s\right) e^{-\frac{\alpha^2}{4}} \cdot \|f\|$$

for all $s \in \mathbb{C}$.

Proof The point is that not only does $\mathrm{Trace}(f U_\alpha H^{-s})$ meromorphically extend to $\mathbb{C}$, but also the formula (10.74) defines $\phi_{f,\alpha}(s)$ by a direct integral formula valid for all $s \in \mathbb{C}$.

As shown above, for a suitable family of states $\mu_{\alpha,t}$ on $C_u(\mathbb{R})$, and a constant which we omit,

$$\Gamma(s) \cdot \mathrm{Tr}(f U_\alpha H^{-s}) \sim \int_0^\infty t^{s-1} \operatorname{csch} t \exp\left(-\frac{\alpha^2}{4} \coth t\right) \mu_{\alpha,t}(f)\, dt$$

$$= \int_0^1 t^{s-1} \operatorname{csch} t \exp\left(-\frac{\alpha^2}{4} \coth t\right) \mu_{\alpha,t}(f)\, dt$$

(10.77)
$$+ \int_1^\infty t^{s-1} \operatorname{csch} t \exp\left(-\frac{\alpha^2}{4} \coth t\right) \mu_{\alpha,t}(f)\, dt = \zeta_1(s) + \zeta_2(s).$$

Consider first $\zeta_1(s)$. Since $\frac{\tanh t}{t}$ and $\frac{\sinh t}{t}$ are bounded on $[0, 1]$, we can bound the integrand of $\zeta_1(s)$ by

$$t^{s-2} e^{-\beta/t} \|f\|, \quad \beta = \frac{\alpha^2}{4}.$$

A change of variables gives

$$\int_0^1 t^{s-2} e^{-\beta/t}\, dt = \int_1^\infty t^{-s} e^{-\beta t}\, dt.$$

If $A_s = \int_0^\infty t^{-s} e^{-\lambda t}\, dt$, then $A_s = \frac{e^{-\beta}}{\beta} + \frac{1}{\beta} A_{s+1}$, by integration by parts, and it follows that $|\int_0^1 t^{s-2} e^{-\beta t}\, dt| \leq \mathrm{const}.\beta^{-\mathrm{Re}(s)} e^{-\beta}$ where the constant does not depend on β or s and hence that

$$|\zeta_1(s)| \le C_s' \cdot \|f\| \cdot \alpha^{-2\mathrm{Re}(s)} e^{-\frac{\alpha^2}{4}}.$$

We can bound $\zeta_2(s)$ as follows:

$$\left| \int_1^\infty t^{s-1} \operatorname{csch} t \exp\left(-\frac{\alpha^2}{4} \coth t \right) \mu_{\alpha,t}(f)dt \right|$$

(10.78)
$$\le \|f\| e^{-\frac{\alpha^2}{4}} \cdot \int_1^\infty t^{s-1} \operatorname{csch} t \, dt = C_s'' e^{-\frac{\alpha^2}{4}} \|f\|.$$

$\square$

Lemma 10.11.13 assists with the meromorphic extension problem for $\mathrm{Trace}(aH^{-s})$, with $a = \sum_{\alpha \in \Gamma} f_\alpha U_\alpha$ an element of a crossed product $C_u(\mathbb{R}) \rtimes \Gamma$, where $\Gamma \subset \mathbb{R}_d$ is a subgroup.

The easiest case is that of a cyclic subgroup, and this produces a very strong result.

If $f \in C_u(\mathbb{R}) \rtimes \Gamma$ for $\Gamma \subset \mathbb{R}$ a subgroup, let f_0 be the coefficient of f at the identity $0 \in \Gamma$.

Let $\mathcal{D}^\infty$ denote the subspace of $\mathcal{D} \subset C_u(\mathbb{R})$ of f such that $\mathrm{Trace}(fH^{-s})$ meromorphically extends to $\mathbb{C}$ with a simple pole at $s = 1$.

Theorem 10.11.14 *Let $a \in C_u(\mathbb{R}) \rtimes_\hbar \mathbb{Z}$, where $\hbar \in \mathbb{R}$ is nonzero, $a = \sum_n f_n[n]$. Then, if $f_0 \in \mathcal{D}^\infty$, then $\mathrm{Trace}(aH^{-s})$, $\mathrm{Re}(s) > 1$, meromorphically extends to $\mathbb{C}$, with a simple pole at $s = 1$, and*

$$\mathrm{Res}_{s=1} \mathrm{Trace}(aH^{-s}) = \frac{1}{2} \mu_u(f_0),$$

where μ_u is the uniform mean of f_0 (10.64).

Proof Suppose first that f has expansion $f = \sum f_n U_{n\hbar}$ with $f_0 = 0$. Then

$$\Gamma(s) \cdot \mathrm{Trace}(fH^{-s}) = \sum_n \mathrm{Trace}(f_n U_{n\hbar} H^{-s}) = \sum_n \phi_n(s),$$

where $\phi_n(s)$ abbreviates $\phi_{f_n, n\hbar}(s)$ of Lemma 10.11.13. The series converges absolutely and uniformly on compact subsets of $\mathbb{C}$ because of the bound

$$|\phi_n(s)| \le \left(C_s' \hbar^{-2\mathrm{Re}(s)} n^{-2\mathrm{Re}(s)} + C_s'' \right) e^{-(\frac{\hbar^2}{4})n^2}.$$

due to the lemma, shows that $\phi_n \to 0$ exponentially fast as $n \to \pm\infty$.

In the general case, $f = f - f_0$, $\mathrm{Trace}((f - f_0)H^{-s})$ extends to an entire function for arbitrary $f \in C_u(\mathbb{R})$, and $\mathrm{Trace}(f_0 H^{-s})$ to a meromorphic function with the stated pole structure if $f_0 \in \mathcal{D}^\infty$ by definition, see Theorem 10.64 for the equivalent condition to being in $\mathcal{D}$.

□

Corollary 10.11.15 *If $a \in A_\hbar := C(\mathbb{T}) \rtimes_\hbar \mathbb{Z}$, then $\mathrm{Trace}(\pi_\hbar(a)H^{-s})$ meromorphically extends to $\mathbb{C}$ with a simple pole at $s = 1$ and*

$$\mathrm{Res}_{s=1}\mathrm{Tr}(aH^{-s}) = \frac{1}{2}\tau(a),$$

where τ is the standard unital trace on $A_\hbar$.

The Heisenberg cycle Definition 10.10.9 over the irrational rotation algebra $A_\hbar$ defines a regular spectral cycle over $A_\hbar^\infty \subset A_\hbar$ with the meromorphic continuation property over the whole C*-algebra $A_\hbar$.

Definition 10.11.16 A finitely generated subgroup $\Lambda \subset \mathbb{R}$ with word length function $|\cdot|_\Gamma$ satisfies a *Diophantine property* if

$$|\alpha| \geq C\,|\alpha|_\Gamma^{-\gamma}$$

for some $\gamma > 0$ and $C > 0$.

Definition 10.11.17 Let $B_\Gamma := C_u(\mathbb{R}) \rtimes \Gamma$, where $\Gamma \subset \mathbb{R}$ is a countable subgroup. Then B_Γ^∞ denotes the completion of the (twisted) group algebra $C_u^\infty(\mathbb{R})[\Gamma]$ with respect to the family of semi-norms $p_{s,m}(f) = \sum_{\alpha\in\Gamma}\|f_\gamma^{(m)}\|\cdot|\gamma|_\Gamma^s$.

Remark 10.11.18 B_Γ^∞ consists of operators in $C_u(\mathbb{R}) \rtimes \Gamma$ whose expansions $f = \sum_{\alpha\in\Gamma} f_\alpha U_\alpha$ have rapid decay in the sense that

(10.79)
$$\sum_{\alpha\in\Gamma}\|f_\alpha^{(m)}\|\cdot|\alpha|_\Gamma^n < \infty,\quad \forall m, n \geq 0.$$

It is not difficult to prove that B_Γ^∞ is closed under holomorphic functional calculus.

Theorem 10.11.19 *Suppose $\Gamma \subset \mathbb{R}$ has a Diophantine property as in Definition 10.11.16, let $f \in B_\Gamma^\infty$, and assume $f_0 \in \mathcal{D}$. Then $\mathrm{Trace}(fH^{-s})$ meromorphically extends to $\mathbb{C}$ with a simple pole at $s = 1$ and $\mathrm{Res}_{s=1}(f) = \frac{1}{2}\mu_u(f)$.*

Proof Write $f = \sum_{\alpha\in\Gamma} f_\alpha U_\alpha \in B_\Gamma^\infty \subset \mathbb{B}(L^2(\mathbb{R}))$, and assume $f_0 = 0$. It suffices to prove that $\mathrm{Trace}(fH^{-s})$ extends to an analytic function on $\mathbb{C}$. This equals

$$\sum_{\alpha\in\Gamma}\mathrm{Trace}(f_\alpha U_\alpha H^{-s}) = \sum_{\alpha\in\Gamma}\phi_{f_\alpha,\alpha}(s),$$

where $\phi_{f,\alpha}$ is as in Lemma 10.11.13, and it suffices to show that this is an absolutely summable sequence of analytic functions, uniformly on compact subsets of $\mathbb{C}$.) Shorten notation $\phi_\alpha := \mathrm{Trace}(f_\alpha U_\alpha H^{-s})$. By the same lemma,

$$(10.80) \qquad |\phi_\alpha(s)| \le \left(C'_s \alpha^{-2\text{Re}(s)} + C''_s\right) e^{-\frac{|\alpha|^2}{4}} \cdot \|f_\alpha\|$$

for all $s \in \mathbb{C}$. Since there are potentially infinitely many α with a small absolute value, the exponential term is no longer of any use, and we discard it, obtaining a polynomial bound for $\phi_\alpha(s)$ of order $|\alpha|^\mu$ for $\mu = -2\text{Re}(s) \in \mathbb{R}$. Since Γ is finitely generated, There exists a constant C_Γ such that $|\alpha| \le C_\Gamma \cdot |\alpha|_\Gamma$ for all $\alpha \in \Gamma$. Combining with the Diophantine assumption gives that

$$C|\alpha|_\Gamma^{-\gamma} \le |\alpha| \le C'|\alpha|_\Gamma.$$

If $\mu \ge 0$, we get

$$C''|\alpha|_\Gamma^{-\mu\gamma} \le |\alpha|^\mu \le C'''|\alpha|_\Gamma^\mu.$$

Hence

$$\sum_{\alpha \in \Gamma} \|f_\alpha\| \cdot |\alpha|^\mu \le C \sum_{\alpha \in \Gamma} \|f_\alpha\| \cdot |\alpha|_\Gamma^\mu,$$

and the last term is finite by (10.79).

If $\mu < 0$, then we use the bound

$$|\alpha|^\mu \le \text{const.}|\alpha|_\Gamma^{-\mu\gamma}.$$

Again, $\sum_{\alpha \in \Gamma} \|f_\alpha\| \cdot |\alpha|_\Gamma^{-\mu\gamma}$ is finite by assumption on f (10.79).

$\square$

10.12   The Local Index Formula for the Heisenberg Cycles

In the paper [40], Alain Connes described an invariant of a finitely generated projective module over $A_\hbar$, generalizing the first Chern number of a complex vector bundle over $\mathbb{T}^2$. This construction is one of the key motivating examples in Noncommutative Geometry.

Connes' construction is the following. Let A be any C*-algebra endowed with a pair α and β of commuting flows, inducing an action of $\mathbb{R}^2$ by automorphisms with (s, t) acting by $\alpha_s \circ \beta_t$.

Let $\delta_i : A \to A$ be the densely defined derivations

$$\delta_1(a) := \lim_{t \to 0} \frac{\alpha_t(a) - a}{t}, \quad \delta_2(a) := \lim_{t \to 0} \frac{\beta_t(a) - a}{t}, \quad a \in A^\infty,$$

where $A^\infty = \cap_{n,m} \mathrm{dom}(\delta^n) \cap \mathrm{dom}(\delta^m)$, the *-subalgebra of elements such that $(s, t) \mapsto \alpha_s(\beta_t(a))$ is smooth.

Let $\tau \colon A \to \mathbb{C}$ be an $\mathbb{R}^2$-invariant tracial state.

Definition 10.12.1 In the above notation, Connes' invariant of an f.g.p. module eA, where e is a projection in A^∞, is given by

$$c_1(e) := \frac{1}{2\pi i} \tau \left(e[\delta_1(e), \delta_2(e)] \right).$$

We call $\theta(e) := e[\delta_1(e), \delta_2(e)]$ the *curvature* of e and $c_1(e)$ the *first Chern number of* e.

The curvature of e is an A^∞-module endomorphism of the f.g.p. module eA^∞.

Theorem 10.12.2 (Connes, [40])) *The number*

$$c_1(e) := \frac{1}{2\pi i} \tau \left(e[\delta_1(e), \delta_2(e)] \right)$$

only depends on the equivalence class of e in $\mathrm{K}_0(A)$.

Moreover, $c_1(1) = 0$, $c_1(\mathcal{E} \oplus \mathcal{E}') = c_1(\mathcal{E}) + c_1(\mathcal{E}')$, *and c_1 thus determines a group homomorphism* $\mathrm{K}_0(A) \to \mathbb{R}$.

It is reasonable to use the term "curvature" here, for suppose X is a manifold and $e \colon X \to M_n(\mathbb{C})$ is a smooth, projection-valued function, with $E \to X$ the corresponding complex vector bundle. Then the curvature of the Grassmann connection on E is given by the $\mathrm{End}(E)$-valued 2-form

$$\theta = e \cdot de \wedge de.$$

If $X = \mathbb{T}^2$ with the standard $\mathbb{R}^2$-action and $n = 1$, then this is the same as the curvature defined by Connes because the derivations δ_i correspond to differentiation in the two coordinate directions.

Example 10.12.3 Let $\hbar \in \mathbb{R}$ and $A_\hbar = C(\mathbb{T}) \rtimes_h \mathbb{Z}$ be the corresponding rotation algebra, with $u \in A_\hbar$ the generator of the $\mathbb{Z}$-action. Then the $\mathbb{R}^2$-action with $\alpha_t(f) = f(x - t)$, $\alpha_t(u) = u$; $\beta_t(u^n) = e^{int} u^n$, $\beta_t(f) = f$, gives rise to the derivations

$$\delta_1 \left(\sum_n f_n[n] \right) = \sum_n f_n'[n], \quad \delta_2 \left(\sum_n f_n[n] \right) = \sum_n in \cdot f_n[n],$$

which up to scale we have already discussed (see, e.g., the discussion before Proposition 10.10.12).

Integrality

In the case of $A = C(\mathbb{T}^2)$, with the standard $\mathbb{T}^2$-action, the first Chern number Definition 10.12.1 of a smooth projection-valued function $e \colon \mathbb{T}^2 \to M_n(\mathbb{C})$ is an *integer*. This is a standard result of topology based on the fact that the first Chern class (and all Chern classes) can be defined using cohomology with *integer coefficients*.

But it also follows from the Atiyah–Singer Index Theorem, which states that

$$c_1(E) = \mathrm{Index}(\bar{\partial}_E),$$

the index of the Dolbeault operator $\begin{bmatrix} 0 & \frac{\partial}{\partial x} - \frac{\partial}{\partial y} \\ \frac{\partial}{\partial x} + \frac{\partial}{\partial y} & 0 \end{bmatrix}$ twisted by E.

The following computation, due to Connes [42], shows that the Chern number, at least of the Rieffel projection, retains this integrality property in the case of the noncommutative rotation algebras $A_\hbar$, with their standard $\mathbb{T}^2$-actions.

Theorem 10.12.4 *Let $p_\hbar \in A_\hbar = C(\mathbb{T}) \rtimes_\hbar \mathbb{Z}$ be the Rieffel projection (Exercise 6.6.4 of Chapter 4). Then*

$$c_1(p_\hbar) = +1.$$

Proof For brevity, for $f \in C(\mathbb{T})$ understood as always as a 2π-periodic function on $\mathbb{R}$, let $f^\hbar(x) := f(x - h)$ denote the action.

The Rieffel projection is given by

$$p_\hbar = f + gu + g^{-\hbar}u^*,$$

where f and g are suitably chosen functions. For $a \in (0, 1)$ and $\epsilon > 0$ small, f equals zero on $[0, a]$ and on $[a + \hbar + \epsilon, 2\pi]$, and $f = 1$ on $[a + \epsilon, a + \hbar + \epsilon]$. We design f so that $f(x) + f(x + \hbar) = 1$. We set $g = \sqrt{f - f^2}$ on $[a + \hbar, a + \epsilon + \hbar]$ and is zero otherwise. We then have to compute $c_1(e) = \frac{1}{2\pi i}\tau([\delta_1(p_\hbar), \delta_2(p_\hbar)])$. We first compute

(10.81)
$$\frac{1}{2\pi i}\tau([\delta_1(p_\hbar), \delta_2(p_\hbar)]) = u^*(gf' - gf'^\hbar) - 2\left((gg')^{-\hbar} - gg'\right) + \left(g^2(f' - f'^\hbar)\right)^{-\hbar}.$$

Multiplying this on the left by $p_\hbar$ produces a terrific mess, but we are only interested in its trace, so the only part that is relevant is

(10.82) $$g^2(f' - f'^\hbar) + 2f\left((gg')^{-\hbar} - gg'\right) + \left(g^2(f' - f'^\hbar)\right)^{-\hbar},$$

which we want to integrate over $\mathbb{T}$.

Set $u = g^2$, $v = f - f^\hbar$. The integral is given by

(10.83)
$$\int uv' + f\left(u'^{-\hbar} - u'\right) + (uv')^{-\hbar}.$$

The middle term is

$$\int f u'^{-h} - \int f u' = \int f^h u' - \int f u' = -\int vu'.$$

Hence we are reduced to computing

$$\int uv' - vu' + (uv')^{-h} = \int 2uv' - vu' = 3\int uv',$$

by integration by parts. Next, since $f^{\hbar} = 1 - f$ on supp(g), we can replace $f' - f'^{\hbar}$ by $-2f'$ and get

$$-6\int f'(f - f^2) = -6\int f'f + 6\int f'f^2$$

(10.84)
$$= -3\int (f^2)' + 2\int (f^3)' = 3 - 2 = 1,$$

where the integration is understood to be restricted to the support of g.

This completes the calculation.   □

We now restate the Local Index Theorem, for two-dimensional spectral cycles.

Theorem 10.12.5 *Let* (H, π, D) *be an even, two-dimensional regular spectral cycle over* $A^{\infty} \subset A$ *with the meromorphic continuation property (over* A^{∞}*).*

Let $[D] \in KK_0(A, \mathbb{C})$ *be the class of the triple. Let* $\Delta := D^2$*, and let* ϵ *be the grading operator on* H*. Let* $\Delta_K = \Delta + \mathrm{pr}_{\ker D}$*, which is invertible.*

Define functionals

* $\psi_0 \colon A \to \mathbb{C}$,

$$\psi_0(a) := \mathrm{Res}_{s=0} \, \Gamma(s) \cdot \mathrm{Tr}(\epsilon a(\Delta + K)^{-s}),$$

and
* $\psi_2 \colon A \otimes A \otimes A \to \mathbb{C}$,

$$\psi_2(a^0, a^1, a^2) := \frac{1}{2} \mathrm{Res}_{s=1}(\epsilon a^0 [D, a^1][D, a^2]\Delta^{-s}).$$

Then if $e \in A^{\infty}$ *is a projection, then*

$$\langle [e], [D] \rangle = \psi_0(e) - \psi_2\left(e - \frac{1}{2}, e, e\right),$$

where $\langle [e], [D] \rangle \in \mathbb{Z}$ *is the pairing between the* $K_0(A)$*-class* $[e]$ *and the* $KK_0(A, \mathbb{C})$
class $[D]$.

The following lemma shows that in the case of the Heisenberg cycles, the zero-dimensional part of the Chern character formula of Theorem 10.12.5 comes from taking the pole at $s = 1$ of the zeta function Trace(aH^{-s}) discussed in the previous section.

Lemma 10.12.6 *If* $a \in \mathbb{B}(L^2(\mathbb{R})$ *and* Re$(s)1$, *then*

$$\Gamma(s)\mathrm{Tr}(\epsilon a(\Delta + \mathrm{proj}_{\ker D})^{-s}) = 2\mathrm{Trace}(aH^{-s-1}) + \psi(s),$$

where $\psi(s)$ *extends to an entire function. In particular, if* Ψ_0 *is the functional b) in Theorem 10.12.5 and if* Trace(aH^{-s}) *meromorphically extends to* $\mathbb{C}$, *then*

$$\Psi_0(a) = \mathrm{Res}\, \mathrm{Tr}(a) := 2\mathrm{Res}_{s=1}\mathrm{Trace}(aH^{-s}).$$

Proof We refer to Theorem 10.12.5. The Hilbert space for the Heisenberg triple is the direct sum of two copies of $L^2(\mathbb{R})$, and $D = \begin{bmatrix} 0 & x - d/dx \\ x + d/dx & 0 \end{bmatrix}$. The kernel of D is the same as the kernel of $x + d/dx$ and is spanned by the ground state $\psi_0(x) = \pi^{-\frac{1}{4}}e^{-\frac{x^2}{2}}$, and $\Delta = \begin{bmatrix} H - 1 & 0 \\ 0 & H + 1 \end{bmatrix}$, where H is the harmonic oscillator. If $p = \mathrm{proj}_{\ker D}$, then $\Delta + p = \begin{bmatrix} H - 1 + p & 0 \\ 0 & H + 1 \end{bmatrix}$. The first copy of the Hilbert space $L^2(\mathbb{R})$ is even in the grading, and the second is odd and so the meromorphic function whose pole at $s = 0$ gives $\Psi_0(a)$

(10.85)
$$\Gamma(s)\mathrm{Trace}(\epsilon a(\Delta + p)^{-s}) = \Gamma(s) \cdot \mathrm{Tr}(a(H - 1 + p)^{-s}) - \Gamma(s) \cdot \mathrm{Tr}(a(H + 1)^{-s}),$$

for $a \in C_u(\mathbb{R})[\mathbb{R}_d]$. Applying Mellin transform and small calculation gives

$$= \int_0^\infty t^{s-1} \sinh t \cdot \mathrm{Trace}(ae^{-tH})dt + E(s),$$

where

$$E(s) = \int_0^\infty t^{s-1}e^t \cdot \mathrm{Trace}(e^{-tH} - e^{-t(H+p)})dt = \mathrm{Trace}(ap) \cdot \int_0^\infty t^{s-1}(1 - e^{-t})dt,$$

giving that $E(s)$ extends meromorphically with a simple pole at $s = -1$, and in particular, $E(s)$ is analytic for Re$(s) > -1$. Hence

$$\Psi_0(a) = \text{Res}_{s=0} \int_0^\infty t^{s-1} \sinh t \, \text{Trace}(ae^{-tH})dt.$$

Since $\sinh t \sim t$ as $t \to 0$,

(10.86)

$$\Psi_0(a) = \text{Res}_{s=0} \int_0^\infty t^s \text{Trace}(ae^{-tH})dt = \text{Res}_{s=1} \int_0^\infty t^{s-1}\text{Trace}(ae^{-tH})dt$$

$$= \text{Res}_{s=1}\text{Trace}(aH^{-s}).$$

$\square$

We now apply the Local Index formula to the general Heisenberg cycles associated with flows which are cohomologically trivial in the sense that the cohomological equation $Xu = f$ is smoothly solvable for arbitrary smooth f of zero mean.

Let $B = C(M) \rtimes \Lambda$, for a flow α on M and $\Lambda \subset \mathbb{R}$ a Diophantine subgroup (Definition 10.11.16). Let μ be an α-invariant measure, X generate the flow, and assume that $Xu = f$ is smoothly solvable for any smooth f such that $\int_M f d\mu = 0$. Fix $p \in M$, and let $\pi : C(M) \rtimes \Lambda \to \mathbb{B}(L^2(\mathbb{R}))$ be the corresponding representation, with $\pi(f) = f_p$, $f_p(t) = f(\alpha_t(p))$.

We have shown that $B^\infty \subset B$ has the property that $\text{Trace}(\pi(b)H^{-s})$ has the meromorphic extension property and that

$$\text{Res}_{s=1}(\pi(b)H^{-s}) = \tau_\mu(b),$$

where $\tau_\mu : B \to \mathbb{C}$ is the trace induced by μ, and $b \in B^\infty$.

Let $\delta_1^\alpha, \delta_2^\alpha$ be the derivations of B^∞ defined

$$\delta_1^\alpha(f) = X(f), \quad \delta_1^\alpha(U_\alpha) = 0, \qquad \delta_2^\alpha(f) = 0, \quad \delta_2^\alpha(U_\alpha) = \alpha.$$

Lemma 10.12.7 *In the above notation, the functional Ψ_2 of Theorem 10.12.5 b) is given on B^∞ by*

(10.87) $$\Psi_2(b^0, b^1, b^2) = \tau_\mu \left(a^0 \delta_1^\alpha(a^1)\delta_2^\alpha(a^2) - a^0 \delta_2^\alpha(a^1)\delta_1^\alpha(a^2) \right)$$

for all $b_0, b_1, b_2 \in B^\infty$.

Proof Note that

$$\delta_1(b) = \left[\pi(b), \frac{d}{dx} \right], \qquad \delta_2(b) = [\pi(b), x].$$

Expand $[D, a^1][D, a^2]$ as a block matrix

$$\left[D, b^1\right]\left[D, b^2\right]$$
$$= \begin{bmatrix} \left[x - d/dx, b^1\right]\left[x + d/dx, b^2\right] & 0 \\ 0 & \left[x + d/dx, b^1\right]\left[x - d/dx, b^2\right] \end{bmatrix}.$$

We deduce that

$$\frac{1}{2}\mathrm{Res}_{s=1}\mathrm{Tr}\left(\varepsilon b^0 \left[D, b^1\right]\left[D, b^2\right] H^{-s}\right)$$

$$= \mathrm{Res}_{s=1}\left(b^0 \left[x, b^1\right]\left[d/dx, a^2\right] H^{-s}\right)$$

$$- \mathrm{Res}_{s=1}\mathrm{Tr}\left(b^0 \left[d/dx, b^1\right]\left[x, b^2\right] H^{-s}\right)$$

(10.88)
$$= \tau_\mu \left(b^0 \delta_1(b^1)\delta_2(b^2) - b^0 \delta_2(b^1)\delta_1(b^2)\right).$$

□

We have thus proved the following.

Theorem 10.12.8 *Let $B = C(M) \rtimes \Lambda$, for a smooth flow α on a compact manifold M. Let $\Lambda \subset \mathbb{R}$ be a Diophantine subgroup. Let μ be an α-invariant measure, X generate the flow, and assume that $Xu = f$ is smoothly solvable for any smooth f such that $\int_M f d\mu = 0$.*

Then the Chern character of the Heisenberg cycle determined by a point $p \in M$ is given by $\tau_\mu - \tau_2$, where τ_μ is the trace on B determined by μ, and τ_2 is the cyclic 2-cocyle for B^∞ given by

$$\tau_2(b^0, b^1, b^2) = \tau_\mu \left(b^0 \delta_1^\alpha(b^1)\delta_2^\alpha(b^2) - b^0 \delta_2^\alpha(b^1)\delta_1^\alpha(b^2)\right)$$

on B^∞.

Corollary 10.12.9 *Let $\hbar \in \mathbb{R}$ and $A_\hbar := C(\mathbb{T}) \rtimes_\hbar \mathbb{Z}$ be the corresponding rotation algebra.*

Let $[D_\hbar]$ be the class of the Heisenberg cycle (Definition 10.10.9)

$$\left(L^2(\mathbb{R}) \oplus L^2(\mathbb{R}), \pi_\hbar, \ D := \begin{bmatrix} 0 & x - d/dx \\ x + d/dx & 0 \end{bmatrix}\right).$$

Then the Chern character of $[D_\hbar]$ is given by $\tau - \hbar\tau_2$, where

$$\tau_2(a^0, a^1, a^2) = \tau \left(a^0 \delta_1(a^1)\delta_2(a^2) - a^0 \delta_2(a^1)\delta_1(a^2)\right).$$

Corollary 10.12.10 *Let $e \in A_{\hbar}^{\infty}$ be a projection and $[e] \in K_0(A_{\hbar})$ its class. Then*

$$\langle [e], [D_{\hbar}] \rangle = \tau(e) - \hbar \cdot c_1(e),$$

where $c_1(e)$ is the first Chern number of e (Theorem 10.12.2).
In particular,

$$\langle [p_{\hbar}], [D_{\hbar}] \rangle = \lfloor \hbar \rfloor,$$

where $\lfloor \hbar \rfloor$ is the greatest integer $< \hbar$.

To round off the discussion, we will also compute the index data for another spectral cycle, which has been intensively studied.

Definition 10.12.11 The *Dirac–Dolbeault* spectral cycle for $A_{\hbar}^{\infty} \subset A_{\hbar}$ is defined as follows. The Hilbert space is $L^2(\mathbb{T}^2) \oplus L^2(\mathbb{T}^2)$, evenly graded. The operator is

$$\bar{\partial} := \begin{bmatrix} 0 & \frac{\partial}{\partial x} - i\frac{\partial}{\partial y} \\ \frac{\partial}{\partial x} + i\frac{\partial}{\partial y} & 0 \end{bmatrix}.$$

The representation is two copies of the representation $\lambda \colon A_{\hbar} \to \mathbb{B}\left(L^2(\mathbb{T}^2)\right)$, which is specified by the covariant pair

$$(\lambda(f)\xi)(x,y) = f(x)\xi(x,y), \quad (\pi(n)\xi)(x,y) = e^{2\pi i n y}\xi(x - n\hbar, y).$$

Exercise 10.12.12 The Dolbeault cycle is two-dimensional and is regular with the meromorphic extension property over $A_{\hbar}^{\infty}$.

Proposition 10.12.13 *Let $[\bar{\partial}] \in KK_0(A_{\hbar}, \mathbb{C})$ be the class of the Dolbeault cycle. Then if $e \in A_{\hbar}^{\infty}$ is a projection, then*

$$\langle [\bar{\partial}], [e] \rangle = c_1(e).$$

Remark 10.12.14 The previous corollary implies the *integrality* of the first Chern number, for *any* projection $e \in A_{\hbar}$.

The proof of Proposition 10.12.13 is left as an exercise: Note that for the Dolbeault operator, one has

$$\bar{\partial}^2 = \begin{bmatrix} \frac{\partial^2}{\partial x^2} + \frac{\partial^2}{\partial y^2} & 0 \\ 0 & \frac{\partial^2}{\partial x^2} + \frac{\partial^2}{\partial y^2} \end{bmatrix},$$

and the kernel of $\bar{\partial}$ is two-dimensional: spanned by a copy of the constant functions in the first $L^2(\mathbb{T}^2)$ and by a copy of the constant functions in the second copy as well.

Cancellation implies that the zeroth functional Ψ_0 associated with this spectral cycle is zero.

Finally, we note that we may draw the following strong corollary, which only depends on the spectral triples we have constructed above, and the computation of their index pairings by the Local Index Formula (and not on computation of the K-theory of $A_\hbar$, which is a deeper result requiring a strong form of Bott Periodicity.)

Corollary 10.12.15 *Suppose $\hbar \in \mathbb{R}$ is nonzero. Then if $\tau \colon A_\hbar \to \mathbb{C}$ is the trace, $\tau_* \colon K_0(A_\hbar) \to \mathbb{R}$ the induced group homomorphism, then*

$$\tau_* \left(K_0(A_\hbar) \right) = \mathbb{Z} + \hbar\mathbb{Z} \subset \mathbb{R}.$$

Proof If $e \in A_\hbar^\infty$ is a projection, then application of our results above gives that $\tau(e) + \hbar c_1(e)$ is an integer. On the other hand, $c_1(e)$ is an integer. This implies $\tau(e) = m + n\hbar$ for a pair of integers m, n. Finally, $A_\hbar^\infty$ is dense and holomorphically closed in $A_\hbar$, so any projection in $A_\hbar$ is represented by a projection in $A_\hbar^\infty$.

$\square$

Chapter 11
An Introduction to KK-Theory

KK-theory is one of the most important achievements of the field of Noncommutative Geometry. KK-theory was invented by Kasparov [109–111], motivated by ideas of Atiyah [9], and was further developed by Connes and Skandalis [51], who introduced an axiomatic approach to the intersection product (the composition operation in the category KK) and produced applications to families and foliation index theorems. The article [18] gives an important description of KK-theory in terms of unbounded operators.

KK-theory gives a language in which it is possible to formulate disparate problems and theorems in geometry and topology, ranging from questions about positive scalar curvature metrics on smooth manifolds, or homotopy invariance conjectures in topology to classification programs in dynamical systems, the topology of orbifolds or of coarse geometric spaces, to the representation theory of Lie groups. KK-theory provides a unified framework for studying K-theory and the many important variants of it used in different contexts: equivariant K-theory [147], coarse K-theory [75], twisted K-theory [62, 143], groupoid-equivariant K-theory [154], and K-theory with $\mathbb{R}$, $\mathbb{R}/\mathbb{Z}$, or $\mathbb{Z}/k$ coefficients [60, 61], and [5, 6]. Many of these variants are important in physics, for example, twisted K-theory is used in connection with String Theory (see, e.g., [121, 144]), while KK-theory coupled with the tools of spectral cycles and Connes' Chern character yields new insight into parts of solid state physics like the Quantum Hall Effect (see [23, 47]).

Some of the deepest mathematical results about KK-theory are in effect attempts at equivariant generalizations of Bott Periodicity, called the Baum–Connes Conjecture (see [19] for an early version). The eventual formulation of the conjecture appeared in [20]. The conjecture concerns the K-theory of crossed products $A \rtimes G$, for a G-C*-algebra A, with G a locally compact group (or groupoid), and asserts, roughly speaking, that for purposes of computing K-theory of the crossed product, $A \rtimes G$ may be replaced by $P \otimes A \rtimes G$ for a specific (up to G-equivariant homotopy) *proper G-C*-algebra P* (see [122]), for which the K-theory may always in principal be computed by repeated excision arguments, because P is proper. The best results

© Springer Nature Switzerland AG 2024
H. Emerson, *An Introduction to C*-Algebras and Noncommutative Geometry*,
Birkhäuser Advanced Texts Basler Lehrbücher,
https://doi.org/10.1007/978-3-031-59850-0_11

to date are the Higson–Kasparov Theorem (see [99]), proving the (strongest form of the) conjecture for amenable groups, Tu's generalization of it to amenable groupoids [154], and Lafforgue's work [117] proving the conjecture for uniform lattices in $SL_3(\mathbb{R})$, among other things, using the full technical power of KK-theory, adapted to Banach algebras. The conjecture is now known to be false in general [96], but it is nonetheless true in many cases, and various proof techniques (the Dirac–dual-Dirac method) developed to tackle it remain powerful tools.

The category of C*-algebras and *-homomorphisms has a tensor product operation: the (spatial) tensor product of two C*-algebras is a C*-algebra, and the tensor product of a pair of *-homomorphisms is a *-homomorphism. This structure holds in KK as well, making it a symmetric monoidal category. In such categories there is a fairly standard notion of duality (see [72]). Two C*-algebras are *dual* in KK if left tensoring by A, a functor KK $\to$ KK, is left adjoint to left tensoring by B. The Baum–Connes Conjecture for a discrete group Γ with compact classifying space $B\Gamma$ is, roughly speaking, a conjectured KK duality between $C(B\Gamma)$ and $C^*(\Gamma)$ (see [73]). *Self*-duality for C*-algebras is quite special. The C*-algebra of continuous functions on a spin^c-manifold is a *self*-dual C*-algebra, and the duality is induced by the Dirac operator, a theorem due to Kasparov [111]. Self-duality almost characterizes smooth manifolds among compact spaces, which suggests that if there were noncommutative examples, they might be deserving of the term "Noncommutative Manifolds" (see [48] for a discussion of this concept). Connes produced the first example of such a duality: the irrational rotation algebra $A_\hbar$ [47, 48], with a duality induced by the Dirac–Dolbeault operator discussed in the previous chapter (see [65] for a recent treatment of this example). Sometime following this the paper [104] proved that the Cuntz–Krieger algebras O_A and O_{A^t} are KK-dual, following this with a duality between the stable and unstable Ruelle algebras of a Smale space in [105], and in [71] it is proved that the crossed products $C(\partial G) \rtimes G$ of boundary actions of Gromov hyperbolic groups are self-dual in KK. None of these examples from dynamics have anything to do with manifolds, which does indicate that there do exist genuinely new manifold-like structures (like Poincaré duality) in the world of noncommutative C*-algebras.

In this chapter we give a basic overview of KK-theory and its main properties, concluding with a proof of Bott Periodicity and one of its equivariant generalizations, and a brief discussion of duality. Our main goal is to illustrate the power of the axiomatic description of the intersection product in computing with concrete examples. The reader who has read the rest of this book should hopefully find the definitions and examples of KK quite natural.

11.1   Basic Definitions of KK

Kasparov defines the $\mathbb{Z}/2$-graded bivariant groups $KK_*(A, B)$, where A and B are C*-algebras, in a very similar way to the way in which we have defined analytic K-homology. The definition goes essentially unchanged, except that H is replaced by a right Hilbert B-*module*.

Definition 11.1.1 Let A and B be C*-algebras. A *Fredholm A–B-bimodule* is a triple $(\mathcal{E}, \pi, F)$, where

a) $\mathcal{E}$ is a right Hilbert B-module.
b) $\pi : A \to \mathbb{B}(\mathcal{E})$ is a *-homomorphism, i.e., a representation of A by adjointable Hilbert B-module operators on $\mathcal{E}$.
c) $F \in \mathbb{B}(\mathcal{E})$ is a self-adjoint Hilbert B-module operator, satisfying

$$(11.1) \qquad \pi(a) \cdot (F^2 - 1), \quad [\pi(a), F] \in \mathcal{K}(\mathcal{E})$$

for all $a \in A$.

The bimodule is *even* if it carries the additional data of a $\mathbb{Z}/2$ grading

$$\mathcal{E} = \mathcal{E}^+ \oplus \mathcal{E}^-,$$

on $\mathcal{E}$, into orthogonal B-submodules, with respect to which elements of A act as even (grading-preserving) operators, and the operator F acts as an odd (grading-reversing) operator.

Otherwise, the bimodule is *odd*.

Clearly, a Fredholm A-$\mathbb{C}$-bimodule is the same as a Fredholm module, as in the previous section.

For any A, B, the triple $(0, 0, 0)$, consisting of the zero Hilbert B-module, understood as a $\mathbb{Z}/2$-graded bimodule, in the only possible way, the zero representation of A, and the zero operator, is a Fredholm A–B-bimodule.

The triple $(A, \mathrm{id}_A, 0)$, where A acts on the left on the Hilbert module A by left multiplication and where the grading is $A = A^+$ (in other words, $A^- = \{0\}$, is an even Fredholm A-A-bimodule, for any A. (Its class in $\mathrm{KK}_0(A, A)$ will be that of the identity morphism.)

There is an obvious definition of unitary isomorphism of such bimodules, and one can clearly take the direct sum of two of them. Let $\mathcal{E}_i(A, B)$ denote the corresponding semigroup of unitary isomorphism classes of even (if $i = 0$) and odd ($i = 1$) Fredholm A–B bimodules. It has a certain natural functoriality. If $\beta : B \to B'$ is a *-homomorphism, then form the right Hilbert B'-module

$$\mathcal{E} \otimes_B B',$$

using the tensor product of Hilbert modules: that of the right Hilbert B-module $\mathcal{E}$, and the right Hilbert B'-module B', over the representation $B \to B' \subset \mathcal{M}(B) = \mathbb{B}(B)$. If $a \in A$, let it act on $\mathcal{E} \otimes_B B'$ by $\pi(a) \otimes 1$ and form the operator $F \otimes 1$. The corresponding triple is a Fredholm A-B'-bimodule $\beta_\sharp(\mathcal{E}, \pi, F)$. It is even or odd according as the original one was.

Thus β determines a semigroup homomorphism

$$(11.2) \qquad \beta_\sharp : \mathcal{E}(A, B) \to \mathcal{E}(A, B').$$

It is even more straightforward that if $\alpha : A' \to A$ is a *-homomorphism, then it induces a semigroup homomorphism

(11.3) $\alpha^{\sharp} : \mathcal{E}(A, B) \to \mathcal{E}(A', B),$

simply by replacing the representation π in an A–B-bimodule by $\pi \circ \alpha$.

If a Fredholm A–B-bimodule $(\mathcal{E}, \pi, F)$ has the property that all the terms in (11.29) are zero, not just compact, then we say it is *degenerate*.

As a particular case of the (forward) functoriality of the $\mathcal{E}(A, B)$ semigroups, note that the point evaluations at $t = 0$ and $t = 1$ give two *-homomorphisms

$$\epsilon_0, \epsilon_1 : C([0, 1]) \to \mathbb{C}$$

and then semigroup homomorphisms

(11.4) $(\epsilon_0)_*, \ (\epsilon_1)_* : \mathcal{E}(A, \ C([0, 1]) \otimes B) \to \mathcal{E}(A, B)$

for any A, B.

Definition 11.1.2 Two Fredhom A–B-bimodules are *homotopic* if they are unitarily isomorphic to the endpoints $(\epsilon_i)_*(\mathcal{E}, \pi, F)$ of a Fredholm A-$C([0, 1]) \otimes B$-bimodule $(\mathcal{E}, \pi, F)$.

Lemma 11.1.3 *If $(\mathcal{E}, \pi, F)$ is a degenerate Fredholm A–B-bimodule, then $(\mathcal{E}, \pi, F)$ is homotopic to the zero bimodule $(0, 0, 0)$.*

Proof The homotopy uses, in the even case, the $\mathbb{Z}/2$-graded right Hilbert $B \otimes C(I)$-module $C_0([0, 1), \mathcal{E})$, graded by the grading on $\mathcal{E}$. The representation of A is by

$$\big(\tilde{\pi}(a)\xi\big)(t) := \pi(a)\xi(t)$$

and the operator

$$(\tilde{F}\xi)(t) := F\big(\xi(t)\big).$$

The triple $\big(C_0([0, 1), \mathcal{E}), \tilde{\pi}, \tilde{F}\big)$ is a Fredholm bimodule because the operators

$$\tilde{\pi}(a) \cdot (\tilde{F}^2 - 1), \quad [\tilde{\pi}(a), \tilde{F}]$$

are actually zero and hence compact (which would not be the case if we merely assumed that $\pi(a) \cdot (F^2 - 1)$ and $[\pi(a), F]$ were merely *compact*).

The endpoints of our homotopy clearly are, respectively, the zero bimodule and our degenerate one, proving the result. $\square$

Exercise 11.1.4 Show that if $\beta_1, \beta_2 : B \to B'$ are homotopic *-homomorphisms, then

$$(\beta_1)_*(\mathcal{E}, \pi, F)$$

is homotopic to

$$(\beta_1)_*(\mathcal{E}, \pi, F)$$

for any Fredholm A-B-bimodule $(\mathcal{E}, \pi, F)$.

Definition 11.1.5 Let A and B be C*-algebras.

Then $KK_0(A, B)$ is the quotient of the semigroup $\mathcal{E}_0(A, B)$ of $\mathbb{Z}/2$-graded (i.e., even) Fredholm A-B-bimodules, by the equivalence relation of homotopy.

$KK_1(A, B)$ is defined in exactly the same way, using odd bimodules.

A *Kasparov morphism* $A \to B$ is an element of $KK_*(A, B) := KK_0(A, B) \oplus KK_1(A, B)$.

Remark 11.1.6 In order to immediately correct an apparent conflict of notation, we point out the following. It is possible to define the equivalence relation(s) on cycles determining KK as we did with K-homology in Section 10.1 of Chapter 10, by using the equivalence relation on cycles generated by addition of degenerates and operator homotopy (a homotopy $\{F_t\}_{t\in[0,1]}$ of operators in the norm topology, but where none of the other data varies).

In the end it turns out that these two approaches agree. Thus, two cycles are homotopic if and only if, and addition of degenerates, they become operator homotopic. This somewhat remarkable result is due to G. Skandalis [149].

Lemma 11.1.7 *With the direct sum operation,* $KK_i(A, B)$ *is a group,* $i = 0, 1$. *A *-homomorphism* $\alpha: A \to B$ *defines a grading-preserving group homomorphism*

$$\alpha_*: KK_*(D, A) \to KK_*(D, B),$$

for any D, and similarly a group homomorphism

$$\alpha^*: KK_*(B, D) \to KK_*(A, D),$$

for any D.

Proof Let $(\mathcal{E}, \pi, F)$ be an even Fredholm A-B-bimodule.

Consider the triple $(-\mathcal{E}, \pi, -F)$, where $-\mathcal{E}$ denotes $\mathcal{E}$ but with the *opposite* grading. The sum of $(\mathcal{E}, \pi, F)$ and $(-\mathcal{E}, \pi, -F)$ is

$$(\mathcal{E} \oplus -\mathcal{E}, \pi \oplus \pi, F \oplus -F).$$

Now let

$$\tilde{F}_t := \begin{bmatrix} \cos t \cdot F & \sin t \\ \sin t & -\cos t \cdot F \end{bmatrix} \in \mathbb{B}(\mathcal{E} \oplus -\mathcal{E}).$$

Using the operators $\tilde{F}_t$, we get a homotopy between $(\mathcal{E} \oplus -\mathcal{E}, \pi \oplus \pi, F \oplus -F)$ and the degenerate bimodule

$$\left(\mathcal{E} \oplus -\mathcal{E}, \pi \oplus \pi, \begin{bmatrix} 0 & 1 \\ 1 & 0 \end{bmatrix}\right).$$

For odd Fredholm bimodules, we use exactly the same method, but dropping any discussion of gradings. The additive inverse of a triple $(\mathcal{E}, \pi, F)$, where $\mathcal{E}$ is an odd A-B-module, is obtained by simply replacing F by $-F$; the same operator homotopy and argument above shows that summing these two cycles gives a homotopy to a degenerate.

This shows that $KK_*(A, B)$ is a group. The functoriality statements follow from the maps on cycles given in (11.2) and (11.3).

$\square$

Example 11.1.8 A *-homomorphism $\alpha \colon A \to B$ determines an element of $\mathcal{E}_0(A, B)$ by setting $\mathcal{E}^+ = B$, $\mathcal{E}^- = 0$, $\pi := \alpha \colon A \to B = \mathcal{K}(\mathcal{E})$, and $F := 0$. The corresponding degree-zero Kasparov morphism is denoted as

$$[\alpha] \in KK_0(A, B).$$

Exercise 11.1.9 Suppose $\alpha, \beta \colon A \to B$ are homotopic *-homomorphisms. Prove that $[\alpha] = [\beta] \in KK_0(A, B)$.

More generally, if $\mathcal{E}$ is a Hilbert B-module and $\pi \colon A \to \mathcal{K}(\mathcal{E})$ is a representation of A as *compact operators* on B, then we can assign the grading $\mathcal{E}^+ := \mathcal{E}$, $\mathcal{E}^- := \{0\}$ and set $F := 0$; then we obtain a cycle $(\mathcal{E}, \pi, 0) \in \mathcal{E}_0(A, B)$, because the terms in (11.29) are all compact.

This situation applies in particular if $\mathcal{E}$ is the underlying right B-module of a *strong Morita A-B-bimodule*, in which π is the left action. Combining these observations gives the following:

Proposition 11.1.10 *A Morita correspondence from A to B in the sense of Definition 6.2.5 determines a Kasparov morphism*

$$[\mathcal{E}] \in KK_0(A, B).$$

Exercise 11.1.11 Suppose that (B, π, F) is a cycle for $KK_1(A, B)$, where the right Hilbert B-module is B. Thus, $\pi \colon A \to \mathbb{B}(B) = \mathcal{M}(B)$ is a *-homomorphism, and F is a self-adjoint multiplier of B, such that $\pi(a) \cdot (F^2 - 1)$, $[\pi(a), F] \in B$, for all $a \in A$. Let α be the class of our cycle.

Let $\sigma \colon B \to B$ be an automorphism and $[\sigma] \in KK_0(B, B)$ its class. Show that $\alpha \otimes_B [\sigma]$ is represented by the triple $(B, \sigma \circ \pi, \sigma(F))$.

Exercise 11.1.12 If χ is a normalizing function, then the triple $(C_0(\mathbb{R}), 1, \chi))$ is a cycle for $KK_1(\mathbb{C}, C_0(\mathbb{R}))$ (it represents the Bott element). Let $\beta \in KK_1(\mathbb{C}, C_0(\mathbb{R}))$

be its class. Let $\tau\colon C_0(\mathbb{R}) \to C_0(\mathbb{R})$ be $\tau(f)(x) = f(-x)$ and $[\tau] \in$ $\mathrm{KK}_0(C_0(\mathbb{R}), C_0(\mathbb{R}))$ its class. Show that

$$\beta \otimes_{C_0(\mathbb{R})} [\tau] = -\beta \in \mathrm{KK}_1(\mathbb{C}, C_0(\mathbb{R})).$$

Kasparov morphisms also appear naturally in connection with K-theory of noncompact spaces. Let (E^+, E^-, u) be a K-theory triple for X. Thus, $E^\pm$ are complex vector bundles over X, and u is a bundle map $E^+ \to E^-$ which is invertible of a compact subset of X.

We can put Hermitian metrics on $E^\pm$, making the spaces of C_0-sections $C_0(X, E^\pm)$ into right Hilbert $C_0(X)$-modules $\mathcal{E}^\pm$. Form the $\mathbb{Z}/2$-graded Hilbert $C_0(X)$-module $\mathcal{E} := \mathcal{E}^+ \oplus \mathcal{E}^-$. We can also assume without loss of generality (by a simple homotopy) that u is unitary of a compact set. Hence the $C_0(X)$-module operator

$$F := \begin{bmatrix} 0 & u* \\ u & 0 \end{bmatrix},$$

acting on $\mathcal{E}$, is self-adjoint, and $F^2 - 1$ is compact (because $F^2 - 1$ vanishes of a compact set).

The triple $(\mathcal{E}, 1, F)$ is a Fredholm $\mathbb{C}$-$C_0(X)$-bimodule and gives a corresponding element of $\mathrm{KK}_0\big(\mathbb{C}, C_0(X)\big)$.

A particular case of a triple is when X is actually compact, and then the bundle map $E^+ \to E^-$ can be taken to be the zero map. The procedure above produces a corresponding Kasparov morphism. These observations show that one has a natural map

$$\mathrm{K}^0(X) \to \mathrm{KK}_0(\mathbb{C}, C_0(X))$$

for any locally compact space X. It can be shown to be an isomorphism.

More generally, let A be a C*-algebra (perhaps not unital) and (p, q, u) be a *relative triple* (Definition 8.4.1) for (A^+, A), where A^+ is its unitization. So $p, q, u \in M_n(A^+)$, and $uu^* - 1$, $u^*u - 1$ and $upu^* - q$ all lie in the ideal $M_n(A)$ of $M_n(A^+)$, for some n. Such triples are by definition cycles for $\mathrm{K}_0(A)$.

Now any $a \in M_n(A^+)$ is in the multiplier algebra of $M_n(A)$ and hence defines a Hilbert A-module map

$$a\colon A^n \to A^n$$

by multiplication, and if a is actually in the ideal $M_n(A) \subset M_n(A^+)$, then a acts as a compact operator on A^n.

Proposition 11.1.13 *If (p, q, u) is a relative triple for A^+ over A, and*

$$w := qup\colon pA^n \to qA^n,$$

then the triple

$$\left(pA^n \oplus qA^n, 1, F := \begin{bmatrix} 0 & w^* \\ w & 0 \end{bmatrix} \right)$$

is a cycle for $\mathrm{KK}_0(\mathbb{C}, A)$.

Proof The assumption implies that

$$u^* q u = p \bmod A.$$

Write

$$u^* q u = p + s, \quad s \in A.$$

We have

$$w^* w = (qup)^* qup = pu^* qup = p(p + s)p = p + psp,$$

which is a perturbation of p by an element of $pM_n(A)p$ and hence by a compact operator on pA^n.

Therefore $w^* w - 1 \in \mathcal{K}(pA^n)$, and similarly $ww^* - 1 \in \mathcal{K}(pA^n)$. The result follows.   □

Exercise 11.1.14 Show that unitarily equivalent relative triples map under the above construction to unitarily equivalent KK-cycles and that degenerate triples map to degenerate KK-cycles.

Theorem 11.1.15 *If A is a σ-unital C*-algebra, then the map on cycles defined by Proposition 11.1.13 descends to a group isomorphism*

$$\mathrm{K}_0(A) \to \mathrm{KK}_0(\mathbb{C}, A).$$

We will sketch the proof of the theorem modulo two important technical theorems.

1. *The Stabilization Theorem*
   The *Stabilization Theorem* for Hilbert modules asserts that if A is a σ-unital C*-algebra, then any right Hilbert A-module $\mathcal{E}$ is a direct summand (in the sense that it is orthogonally complemented) in the standard Hilbert A-module $A \otimes l^2$. To be explicit, $\mathcal{E} \oplus \mathcal{F} \cong A \otimes l^2$ for some Hilbert A-module $\mathcal{F}$, where the direct sum is in the category of Hilbert modules, and the isomorphism is a unitary isomorphism of Hilbert modules. See [108] (or the book [26]) for proofs of the Stabilization Theorem.

2. *Kuiper Theorem*
   Kuiper's theorem [114] asserts the contractibility of the unitary group of $\mathbb{B}(H)$. The theorem is generalized in [59] to Hilbert modules and implies the following:

Theorem 11.1.16 *Let A be any C*-algebra and $\mathcal{M}^s(A) := \mathcal{M}(A \otimes \mathcal{K})$. Then*

$$K_i(\mathcal{M}^s(A)) = 0, \quad i = 1, 2.$$

The theorem is what is needed to prove Theorem 11.1.15 in the following manner:

Proof of Theorem 11.1.15 Let

$$\Phi \colon K_0(A) \to KK_0(\mathbb{C}, A)$$

be the map constructed using relative triples, above. We construct a map

$$\Psi \colon KK_0(\mathbb{C}, A) \to K_0(A)$$

inverting Φ. Consider the exact sequence of C*-algebras

$$(11.5) \qquad 0 \to A \otimes \mathcal{K} \to \mathcal{M}^s(A) \to Q^s(A) \to 0,$$

where $\mathcal{M}^s(A) := \mathcal{M}(A \otimes \mathcal{K})$ and $Q^s(A) = \mathcal{M}^s(A)/A \otimes \mathcal{K}$. The associated long exact sequence in K-theory of Theorem 8.5.6 gives an exact sequence

$$\cdots K_2(Q^s(A)) \overset{\delta}{\to} K_1(A) \to K_1(\mathcal{M}^s(A))$$

$$(11.6) \qquad \to K_1(Q^s(A)) \overset{\delta}{\to} K_0(A) \to K_0(\mathcal{M}^s(A)) \to K_0(Q^s(A)),$$

and by Kuiper's theorem, $K_i(\mathcal{M}^s(A)) = 0$, $i = 0, 1$, so

$$\delta \colon K_1(Q^s(A)) \to K_0(A)$$

is an isomorphism. So it remains to show that

$$KK_0(\mathbb{C}, A) \cong K_1(Q^s(A)).$$

Suppose that $(\mathcal{E}, \pi, F)$ is a cycle for $KK_0(\mathbb{C}, A)$. We may assume that $\pi \colon \mathbb{C} \to \mathbb{B}(\mathcal{E})$ is unital and otherwise replace the cycle by the one obtained by compressing everything by the projection $\pi(1)$. The original cycle is then the direct sum of this one and a degenerate cycle. So we may assume that the representation involved in our cycle is unital, so the cycle has the form $(\mathcal{E}, 1, F)$.

Next, using the Stabilization Theorem, there exist Hilbert A-modules $\mathcal{F}^\pm$ such that

$$\mathcal{E}^\pm \oplus \mathcal{F}^\pm \cong A \otimes l^2,$$

and we may assume in addition, without loss of generality, that $\mathcal{F}^+ \cong \mathcal{F}^-$, by a unitary w. Let $\mathcal{F} = \mathcal{F}^+ \oplus \mathcal{F}^-$.

We now modify our cycle $(\mathcal{E}, 1, F)$ by adding to it the degenerate cycle

$$\left(\mathcal{F}, 1, \begin{bmatrix} 0 & w^* \\ w & 0 \end{bmatrix} \right).$$

This results in a cycle for $KK_0(\mathbb{C}, A)$ in which the only remaining variable is the operator F, which has the form

$$F = \begin{bmatrix} 0 & u \\ u & 0 \end{bmatrix},$$

where

$$u \in \mathbb{B}(A \otimes l^2) \cong \mathcal{M}^s(A),$$

is an *essential unitary*. In particular, u defines a class

$$[u] \in K_1(Q^s(A)).$$

We let Ψ be defined on the cycle we started with by

$$\delta([u]) \in K_0(A),$$

where δ is the connecting map in the exact sequence (11.6).

Notice that if the cycle, reduced to one of the form,

$$\left(A \otimes l^2 \oplus A \otimes l^2, 1, \begin{bmatrix} 0 & u^* \\ u & 0 \end{bmatrix} \right)$$

is degenerate, then u is actually unitary in $\mathcal{M}^s(A)$, and hence by the exact sequence (11.6),

$$\delta([u]) = 0.$$

Finally, if two Kasparov cycles are operator homotopic, then they determine homotopic unitaries in $Q^s(A)$, as we leave it to the reader to check. Since the equivalence relation defining KK can be taken to be operator homotopy and addition of degenerates, we obtain a well-defined map

$$KK_0(\mathbb{C}, A) \to K_0(A).$$

The fact that it inverts Φ is left to the assiduous reader.   □

There is also an isomorphism $K_1(A) \cong KK_1(\mathbb{C}, A)$ for any A. Indeed, arguing as in the proof above, one shows that $KK_1(\mathbb{C}, A) \cong K_0(Q^s(A))$. Indeed, if $(\mathcal{E}, \pi, F)$ is an odd Fredholm $\mathbb{C}$-A-bimodule, with $\mathcal{E} \cong A \otimes l^2$, then F defines a self-adjoint contraction in $\mathcal{M}^s(A)$ such that $F^2 - 1 \in A \otimes \mathcal{K}$. If $P = \frac{F+1}{2}$, then P is self-adjoint and $P^2 - P \in A \otimes \mathcal{K}$, whence P defines a projection in $Q^s(A)$. This determines a map $KK_1(\mathbb{C}, A) \to K_0(Q^s(A))$, and applying the exponential map $K_0(Q^s(A)) \xrightarrow{\cong} K_1(A)$ gives an isomorphism $KK_1(\mathbb{C}, A) \cong K_1(A)$. The inverse map is more tricky to compute, if $K_1(A)$ is described in terms of unitaries in matrix algebras over A, as to find a KK cycle mapping to the class of a unitary involves lifting the unitary under the exponential map.

See the Exercise below. We state the result:

Corollary 11.1.17 *For any separable C*-algebra A, $K_i(A) \cong KK_i(\mathbb{C}, A)$, $i = 1, 2$.*

Exercise 11.1.18 If $(\mathcal{E}, \pi, F)$ is an *odd* Fredholm A-B bimodule with $F = 0$, then $(\mathcal{E}, \pi, F)$ is equivalent to the zero bimodule. Why is this not also true if the bimodule is *even*?

Example 11.1.19 Identify $C(\mathbb{T})$ with $\{f \in C([0, 1]) \mid f(0) = f(1)\}$. Let $\mathcal{E} \subset C(\mathbb{T})$ be the ideal $C_0((0, 1))$, a right Hilbert $C(\mathbb{T})$-module. Let $F \in \mathbb{B}(\mathcal{E})$ be the operator $\mathcal{E} \to \mathcal{E}$ of multiplication by the function $f(t) = t$. Then F is self-adjoint and $F^2 - 1 \in \mathcal{K}(\mathcal{E})$, so $(\mathcal{E}, \pi, F)$ is a cycle for $KK_1(\mathbb{C}, C(\mathbb{T}) \cong K_0(Q^s(C(\mathbb{T})))$, where π is the scalar multiplication representation of $\mathbb{C}$ on $\mathcal{E}$. The class of this cycle maps under the exponential map to the class $[z] \in C(\mathbb{T})$ of the unitary $z \in C(\mathbb{T})$.

11.2 Category Structure of KK

The most important result about KK-theory is that it has a category structure: Kasparov morphisms can be composed in a manner extending, roughly speaking composition of *-homomorphisms. From this categorical point of view, the index pairing between K-theory $K_*(A) \cong KK_*(\mathbb{C}, A)$ and K-homology $KK_*(A, \mathbb{C})$ discussed in Section 10.1 is nothing but composition of morphisms: A morphism $\mathbb{C} \to A$ and a morphism $A \to \mathbb{C}$ compose to give a morphism $\mathbb{C} \to \mathbb{C}$, equivalently, an integer, since $KK_*(\mathbb{C}, \mathbb{C}) = \mathbb{Z}$.

For any A separable, let $1_A \in KK_0(A, A)$ be the class of the Fredholm A-A-bimodule $(A, \text{id}, 0)$, where $\text{id} \colon A \to \mathcal{K}(A) = A$ is the representation of A as left multipliers.

Theorem 11.2.1 *For any A, B, there is a bilinear pairing*

$$(11.7) \qquad KK_*(A, B) \times KK_*(B, C) \to KK_*(A, C)$$

mapping a pair of morphisms $f \in KK_i(A, B)$ and $g \in KK_j(B, C)$ to a morphism

$$f \hat{\otimes}_B g \in \mathrm{KK}_{i+j}(A, C),$$

a) *The Kasparov product gives* KK *the structure of a* $\mathbb{Z}/2$-*graded category, with objects (separable)* C*-algebras and morphisms* $A \to B$ *the elements of the* $\mathbb{Z}/2$-*graded abelian group* $\mathrm{KK}_*(A, B)$. *For any* A, *the element* $1_A \in \mathrm{KK}_0(A, A)$ *defined above acts as the identity morphism from* A *to* A.

b) *If* $\alpha \colon A \to A'$ *is a* *-homomorphism and* $[\alpha] \in \mathrm{KK}_0(A, A')$ *its class, then* $\alpha^*(f) = [\alpha] \hat{\otimes}_{A'} f \in \mathrm{KK}_*(A, B)$ *for any* $f \in \mathrm{KK}_*(A', B)$. *Similarly* $\alpha_*(g) = g \hat{\otimes}_A [\alpha] \in \mathrm{KK}_*(B, A')$ *for any* $g \in \mathrm{KK}_*(B, A)$.

c) *If* $\alpha \colon C \to C'$ *is a* *-homomorphism and* $f \in \mathrm{KK}_*(A, B)$, $g \in \mathrm{KK}_*(B, C)$, *then*

$$\alpha_*(f \hat{\otimes}_B g) = f \hat{\otimes}_B \alpha_*(g).$$

d) *If* $\alpha \colon A \to B$ *is a* *-homomorphism* $[\alpha] \in \mathrm{KK}_0(A, B)$ *the class defined in Example 11.1.8, then mapping* α *to* $[\alpha]$ *determines a functor from the category of separable* C*-algebras and *-homomorphisms, to the category* KK.

Let $\mathbf{C}^* - \mathbf{alg}$ be the category of separable C*-algebras and *-homomorphisms. Let

$$\mathrm{C} \colon \mathbf{C}^* - \mathbf{alg} \to \mathrm{KK}$$

be the canonical functor discussed above. Exercise 11.1.9 shows that C is homotopy invariant. The following shows that the KK functor is *stable*. It is easily proved, and we establish this *Morita invariance* of KK in the next section.

Proposition 11.2.2 (Stability) *If* $p \in \mathcal{K}$ *is a rank-one operator on a Hilbert space, then the* *-homomorphism* $e \colon A \to A \otimes \mathcal{K}$, $e(a) := a \otimes p$, *induces an isomorphism*

$$e_* \colon \mathrm{KK}_*(D, A) \cong \mathrm{KK}_*(D, A \otimes \mathcal{K}),$$

for any D, *and an isomorphism*

$$e^* \colon \mathrm{KK}_*(A \otimes \mathcal{K}, D) \to \mathrm{KK}_*(A, D),$$

for any D.

A functor $\mathrm{F} \colon \mathbf{C} * - \mathbf{alg} \to \mathrm{Ab}$ from the C*-algebra category to the category of abelian groups is *split exact* if the following holds. If $0 \xrightarrow{j} D \xrightarrow{\pi} D/J \to 0$ is an exact sequence of separable C*-algebras which is split by a *-homomorphism $s \colon D/J \to D$, then $0 \xrightarrow{F(j)} D \xrightarrow{F(\pi)} D/J \to 0$ is a split exact sequence of abelian groups.

The following is proved in [95], see also [26].

Proposition 11.2.3 *Let* $0 \xrightarrow{j} D \xrightarrow{\pi} D/J \to 0$ *be an exact sequence of separable C*-algebras, which is split by a *-homomorphism* $s: D/J \to D$.

Then, the following sequences of abelian groups are split exact for any A, B:

$$0 \to KK_*(A, J) \xrightarrow{j_*} KK_*(A, D) \xrightarrow{\pi_*} KK_*(A, D/J) \to 0,$$

and

$$0 \to KK_*(D/J, B) \xrightarrow{\pi^*} KK_*(D, B) \xrightarrow{j^*} KK_*(J, B) \to 0.$$

The splittings are given by s_* *and* s^*, *respectively.*

In the paper [95], N. Higson proved the following remarkable result. It shows that KK-theory has an extremely strong uniqueness property.

Let $\mathbf{C^* - alg}$ be the category of separable C*-algebras and *-homomorphisms. Let

$$C: \mathbf{C^* - alg} \to KK$$

be the canonical functor discussed above.

Theorem 11.2.4 *Let* $F: \mathbf{C^* - alg} \to A$ *be any functor from the category of separable C*-algebras to an additive category. Assume that for every* $X \in \mathrm{Obj}(A)$, *the functor* $\mathrm{Hom}_A (X, F(\cdot))$ *is a homotopy-invariant, stable, and split exact functor into abelian groups.*

Then F *factors through* KK: *There exists a functor* $\widehat{F}: KK \to A$ *such that* $\widehat{F} \circ C = F$.

The category of C*-algebras has a tensor product operation, which extends to KK by way of the *external product* operation.

This gives KK the additional structure of a *symmetric monoidal* category, which we now explain.

Let $\alpha: A \to B$ be a *-homomorphism and D any other C*-algebra. Then $\alpha \otimes 1_D: A \otimes D \to B \otimes D$ and $1_D \otimes \alpha: D \otimes A \to D \otimes B$ are *-homomorphisms. The tensor product of two *-homomorphisms $\alpha_1: A_1 \to B_1$ and $\alpha_2: A_2 \to B_2$ is the *-homomorphism

$$\alpha_1 \otimes \alpha_2: A_1 \otimes A_2 \to B_1 \otimes B_2,$$

obtained by the composition

$$A_1 \otimes A_2 \xrightarrow{\alpha_1 \otimes 1_{A_2}} B_1 \otimes A_2 \xrightarrow{1_{B_1} \otimes \alpha_2} B_1 \otimes B_2.$$

Exercise 11.2.5 Suppose $\psi \in KK_*(A, B)$ is a morphism represented by the Fredholm A-B-bimodule $(\mathcal{E}, \pi, F)$. If D is any C*-algebra, then the triple $(\mathcal{E} \otimes_{\mathbb{C}}$

$D, \pi \otimes 1_D, F \otimes 1_D)$ is a Fredholm $A \otimes D$-$B \otimes D$-bimodule. Denote its class in $KK_*(A \otimes D, B \otimes D)$ by $\psi \hat{\otimes}_{\mathbb{C}} 1_D$.

Verify that the map

$$KK_*(A, B) \to KK_*(A \otimes D, B \otimes D), \quad f \mapsto f \hat{\otimes}_{\mathbb{C}} 1_D$$

is a well-defined group homomorphism.

Similarly, $f \mapsto 1_D \hat{\otimes}_{\mathbb{C}} f$ defines a group homomorphism $KK_*(A, B) \to KK_*(D \otimes A, D \otimes B)$.

Based on the Exercise and the existence of the Kasparov product, we may define the external product of two Kasparov morphisms as follows:

Definition 11.2.6 Let $\alpha_1 \in KK_*(A_1, B_1)$, $\alpha_2 \in KK_*(A_2, B_2)$. Their *external product* is the Kasparov morphism $\alpha_1 \hat{\otimes}_{\mathbb{C}} \alpha_2$ given by the Kasparov composition

$$\left(\alpha_1 \hat{\otimes}_{\mathbb{C}} 1_{A_2}\right) \otimes_{B_1 \otimes A_2} \left(1_{B_1} \hat{\otimes}_{\mathbb{C}} \alpha_2\right) \in KK_*(A_1 \otimes A_2, B_1 \otimes B_2).$$

Theorem 11.2.7 *If $\alpha_1 \in KK_i(A_1, B_1)$ and $\alpha_2 \in KK_j(A_2, B_2)$, $\sigma: A_1 \otimes A_2 \to A_2 \otimes A_1$, and $\tau: B_1 \otimes B_2 \to B_2 \otimes B_1$ are the flips, then*

$$\alpha_1 \hat{\otimes}_{\mathbb{C}} \alpha_2 = (-1)^{ij} \sigma^* \tau_* (\alpha_2 \hat{\otimes}_{\mathbb{C}} \alpha_1).$$

Thus, the external product is graded commutative.

If $\alpha_3 \in KK_k(A_3, B_3)$, then $(\alpha_1 \hat{\otimes}_{\mathbb{C}} \alpha_2) \hat{\otimes}_{\mathbb{C}} \alpha_3 = \alpha_1 \hat{\otimes}_{\mathbb{C}} (\alpha_2 \hat{\otimes}_{\mathbb{C}} \alpha_3)$. The external product is associative.

The identity $1_{\mathbb{C}} \in KK_0(\mathbb{C}, \mathbb{C})$ acts as a unit under the external product operation; $x \hat{\otimes}_{\mathbb{C}} 1_{\mathbb{C}} = 1_{\mathbb{C}} \hat{\otimes}_{\mathbb{C}} x = x$.

Let A be a C*-algebra, and for any X, Y, let $\tau_A: KK_*(X, Y) \to KK_*(X \otimes A, Y \otimes A)$ be the map $\tau_A(f) := f \hat{\otimes}_{\mathbb{C}} 1_A$. Then $\tau_A(x \hat{\otimes}_{\mathbb{C}} y) = \tau_A(f) \hat{\otimes}_A \tau_A(y)$ for any $x \in KK_*(X, X')$, $y \in KK_*(Y, Y')$.

An analogous statement holds for τ^A, with $\tau^A(f) := 1_A \hat{\otimes}_{\mathbb{C}} f$.

See [111] for the proof.

Exercise 11.2.8 Let A, A', B be separable C*-algebras and $f \in KK_*(A, A')$. Let $\sigma: A \otimes B \to B \otimes A$ and $\tau: B \otimes A' \to A' \otimes B$ be the flips. Show that $\sigma^*(f \hat{\otimes}_{\mathbb{C}} 1_B) = \tau_*(1_B \hat{\otimes}_{\mathbb{C}} f)$.

Exercise 11.2.9 Let A be a C*-algebra and τ_A be the functor $KK \to KK$ which on objects maps D to $D \otimes A$ and on morphisms maps $f \in KK(D_1, D_2)$ to $f \hat{\otimes} 1_A \in KK_*(D_1 \otimes A, D_2 \otimes A)$. Show that $\tau_A(x \hat{\otimes}_{\mathbb{C}} y) = \tau_A(f) \hat{\otimes}_A \tau_A(y)$ for any $x \in KK_*(X, X')$, $y \in KK_*(Y, Y')$.

Exercise 11.2.10 Suppose $a \in KK_*(\mathbb{C}, A)$, $b \in KK_*(\mathbb{C}, B)$, $c = a \hat{\otimes}_{\mathbb{C}} b \in KK_*(\mathbb{C}, A \otimes B)$, $f \in KK_*(A, A')$, and $g \in KK_*(B, B')$ are all homogeneous elements with respect to the grading. Show that

$$c \hat{\otimes}_{A \otimes B} (f \hat{\otimes}_{\mathbb{C}} g) = (-1)^{\partial b \partial f} (a \hat{\otimes}_A f) \hat{\otimes}_{\mathbb{C}} (b \hat{\otimes}_B g) \in \mathrm{KK}_*(\mathbb{C}, A' \otimes B').$$

Kasparov combines the Kasparov product and the external product (see [111]), mainly for purposes of notation, to obtain the following "cup-cap" product operation in KK.

Definition 11.2.11 The cup-cap product

$$(11.8) \quad \mathrm{KK}_*(A_1, B_1 \otimes D) \times \mathrm{KK}_*(D \otimes A_2, B_2) \to \mathrm{KK}_*(A_1 \otimes A_2, B_1 \otimes B_2)$$

is defined

$$\alpha_1 \hat{\otimes}_D \alpha_2 := \left(\alpha_1 \hat{\otimes}_{\mathbb{C}} 1_{A_2} \right) \hat{\otimes}_{B_1 \otimes D \otimes A_2} \left(1_{B_1} \hat{\otimes}_{\mathbb{C}} \alpha_2 \right).$$

We occasionally use this notation. This general cup-cap product is clearly associative and bilinear, since it is built from the external product and Kasparov composition.

As an application of these operations, we now describe the process of twisting an elliptic operator by a vector bundle (see Section 9.5) in KK-theoretic terms.

If D is an elliptic order 1 differential (odd) operator on sections of a $\mathbb{Z}/2$-graded bundle $S \to X$, where X is a smooth compact manifold, then D defines a class $[D] \in \mathrm{KK}_0(C(M), \mathbb{C})$. If $E \to X$ is a complex vector bundle, one can construct the twisted operator D_E, which is again elliptic order 1 and also defines a class $[D_E] \in \mathrm{KK}_0(C(X), \mathbb{C})$. What is the relationship between these classes?

Let $\delta \colon X \to X \times X$ the diagonal map, then δ defines the multiplication *-homomorphism $C(X) \otimes C(X) \to C(X)$ and $\delta_* \colon \mathrm{KK}_*(C(X), \mathbb{C}) \to \mathrm{KK}_*(C(X \times X), \mathbb{C})$. Set

$$\Delta := \delta_*([D]) \in \mathrm{KK}_0(C(X) \otimes C(X), \mathbb{C}).$$

We will verify in the next section, once we have defined the Kasparov composition, that the class $[D_E] \in \mathrm{KK}_0(C(X), \mathbb{C})$ of D twisted by a complex vector bundle $E \to X$ satisfies

$$[D_E] = [E] \hat{\otimes}_{C(M)} \Delta \in \mathrm{KK}_0(C(M), \mathbb{C}),$$

where we are using the notation of Definition 11.2.11. Thus, $[D_E]$ is the image under a map

$$(11.9) \quad \Delta \cap \colon \mathrm{KK}_0(\mathbb{C}, C(X)) \to \mathrm{KK}_0(C(X), \mathbb{C}), \quad \Delta \cap x := x \hat{\otimes}_{C(M)} \Delta.$$

Such maps, using both the composition operation and the external product operation, are of great importance in KK-theory. One reason is that if X is spin^c and D is the Dirac operator, then $\Delta \cap$ *is an isomorphism*. This means that the K-homology of

$C(X)$ is generated by the single class of the Dirac operator, as a module under the twisting action of bundles.

Exercise 11.2.12 Let X be any locally compact space. Using external products and the diagonal map $\delta : X \to X \times X$,

a) Describe the ring structure on $K^*(X) = KK_*(\mathbb{C}, C_0(X))$ in KK terms.
b) Show that $KK_*(C_0(X), \mathbb{C})$ is a module over the ring $KK_*(\mathbb{C}, C_0(X)) = K^*(X)$.

The above discussion fits into a rather general notion of *duality* for C*-algebras, of which there are a number of interesting examples.

Definition 11.2.13 Separable C*-algebras A and B are *dual* in KK, with a dimension shift of n, if there exist classes

$$\Delta \in KK_n(A \otimes B, \mathbb{C}), \qquad \widehat{\Delta} \in KK_n(\mathbb{C}, A \otimes B),$$

such that, in the notation of Definition 11.2.11, satisfying

$$\sigma_*(\widehat{\Delta}) \hat{\otimes}_A \Delta = 1_B, \qquad \widehat{\Delta} \hat{\otimes}_B \sigma^*(\Delta) = (-1)^n 1_A,$$

where $\sigma : A \otimes B \to A \otimes B$ is the flip.

Proposition 11.2.14 *Given A, B dual as in Definition 11.2.13, the map*

$$K_*(A) = KK_*(\mathbb{C}, A) \to KK_{*+n}(B, \mathbb{C}), \qquad x \in KK_*(\mathbb{C}, A) \mapsto x \hat{\otimes}_A \Delta$$

is an isomorphism with inverse of the map

$$KK_*(B, \mathbb{C}) \to KK_{*+n}(\mathbb{C}, A), \qquad y \mapsto \widehat{\Delta} \hat{\otimes}_B y.$$

We leave the proof to the reader, and it is an excellent exercise. (See [72].)

We say a C*-algebra exhibits *Poincaré duality* if A is self-dual.

Theorem 11.2.15 *Let X be a compact* spin^c*-manifold and $[D] \in KK_n(C(X), \mathbb{C})$ the class of the Dirac operator on X. Let $m : C(X \times X) \to C(X)$ be the multiplication homomorphism, Gelfand dual to $\delta : X \to X \times X$. Let ν be the normal bundle to the smooth immersion δ, then ν carries a canonical K-orientation. Let $\xi \in K^{-n}(\nu)$ be the corresponding Thom class and $\varphi : \nu \to X \times X$ the tubular neighborhood embedding. Set $\widehat{\Delta} := \varphi!(\xi) \in KK_n(\mathbb{C}, C(X) \otimes C(X))$ and $\Delta := m^*([D]) \in KK_n(C(X) \otimes C(X), \mathbb{C})$. Then Δ and $\widehat{\Delta}$ induce an n-dimensional duality between $C(X)$ and $C(X)$.*

The C*-algebra $C(X)$ has Poincaré duality in KK-theory.

In [71] the following result is proved.

Theorem 11.2.16 *Let G be a torsion-free Gromov hyperbolic group ∂G its Gromov boundary. Then $C(\partial G) \rtimes G$ exhibits Poincaré duality in KK-theory. It is induced by* *the class* $\Delta \in KK_1(C(\partial G) \rtimes G \otimes C(\partial G) \rtimes G, \mathbb{C})$ *of the B.D.F. cycle of Corollary 10.5.18.*

Exercise 11.2.17 Let G be a finite group. Let λ be the left regular representation of G on $l^2 G$ and ρ the right regular representation. They commute and determine a *-homomorphism $\gamma : C^*(G) \otimes C^*(G) \rightarrow \mathcal{K}(l^2 G)$. Prove that $\Delta := [\gamma] \in KK_0(C^*(G) \otimes C^*(G), \mathbb{C})$ induces a Poincaré duality for $C^*(G)$. Describe the corresponding class $\widehat{\Delta}$ as that of a suitable projection in $C^*(G) \otimes C^*(G)$.

Remark 11.2.18 The famous *Baum–Connes* assembly map for discrete groups with compact classifying space BG also has the form of a duality, albeit not *Poincaré* duality. Let X be a locally compact and G-compact model for its classifying space $\mathcal{E}G$ for proper actions and $\mathcal{E}_{G,X}$ the Mischenko module of Example 5.4.11.

Then $[\mathcal{E}_{G,X}] \in KK_0(\mathbb{C}, C^*(G) \otimes C(BG))$ determines a duality map

$$KK_*(C(BG) \otimes D_1, D_2) \rightarrow KK_*)(D_1, C^*(G) \otimes D_2),$$

conjectured to be an isomorphism in general.

The Baum–Connes conjecture has been verified for the classes of amenable groups and hyperbolic groups, see the discussion in the Overview to this chapter.

For other examples of KK-duality in connection with hyperbolic dynamics, see [104] and [105].

11.3 The Axiomatic Approach to the Kasparov Product

What is of principal interest in KK-theory is the calculation of specific Kasparov compositions in geometric examples. This is made possible by an axiomatic description of the Kasparov composition due to Connes and Skandalis [51]. The *existence* of at least one cycle satisfying the axioms is a hard technical theorem due to Kasparov, but the proof is not constructive so is not overly helpful in dealing with concrete situations. We have (therefore) omitted any discussion of the proof of Theorem 11.5.5 in this book, but instead will focus on the axiomatic approach and how to use it.

Theorem 11.3.1 *Let $(\mathcal{E}_1, \pi_1, F_1)$ be a cycle for $KK_1(A, B)$, defining a class x, and $(\mathcal{E}_2, \pi_2, F_2)$ a cycle for $KK_1(B, C)$, with class y. Let*

$$\mathcal{E} := \mathcal{E}_1 \otimes_B \mathcal{E}_2,$$

*the Hilbert module tensor product over the *-homomorphism $\pi : B \rightarrow \mathbb{B}(\mathcal{E}_2)$, and a right Hilbert C-module.*

Let

$$\pi : A \to \mathbb{B}(\mathcal{E}), \quad \pi(a) = \pi_1(a) \otimes 1_{\mathcal{E}_2}.$$

In addition, assume that $u \in \mathbb{B}(\mathcal{E})$ is a bounded operator satisfying the following conditions:

a) $\pi(a) \cdot (u^*u - 1)$, $\pi(a) \cdot (uu^* - 1)$, $[\pi(a), u] \in \mathcal{K}(\mathcal{E})$ *for all $a \in A$.*
b) *For all $\xi \in \mathcal{E}_1$, the operators*

$$(11.10) \quad i\, T_\xi \circ F_2 - u \circ T_\xi \in \mathbb{B}(\mathcal{E}_2, \mathcal{E}), \quad -i\, T_\xi \circ F_2 - u^* \circ T_\xi \in \mathbb{B}(\mathcal{E}_2, \mathcal{E})$$

are compact operators, where $T_\xi : \mathcal{E}_2 \to \mathcal{E}$ is the operator

$$T_\xi(\eta) := \xi \otimes \eta.$$

c) *For any $a \in A$, the operators*

$$(11.11) \quad \pi(a) \cdot (F_1 u + u^* F_1) \cdot \pi(a^*), \quad \pi(a) \cdot (F_1 u^* + u F_1) \cdot \pi(a^*)$$

are positive in the C*-algebra $\mathbb{B}(\mathcal{E})/\mathcal{K}(\mathcal{E})$.

Then the Kasparov composition

$$x \hat{\otimes}_B y \in \mathrm{KK}_0(A, C)$$

is represented by the triple $(\mathcal{E} \oplus \mathcal{E}, \pi \oplus \pi, F := \begin{bmatrix} 0 & u^* \\ u & 0 \end{bmatrix})$.

The axioms present certain relationships between the operators F_1, F_2 and the operator F. These relationships guarantee that any two F's satisfying the axioms are actually operator homotopic, as we show below.

The axioms for the intersection product may be rephrased as follows. Firstly, the conditions on u when phrased in terms of $F \in \mathbb{B}(\mathcal{E} \oplus \mathcal{E})$ assert that

$$(11.12) \qquad \pi(a) \cdot (F^2 - 1), \quad [\pi(a), F] \in \mathcal{K}(\mathcal{E} \oplus \mathcal{E}),$$

where here π denotes the direct sum of two copies of the original representation. Let

$$\tilde{F}_2 = \begin{bmatrix} 0 & -i F_2 \\ i F_2 & 0 \end{bmatrix} \in \mathbb{B}(\mathcal{E}_2 \oplus \mathcal{E}_2).$$

For $\xi \in \mathcal{E}_1$, let $\tilde{T}_\xi : \mathcal{E}_2 \oplus \mathcal{E}_2 \to \mathcal{E} \oplus \mathcal{E}$ be the direct sum of two copies of T_ξ. Then the connection condition b) asserts that

$$(11.13) \qquad\qquad \tilde{T}_\xi \cdot \tilde{F}_2 - F \cdot \tilde{T}_\xi$$

is a compact operator. Finally, the alignment condition can be written as follows:
Let

$$\tilde{F}_1 = \begin{bmatrix} 0 & F_1 \\ F_1 & 0 \end{bmatrix}.$$

The alignment condition c) says in this notation

(11.14) $\qquad \pi(a) \cdot [\tilde{F}_1, F]_s \cdot \pi(a)^* \geq 0 \quad \mod \mathcal{K}(\mathcal{E} \oplus \mathcal{E}),$

where $[\cdot, \cdot]_s$ denotes the *graded commutator*

$$[A, B]_s := AB + BA$$

of two (odd) operators on a graded space.

Proposition 11.3.2 *Let F and F' be two self-adjoint, odd operators on $\mathcal{E} \oplus \mathcal{E}$ which satisfy conditions (11.12), (11.13), and (11.14). Then there is a path of self-adjoint odd operators F_t between them, also satisfying the axioms.*

See [149] for the straightforward proof.

Although we will be studying mainly intersection products of KK_1-classes, we state the version of the axioms for a KK_0-pairing as well.

Theorem 11.3.3 *Let $(\mathcal{E}_1, \pi_1, F_1)$ be a cycle for $KK_0(A, B)$, defining a class x, and $(\mathcal{E}_2, \pi_2, F_2)$ a cycle for $KK_0(B, C)$, with class y. Let*

$$\mathcal{E} := \mathcal{E}_1 \otimes_B \mathcal{E}_2,$$

*the graded Hilbert module tensor product over the *-homomorphism $\pi: B \to \mathbb{B}(\mathcal{E}_2)$, and a right Hilbert C-module.*

   Let

$$\pi: A \to \mathbb{B}(\mathcal{E}), \quad \pi(a) = \pi_1(a) \otimes 1_{\mathcal{E}_2},$$

and use the same letter for the diagonal representation on the $\mathbb{Z}/2$-graded $\mathcal{E} \oplus \mathcal{E}$.

   Suppose that $F \in \mathbb{B}(\mathcal{E} \oplus \mathcal{E})$ is an odd, bounded operator satisfying the following conditions:

a) $\pi(a) \cdot (F^2 - 1)$, $[\pi(a), F] \in \mathcal{K}(\mathcal{E})$ *for all $a \in A$.*
b) *For all $\xi \in \mathcal{E}_1$,*

(11.15) $\qquad T_\xi \circ F_2 - F \circ T_\xi \in \mathbb{B}(\mathcal{E})$

   is a compact operator, where $T_\xi: \mathcal{E}_2 \to \mathcal{E}$ is the operator

$$T_\xi(\eta) := \xi \otimes \eta.$$

c) *For any $a \in A$, the operators*

(11.16) $$\pi(a) \cdot [F_1, F]_s \pi(a)^*$$

are positive *in the C*-algebra $\mathbb{B}(\mathcal{E})/\mathcal{K}(\mathcal{E})$, where $[\cdot, \cdot]_s$ denotes the graded commutator $F_1 F + F F_1$.*

Then the intersection product

$$x \hat{\otimes}_B y \in KK_0(A, C)$$

is represented by the triple $(\mathcal{E} \oplus \mathcal{E}, \pi \oplus \pi, F))$.

Example 11.3.4 If $\mathcal{E}$ is a Morita A-B-equivalence bimodule and $\lambda \colon A \to \mathbb{B}(\mathcal{E})$ is the left action of A, we give $\mathcal{E}$ the $\mathbb{Z}/2$-grading with $\mathcal{E}^+ = \mathcal{E}$, $\mathcal{E}^- = \{0\}$.

Then the triple $(\mathcal{E}, \lambda, 0)$ defines a Fredholm A-B-bimodule, because by Proposition 6.1.11, the left action of A is by compact operators.

We define a "conjugate" bimodule as follows. Let $\mathcal{E}^*$ be $\mathcal{E}$ as an additive group, but with the conjugate $\mathbb{C}$-multiplication $\lambda x := \bar{\lambda} x$, making it a $\mathbb{C}$-vector space.

Denote elements of $\mathcal{E}^*$ by $\bar{x}$ (where $x \in \mathcal{E}$).

Then $\mathcal{E}^*$ together with the B-A-bimodule structure

$$b \bar{x} a := \overline{a^* x b^*}$$

and inner products

$$_B\langle \bar{x}, \bar{y} \rangle := \langle x, y \rangle_B, \quad \langle \bar{x}, \bar{y}, \rangle_A :=_A \langle x, y \rangle$$

is a strong Morita equivalence B-A-bimodule.

Theorem 11.3.5 *The class $[\mathcal{E}] \in KK_0(A, B)$ of an A-B Morita equivalence bimodule is an invertible. The inverse is the class of the conjugate bimodule $[\mathcal{E}^*]$.*

Proof The proof is simply the observation that

$$\mathcal{E}^* \otimes_A \mathcal{E} \cong B, \qquad \mathcal{E} \otimes_B \mathcal{E}^* \cong A,$$

as right Hilbert B bimodules, A-modules, respectively. As we have noted already, the bimodule A itself defines the identity morphism in $KK_0(A, A)]$. So this proves that $[\mathcal{E}^*] \otimes_A [\mathcal{E}] = 1_B \in KK_0(B, B)$.

To see why that $\mathcal{E}^* \otimes_A \mathcal{E} \cong B$, recall that the tensor product $\mathcal{E}^* \otimes_A \mathcal{E}$ is defined as the completion of the algebraic tensor product over $\mathbb{C}$ with respect to the Hermitian B-valued form

(11.17) $\langle \bar{x}_1 \otimes y_1, \bar{x}_2 \otimes y_2 \rangle_B := \langle y_1, \langle \bar{x}_1, \bar{x}_2 \rangle_A \cdot y_2 \rangle_B. = \langle y_1, {}_A\langle x_1, x_2 \rangle y_2 \rangle_B.$

Let

$$U : \mathcal{E}^* \otimes_A \mathcal{E} \to B, \quad U(x^* \otimes y) := \langle x, y \rangle_B.$$

Then U is a well-defined B-bimodule map: To see that it is a bimodule map compute

$$U\left(b_1 \cdot (\bar{x} \otimes y)b_2\right) = U\left(\overline{xb_1^*} \otimes yb_2\right) = \langle xb_1^*, yb_2 \rangle_B = b_1^* \langle x, y \rangle_B b_2.$$

Finally,

$$U(\bar{x}_1 \otimes y_1)^* U(\bar{x}_2 \otimes y_2) = \langle y_1, x_1 \rangle_B \cdot \langle x_2, y_2 \rangle_B = \langle y_1, x_1 \langle x_2, y_2 \rangle_B \rangle_B$$

$$(11.18) \qquad\qquad\qquad = \langle y_1, {}_A\langle x_1, x_2 \rangle \cdot y_2 \rangle_B,$$

which agrees with (11.17). Hence U is an isometry and is clearly surjective so is an isomorphism of right Hilbert B-modules.   □

Example 11.3.6 Let D be the Dirac operator on sections of a spinor bundle $S \to X$. So, for some connection ∇^S on S and some Clifford module structure c on S, D acts on smooth sections of S by

$$(Ds)(x) = \sum_i c(e_i)(\nabla_i s)(x), \quad s \in \Gamma^\infty(S),$$

with (e_i) a local frame for TX. From D we obtain a class

$$[D] \in KK_0(C(X), \mathbb{C}), \quad [D] = \text{class of the triple } \left(L^2(S), \pi, F := \chi(D)\right).$$

Now choose any Hermitian metric and compatible connection ∇^E on E. Let $\nabla := \nabla^S \otimes 1 + 1 \otimes \nabla^E$, the tensor product connection. Set

$$(11.19) \qquad c_E(x, \xi) := c(x, \xi) \otimes 1_{E_x} : S_x \otimes E_x \to S_x \otimes E_x;$$

this gives $S \otimes E$ the structure of a Clifford module over TX. With respect to a local orthonormal frame $e_1, \ldots e_n$, we set

$$(D_E)s(x) = \sum_i c_E(e_i)(\nabla_i s)(x), \quad s \in \Gamma^\infty(S \otimes E).$$

The ellipticity of D_E implies that $(1 + D_E^2)^{-1}$ is compact, and D_E commutes mod bounded operators with multiplications by smooth functions $f \in C^\infty(X)$, so we get a class

$$[D_E] \in KK_0(C(X), \mathbb{C}), \quad [D_E] = \left[\left(L^2(S \otimes E), \pi, F_E := \chi(D_E)\right)\right].$$

To the bundle E, we associate the Kasparov triple

$$[[E]] := [(\Gamma(E), \rho, 0)] \in KK_0 (C(X), C(X)),$$

where ρ denotes the multiplication action of $C(X)$ on $\Gamma(E)$.

Then the twisting procedure may be translated into the KK-theory framework by the following:

Proposition 11.3.7 *In the above notation,*

$$[[E]] \otimes_{C(X)} [D] = [D_E] \in KK_0(C(X), \mathbb{C}).$$

Furthermore,

$$\text{Index}([D_E]) = [E] \otimes_{C(X)} [D] \in KK_0(\mathbb{C}, \mathbb{C}) = \mathbb{Z}.$$

Proof We leave the second statement as an exercise. To check the first, since the triple defining $[[E]]$ has the zero operator, it is not needed to check alignment, and we only need to check that the operator $F_E := \chi(D_E)$ satisfies the connection condition b) of Theorem 11.3.3. We show first that

$$T_\xi D - D_E T_\xi$$

is bounded, for $\xi \in \Gamma^\infty(E)$ a smooth section of E. Let s be a smooth section of S. Let(e_i) be a local orthonormal frame for TX. Let c_E be as in (11.19). Then

$$
\begin{aligned}
\left(T_\xi D - D_E T_\xi\right)(s)(x) \\
= \sum_i c(e_i)(\nabla_i^S s)(x) \otimes \xi(x) - \sum_i c_E(e_i)\nabla_i(\xi \otimes s)(x) \\
= \sum_i c(e_i)(\nabla_i^S s)(x) \otimes \xi(x) - \sum_i c(e_i)(\nabla_i^E \xi)(x) \otimes s(x) \\
- \sum_i \xi(x) \otimes c(e_i)(\nabla_i^S s)(x)
\end{aligned}
$$

(11.20)
$$= -\sum_i c(e_i)s(x) \otimes \nabla_i^E(\xi)(x).$$

Since $\xi \in \Gamma^\infty(E)$ is fixed, this is (clearly) a bounded operator on spinor sections s.

Exercise 11.3.8 Use the integral formula (9.30) to prove that boundedness of the operators $T_\xi D - D_E T_\xi$ in the above argument implies compactness of the operators $T_\xi F - F_E T_\xi$.   □

Example 11.3.9 Let $X = \mathbb{T}^2$, the 2-torus, which we regard as $\mathbb{R}/\mathbb{Z}$ and use notation like $x, y, \ldots$ for points in it.

The Dirac–Dolbeault operator

$$\bar{\partial} = \begin{bmatrix} 0 & \frac{\partial}{\partial z} \\ \frac{\partial}{\partial \bar{z}} & 0 \end{bmatrix},$$

where $\frac{\partial}{\partial \bar{z}} = \frac{\partial}{\partial x} + i\frac{\partial}{\partial y}$, acts on sections of the spinor bundle coming from the complex structure. The bundle is trivial, isomorphic to $\mathbb{T}^2 \times \mathbb{C}^2$, so that the spinor grading corresponds to grading the first factor of $\mathbb{C}^2$ even and the second odd. An exercise in Clifford algebras shows that the corresponding Dirac operator is given by the above matrix.

Under Fourier transform, $L^2(\mathbb{T}^2) \cong l^2(\mathbb{Z}^2)$ corresponds to the standard basis $z_1^a z_2^b \in C(\mathbb{T}^2) \subset L^2(\mathbb{T}^2)$. These are eigenvectors for $\frac{\partial}{\partial \bar{z}}$, as one checks

$$\frac{\partial}{\partial \bar{z}}(z_1^a z_2^b) = \left(\frac{\partial}{\partial x} + i\frac{\partial}{\partial y} \right)(z_1^a z_2^b) = 2\pi i(a - ib)z_1^a z_2^b.$$

Therefore, up to unitary equivalence, the Dirac cycle for $\mathbb{T}^2$ is the triple

$$\left(l^2(\mathbb{Z}^2) \oplus l^2(\mathbb{Z}^2), \pi, \bar{\partial} := \begin{bmatrix} 0 & M_{-2\pi i(a+ib)} \\ M_{2\pi i(a-ib)} & 0 \end{bmatrix} \right),$$

where $M_{2\pi i(a-ib)}$ is the diagonal operator in the standard basis with entries $2\pi i(a - ib)$. The representation $\pi : C(\mathbb{T}^2) \to \mathbb{B}(l^2(\mathbb{Z}^2))$ involves a Fourier transform. It is defined by

$$\pi(f) = \lambda(\hat{f}),$$

where $\lambda : C^*(\mathbb{Z}^2) \to \mathbb{B}(l^2(\mathbb{Z}^2))$ is the regular representation.

Let $[\bar{\partial}]$ denote the class in $\mathrm{KK}_0(C(\mathbb{T}^2), \mathbb{C})$ of this (spectral) cycle.

Note that the kernel of $\frac{\partial}{\partial \bar{z}}$ is the holomorphic functions on $\mathbb{T}^2$ and its cokernel the anti-holomorphic functions, and each space is one-dimensional, consisting of constants. However, this fact is also obvious from the diagonalized picture of it and for the kernel of $a - ib$ is one-dimensional, happening when $a = b = 0$, and likewise for $a + ib$.

We are interested in the result of twisting the Dirac–Dolbeault operator by the Poincaré bundle $\mathcal{P}$(see Exercise 5.2.13) over $\mathbb{T}^2$, whose right Hilbert $C(\mathbb{T}^2)$-module of sections is

$$\Gamma(\mathcal{P}) = \{ f \in C(\mathbb{R}^2) \mid f(x + 1, y) = e^{-2\pi iy} f(x, y), \quad \forall x, y, \in \mathbb{R}, n \in \mathbb{Z} \}.$$

To "find" elements in $\Gamma(\mathcal{P})$, let $f \in C_c(\mathbb{R})$. Then the series

$$\hat{f}(x, y) = \sum_{n \in \mathbb{Z}} f(x + n) e^{2\pi i n y}$$

is, for each point x, finite, and $\hat{f}$ defines an element of $\mathcal{P}$.

Exercise 11.3.10 Let $L^2(\mathcal{P})$ denote the Hilbert module tensor product $\Gamma(\mathcal{P}) \otimes_{C(\mathbb{T}^2)} L^2(\mathbb{T}^2)$. Prove that $f \mapsto \hat{f}$ given above extends to a unitary isomorphism

$$L^2(\mathcal{P}) \cong L^2(\mathbb{R}).$$

The action of $C(\mathbb{T}^2)$ on (the left) of $\Gamma(\mathcal{P})$ by fiberwise multiplication determines an action on $L^2(\mathcal{P})$: What representation ρ of $C(\mathbb{T}^2)$ on $L^2(\mathbb{R})$ corresponds to it? Show that under this isomorphism the operator

$$D := \begin{bmatrix} 0 & x - \frac{d}{dx} \\ x + \frac{d}{dx} & 0 \end{bmatrix}$$

on $L^2(\mathbb{R}) \oplus L^2(\mathbb{R})$ satisfies the connection axiom for the triple $\big(L^2(\mathbb{R}) \oplus L^2(\mathbb{R}), \rho, D\big)$ to represent the Kasparov product $[[\mathcal{P}]] \otimes_{C(\mathbb{T}^2)} [\bar{\partial}] \in \mathrm{KK}_0(C(\mathbb{T}^2), \mathbb{C})$.

Example 11.3.11 We now study an example of an *external product*.

Proposition 11.3.12 *Let* $x \in \mathrm{KK}_1(\mathbb{C}, C_0(\mathbb{R}))$ *be the class of the odd Fredholm* $\mathbb{C}$-$C_0(\mathbb{R})$ *bimodule* $(C_0(\mathbb{R}), 1, \chi)$, *where* χ *is a normalizing function, acting as a (self-adjoint) multiplier of* $C_0(\mathbb{R})$.
Let $\beta_{\mathbb{R}^2} \in \mathrm{KK}_0(\mathbb{C}, C_0(\mathbb{R}^2))$ *be the Bott class. Then*

$$x \hat{\otimes}_{\mathbb{C}} x = \beta_{\mathbb{R}^2} \in \mathrm{KK}_0(\mathbb{C}, C_0(\mathbb{R}^2)).$$

Proof For purposes of the proof, we will represent the Bott class by the (even) Fredholm $\mathbb{C}$-$C_0(\mathbb{R}^2)$ bimodule

$$\left(C_0(\mathbb{R}^2) \oplus C_0(\mathbb{R}^2), 1, \frac{1}{\sqrt{1 + x^2 + y^2}} \begin{bmatrix} 0 & x - iy \\ x + iy & 0 \end{bmatrix} \right).$$

We wish to compute the image of (x, x) under the external map

$$\mathrm{KK}_1\big(\mathbb{C}, C_0(\mathbb{R})\big) \times \mathrm{KK}_1\big(\mathbb{C}, C_0(\mathbb{R})\big) \to \mathrm{KK}_0\big(\mathbb{C}, C_0(\mathbb{R}^2)\big).$$

By definition of the external product, which is defined in terms of the *internal* product,

(11.21) $x \otimes_{\mathbb{C}} x = x \otimes_{C_0(\mathbb{R})} (1_{C_0(\mathbb{R})} \otimes_{\mathbb{C}} x).$

The class $1_{C_0(\mathbb{R})} \otimes x \in \mathrm{KK}_1(C_0(\mathbb{R}), C_0(\mathbb{R}^2))$ is represented by the cycle

$$(C_0(\mathbb{R}^2), \mu_1, 1 \otimes \chi),$$

where $\mu_1(f)(x, y) = f(x)$.

The intersection product, and therefore the external product, is represented by a cycle with $\mathbb{Z}/2$-graded module the direct sum of two copies of

$$C_0(\mathbb{R}) \otimes_{C_0(\mathbb{R})} C_0(\mathbb{R}^2) \cong C_0(\mathbb{R}^2).$$

The operator will be of the form

$$F = \begin{bmatrix} 0 & u^* \\ u & 0 \end{bmatrix},$$

where u is going to be a suitable multiplier of $C_0(\mathbb{R}^2)$.

We set

(11.22)
$$u(x, y) := \frac{x + iy}{\sqrt{1 + x^2 + y^2}}.$$

We show that it satisfies the axioms for the intersection product (11.21) stated in Theorem 11.3.1.

We refer to some of the notation of the theorem. The module $\mathcal{E}_1$ is $C_0(\mathbb{R})$. The module $\mathcal{E}_2$ is $C_0(\mathbb{R}) \otimes C_0(\mathbb{R}) \cong C_0(\mathbb{R}^2)$. The connection axiom (11.10) refers to the operator denoted T_ξ there, for $\xi \in \mathcal{E}_1 = C_0(\mathbb{R})$. We can assume ξ is compactly supported.

The Hilbert $C_0(\mathbb{R}^2)$-module operator T_ξ maps $\mathcal{E}_2 = C_0(\mathbb{R}^2)$ to the tensor product $\mathcal{E}_1 \otimes_{C_0(\mathbb{R})} (C_0(\mathbb{R}) \otimes \mathcal{E}_2)$ which is isomorphic to $C_0(\mathbb{R}^2)$ as a $C_0(\mathbb{R}^2)$-Hilbert module. With this identification, $T_\xi : C_0(\mathbb{R}^2) \to C_0(\mathbb{R}^2)$ is the multiplier by the bounded function $\xi \otimes 1$.

Similarly, the operator $i T_\xi F_2 - u T_\xi$ identifies with a right $C_0(\mathbb{R}^2)$-module operator $C_0(\mathbb{R}^2) \to C_0(\mathbb{R}^2)$, and it is multiplication by the bounded function

(11.23)
$$\xi(x) \cdot (i\chi(y) - u(x, y)).$$

The axiom requires that (11.23) vanishes as $(x, y) \to \infty$, since this is equivalent to compactness of the corresponding Hilbert module operator. Since the support of ξ is assumed compact, the requirement then is that $u(x, y) - i\chi(y) \to 0$ as $y \to \infty$ and x remains within a compact set, which is clearly true in the present case where the statement is that if C is a constant, then

$$u(x, y) - i\chi(y) = \frac{x + iy}{\sqrt{1 + x^2 + y^2}} - \frac{iy}{\sqrt{1 + y^2}} \to 0, \quad \text{as } y \to \infty, \ |x| \le C,$$

which is clear by direct computation.

The connection axiom is therefore satisfied.

For the positivity or alignment condition (11.11), we need to verify that

$$(F_1 \otimes 1)\, u + u^* \,(F_1 \otimes 1) \geq 0 \ \text{ mod compacts.}$$

On the right Hilbert $C_0(\mathbb{R}^2)$-module $C_0(\mathbb{R}^2)$, $F_1 \otimes 1$ acts by $\chi \otimes 1$, which using the standard normalizing χ is $(\chi \otimes 1)(x, y) = \frac{x}{\sqrt{1+x^2}}$. We therefore get that

$$(F_1 \otimes 1)\, u + u^* \,(F_1 \otimes 1) = 2\chi \cdot \mathrm{Re}(u)$$

from which we see that the alignment condition is satisfied if $\mathrm{Re}(u)$ is ≥ 0 on $[0, \infty)$ and ≤ 0 on $(-\infty, 0]$. Evidently this is satisfied by u as in (11.22).

Remark 11.3.13 The word "alignment" is suggested by the requirement here that the real part of u should be positive (respectively, negative) on $\mathbb{R}_\pm \times \mathbb{R}$, that is, in the same places where $\chi \otimes 1$ is positive (negative). $\qquad\square$

Note that since $x \in \mathrm{KK}_1(\mathbb{C}, C_0(\mathbb{R}))$ has degree 1, the graded commutativity of the external product implies that

$$x \otimes_\mathbb{C} x = -\sigma_*(x \otimes_\mathbb{C} x) \in \mathrm{KK}_0(\mathbb{C}, C_0(\mathbb{R}^2)),$$

where $\sigma : C_0(\mathbb{R}^2) \to C_0(\mathbb{R}^2)$ is induced by the coordinate flip $\mathbb{R}^2 \to \mathbb{R}^2$. As $x \otimes_\mathbb{C} x = \beta_{\mathbb{R}^2}$ is the Bott element, by the result above, this is equivalent to

$$(11.24) \qquad\qquad \sigma_*(\beta_{\mathbb{R}^2}) = -\beta_{\mathbb{R}^2}.$$

Exercise 11.3.14 Prove (11.24) directly using the definition of $\beta_{\mathbb{R}^2} \in \mathrm{KK}_0(\mathbb{C}, C_0(\mathbb{R}^2))$ presented above. The map σ flips the coordinates, $\sigma(x, y) = (y, x)$.

We close with a discussion of the compatibility of the K-homology/K-theory pairings defined in Section 10.1 with the Kasparov composition.

Theorem 11.3.15 *The group isomorphism $K_0(A) \to \mathrm{KK}_0(\mathbb{C}, A)$ identifies the pairing $K_0(A) \times \mathrm{KK}_0(A, \mathbb{C}) \to \mathbb{Z}$ of Section 10.1 with the combination of the Kasparov composition*

$$\mathrm{KK}_0(\mathbb{C}, A) \times \mathrm{KK}_0(A, \mathbb{C}) \to \mathrm{KK}_0(\mathbb{C}, \mathbb{C})$$

and the index isomorphism $\mathrm{KK}_0(\mathbb{C}, \mathbb{C}) \cong \mathbb{Z}$.

Proof Let A be unital for simplicity, let $p \in A$ be a projection, and $\mathcal{E}_1 = pA$. Suppose that $(H = H^+ \oplus H^-, \pi, F)$ is a cycle for $\mathrm{KK}_0(A, \mathbb{C})$, and let α be its class. Then $\mathcal{E}_A H \cong pH$, a $\mathbb{Z}/2$-graded Hilbert space, and $pFp \in \mathbb{B}(pH)$ is an odd operator on pH, and the pairing of $[p]$ and α defined in Section 10.1 is given by the Fredholm index of pF_+p, where F_+ is the restriction of F to H_+. All that needs to be proved is that pFp satisfies the connection condition of the Kasparov product, which we leave as an exercise.

□

To deal with the odd pairing, let $A = C(\mathbb{T})$, identify $\mathbb{T}$ with $I := (0, 1)$ with the endpoints identified, and note $C_0(I)$ is then an ideal. The function $\chi(t) = t$ on I is a self-adjoint multiplier of $C_0(I)$ and $\chi^2 - 1 \in C_0(I)$, so we have a right Hilbert $C(\mathbb{T})$-module $C_0(I)$ and a Fredholm $\mathbb{C}$-$C(\mathbb{T})$-bimodule $(C_0(I), 1, \chi)$.

Lemma 11.3.16 *The class of the triple* $(C_0(I), 1, \chi)$ *in* $\mathrm{KK}_1(\mathbb{C}, C(\mathbb{T}))$ *equals the class of the unitary generator* $[z] \in C(\mathbb{T})$ *under the identification* $\mathrm{KK}_1(\mathbb{C}, C(\mathbb{T})) \cong \mathrm{K}_1(C(\mathbb{T}))$.

Now let $u \in A$ be a unitary in a unital C*-algebra. Then $\mathrm{Spec}(u) \subset \mathbb{T} = I^+$, the one-point compactification of $I = (0, 1)$. Functional calculus defines a *-homomorphism $C(\mathbb{T}) \to A$ and then restricts to a *-homomorphism $\alpha_u \colon C_0(I) \to A$, which pushes the triple $(C_0(I), 1, \chi)$ for $C(\mathbb{T})$ to one for A, given by the triple $(C_0(I) \otimes_{C_0(I)} A, 1, \chi \otimes 1)$.

This procedure defines a group homomorphism $\mathrm{K}_1(A) \to \mathrm{KK}_1(\mathbb{C}, A)$ which agrees with the homomorphism $\mathrm{K}_0(A \otimes C_0(I)) \to \mathrm{KK}_0(\mathbb{C}, A \otimes C_0(I))$ of Theorem 11.1.15 when combined with the identification of $\mathrm{K}_0(A \otimes C_0(I))$ with the unitary picture of $\mathrm{K}_1(A)$. We get the following theorem:

Theorem 11.3.17 *The group isomorphism* $\mathrm{K}_1(A) \to \mathrm{KK}_1(\mathbb{C}, A)$ *defined above identifies the pairing* $\mathrm{K}_1(A) \times \mathrm{KK}_1(A, \mathbb{C}) \to \mathbb{Z}$ *of Section 10.1 with the combination of the Kasparov composition*

$$\mathrm{KK}_1(\mathbb{C}, A) \times \mathrm{KK}_1(A, \mathbb{C}) \to \mathrm{KK}_0(\mathbb{C}, \mathbb{C})$$

and the index isomorphism $\mathrm{KK}_0(\mathbb{C}, \mathbb{C}) \cong \mathbb{Z}$.

Exercise 11.3.18 Let A be unital and $J \subset A$ an ideal. Suppose $\chi \in A$ is self-adjoint and $\chi^2 - 1 \in J$. Then the Fredholm $\mathbb{C}$-J-bimodule $(J, 1, \chi)$ defines a class $[\chi] \in \mathrm{KK}_1(\mathbb{C}, J)$.

a) Show that $\frac{\chi+1}{2}$ projects to a projection $p \in A/J$ and that $\delta([p]) = [\chi] \in \mathrm{K}_1(J)$, where $\delta \colon \mathrm{K}_0(A/J) \to \mathrm{K}_1(J)$ is the connecting map (the exponential map) in the 6-term exact sequence for $J \subset A$.
b) If $i \colon J \to A$ is the inclusion, show that $i_*([\chi]) = 0 \in \mathrm{KK}_1(\mathbb{C}, A)$.
c) Why does this not show that the triple of Lemma 11.3.16 defines the zero element of $C(\mathbb{T})$?

11.4   The Bott Periodicity Theorem in KK-Theory

In this section we compute the single most important example of a Kasparov composition—one which results in an important type of proof of Bott Periodicity.

The beauty of this Dirac–Schrödinger proof is that it is built in a way that reflects the geometry of $\mathbb{R}$. These strong geometric features lead to an equivariant version of the Periodicity Theorem of fundamental importance for computing, for example, K-theory of crossed products by the integers $\mathbb{Z}$.

Let $\chi : \mathbb{R} \to [-1, 1]$ be a normalizing function, which we generally take $\chi(x) = \frac{x}{\sqrt{1+x^2}}$, but most of the discussion below applies to any normalizing function. Let

$$F_1 \in \mathbb{B}(\mathcal{E}_1), \quad (F_1\xi)(t) = \chi(t) \cdot \xi(t),$$

that is, F_1 is multiplication by the bounded continuous function χ. Clearly $F_1^2 - 1$ is compact, since it is multiplication by the C_0-function $\chi(t)^2 - 1$. We obtain a cycle $(\mathcal{E}_1, 1, F_1)$ for $\mathrm{KK}_1(\mathbb{C}, C_0(\mathbb{R}))$.

Definition 11.4.1 Let

$$x \in \mathrm{KK}_1(\mathbb{C}, C_0(\mathbb{R}))$$

be the class of the odd Fredholm $\mathbb{C}$-$C_0(\mathbb{R})$-bimodule $(C_0(\mathbb{R}), 1, F_1)$. We will call it the *Bott morphism*.

Next, let $\mathcal{E}_2 := L^2(\mathbb{R})$, and D_2 the self-adjoint extension of the densely defined unbounded operator $-i\frac{d}{dx}$, on $L^2(\mathbb{R})$. Let $F_2 \in \mathbb{B}(L^2\mathbb{R})$ be $\chi(D_2)$ (note it is the Fourier conjugate of F_1), so that

$$F_2 = \chi(D_2).$$

If $f \in C_c^\infty(\mathbb{R})$, then

$$[f, D_2] = -if',$$

and in particular, the commutator is bounded for smooth and compactly supported functions. Furthermore, $\rho \cdot (1 + D_2^2)^{-1}$ is a compact operator for any $\rho \in C_c^\infty(\mathbb{R})$. It follows that

$$[f, D_2(1 + D_2^2)^{-\frac{1}{2}}] \in \mathcal{K}(L^2\mathbb{R})$$

for any $f \in C_c^\infty(\mathbb{R})$ and hence for all $f \in C_0(\mathbb{R})$. In fact, all of these statements can be checked directly by taking Fourier transforms.

Therefore $(L^2\mathbb{R}, \pi, D_2)$ is an odd, Fredholm $C_0(\mathbb{R})$-$\mathbb{C}$-bimodule.

Definition 11.4.2 Let

$$y \in \mathrm{KK}_1(C_0(\mathbb{R}), \mathbb{C})$$

the class of the cycle $(L^2(\mathbb{R}), \pi, F_2)$, where $\pi : C_0(\mathbb{R}) \to \mathbb{B}(L^2(\mathbb{R}))$ is the representation by multiplication operators. We will call it the *Dirac morphism*.

Finally, we let $[x + \frac{d}{dx}] \in KK_0(\mathbb{C}, \mathbb{C})$ be the class of the spectral Fredholm module over $\mathbb{C}$ given by $\left(L^2(\mathbb{R}) \oplus L^2(\mathbb{R}), 1, D := \begin{bmatrix} 0 & x - \frac{d}{dx} \\ x + \frac{d}{dx} & 0 \end{bmatrix} \right)$. The corresponding Fredholm module has operator

$$F := \chi(D) = \begin{bmatrix} 0 & A(H+2)^{-\frac{1}{2}} \\ AH^{-\frac{1}{2}} & 0 \end{bmatrix},$$

see Definition 10.10.9, $A = x + \frac{d}{dx}$, and $H = -\frac{d^2}{dx^2} + x^2$ is the harmonic oscillator.

Lemma 11.4.3 *The equality*

$$x \hat{\otimes}_{C_0(\mathbb{R})} y = [x + \frac{d}{dx}] \in KK_0(\mathbb{C}, \mathbb{C})$$

holds.

Proof The tensor product of Hilbert modules for the intersection product is

$$\mathcal{E} = \mathcal{E}_1 \otimes_{C_0(\mathbb{R})} \mathcal{E}_2 = C_0(\mathbb{R}) \otimes_{C_0(\mathbb{R})} L^2(\mathbb{R}) \cong L^2(\mathbb{R}).$$

So the intersection product will be represented by a Fredholm module of the form

$$(L^2(\mathbb{R}) \oplus L^2(\mathbb{R}), 1, F),$$

where F is a suitable odd, self-adjoint operator. We need to check that

$$F = \chi(D) = \begin{bmatrix} 0 & A(H+2)^{-\frac{1}{2}} \\ AH^{-\frac{1}{2}} & 0 \end{bmatrix}$$

satisfies the axioms for the Kasparov product.

The Fredholm condition has already been verified: $\chi(D)$ is Fredholm, as we showed in the previous section. Thus, $(L^2(\mathbb{R}) \oplus L^2(\mathbb{R}), F)$ is a cycle for $KK_0(\mathbb{C}, \mathbb{C})$.

We first discuss the connection condition b), which states that

$$i \, T_\xi \, F_2 - u \, T_\xi, \quad -i T_\xi \, F_2 - u^* T_\xi$$

are compact operators, where $u = AH^{-\frac{1}{2}}$ and $\xi \in C_0(\mathbb{R})$. We may assume that ξ is smooth and compactly supported. The operator T_ξ is multiplication by ξ on $L^2(\mathbb{R})$. It follows that the connection condition boils down to

(11.25) $i F_2 \xi - \xi w$

is a compact operator, for any $\xi \in C_c^{\infty}(\mathbb{R})$, where $F_2 = \chi(-i\frac{d}{dx})$ as before. Since T_ξ commutes mod $\mathcal{K}(L^2\mathbb{R})$ with F_2 and with u, we are reduced therefore to showing that

(11.26) $(iF_2 - u)\xi$

is a compact operator. Let $D_2' := \begin{bmatrix} 0 & -\frac{d}{dx} \\ \frac{d}{dx} & 0 \end{bmatrix}$. Then D_2' is self-adjoint and

$$(D_2')^2 = \begin{bmatrix} -\frac{d^2}{dx^2} & 0 \\ 0 & -\frac{d^2}{dx^2} \end{bmatrix}.$$

Therefore

$$\chi(D_2') = \begin{bmatrix} 0 & -\frac{d}{dx}(1-\frac{d^2}{dx^2})^{-\frac{1}{2}} \\ \frac{d}{dx}(1-\frac{d^2}{dx^2})^{-\frac{1}{2}} & 0 \end{bmatrix} = \begin{bmatrix} 0 & -iF_2 \\ iF_2 & 0 \end{bmatrix}.$$

We have

$$\begin{bmatrix} 0 & -iF_2 - u^* \\ iF_2 - u & 0 \end{bmatrix} = \begin{bmatrix} 0 & -iF_2 \\ iF_2 & 0 \end{bmatrix} - \begin{bmatrix} 0 & u^* \\ u & 0 \end{bmatrix} = \chi(D_2') - \chi(D).$$

Now if $\xi \in C_c(\mathbb{R})$, then $M_\xi(D_2' - D)$ is bounded. By Lemma 9.5.7,

$$\big(\chi(D_2') - \chi(D)\big) M_\xi$$

is compact. This verifies the connection axiom.

We now verify the alignment condition. We will need the following lemma.

Let C be the downward-oriented vertical line $\mathrm{Re}(s) = \frac{1}{2}$ in the complex plane. The contour misses the spectrum of H, and we have for any $s \in \mathbb{C}$ with $\mathfrak{R}(s) > 0$,

$$H^{-s} = \frac{1}{2\pi i} \int_C \lambda^{-s} (\lambda - H)^{-1} d\lambda.$$

The integrand is a function valued in $\mathcal{K}(L^2\mathbb{R})$. Since

$$\|(\lambda - H)^{-1}\| = O(|\lambda|^{-1}), \quad |\lambda| \to \infty,$$

we get that $\|\lambda^{-s}(\lambda - H)^{-1}\| = O(|\lambda|^{-s-1})$, and hence the integral converges absolutely if $\mathrm{Re}(s) > 0$.

If T is, say, an operator on the Schwartz space of $\mathbb{R}$, then

$$[T, (\lambda - H)^{-1}] = (\lambda - H)^{-1} \cdot [T, H] \cdot (\lambda - H)^{-1}$$

by algebra, and hence we get

$$(11.27) \qquad [T, H^s] = \frac{1}{2\pi i} \int_C \lambda^{-s} (\lambda - H)^{-1} \cdot [T, H] \cdot (\lambda - H)^{-1} d\lambda.$$

Lemma 11.4.4 *If $f \in C^\infty(\mathbb{R})$, f' and f'' are both bounded, and $\mathrm{Re}(s) > 0$, then $[f, H^{-s}]H^{\frac{1}{2}}$ is bounded.*

Proof Since $AH^{-\frac{1}{2}}$ and $A^* H^{-\frac{1}{2}}$ are bounded, it follows that $\frac{d}{dx} H^{-\frac{1}{2}}$ is bounded. From this we get $[f, H]H^{-\frac{1}{2}} = (f'' + f' \frac{d}{dx})H^{-\frac{1}{2}}$ is bounded. It follows that $[f, H](\lambda - H)^{-1} H^{\frac{1}{2}}$ is bounded uniformly in $\lambda \in C$. By Cauchy's formula,

$$[f, H^{-s}]H^{\frac{1}{2}} = \frac{1}{2\pi i} \int_C \lambda^{-s} (\lambda - H)^{-1} [f, H](\lambda - H)^{-1} H^{\frac{1}{2}} \, d\lambda.$$

This is an absolutely convergent integral of bounded operators, so is bounded. □

We need to show that

$$F_1 u + u^* F_1 \geq 0 \quad \text{mod compact operators.}$$

In this case, F_1 is the multiplication operator χ. Now as $\begin{bmatrix} \chi & 0 \\ 0 & \chi \end{bmatrix}$ commutes with D modulo bounded operators, it follows that the same matrix commutes with $\chi(D)$ modulo compact operators, and it follows that χ commutes with $u = AH^{-\frac{1}{2}}$ and $A^* H^{-\frac{1}{2}}$ mod compact operators. Thus

$$\chi u + u^* \chi \sim \chi \left(u + u^* \right)$$

$$= \chi \left(AH^{-\frac{1}{2}} - H^{-\frac{1}{2}} A^* \right) \sim \chi \left(AH^{-\frac{1}{2}} - A^* H^{-\frac{1}{2}} \right) = 2\chi x H^{-\frac{1}{2}} = 2f H^{-\frac{1}{2}},$$

where $f(x) = x^2 (1 + x^2)^{-\frac{1}{2}}$.

So since $f^{\frac{1}{2}} H^{-1} f^{\frac{1}{2}} \geq 0$, it suffices to show that

$$R := fH^{-\frac{1}{2}} - f^{\frac{1}{2}} H^{-1} f^{\frac{1}{2}} = f^{\frac{1}{2}} [f^{\frac{1}{2}}, H^{-\frac{1}{2}}]$$

is compact, equivalently, that $R^* R$ is compact. We have

$$R^* R = [f^{\frac{1}{2}}, H^{-\frac{1}{2}}] f [f^{\frac{1}{2}}, H^{-\frac{1}{2}}].$$

Since the first and second derivatives of $f^{\frac{1}{2}}$ are both bounded, Lemma 11.4.4 applies and gives that the commutator $[f^{\frac{1}{2}}, H^{-\frac{1}{2}}]H^{\frac{1}{2}}$ is bounded, and hence

$[f^{\frac{1}{2}}, H^{-\frac{1}{2}}]$ is compact. On the other hand, f is a bounded perturbation of x, from which we deduce that $[f^{\frac{1}{2}}, H^{-\frac{1}{2}}](f - x)[f^{\frac{1}{2}}, H^{-\frac{1}{2}}]$ is compact. So it remains to show $[f^{\frac{1}{2}}, H^{-\frac{1}{2}}]x[f^{\frac{1}{2}}, H^{-\frac{1}{2}}]$ is compact. But $xH^{-\frac{1}{2}} = (A + A^*)H^{-\frac{1}{2}} =: T$ is bounded. We get

$$[f^{\frac{1}{2}}, H^{-\frac{1}{2}}]x[f^{\frac{1}{2}}, H^{-\frac{1}{2}}] = [f^{\frac{1}{2}}, H^{-\frac{1}{2}}]T H^{\frac{1}{2}}[f^{\frac{1}{2}}, H^{-\frac{1}{2}}]$$

is compact and $H^{\frac{1}{2}}[f^{\frac{1}{2}}, H^{-\frac{1}{2}}]$ is bounded by Lemma 11.4.4 and $[f^{\frac{1}{2}}, H^{-\frac{1}{2}}]$ is compact, so R^*R, and hence R, is compact as claimed. $\square$

Corollary 11.4.5 *We have*

$$x \otimes_{C_0(\mathbb{R})} y = 1_{\mathbb{C}} \in KK_0(\mathbb{C}, \mathbb{C}).$$

Proof The operator $A = x + \frac{d}{dx}$ acts by a weighted left-shift in the basis described above, and it follows that

$$\ker(A) = \mathbb{C}\psi_0,$$

while

$$\ker(A^*) = \{0\}.$$

In particular,

$$\mathrm{Index}(D) = \ker(A) - \ker(A^*) = 1.$$

Since the Fredholm index parameterizes $KK_0(\mathbb{C}, \mathbb{C})$, this concludes the proof. $\square$

As with Atiyah's arguments in topological K-theory, a simple rotation trick proves that x and y are actually two-sided inverses of each other in KK:

Lemma 11.4.6 *With x, y the Bott and Dirac morphisms,*

$$y \hat{\otimes}_{\mathbb{C}} x = 1_{C_0(\mathbb{R})} \in KK_0(C_0(\mathbb{R}), C_0(\mathbb{R})).$$

We will use the following:

Lemma 11.4.7 *Let $x \in KK_1(\mathbb{C}, C_0(\mathbb{R})$ be the Bott element. Then*

$$1_{C_0(\mathbb{R})} \otimes x = -x \otimes 1_{C_0(\mathbb{R})} \in KK_1(C_0(\mathbb{R}), C_0(\mathbb{R}^2)).$$

Proof Let $\sigma : C_0(\mathbb{R}^2) \to C_0(\mathbb{R}^2)$ be the map

$$\sigma(f)(x, y) = f(y, -x).$$

By Exercise 11.1.11, the class $(1_{C_0(\mathbb{R})} \otimes x) \otimes_{C_0(\mathbb{R}^2)} [\sigma]$ is represented by the isomorphism class of the following triple. The right $C_0(\mathbb{R}^2)$ module is $C_0(\mathbb{R}^2)$. The left action of $C_0(\mathbb{R})$ is given by letting f act by multiplication by the function $(x, y) \mapsto f(y)$. The operator is given by the multiplier $(x, y) \mapsto \chi(-x)$. Since $\chi(-x) = -\chi(x)$, these computations show that σ maps the cycle for $1_{C_0(\mathbb{R})} \otimes x$ to a cycle identical to that for $x \otimes 1_{C_0(\mathbb{R})}$, except that the operator $\chi \otimes 1$ for the latter cycle has been replacing by $-\chi \otimes 1$. Since replacing the operator by its negative makes the additive inverse in KK_1, we have shown that

$$(1_{C_0(\mathbb{R})} \otimes x) \otimes_{C_0(\mathbb{R}^2)} [\sigma] = -x \otimes 1_{C_0(\mathbb{R})}.$$

The conclusion follows from observing that σ is homotopic to the identity automorphism of $C_0(\mathbb{R}^2)$, since it is induced by a rotation of the plane. $\qquad\square$

Proof of Lemma 11.4.6 The external product $\hat{\otimes}_{\mathbb{C}}$ over $\mathbb{C}$ is graded commutative. Hence

$$y\hat{\otimes}_{\mathbb{C}}x = -x\hat{\otimes}_{\mathbb{C}} y.$$

By definition,

(11.28) $\qquad\qquad - x\hat{\otimes}_{\mathbb{C}}y = -(x\hat{\otimes}1_{C_0(\mathbb{R})}) \otimes_{C_0(\mathbb{R}^2)} (1_{C_0(\mathbb{R})}\hat{\otimes}y).$

By the lemma, $x \otimes 1_{C_0(\mathbb{R})} = - 1_{C_0(\mathbb{R})} \otimes x$. Substituting into the above, we arrive at

$$= +(1_{C_0(\mathbb{R})} \otimes x) \otimes_{C_0(\mathbb{R}^2)} (1_{C_0(\mathbb{R})} \otimes y) = 1_{C_0(\mathbb{R})} \otimes (x \otimes_{C_0(\mathbb{R})} y).$$

The result follows from $x \otimes_{C_0(\mathbb{R})} y = 1_{\mathbb{C}}$. $\qquad\square$

We have established the following.

Theorem 11.4.8 *The Bott and Dirac morphisms x and y are KK_1-equivalences. In particular, $A \otimes C_0(\mathbb{R})$ is KK_1-equivalent to A, for any separable C*-algebra A.*

Proof We have already shown that

$$y \otimes_{\mathbb{C}} x = 1_{C_0(\mathbb{R})}, \quad x \otimes_{C_0(\mathbb{R})} y = 1_{\mathbb{C}}.$$

It follows that $x\hat{\otimes}1_A \in \text{KK}_1(A, C_0(\mathbb{R}) \otimes A)$ is a KK_1-equivalence for any A, with inverse $y\hat{\otimes}1_A$, by Theorem 11.2.7 (in the notation of the theorem, $x\hat{\otimes}_{\mathbb{C}}1_A = \tau_A(x)$). $\qquad\square$

Exercise 11.4.9 Prove that if $\tau \colon C_0(\mathbb{R}) \to C_0(\mathbb{R})$ is $\tau(f)(x) = f(-x)$, then

$$[\tau] = -1_{C_0(\mathbb{R})} \in \text{KK}_0(C_0(\mathbb{R}), C_0(\mathbb{R})).$$

11.5   Equivariant Bott Periodicity and the K-Theory of Crossed Products

In this book we have given two proofs of Bott Periodicity: the Toeplitz proof and the KK-proof. There are several other well-known proofs. Atiyah has proved Bott Periodicity using elementary linear algebra applied to matrix-valued Laurent polynomials on the circle. Joachim Cuntz has given an extremely general argument that Bott Periodicity is forced by basic properties of the K-theory functor (half-exactness and stability).

The merit, however, of the "Dirac–Schrödinger proof" we have given in the previous section is that it is based on the geometry and analysis of the real line $\mathbb{R}$ and is, in a certain sense, *essentially translation-invariant*, a feature not possessed by the other proofs alluded to above. More precisely, the cycles we constructed in the previous section determine cycles in *equivariant* KK-theory, and the equivariant version of the Bott Periodicity theorem still holds for subgroups of $\mathbb{R}$.

This "equivariant" KK-theory is a theory defined on G-C*-algebras, for G a locally compact group, by making several simple and fairly obvious changes to the definitions, to make them equivariant. G-equivariant KK-theory is equipped with a *descent* construction, which is a functor from the category of G-C*-algebras to the category of C*-algebras mapping an object A to the crossed product $A \rtimes G$.

Combining the equivariant Bott Periodicity Theorem with descent results in a number of very strong statements relating to the K-theory of crossed products by subgroups $G \subset \mathbb{R}$.

Equivariant KK-theory contains many special cases of interest, like equivariant K-theory and K-homology—and representation theory. In equivariant KK-theory for *compact* groups, for example, the ring $\mathrm{KK}_*^G(\mathbb{C}, \mathbb{C})$ is (supported in degree 0 and) isomorphic to the representation ring $R(G)$ of the group. While KK-theory morphisms form a group, that is, a $\mathbb{Z} = \mathrm{KK}_0(\mathbb{C}, \mathbb{C})$-module, KK^G-morphisms form a module over the ring $\mathrm{KK}_0^G(\mathbb{C}, \mathbb{C}) \cong R(G)$. So the representation ring acts as "scalars" in this theory, as the integers did in ordinary KK. For compact groups, $\mathrm{KK}_1^G(\mathbb{C}, \mathbb{C}) = \{0\}$ (the duals of compact groups are discrete noncommutative spaces), but for noncompact groups this is no longer the case, making things slightly more complicated, but there is still a module structure of KK_*^G over $\mathrm{KK}_*^G(\mathbb{C}, \mathbb{C})$.

Let G be a locally compact group and B be a G-C*-algebra, that is, a C*-algebra with a strongly continuous action of G on B by C*-algebra automorphisms.

Definition 11.5.1   A *G-equivariant Hilbert B-module* $\mathcal{E}$ is a right Hilbert B-module, together with a $\mathbb{C}$-linear action of G on $\mathcal{E}$, satisfying

$$g(\xi b) = g(\xi)g(b), \quad \forall \xi \in \mathcal{E}, b \in B, \quad \langle g(\xi), g(\eta) \rangle = g(\langle \xi, \eta \rangle), \quad \xi, \eta \in \mathcal{E}.$$

Note that B itself is a G-equivariant right Hilbert B-module, with the given action. We emphasize that in a G-equivariant Hilbert B-module $g \in G$ does not act by a right Hilbert B-module map, unless the action on B is trivial: We have $g(\xi b) = g(\xi)g(b)$.

Example 11.5.2 Let X be a compact space with an action of G, and let $\pi : V \to X$ be a G-equivariant Hermitian vector bundle over X (Section 6.3). Assume that the G-action preserves the Hermitian metric on the bundle in the sense that

$$\langle g(v_1), g(v_2) \rangle_{g(x)} = \langle v_1, v_2 \rangle_x, \quad \forall x \in X, \ v_1, v_2 \in V_x.$$

Then the module $\Gamma(X, V)$ of continuous sections of V, endowed with its canonical $C(X)$-valued inner product from the metric, has the structure of a G-equivariant Hilbert $C(X)$-module by $g(s)(x) = g\left(s(g^{-1}x)\right)$, for a section $s \in \Gamma(X, V)$.

Definition 11.5.3 Let G be a locally compact group and A and B be G-C*-algebras. A *G-equivariant Fredholm A-B-bimodule* is a triple $(\mathcal{E}, \pi, F)$, where

a) $\mathcal{E}$ is a G-equivariant right Hilbert B-module.
b) $\pi : A \to \mathbb{B}(\mathcal{E})$ is a G-equivariant *-homomorphism, i.e., a representation of A by adjointable Hilbert B-module operators on $\mathcal{E}$ such that

$$g\left(\pi(a)\xi\right) = \pi\left(g(a)\right) g(\xi), \quad \forall a \in A, \ \xi \in \mathcal{E}, \ g \in G.$$

c) $F \in \mathbb{B}(\mathcal{E})$ is a self-adjoint Hilbert B-module operator, satisfying

$$(11.29) \qquad \pi(a) \cdot (F^2 - 1), \quad [\pi(a), F] \quad \pi(a) \cdot (g(F) - F) \in \mathcal{K}(\mathcal{E})$$

for all $a \in A$, $g \in G$.

The bimodule is *even* if it has a $\mathbb{Z}/2$ grading

$$\mathcal{E} = \mathcal{E}^+ \oplus \mathcal{E}^-,$$

on $\mathcal{E}$, into orthogonal B-submodules, with respect to which elements of A act as even (grading-preserving) operators, the operator F acts as an odd (grading-reversing) operator, and G acts by even operators.

For an *odd* bimodule, we drop any mention of gradings.

There is an obvious notion of unitary isomorphism of such triples. Homotopy is defined in the same way as before, using the C*-algebra $C_0([0, 1))$, viewed now as a G-C*-algebra with trivial G-action.

Definition 11.5.4 For any pair of G-C*-algebras A and B, let $\mathrm{KK}_*^G(A, B)$ be the quotient by homotopy of the set of (isomorphism classes) of G-equivariant Fredholm A-B-bimodules.

Equivariant KK-theory has the same basic formal properties as the nonequivariant version. If A is any G-C*-algebra, the Fredholm A-A-bimodule A, with the given action, the left multiplication operation of A on itself, and the zero operator F, defines a cycle whose class is denoted $1_A \in \mathrm{KK}_0(A, A)$.

Theorem 11.5.5 *Let G be a locally compact group. For any G-C*-algebras A, B there is a bilinear pairing (the Kasparov composition)*

$$(11.30) \qquad KK_*^G(A, B) \times KK_*^G(B, C) \to KK_*^G(A, C)$$

mapping a pair of morphisms $f \in KK_i^G(A, B)$ *and* $g \in KK_j^G(B, C)$ *to a morphism*

$$f \hat\otimes_B g \in KK_{i+j}^G(A, C),$$

a) *The Kasparov composition gives* KK^G *the structure of a* $\mathbb{Z}/2$*-graded category, with objects (separable) G-C*-algebras, and morphisms* $A \to B$ *the elements of the* $\mathbb{Z}/2$*-graded abelian group* $KK_*^G(A, B)$. *For any A, the element* $1_A \in KK_0^G(A, A)$ *defined above acts as the identity morphism from A to A.*
b) *If* $\alpha: A \to A'$ *is a *-G-equivariant *-homomorphism,* $[\alpha] \in KK_0^G(A, A')$ *its class, then* $\alpha^*(f) = [\alpha] \hat\otimes_{A'} f \in KK_*^G(A, B)$ *for any* $f \in KK_*^G(A', B)$. *Similarly* $\alpha_*(g) = g \hat\otimes_A [\alpha] \in KK_*^G(B, A')$ *for any* $g \in KK_*^G(B, A)$.
c) *If* $\alpha: C \to C'$ *is a G-equivariant *-homomorphism and* $f \in KK_*^G(A, B)$, $g \in KK_*^G(B, C)$, *then*

$$\alpha_*(f \hat\otimes_B g) = f \hat\otimes_B \alpha_*(g).$$

The external product $\hat\otimes_\mathbb{C}$ also extends to KK^G with the same formal properties and relation to the Kasparov composition. We omit the statements.

For purposes of studying K-theory of crossed products, the key point about equivariant KK-theory is that it comes equipped with a *descent map*.

The crossed product construction may be thought of as a *functor*, from the category of G-C*-algebras to the category of C*-algebras, sending a G-C*-algebra A to the C*-algebra $A \rtimes G$, and a G-equivariant *-homomorphism $\alpha: A \to B$ to the corresponding integrated form $\alpha_G: A \rtimes G \to B \rtimes G$. Descent extends this functor to a functor $KK^G \to KK$, agreeing with the one just described on objects and on the the KK^G-morphisms provided by equivariant *-homomorphisms.

The descent construction is based on the following observations about G-equivariant Hilbert modules. Firstly, let A be a G-C*-algebra. Then A is a G-equivariant right Hilbert A-module as well. The algebra multiplication in $A \rtimes G$ extends the twisted convolution operation on $C_c(G, A)$ given by

$$(a * b)(t) = \int_G a(s) s \left(b(s^{-1}t) \right) ds,$$

with ds Haar measure. The adjoint is given by $a^*(s) = s^{-1} \left(a(s^{-1})^* \right)$ times a possible factor of the modular function of the group if it is not unimodular. This cancels with the change of variables $s \to s^{-1}$ giving the integral expression

$$(a^* * b)(t) = \int_G s^{-1} \left(a(s)^* b(st) \right) ds$$

describes the inner product $\langle a, b \rangle$ on $A \rtimes G$ when viewing the latter as a right Hilbert $A \rtimes G$-module.

The formula extends to any G-equivariant Hilbert A-module (see [111] for further details).

Lemma 11.5.6 *Let A be a G-C^*-algebra and $\mathcal{E}$ a G-equivariant right Hilbert A-module. On $C_c(G, \mathcal{E})$ define the right $A \rtimes G$-valued inner product*

$$\langle \xi_1, \xi_2 \rangle := \int_G s^{-1} \left(\langle \xi_1(s), \xi_2(st) \rangle \right) ds$$

and right $C_c(G, A)$-module structure by

$$(\xi a)(t) := \int_G \xi(s) s \left(a(s^{-1}t) \right) ds.$$

Then the resulting completion is a right Hilbert $A \rtimes G$-module, denoted $\mathcal{E} \rtimes G$.

If $(\mathcal{E}, \pi, F)$ is a G-equivariant Fredholm A-B-bimodule, then the construction above produces a right $B \rtimes G$-Hilbert module $\mathcal{E} \rtimes G$. On the other hand, if $T \in \mathbb{B}(\mathcal{E})$, then T acts by an adjointable right $\mathcal{E} \rtimes G$-module operator on $\mathcal{E} \rtimes G$ by extending the formula $(T\xi)(t) = T(\xi(t))$. The group G acts by module operators by $(g\xi)((t) = \xi(g^{-1}t)$. This defines a covariant pair and consequent *-homomorphism $\mathbb{B}(\mathcal{E}) \rtimes G \to \mathbb{B}(\mathcal{E} \rtimes G)$. So F determines an operator F_G on $\mathcal{E} \rtimes G$. Finally, the G-equivariant *-homomorphism $\pi \colon A \to \mathbb{B}(\mathcal{E})$ determines a *-homomorphism $\pi_G \colon A \rtimes G \to \mathbb{B}(\mathcal{E}) \rtimes G \subset \mathbb{B}(\mathcal{E} \rtimes G)$. It is a routine exercise to show that $(\mathcal{E} \rtimes G, \pi_G, F_G)$ defines a Fredholm $A \rtimes G$-$B \rtimes G$ bimodule.

Theorem 11.5.7 *The construction above determines a functor $j_G \colon KK^G \to KK$, called* descent, *which on objects maps a G-C^*-algebra A to $A \rtimes G$, and which extends the crossed product functor.*

We make the following final point about the machinery of equivariant KK-theory: The axiomatic approach to the Kasparov composition is essentially exactly the same as in the nonequivariant case except for some small obvious modifications. We now proceed to use these tools to prove the equivariant Bott Periodicity theorem for subgroups $G \subset \mathbb{R}$.

The translation action of $\mathbb{R}$ on $\mathbb{R}$ gives $C_0(\mathbb{R})$ the structure of an $\mathbb{R}$-C^*-algebra or G-C^*-algebra for any countable subgroup $G \subset \mathbb{R}$ or $G = \mathbb{R}$ itself with the standard topology.

Lemma 11.5.8 *Let $G = \mathbb{R}$ or G be a countable subgroup of $\mathbb{R}$ (with the discrete topology), and let χ be a normalizing function.*

a) *The cycle $(C_0(\mathbb{R}), 1, \chi)$, in which $C_0(\mathbb{R})$ is regarded as a G-equivariant right Hilbert $C_0(\mathbb{R})$-module, is an odd G-equivariant Fredholm $\mathbb{C}$-$C_0(\mathbb{R})$-bimodule.*

b) *Give $L^2(\mathbb{R})$ the unitary action of G induced by the translation action on $\mathbb{R}$. The $\frac{d}{dx}$ commutes with G with this action, and hence $F := \chi(-i\frac{d}{dx})$ commutes with*

G as well. The triple $(L^2(\mathbb{R}), M, F)$ *is an odd G-equivariant Fredholm* $C_0(\mathbb{R})$-
$\mathbb{C}$ *bimodule.*

We let

$$\alpha_G \in \text{KK}_1^G(\mathbb{C}, C_0(\mathbb{R})), \beta_G \in \text{KK}_1^G(C_0(\mathbb{R}), \mathbb{C})$$

be the classes of the cycles in a), b), respectively.

Proof The only additional comment regards a): For $s \in \mathbb{R}$ fixed, note that $\chi(x +$
$s) - \chi(x) \to 0$ uniformly in x, for fixed $s \in \mathbb{R}$, So $s(\chi) - \chi$ is a compact multiplier
of $C_0(\mathbb{R})$.   □

Now consider the cycle $\left(L^2(\mathbb{R}) \oplus L^2(\mathbb{R}), 1, F := \chi(D)\right)$ for $\text{KK}_0^G(\mathbb{C}, \mathbb{C})$, where

$$D = \begin{bmatrix} 0 & x - \frac{d}{dx} \\ x + \frac{d}{dx} & 0 \end{bmatrix},$$

where G acts on $L^2(\mathbb{R})$ by the translation action, as in the definition of β_G above.
Note that if $s \in \mathbb{R}$, then $s(D) - D = \begin{bmatrix} 0 & s \\ s & 0 \end{bmatrix}$, which is bounded. Applying
Lemma 9.5.10 gives that $s(F) - F$ is compact.
    We let

$$\gamma_G \in \text{KK}_0^G(\mathbb{C}, \mathbb{C})$$

be the class of this cycle.

Lemma 11.5.9 *In reference to the three examples above,*

$$\alpha_G \otimes_{C_0(\mathbb{R})} \beta_G = \gamma_G \in \text{KK}_0^G(\mathbb{C}, \mathbb{C}).$$

The proof is exactly the same as in the nonequivariant case: One verifies that the
Dirac–Schrödinger triple satisfies the axioms for the Kasparov product.

Theorem 11.5.10 α_G *and* β_G *are inverse* KK_1^G-*equivalences, that is,*

$$\alpha \otimes_{C_0(\mathbb{R})} \beta = 1_{\mathbb{C}}, \quad \beta \otimes_{\mathbb{C}} \alpha = 1_{C_0(\mathbb{R})}.$$

Moreover, if A is any G-C*-algebra, then

$$\alpha \otimes 1_A \in \text{KK}_1^G(A, C_0(\mathbb{R}) \otimes A), \quad \beta \otimes 1_A \in \text{KK}_1^G(C_0(\mathbb{R}) \otimes A, A)$$

are inverse KK_1^G-*equivalences.*

Proof By linearly deforming the G action on $\mathbb{R}$ to the trivial action (let g act at
time t by translation by tg), we obtain a homotopy between the cycle for γ_G defined

above and a cycle which is identical except insofar as the G-action is trivial. Since the Fredholm index of the Dirac–Schrödinger cycle is $+1$, $\gamma_G = 1_{\mathbb{C}} \in \mathrm{KK}_0^G(\mathbb{C}, \mathbb{C})$.

The second statement follows from the general mechanics of KK. □

Consider the morphisms α_G and β_G of Theorem 11.5.10. If A is any G-C*-algebra, we get morphisms $\alpha_G \otimes 1_A \in \mathrm{KK}_1^G(A, C_0(\mathbb{R}, A))$ and $\beta_G \otimes 1_A \in \mathrm{KK}_1(C_0(\mathbb{R}, A), A)$. Here $C_0(\mathbb{R}, A)$ carries the diagonal action of G.

Applying the descent functor $j\colon \mathrm{KK}^G \to \mathrm{KK}$ to these morphisms then gives morphisms $j(\beta_G) \in \mathrm{KK}_1(C_0(\mathbb{R}, A) \rtimes G, A \rtimes G)$ and $j(\alpha_G) \in \mathrm{KK}_1(A \rtimes G, C_0(\mathbb{R}, A) \rtimes G)$. They are inverse KK_1-equivalences, since descent is a functor.

Corollary 11.5.11 *If $G \subset \mathbb{R}$ is $\mathbb{R}$ with the usual topology, or any countable subgroup, and A is a G-C*-algebra, then $C_0(\mathbb{R}, A) \rtimes G$ and $A \rtimes G$ are KK_1-equivalent.*

The power of this theorem is quite simple to see. Suppose that $A = C_0(X)$, for a locally compact $\mathbb{Z}$-space X. Then the diagonal action of $\mathbb{Z}$ on $\mathbb{R} \times X$ is a proper action. In fact it is a free and proper action, and therefore

$$C_0(\mathbb{R} \times X) \rtimes \mathbb{Z} \sim C_0(\mathbb{R} \times_{\mathbb{Z}} X),$$

where $\mathbb{R} \times_{\mathbb{Z}} X$ is the quotient of $\mathbb{R} \times X$ by the diagonal group action, and $\sim$ is strong Morita equivalence.

If $G = \mathbb{R}$, then the same remarks apply, and since $\mathbb{R} \times_{\mathbb{R}} X \cong X$, we obtain the following corollary:

Corollary 11.5.12 *Let X be a G space, where G is $\mathbb{R}$ or $\mathbb{Z}$ and X be a G-space.*

a) *If $G = \mathbb{Z}$, then $C(X) \rtimes \mathbb{Z}$ is KK_1-equivalent to $C(\mathbb{R} \times_{\mathbb{Z}} X)$.*
b) *If $G = \mathbb{R}$, then $C(X) \rtimes \mathbb{R}$ is KK_1-equivalent to A.*

The second statement is called the *Connes–Thom Isomorphism*.

An integer action on X is determined by where the generator 1 goes. Say that two $\mathbb{Z}$ actions are *isotopic* if the corresponding pair of homeomorphisms can be connected to each other by a continuous path of homeomorphisms.

It is not at all obvious that the K-theory $\mathrm{K}_*(C_0(X) \rtimes \mathbb{Z})$ of the crossed product only depends on the isotopy class of the action, although one would imagine this must be true. It follows, however, from equivariant Bott Periodicity.

Corollary 11.5.13 *If two $\mathbb{Z}$-actions are isotopic, then the K-theory groups of the corresponding crossed products $\mathrm{K}_*(C_0(X) \rtimes \mathbb{Z})$ are isomorphic.*

Proof It is obvious that isotopic $\mathbb{Z}$-actions lead to homeomorphic mapping cylinders $\mathbb{R} \times_{\mathbb{Z}} X$. The result follows from ordinary homotopy-invariance of K-theory. □

For the case of the irrational rotation algebra $A_\hbar := C(\mathbb{T}) \rtimes \mathbb{Z}$, defined by letting the homeomorphism be rotation by an irrational angle $\hbar \in \mathbb{R}/\mathbb{Z}$, we obtain the following corollary:

Corollary 11.5.14 $K_0(A_\hbar) \cong \mathbb{Z} \oplus \mathbb{Z}$, *with generators the class* $[1_{A_\hbar}]$ *of the unit in* $A_\hbar$, *and the class* $[p_\hbar]$ *of the Rieffel projection, and* $K_1(A_\hbar) \cong \mathbb{Z} \oplus \mathbb{Z}$ *with generators the class* $[z]$ *of the unitary complex coordinate on* $\mathbb{T}$, *and the class* $[u]$ *of the generator* $u \in C(\mathbb{T}) \rtimes \mathbb{Z}$ *of* $\mathbb{Z}$ *in the crossed product.*

Proof The crossed product $C(\mathbb{T}) \rtimes \mathbb{Z}$ is KK_1-equivalent to the mapping cylinder $\mathbb{R} \times_\mathbb{Z} \mathbb{T}$, which, as we have noted, is naturally homeomorphic to the ordinary 2-torus $\mathbb{T}^2$ (note that the homeomorphism uses, precisely, the obvious isotopy between rotation by $\hbar$ and the identity). Since $K_0(\mathbb{T}^2) \cong \mathbb{Z}^2$ and $K^1(\mathbb{T}^2) \cong \mathbb{Z}^2$, we get $K_1(A_\hbar) \cong \mathbb{Z}^2$ and $K_0(A_\hbar) \cong \mathbb{Z}^2$ follow, as abstract groups. Verifying the assertions about the generators is left to the reader, using the various tools explained in this book.   □

Corollary 11.5.15 *Suppose* $\phi \colon X \to X$ *is a minimal homeomorphism of the Cantor set* X. *Then* $K_0(C(X) \rtimes \mathbb{Z})$ *is naturally isomorphic to the cokernel of the abelian group homomorphism*

$$\mathrm{id} - \phi^* \colon C(X, \mathbb{Z}) \to C(X, \mathbb{Z}),$$

that is,

$$K_0(C(X) \rtimes \mathbb{Z}) \cong C(X, \mathbb{Z})/\mathrm{ran}(\mathrm{id} - \phi^*),$$

with $\phi^*(f) := f \circ \phi$, *and* $C(X, \mathbb{Z})$ *the group of integer-valued, continuous functions on* X.

And $K_1(C(X) \rtimes \mathbb{Z}) \cong \mathbb{Z}$, *with generator the class* $[u]$ *of the unitary* $u \in C(X) \rtimes \mathbb{Z}$ *generating the action.*

This is an immediate corollary of the more general sequence, called the *Pimsner–Voiculescu sequence.*

Corollary 11.5.16 *For any* $\mathbb{Z}$*-action on a* C^**-algebra* A, *there is a cyclic 6-term exact sequence of the form*

$K_0(A) \xrightarrow{\mathrm{id}-\alpha_*} K_0(A) \xrightarrow{i_*} K_0(A \rtimes \mathbb{Z})$

$\delta \uparrow \qquad\qquad\qquad\qquad\qquad \downarrow \delta$

$(11.31) \qquad K_1(A \rtimes \mathbb{Z}) \xleftarrow{i_*} K_1(A) \xleftarrow{\mathrm{id}-\alpha_*} K_1(A)$

where $\alpha \colon A \to A$ *is the automorphism generating the action, and* $i \colon A \to A \rtimes \mathbb{Z}$ *is the inclusion.*

The sequence is natural with respect to $\mathbb{Z}$-equivariant *-homomorphisms.

Proof of Corollary 11.5.15 This is an application of Corollary 11.5.16 in the case $A = C(X)$. In this case, $K_1(A)$ is the zero group, and hence the Pimsner–Voiculescu sequence reduces to a sequence of the form

(11.32)

$$0 \to \mathrm{K}_1(C(X) \rtimes \mathbb{Z}) \xrightarrow{\delta} \mathrm{K}^0(X) \xrightarrow{\mathrm{id}-\alpha_*} \mathrm{K}^0(X) \xrightarrow{i_*} \mathrm{K}_0(C(X) \rtimes \mathbb{Z}) \to 0.$$

Since $\mathrm{K}^0(X) \cong C(X, \mathbb{Z})$ as abelian groups, it follows that

$$\mathrm{K}_0(C(X) \rtimes \mathbb{Z}) \cong C(X, \mathbb{Z})/(\mathrm{id} - \alpha_*)C(X, \mathbb{Z}),$$

that is, $\mathrm{K}_0(C(X) \rtimes \mathbb{Z})$ is isomorphic to the cokernel of $\mathrm{id} - \alpha_*$ acting on $C(X, \mathbb{Z})$. Furthermore, if the $\mathbb{Z}$-action is minimal, then no continuous, complex-valued function $f \colon X \to \mathbb{C}$ can satisfy

$$f \circ \alpha = f,$$

unless it is constant. Hence $\ker(\mathrm{id} - \alpha_*)$ consists of the subgroup of constant functions in $C(X, \mathbb{Z})$ and thus is infinite cyclic. Hence $\mathrm{K}_1(C(X) \rtimes \mathbb{Z})$ is also infinite cyclic, and we leave it as an exercise to verify that the class $[u]$ described in the statement is a generator. □

As our final example, suppose that $\Gamma \subset \mathbf{PSL}_2(\mathbb{R})$ is a uniform lattice, acting on the hyperbolic plane $\mathbb{H}^2$ (by Möbius transformations) and on its boundary $\partial\mathbb{H}$ with $\Gamma\backslash\mathbb{H}^2 = M$ a compact genus g surface, $g \geq 2$. See Section 6.7. We proved there that $C(\partial\mathbb{H}^2) \rtimes \Gamma$ is Morita equivalent to $C(SM) \rtimes B$, where B is the Borel upper triangular subgroup $B = \{\begin{bmatrix} a & b \\ 0 & a^{-1} \end{bmatrix} \mid a, b \in \mathbb{R}, \ a \neq 0\}$ of $\mathbf{PSL}_2(\mathbb{R})$. The subgroup B is a semi-direct product $B \cong \mathbb{R} \rtimes \mathbb{R}$, with the two copies of $\mathbb{R}$ acting by geodesic and horocycle flow on the sphere bundle SM. It follows that the crossed product $C(SM) \rtimes B$ can be written as an iterated crossed product $(C(SM) \rtimes_h \mathbb{R}) \rtimes_g \mathbb{R}$. Two applications of the Connes–Thom Isomorphism Corollary 11.5.12 b) give a KK_0-equivalence between $C(\partial\mathbb{H}^2) \rtimes \Gamma$ and $C(SM)$ and the following result:

Corollary 11.5.17 *Let* $\Gamma \subset \mathbf{PSL}_2(\mathbb{R})$ *be a uniform lattice, acting on the boundary* $\partial\mathbb{H}$ *of the hyperbolic plane. Then* $C(\partial\mathbb{H}^2) \rtimes \Gamma$ *is* KK_0*-equivalent to* $C(SM)$, *where* $M = \Gamma\backslash\mathbb{H}^2$ *and* SM *is the sphere bundle to this Riemann surface.*

An interesting corollary of this is that the class $[1_{C(\partial\mathbb{H}^2)\rtimes\Gamma}] \in \mathrm{K}_0\left(C(\partial\mathbb{H}^2) \rtimes \Gamma\right)$ of the unit of this C*-algebra is torsion of order equal to $2g - 2$, the Euler characteristic of M. This follows from computation of $\mathrm{K}^*(SM)$ (and the isomorphism with $\mathrm{K}_*\left(C(\partial\mathbb{H}^2) \rtimes \Gamma\right)$). The K-theory of sphere bundles SM is described by a "Gysin sequence" (see [34] for cohomology) which in particular implies the torsion result above.

Actually, a rather close analog of the Gysin sequence holds for *general* hyperbolic groups G acting on their boundaries holds. This is shown in [74]. This piece of K-theory and other considerations suggest an analogy between the cross products $C(\partial G) \rtimes G$ and the C*-algebras of sphere bundles SM (of compact Riemannian manifolds), in which the C*-algebra inclusion $C^*(G) \to C(\partial G) \rtimes G$ plays the role of the inclusion $C(M) \to C(SM)$ by Gelfand dualizing the projection $SM \to M$. Thus, one can think of $C(\partial G) \rtimes G$ as a kind of "noncommutative sphere bundle" over the noncommutative space $\widehat{G}$.

Exercise 11.5.18 If A is a unital, *commutative* C*-algebra, then the class $[1_A] \in K_0(A)$ of the unit is never torsion and never zero.

Bibliography

1. S. Adams: *Boundary amenability for word hyperbolic groups and an application to smooth dynamics of simple groups*, Topology 33 (1994), no. 4, 765–783.
2. J.F. Adams: *Vector fields on spheres*, Ann. of Math. (3) Vol. 75, no. 3 (1962), pp. 603–632.
3. J.F. Adams: *On the non-existence of elements of Hopf invariant one*, Ann. of Math. (2) 72 (1960), 20–104.
4. C. Anantharaman-Delaroche: *Purely infinite C*-algebras arising from dynamical systems*, Bull. Soc. Math. France 125, no. 21 (1997), 99–225.
5. P. Antonini, S. Azzali, Sara, G. Skandalis: *Flat bundles, von Neumann algebras and K-theory with R/Z-coefficients*, J. K-Theory 13 (2014), no. 2, 275–303.
6. P. Antonini, Paolo; S. Azzali, G. Skandalis *Bivariant K-theory with R/Z-coefficients and rho classes of unitary representations*, J. Funct. Anal. 270 (2016), no. 1, 447–481.
7. W.B. Arveson: *An invitation to C*-algebras*, Graduate Texts in Mathematics 39, Springer-Verlag, Berlin, New York (1976).
8. M.F. Atiyah *K-theory*, Lecture notes by D. W. Anderson (1967), W. A. Benjamin, Inc., New York-Amsterdam.
9. M.F. Atiyah: *Bott periodicity and the index of elliptic operators*, Quart. J. Math. Oxford Ser. (2) (1968), 113–140.
10. M.F. Atiyah, R. Bott, V.K. Patodi: *On the heat equation and the index theorem*, Invent. Math. 19 (1973), 279–330.
11. M.F. Atiyah, R. Bott, A. Shapiro: *Clifford modules*, Topology 3 (1964), no. 1, 3–38.
12. M.F. Atiyah, I.M. Singer: *The Index of Elliptic Operators on Compact Manifolds*, Bull. Amer. Math. Soc., 69 (3) (1963), 42–433.
13. M. F. Atiyah, I. M. Singer: *The index of elliptic operators. I*, Ann. of Math. (2) 87 (1968), 484–530.
14. M. F. Atiyah, I. M. Singer: *The index of elliptic operators II*, Ann. of Math. (2) 87 (1968), 531–535.
15. M. F. Atiyah, I. M. Singer: *The index of elliptic operators III*, Ann. of Math. (2) 87 (1968), 536- 604.
16. M. F. Atiyah, I. M. Singer: *The index of elliptic operators V*, Ann. of Math. (2) 87 (1971), 139–149.
17. M. F. Atiyah, I. M. Singer: *The index of elliptic operators IV*, Ann. of Math. (2) 93 (1971), 119–138.
18. S. Baaj, P. Julg: *Théorie bivariante de Kasparov et opérateurs non bornés dans les C*-modules hilbertiens*, C. R. Acad. Sci. Paris Sér. I Math. 296 (1983), no. 21, 875–878.

© Springer Nature Switzerland AG 2024
H. Emerson, *An Introduction to C*-Algebras and Noncommutative Geometry*,
Birkhäuser Advanced Texts Basler Lehrbücher,
https://doi.org/10.1007/978-3-031-59850-0

19. P. Baum, A. Connes: *K -theory for discrete groups*, Operator algebras and applications, Vol. 1, 1–20, London Math. Soc. Lecture Note Ser., 135, Cambridge Univ. Press, Cambridge, 1988.

20. P. Baum, A. Connes, N. Higson: *classifying space for proper actions and K-theory of group C*-algebras*, C*-algebras: 1943–1993 (San Antonio, TX, 1993), 240–291, Contemp. Math., 167, Amer. Math. Soc., Providence, RI, 1994.

21. P. Baum, Paul, N. Higson, T. Schick: *A geometric description of equivariant K-homology for proper actions*, Quanta of maths, 1–22, Clay Math. Proc., 11, Amer. Math. Soc., Providence, RI, 2010.

22. J. Bellissard: *C*-algebras in solid state physics*, 2D electrons in a uniform magnetic field. Operator algebras and applications, Vol. 2, 49–76, London Math. Soc. Lecture Note Ser., 136, Cambridge Univ. Press, Cambridge, 1988.

23. J. Bellisard: *Ordinary quantum Hall effect and noncommutative cohomology*, Localization in disordered systems (Bad Schandau, 1986), 61–74, Teubner-Texte Phys., 16, Teubner, Leipzig, 1988.

24. M. Berger, P. Gauduchon, E. Mazet: *Le spectre d'une variété riemannienne*, Lecture Notes in Mathematics (in French), 194 (1971), Berlin-New York: Springer-Verlag.

25. N. Berline, E. Getzler, M. Vergne: *Heat kernels and Dirac operators*, Grundlehren der Mathematischen Wissenschaften 298 (1992), Springer-Verlag, Berlin.

26. B. Blackadar: *K-theory for Operator Algebras*, 2nd edition, (1998) MSRI Publications, vol. 5.

27. N.P. Brown, N. Ozawa: *C*-algebras and finite-dimensional approximations*, Graduate Studies in Mathematics, 88. American Mathematical Society, Providence, RI, 2008.

28. J.-B. Bost, A. Connes: *Hecke Algebras, Type III factors and Phase transitions with Spontaneous Symmetry Breaking in Number Theory*, Selecta. Mathematica, New Series, Vol 1, no. 3 (1995), 11–56.

29. L. Brown, P. Green, M. Rieffel: *Stable isomorphism and strong Morita equivalence of C*-algebras*, Pacific J. Math. 71, no. 2 (1977), 349–363.

30. R. Bott: *An application of the Morse theory to the topology of Lie-groups,* Bulletin de la Société Mathématique de France, 84 (1956), 251–281

31. R. Bott: *The stable homotopy of the classical groups*, Proceedings of the National Academy of Sciences of the United States of America, 43 (10): (1957) 933–5.

32. R. Bott: *The stable homotopy of the classical groups* , Annals of Mathematics, Second Series, 70 (2) (1959), 313–337.

33. R. Bott: *The periodicity theorem for the classical groups and some of its applications*, Advances in Mathematics, 4 (3), (1970), 353–411.

34. R. Bott, L.W. Tu: *Differential forms in algebraic topology*, Graduate Texts in Mathematics, 82. Springer-Verlag, New York-Berlin, 1982.

35. O. Bratteli: *Inductive limits of finite dimensional C*-algebras*, Trans. Amer. Math. Soc., 171 (1972) 195–234.

36. L. Brown: *Stable isomorphism of hereditary subalgebras of C*-algebras*, Pacific J. Math. 71, no. 2 (1977), 335–348.

37. A. Chamseddine, A. Connes: *Universal formula for noncommutative geometry actions: unification of gravity and the Standard Model*, Phys. Rev. Lett. 77, (1996).

38. A. Chamseddine, A. Connes: *The Spectral action principle*, Comm. Math. Phys. Vol.186 (1997), 731–750.

39. A. Chamseddine, A. Connes: *Inner fluctuations of the spectral action*, J. Geom. Phys. 57, no. 1 (2006), 1–21.

40. A. Connes: *C*-algèbras et géométrie différentielle*. C.R. Acad. Sci. Paris, Ser. A-B (1980), 599–604.

41. A. Connes: *Noncomutative Differential Geometry*, Publ. Math. IHES 39 (985), 257–360.

42. A. Connes: *A survey of foliations and operator algebras*, Operator algebras and applications, Part I (Kingston, Ont., 1980), Amer. Math. Soc., Providence, R.I. 1982, vol. 38, 521–628.

43. A. Connes: *Cyclic cohomology and the transverse fundamental class of a foliation*, Geometric methods in operator algebras (Kyoto, 1983), Pitman Res. Notes Math. Ser. 123 (1986), 52–144.

44. A. Connes: *An analogue of the Thom isomorphism for crossed products of a C*-algebra by an action of R*. Adv. in Math. 39 (1981), no. 1, 31–55.

45. A. Connes: *Noncommutative differential geometry*, Inst. Hautes Études Sci. Publ. Math. No. 62 (1985), 257–360.

46. A. Connes: *Compact metric spaces, Fredholm modules, and hyperfiniteness*, Ergodic Theory Dynam. Systems 9 (1989), no. 2, 207–220.

47. A. Connes: *Noncommutative Geometry*, Academic Press, Inc., (1994) San Diego, CA.

48. A. Connes: *Gravity coupled with matter and the foundation of non-commutative geometry*, Comm. Math. Phys. 182 (1996), no. 1, 155–176.

49. A. Connes: *Essay on physics and noncommutative geometry*, The interface of mathematics and particle physics, Oxford (1988), 9–48, Inst. Math. Appl. Conf. Ser. New Ser., 24, Oxford Univ. Press, New York, 1990.

50. A. Connes, A. H. Chamseddine: *Why the standard model*. J. Geom. Phys. 58 (2008), no. 1, 38–47.

51. A. Connes, G. Skandalis: *The longitudinal index theorem for foliations*, Publ. Res. Inst. Math. Sci. 20 (1984), no. 6, 1139–1183.

52. A. Connes, H. Moscovici: *The local index formula in noncommutative geometry*, Geom. Funct. Anal. 5 (1995), no. 2, 174–243.

53. A. Connes, M. Marcolli: *From physics to number theory via noncommutative geometry*, Frontiers in number theory, physics and geometry 1, Springer, Berlin (2006), 269–347.

54. J.B. Conway: *A Course in Functional Analysis*, Second edition., Grad. Texts Math. 96, Springer-Verlag, Berlin, New York (2007).

55. J. Cuntz: *Simple C*-algebras generated by isometries*, Comm. Math. Phys. 57 (1977), no. 2, 173–185.

56. J. Cuntz: *K-theory and C*-algebras*, in Algebraic K-theory, number theory, geometry and analysis (Bielefeld 1982), 55–79. Lecture Notes in Math. 1046 (1984), Springer, Berlin.

57. J. Cuntz; C. Deninger; M. Laca: *C*-algebras of Toeplitz type associated with algebraic number fields*, Math. Ann. 355 (2013), no. 4, 1383–1423.

58. J. Cuntz; W. Krieger: *A class of C*-algebras and topological Markov chains*, Invent. Math. 56 (1980), no. 3, 251–268.

59. J. Cuntz, N. Higson: *Kuiper's theorem for Hilbert modules*, Contemp. Math. 62 (1967), 429–435.

60. R.J. Deeley: *Geometric K-homology with coefficients I: Z/kZ-cycles and Bockstein sequence*, J. K-Theory 9 (2012), no. 3, 537–564.

61. R.J. Deeley: *Geometric K-homology with coefficients II: The analytic theory and isomorphism*. J. K-Theory 12 (2013), no. 2, 235–256.

62. P. Donovan, M. Karoubi: *Graded Brauer groups and K-theory with local coefficients*, Inst. Hautes Études Sci. Publ. Math. No. 38 (1970), 5–25.

63. C. Delaroche, J. Renault: *Amenable groupoids*, Groupoids in analysis, geometry, and physics (Boulder, CO, 1999), 35–46, Contemp. Math., 282, Amer. Math. Soc., Providence, RI, 2001.

64. R.G. Douglas: *C*-algebra extensions and K-homology*, Annals of Mathematics Studies, No. 95 Princeton University Press, Princeton, N.J.; University of Tokyo Press, Tokyo, 1980.

65. A. Duwenig, H. Emerson: *Transversals, duality, and irrational rotation*, Trans. Amer. Math. Soc. Ser. B 7 (2020), 254–289.

66. S. Echterhoff, D.P. Williams: *The Mackey machine for crossed products: inducing primitive ideals*, Group representations, ergodic theory, and mathematical physics: a tribute to George W. Mackey, 129–136, Contemp. Math., 449, Amer. Math. Soc., Providence, RI, 2008.

67. S. Echterhoff, H. Emerson: *Structure and K-theory of crossed products by proper actions*, Expo. Math. 29 (2011), no. 3, 300–344.

68. E. Effros: *Why the Circle is Connected: An Introduction to Quantized Topology*, Math. Intelligencer 11 (1989), no. 1, 27–34.

69. G.A. Elliot: *On the classification of inductive limits of sequences of semisimple finite-dimensional algebras*, J. Algebra, 38 (1) (1976) 29–44.
70. G. Elliot, G. Gong, L. Li: *On the classification of simple inductive limit C*-algebras, II: The isomorphism theorem*. Invent. Math. 168 (2007), no. 2, 249–320.
71. H. Emerson: *Noncommutative Poincaré duality for boundary actions of hyperbolic groups*, J. Reine Angew. Math. 564 (2003), 1–33.
72. H. Emerson: *Lefschetz numbers for C*-algebras*, Canad. Math. Bull. 54 (2011), no. 1, 82–99.
73. H. Emerson, R. Meyer: *Dualities in equivariant Kasparov theory*, New York J. Math. 16 (2010), 245–313.
74. H. Emerson, R. Meyer: *Euler characteristics and Gysin sequences for group actions on boundaries*, Math. Ann. 334 (2006), no. 4, 853–904.
75. H. Emerson, R. Meyer: *Dualizing the coarse assembly map*, J. Inst. Math. Jussieu 5 (2006), no. 2, 161–186.
76. H. Emerson, S. Echterhoff, H.-J. Kim: *KK-theoretic duality for proper twisted actions*. Math. Ann. 340 (2008), no. 4, 839–873.
77. H. Emerson, R. Meyer: *Equivariant representable K-theory*, J. Topology. (2009), no. 2 123–156.
78. H. Emerson, R. Meyer: *Equivariant embedding theorems and topological index maps*, Adv. Math. 225 (2010), 2840–2882
79. H. Emerson, R. Meyer: *Bivariant K-theory via correspondences*, Adv. Math. 225 (2010), 2883–2919.
80. H. Emerson, I. Dell'Ambrolio, I., R. Meyer: *A equivariant Lefschetz fixed-point formula for correspondences*. Doc. Math. 19 141–194.
81. H. Emerson, B. Nica: *K-homological finiteness and hyperbolic groups*, J. Reine Angew. Math. 745 (2018), 189–229.
82. T. Fack, G. Skandalis, *L'analogue de l'isomorphisme de Thom pour les groups de Kasparov*, C.R. Acad. Sci. Paris (A-B) 291 no. 10 (1980), 579–581.
83. G. B. Folland: *Real Analysis, Modern Techniques and Applications*, 2nd. ed., A Wiley-Interscience Publication (1999).
84. J.M. Gracia-Bondia, Joseph C. Varilly, Hector Figueroa: *Elements of Noncommutative Geometry*, Birkhäuser Advanced Texts (2001), Birkäuser Boston.
85. I. M. Gelfand: *On elliptic equations*, Russ. Math. Surv., 15 (3) (1960), 11–123.
86. I. M. Gelfand, M. A. Naimark: *On the imbedding of normed rings into the ring of operators on a Hilbert space*. Mat. Sbornik. 12 (2) (1943), 197–217.
87. E. Getzler: *The odd Chern character in cyclic homology and spectral flow*, Topology, 32(3) (1993) 489–507.
88. E. Getzler and A. Szenes: *On the Chern character of a theta-summable Fredholm module*, J. Funct. Anal., 84(2) (1989) 343–357.
89. T. Giordano, H. Matui, I.F. Putnam, C.F. Skau: *Orbit equivalence for Cantor minimal Zd-systems*, Invent. Math. 179 (2010), no. 1, 119–158.
90. M. Grensing: *Universal cycles and homological invariants of locally convex algebras*, J. Funct. Anal. 263 (2012), no. 8, 2170–2204.
91. M. Gromov: *Hyperbolic groups*, in *Essays in group theory* (Publ. MSRI 8, Springer 1987), 75–263.
92. M. Gromov, H.B. Lawson Jr.: *Positive scalar curvature and the Dirac operator on complete Riemannian manifolds*, Inst. Hautes Etudes Sci. Publ. Math., 58 (1983), 83–196.
93. É. Ghys, P. de la Harpe: *Sur les groupes hyperboliques d'après Mikhael Gromov*, Progress in Mathematics 83, Birkhäuser 1990.
94. E. Guentner, J. Kaminker: *Exactness and the Novikov conjecture*, Topology 41 (2002), no. 2, 411–418.
95. N. Higson: *A characterization of KK-theory* Pacific J. Math 126 (1987) no. 2, 253–276.
96. N. Higson, V. Lafforgue, V., G. Skandalis: *Counterexamples to the Baum-Connes conjecture*, Geom. Funct. Anal. 12 (2002), no. 2, 330–354.

97. N. Higson, J. Roe: *Amenable group actions and the Novikov conjecture*, J. Reine Angew. Math. 519 (2000), 143–153.

98. N. Higson, J. Roe: *Analytic K-homology*, Oxford Mathematical Monographs. Oxford Science Publications. Oxford University Press, Oxford, 2000.

99. N. Higson, G. G. Kasparov: *E-theory and KK-theory for groups which act properly and isometrically on Hilbert space*, Invent. Math. 144 (2001), no. 1, 23–74.

100. N. Higson: *The residue index theorem of Connes and Moscovici*, Surveys in noncommutative geometry, 71–126, Clay Math. Proc., 6, Amer. Math. Soc., Providence, RI, 2006.

101. E. Hopf: *Fuchsian groups and ergodic theory*, Trans. Amer. Math. Soc. 39 (1936), no. 2, 299–314.

102. P. Julg, A. Valette: *Group actions on trees and K-amenability*, Operator algebras and their connections with topology and ergodic theory (Buşteni, 1983), 289–296, Lecture Notes in Math., 1132, Springer, Berlin, 1985.

103. P. Julg: *K-théorie équivariante et produits croises*, C. R. Acad. Sci., Paris, Sér. I, 292:629–632, 1981.

104. J. Kaminker, I.P. Putnam: *K -theoretic duality of shifts of finite type*, Comm. Math. Phys. 187 (1997), no. 3, 509–522.

105. J. Kaminker, I.F. Putnam, M.F. Whittaker: *K-theoretic duality for hyperbolic dynamical systems*, J. Reine Angew. Math. 730 (2017), 263–299.

106. M. Kac: *Can one hear the shape of a drum?* American Mathematical Monthly, 73 (1966), 1–23.

107. M. Karoubi: *K-theory: An Introduction*, Springer-Verlag, series *Classics in Mathematics* (1978).

108. G.G. Kasparov: *Hilbert modules: Theorems of Stinespring and Voiculescu*, J. Op. Theory 4 no. 1 (1980), 133–150.

109. G. G. Kasparov: *Topological invariants of elliptic operators: I. K-homology*, Izv. Akad. Nauk SSSR Ser. Mat. 39 (1975), no. 4, 796–838.

110. G.G. Kasparov: *The operator K-functor and extensions of C* -algebras*, Izv. Akad. Nauk SSSR Ser. Mat. 44 (1980), no. 3, 571–636, 719.

111. G. Kasparov: *Equivariant KK-theory and the Novikov Conjecture*, Invent. Math. 91 (1988), 147–201.

112. A. Katok, Hasselblatt, Boris: *An Introduction to the Modern Theory of Dynamical Systems*, With a supplementary chapter by Katok and Leonardo Mendoza. Encyclopedia of Mathematics and its Applications, 54. Cambridge University Press, Cambridge, 1995.

113. A. Kitaev: *Periodic table for topological insulators and superconductors* , AIP Conference Proceedings, 1134 (1) (2009) 22–30.

114. N. Kuiper: *The homotopy type of the unitary group of Hilbert space*. Topology. 3 (1) (1965), 19–30.

115. M. Laca, M. van Frankenhuijsen: *Phase transitions with spontaneous symmetry breaking on Hecke C*-algebras from number fields*, Noncommutative geometry and number theory, 205–216, Aspects Math., E37, Friedr. Vieweg, Wiesbaden, 2006.

116. E. C. Lance: *Tensor products and nuclear C*-algebras*, Operator algebras and applications, Part I (Kingston, Ont., 1980), Proc. Sympos. Pure Math., 38, (1982), Providence, R.I.: Amer. Math. Soc., 379–399.

117. V. Lafforgue: Lafforgue: *K-théorie bivariante pour les algèbres de Banach et conjecture de Baum-Connes*. (French) [Bivariant K-theory for Banach algebras and the Baum-Connes conjecture] Invent. Math. 149 (2002), no. 1, 1–95.

118. H. B. Lawson, M.-L. Michelsohn: *Spin geometry*, Princeton Mathematical Series, 38 (1989) Princeton University Press, Princeton, NJ, . xii+427 pp. ISBN: 0-691-08542-0.

119. J.-L. Loday: *Cyclic homology*, Appendix E by María O. Ronco. Second edition, Grundlehren der Mathematischen Wissenschaften, 301. Springer-Verlag, Berlin, 1998.

120. W. Lück, B. Oliver: *The completion theorem in K-theory for proper actions of a discrete group*, Topology 40 (2001), no. 3, 585–616.

121. V. Mathai, J. Rosenberg, Jonathan: *T-duality for circle bundles via noncommutative geometry*, Adv. Theor. Math. Phys. 18 (2014), no. 6, 1437–1462.
122. R. Meyer, R. Nest: *The Baum-Connes conjecture via localisation of categories*, Topology 45 (2006), no. 2, 209–259.
123. F.G. Mehler: "Über die Entwicklung einer Function von beliebig vielen Variabeln nach Laplaceschen Functionen höherer Ordnung", J. Reine Angew. Math. 66. 161–176.
124. J. Milnor: *Morse theory*, Based on lecture notes by M. Spivak and R. Wells, Annals of Mathematics Studies, No. 51 Princeton University Press, Princeton, N.J. 1963.
125. P. Muhly, J. Renault, D. Williams: *Equivalence and isomophism for groupoid C*-algebras*, J. Operator Theory 17 (1987), 3–22.
126. G.J. Murphy: *C*-algebras and Operator Theory*, Academic Press, Inc., Boston, MA, 1990.
127. G. Pedersen: *C*-algebras and their automorphism groups*, Second edition of [MR0548006]. Edited and with a preface by Søren Eilers and Dorte Olesen. Pure and Applied Mathematics (Amsterdam). Academic Press, London, 2018.
128. R. Ponge: *A new short proof of the Local Index formula and some of its applications*, Comm. Math. Phys. 241 (2003), no. 2–3, 215–234.
129. R. Ponge, H. Wang: *Noncommutative geometry and conformal geometry*, I. Local index formula and conformal invariants. J. Noncommut. Geom. 12 (2018), no. 4, 1573–1639.
130. M. Puschnigg: *Finitely summable Fredholm modules over higher rank groups and lattices*, J. K-Theory 8 (2011), no. 2, 223–239.
131. I.F. Putnam: *The C*-algebras associated with minimal homeomorphisms of the Cantor set*, Pacific J. Math. 136 (1989), no. 2, 329–353.
132. I.F. Putnam, J. Spielberg: *The structure of C*-algebras associated with hyperbolic dynamical systems*, J. Funct. Anal. 163 (1999), no. 2, 279–299.
133. D. Quillen:*Projective modules over polynomial rings*, Inventiones Mathematicae, 36 (1) (1976), 167–171.
134. I. Raeburn, D.P. Williams: *Morita Equivalence and Continuous-Trace C*-algebras*, Mathematical Surveys and Monographs, 60. American Mathematical Society, Providence, RI, 1998.
135. J. Renault: *A groupoid approach to C*-algebras*, Lect. Notes Math. 793, Springer, Berlin (1980).
136. M. Rieffel: *Morita equivalence for C*-algebras and W*-algebras*, Journal of Applied Algebra 5, no. 1, 51–96.
137. M. A. Rieffel: *Morita equivalence for operator algebras*, Operator algebras and applications, Part I, Kingston, Ont. (1980), 285–298, Proc. Sympos. Pure Math., 38, Amer. Math. Soc., Providence, R.I., (1982).
138. M. A. Rieffel: *Induced representations of C*-algebras*, Adv. Math. 13 (1974) 176–257.
139. M.A. Rieffel: *Applications of strong Morita equivalence to transformation group C*-algebras,* Proc. Symp. Pure Math. 38 (1982) 299–310.
140. B. Riemann:*Theorie der Abel'schen Functionen,* J. Reine Angew. Math.1857 (54) (1857), 11–155.
141. G. Roch: *Über die Anzahl der willkurlichen Constanten in algebraischen Functionen,* J. Reine Angew. Math. 64 (1865), 37–376.
142. J. Roe: *Elliptic operators, topology and asymptotic methods*, Second edition. Pitman Research Notes in Mathematics Series, 395. Longman, Harlow, 1998.
143. J. Rosenberg: *Continuous-Trace Algebras from the Bundle Theoretic Point of View*, Journal of the Australian Mathematical Society, Series A, 47 (3) (1989), 36–381.
144. J. Rosenberg: *Topology, C*-algebras, and string duality*, CBMS Regional Conference Series in Mathematics, 111. Published for the Conference Board of the Mathematical Sciences, Washington, DC; by the American Mathematical Society, Providence, RI, 2009.
145. S. Rosenberg: *The Laplacian on a Riemannian manifold. An introduction to analysis on manifolds*. London Mathematical Society Student Texts, 31. Cambridge University Press, Cambridge, 1997.
146. J. Rosenberg: *Algebraic K-theory and its applications*, Graduate Texts in Mathematics, 147. Springer-Verlag, New York, 1994.

147. G. Segal: *Equivariant K-theory*, Inst. Hautes Études Sci. Publ. Math. No. 34 (1968), 129–151.
148. C. Series: *Symbolic dynamics for geodesic flows*, Proceedings of the International Congress of Mathematicians, Vol. 1, 2 (Berkeley, Calif., 1986), 1210–1215, Amer. Math. Soc., Providence, RI, 1987.
149. G. Skandalis, *Some Remarks on Kasparov Theory*, Journal of Functional Analysis 56 (1984), 337–447.
150. J. Spielberg: *Cuntz-Krieger algebras associated with Fuchsian groups*, Ergodic Theory and Dynamical Systems, 13 (1993) no. 3, 581–595.
151. D. Sullivan: *On the ergodic theory at infinity of an arbitrary discrete group of hyperbolic motions*, Ann. Math. Stud. 97 (1981), 465–496.
152. R. Swan: *Vector Bundles and Projective Modules*, Transactions of the American Mathematical Society, 105 (2) (1962) 264–277.
153. M. Takesaki, *On the cross-norm of the direct product of C*-algebras*, The Tohoku Mathematical Journal, Second Series, 16 (1964), 111–122.
154. J.-L. Tu: *La conjecture de Baum-Connes pour les feuilletages moyennables*, K-Theory 17 (1999), no. 3, 215–264
155. R. O. Wells Jr.: *Differential analysis on complex manifolds* 65 Graduate Texts in Mathematics, Springer, New York, (2008).
156. E. Edward: *D-Branes and K-Theory*, Journal of High Energy Physics, 9812 (19) (1998),1126–6708.
157. J. Väisälä: *Gromov hyperbolic spaces*, Expo. Math. 23 (2005), no. 3, 187–231.
158. R.J. Zimmer: *Ergodic theory and semisimple groups*, Monographs in Mathematics, 81. Birkhäuser Verlag, Basel, 1984.